ONE WEEK LOAN

FORENSIC ENTOMOLOGY

The Utility of Arthropods
in Legal Investigations

SECOND EDITION

FORENSIC ENTOMOLOGY

The Utility of Arthropods in Legal Investigations

SECOND EDITION

Edited by

Jason H. Byrd & James L. Castner

CRC Press
Taylor & Francis Group
Boca Raton   London   New York

CRC Press is an imprint of the
Taylor & Francis Group, an **informa** business

Cover photo courtesy of Joseph Berger, University of Georgia; bugwood.org. Credited authors hold the copyright for their photographs and illustrations.

CRC Press
Taylor & Francis Group
6000 Broken Sound Parkway NW, Suite 300
Boca Raton, FL 33487-2742

© 2010 by Taylor and Francis Group, LLC
CRC Press is an imprint of Taylor & Francis Group, an Informa business

No claim to original U.S. Government works

Printed & bound in Singapore by Markono Print Media Pte Ltd
10 9 8 7 6 5 4 3 2 1

International Standard Book Number: 978-0-8493-9215-3 (Hardback)

Library of Congress Cataloging-in-Publication Data

Forensic entomology : the utility of arthropods in legal investigations / edited by Jason H. Byrd and James L. Castner. -- 2nd ed.
    p. cm.
Includes bibliographical references and index.
ISBN-13: 978-0-8493-9215-3 (alk. paper)
ISBN-10: 0-8493-9215-2 (alk. paper)
  1. Forensic entomology. 2. Arthropoda--Miscellanea. 3. Postmortem changes. 4. Death--Time of. I. Byrd, Jason H. II. Castner, James L. III. Title.

RA1063.45.F67 2009
614'.17--dc22
                                                  2008035462

**Visit the Taylor & Francis Web site at
http://www.taylorandfrancis.com**

**and the CRC Press Web site at
http://www.crcpress.com**

In Memoriam

Chester Lamar Meek

January 16, 1944–June 27, 2002

Photograph courtesy of *The Advocate*.

C. Lamar Meek

1944–2002

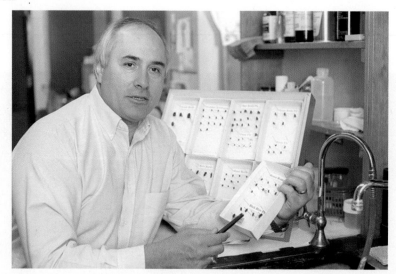

Photo courtesy of *LSU Today*

In 1975, with his newly minted PhD in entomology from Texas A&M University, Dr. Lamar Meek began his career as an assistant professor at Louisiana State University. By 1983 he was a full professor, with his research focusing mainly on mosquito control and the diseases mosquitoes may carry, such as encephalitis. Over the years he became renowned in his field. Cooperating with like researchers from across the country, he authored or coauthored more than 125 papers on medical entomology topics that included biology, ecology, and chemical and biologic resistance to insecticides, as well as many others. However, by the mid-1980s Dr. Meek had found another interest: forensic entomology. As a pioneer in forensic entomology who influenced countless entomology and anthropology graduate students, he was a legend to his followers. His graduate-level forensic entomology course was always full, even as he challenged his students to immerse themselves in data collection, a task they did not fully appreciate until he charged them with sampling larvae from all natural and artificial orifices of the carcasses he used for class.

This kind and gentle man, whom I always called Lamar, never raised his voice once to anyone in my presence in all the 17 years I knew him. Throughout those 17 years I frequently would seek his advice in trying to determine time since death for countless forensic cases. Lamar cautioned others that his work with insects in such settings was to figure out the biological clock, not necessarily the time since death. That clock represented the time of insect activity on a body, or the period of time the body had been available to the insects. Sometimes we would go to a scene together; sometimes he would simply meet me there. He was always considerate and respectful of the dead, always asking my opinion, and always educating. The first time in my lab when he saw the larvae of a hairy maggot blow fly, years before that particular species had made its way north to Tennessee and beyond, he commented, "Oh, it's here now," as he gingerly pointed out the spiny little maggot that reminded me of a caterpillar in my naïveté. Lamar asked me if I knew what it was—I had not a clue. I knew it looked different from the other larvae we plopped into jars of alcohol or placed in canisters for subsequent rearing. I knew I did not want to pick it up. Soon, as he explained, I knew it was one of those hitchhikers of the universe who unwittingly had made its way to North America, not unlike the fire ant and kudzu before it. He further explained how important it was to record the hairy maggot's presence at a death scene because it could skew the estimate of time since death by cannibalizing the other species' larvae.

It was Lamar who showed me how insect succession might be delayed on a corpse in the trunk of a car, but how ingenious the blow flies were ultimately at getting into the vehicle, laying their eggs, going through the various "instars," and burrowing under the carpet in the rear of the car to complete their development during the pupation stage. It was Lamar who determined that the dark, crusty, lifeless maggot that appeared to be burned when I found it on a human body in a burned-out car really was just completely enveloped in dried human body fluids. If it had been burned, we both knew it could suggest that the car had not been set on fire until sometime after death, after the larvae had time to be present on the body.

Over the years, we worked side by side on cases as he focused in one area of forensics and I in another in a science that ultimately would capture the attention of a worldwide audience. In forensic entomology he published more than twenty papers and a book chapter on necrophilous arthropods in death investigations of both humans and wildlife species. Though he enjoyed forensic entomology, he continued his teaching and research in medical and veterinary entomology. His awards and honors in that area were numerous and included the 1991 American Mosquito Control Association Memorial Lectureship Award, the 1989 Hathaway-Ridder Distinguished Achievement Award from the Louisiana Mosquito Control Association, the 1987 Outstanding Award in Medical-Veterinary Entomology from the American Registry of Professional Entomologists, and the 1986 Meritorious Service Award from the American Mosquito Control Association (AMCA). He was also a member of various local, regional, and national professional societies, which included the Louisiana Mosquito Control Association (LMCA), the Louisiana Entomological Society, the Entomological Society of America, the Texas Mosquito Control Association, the American Registry of Professional Entomologists, and the Society of Vector Ecology. He served on the governing board of the Entomology Society of America, on the board of directors of the LMCA, as treasurer of AMCA, and as past president of both the LMCA and the Louisiana Entomology Society.

Lamar was loved and admired by everyone who knew him, especially his students. He graduated nine MS students and seven PhDs in his all too brief career. Today, they continue his legacy as they work in a variety of research and applied positions. Lamar died on June 27, 2002, from a heart attack he experienced while collecting mosquito control data in Mississippi. He was 58. To continue the legacy of Dr. Meek and his dedication to forensic and medical and veterinary entomology, Louisiana State University and family members have established a scholarship in his name, the Lamar Meek Graduate Student Memorial Fund. His family, students, and many others, like me, remember him with great respect and love, and sorely miss his counsel.

Mary H. Manhein
Director
Forensic Anthropology and Computer Enhancement Services Laboratory
Louisiana State University

Foreword

Insects and arachnids are found in almost all conceivable habitats worldwide. They are abundant, widely distributed, and occupy all places of human inhabitation on planet Earth, and as such they are often associated with our daily activities. Insects are the largest group of organisms: one-half to almost two-thirds of global biodiversity estimated to be 10 million species (<1.75 million described). Having survived so long, over 300 million years, insects are highly adaptable to changing environment and have adapted to diverse habitats and environmental conditions on land, water, and soil. They are invaluable ecological partners in sustaining our life-support system on the planet Earth. Since insects inhabit almost everywhere humans live, they are closely associated with humans and human enterprises. As a result, insects and arachnids often become invaluable evidence to crime, providing important information for forensic investigation. They often are living witness to crime or the cause of litigation, to which forensic entomologists apply their knowledge about specific species or groups of species to help solve crimes. In other words, forensic entomology is a branch of entomological sciences that offers an important specialty discipline for the forensic sciences at large.

Forensic entomology refers to the science of insects and related arthropods (specifically arachnids, mites, ticks, scorpions, and spiders), which is applied to help solve litigation in civil and criminal cases. It not very surprising to find out that our wise ancestors used insects to help determine the cause of death, help establish the time of death, and make conclusions as to the trauma suffered by victims of violent crime. When I started to involve myself in forensic work, I found that many experienced entomologists often engaged in litigation to help solve civil and criminal cases that involved entomological evidence. In recent years, an entomological specialty known as forensic entomology has often related more closely to criminal cases such as homicides and rapes, and thus is more commonly identified with criminal investigation. This trend culminated in the formation of the North American Forensic Entomology Association (NAFEA) in 2005 to provide a scientific and professional forum for forensic entomology.

At this juncture, forensic entomology as a science and a profession faces many challenges as it attracts the public interest and also offers an opportunity for expanding research related to forensically important insects and arachnids. This makes for a unique contribution to science education as our knowledge base about many insect species is quite limited. Forensic entomology as a discipline in the forensic science community is also expanding, with a few entomologists leading in forensic science education. Forensic entomology must be based on entomology and an application of entomological knowledge. However, the professional focus in forensic entomology has not been directed to expansion of entomological research pertaining to the deficient areas of forensic entomology as a forensic science.

The first edition of this multiauthored book, *Forensic Entomology: The Utility of Arthropods in Legal Investigation*, brought together for the first time the results of research on insects of forensic importance for the last quarter of the twentieth century. This work offered the forensic science community up-to-date information on medicocriminal aspects of forensic entomology. As the first edition of this volume effectively served the forensic science community, this second edition will again present updated information based on research done since the dawn of the third millennium.

Ke Chung Kim, PhD
Diplomate
American Board of Forensic Entomology
Pennsylvania State University

Preface

Forensic entomology is not a new science. It has been around since the 13th century, first recorded in China. However, as a science it began to gain real momentum in the United States in the late 1960s when it became widely evident that carrion insects possessed predicable growth rates and behavioral patterns. Since that time, it has been determined that swine carcasses make suitable research subjects that closely mimic human decomposition rates, and research and experimentation has occurred at a dizzying pace. Forensic entomology is now a familiar term to almost all medico-legal death investigators, crime scene analysts, medical examiners and coroners. Today, when entomological evidence is present at death scenes; it is almost always collected for analysis. This is an exciting time to be involved in forensic entomology.

With the increased attention focused upon our science, forensic entomologists carry the burden of providing adequate research methods and protocols for the collection, preservation and analysis of insect evidence. Many of the questions posed in forensic entomology have yet to be answered, and we are faced with the challenges that an emerging science invariably confronts. We must ensure that entomological evidence is reliable and repeatable in an environment that routinely deals with natural biological variation. Additionally, we must ensure that the terminology used is clear and succinct. In casual conversation it is often said that the forensic entomologist determines the postmortem interval. In fact, the entomologist determines the period of time for which insects are active on the tissues of the body (human or animal). In addition to this estimation of the colonization time, the entomologist estimates the length of time required for the adult female flies to arrive at the remains. Together, these time intervals equal the period of insect activity. This time period may, or may not, closely approximate the entire postmortem interval. Generally speaking, the period of insect activity is usually less than the postmortem interval. It is up to the individual forensic entomologist to determine the overall difference between these two related time periods, utilizing weather conditions and any other relevant data available.

The concept of this book formed in 1995 when I was teaching a graduate course in medical and veterinary entomology in the Department of Entomology and Nematology at the University of Florida. During that course, I was asked about the feasibility of teaching a course in forensic entomology. Three things, which eventually provided the impetus to organize this work, immediately occurred to me. The first was that a full course in forensic entomology existed only at two universities. The second was that a suitable textbook did not exist for a course of such nature. And finally, most taxonomic keys for identification were too difficult for the introductory student to navigate. Humans are visual learners, and in order to make a forensic entomology text palatable to many people, it had to visual.

A fellow entomologist and scientific photographer, Dr. James Castner, had a keen interest in carrion insects and was surprised at the void in photographic documentation for these insects. The next three years were spent collecting and photographing carrion inhabiting insects. During this time, forensic entomology became an area in which research funding became available, and the need for the current literature to be compiled into a single source was evident.

This work has been a continuing effort to document the level of professional development that exists within the field of forensic entomology. It contains definitive research by some of the leading forensic scientists in the United States and from around the world. The first edition was written so that other forensic scientists, law enforcement personnel, as well as medical and legal professionals could gain a basic understanding of forensic entomology and how it is applied. The increased understanding of this discipline by those outside of the field will result in a higher probability they will seek the expert consultation of a forensic entomologist when needed. The editors hope that this work facilitates the proper application of insect-related evidence within our legal system. The field of forensic entomology has been tasked to produce the scientific data necessary to prove its rightful place within the forensic sciences. Fortunately, both practicing and research entomologists have been able to publish a wide array of scientific literature to detail the scope and applications of forensic entomology. Their work is the basis for forensic entomology being a widely accepted discipline in the forensic sciences, and it is our honor to showcase some of this research.

Jason H. Byrd
James L. Castner

The Editors

Dr. Jason H. Byrd, PhD, D-ABFE, is a board-certified forensic entomologist and diplomate of the American Board of Forensic Entomology. He is the current vice president of the American Board of Forensic Entomology, and the current president of the North American Forensic Entomology Association. He is the first person to be elected president of both professional North American forensic entomology associations. Dr. Byrd is a bureau chief with the Florida Division of Agriculture and Consumer Services, and he serves as the associate director of the William R. Maples Center for Forensic Medicine, University of Florida College of Medicine. At the University of Florida, he instructs courses in forensic science at the University of Florida's nationally recognized Hume Honors College. He is also a faculty member of the Virginia Institute of Forensic Science and Medicine.

Outside of academics Dr. Byrd serves as an administrative officer within the National Disaster Medical System, Disaster Mortuary Operational Response Team, Region IV. He also serves as the logistics chief for the Florida Emergency Mortuary Operations Response System. Currently he serves as a subject editor for the *Journal of Medical Entomology*. He has published numerous scientific articles on the use and application of entomological evidence in legal investigations. Dr. Byrd has combined his formal academic training in entomology and forensic science to serve as a consultant and educator in both criminal and civil legal investigations throughout the United States and internationally. Dr. Byrd specializes in the education of law enforcement officials, medical examiners, coroners, attorneys, and other death investigators on the use and applicability of arthropods in legal investigations. His research efforts have focused on the development and behavior of insects that have forensic importance, and he has over 15 years experience in the collection and analysis of entomological evidence. Dr. Byrd is a fellow of the American Academy of Forensic Sciences.

James L. Castner, PhD, is an entomologist-photographer-writer with special interests in tropical biology, cultural anthropology, and human artifacts that are considered primitive tribal art. His undergraduate work was done at Rutgers, followed by advanced degrees at the University of Florida. He is currently an adjunct professor in the Biology Department at Pittsburg State University in Pittsburg, Kansas.

Combining photographic expertise with an academic background in entomology, biology, and Spanish, Dr. Castner has endeavored to create and publish works that bring scientific topics to both a professional and general audience. His photographs have appeared in a variety of books and magazines, including almost every college-level biology textbook. His favorite topics are related to the insect world and the rainforest. Some of his writing and photo credits include: *National Geographic*, *Natural History*, *International Wildlife*, *Ranger Rick*, and *Kids Discover*. His book credits include: *Shrunken Heads*, *Amazon Insects*, and *Medicinal and Useful Plants of the Upper Amazon*.

In 1997, Dr. Castner left his academic position to pursue writing and the development of educational materials full-time. His *Photographic Atlas of Entomology* and *Photographic*

Atlas of Botany are used by colleges and universities throughout the country. He has been actively involved as an educator of secondary school students and their teachers for many years, often acting as a workshop leader or instructor of field courses. He designs and leads natural history tours for teachers, students, and naturalists to the Amazon Basin and often serves as a consultant in many capacities. He has traveled and photographed throughout South and Central America, including over fifty trips to Peru to study tropical insect biodiversity. He is an authority on the authentication of shrunken heads, or *tsantsas*, and in 2007 published a novel on this topic called *The Tsantsa Homicides*.

Acknowledgments

The editors of this work extend our thanks and deep appreciation to the contributors of this work. It is their research and professional efforts that make volumes like this possible. The compilation of such an impressive amount of research and knowledge within a single book project undoubtedly serves to bring recognition and acceptance of forensic entomology into the mainstream of forensic science. Such works also educate many others who do not have access to scientific journals about the topic of forensic entomology, and we applaud the manner in which the contributors have constructed their various chapters so that a complex discipline is easily understandable to most audiences. We also thank Nikita Vikhrev, Cor Zonneveld, Susan Ellis, Jay Cossey, Joseph Berger, Paul Beuk, and Steve Graser for the use of their excellent photos of many insect species of forensic importance.

Contributors

Gail S. Anderson*
School of Criminology
Simon Fraser University
Burnaby, British Columbia, Canada

Mark Benecke*
International Forensic Research and Consulting
Cologne, Germany

Henk R. Braig
School of Biological Sciences
Bangor University
Bangor, United Kingdom

Rosemary Brown
Crime Scene Unit
Tallahassee Police Department
Tallahassee, Florida

Jason H. Byrd*
William R. Maples Center for Forensic Medicine
Department of Pathology, Immunology and
   Laboratory Medicine
University of Florida College of Medicine
Gainesville, Florida

James L. Castner*
Department of Biology
Pittsburg State University
Pittsburg, Kansas

Ian Dadour
Centre for Forensic Science
University of Western Australia
Crawley, Western Australia, Australia

Shari L. Forbes*
Faculty of Science
University of Ontario Institute of Technology
Oshawa, Ontario, Canada

Alison Galloway*
Department of Anthropology
University of Santa Cruz
Santa Cruz, California

M. L. Goff*
Forensic Sciences Program
Chaminade University of Honolulu
Honolulu, Hawaii

Robert D. Hall*
Office of Research
University of Missouri
Columbia, Missouri

Neal H. Haskell
Department of Biological Sciences
Saint Joseph's College
Rensselaer, Indiana

Roger Hawkes
Crime Scene Unit
Tallahassee Police Department
Tallahassee, Florida

Leon G. Higley*
Department of Entomology
University of Nebraska-Lincoln
Lincoln, Nebraska

Timothy Huntington
Natural Science Department
Concordia University, Nebraska
Seward, Nebraska

K. C. Kim
Department of Entomology
Pennsylvania State University
University Park, Pennsylvania

K. Lane Tabor Kreitlow*
North Carolina Department of Agriculture and
   Consumer Services
Plant Industry Division
Raleigh, North Carolina

Lynn R. LaMotte
Department of Experimental Statistics
Louisiana State University
Baton Rouge, Louisiana

* Denotes senior author.

Wayne D. Lord
Forensic Science Institute and Department of
  Biology
University of Central Oklahoma
Edmond, Oklahoma

Richard W. Merritt*
Department of Entomology and Fisheries and
  Wildlife
Michigan State University
East Lansing, Michigan

M. Anderson Parker*
City-County Bureau of Investigation
Wake County Public Safety Building
Raleigh, North Carolina

M. Alejandra Perotti*
School of Biological Sciences
University of Reading
Reading, United Kingdom

John R. Scala*
WGAL-TV
Lancaster, Pennsylvania

Jamie R. Stevens
Hatherly Laboratories
School of Biosciences
University of Exeter
Exeter, United Kingdom

Justin Talley
Department of Entomology and Plant Pathology
Oklahoma State University
Stillwater, Oklahoma

Jeffrey K. Tomberlin*
Department of Entomology
Texas A&M University System
College Station, Texas

Sherah L. VanLaerhoven*
Forensic Sciences Program
University of Windsor
Windsor, Ontario, Canada

John R. Wallace
Department of Biology
Millersville University
Millersville, Pennsylvania

Heather A. Walsh-Haney
Division of Justice Studies
College of Professional Studies
Florida Gulf Coast University
Ft. Myers, Florida

Jeffrey D. Wells*
Department of Biology
West Virginia University
Morgantown, West Virginia

Terry Whitworth*
Whitworth Pest Solutions
Puyallup, Washington

* Denotes senior author.

Table of Contents

Dedication

This volume is dedicated to Hugh David Byrd. It is because of you that I understand fortitude. To Evelyn, Justin, and Juli, you have provided a lifetime of support and inspiration for which I will forever be thankful. To Heather, your seemingly infinite patience has helped me complete this and many other personal and professional projects. Most importantly, you have taught me that life would not be a picnic without the ants.

—JHB

For my father, Chief of Detectives (Ret.) Franklyn Castner, who devoted his professional life to law enforcement. Thank you for everything.

—JLC

Introduction: Perceptions and Status of Forensic Entomology

ROBERT D. HALL
TIMOTHY E. HUNTINGTON

Contents

Introduction

Forensic entomology is the broad field where arthropod science and the judicial system interact. It has been subdivided into three principal areas focused on those issues most often litigated (Lord and Stevenson, 1986). Therefore, urban entomology concentrates mainly on controversies involving termites, cockroaches, and other insect problems accruing to the human environment. The name *urban* is somewhat a misnomer, in that private or public nuisance actions involving pest insects, such as flies emanating from livestock or similar facilities, are categorized under this heading. There is currently much litigation involving insect nuisance as it relates to agricultural endeavors, especially cattle feedlots, poultry houses, and "corporate" hog facilities. That numerous lawsuits relate to termite damage, termite extermination, and the effect of termites and related structural pests on the value of real estate should not come as a surprise. However, other litigation under the urban rubric is not as easily foreseen by the uninitiated. Patients in hospitals and nursing homes occasionally suffer myiasis (infestation by fly larvae), and this usually results in actions claiming neglect. The neglect of pets is another area where the duration of myiasis may be used to help prove neglect (Anderson and Huitson, 2004). Negligence actions against mortuaries may result from maggot-infested corpses. To illustrate how diverse the field can be, the authenticity of figurines and other artifacts (Figure I.1a–d) from west Mexican shaft tombs has been verified by insect evidence (Figures I.2a and b, and I.3a and b) (Pickering et al., 1998).

The area of stored products entomology involves disputes over arthropods and arthropod parts in food and other products. Insect debris in breakfast cereal, caterpillars in cans of vegetables, and fly maggots in sandwiches from fast-food restaurants are good examples of commonly litigated cases in the stored products area (Figure I.4). Occasionally, a consumer will

(a)

(b)

(c)

(d)

Figure I.1 (a) This olla form pot in the Colima red style represents a shallow basket of fruit with a smaller olla in the middle. It is believed that these vessels were filled with drink offerings, and they were commonly placed in the shaft tombs approximately 2000 years ago. (b) This small standing figurine is characteristic of the Tuxcaceuso-Ortices style. Manganese stains can be seen over parts of the figure, some of which included mineralized puparia. (c) Solid ceramic figurine of a mother and child. On this figure, two widely separated impressions of puparia were discovered. (d) A detail of the back of a small solid ceramic figurine. A cluster of insect puparia that have been mineralized by manganese deposits can be seen in the middle of the photograph (black deposits), while unmineralized impressions can be seen adjacent and in the upper right. (Photos courtesy of Robert B. Pickering, Photo Archives, Denver Museum of Natural History.)

(a)   (b)

Figure I.2 Photomicrographs of a single insect puparium (a) and a grouping of puparia (b) that have been mineralized by manganese (black deposits) on Colima dog figurines discovered in west Mexican shaft tombs. (Photos courtesy of Ephraim A. Cuevas, Photo Archives, Denver Museum of Natural History.)

attempt to defraud a restaurant or other business by "planting" insects or insect parts in products purchased beforehand. The resolution of such cases requires a forensic entomologist.

The focus of the present text, the area often called medicolegal entomology or forensic medical entomology, is now commonly known as medicocriminal entomology because of its emphasis on the utility of arthropod evidence in solving crimes, most often crimes of violence (Hall, 1990). Medicocriminal entomology usually involves all of those elements necessary to produce a fascinating story. It contains the intrigue surrounding human death, the decay process with its grisly aspects, the detective work necessary to bring perpetrators to trial, the adversarial criminal justice system with its arcane terminology, seeming inconsistencies, and the drama of the courtroom. Add to this the application of an impartial biological science and it is not surprising that medicocriminal entomology has been embraced by a broad spectrum of individuals. Those individuals include consumers (criminalists and detectives), advocates (prosecuting and defense attorneys), and future practitioners, such as students in many colleges and universities. It also has been used to

(a)   (b)

Figure I.3 (a) Photomicrographs of an unmineralized single insect puparium from the outer surface of a pitcher discovered in a west Mexican shaft tomb. (b) An unmineralized grouping of insect puparia on the inside of a broken pot discovered in a west Mexican shaft tomb. (Photos courtesy of Ephraim A. Cuevas, Photo Archives, Denver Museum of Natural History.)

Figure I.4 Indian meal moth larvae (*Plodia interpunctella*) (Lepidoptera: Pyralidae) infesting a bite-sized chocolate candy. Often the civil litigation surrounding insect contamination of food products requires a forensic entomologist. (Photo courtesy of Dr. Jason H. Byrd.)

resolve questions about the death of animals other than humans, including livestock, pets, or protected species such as bears (Figure I.5) (see Anderson, 1998).

Although medicocriminal entomology may involve deliberate homicide or assault using insects, cases of unexplained sudden death (such as anaphylaxis from bee stings), or causation of traffic accidents (e.g., inattention to driving during frantic attempts to evade a wasp inside an automobile), the typical questions posed to the medicocriminal entomologist involve estimates of the time a decedent has been dead (the postmortem interval [PMI]) and, less frequently, the place (situs) where death occurred (Hall, 1990). For the

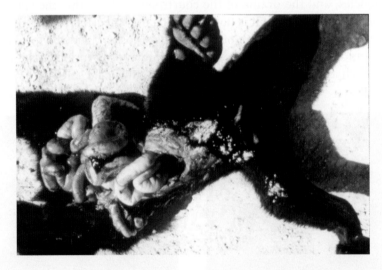

Figure I.5 Forensic entomologists are often involved in cases of wildlife poaching concerning protected species in an effort to resolve questions as to the time of death, as with this black bear cub found in the parking lot of a garbage dump. Recently, forensic entomologists have teamed with police and wildlife officials in Canada to solve cases in which bears have been poached to satisfy the market for their gall bladders, which are sold as supposed aphrodisiacs. (Photograph taken by W. Campbell, RCMP, Winnipeg, Manitoba, FIS. With permission of the Royal Canadian Mounted Police.)

former determination, two principal approaches are employed. The first involves application of the temperature-dependent development of insects (typically flies), and the second recognizes that a generally predictable succession of arthropods often facilitates decomposition of organic matter, including human corpses or animal cadavers, and that, by recognizing how a corpse's fauna relate to such patterns of colonization, an estimate of PMI may be made.

Because of its association with flies, which are of broad importance to human and animal health, and the ecology of decomposition, the field of medicocriminal entomology is rightly recognized as a specialty within medical entomology itself. The province of medical entomology, preventive medicine, in its broadest form involves the study of mosquitoes, ticks, mites, and all other arthropod species that can damage human or animal health directly or vector pathogenic organisms that in turn cause disease. Thus, medical entomology is an important biomedical science, and the medical entomologist is no stranger to the medical community. For many years, medical entomologists have worked alongside physicians and veterinarians in fields including infectious diseases, pathology, and dermatology, elucidating the various direct and indirect ways that insects and their allies impact human and animal health.

The aforementioned phenomenon called myiasis is commonly noted in veterinary medicine and can occur in humans even in the West's sanitized society. Those species of flies producing facultative, or secondary, myiasis have long been recognized as major players in the decomposer biotic community. The natural ecological cycles returning dead plants and animals into raw material for future life frequently involve insects. In particular, the decomposition of human or livestock feces, or the physical remains of humans or other animals, depends heavily on insect involvement. The nuisance factors associated with flies from rural livestock facilities (Thomas and Skoda, 1993) or the beneficial effects of fly maggots in wounds (see Graner, 1997) directly included medical entomologists in these health-related issues. Therefore, it is not surprising that most medicocriminal entomologists currently operate under the broad rubric of medical entomology, which includes veterinary applications.

Medicocriminal Entomology as a Profession

Most modern progress in medical entomology has been recorded during the past century, especially since Pasteur's articulation of the "germ theory" in the latter part of the 1800s, and this progress correlates well with similar advances in medicocriminal entomology. The medicocriminal entomologist thus finds his or her professional status currently analogous to that of the medical entomologist in the broad sense. The status of medicocriminal entomology as a profession, therefore, can be examined most efficiently in the context of a biomedical career track.

Although the data are now a few decades old, a survey of forensic entomologists worldwide provided an interesting perspective (Lord and Stevenson, 1986). Of sixty-two entomologists responding, thirty-three indicated that they were associated solely with the medicocriminal specialty, and five more were involved with both medicocriminal and other forensic entomology specialties. Most held full-time positions with universities or governmental agencies, and about 35% indicated involvement with consulting work. Fewer than 40% of respondents were members of the American Registry of Professional

Entomologists (ARPE), the professional association of American entomologists, which has since been amalgamated into the Entomological Society of America as the Board Certified Entomologist (BCE) program. The majority of entomologists responding to the survey held the PhD degree or equivalent, with the remainder possessing MS or occasionally MD degrees.

Bottlenecking

When compared to licensed professions, medicocriminal entomology will be seen to suffer from several factors that may be viewed as shortcomings. The most important of these will center on the issues of quality assurance and career opportunity. Every career track exhibits some form of "bottleneck," perhaps known better to students or other aspirants as the "weeding-out point." Once these bottlenecks are identified, it becomes more straightforward to succeed in the ultimate goal of career progression. In the professional curricula of medicine, veterinary medicine, and law, the initial bottleneck is highly visible; it is at the point of admission to the appropriate institution.

Although the well-known principle of supply and demand causes fluctuations, virtually every professional curriculum boasts many better-qualified applicants than are admitted annually. Most medical and law schools currently admit about 10% of applicants; those that are admitted usually succeed and progress on to graduation and subsequent professional careers. In contrast, most graduate curricula in the life sciences—and entomology is no exception—admit a higher proportion of qualified students who apply, even though all admitted might not receive financial support. What this means in practice is that a student with a reasonable undergraduate academic record will likely be accepted into a postgraduate program in the life sciences, including entomology.

The underlying reasons for this include, but are not limited to, the trend to measure productivity of academic curricula by easily calculated numbers such as students enrolled, students graduated, and total student credit hours taught. A major pressure on most graduate life science faculties is to recruit and retain students. Coupled with this is the dependence of university research on the economical labor provided by highly motivated graduate students. After even a short period of enrollment, such students become quite valuable to both departments and advisors, in that they may have yet-unfinished research projects, done at considerable investment of resources, that will yield scientific publications (another easily counted measure of productivity) in the future. The pressure, therefore, trends toward retaining students who are making satisfactory progress.

Most of these students eventually earn their graduate degree. The points of admission and academic success, therefore, are not the principal bottlenecks in the entomology graduate career track. The major bottleneck occurs at the point of competition for the few professional positions available. A desirable faculty, governmental, or industry job may generate a hundred or more applications, and virtually all applicants will possess the requisite terminal degree. What separates the successful candidate from the rest is the remainder of his or her portfolio, including publications, involvement with grants (overhead-producing extramural support is another easily evaluated number), and ancillary professional involvement.

The opportunities for this professional involvement vary widely by university, program, and even academic advisor. Additionally, proper training in medicocriminal entomology is limited by the relatively low number of competent medicocriminal entomologists. For the student seeking a professional career in medicocriminal entomology, the mentorship of a practicing expert in the field is invaluable. This additional bottleneck caused by the limited availability of competent mentors, combined with the recent increase in interest of all things dealing with forensic science, has led to the inception of programs in forensic entomology at universities, large and small, that have very little of the expertise that is needed for such an undertaking. The advent of learning through distance education being offered by many different colleges and universities adds yet another twist to this bottle-neck. Although a master's degree in entomology may be obtained without setting foot in a proper laboratory, one could hardly feel comfortable having evidence analyzed by someone so trained.

The bottleneck at the point of job competition is made more compelling when one notes that few professional positions exist for entomologists—and indeed most other life scientists—outside the "institution." Whether the institution in question is a college, university, government agency, or private corporation, the point is the same: there is little professional life for the entomologist outside the framework of the employing entity. This is in direct contrast to the professional career tracks in medicine (including dentistry), veterinary medicine, and law. Those in the latter fields indeed have opportunities to compete for faculty, government, and industry positions, but they also have an option not typically available to entomologists: they can operate independently under their state-issued professional license. The "licensed" pest-control industry provides less opportunity for the graduate entomologist than one would think, because most of the actual control practice is conducted by technicians.

Taking all the above into account, one unexploited career path might include the MD degree and a subsequent residency in forensic pathology at an institution also offering a graduate program in forensic entomology. A collateral or consecutive doctoral degree in the latter area would give such an individual the requisite entomological credentials along with ready employability as a forensic pathologist.

Licensing and Quality Assurance

The issue of, and controversy about, professional licensing is not new to entomology, having been debated most enthusiastically since the advent of organic insecticides produced an easy analogy to prescription drugs in medicine (Hall and Hall, 1986). However, entomologists have been remarkably resistant to any sort of licensing effort (Perkins, 1982; Hall and Hall, 1986). There is a second "weeding-out point" in the professional career tracks of physicians, veterinarians, and attorneys. This is the state licensing examination, which must be passed before one can practice medicine, veterinary medicine, or law. Far from being perfunctory or "ritual" exams, these tests serve as a major hurdle, and by no means do all candidates pass, even though at the point of examination they may hold the terminal academic degree. As such professionals further specialize, there exists a vast array of board examinations in the medical field. The purpose for these excruciating tests is simple: quality assurance. As a society, we have agreed that we want those practicing medicine, veterinary medicine, and law to be fully qualified and to meet objective educational and performance minima. Curiously, we have not insisted on similar standards for other career fields, including medicocriminal entomology.

Perhaps the field of entomology is sufficiently arcane in and of itself to give anyone associated with it the aura of "expert," but in practice this is far from the truth. The fact that an individual possesses an earned PhD with a major in entomology is no indication of competence in the medicocriminal field. Unfortunately, it is often no indication of general competence in entomology itself. The American Board of Forensic Entomology (ABFE) was formed to serve as a quality assurance tool in medicocriminal entomology work. Similar to other such boards associated with the American Academy of Forensic Sciences, the ABFE has educational and performance criteria that must be met before an applicant can be certified as a member or diplomate. An earned PhD (MS for member status) with a major in medical entomology is the educational minimum (similar to the general criterion for faculty appointments at the assistant professor level or higher), and applicants must demonstrate involvement with and contributions to both research and case work in medicocriminal entomology. Peer recommendations are solicited, and applicants are screened by committee and approved by the board to sit for the certification examination. As with other professional certification bodies, a goal is to have annual continuing educational requirements for retention of certification. Because the ultimate application of medicocriminal entomology is in the context of the adversarial legal system, the ability to qualify as an expert witness is necessary to fulfill this professional function. As detailed in Chapter 14, expert testimony can be excluded if it can be shown that the witness offering it is unqualified. As more trial judges assume the role of gatekeeper under *Daubert* standards, and as more pretrial motions to exclude are filed, there will assuredly be closer scrutiny of the entomologist's professional credentials. Membership in the ABFE is intended to reflect the professional status of fully qualified medicocriminal entomologists, and the membership list of that organization is a good place to start when such expertise is needed. A current list of ABFE members is available on the Internet.

Development of Medicocriminal Entomology

The earliest record of medicocriminal entomology comes from thirteenth-century China. The Chinese criminalist Sung Tz'u was recorded in the seminal *Washing Away of Wrongs* (translated by McKnight, 1981) to have investigated a murder by slashing in a village. When all villagers were required to bring their scythes to one spot, he noted that flies congregated on only one, ostensibly because of minute traces of blood and other tissue. Confronted by this evidence, the guilty villager reportedly confessed. In addition to concluding this earliest known case, Sung Tz'u also recorded the rapid appearance of maggots on a decedent during the warm season, and the potential utility of maggot infestation in recognizing antemortem wounds.

It was not until the mid-1800s that medicocriminal entomology saw recorded use in the West. Bergeret (1855) investigated the death of an infant near Paris, France, where the corpse had been discovered behind a mantle. By evaluating the insect fauna on the mummified remains, he concluded—perhaps incorrectly (see Greenberg, 1991)—that the baby had been dead about 2 years, thus exonerating the current inhabitants of the house, who had resided there a shorter time. Despite any putative errors, this case represents the first application of insect succession data in forensic entomology and paved the way for later studies cited in Nuorteva (1977), Keh (1985), Smith (1986), and Catts and Goff (1992). J.

P. Megnin is usually credited with focusing Western attention on the forensic utility of entomology, especially with publication of his famous capstone work, *Fauna of Cadavers* (Megnin, 1894). The eight stages of human decomposition described therein, which were followed by Leclercq (1969) and Easton and Smith (1970), and the insects associated with them, have perhaps served as much as an obstacle to understanding the entomology-associated decay process as illuminative of it (see Greenberg, 1991). An analysis of eleven insect faunal succession studies suggested that the phenomenon represented a continuum of decay rather than the tidy "seres" described by Megnin (Schoenly and Reid, 1987). This area continues as an active field of research in forensic entomology, and computer technology has been incorporated (Byrd, 1998; Schoenly et al., 1992). Perhaps the most frequent forensic use of successional insect colonization of corpses and cadavers has been in the Hawaiian Islands, likely because of the climatic stability and faunal predictability found there (see Goff et al., 1986, 1988; Goff and Flynn, 1991). Researchers in this geographic area also noted the possible effects of certain illicit drugs on fly development (Goff et al., 1989, 1991). Goff and Lord provide an extensive treatment of entomotoxicology in Chapter 12.

The foundation necessary for reliable application of forensic entomology was laid during the first half of the twentieth century by taxonomists interested in those insect species of medicocriminal importance. These include two principal families: the Sarcophagidae (or "flesh flies," which is not descriptive because most sarcophagid species are not necrophilous) and the Calliphoridae ("bottle flies" or "blow flies"). The latter family contains the species of greatest medicocriminal interest. When J. M. Aldrich applied Boettcher's concepts regarding distinctive insect male genitalia to specific determinations of the flesh flies (Aldrich, 1916), he for the first time enabled reliable species identification of this forensically significant family. *The Blowflies of North America* (Hall, 1948) followed Aldrich's lead, was the first monographic treatment of Nearctic Calliphoridae, and laid the foundation for modern work in North American medicocriminal entomology. Hall chose this family at least in part as a response to requests by the FBI and Washington Metropolitan Police regarding the utility of insect evidence (Hall and Townsend, 1977). This monograph remains the standard reference work on American blow flies, although changes in nomenclature have been proposed (Shewell, 1987; Whitworth, 2006) and considerable work has been accomplished on the immature forms (e.g., Knipling, 1936, 1939; Liu and Greenberg, 1989; Greenberg and Singh, 1995).

Early research on immatures proceeded by collecting living female flies and garnering their eggs or larvae, which were then divided into subsamples with some preserved and described, and others reared to adults. When adult specimens included males that could be identified by their genitalia, the circle was complete. More recent work has employed scanning electron microscopy (Liu and Greenberg, 1989; Wells et al., 1999; Sukontason et al., 2003), and as continually refined tools are employed in taxonomic research, it can be anticipated that more accurate identifications will be possible. In the future, molecular techniques employing DNA sequencing and polymerase chain reaction may facilitate identification of the immature stages of flies, including eggs and early instars, or confirm visual identifications based on morphological features (for example, see Gleeson and Sarre, 1997; Roehrdanz and Johnson, 1996; Sperling et al., 1994; Stevens and Wall, 1997; Wallman and Adams, 1997; Malgorn and Coquoz, 1999; Wells and Sperling, 1999, 2001; Wells et al., 2001; Harvey et al., 2003). Sophisticated statistical analyses have been used to facilitate understanding of the relationships between blow fly species (Stevens and Wall, 1996).

In addition, several species of blow flies, heretofore confined to the Old World, have recently become established in the Americas (Baumgartner and Greenberg, 1984; Richard and Ahrens, 1983; Wells et al., 1999). Of special interest is *Chrysomya rufifacies*, the invading hairy maggot blow fly, first noted in the Gulf Coast region (Martin et al., 1996), but which has established itself throughout the central United States (Figarola and Skoda, 1998; Shahid et al., 2000) and has been collected as far north as East Lansing, Michigan (R. W. Merritt, personal communication). Another exotic species of blow fly to recently establish itself in North America is *Chrysomya megacephala* (Greenberg, 1988; Tomberlin et al., 2001; Tenorio et al., 2003; Wells, 1991). Although this species has not achieved the same distribution status as *C. rufifacies* (Whitworth, 2006), its appearance in the United States should be noted by practicing medicocriminal entomologists. Attention has been paid to the potential effect of such exotic species on the indigenous blow fly fauna (Wells and Greenberg, 1992).

As mentioned earlier, the principal methodology used in medicocriminal entomology is application of the temperature-dependent development of insects, especially flies, in estimating a decedent's PMI. Other things being equal, insects (as poikilothermic animals) develop faster as temperatures increase over a thermal minimum threshold up to a lethal maximum (described by the well-known S-shaped biological curve) (Andrewartha and Birch, 1973; Davidson, 1944). Laboratory rearing of forensically important species at constant temperature (Kamal, 1958) permits calculation of the number of accumulated thermal units (such as degree-days or degree-hours) necessary for the insect to progress from the stage deposited on the decedent to the stage collected. Granted that the season is amenable to fly development, and that flies had access to the decedent, the time required for such development under appropriate field conditions constitutes the minimum PMI. Confounding factors include delayed oviposition because the decedent was protected from flies (for example, in an automobile trunk, inside a house, or wrapped in plastic) or any phenomenon that might alter the thermal regimen, such as "maggot mass" effects (Greenberg, 1991). Because retrospective and generally remote temperature data are used for most medicocriminal entomology analyses, there are many other factors that may complicate the analysis.

Although a wide variety of food substrates have been used for this sort of research, especially beef or pork liver (Hall, 1948) or small animals (e.g., Payne, 1965), the ultimate medicocriminal application is human tissue. Establishment of the Anthropological Research Facility (ARF) in Knoxville, Tennessee, permitted research on insect colonization and succession on human corpses (Figure I.6a and b) (Rodriguez and Bass, 1983) and validated the use of swine cadavers as surrogates in other areas where research on human remains is illegal. One objection voiced to ARF-generated data was theoretical "faunal enrichment" at that site, where "local populations of carrion insects are concentrated and abundant" (Catts and Goff, 1992). Comparisons of decomposition measured on replicate swine carcasses at the ARF with that at remote sites in the Knoxville area failed to substantiate differences in decay rates (Shahid et al., 2003; Schoenly et al., 2005). Similarly, research on oviposition habits of blow flies has been conducted to test the general assertion that these insects usually do not lay eggs after dark (Greenberg, 1990; Hall, 1948; Nuorteva, 1977; Tessmer et al., 1995). Initial results in Missouri confirm that blow fly egg-laying ceases during hours of darkness in rural sites with no artificial lighting (C. Hempel, personal communication). The effect of time since death on resultant attractiveness of a carcass to blow and flesh flies also has been evaluated by Hall and Doisy (1993).

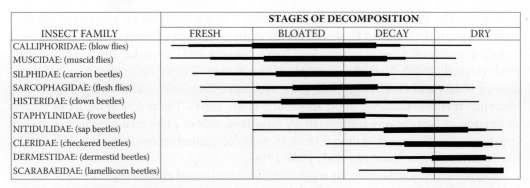

INSECT FAMILY	STAGES OF DECOMPOSITION			
	FRESH	BLOATED	DECAY	DRY
CALLIPHORIDAE: (blow flies)				
MUSCIDAE: (muscid flies)				
SILPHIDAE: (carrion beetles)				
SARCOPHAGIDAE: (flesh flies)				
HISTERIDAE: (clown beetles)				
STAPHYLINIDAE: (rove beetles)				
NITIDULIDAE: (sap beetles)				
CLERIDAE: (checkered beetles)				
DERMESTIDAE: (dermestid beetles)				
SCARABAEIDAE: (lamellicorn beetles)				

* Each stage of decomposition is given the same amount of space in this table.

——— Indicates a small number of individuals present.
▬▬▬ Indicates a moderate number of individuals present.
███ Indicates a large number of individuals present.

(a)

INSECT FAMILY	STAGES OF DECOMPOSITION			
	FRESH	BLOATED	DECAY	DRY
CALLIPHORIDAE: (blow flies)				
MUSCIDAE: (muscid flies)				
SILPHIDAE: (carrion beetles)				
SARCOPHAGIDAE: (flesh flies)				
STAPHYLINIDAE: (rove beetles)				
DERMESTIDAE: (dermestid beetles)				
SCARABAEIDAE: (lamellicorn beetles)				

* Each stage of decomposition is given the same amount of space in this table.

——— Indicates a small number of individuals present.
▬▬▬ Indicates a moderate number of individuals present.
███ Indicates a large number of individuals present.

(b)

Figure I.6 (a) Succession of adult arthropods on human cadavers in east Tennessee (during spring and summer). (b) Succession of arthropod larvae on human cadavers in east Tennessee (during spring and summer). (Adapted from Rodriguez and Bass, 1983. © ASTM. With permission.)

Current Perceptions

The expanding literature in medicocriminal entomology is still small by comparison to many other fields of science. A bibliography published in 1985 listed only 329 references through 1983 (Vincent et al., 1985). In a bibliography related to the broad topic of recovering evidence from outdoor settings, Hochrein (2004) included approximately 800 publications from 1834 to 2004 that were related to forensic entomology. A review of forensic entomology literature was also compiled by Tomberlin and Byrd (2003), listing 112 references from 1999 to 2003. Similarly, Huntington and Baxendale (2005) composed a bibliography of forensic entomology from the year 2004, yielding 79 papers. To supplement the one textbook on the subject (Smith, 1986), a guide was published to facilitate acquisition of entomological evidence by crime scene personnel (Catts and Haskell, 1990). Because of the increased use of medicocriminal entomology evidence in litigation, the subject was included in the 1995 iteration of *Forensic Sciences*, which contains an analysis of the substantive literature (Hall and Haskell, 1995), and many current editions of medicolegal texts include chapters on forensic entomology (see Anderson and Cervenka, 2002; Haskell et al., 1997; Haskell, 2006). There are multiple recent review articles on medicocriminal entomology (Catts and Goff, 1992; Keh, 1985; Amendt et al., 2000, 2004) along with several historical reviews (Benecke, 2001; Klotzbach et al., 2004). The practical utility of insects as forensic indicators was recognized by Nuorteva, whose published case histories were largely responsible for rekindling interest in medicocriminal entomology during the 1970s (see Nuorteva, 1974; Nuorteva et al., 1967, 1974).

A commonly voiced concern regarding medicocriminal entomology is that it seldom links any particular suspect or defendant with a crime, providing instead mainly inferential data on postmortem interval. Occasionally, however, the science is able to provide incriminating evidence in the former sense. In one case, the presence of chigger bites on a murder suspect was used to link him to the crime scene (Lord, 1990). Molecular technology may in the future permit analysis of arthropod gut contents, with ingested tissue such as blood (from blood-feeding arthropods) or semen (in maggot intestinal tracts) possibly linking a suspect to a particular decedent or locale (see Introna et al., 1999). Recent research has brought the possibility to extract gunshot residue (GSR) from the gut contents of feeding maggots (Roeterdink et al., 2004).

Conclusions

Medicocriminal entomology is one of three areas in the broad field of entomology that routinely is involved in forensic applications. Although litigation involving urban entomology and stored products entomology typically occurs, it is the area of medicocriminal entomology that is directly utilized by law enforcement agencies in death investigations, and is what most people think of when they hear the term *forensic entomology*. Insect evidence can be paramount in establishing an accurate postmortem interval for a decedent, as well as providing additional information to those investigators who are capable of deciphering the entomological clues.

Insect attraction to and interaction with human remains has been known, and even used, for centuries, yet medicocriminal entomology is still considered to be in its infancy. The scientific literature available on this topic, although constantly growing, remains small when compared to the areas of entomology that deal with agriculture or disease vectors. Likewise, the number of qualified practicing forensic entomologists capable of fully utilizing insect evidence is currently very small.

Medicocriminal entomology has reached an exciting stage in its evolution as testimony based on the interpretation of insect evidence is now routinely provided in court by expert witnesses. The establishment of the ABFE is an attempt to provide courts and others with quality assurance regarding the individuals presenting such testimony. The increased acceptance and recognition of medicocriminal entomology as a forensic discipline, coupled with the increased reliance of courts on biological evidence, shall continue to present increased opportunities for qualified forensic entomologists. The availability of an accurate PMI can be responsible for the overall direction of an investigation, and the interpretation of entomological evidence may eventually be the deciding factor in the determination of guilt or innocence in a court of law.

Acknowledgments

We thank Bob Pickering, deputy director for Collections and Education, Buffalo Bill Historical Center; Ephraim Cuevas, Cuevas Chemical Consulting; and the Photo Archives of the Denver Museum of Natural History for providing the illustrations of the west Mexican tomb artifacts and photomicrographs of insect puparia impressions on the artifacts.

References

Aldrich, J. M. 1916. *Sarcophaga and allies*. Lafayette, IN: Thomas Say Foundation.

Amendt, J., R. Krettek, C. Niess, R. Zehner, and H. Bratzke. 2000. Forensic entomology in Germany. *Foren. Sci. Int.* 113:309–14.

Amendt, J., R. Krettek, and R. Zehner. 2004. Forensic entomology. *Naturwissenschaften* 91:51–65.

Anderson, G. S. 1998. Wildlife forensics: Using forensic entomology in the investigation of illegally killed black bears. *Proc. Am. Acad. Foren. Sci.* 4:45.

Anderson, G. S., and V. J. Cervenka. 2002. Insects associated with the body: Their use and analyses. In Haglund, W. D., and M. H. Sorg, Eds., *Advances in forensic taphonomy: Method, theory, and archaeological perspectives*. Boca Raton, FL: CRC Press LLC, 173–200.

Anderson, G. S., and N. R. Huitson. 2004. Myiasis in pet animals in British Columbia: The potential of forensic entomology for determining duration of possible neglect. *Can. Vet. J.* 45:993–98.

Andrewartha, H. G., and L. C. Birch. 1973. The history of insect ecology. In Smith, R. F., T. E. Mittler, and C. N. Smith, Eds., *History of entomology*. Palo Alto, CA: Annual Reviews, 229–66.

Baumgartner, D. L., and B. Greenberg. 1984. The genus *Chrysomya* (Diptera: Calliphoridae) in the New World. *J. Med. Entomol.* 21:105–13.

Benecke, M. 2001. A brief history of forensic entomology. *Foren. Sci. Int.* 120:2–14.

Bergeret, M. 1855. Infanticide, momification du cadavre. Decouverte du cadavre d'un enfant nouveau-ne dans une dheminee ou il setait momifie. Determination de l'epoque de la naissance par la presence de nymphes et de larves d'insectes dans le cadavre et par l'etude de leurs metamorphoses. *Ann. Hyg. Legal Med.* 4:442–52.

Byrd, J. H. 1998. Temperature dependent development and computer modeling of insect growth: Its application to forensic entomology. PhD dissertation, Department of Entomology and Nematology, University of Florida, Gainesville.

Catts, E. P., and M. L. Goff. 1992. Forensic entomology in criminal investigations. *Ann. Rev. Entomol.* 37:253–72.

Catts, E. P., and N. H. Haskell, Eds. 1990. *Entomology and death: A procedural guide*. Clemson, SC: Joyce's Print Shop.

Davidson, J. 1944. On the relationship between temperature and rate of development of insects at constant temperature. *J. An. Ecol.* 13:26–38.

Easton, A. M., and K. V. G. Smith. 1970. The entomology of the cadaver. *Med. Sci. Law* 10:208–15.

Figarola, J. L. M., and S. R. Skoda. 1998. *Chrysomya rufifacies* (Macquart) (Diptera: Calliphoridae) in Nebraska. *J. Entomol. Sci.* 33:319–21.

Gleeson, D. M., and S. Sarre. 1997. Mitochondrial DNA variability and geographic origin of the sheep blowfly, *Lucilia cuprina* (Diptera: Calliphoridae) in New Zealand. *Bull. Entomol. Res.* 87:265–72.

Goff, M. L., W. A. Brown, K. A. Hewadikaram, and A. I. Omori. 1991. Effect of heroin in decomposing tissues on the development rate of *Boettcherisca peregrina* (Diptera: Sarcophagidae) and implications to the estimation of postmortem intervals using arthropod developmental patterns. *J. Foren. Sci.* 36:537–42.

Goff, M. L., M. Early, C. B. Odom, and K. Tullis. 1986. A preliminary checklist of arthropods associated with exposed carrion in the Hawaiian Islands. *Proc. Hawaiian Entomol. Soc.* 26:53–57.

Goff, M. L., and M. M. Flynn. 1991. Determination of postmortem interval by arthropod succession: A case study from the Hawaiian Islands. *J. Foren. Sci.* 36:607–14.

Goff, M. L., A. I. Omori, and J. R. Goodbrod. 1989. Effect of cocaine in tissues on the development rate of *Boettcherisca peregrina* (Diptera: Sarcophagidae). *J. Med. Entomol.* 26:91–93.

Goff, M. L., A. I. Omori, and K. Gunatiklake. 1988. Estimation of postmortem interval by arthropod succession: Three case studies from the Hawaiian Islands. *Am. J. Foren. Med. Pathol.* 9:220–25.

Graner, J. L. 1997. S. K. Livingston and the maggot therapy of wounds. *Mil. Med.* 62:296–300.

Greenberg, B. 1988. *Chrysomya megacephala* (F.) (Diptera: Calliphoridae) collected in North America and notes on *Chrysomya* species present in the New World. *J. Med. Entomol.* 25:199–200.

Greenberg, B. 1990. Blow fly nocturnal oviposition behavior. *J. Med. Entomol.* 27:807–10.

Greenberg, B. 1991. Flies as forensic indicators. *J. Med. Entomol.* 28:565–77.

Greenberg, B., and D. Singh. 1995. Species identification of calliphorid (Diptera) eggs. *J. Med. Entomol.* 32:21–26.

Hall, D. G. 1948. The *blowflies of North America*. Baltimore: Thomas Say Foundation.

Hall, D. G., and R. D. Hall. 1986. Entomologists, taxonomists, public opinion, and professionalism. *Bull. Entomol. Soc. Am.* 32:8–21.

Hall, R. D. 1990. Medicocriminal entomology. In Catts, E. P., and N. H. Haskell, Eds., *Entomology and death: A procedural guide*. Clemson, SC: Joyce's Print Shop, 1–8.

Hall, R. D., and K. E. Doisy. 1993. Length of time after death: Effect on attraction and oviposition or larviposition of midsummer blow flies (Diptera: Calliphoridae) and flesh flies (Diptera: Sarcophagidae) of medicolegal importance in Missouri. *Ann. Entomol. Soc. Am.* 86:589–93.

Hall, R. D., and N. H. Haskell. 1995. Forensic entomology: Applications in medicolegal investigations. In Wecht, C., Ed., *Forensic sciences*. New York: Matthew Bender.

Hall, R. D., and L. H. Townsend, Jr. 1977. *The blow flies of Virginia (Diptera: Calliphoridae)*. The Insects of Virginia, No. 11. Virginia Polytechnic Institute and State University Research Bulletin, No. 123.

Harvey, M. L., M. W. Mansell, M. H. Villet, and I. R. Dadour. 2003. Molecular identification of some forensically important blowflies of southern Africa and Australia. *Med. Vet. Entomol.* 17:363–69.

Haskell, N. H. 2006. Time of death and changes after death: Forensic entomology. In Spitz, W. U., and D. J. Spitz, Eds., *Spitz and Fisher's medicolegal investigation of death: Guidelines for the application of pathology to crime investigation*. 4th ed. Springfield, IL: Charles C. Thomas, 149–73.

Haskell, N. H., R. D. Hall, V. J. Cervenka, and M. A. Clark. 1997. On the body: Insects' life stage presence, their postmortem artifacts. In Haglund, W. D., and M. H. Sorg, Eds., *Forensic taphonomy: The postmortem fate of human remains*. Boca Raton, FL: CRC Press LLC, 415–48.

Hochrein, M. J. 2004. *A bibliography related to crime scene interpretation with emphases in geotaphonomic and forensic archaeological field techniques*. 11th ed. Washington, DC: U.S. Department of Justice, FBI, FBI Print Shop.

Huntington, T. E., and F. P. Baxendale. 2005. A year in review: Forensic entomology research published in 2004. In *Third Annual North American Forensic Entomology Conference*, Orlando, FL, July 20–22.

Introna, F., J. D. Wells, D. V. Giancarlo, C. P. Campobasso, and F. A. H. Sperling. 1999. Human and other animal mtDNA analysis from maggots. Presented at Annual Meeting of the American Academy of Forensic Sciences, Orlando, FL, Paper G75.

Kamal, A. S. 1958. Comparative study of thirteen species of sarcosaprophagous Calliphoridae and Sarcophagidae (Diptera). I. Bionomics. *Ann. Entomol. Soc. Am.* 51:261–70.

Keh, B. 1985. Scope and applications of forensic entomology. *Ann. Rev. Entomol.* 30:137–54.

Klotzbach, H., R. Krettek, H. Bratzke, K. Püschel, R. Zehner, and J. Amendt. 2004. The history of forensic entomology in German-speaking countries. *Foren. Sci. Int.* 144:259–63.

Knipling, E. F. 1936. Some specific taxonomic characters of common Lucilia larvae–Calliphorinae–Diptera. *Iowa State Coll. J. Sci.* 10:275–93.

Knipling, E. F. 1939. A key for blowfly larvae concerned in wound and cutaneous myiasis. *Ann. Entomol. Soc. Am.* 32:376–83.

Leclercq, M. 1969. Entomological parasitology: The relations between entomology and the medical sciences. In *Modern trends in physiological sciences*. Monograph in Pure and Applied Biology. Oxford: Pergamon Press, 128–142.

Liu, D., and B. Greenberg. 1989. Immature stages of some flies of forensic importance. *Ann. Entomol. Soc. Am.* 82:80–93.

Lord, W. D. 1990. Case histories of the use of insects in investigations. In Catts, E. P., and N. H. Haskell, Eds., *Entomology and death: A procedural guide*. Clemson, SC: Joyce's Print Shop, 9–37.

Lord, W. D., and J. R. Stevenson. 1986. *Directory of forensic entomologists*. 2nd ed. Washington, DC: Defense Pest Management Information Analysis Center, Walter Reed Army Medical Center.

Malgorn, Y., and R. Coquoz. 1999. DNA typing for identification of some species of Calliphoridae: An interest in forensic entomology. *Foren. Sci. Int.* 102:111–19.

Martin, C. S., C. E. Carlton, and C. L. Meek. 1996. New distribution records for the hairy maggot blow fly *Chrysomya rufifacies*. *Southwest. Entomol.* 21:477–78.

McKnight, B. E. 1981. *The washing away of wrongs: Forensic medicine in thirteenth century China*. Ann Arbor: University of Michigan.

Megnin, J. P. 1894. *La faune des cadavres: Application de l'entomologie a la medicine legale*. Encyclopedie Scientifique des Aide-Memoires. Paris: Masson et Gauthier-Villaars.

Nuorteva, P. 1974. Age determination of a blood stain in a decaying shirt by entomological means. *Foren. Sci.* 3:89–94.

Nuorteva, P. 1977. Sarcosaprophagous insects as forensic indicators. In Tedeschi, C. G., W. G. Eckert, and L. G. Tedeschi, Eds., *Forensic medicine: A study in trauma and environmental hazards*. Philadelphia: Saunders.

Nuorteva, P., M. Isokoski, and K. Laiho. 1967. Studies on the possibilities of using blow flies (Dipt.) as medicolegal indicators in Finland. I. Report of four indoor cases from the city of Helsinki. *Ann. Entomol. Soc. Finl.* 33:217–225.

Nuorteva, P., H. Schumann, M. Isokoski, and K. Laiho. 1974. Studies on the possibilities of using blowflies (Dipt., Calliphoridae) as medicolegal indicators in Finland. II. Four cases where species identification was performed from larvae. *Ann. Entomol. Soc. Finl.* 40:70–74.

Payne, J. A. 1965. A summer carrion study of the baby pig *Sus scrofa* Linnaeus. *Ecology* 46:592–602.

Perkins, J. H. 1982. *Insects, experts, and the insecticide crisis*. New York: Plenum Press.

Pickering, R. B., J. Ramos, N. H. Haskell, and R. Hall. 1998. El significado de las cubiertas de crisalidas de insectos que aparecen en las figurillas del occidente de Mexico. In *El occidente de Mexico: Arqueologia, historia y medio ambiente, Perspectivas regionales*. Guadalajara, Mexico: Actas del IV Coloquio de Occidentalistas. Universidad de Guadalajara, Instituto Frances de Investigacion Ceintifica para el Desarrollo en Cooperacion.

Richard, R. D., and E. H. Ahrens. 1983. New distribution record for the recently introduced blow fly *Chrysomya rufifacies* (Macquart) in North America. *Southwest. Entomol.* 8:216–18.

Rodriguez, W. C., and W. M. Bass. 1983. Insect activity and its relationship to decay rates of human cadavers in east Tennessee. *J. Foren. Sci.* 28:423–32.

Roehrdanz, R. L., and D. A. Johnson. 1996. Mitochondrial NDA restriction site map of *Cochliomyia macellaria* (Diptera: Calliphoridae). *J. Med. Entomol.* 33:863–65.

Roeterdink, E. M., I. R. Dadour, and R. J. Watling. 2004. Extraction of gunshot residues from the larvae of the forensically important blowfly *Calliphora dubia* (Macquart) (Diptera: Calliphoridae). *Int. J. Leg. Med.* 118:63–70.

Schoenly, K., M. L. Goff, and M. Early. 1992. A BASIC algorithm for calculating the postmortem interval from arthropod successional data. *J. Foren. Sci.* 37:808–23.

Schoenly, K., and W. Reid. 1987. Dynamics of heterotrophic succession in carrion arthropod assemblages: Discrete seres or a continuum of change? *Oecologia* 73:192–202.

Schoenly, K., S. A. Shahid, N. H. Haskell, and R. D. Hall. 2005. Does carcass enrichment alter community structure of predaceous and parasitic arthropods? A second test of the arthropod saturation hypothesis at the Anthropology Research Facility in Knoxville, Tennessee. *J. Foren. Sci.* 50:134–42.

Shahid, S. A., R. D. Hall, N. H. Haskell, and R. W. Merritt. 2000. *Chrysomya rufifacies* (Macquart) (Diptera: Calliphoridae) established in the vicinity of Knoxville, Tennessee, USA. *J. Foren. Sci.* 45:896–97.

Shahid, S. A., K. Shoenly, N. H. Haskell, R. D. Hall, and W. Zhang. 2003. Carcass enrichment does not alter decay rates or arthropod community structure: A test of the arthropod saturation hypothesis at the Anthropology Research Facility in Knoxville, Tennessee. *J. Med. Entomol.* 40:559–69.

Shewell, G. E. 1987. Calliphoridae. In McAlpine, J. F., et al., Eds., *Manual of Nearctic Diptera*. Vol. 2, monograph 28. Ottawa, Canada: Research Branch, Agriculture Canada.

Smith, K. G. V. 1986. *A manual of forensic entomology*. London: British Museum (Natural History).

Sperling, F. A. H., G. S. Anderson, and D. A. Hickey. 1994. A DNA-based approach to the identification of insect species used for postmortem interval estimation. *J. Foren. Sci.* 39:418–27.

Stevens, J., and R. Wall. 1996. Classification of the genus *Lucilia* (Diptera: Calliphoridae): A preliminary parsimony analysis. *J. Nat. Hist.* 30:1087–94.

Stevens, J., and R. Wall. 1997. Genetic variation in populations of the blowflies, *Lucilia cuprina* and *Lucilia sericata* (Diptera: Calliphoridae). Random amplified polymorphic DNA analysis and mitochondrial DNA sequences. *Biochem. System. Ecol.* 25:81–97.

Sukontason, K. L., K. Sukontason, S. Lertthamnongtham, B. Kuntalue, N. Thijuk, R. C. Vogtsberger, and J. K. Olson. 2003. Surface ultrastructure of *Chrysomya rufifacies* (Macquart) larvae (Diptera: Calliphoridae). *J. Med. Entomol.* 40:259–67.

Tenorio, F. M., J. K. Olson, and C. J. Coates. 2003. Decomposition studies, with a catalog and description of forensically important blow flies (Diptera: Calliphoridae) in central Texas. *Southwes. Entomol.* 28:37–45.

Tessmer, J. W., C. L. Meek, and V. I. Wright. 1995. Circadian patterns of oviposition by necrophilous flies (Diptera: Calliphoridae) in southern Louisiana. *Southwest. Entomol.* 20:439–45.

Thomas, G. D., and S. R. Skoda, Eds. 1993. Rural flies in the urban environment? University of Nebraska Research Bulletin 317. *North Cent. Reg. Res. Pub.* 335:97.

Tomberlin, J. K., and J. Byrd. 2003. State of forensic entomology research in the U.S. and abroad. In *North American Forensic Entomology Conference*, Las Vegas, NV, August 7–9.

Tomberlin, J. K., W. K. Reeves, and D. C. Sheppard. 2001. First record of *Chrysomya megacephala* (Diptera: Calliphoridae) in Georgia, U.S.A. *Florida Entomol.* 84:300–1.

Vincent, C. D., K. McE. Kevan, M. Leclercq, and C. L. Meek. 1985. A bibliography of forensic entomology. *J. Med. Entomol.* 22:212–19.

Wallman, J. F., and M. Adams. 1997. Molecular systematics of Australian carrion-breeding blowflies of the genus *Calliphora* (Diptera: Calliphoridae). *Aust. J. Zool.* 45:337–56.

Wells, J. D. 1991. *Chrysomya megacephala* (Diptera: Calliphoridae) has reached the continental United States: Review of its biology, pest status, and spread around the world. *J. Med. Entomol.* 28:471–73.

Wells, J. D., J. H. Byrd, and Tarek I. Tantawi. 1999. Key to the third instar Chrysomyinae (Diptera: Calliphoridae) from carrion in the continental USA. *J. Med. Entomol.* 36:638–641.

Wells, J. D., and B. Greenberg. 1992. Interaction between *Chrysomya rufifacies* and *Cochliomyia macellaria* (Diptera: Calliphoridae): The possible consequences of an invasion. *Bull. Entomol. Res.* 82:33–137.

Wells, J. D., T. Pape, and F. A. H. Sperling. 2001. DNA-based identification and molecular systematics of forensically important Sarcophagidae (Diptera). *J. Foren. Sci.* 46:1098–102.

Wells, J. D., and F. A. H. Sperling. 1999. Molecular phylogeny of *Chrysomya albiceps* and *C. rufifacies* (Diptera: Calliphoridae). *J. Med. Entomol.* 36:222–26.

Wells, J. D., and F. A. H. Sperling. 2001. DNA-based identification of forensically important Chrysomyinae (Diptera: Calliphoridae). *Forensic Sci. Int.* 3065:110–115.

Whitworth, T. 2006. Keys to the genera and species of blow flies (Diptera: Calliphoridae) of America north of Mexico. *Proc. Entomol. Soc. Wash.* 108:689–725.

General Entomology and Insect Biology

1

JAMES L. CASTNER

Contents

Introduction

Arthropods are invertebrate animals having an exoskeleton and jointed legs, which includes the insects, arachnids, centipedes, millipedes, and crustaceans. One of the largest groups of arthropods, the insects, differs from the rest by having six legs and a body divided into three distinct regions. The study of insects is termed entomology. Following best practices in crime scene protocol, the forensic science team should have a consulting entomologist (a biologist who studies insects) with forensic specialization as a member. Such a specialist should be consulted for the accurate identification and interpretation of insect evidence. Although an entomologist should be consulted, a degree of familiarity with basic insect anatomy, development, and behavior can be extremely useful to all death scene investigators in order to understand the ecological roles insects play and to fully appreciate their forensic value in investigations. For example, the pupal shells of flies can easily be overlooked or mistaken for rodent droppings at a crime scene. Postmortem feeding by fire ants or cockroaches on a body may damage the tissue and produce artifacts similar in appearance to chemical scarring, abrasions, contusions, or antemortem trauma to the untrained eye. Knowledge of general entomology and basic insect biology is essential for the accurate interpretation of insect evidence. The ability to recognize which arthropods need to be collected as forensic evidence at a death scene is an invaluable skill and shall be dealt with in Chapter 2.

Insects are the most numerous and diverse organisms on the planet (Figure 1.1). While less than a million species have been described and named, research indicates that as many as 3 to 30 million may actually exist. They are found in almost all terrestrial habitats and in most aquatic ones as well, except for salt water. As a group, insects have evolved the presence of wings, a feature that distinguishes them from all other invertebrates. This enables them to travel considerable distances when foraging for food or attempting to locate a suitable habitat for laying their eggs. This is an extremely important factor in species of forensic importance that must quickly locate and utilize temporary resources such as carrion, including human remains.

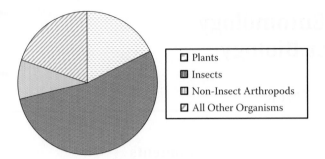

Figure 1.1 Insects comprise more than half of all the living species on Earth, exceeding all other groups of organisms combined. (Figure courtesy of Dr. Jason H. Byrd.)

Two major groups of insects are predictably attracted to cadavers and provide the majority of information derived from forensic investigations. They are the beetles (order Coleoptera) and the flies (order Diptera). Both groups are diverse, although the beetles have far more known species than any animal group, insect or otherwise. There are approximately 300,000 described species in the world, of which 30,000 species occur in North America. Estimates of total beetle species in the world are at least 750,000, and they currently compose over one-third of all known insects. Flies are much less numerous, with only 86,000 described species worldwide, of which 16,300 occur in North America.

Many other types of arthropods are found in association with bodies, but these are typically opportunistic feeders taking advantage of the circumstances. They are made up of both scavengers feeding on decaying material and predators feeding on the species that have been attracted to and colonized the carrion. The beetles and flies have distinctive external structures and develop in significantly different ways. These body features and developmental cycles shall be discussed in the remainder of this chapter.

External Anatomy

When we look at insects, we see an almost bewildering array of shapes, sizes, and colors. However, all insects have evolved from a common ancestral form and retain certain diagnostic features about their external anatomy. Scientists believe that the insect predecessor was elongate, roughly cylindrical, and segmented with paired appendages on each segment. Through time, certain segments grouped together into functional regions on the insect's body. Three such regions have resulted. They are represented on the insect body by the head, thorax, and abdomen.

The body wall of an insect is called its exoskeleton, and serves two functions. One is providing points of attachment for the muscles, and the other is protection by means of the hard outer layer of the exoskeleton, called the cuticle. This durable cuticle results in the bodies of insects remaining in the environment for extended periods of time. In this way, an insect can serve as forensic evidence long after the organism itself is dead. Fragments and parts of dead insects have been used successfully to link suspects to a crime scene or victim in numerous cases.

The insect skin is not one continuous hard shell, but rather is composed of a number of hardened plates. These plates (or sclerites) are separated from one another by seams or sutures, and by larger membranous areas, such as between body segments (Figure 1.2a–c).

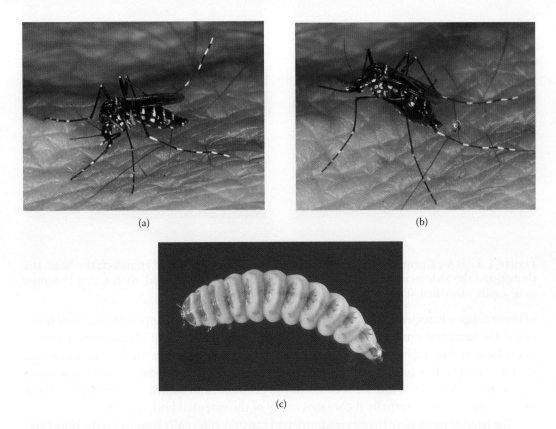

(a)

(b)

(c)

Figure 1.2 (a) The bodies of insects must have the flexibility to expand. The membranous portion of this mosquito's abdomen is visible as a narrow region between the white spots on the top and bottom portions of the body. (b) The same mosquito after finishing its blood meal shows a greatly distended abdomen where the membranous area has expanded. (c) Hardened plates or sclerites are clearly visible as small darker areas with various shapes on the back of this carrion beetle larva. (Photos courtesy of Dr. James L. Castner.)

The degree of hardness of an insect's body varies. Extremely hard areas of the insect, such as the jaws, or the wing covers in most beetles, are said to be heavily sclerotized. The more lightly sclerotized or membranous areas of the body permit the insect a degree of flexibility, greater range of movement, and to expand while feeding in some species.

A species' biology and behavior will greatly affect its morphology, including whether it is hard or soft bodied. The exoskeleton of a scarab beetle can be incredibly hard, making its body difficult to penetrate, affording excellent protection from other insect predators. Yet insects that live in protected situations (such as internal parasites), or whose immature forms receive adult care (such as with bees, wasps, and ants), may be completely soft and vulnerable. Even the body of the scarab beetle is soft and unprotected when it is an immature grub underground. Thus, a great difference may be observed from one life stage of an insect to the next.

Body Regions

As previously mentioned, through the course of evolution the segments of insects have fused to form three body regions: the head, thorax, and abdomen (Figure 1.3a and b). Each

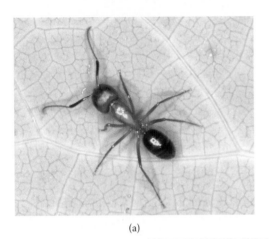

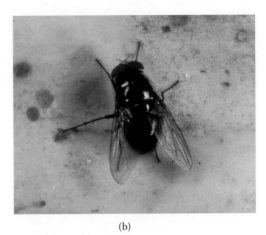

(a)   (b)

Figure 1.3 (a) A carpenter ant clearly shows the three body regions of an insect: the head, the thorax, and the abdomen. (b) A dorsal view of a blow fly allows the head, thorax, and abdomen to be easily identified. (Photos courtesy of Dr. James L. Castner.)

of these regions has specialized external and internal structures that perform certain functions. The head is the main area of sensory perception and the point of ingestion. The thorax is located directly behind the head and contains the segments with the legs and wings on adult insects. It is primarily responsible for locomotion. The abdomen is the hindmost region and is behind the thorax. It contains the genitalia, and sometimes other specialized external structures. Internally it contains many of the essential body systems.

The head of most adult insects is a hardened capsule. Internally it contains the brain and a framework of structural supports for the musculature of the mandibles and mouthparts. Its most noticeable external features are the eyes, antennae, and mouthparts. All of these are important characteristics used in the identification of insects, and sometimes in the determination of sex.

Insects have evolved various mechanisms for processing the light information from the environment in which they live. In predatory insects, as in the beetles and adult flies that are associated with carrion, vision may be extremely acute. In other species that spend the majority of their life in the absence of light (such as underground beetle grubs) or that live immersed in their food source (such as the maggots of carrion flies), vision may be totally lacking. One of the most prominent features of the insect head is a pair of multifaceted compound eyes. In predators such as mantids or highly agile flying insects like dragonflies and horseflies, the compound eyes may make up the majority of the head (Figure 1.4a and b). The compound eye itself is made up of facets, which may number from only a few to as many as several thousand. The shape of the compound eyes, their location on the head, and whether or not they touch are all characters sometimes used in identification.

There may also be zero, one, two, or three simple eyes found on the head of adult insects. These ocelli are composed merely of a single facet and can only detect changes in light intensity. Externally they appear like small jewels embedded in the surface of the insect head, often amber in color. Simple eyes or photoreceptors are found on the head of certain larval insects, as well as on the adults of more primitive insect groups.

The antennae or "feelers" are often the most noticeable appendages on the insect head. There is a great variety of shapes and sizes, from the long thread-like antennae of katydids

(a)   (b)

Figure 1.4 (a) Compound eyes may make up the majority of the head in flying insects, such as this horsefly. The individual facets are barely discernible at this level of magnification. (b) The large red compound eyes of this oriental latrine fly, which possess unequal facets, are a helpful character in field identification. (Photos courtesy of Dr. James L. Castner.)

that may be twice the body length, to the tiny bristle-like antennae of dragonflies that could be easily overlooked. The type of antennae is one of the key taxonomic features that enable us to identify insects down to the family level.

Insect antennae are sensory organs and they are covered with chemical receptors that allow the insect to evaluate its environment. While all antennae serve the same basic function, a diversity of physical appearances and shapes have evolved throughout the insect world. For example, the antennae of beetles are usually long and obvious, composed of ten to twenty individual segments (Figure 1.5). In the carrion beetles (family Silphidae), the tip of the antenna is enlarged into a broadened club. This antenna type is called clavate (Figure 1.6). A variation on this clubbed antenna is also found in the scarab beetles (family Scarabaeidae), another group often collected at remains. The terminal antenna segments that form the "club" of scarab beetles are enlarged parallel plates that stick out perpendicular to the rest of the antenna. This antenna type is called lamellate (Figure 1.7).

Flies have very different antennae from beetles. Three of the most important groups of flies (blow flies, flesh flies, and house flies) attracted to human remains have an antenna type

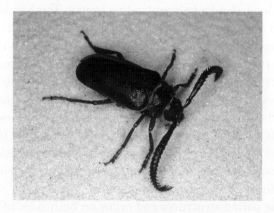

Figure 1.5 Longhorned woodboring beetles (family Cerambycidae) get their common name from their extremely long antennae. (Photo courtesy of Dr. James L. Castner.)

Figure 1.6 The expanded terminal segments of a carrion beetle's antennae form a club. This type of antenna is termed clavate. (Photo courtesy of Dr. James L. Castner.)

Figure 1.7 The lamellate antennae of this dung beetle have flattened parallel segments at the tip. These segments are shown widely spread in this photo, but they can also be held tightly together. (Photo courtesy of Dr. James L. Castner.)

called aristate (Figure 1.8). This antenna is composed of three large, wide segments, the outermost of which has a long, slender hair protruding from it. This hair is called the arista. The antennae of these flies are much smaller in proportion than the beetle antennae discussed. They are difficult to see with the naked eye and must be viewed with a microscope. The twelve most common types of insect antennae are listed and described in the appendix.

The other significant features of the head are the mouthparts. In some species that do not feed as adults, they may be vestigial or absent. However, most adult insects will have distinctive mouthparts whose shape is indicative of the type of food they consume. The chewing mouthtype is the most primitive and the most commonly observed in the insect world (Figure 1.9). It is found on nearly all the adult beetles of forensic importance, as well as the majority of larval insects associated with carrion. The chewing mouthtype is also called mandibulate, due to the presence of mandibles.

The mandibles are paired structures and occur on both sides of the mouth and meet in the middle. They are responsible for tearing or biting off chunks of food and grinding or chewing it to a consistency that the insect can ingest. The darkest portions of the

Figure 1.8 Aristate antennae are found on many flies of forensic importance. The name derives from a hair called an arista, which is found on the last antennal segment. The arista shown above is lined with many smaller hairs, and therefore termed plumose. (Photo courtesy of Dr. James L. Castner.)

Figure 1.9 The large mandibles of this staphylinid beetle are indicative of the chewing insect mouthtype. (Photo courtesy of Dr. James L. Castner.)

mandibles are the hardest or most heavily sclerotized. Looking at a mandible under the microscope will show areas that are sharp for cutting and others that are blunt for grinding. These areas are comparable to the incisors and molars of humans.

Many groups of insects have chewing mouthparts similar to those described above. In some groups, however, modifications of the general design have occurred. Fly larvae (commonly called maggots) have paired, hardened mouth hooks that tear and cut into the carrion upon which they feed (Figure 1.10a and b). These mouth hooks often have a distinctive shape and can be used in the identification of fly larvae to species. Adaptations that allowed one insect group to exploit a food source not used by the majority of other groups would have decreased competition and given a survival advantage. Thus, we see a variety of mouthtypes that enable insects to feed on different kinds of food in vastly different ways.

Many insects feed on liquid rather than solid food. The nourishing liquid may take the form of nectar from flowers, or plant juices withdrawn directly from plugging in to the plant's vascular system, or even blood that is sucked out of an animal's vascular system. A piercing-sucking mouthtype is often responsible for the removal of a liquid food source, and the mouthparts themselves are commonly referred to as a beak (Figure 1.11). These beaks

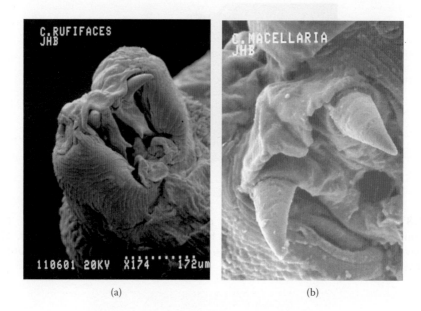

(a)   (b)

Figure 1.10 (a) The tiny mouth hooks of this hairy maggot blow fly larva provide it with the tools necessary for macerating flesh. This scanning electron micrograph shows the head of the maggot 80× life size. (b) The paired mouth hooks of maggots are distinct and can be used with other larval characters to make a species identification. This scanning electron micrograph shows the mouth hooks at 350× life size. (Photos courtesy of Dr. Jason H. Byrd.)

Figure 1.11 The wheel bug is a common predatory insect. Beneath its head is a stout, curving, three-segmented beak that exemplifies a piercing-sucking mouthtype. (Photo courtesy of Dr. James L. Castner.)

Figure 1.12 (a) The sponging mouthparts of a fly are usually kept retracted and close to the head unless it is actually feeding. Here, the mouthparts of an oriental latrine fly are extended. (b) A bronzebottle fly extends its proboscis to sponge up liquid on the surface of a rotting fruit. (Photos courtesy of Dr. James L. Castner.)

can be formidable, as evidenced by the predatory assassin bug that feeds on other insects. Such bugs are sometimes found preying on the insects that have colonized a body. An intermediate-sized beak that most of us are all too familiar with is found on mosquitoes.

The majority of species of flies attracted to human remains have a sponging mouthtype (Figure 1.12a and b). Instead of a piercing beak, the mouthparts have evolved into a mechanism that permits the noninvasive sucking of liquids. In most sponging mouthtypes, the tip consists of a fleshy lobe or lobes that are crossed with grooves. Liquid is drawn into these grooves via capillary action and then on up into a food channel. The secretion of saliva by the fly onto the food material helps to break it down and liquefy it so that it can be sucked up easily.

Other mouthtypes exist and are often adapted for special diets. Variations on the sponging mouthtype are found in a number of fly species that suck blood. Butterflies and moths have a tongue that is a long, coiled tube that functions like a straw (Figure 1.13). The

Figure 1.13 The tongue of butterflies and moths can be uncoiled and extended to reach the nectar in flowers. The coiled mouthparts of this heliconian butterfly are covered with pollen. (Photo courtesy of Dr. James L. Castner.)

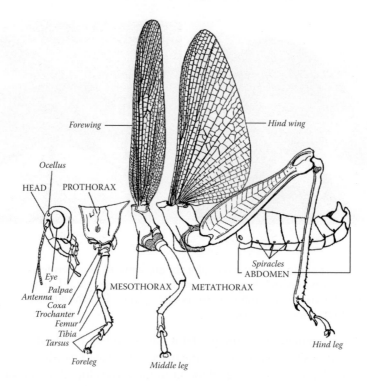

Figure 1.14 This diagrammatic illustration shows the prothorax, mesothorax, and metathroax of an insect. (Figure courtesy of the U.S. Department of Agriculture.)

honeybee has an unusual combination of chewing-lapping mouthtypes. It can suck nectar through its tongue, but also has mandibles that are used for molding wax and chewing its way out of its pupal case. The larval stages of lacewings and antlions have long, grooved, sickle-like jaws. They impale their victims and feed on the body juices that flow down the grooved mandibles.

Directly behind the head, we encounter a region composed of three fused segments and called the thorax. The obvious external structures (legs and wings) found associated with the thorax of most adult insects make it clear that it is the center of locomotion. The individual segments making up the thorax, from front to back, have been named the prothorax, mesothorax, and metathorax. The prefixes *pro-*, *meso-*, and *meta-* are commonly used when discussing thoracic structures and appendages to specifically identify the segment on which they are located (Figure 1.14). Therefore, a leg on the second thoracic segment would be a mesothoracic leg. A wing on the third and last thoracic segment would be a metathoracic wing.

The thoracic region is usually heavily sclerotized to provide support and bracing for the movement of the legs and wings. The top portion of each thoracic segment is called a notum. Using the same convention as earlier, the pronotum is often prominent and conspicuous, and in insects like beetles and cockroaches it may cover the entire top part of the thorax.

Most adult insects have six legs, which is why the current class Insecta is sometimes listed by its older name of class Hexapoda. One pair of legs originates from each thoracic segment. The typical adult insect leg has five major parts or segments. The femur and tibia are usually the longest and most evident portions. The terminal portion of the leg is the tarsus, which may be composed of from one to five segments. The last tarsal segment often has a tarsal claw.

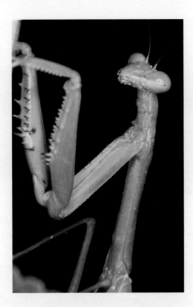

Figure 1.15 The predatory praying mantis catches its prey by grasping. The raptorial front legs are covered with sharp spines to make such captures easier. (Photo courtesy of Dr. James L. Castner.)

The legs of an insect living underground that contends with moving through a heavy medium such as soil or sand will show morphological adaptations extremely different from species that run on the soil surface. Insects living in an aquatic environment will exhibit anatomical structures that facilitate living in or under water. While the insect body as a whole will reflect modifications for living within a certain physical habitat, some of the most radical and obvious morphological changes have occurred to the legs.

Raptorial legs are characterized by sharp teeth or spines that impale and cling to prey organisms. The praying mantis is probably the most commonly recognized insect with raptorial legs (Figure 1.15), but variations on this leg type are seen throughout the insect world. Many hunting species have evolved legs for grasping prey, including assassin bugs, ambush bugs, diving beetles, and dragonflies. Life underground or beneath the leaf litter is termed fossorial, as is the leg type evolved for locomotion and movement underground. The front or prothoracic legs are usually the most affected. Shoveling through the soil requires wide, flattened structures for scooping and moving earth (Figure 1.16). Some of the larger beetles found with human remains, such as species of sexton and scarab beetles, have exactly these types of adaptations on their front legs. Other insect leg types include cursorial (evidenced by cockroaches and used for running), saltatorial (thick, muscular, and used for jumping, as in grasshoppers), and natatorial (flattened and used for swimming) (Figure 1.17).

The final structures associated with the thorax that we shall discuss are the wings. An important feature in the success of insects as a group, functional wings are almost always found only on adult insects. However, not all insects have wings. Some of the most primitive groups are apterous, or wingless. Other more advanced groups of insects have "lost" their wings evolutionarily as a survival adaptation to a specialized way of life. For example, both the fleas and the sucking lice are apterous, probably because wings would be a hindrance to their movements among the fur and between the hairs of their hosts. Most winged insects have two pairs of wings. Some only have one pair of wings (flies), while in

Figure 1.16 The front leg of a mole cricket possesses large digging claws well adapted for a subterranean existence. (Photo courtesy of Dr. James L. Castner.)

Figure 1.17 The hind leg of this aquatic beetle is covered with long, stiff hairs. These increase its surface area and allow the leg to function as an oar in propelling the beetle through the water. (Photo courtesy of Dr. James L. Castner.)

others winged individuals serve a reproductive purpose and are produced only at certain times of the year (ants and termites).

Dark thickened lines appear to radiate out and crisscross throughout the wing (Figure 1.18a and 8b). These wing veins are very important characters for identification in some groups. In other groups, venation is greatly reduced or almost nonexistent. Veins connect to the insect body and circulatory system. The movement of blood through these veins is essential in expanding the wings to their full size after an insect has shed its skin for the last time.

Adult flies are unique in having a single pair of wings. The hindwings are absent, and each has been replaced by a small, knob-like structure called a haltere (Figure 1.19). In beetles, the forewings have undergone extreme modification into hard, shell-like covers that protect the membranous hindwings that remain folded beneath them (except during flight). These hardened forewings are called elytra, and their color, shape, and texture can be very important in species identification (Figure 1.20).

The abdomen is the third and final region and forms the posterior portion of the insect's body. It contains organs whose functions deal mainly with reproduction, digestion, circulation, and respiration. There are typically eleven abdominal segments, although a reduction in number has occurred in some groups. Each segment appears to be divided

(a)   (b)

Figure 1.18 (a) The wings of butterflies are usually covered with scales that obscure the venation. This clear-winged butterfly shows the underlying wing veins. (b) Wing veins are easily visible on the wings of a dragonfly. This reticulate venation strengthens the wings in these active flyers. (Photos courtesy of Dr. Jason H. Byrd.)

Figure 1.19 This papaya fruit fly is positioned so that its left haltere is clearly visible as a small yellow knob between the base of the wing and the abdomen. (Photo courtesy of Dr. James L. Castner.)

Figure 1.20 A scarab beetle shows the hard, brown wing covers called elytra, which normally conceal the larger membranous rear wings used for flying. (Photo courtesy of Dr. James L. Castner.)

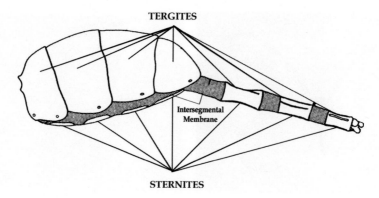

TERGITES

Intersegmental Membrane

STERNITES

Figure 1.21 The flexible intersegmental membrane between the insect's sclerites allows for flexibility and movement. (Illustration courtesy of Contemporary Publishing Company of Raleigh, Inc. With permission.)

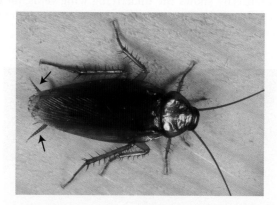

Figure 1.22 A pair of segmented cerci protrude from the tip of the abdomen of this American cockroach, looking like small thick antennae. (Photo courtesy of Dr. James L. Castner.)

into a top and bottom plate with a membranous area between. This membranous intersegmental area permits flexibility and expansion or extension of the body (Figure 1.21).

One of the most obvious abdominal features in most insects occurs at the tip of the abdomen. These are the paired, feeler-like structures called cerci (Figure 1.22). The cerci vary in size among different groups of insects, and sometimes are not present. They appear to be sensory organs that are useful in detecting vibrations and disturbances in air currents. However, in some insects, like the earwigs, they have evolved into hardened pincers that are used for both defense and prey capture (Figure 1.23). The cerci of silverfish are extremely long and tail-like.

Another feature found at the tip of the abdomen, but only in female insects, is the ovipositor. The female uses this structure when laying her eggs. It allows for careful and exact placement of the eggs, and sometimes provides the mechanical means for introducing the eggs into a specific material such as the ground or a plant stem. It may be large and obvious, as in some crickets, or it may be withdrawn and hidden within the body, as in female house flies and blow flies (Figure 1.24a). The latter type often has segments that telescope one within the other. Some insects have extremely long ovipositors used to deposit eggs in other insects, plants, or soil (Figure 1.24b and c). Other wasps and bees have ovipositors that are modified for defense. These ovipositors are called stingers

Figure 1.23 The cerci found at the tip of an earwig's abdomen are pincer-like. They are used both in defense and to help catch food items. (Photo courtesy of Dr. James L. Castner.)

(a)   (b)

(c)

Figure 1.24 (a) Greenbottle flies add eggs to an existing mass. The narrowing tips of their abdomens serve as ovipositors. (b) The ovipositor of a field cricket is used to place her eggs in the soil. It is long and needle-like, extending straight back from the abdomen between the long slender cerci. (c) The female papaya fruit fly uses her ovipositor to penetrate the skin of fruits to deposit her eggs in the center. (Photos courtesy of Dr. James L. Castner.)

(a)   (b)

Figure 1.25 (a) The spiracles of the tobacco hornworm caterpillar are visible here as eight white-rimmed oval black spots along the side of the body. (b) A close-up view of the spiracles of a tobacco hornworm caterpillar. (Photos courtesy of Dr. James L. Castner.)

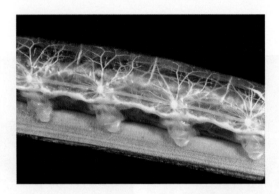

Figure 1.26 The white, thread-like tracheae of a skipper butterfly caterpillar are clearly visible through its nearly transparent skin. They branch out from the oval white spiracle seen above each leg. (Photo courtesy of Dr. James L. Castner.)

and can be extruded to puncture intruders, often with the injection of venom. Female honeybees, bumblebees, yellowjackets, paper wasps, and hornets protect themselves and their nests in this way.

A series of small oval holes, one pair per segment, may also be present along each side of the abdomen. They are easily visible on most caterpillars and are called spiracles (Figure 1.25a and b). They represent the external openings of the insect's respiratory system. The spiracles connect to tiny tubes that get progressively smaller as they ramify throughout the insect's body (Figure 1.26). By coordinating muscular movement with the opening and closing of the spiracles, the insect can effectively ventilate air through its body. Maggots have a thoracic spiracle on each side of the body as well as a pair of posterior spiracles at the tip of the abdomen (Figure 1.27a and b). The shape and coloration of these structures are useful in determining the species.

Copulatory structures are also present at the tip of the abdomen. In male insects, the penis or aedeagus is often withdrawn into the body. Accessory structures such as claspers, used to help hold the female during mating, are also often present. In most insects sperm transfer is direct from the male to the female during copulation or mating.

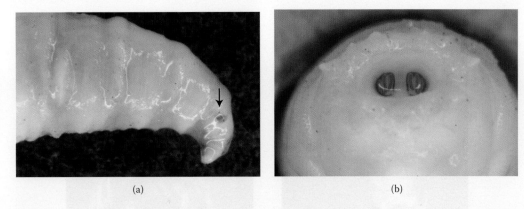

(a)   (b)

Figure 1.27 (a) The thoracic spiracle of this flesh fly maggot is visible as a darkened spot near the larva's head on the right. (b) The two posterior spiracles of fly larvae are spaced close together at the tip of the abdomen such as in this flesh fly maggot. (Photos courtesy of Dr. James L. Castner.)

Insect Growth and Development

Insects grow by passing through a series of stages, starting with an egg and ending with the adult. The physical appearance of the insect in these stages and the time spent within each vary with the species and with the environmental conditions that are present. For example, the developmental period is usually shortened as temperature increases, which is why accurate climatic data are of utmost importance in the calculation and estimation of the postmortem interval based on insect evidence. The process of undergoing physical changes from one life stage to the next is known as metamorphosis. This is accomplished by means of the insect "shedding its skin," or undergoing ecdysis, at certain times as it grows. The old shed skin that is left behind is called an exuvium (Figure 1.28a and b). The time period spent in any particular life stage is referred to as a stadium. Finally, the insect itself may also be called an instar, especially during larval development. For example, a second larval instar has shed its skin once to proceed to its second stage since hatching from an egg.

The simplest form of development occurs when the only differences between the immature and adult forms are size and sexual maturity. Very primitive insects, such as silverfish and springtails, show this development, which is called ametabolous. The young hatch from an egg, and they are essentially a small version of the adult, since there is no outward morphological difference between immatures and adults except size. The insects that undergo ametabolous development are all wingless.

A slightly more complex form of growth is hemimetabolous development, or gradual metamorphosis. We see this growth pattern in the true bugs, such as the assassin bugs. With this type of development, eggs are laid and hatch out into immature forms that are called nymphs. These nymphs are somewhat similar to the adults, but have no wings. With each shedding of the skin, the nymph becomes a little bigger. After ecdysis occurs several times, small wing pads begin to appear on the outside of the nymph. These pads get larger after the skin is shed each time, until the adult stage is reached. At this point, the wings have developed to full size. This development is also seen in crickets, grasshoppers, katydids, and cockroaches.

The most involved and complex growth pattern is called holometabolous, or complete metamorphosis. Most people are probably familiar with the life cycle of a butterfly, which

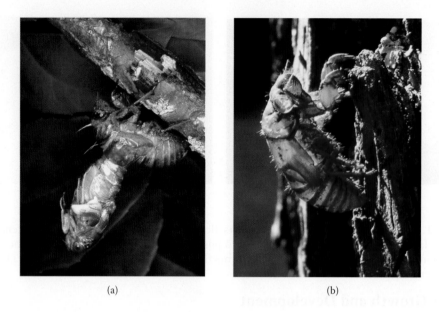

(a)   (b)

Figure 1.28 (a) A cicada sheds its nymphal skin during the process of molting. (b) The shed skin (exuvium) of a cicada. (Photos courtesy of Dr. James L. Castner.)

starts with an egg, progresses through caterpillar stages, forms a pupa, then eventually emerges as a butterfly. The same type of development is also seen in the two groups of insects with greatest value as forensic indicators of postmortem interval. Beetles undergo complete metamorphosis, but their egg and pupal stages are often hidden in protected areas. The larvae are often underground or under the body in contact with the exudate-soaked ground beneath and not readily visible. The same distinct four stages of egg, larva, pupa, and adult are part of each beetle life cycle (Figure 1.29). In some cases, such as the rove beetles (family Staphylinidae), the larvae are highly mobile and capable of foraging on their own for food. In some species of carrion beetles (family Silphidae), parental care has evolved and the adult beetle regurgitates food into the mouths of the young larvae until they are mature. Beetles usually pass the pupal stage underground or in a protected cavity or area. They tend to be white or yellowish in color, becoming increasingly dark as

Figure 1.29 The stages in the life cycle of a carrion beetle. From left to right: a young larva, an older larva, a pupa, an adult. The egg is not shown. (Photo courtesy of Dr. James L. Castner.)

Figure 1.30 The exarate pupa of a sexton beetle demonstrates how the legs and wings develop externally to the rest of the body. (Photo courtesy of Dr. James L. Castner.)

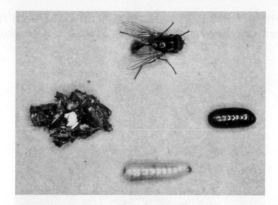

Figure 1.31 The life stages of a fly consist of, clockwise from right, the egg, larva or maggot, pupa, and adult. (Photo courtesy of Dr. James L. Castner.)

they mature. The legs and appendages are easily visible and free from the rest of the body, characteristic of the pupal type referred to as exarate (Figure 1.30).

Flies also undergo holometabolous development, but with a distinct variation. Although it is still complete metamorphosis, the chain of events and appearance of the life stages are quite different (Figure 1.31). Those families of flies that are both most commonly encountered and most useful as forensic evidence are the Calliphoridae (blow flies), Sarcophagidae (flesh flies), and Muscidae (house flies). All are among the most evolutionarily advanced of the flies and belong to the suborder Cyclorrapha.

These flies begin life as eggs that are usually laid in large numbers on carrion, feces, or decaying material. In the flesh flies, the eggs are retained by the female until hatching. The small, first-instar larvae are then deposited directly on the food source. The young or maggots pass through three larva stages and are then ready to pupate. At this point, the mature maggots usually migrate away from the remains or food source to pupate in the soil (Figure 1.32). Sometimes, however, fly puparia can be found in the clothing of the deceased (Figure 1.33) or under carpets or in furniture when in a house environment.

Blow flies, flesh flies, and house flies all molt or shed their skin to form an exarate pupa, but they do it inside the old skin of the mature larva, which then shrinks and hardens. This outer skin is called the puparium, and this pupal type is referred to as coarctate (Figure 1.34). The puparium darkens with time to brown, reddish brown, or almost black, depending on the species (Figure 1.35). Morphological characteristics of the larvae are

Figure 1.32 Maggots pupate in the soil, typically after migrating away from their food source. These puparia were uncovered by removing the soil surface debris. (Photo courtesy of Dr. James L. Castner.)

Figure 1.33 Under certain circumstances fly puparia may be found attached or embedded in the clothes of the decedent. (Photo courtesy of Dr. James L. Castner.)

Figure 1.34 The coarctate pupae of many flies of forensic importance form when the last larval skin shrinks and hardens such as in these flesh fly (left) and blow fly (right) puparia. (Photo courtesy of Dr. James L. Castner.)

Figure 1.35 As most fly puparia mature, they darken and harden. In this image, the youngest flesh fly puparium is at the left and the oldest is at the right. (Photo courtesy of Dr. James L. Castner.

retained on the outer surface of the puaparium and are often distinctive enough to permit a species identification to be made.

Insects in aquatic environments also undergo gradual or complete metamorphosis. For example, mayflies (order Ephemeroptera) have hemimetabolous development, just as true bugs do. The immature forms of aquatic insects with gradual metamorphosis are called naiads rather than nymphs. Caddisflies (order Trichoptera) are aquatic insects that are moth-like as adults and undergo complete metamorphosis. Their immature forms are still referred to as larvae.

Conclusions

Insects and other arthropods are found in almost every conceivable type of habitat. As a result, human remains are often found colonized by carrion-feeding invertebrates or in association with species that are scavengers or opportunistic feeders. The ability of true carrion-feeding insects to rapidly locate a body enables them to colonize remains that have even been wrapped, buried, or otherwise protected. The ubiquitous nature of insects makes their eventual appearance at a death scene a near certainty.

To take advantage of the potential forensic value of arthropods, evidence must be systematically collected and processed. For an investigator or crime scene technician to collect such material, he or she must first know what to look for. A basic understanding of insect biology and anatomy, especially with regards to the flies and beetles, shall facilitate the search, recognition, and collection of insect specimens for evidence. Recognition of insects and other arthropods to a greater taxonomic level, such as that of family or species, is covered in Chapter 2.

Suggested Readings

Castner, J. L. 2000. *Photographic atlas of entomology and guide to insect identification.* Gainesville, FL: Feline Press.
Elzinga, R. J. 2004. *Fundamentals of entomology.* 6th ed. Upper Saddle River, NJ: Prentice Hall.

Evans, H. E. 1984. *Insect biology.* Reading, MA: Addison-Wesley Publishing Co.

Gillott, C. 2005. *Entomology.* 3rd ed. The Netherlands: Springer.

Gullan, P. J., and P. S. Cranston. 2004. *The insects.* 3rd ed. Hoboken, NJ: Wiley-Blackwell.

Johnson, N. F., and C. A. Triplehorn. 2005. *Borror and DeLong's An introduction to the study of insects.* 7th ed. St. Paul, MN: Brooks Cole.

Romoser, W. S., and J. G. Stoffolano. 1997. *The science of entomology.* 4th ed. Dubuque, IA: William C. Brown.

Ross, H. H., C. A. Ross, and J. R. P. Ross. 1991. *A textbook of entomology.* 4th ed. Malabar, FL: Krieger Publishing Co.

Appendix: Insect Antennae

Recognition of the different types of antennae is essential for identifying insects. A brief description is given below for each of the major antenna types, along with selected insect groups that serve as examples.

Aristate: A distinctive three-segmented antenna found on certain flies where the third segment bears a protruding hair called an **arista**. Blow flies, house flies, flesh flies, and tachinid flies all have aristate antennae.

Capitate: The tip of the antenna is enlarged into a rounded knob. Examples are found on butterflies, antlions, and owlflies.

Clavate: The tip of the antenna is enlarged into a broadened club. Carrion beetles and carpet beetles both have distinctly clubbed antennae.

Filiform: Thread-like or hair-like. Composed of a series of cylindrical or flattened segments. Examples are cockroaches, crickets, grasshoppers, true bugs, and bark lice.

Geniculate: Elbowed. Abruptly bent, such as in a knee joint or elbow joint. Examples of such antennae occur on ants, bees, and weevils.

Lamellate: A form of clubbed antenna where the terminal segments are enlarged parallel plates that stick out perpendicular to the rest of the antenna. These plates are close fitting and may not show any obvious space between them on dead specimens. Scarab beetles typically have lamellate antennae.

Moniliform: Bead-like and composed of a series of rounded segments like a pearl necklace. Examples include termites and some beetles.

Pectinate: Lateral processes stick out from the antenna at regular intervals, like the teeth of a comb. Comb-like. Examples are glow-worms and some fireflies.

Plumose: Feather-like, or with whorls or clumps of hairs. Mosquitoes and silk moths both exhibit plumose antennae, which are more pronounced in the males of each.

Serrate: A combination of roughly triangular segments that give a saw-tooth appearance. Found on click beetles and others.

Setaceous: Slender, bristle-like, and gradually tapering to a tip. Examples include dragonflies, damselflies, and cicadas.

Stylate: Antenna that terminates in a long slender point called a **style**. The style may be hair-like and similar to an arista, but is found at the tip of the antenna rather than projecting from the side. Stylate antennae are found on robber flies and bee flies.

Insects of Forensic Importance

2

JASON H. BYRD
JAMES L. CASTNER

Contents

Introduction

The first studies of faunal succession on decomposing remains were conducted in the mid-1600s. Francesco Redi was one of the first scientists to disprove the theory of spontaneous generation in 1668. He did so by proving that the seemingly spontaneous appearance of "worms" on meat was indeed the larvae of flies that hatched from eggs. On further observation of these worms, he discovered that they developed into flies. In 1894 Mégnin published *La Faune des Cadavers*, which was a thesis on the observations of insects on human corpses. He was also the first individual to develop a method to calculate the age of a dead body by the insects present.

To the casual observer, it would appear that all arthropods inhabiting human remains or animal carcasses share the same ecological resource. However, in order to survive, each species must have its own ecological niche, or habitat, within any particular environment. Therefore, each species of insect developing on a carcass is present because of its unique niche, and within that niche it possesses some adaptive advantage over all other species. Such advantages may include early adult arrival time, accelerated development, predation, or location preference. Therefore, one of the first steps in a forensic entomology case analysis is the proper identification of the insect and arthropod species present. This genus/species identification allows the forensic entomologist to access the correct developmental data and distribution ranges to be applied for the case. An incorrect species determination can lead to potential error in the estimation of the time of colonization (TOC) and estimated period of subsequent insect activity (PIA). Each of these biological time markers can be utilized to estimate a portion of the total postmortem interval (PMI).

With advances in molecular biology and genetic identification techniques, the identification of insects by DNA methods is becoming increasingly common. Genetic analysis does have it advantages in being able to conclusively identify fragmentary remains and species without distinct morphological differences. However, the costs of such laboratory analysis is prohibitive for some agencies, and in many cases, such detailed analysis is not necessary. Morphological examinations can be done quickly and are extremely cost-effective. This publication is not intended to be an exhaustive taxonomic guide to insects of forensic importance. However, it does offer a dichotomous key to the common Calliphoridae found on human remains in North America (see Chapter 19).

In this chapter we present color photographs of the most common species of insects likely to be recovered from human remains. It is our intent to show photographs of the insects so that those unfamiliar with the insect fauna (i.e., crime scene investigators and homicide detectives) can use this work as a tool for the easy and rapid recognition of the insects that would be most important to document and collect from a scene of human or animal death. Familiarity with the insect species that can provide valuable clues to a death investigation will allow for more precise data collection by focusing collection efforts on the species of importance.

It should be noted that some insect species are easily visually identified. Others appear similar and have growth and development rates that can be applied from one to the other with little or no error in the resulting estimation of the time of colonization, or the period of insect activity. However, species exist that are not morphologically distinct or have highly variable morphological characteristics, yet have drastically different growth rates, behaviors, and habitat preferences. Due to these developmental traits and morphological characteristics, it is not possible to make comparisons between insect appearance and developmental physiology. Therefore, it is essential that a certified forensic entomologist or qualified insect taxonomist should always make the species determinations for forensic casework.

While some species of the insects of forensic importance can be readily identified by simply viewing a photograph, such instances are in the minority. This chapter does not encourage forensic investigators to make definitive species identifications, but illustrates general taxonomic characteristics and describes the basic biology of species when it is known. Familiarization with the appearance and habits of selected species will enable forensic investigators to make informed decisions regarding the collection of insect evidence. Therefore, this chapter features color photographs of the most common species of insects and other arthropods of forensic importance in North America. Although this work is not inclusive of every possible species that could be recovered during the processing of a death scene, it does represent the most commonplace and useful fauna. For illustrative purposes, insect-induced postmortem artifacts often confused with perimortem trauma have also been included.

Insects and their arthropod relatives play a diverse role in civil and criminal investigations. The species included in this chapter are ones that are widely distributed and common. As such, much of their developmental history, behavior, and geographic distribution are well known, and thus are frequently utilized in forensic investigations. Since there are situations when any insect can be of forensic importance, it is beyond the scope of this work to include anything but the most common examples of insects as they are used in assisting with postmortem interval determinations. Insect evidence also has other valuable uses beyond aiding in the estimation of a postmortem interval. For instance, a moth larva recovered from a plant seed attached to a blanket used in a rape or sexual assault may provide evidence linking the crime to a particular location. Sometimes entomological evidence is found as broken and fragmented insect parts in the clothing or personal belongings of the suspect or victim. These fragmentary remains are sometimes utilized successfully to create links among the victim, suspect, and scene. Many species of insects could serve in this role, but it is beyond the scope of this work to illustrate fragmentary arthropod remains.

Forensic investigators should become familiar with the appearance of the arthropods associated with human remains, and follow the collection methodologies as described in

Chapter 3. This will enhance their ability to provide the forensic entomologist with scene information and properly collected entomological evidence that will yield the maximum amount of information possible. The use of the scene forms and collection checklists included in Chapter 3 are encouraged. These are useful for the orderly arrangement of scene data and can help ensure that all steps of the collection process are properly completed.

Flies (Order Diptera)

This insect is one of the largest, with over 86,000 known species, which are all commonly known as "flies," with over 16,000 of these species occurring in North America. Although they differ greatly in appearance, they are all characterized by possessing a single pair of wings used for flight, and a second pair of "wings" is reduced to small, knob-like organs called the halters, which are used to stabilize the insect in flight. It is standard nomenclature to incorporate the word *fly* written as a separate word from the descriptive name (i.e., house fly). Whereas with the non-Diptera, *fly* is written together with the descriptive name (i.e., dragonfly). True flies can be found in almost any terrestrial habitat, and have even been collected at sea many miles from any land mass. Flies have large compound eyes with mouthparts of various types (Chapter 1). However, most flies associated with decomposing animal, human, or vegetative matter possess sponging mouthparts. The larvae of all fly species are called maggots, and all are legless. Most are white or cream colored, soft bodied, and many lack a distinctly visible head (Figure 2.1). Most aquatic larvae (such as those of midges, mayflies, and mosquitoes) are slender and have a recognizable head capsule (Figure 2.2). Flies (and beetles) are important scavengers, feeding upon and thus removing decomposing plant and animal material from the environment. Some are predators and parasites of other insect species, and others aid in the pollination of plants (Figure 2.3a and b). The remainder of this section focuses on the species of Diptera that are important decomposers and scavengers.

Figure 2.1 This mature red-tailed flesh fly maggot (*Sarcophaga haemorrhoidalis*) illustrates the general form of most fly larvae. It is cylindrical, lacks appendages, and does not have a distinct head. However, the maggots of some species may be flattened, or have small fleshy projections on the body. (Photo courtesy of Dr. James L. Castner.)

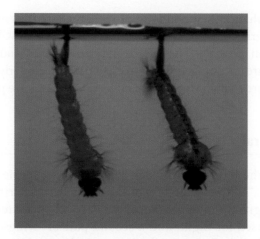

Figure 2.2 The mosquito is one of many insect groups with aquatic larvae. Most of these aquatic larvae have a distinct head, and possess special structures and adaptations for living in an aquatic environment. (Photos courtesy of Dr. James L. Castner.)

(a)   (b)

Figure 2.3 (a) The feather-legged fly (*Trichapoda pennipes*) is a parasite of the green stink bug. It lays an egg on the stink bug, which will eventually kill its host once it hatches. (b) Robber flies catch and feed on other insects. Some, like this Florida species (*Mallophora orcina*), are excellent mimics of bees and wasps. (Photo courtesy of Dr. James L. Castner.)

Blow Flies (Family Calliphoridae)

This insect family contains the familiar metallic blue and green flies commonly seen around refuse and wastes during the warm summer months. This is an extremely large group of medium-sized flies that contains over 1,000 species, and members of this family can be found worldwide. They, along with the sarcophagid and muscid flies, are the most important species that provide information relating to the accurate estimation of the period of insect activity, and thus accounting for a portion of the postmortem interval. The calliphorid flies are attracted to decomposing human tissues, animal carrion, excrement, some vegetative material, and with some species, exploiting open wounds in living humans and animals. This family includes the familiar green bottle flies (genus *Phaenicia*) and blue bottle flies (genus *Calliphora*), as well as the notorious screwworm flies (genus *Cochliomyia*). In addition to their forensic importance,

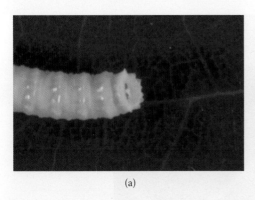
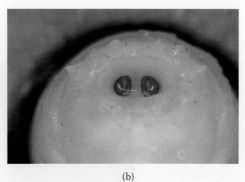

(a)   (b)

Figure 2.4 (a) Fleshy tubercles (seen here as triangular projections on the posterior segment of the lower larvae) are often found on the last segment of fly larvae, and they are useful in the identification of some species. (b) The two dark structures at the posterior end, which the tubercles surround, are the posterior spiracles through which the maggot breathes. (Photos courtesy of Dr. James L. Castner.)

this family is extremely valuable in nutrient recycling and community ecology based on their removal and breakdown of vertebrate carcasses from the environment.

Both male and female adult calliphorids range from 6 to 14 mm in length. The adult size is dependent on species and food availability to the larval stages. The majority of these species are metallic in appearance, with colors ranging from brilliant green or blue to bronze or shiny black. In some species, a covering of fine hairs, powder, or dust masks the bright metallic coloration of the fly epicuticle. This results in only a dull metallic sheen remaining. Adults are characterized by a three-segmented antenna located between and anterior to the pair of compound eyes. This antenna has a hair or arista on the last segment (Figure 1.8), which is plumose or hairy throughout its entire length.

The mature larvae (third instar) of blow flies range from 8 to 23 mm in length and are white or cream colored. The terminal segment of the larval body typically has six or more cone-shaped tubercles about its perimeter (Figure 2.4). This segment also contains the posterior spiracles, which are the primary breathing apparatus of the larva (Figure 1.27b). The posterior spiracles are important morphological indicators for species identification. The slits that form the opening to each trachea within each spiracle slant toward the center of the larva, as opposed to those of the Sarcophagidae, which slant more outward or downward, and those of the Muscidae, which are S shaped or sinuous.

Blow flies are among the first insects to detect and colonize human and animal remains. In experimental studies, calliphorid flies have been recorded arriving at carcasses within minutes of their exposure. Blow flies locate human and animal remains in a two-step process that first consists of chemical detection via receptors on their antenna (olfaction) and a visual search. Once in close proximity to a body or carcass, flies switch from a primarily olfactory-driven search to a visual search. Once they locate the body, they will visually assess the size of the carcass and find the location of natural orifices and any sites of trauma. Flies are often observed walking extensively over the surface of a human body or animal carcass. This behavior aids in the visual survey of the remains, and since the taste receptors of flies are located in sensory organs on their body, legs, and feet, it also assists in the detection of the most suitable oviposition site. Once a suitable oviposition site is found, gravid female flies will begin to deposit their eggs. During this process, the telescoping

Figure. 2.5 Adult *Calliphora coloradensis* with typical light gray thorax with darker longitudinal stripes, and metallic blue thorax. (Photo courtesy of Dr. James L. Castner.)

segments of the tip of the female's abdomen extend to form an ovipositor, which is used for egg laying. Large numbers of eggs are commonly placed in the nose and mouth, as well as other natural body openings that are exposed. Due to the presence of blood (a source of moisture, sugar, and protein), areas of trauma (open wounds) are also selected for egg placement. Thus, subsequent maggot mass formations and patchy, uneven distributions of insect activity and tissue removal can be indicators of perimortem trauma. Therefore, it is good practice to have both a forensic pathologist and a forensic anthropologist closely examine these sites.

References: Arnett and Jacques 1981; Bland and Jacques 1978; Borror and White 1970; Borror et al. 1989; Castner et al. 1995; Hall and Doisy 1993; Hogue 1993; James and Harwood 1969; Liu and Greenberg 1989; Peterson 1979; Shewell 1987.

Calliphora coloradensis Hough

Nearctic in distribution and common from Mexico to Canada, and east to Ontario. Most commonly abundant in Arizona, Colorado, New Mexico, Texas, and Wyoming. This species seems to prefer woodland environments, and is somewhat intolerant of high temperatures, becoming uncommon during summer months throughout most of its range. The thorax is black with five longitudinal dorsal stripes. The abdomen is blue, and may appear to have a coating of whitish pollen under bright lighting.

References: Hall 1948.

Calliphora latifrons Hough *(= Eucalliphora lilaea* Walker)

This species is Nearctic in distribution and common from Mexico to Canada, and can be found westward into Wisconsin and Ontario. It is most common in the Rocky Mountain region from Colorado northward. The thorax and abdomen are a dark metallic blue. However, as with some other species of Calliphoridae, the thorax is marked with dark longitudinal stripes on the dorsum. These stripes do not continue onto the abdomen. The legs are black. This species is commonly collected from excrement and decomposing tissues of vertebrate carcasses.

References: Hall 1948, Townsend 1937.

Figure 2.6 *Calliphora latifrons* is common in the Rocky Mountain area of the United States. This species has many of the typical traits of a blue blow fly, but it also has a distinctly tessellated abdomen. (Photo courtesy of Dr. James L. Castner.)

Figure 2.7 *Calliphora loewi* adults prefer to avoid human settlements and are most commonly found in forested regions. (Photograph courtesy of Nikita Vikhrev.)

Calliphora loewi Enderlein (= *C. morticia* Shannon)

Palaearctic and Nearctic in distribution. Very common in both upper and lower Alpine regions. In North America it is found throughout Alaska and Canada. In Europe it is found in the northern and central forest regions. Also found in Central Asia, Mongolia, and Japan. Throughout its ranges, this species generally avoids urban areas and human activity.

Calliphora vicina Rodineau-Desvoidy (= *C. erythrocephala* Meigen)

Common name: (European blue bottle fly)
Several species of Calliphoridae exist that share the common name of "blue bottle fly." This species has become nearly worldwide in distribution. Within the United States, it is most abundant in the northern half of the country. *C. vicina* is a large fly, usually ranging from 10 to 14 mm long. The head is black in color, with the lower part of the bucca, or "cheeks," appearing red to yellow. The cheeks are black in *C. vomitoria*, a very similar species. The epicuticle of the thorax is black to dark blue-green. However, it is coated with

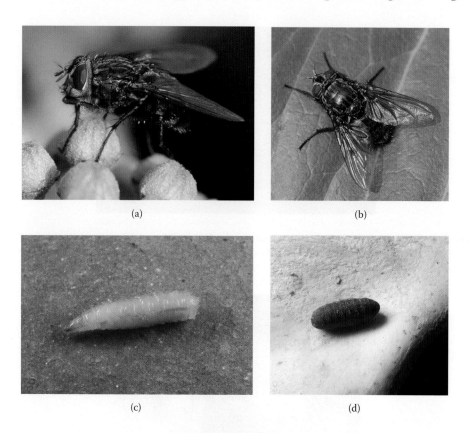

(a)   (b)

(c)   (d)

Figure 2.8 (a) The adult European blue bottle fly (*Calliphora vicina*) with the characteristic yellow-orange bucca, or cheek. (Photo courtesy of Cor Zonneveld, www.corzonneveld.nl.) (b) The body is not brightly metallic, as with the green bottle blow flies, but does have a metallic sheen when examined closely. (Photo courtesy of Cor Zonneveld, www.corzonneveld.nl.) (c) This mature larva shows the typical maggot shape and coloring. The darkened mouth hooks, used for shredding tissue, are visible through the transparent skin of the anterior end. (Photo courtesy of Dr. James L. Castner.) (d) The segmentations of this puparium are artifacts of the larval body segments. (Photo courtesy of Dr. James L. Castner.)

fine hairs and grayish powder, giving it an overall grayish blue to greenish silver appearance. The thorax also has faint dark longitudinal stripes on the dorsal surface between the bases of the wings. The abdomen is a noticeable metallic blue, often with dark blue to black coloration along the posterior margins of the abdominal tergites, which are otherwise patterned with silver. The legs are black, and overall the body appears very bristly.

The adults are attracted to most types of decaying matter and frequent rotting fruit, decaying meat, and feces. However, the larvae are found primarily on carrion. *C. vicina* is extremely common on human corpses throughout the United States and Europe in temperate regions. In the southern United States this is considered a winter species, while a spring and fall species in the temperate zones, and a summer fly in the subpolar regions. This species primarily favors shady situations and urban habitats, where it is often the dominant species on human cadavers. This species is another that has been known to produce myiasis in humans and animals.

References: Erzinclioglu 1985; Greenberg 1971; Hall 1948; Hall and Townsend 1977; James 1947; Nuorteva 1977; Payne 1965; Smith 1986.

Figure 2.9 The adult Holarctic blue bottle fly (*Calliphora vomitoria*) is similar in size and appearance to the European blue bottle fly. The thorax is a dusty blue, with a metallic abdomen. The dark bucca is in contrast to the orange-yellow coloration of *C. vicina*. (Photo courtesy of Cor Zonneveld, www.corzonneveld.nl.)

Calliphora vomitoria (Linnaeus)

Common name: Holarctic blue blow fly

This blow fly species is Holarctic in distribution, as the common name implies. In the United States it is most common from Virginia to California and northward to Alaska. It is also found throughout southern Canada. The Holarctic blue blow fly ranges from 7 to 13 mm in length. Much like *C. vicina*, the thorax is dark blue to black and adorned with four dark longitudinal stripes on the dorsal thorax between the wings. The thorax is covered with fine hairs and appears to have a light gray dusty coating, giving it a silver coloration. Depending on the condition of the specimen, darker blue longitudinal stripes may be very prominent or completely lacking. The abdomen is bright metallic blue and patterned with a silver-gray powder. This pollen-like coating masks much of the metallic nature of the color, but a glint of metallic sheen is still visible when specimens are examined closely. The posterior margin of the abdominal tergites is dark blue to black. The legs of this species are black.

This species is similar in appearance to *C. vicina*, except the head appears almost entirely black. *C. vomitoria* lacks the prominent orange bucca, or "cheek" area, that is characteristic of *C. vicina*, and it possesses only a few red to orange hairs near the posterior margin of the head below the eyes and around the mouthparts. Overall, the body appears very stocky and bristly. This species is common in wooded rural as well as suburban areas, where it prefers shaded locations. The Holarctic blue blow fly is slow flying and makes a loud buzzing sound during flight. The biology is much the same as described for *C. vicina*, but *C. vomitoria* is not as common throughout its range.

References: Erzinclioglu 1985; Greenberg 1971; Hall 1948; Hall and Townsend 1977; James 1947; Smith 1986.

Chrysomya albiceps (Wiedemann)

Common name: Hairy maggot blow fly

This African species is closely related to the tropical Australasian and Oriental *Chrysomya rufifacies*. *C. albiceps* is also a common producer of secondary myiasis in animals, but there are no recorded cases from humans. The hairy maggot blow fly larvae have a series of fleshy projections on each body segment, and like *C. rufifacies*, the larvae are predatory and

(a)   (b)

Figure 2.10 (a) The bucca of *Chrysomya albiceps* is white, in contrast to the orange bucca of *Chrysomya rufifacies*. (b) The fleshy protrusions on the larvae give the common name hairy maggot blow fly. (Photos courtesy of Nikita Vikhrev.)

cannibalistic in the third instar. The adults have a thorax and abdomen that is shiny metallic green. The abdominal tergites each have a dark blue to metallic blackish purple band on the posterior margin. The anterior spiracle is white to light yellow, and the bucca is white.

References: Smith 1986.

Chrysomya megacephala (Fabricius)

Common Name: Oriental latrine fly

This distinctive blow fly is widely distributed throughout the Oriental regions, South Africa, and South America. It is now also well established in the southern United States, particularly Florida, Georgia, Alabama, Mississippi, Louisiana, and Southern Texas. The adults have short stout bodies similar in appearance to *C. rufifacies*, but with a noticeably larger head. The eyes are unusually large and are a very prominent shade of red, making this fly easily recognizable in the field. In the males, the facets of the compound eye change its orderly pattern on the lower third of the ocular surface, making this an easily recognized species.

Both male and female adult flies are attracted to carrion, fruit, sweet foods, and urine and fecal material—hence the common name. Although *C. megacephala* has a pronounced activity peak during the heat of the afternoon, this species is one of the first to become active in the early morning hours and is one of the last to depart carrion at nightfall. Due to their activity period, they are often the first species to arrive at human and animal remains, and thus are often the first species to deposit eggs. Although *C. megacephala* may be the first to colonize decomposing remains, the later arriving *Chrysomya rufifacies* develop more quickly and often displace *C. megacephala* for food resources. Once *C. megacephala* adults have settled on carrion, they are not easily disturbed, and therefore are commonly collected if present even in low numbers. Unlike some species of Calliphoridae, the adults will readily enter dwellings in search of food and suitable oviposition sites. The larvae are primarily carrion feeders, and the adult oriental latrine fly shows a preference for fresh remains. Carrion that has dried or is in advanced stages of decomposition has little attraction for this species. While this species is slowly increasing its range northward in the United States, it has not done so as rapidly as its sister species *C. rufifacies*.

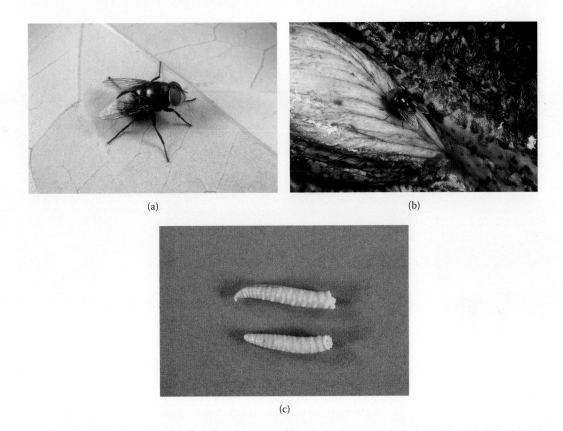

(a)   (b)

(c)

Figure 2.11 (a) The Oriental latrine fly (*Chrysomya megacephala*) is characterized by a large head with extremely large red eyes. The specimen pictured above is a female, which has the characteristic separation between the compound eyes on the top of the head. (b) This adult male has compound eyes that nearly touch on top of the head, which is characteristic of all adult male flies. (c) Typical mature larvae have the body narrowing toward the anterior end (left) and widening toward the posterior end (right). The last larval segment has a ring of fleshy tubercles that are useful identification characteristics in some species. (Photos courtesy of Dr. James L. Castner.)

References: Bohart and Gressitt 1951; Gagne 1981; Greenberg 1971; Hall 1948; Hall and Townsend 1977; James 1947; Prins 1982; Smith 1986; von Zuben et al. 1993.

Chrysomya rufifacies (Macquart)

Common name: Hairy maggot blow fly, hairy sheep maggot

The hairy maggot blow fly is indigenous to the Australian and Oriental regions of the Old World tropics. It was introduced into the continental United States in 1981 and is now well established in the southern tier of the United States from California into North Carolina, with occasional trappings during the summer months as far north as Michigan. This species has drastically expanded its range, and currently ranges into areas previous thought to be inhospitable to this cold-intolerant species. The adults of *C. rufifacies* have stout bodies and are brilliant blue-green in appearance. The terminal edge of the abdominal segments is noticeably tinted a dark blue to black with purple reflections.

Second only to *Chrysomya megacaphala*, the adults of this species are usually the first to arrive on carrion, often within hours after death, throughout its range. Unlike *C.*

(a)

(b)

(c)

Figure 2.12 (a) The hairy maggot blow fly (*Chrysomya rufifacies*) is a species recently introduced in the United States, where it is continuing to expand its range. (Photo courtesy of Dr. Jason H. Byrd.) (b) The mature larva is distinctive because it is covered with fleshy protuberances, vestiges of which remain when the larva forms the puparium. The larva of this species is both predatory and cannibalistic. (Photo courtesy of Dr. James L. Castner.) (c) The puparia are distinctive, with rings of backward-pointing projections on the body segments. (Photo courtesy of Dr. James L. Castner.)

megacephala, this species rarely enters dwellings, and will develop only on carrion and not excrement. The larvae of this species are readily distinguished from other larvae in the family Calliphoridae that commonly occur in the United States by the presence of prominent fleshy protrusions along their body. The larvae are both predacious and cannibalistic, and therefore should be separated from other species when live collections are made for shipment to a forensic entomologist. If the food supply becomes depleted during shipment, the larvae will consume and often totally eliminate other species collected from the carcass. This same behavior is also common under natural conditions. In 1934 it was noted that competition with *C. rufifacies* greatly reduced the population's *Lucilia* species. The larvae will colonize buried remains if the burial site is shallow and a portion of the body becomes exposed during the bloat stage of decomposition. This species continues to rapidly expand its range throughout the United States, and due to its predatory nature, it is likely that forensic entomologists will encounter the hairy maggot blow fly as the sole indicator species with increasing frequency.

References: Baumgartner 1986, 1993; Baumgartner and Greenberg, 1984; Fuller 1934; Gagne 1981; Greenberg 1971, 1988; Hall 1948; Hall and Townsend 1977; James 1947;

Oldroyd and Smith 1973; Richard and Ahrens 1983; Smith 1986; Wells and Greenberg 1992; Zumpt 1965.

Cochliomyia macellaria (*Fabricius*)

Common name: Secondary screwworm fly

This species is very abundant throughout the New World, where it can be found from the American tropics throughout the United States to the Canadian border. However, its population is being greatly reduced in areas where habitat is shared with *C. rufifacies*. In some areas or range overlap, *C. macellaria* is now considered to be scarce.

The adults are easily distinguished from most all other blow flies due to their metallic greenish blue coloration, with three pronounced dark green longitudinal stripes on the dorsal surface of the thorax between the base of the wings. These stripes do not extend onto the abdomen. On close observation, the head appears orange and the legs may vary from a reddish brown to dark brown in color. The larvae have readily visible respiratory tracheae on their posterior end. The tracheae appear as swirling black lines easily visible against the white background of the maggot's body. These characters make this species readily identifiable in the field.

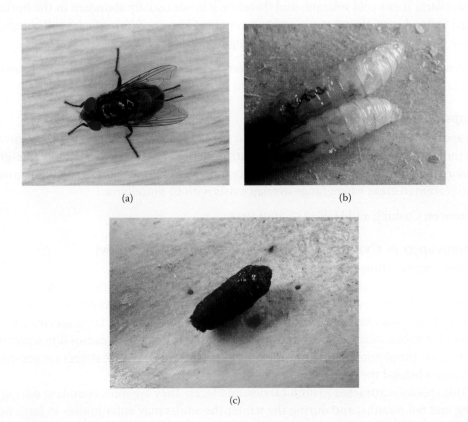

(a)   (b)

(c)

Figure 2.13 (a) The adult secondary screwworm fly (*Cochliomyia macellaria*) is easy to identify by the three dark longitudinal stripes on its blue-green thorax. (b) The tracheae, visible here as swirling black lines within the larvae, are characters that may aid in the field identification of this species. (c) The puparium has the typical cylindrical blow fly shape. They are typically not found in direct association with the remains since the larvae usually undergo a migration of some distance before pupation will begin. (Photos courtesy of Dr. James L. Castner.)

Figure 2.14 *Compsomyiops callipes* appears very similar to *C. macellaria*, which is also metallic blue-green in color, with three dark longitudinal thoracic stripes. (Photo courtesy of Dr. James L. Castner.)

The secondary screwworm fly prefers warm, humid weather. It occurs throughout the southern United States and is most abundant during rainy periods. This species frequents carrion in both sunny and shaded locations, but is rarely recovered from indoor habitats. *C. macellaria* is not cold tolerant, and therefore it is not usually abundant in the northern United States during the winter months. It is not yet known if *C. rufifacies* will completely replace *C. macellaria* within their overlapping ranges.

References: Denno and Cothran 1975; Greenberg 1971; Hall 1948; Hall and Townsend 1977; James 1947; Smith 1986; Rodriguez and Bass 1983; Wells and Greenberg 1992.

Compsomyiops callipes (Bigot)
This species is restricted to the southwestern United States. The adult has a dark metallic blue thorax, with three dark longitudinal stripes. The abdomen is a metallic blue-green in color. This species is similar in appearance to *C. macellaria*. Caution should be used in identification in areas where range overlap exists with *C. macellaria*.

References: Cushing and Hall 1937; Hall 1948.

Cynomyopsis (= Cynomya) cadaverina (Robineau-Desvoidy)
Common name: Shiny blue bottle fly
This species ranges throughout the Nearctic region. It is most commonly found in the northern United States and southern Canada, but can also occasionally be found as far south as Texas and northern Florida. The shiny blue bottle fly is a large species with the thorax blue to blue-black and covered with a silvery powder, which causes it to sometimes appear gray. The abdomen is a shiny metallic blue. Three darker blue stripes are present on the dorsum behind the head.

This species is attracted to both carrion and feces. They are most abundant during the spring and fall months, and during the winter the adults may enter houses in large numbers. The adults are slow flying and easily captured when making collections at the death scene. The larvae are typically found on carrion during the advanced stages of decomposition, and the adults are not usually attracted to fresh remains.

References: Erzinclioglu 1985; Greenberg 1971; Hall 1948; Hall and Townsend 1977; James 1947; Smith 1986.

(a)   (b)

Figure 2.15 (a) The shiny blue bottle fly (*Cynomyopsis cadaverina*) is similar in appearance to other blue bottle flies. It is a large, bristly fly with dark longitudinal stripes on the thorax. (Photo courtesy of Dr. James L. Castner.) (b) The coloration of some adults may be so dark as to appear black under certain lighting conditions. (Photo courtesy of Jim Kalish, University of Nebraska, Lincoln.)

(a)   (b)

Figure 2.16 (a) An adult *Cynomya mortuorum* with the characteristic yellow-orange coloration on the anterior aspect of the head. (b) Typical contrast in coloration between the dull thorax and the hairy, metallic abdomen. (Photos courtesy of Cor Zonneveld, www.corzonneveld.nl.)

Cynomya mortuorum (Linnaeus) (= Calliphora mortuorum)
Common name: Blue bottle fly

Holarctic in distribution, most common in northern Europe and Asia. In North America, it is found in Alaska and Canada. This species has a dull blue-green thorax with a silvery coating, and sometimes appearing as if coated with a golden pollen. The thorax has three dark longitudinal stripes, which may not be readily apparent. The abdomen is hairy, and a dark metallic blue. The facial area and bucca is yellow-orange.

References: Hall 1948; Smith 1986.

Lucilia illustris (Meigen) (= Musca illustris)
Common name: Green bottle fly

Figure 2.17 *Lucilia illustris* is one of several common green bottle flies that are early arrivers at carrion. This species has a shiny metallic green body with black legs. (Photo courtesy of Dr. James L. Castner.)

This species is one of several that have been given the common name of green bottle fly, and *L. illustris* may be one of the most common species of green bottle fly. *L. illustris* is distributed throughout the Holarctic region. In North America this is a very common species that ranges from Mexico to southern Canada and throughout the American Midwest.

The adults are approximately 6 to 8 mm in length. The thorax and abdomen are a shining greenish blue, while the legs are black. This is considered a warm weather species, as it is most abundant in open woodlands during the summer months. The adults are attracted primarily to fresh carrion, but they can occasionally be collected on excrement. Adult females are most active and most commonly oviposit on remains in bright or sunlit locations. The larvae feed on both carrion and fecal material, but they are also most common on carrion.

References: Greenberg 1971; Hall 1948; Hall and Townsend 1977; Smith 1986.

Lucilia silvarum (Meigen, 1826) *(= Bufolucilia silvarum)*

Holarctic in distribution. This species is found throughout North America, Asia, Europe, and northern Africa. In the United States, this is an uncommon species, but it is most frequently collected west of the Rocky Mountains. Throughout its range, it is most common during the warm summer months. However, it has been collected in Virginia, south to Arkansas. The thorax and abdomen are a shining metallic, dark blue. Legs are black. This species has been found as a parasite on live amphibians, and it has been collected from the carcasses of amphibians. It is thought to readily kill amphibians, which it parasitizes. It has been observed to be active early in the morning, and remains active until late in the evening, and in very low light conditions. This behavior could be of tremendous significance in forensic investigations.

Figure 2.18 *Lucilia silvarum* is a relatively uncommon species. Its appearance is distinctive from the green bottles by the dark blue, metallic thorax and abdomen. (Photo courtesy of Dr. James L. Castner.)

Figure 2.19 *Phaenicia cluvia* is a green bottle fly whose shiny body is sometimes a greenish blue. This species is attracted to fresh remains and may be confused with other species of *Phaenicia*. (Photo courtesy of Dr. James L. Castner.)

Phaenicia cluvia (Walker) *(= Lucilia cluvia)*

Common name: Green bottle fly

This fly is Nearctic to Neotropical in distribution, and is most common in the southeastern United States during the late summer and fall. The thorax and abdomens are a bright metallic green with copper reflections. The first segment of the abdomen is dark blue on the dorsal surface. The legs are dark brown to black in color. The biology and habits of this fly are nearly identical to that of *P. coeruleiviridis*. The bucca of the adult is reddish brown with dark hair anterior and dorsally and orange-yellow hair posterior to the bucca. The adults are attracted to both carrion and rotten fruits.

References: Hall 1948.

Phaenicia coeruleiviridis (Macquart) *(= Lucilia coeruleiviridis)*

Common name: Green bottle fly

(a)   (b)

(c)   (d)

Figure 2.20 Female green bottle flies (*Phaenicia coeruleiviridis*) frequently cluster together in surface depressions, creases, natural orifices, or at sites of trauma to deposit their eggs on dead tissues. (a) A batch of recently deposited eggs can be seen on the left. (Photo courtesy of Dr. Jason H. Byrd.) (b) The adult fly is a shiny green, with the femora partially metallic green. (c) Photography of pupae found at a death scene is of critical importance. The color of the pupa is an excellent marker of life stage and extent of development. This is a typical accumulation of puparia after migration has occurred. (d) Mature larvae quickly disperse in search of suitable pupation sites, and documentation of the presence or absence of this stage is essential. (Photo courtesy of Dr. James L. Castner.)

This green bottle fly has a Nearctic distribution and is very common in the southern United States. The thorax and abdomen are a shiny, metallic blue-green, which may be tinted with purple or bronze. The legs are dark brown to black, with the femora a slight metallic green. The adults of this species are attracted to almost all decaying human and animal matter, and this is one of the most common species attracted to fresh carrion. This fly has been the predominant species of blow fly in the southeastern United States during the spring and fall; however, it too has undergone drastic population reductions in areas of range overlap with *C. rufifacies*. However, this species remains active all year during the mild winters in the southeastern United States.

References: Hall 1948; Hall and Townsend 1977.

Phaenicia cuprina (Wiedemann) *(= Lucilia Cuprina Wiedemann) (= Phaenicia pallescens Shannon)*

Common name: Australian sheep maggot, bronze bottle fly

This species of Nearctic distribution is closely related to *Phaenicia sericata*. However, *P. sericata* is more cold tolerant and found most abundantly in northern portions of its range.

(a)

(b)

(c)

(d)

Figure 2.21 (a) The distinct bronze tone overlies the metallic green on the body of the bronze bottle fly (*Phaenicia cuprina*). (Photo courtesy of Dr. Jason H. Byrd.) (b) Egg clusters are frequently deposited on the surface of almost any decomposing animal or human tissue. Eggs are commonly deposited near natural body openings and at sites of trauma, so that the larvae have easy access to the interior of the body. Newly hatched larvae cannot penetrate normally unbroken human skin. (Photo courtesy of Dr. James L. Castner.) (c) *P. cuprina* larvae of different ages feed on the surface of exposed meat. (Photo courtesy of Dr. James L. Castner.) (d) The puparium of *P. cuprina* is generally smooth and cylindrical on the surface. In this specimen, the developing adult is visible through the puparium. (Photo courtesy of Dr. James L. Castner.)

P. cuprina is found most commonly in the southern portions of its range, particularly the southeastern United States. There is debate as to *P. pallescens* being a synonym of *P. cuprina*. Individual specimens collected from North America have morphological differences from those found in Asia, Africa, and Australia. In the United States, *P. cuprina* occurs primarily in the southern region. The bronze bottle fly is 6 to 8 mm in length and is a metallic yellow-green to dull copper in color, which may be tinted with green. Legs are colored, as in *P. sericata*. The adults seem to prefer excrement to carrion, but are commonly attracted to both, as well as to decaying fruit. They often alight on the ground or vegetation in proximity to a food source and take flight readily when disturbed, thus making them difficult to collect. The larvae of this species are most frequently found on carrion, and adults are commonly attracted to fruit.

The bronze bottle fly is abundant from spring through fall and can be collected year-round in Florida. It is often found near dwellings and readily enters homes. This species has been known to produce myiasis (infest living tissue) in both man and animals.

Figure 2.22 *Phaenicia eximia* is a green bottle fly that ranges in color from blue-green to blue-purple as an adult. Although relatively common, it is seldom documented as present at human remains since it is frequently confused with other species in the genus *Phaenicia*. As with most species of *Phaenicia*, it arrives early in the sequence of insect succession on carrion. (Photo courtesy of Dr. James L. Castner.)

References: Arnett and Jacques 1981; Bland and Jacques 1978; Borror and White 1970; Borror et al. 1989; Castner et al. 1995; Greenberg 1971; Hall 1948; Hall and Townsend 1977; Hogue 1993; James and Harwood 1969; Peterson 1979; Smit 1931; Smith 1986.

Phaenicia eximia (Wiedemann) *(= Lucilia eximia)*
Common name: Green bottle fly

This green bottle species is Nearctic and Neotropical in distribution. It is found in the southern United States and throughout Central and South America, where is it possibly the most common species of Calliphoridae. The adult is similar in appearance to *Phaenicia sericata* and *Phaenicia coeruleiviridis*. Adults are bright metallic blue-green to entirely blue or purple with legs that are black to dark brown. Adult flies are attracted to carrion, dung, and decaying fruits, with the larvae developing on the same substances. The adults are often reported to be the first to arrive at carrion in open sunlit conditions, and as a result, their larvae are typically found during the early stages of decomposition. *P. eximia* has also undergone drastic reductions in population due to increased competition from *C. rufifacies*.

References: Hall 1948; Hall and Townsend 1977, Archer and Elgar 2003.

Phaenicia sericata (Meigen)
Common name: Sheep blow fly

Historically, this species has been Holarctic in distribution, but now it is nearly cosmopolitan in range. Worldwide this may be the most common species of Calliphoridae, and it is most abundant in the temperate zone of the northern hemisphere. In North America, it can be collected throughout the United States and Canada, being most common in the western regions.

Adult sheep blow flies are 6 to 9 mm in length. This fly is a brilliant metallic blue-green, yellow-green, green, or golden bronze. The thorax has three prominent transverse grooves on its dorsal surface, and the front femora are black or deep blue, a useful character in identification. The larvae of this species can successfully develop in a wide variety of food substrates, but they are best suited to carrion. Along with other species of the genus, this fly is one of the earliest arriving fly species on human and animal remains. Commonly this fly will deposit eggs on carrion only a few hours after death. The adults prefer carcasses

(a)   (b)

Figure 2.23 (a) The adult sheep blow fly (*Phaenicia sericata*). Wide color variation exists within the species, ranging from green to blue-green to gold-green. (b) Female (left) and male (right) *P. sericata*. Due to its propensity to arrive early to fresh carrion, and its wide distribution, this species is of tremendous forensic importance. (Photos courtesy of Cor Zonneveld, www.corzonneveld.nl.)

located in bright sunshine and open habitats; however, they will seek shaded areas of the body in which to deposit their eggs. There have been reports of this species anticipating death and ovipositing on the wounds of the dying. However, there are also reports that the larvae develop most rapidly on decomposing (not fresh) carrion. The larvae of *P. sericata* have been routinely employed in maggot therapy for the removal of necrotic tissue from wounds with great success.

References: Cragg 1956; Cragg and Cole 1952; Cragg and Hobart 1955; Davis 1928; Greenberg 1971; Hall 1948; Hall and Townsend 1977; Hudson 1914; James 1947; Ratcliffe 1935; Smith 1986; Zumpt 1965.

Phormia regina (Meigen)

Common name: Black blow fly

Holarctic in distribution, the black blow fly can be found throughout the United States, with the exception of southern Florida. The adult flies typically range from 7 to 9 mm in length. Contrary to its common name, this species is actually dark green to olive colored on both the thorax and abdomen. The legs are black. Under some lighting conditions, the entire body may appear shiny black. The anterior thoracic spiracle is surrounded by distinctive orange hairs, which serve readily in the identification of this species.

Typically considered a cold weather fly, it is most abundant throughout most of the United States in the spring and fall. In the northern latitudes of the United States and in southern Canada, it is commonly found throughout the summer months. However, in the southern United States, it is predominant during the winter months, and it is not typically found during the hot summer months. There is a large variation in the adult morphology of this species. This variation includes the size of the eyes, and the number of sternopleural bristles and discal scutellar bristles. The larvae develop mainly in carrion and are well-known myiasis producers. The larvae also have a great deal of variation in

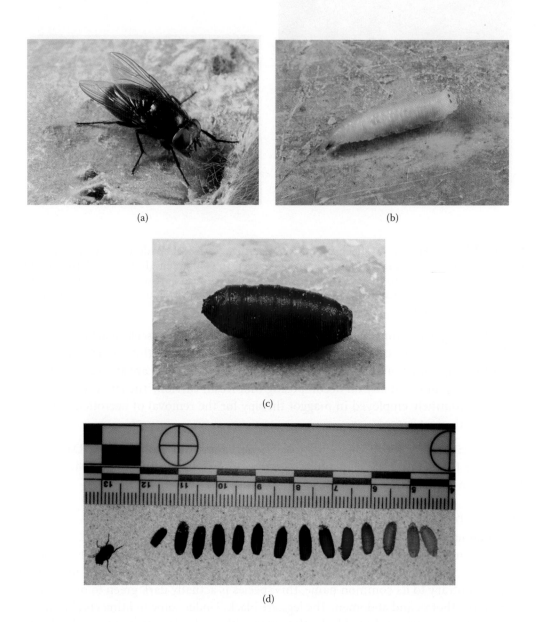

(a)   (b)

(c)

(d)

Figure 2.24 (a) The body of the adult black blow fly (*Phormia regina*) is actually a dark olive green, but it frequently appears black under most lighting conditions. (b) Black mouth hooks and brown posterior spiracles are larval identification characters easily seen on this mature black blow fly maggot. (Photo courtesy of Dr. James L. Castner.) (c) Pupae of this species may be found either on the surface of the remains or in the clothing, since many larvae of this species may not migrate a significant distance. (Photo courtesy of Dr. James L. Castner.) (d) Photography of the pupal stage is important due to the gradual color change as the pupa develops. (Photo courtesy of Dr. Jason. H. Byrd.)

development time. Melvin (1934) found that at 59°F egg hatch took 52 hours, at 104°F just less than 9 hours was required, and at 109°F no egg hatch occurred. Total duration of the larval stage ranges from 4 to 15 days (Bishopp 1915), with a pupal duration of 3 to 13 days, with total development ranging from 10 to 25 days (Bishopp 1915). Like *P. sericata*, this species is used in maggot therapy and historically has been used as a "surgical maggot" to cleanse wounds.

References: Bishopp 1915; Denno and Cothran 1975; Greenberg 1971; Hall 1948; Hall and Townsend 1977; James 1947; Kamal 1958; Melvin 1934; Smith 1986; Zumpt 1965.

Protophormia terraenovae (Robineau-Desvoidy)
(= *Protocalliphora terrae-novae*)
Common name: Bird's nest screwworm fly, Holarctic blow fly
This species of blow fly is Holarctic in distribution and most commonly found in Asia and Europe. In North America is it found most commonly in Alaska and Canada, and in high

(a)

(b)   (c)

Figure 2.25 (a) The Holarctic blow fly (*Protophormia terraenovae*) is most common in the northern United States, and is one of the most cold-adapted species of blow fly. (Photo courtesy of Nikita Vikhrev.) (b) The larva of this species has the spiracular place set into a depression slightly deeper than most Calliphoridae larvae. (Photo courtesy of Dr. James L. Castner.) (c) The puparia are similar to that of other blow flies, and they are often the most abundant species found in the soil surrounding remains in colder, high-altitude geographic regions. (Photo courtesy of Dr. James L. Castner.)

altitudes 3,000+ feet during the summer months in the American West. However, it has been reported from Texas and northern Georgia during the cooler months.

Adult Holarctic blow flies are 7 to 12 mm in length. This species has a dark blue to black body coated with a silver-gray powder. The abdomen is a greenish blue to blue, but with its powdery coating appears tessellated. The legs of this fly are black. Depending on the condition of the specimen, darker black longitudinal stripes may be visible on the dorsal surface of the thorax, starting immediately behind the head. This species is rather large in size and appears more hairy than most of the Calliphoridae (with the exception of *Calliphora*).

Although most common during the spring months, *P. terraenovae* can be collected throughout the summer in the higher elevations of the United States. Cool weather favors development, and this species is the most cold tolerant of all calliphorid species. It is the northern counterpart to *P. regina*, whereas it has much the same behavior, but ranges much farther north. In Canada it is an early spring species and it is most abundant in July. The larvae develop primarily on carrion, but they have been recorded as myiasis producers in livestock and as parasites of birds.

References: Greenberg 1971; Hall 1948; Hall and Townsend 1977; James 1947; Kamal 1958; Nuorteva 1977, 1987; Smith 1986.

Flesh Flies (Family Sarcophagidae)

Flesh flies comprise a large family with over 2,000 species, approximately 327 of which occur in the United States and Canada. Representatives of this family are found throughout the world, with most species occurring either in tropical or warm temperate regions. Adults are common and often locally abundant. They feed on decomposing human and animal tissues, as well as on decomposing vegetation. The adults are often found on flowers, where they are attracted to nectar. They feed on other sweet substances as well, including sap and honeydew. However, the family's Latin name means "flesh eating" and apparently refers to the larvae or maggots that typically feed on some sort of animal material. In addition to carrion, they may also feed on excrement or exposed meats. They have been known to cause myiasis and may be involved in the mechanical transmission of diseases. Many species of sarcophagid or flesh flies are parasitic on other insects, especially bees and wasps. At least one species is beneficial to man, however, and serves as a major natural control of the forest tent caterpillar.

Flesh flies are medium sized and range in length from 2 to 14 mm. The adults commonly have gray and black longitudinal stripes on the thorax and have a tessellated (checkerboard) pattern on the abdomen. Although they are roughly the same size as the blow flies and the bottle flies (family Calliphoridae), flesh fly adults never have a metallic coloration like the others. Also, the arista of the antennae of flesh flies is plumose only at the base, while it is plumose throughout its length in calliphorids. The bodies of sarcophagids tend to be bristly, and the eyes are fairly widely separated in both sexes. In some species the eyes are bright red in color, as are the highly visible genitalia at the tip of the abdomen.

The larvae of flesh flies have the posterior spiracles located in a pit or depression at the tip of the abdomen, which is edged with fleshy tubercles. This character can be used to differentiate between flesh fly and blow fly larvae. Species of sarcophagids are similar to one another in both the adult and larval stages, and are notoriously difficult to identify. Maggots should always be reared to the adult stage to facilitate positive species identification.

Flesh flies are attracted to carrion under most conditions, including sun, shade, dry, wet, indoors, and outdoors. They are frequently found on any decomposing tissues located within an indoor environment. They can be found associated with carcasses throughout both the early and late stages of decomposition. Female flies in this family deposit living first-instar larvae on decomposing remains. They do not lay eggs, and thus fly egg masses associated with human remains cannot be attributed to sarcophagids. The time period necessary for egg development must also be eliminated when calculating a PMI based on flesh fly evidence.

Flies of the genus *Sarcophaga* arrive concurrently with, or slightly after, the blow flies on human remains. They are known to fly under inclement conditions that would prevent the flight of most other flies, and therefore may actually be the first species to arrive on carrion. The adults commonly enter indoor habitats, and thus the larvae are often recovered from human remains located in homes. This is the predominant fly recovered on human bodies located in indoor habitats during the summer months in the southeastern United States.

References: Aldrich 1916; Arnett and Jacques 1981; Bland and Jacques 1978; Borror and White 1970; Borror et al. 1989; Castner et al. 1995; Greenberg 1971; Hall and Doisy 1993; Hogue 1993; James 1947; James and Harwood 1969; Knipling 1936; Payne 1965; Peterson 1979; Shewell 1987; Smith 1956, 1975, 1986.

Sarcophaga bullata (Parker)

This species occurs only in the Nearctic region, or that part of North America north of Mexico. Although it can be found from Canada south to Florida, and west to California and Washington, it is most common in the southern United States. Adults of *S. bullata* range from 8 to 14 mm in length. They appear very much like the adults of *S. haemorrhoidalis*, to which they are closely related. They are also very similar in behavior and habitat preferences.

References: Aldrich 1916; James 1947.

Sarcophaga haemorrhoidalis (Fallen)

Common name: Red-tailed flesh fly

The red-tailed flesh fly is cosmopolitan in distribution, occurring worldwide. Investigators may find it associated with human remains throughout the United States and Canada. It is a large species that ranges from 8 to 14 mm in size. This fly appears very similar to flies in the family Muscidae. However, it is typically larger, and its abdomen terminates in a red tip, which is actually the external genitalia. Although named the red-tailed flesh fly, it shares this characteristic with other sarcophagid species as well. The body is black, but covered in whitish powder that gives it an ash-gray appearance. It has three dark longitudinal stripes on the thorax between the wing bases. These stripes do not continue on the abdomen, which has a black and gray tessellated (checkerboard) pattern.

The adults of this species are attracted to both excrement and carrion. Females commonly enter dwellings to deposit their larvae, which are frequently found on human corpses located indoors, particularly during the summer months. Eggs are not laid, as they hatch within the body of the female. Therefore, it is the living first-instar larva that is deposited on carrion or excrement. This species can infect the wounds or digestive tracts of living

(a)

(b)

(c)

Figure 2.26 (a) Dorsal view of the red-tailed flesh fly (*Sarcophaga bullata*) showing the red eyes and the red tip of the abdomen, which is actually the external genitalia. The adults are characterized by having a gray and black body, with stripes on the thorax and a checkerboard pattern on the abdomen. (b) Dorsal view of a mature maggot showing the darkened mouth hooks at the anterior end (bottom) and the fleshy tubercles at the posterior end (top). (c) Illustration of the color change of the pupal stage. The ages of the pupae vary by only a couple of hours. Photography greatly aids in establishing the age of the pupal stage. (Photo courtesy of Dr. James L. Castner.)

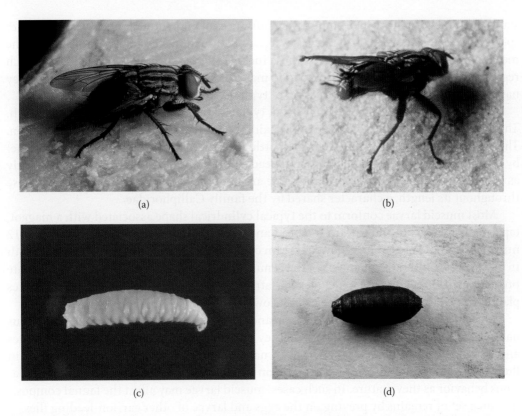

(a)   (b)

(c)   (d)

Figure 2.27 (a) The large-size, bright red eyes, and the noticeably red genitalia make the red-tailed flesh fly (*Sarcophaga haemorrhoidalis*) easily recognizable in the field. (b) This view of the adult fly clearly shows the red tip of the abdomen. (c) Lateral view of a mature maggot showing the typical maggot body form. Note the darker spot at the anterior end (right), which is the thoracic spiracle. (d) The fully formed pupae of the red-tailed flesh fly are larger than those of the blow flies, and they are usually found away from the remains since the larvae undergo active migration. (Photo courtesy of Dr. James L. Castner.)

animals, particularly mammals. *S. haemorrhoidalis* has been recorded as even completing development entirely within the human digestive tract.

References: Aldrich 1916; Greenberg 1971; James 1947; Smith 1975, 1986; Zumpt 1952.

Muscid Flies (Family Muscidae)

Muscid flies belong to a large family that is worldwide in distribution, with over 700 species in North America alone. Many species are ubiquitous and synanthropic (found closely associated with man). This habit has often resulted in the unintentional transportation and introduction of species to areas outside their natural ranges. The propensity with which muscid flies are found in domestic situations contributes to both their medical and forensic importance. Common members of this family are the house fly, stable fly, horn fly, and latrine fly. The tsetse fly, which vectors African sleeping sickness, is also a muscid fly.

The biology and habits of muscid flies are quite varied, as the adults may feed on decaying plant and animal material, dung or excrement, pollen, or even blood. In the latter case, stable flies and horn flies can become severe nuisances of both humans and livestock. Other species that breed and feed on garbage, sewage, and human waste can be responsible for the mechanical transmission of diseases. These include species such as the house flies, which regurgitate digestive fluid directly onto food materials as part of their feeding process. They have been implicated in the transmission of typhoid, anthrax, dysentery, and yaws.

Muscid flies are small to medium sized, typically ranging from 3 to 10 mm in length. They tend to be dull gray to dark in color, although a few species have a metallic sheen. These species can be separated from the bottle flies and blow flies (family Calliphoridae) by taxonomic characters of the wings and head. Adult muscids are generally not as bristly as the sarcophagid and calliphorid flies. The arista of the antennae of muscids is plumose throughout its length, a character shared by the family Calliphoridae.

Most muscid larvae conform to the typical cylindrical shape associated with a maggot, tapering from the tail end toward the head and mouth. Mature larvae range from 5 to 12 mm in length and are white, yellow, or cream colored. The surface of the maggot is smooth in most species, although members of the genus *Fannia* are flattened and have many ornate projections. This genus, which includes the latrine fly and the lesser house fly, is sometimes placed in the family Anthomyiidae or in its own family (the Faniidae).

Muscid flies are of great forensic importance due to their wide distribution, ubiquitous nature, and close association with man. They tend to arrive at bodies after the flesh flies and blow flies. They often lay their eggs in natural body openings, at wound sites, or in fluid-soaked clothing. Larvae feed directly on carrion, but in some species exhibit predacious behavior as they mature. In such cases, muscid larvae may affect the faunal composition on a set of remains by preying on the eggs and larvae of other carrion-feeding flies.

References: Arnett and Jacques 1981; Bland and Jacques 1978; Borror and White 1970; Borror et al. 1989; Castner et al. 1995; Hogue 1993; Huckett and Vockeroth 1987; James and Harwood 1969; Peterson 1979.

Fannia canicularis (Linnaeus)

Common name: Lesser house fly

The lesser house fly (also called the little house fly) is cosmopolitan in distribution. This species is a small, slender fly usually attaining a length of 6 to 7 mm. It has a black to dark brown thorax and an abdomen coated with silvery-gray powder. The legs appear yellowish in color. The thorax has three longitudinal brown stripes that do not continue onto the abdomen.

The lesser house fly is most abundant during warm summer months, and it is the species of *Fannia* most commonly involved in myiasis. They are most often found on excrement, decomposing vegetable matter (such as fruits), dairy products, and materials saturated with urine. They are often recovered on decomposing human remains when excrement or gut contents have been exposed. The adults are one of the most common species of Muscidae found in homes in some parts of the United States.

The maggots develop well on decomposing substances that are liquefying. As an adaptation to this type of a habitat, the larvae are flattened with branched protuberances that apparently aid in flotation in semiliquid environments. The larvae of *F. canicularis* are similar in appearance to those of *F. scalaris*, but the protuberances are not as branched. They can also grow longer than *F. scalaris* (the latrine fly). The larvae will migrate to drier habitats in order to pupate. The adults are prone to enter indoor habitats where they may be seen hovering in mid-air or flying about in an erratic manner. They are often considered

(a)

(b)

(c)

(d)

Figure 2.28 (a) The lesser house fly (*Fannia canicularis*) is a common pest in chicken houses, and can be found at decomposing remains when excrement or exposed gut contents are present. (Photograph courtesy of Nikita Vikhrev.) (b) The pupa has rings of projections on each segment, which are remnants of the fleshy projections on the larval body. (Photo courtesy of Dr. James L. Castner.) (c) Long, backward-pointing projections that adorn the larvae are an adaptation to living in a semiliquid environment. (Photo courtesy of Dr. James L. Castner.) (d) A group of larvae on artificial medium. Due to their color, shape, small size, and occurrence in relatively low numbers, these larvae often go unnoticed by the investigator. (Photo courtesy of Dr. James L. Castner.)

nuisances, as they can aggregate in relatively large numbers indoors during certain periods of the year. The adults are strongly attracted to both carrion and excrement.

References: Greenberg 1971; James 1947; Nuorteva 1974; Payne 1965; Smith 1986; Wasti 1972.

Fannia scalaris (Fabricius)
Common name: Latrine fly

The latrine fly is worldwide or cosmopolitan in distribution. It is very similar to the lesser house fly in appearance. Adults are typically 6 to 7 mm long. This species is a black fly with the thorax and abdomen having a dusty silver gray coat that makes it appear superficially like a small house fly. The larvae are small and flattened, usually reaching a length of no more than 6 mm. The immatures possess a pair of lateral processes (tubercles) on each body segment. These protuberances are feathered and may aid in flotation in their preferred habitat of semiliquid fecal masses.

(a)

(b)

(c)

Figure 2.29 (a) The adult latrine fly (*Fannia scalaris*) does not enter indoor locations to deposit its eggs as commonly as its close relative the lesser house fly. (b) The flattened larva has feathery projections that are more plumose than those found on the larva of the lesser house fly. It is thought that these projections may aid in flotation, an adaptation to life in their semiliquid environment. (c) The hardened brown puparium has the same noticeable projections found in the larval form. (Photo courtesy of Dr. James L. Castner.)

Eggs are typically laid in human or animal dung, as well as decaying vegetable material. The larvae can develop in anything from dung and carrion to the nests of birds and other insects to human cadavers. The latter is especially true when excrement or exposed gut contents are present. The larvae of this species are likely to be encountered on urine-soaked clothes and on other such soiled materials. This species is not as prone to enter indoor habitats as is *F. canicularis*. The presence of the latrine fly indoors is usually indicative of unsanitary conditions.

References: Greenberg 1971; James 1947; Nuorteva 1974; Smith 1986.

(a)   (b)

(c)

Figure 2.30 (a) The dark, shiny adult of the bronze dump fly (*Hydrotaea aenescens*) tends to have a gold or bronze tint. This species readily enters dwelling to colonize remains. (b) The larvae are predatory and develop in both carrion and excrement. (c) The puparia are generally smaller than that of the blow flies, but have the same characteristic appearance. (Photo courtesy of Dr. James L. Castner.)

Hydrotaea (= Ophyra) aenescens (Wiedemann)

Common name: Bronze dump fly

In the United States, the bronze dump fly is distributed from Oregon to Arizona, Illinois to Florida, and along the Atlantic Coast. It is also found throughout Central and South America. This is a small, shining, blue-black species with a bronze tint. Like *H. leucostoma*, it readily enters dwellings seeking suitable oviposition sites and food sources. Although the adults are attracted to both carrion and excrement, the maggots develop primarily in feces. The larvae of *H. aenescens* are predatory. They may be recovered on human cadavers during the late or active decay stage. Larvae are often found in body exudates that have soaked into the soil beneath remains, and are commonly associated with exposed gut contents. The bronze dump fly is most abundant during the summer months in the United States.

References: Greenberg 1971; James 1947; Smith 1986.

Hydrotaea (=Ophyra) leucostoma (Wiedemann)

Common name: Black dump fly

The black dump fly is found throughout the United States. Although Holarctic in distribution, it is restricted to the area north of Mexico in North America. This is a species with shining blue-black to black adults. The flies are attracted to both carrion and excrement,

(a)

(b)

(c)

Figure 2.31 (a) Lateral view of the adult black dump fly (*Hydrotaea leucostoma*) showing the shiny jet black body and legs. This species as well as the bronze dump fly are most attracted to excrement. (b) *Hydrotaea* are small and slender when compared to most other Muscidae larvae, and are smaller than all Calliphoridae larva. (c) The younger puparium (left) is lighter in color and reddish brown, while the older puparium (right) is dark brown. (Photo courtesy of Dr. James L. Castner.)

where they frequently deposit their eggs. The adults may be noticed hovering either above or near decomposing carcasses on days with little to no wind. The larvae develop primarily in feces, and are predatory in nature. They frequently attack the larvae of *M. domestica* when they occur together. The larvae usually appear during the late or active decay stages on human cadavers and are often found in the fluid-soaked soil underneath the body. They are most abundant during the summer months. This is one of several species that is frequently recovered from remains found in indoor habitats.

References: Greenberg 1971; James 1947; Smith 1986.

Musca domestica (Linnaeus)

Common name: House fly

The house fly is worldwide in distribution. Although it is found throughout the United States and Canada, it may not be the predominant fly species in all geographical areas. The adults are 6 to 7 mm in length. They are gray in color with black longitudinal stripes on the thorax that continue onto the abdomen. This species is almost never found away from man and his dwellings, and it is a major nuisance due to its movements from garbage, carrion, and excrement to

(a)

(b)   (c)

Figure 2.32 (a) Adult house fly (*Musca domestica*) is found in urban environments associated with human activity throughout the world. (Photograph courtesy of Nikita Vikhrev.) (b) A mature maggot is visibly thicker at the posterior end (left), giving it the typical club-like appearance, as seen here. (Photo courtesy of Dr. James L. Castner.) (c) The puparium is typically smooth, cylindrical, and dark brown when mature. (Photo courtesy of Dr. James L. Castner.)

human food. It is a vector of over a hundred disease-causing pathogens. Both adults and larvae prefer excrement and decomposing vegetable matter. The adults are also attracted to sweet foods and meat, and their larvae can adequately develop in these materials as well.

The adults will readily enter dwellings in order to colonize decomposing remains. They are among the first flies attracted to excrement, and are also attracted to carrion, usually following the blow flies. This species can be found year-round throughout its range, but it is most abundant during the summer months. In temperate zones, the population reaches its peak in spring and late summer. In the tropics and subtropics, house flies are most abundant during the hot summer months. The presence of this species on a fresh corpse is rare, unless excrement is present or gut contents are exposed. This species may be present during active or advanced decomposition.

References: Chapman 1944; Greenberg 1971; James 1947; Smith 1986.

Synthesiomyia nudiseta (Van Der Wulp)

This species is found throughout the tropical and subtropical regions of the world. In the United States it has been collected primarily from California to Texas, and from North Carolina to Florida. This fly is one of the largest muscids, approximately 7 to 10 mm in length. It has a gray color with a tessellated or checkerboard pattern on the abdomen, closely resembling that found on the flesh flies. However, it is easily separated from sarcophagids because the thorax is patterned with four distinct longitudinal stripes instead of three, and the terminal segment of the abdomen appears yellow (or bright orange) instead of red. The antenna and palpi are also yellowish to orange in color. The larvae are commonly found in animal and human feces as well as in decaying vegetable materials, refuse, and garbage. However, they have been reported to prefer carrion as the food source of choice.

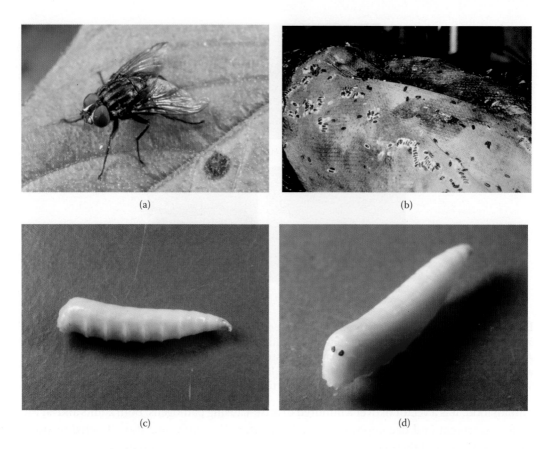

(a)   (b)

(c)   (d)

Figure 2.33 (a) The adult *Synthesiomyia nudiseta* is a muscid fly. Although somewhat similar to a flesh fly in appearance, it has four dark stripes on the thorax rather than three. (b) This is the typical appearance of the puparia on human remains. Note the silky white material produced by the pupating larvae that surrounds the puparia on the clothing. In outdoor environments, the puparia may be covered with small soil particles and have a more cocoon-like appearance. (c) Lateral view of the mature larva showing that the posterior segment is smooth and does not have tubercles on the posterior segment (as in the blow flies or flesh flies). (d) The posterior spiracles are smaller and more widely separated than in most species of maggots, and they are located on the smooth, convex posterior segment. (Photo courtesy of Dr. James L. Castner.)

Although this species have been noted to arrive along with flies in the genus *Sarcophaga* to deposit their eggs, their larvae develop more slowly and pupate with the larvae of later-arriving fly species. The larvae of this species are predacious, and are one of the few species that is noted to consume the larvae of *Chrysomya rufifacies*. The puparia of this species may be difficult to recognize as the larvae often secrete a silky white substance in which the puparium forms. This substance, which is white and cottony in appearance, hardens and becomes an effective cement that serves to hold the pupa in place and attached to its substrate. Additionally, soil particles may also become cemented to the outer surface to help form this protective "cocoon."

References: Greenberg 1971; James 1947; Lord et al. 1992; Rabinovich 1970.

Skipper Flies (Family Piophilidae)

The skipper flies comprise a small family of only sixty-nine species, but are found worldwide, where their greatest diversity is found in temperate regions. Adults are metallic blue or black in color and typically range from 2.5 to 4.5 mm in length. Skipper flies are found in a variety of habitats that may include carrion, human waste, bones, skin, and fur. They are common and usually associated with protein-rich food sources that are dry in nature.

The common name *skipper* comes from an unusual behavior exhibited by the larvae of some species, including the cheese skipper (*Piophila casei*). The larvae will grasp small protrusions on the anal segment with the mouth hooks and suddenly release their grip (Figure 2.34c). This action flings the larva into the air 3 to 4 inches, and laterally over a distance of 6 to 8 inches. This "jumping" behavior is used as an effective "escape" mechanism, and it is also utilized extensively during larval migration. However, they also move in the more traditional creeping manner exhibited by most fly larvae.

Skipper fly larvae are primarily scavengers and feed on the same types of decaying food materials on which the adults are found. One group are specialty feeders on rotting fungi, while in Europe there is a species whose larva is a blood-sucking ectoparasite of young birds. The maggots of skipper flies can be considerably larger than the adults and usually range from 5 to 10 mm in length.

References: Arnett and Jacques 1981; Bland and Jacques 1978; Borror and White 1970; Borror et al. 1989; Castner et al. 1995; Hogue 1993; James and Harwood 1969; McAlpine 1987; Peterson 1979; Simmons 1927).

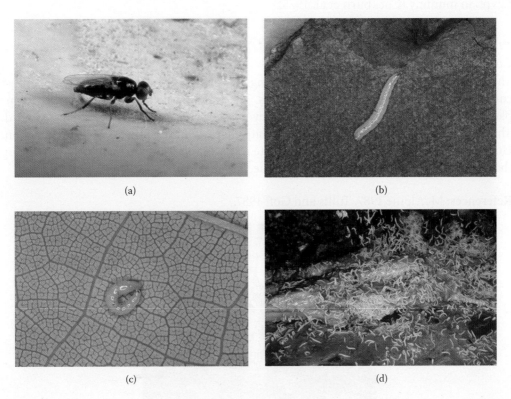

(a)   (b)

(c)   (d)

Figure 2.34 (a) The adult cheese skipper (*Piophila casei*) is a small, fast-moving fly with a shiny black body and black and yellow legs. (b) The larva is also more slender than that of blow flies. (c) Prior to using its unique jumping ability, which is an evasive and locomotory maneuver, the maggot will form itself into a tight loop before springing into the air. (d) Hundreds of larvae often infest human remains, especially if the body is located in a moist and shaded environment. (Photo courtesy of Dr. James L. Castner.)

Piophila casei (Linnaeus)

Common name: Cheese skipper, jumping maggot

Cheese skippers are cosmopolitan in distribution and found throughout the world. They are small (2.5 to 4.5 mm) shining black flies with yellow legs. The adults are found on carrion, excrement, and garbage. Observations show that they prefer darker, shaded areas and hold their wings flat over the back when at rest. Almost any protein-rich food source is a suitable substrate for the developing larvae.

Cheese skippers are a serious pest of the food industry. The adults lay eggs on cheeses, bacon, smoked fish, and cured meats. This preference for some of the same food items used by man has resulted in numerous cases of immature cheese skippers being ingested. Surprisingly, this species is able to survive and pass through the digestive system of humans, and their presence results in a condition referred to as enteric pseudomyiasis.

The larvae are often recovered on human bodies after active decay, as the body begins to dry. Adults have been recovered at remains only 3 to 4 days after death; however, it is important to note that the presence of an adult insect does not necessarily mean that oviposition (colonization) is occurring. Although skipper flies in general seem to prefer a food substrate that is dry, they are often recovered from bodies found in aquatic situations. The puparia of this species have been recovered on human organs inside of a 2,000-year-old Egyptian mummy (Cockburn et al. 1975).

Prochyliza sp. (Walker)

Common name: Waltzing flies

This genus is quite common on carrion during the spring, summer, and fall. They are most commonly found in bright, sunny locations. The waltzing flies are noted for their complex courtship and combat displays, and they possess a characteristically elongated head. The adult can be found at carrion that is relatively early in the process of decomposition. However, the larvae are not evident until late in decomposition and can be found during the dry remains state. As in the Calliphoridae, larvae pupate in the soil.

References: McAlpine 1987, Tullis and Goff 1987.

Figure 2.35 Flies in the genus *Prochyliza* are very small, but have distinctive elongated head capsules. The flies have elaborate courtship and combat displays that sometimes last for several minutes. This behavior is often observed by crime scene analysts while processing the remains. (Photo courtesy of Jay Cossey, www.photographsfromnature.com.)

Dung Flies (Family Scathophagidae)

The dung flies are treated here as a separate family, although other references may include them as a subfamily under the family Muscidae or family Anthomyiidae. The Latin name for this family means "dung eating," and although *Scathophagidae* is most widely used, some authors refer to the family as the *Scatophagidae*. There are over 250 species, almost all of which are found in the Holarctic region. Approximately 150 species occur in temperate North America, with the greatest diversity found in southern Canada.

The most common species of dung flies are red or yellow and densely hairy. Some species are attracted to excrement or decaying plant material, and their larvae are most commonly found in excrement. Other species are dark as adults and have larvae that are leaf miners, pests of flower heads, parasitic, predacious, and aquatic. The larvae of the genus *Scatophaga* feed on dung, but are also often found on decaying seaweed.

References: Arnett and Jacques 1981; Bland and Jacques 1978; Borror and White 1970; Borror et al. 1989; Castner et al. 1995; Hogue 1993; James and Harwood 1969; Peterson 1979; Vockeroth 1987.

Scatophaga stercoraria Linnaeus
Common name: Common dung fly

The common dung fly is found throughout the United States and Canada. The adult is slender and reaches a length of 7 to 10 mm. Although the body itself is black, it is covered with a dense coat of yellow or yellow-brown hair, thus making the entire insect appear bright yellow. The thorax has several long, black hairs protruding from it, and the wings are typically held flat over the back when at rest. The adults are present from spring until late fall. The larvae occur in a variety of habitats, but can be found on cadavers when excrement is present or the gut contents have become exposed.

Figure 2.36 The adult of the common dung fly (*Scatophaga stercoraria*) is a medium-sized fly whose body and legs are covered with long yellow hairs. They are attracted to decomposing remains when excrement is present. (Photo courtesy of Steve Graser, www.BeautyOfNature. net.)

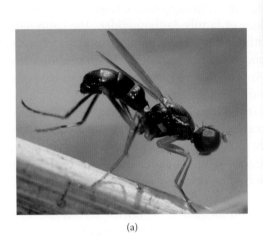

(a)   (b)

Figure 2.37 (a) A typical black scavenger fly (*Sepsis* sp.) has a rounded head and a constriction between the thorax and the abdomen. They also have the behavioral habit of waving their wings as they walk. (b) An additional identification characteristic seen on many species is a black spot near the tip of each wing. (Photos courtesy of Nikita Vikhrev.)

Black Scavenger Flies (Family Sepsidae)

Black scavenger flies or sepsids are worldwide in distribution and represented by at least 240 species. The adults are small, shining black, purple, or red flies that are usually no more than 3.5 mm in length. They have a characteristic shape due to a head that is noticeably rounded and a constriction or narrowing of the abdomen at the base. Despite their small size, they are easily identified by the behavioral characteristic of flicking their wings outward as they walk. This habit of wing waving has also given them the common name of waggle flies. These flies often occur in large numbers and are very common on dung.

Sepsid fly larvae develop in a variety of decomposing organic matter, including carrion and excrement. Some species are found in decaying seaweed. Mature larvae are small, ranging from 3 to 6 mm in length.

References: Arnett and Jacques 1981; Bland and Jacques 1978; Borror and White 1970; Borror et al. 1989; Castner et al. 1995; Hogue 1993; James and Harwood 1969; Streyskal 1987; Peterson 1979.

Sepsis sp.

Common name: Black scavenger flies

Black scavenger flies of the genus *Sepsis* are worldwide in distribution. They are commonly found throughout the United States and Canada. The adults are shiny black and show all the other features described previously under the family Sepsidae. In addition, adult species of *Sepsis* also have a dark spot near the wing tip along the front margin.

These flies are attracted to carrion, excrement, and other types of decaying matter. They are sometimes mistaken for ants due to their small size and narrow "waist" (constricted abdomen). Black scavenger flies can be abundant on carcasses, especially during the advanced stages of decay, as well as on excrement and other types of decaying matter.

References: Greenberg 1971; James 1947; Smith 1975, 1986.

Small Dung Flies, Minute Scavenger Flies (Family Sphaeroceridae)

The small dung flies are a group with cosmopolitan distribution and over 240 species in North America. They can be found throughout the United States and Canada. They are small flies, ranging from 1 to 5 mm in length, and typically a dull black or brown color. Taxonomic characters of the hind legs and the wing venation are used in their identification to the family level.

Sphaerocerid fly adults and larvae are found in association with dung and excrement. They are a common part of the insect fauna that inhabits cow dung, and the transport of which is probably largely responsible for the spread of certain species. They are often found with the larger dung flies (family Scatophagidae) and the black scavenger flies (family Sepsidae).

In addition to animal waste, the small dung flies may also be found on carrion, refuse, seaweed, fungi, and other types of decaying organic matter. The presence of excrement or soiled clothing on human remains probably greatly increases the chance of encountering small dung flies in association with human remains. Unfortunately, little research has been done on the developmental times of these flies.

References: Arnett and Jacques 1981; Bland and Jacques 1978; Borror and White 1970; Borror et al. 1989; Castner et al. 1995; Hogue 1993; James and Harwood 1969; Marshall and Richards 1987; Peterson 1979.

Poecilosomella angulata (Thomson)
Common name: Small dung fly

Figure 2.38 The adult of the minute scavenger fly (*Poecilosomella angulata*) is only 2 to 3 mm long, and rarely recovered at human remains unless excrement is present. (Photo courtesy of Dr. James L. Castner.)

This species is small, only 2 to 3 mm long. It is not often recovered from human remains unless excrement or excrement-soiled clothing is present. On occasion, they can also be attracted when gut contents are exposed or when the soil has become saturated with gut contents. They are common at very large vertebrate carcasses (i.e., bovines) where large amounts of gut contents are exposed during the process of decomposition. This species has been responsible for cases of intestinal myiasis.

References: Greenberg 1971, McAlpine 1987.

Soldier Flies (Family Stratiomyidae)

Soldier flies encompass more than 250 described species that range in size from 5 to 20 mm. The color of adults varies, although many of the most common species are wasp-like in appearance. The antennae of soldier flies consist of three segments, the last of which is either elongate or rounded, with a long hair protruding from it. Adults are often found at flowers and occur near woods and on vegetation in wet areas, as well as on almost any decomposing plant or animal tissue. The common name of this family derives from the spines found on the adults of certain species, which led some observers to consider them "armed" like soldiers.

Mature larvae vary greatly in length, from 4 to 40 mm, depending upon the species. Some are purely aquatic, while others are found in terrestrial habitats. The terrestrial larvae usually develop in decomposing plant or animal material, although in some cases they may be predacious. Rotting wood, compost piles, and excrement are typical habitats. The aquatic forms can sometimes be found in surprisingly hostile environments such as hot springs or extremely saline waters. Soldier fly larvae are sometimes ingested through the consumption of infested fruit, subsequently appearing in stools. Several cases of such enteric pseudomyiasis have been reported.

References: Arnett and Jacques 1981; Bland and Jacques 1978; Borror and White 1970; Borror et al. 1989; Castner et al. 1995; Hogue 1993; James 1981; James and Harwood 1969; Peterson 1979.

Hermetia illucens (Linnaeus)

Common name: Black soldier fly

This species is Neotropical in origin, but now found in almost all temperate and tropical areas of the world, where it has probably been transported via contaminated food. The black soldier fly is found throughout the United States and Canada. It is very common in the southeastern United States, and least common in the Northwest. Adults are very wasp-like in appearance and are even sometimes referred to as the wasp fly.

Adult flies are approximately 15 to 20 mm long. The females are entirely blue-black, while the males have an abdomen that is more brown in color. The tips of the legs of both sexes are white. The wings appear smokey-black and are held flat over the back (wasp-like) when at rest. The abdomen is elongate and narrow at the base, with the first two segments bearing translucent areas. This feature contributes to their narrow waist appearance and aids in the resemblance of a wasp. In addition, the flies will also behave in a wasp-like manner, moving their antennae rapidly, running about in an agitated manner, and sometimes even pretending to sting.

The larvae of the black soldier fly are broad, flattened, and distinctly segmented with a narrow prominent "head." Each body segment has a row of broad stout hairs. Larvae are

(a)   (b)

Figure 2.39 (a) The adult black soldier fly (*Hermetia illucens*) is extremely wasp-like in appearance and behavior. It can be found at remains throughout the early and advanced stages of decay. (b) The large larva is flattened, distinctly segmented, and has a pronounced head capsule. Although easily noticeable, they are extremely slow moving, causing many investigators to assume they are dead. (Photo courtesy of Dr. James L. Castner.)

gray to brown in color and have a hard, thick surface. The body has a cobblestone texture when viewed under magnification, and this type of surface and texture is described as shagreened. An outer layer of calcium carbonate deposited on the skin, and serving as protection from desiccation during dry periods, causes their unusual external appearance. The larvae develop on many different types of decaying organic material. Under some circumstances, the black soldier fly is considered a beneficial species because it suppresses house fly populations through competition for the same food source. Larvae are commonly found on rotting fruits and vegetables, in compost and refuse, in excrement, and on carrion. When associated with human remains, the larvae usually occur during the advanced to dry stages of decay.

References: Bohart and Gressitt 1951; Dunn 1916; Greenberg 1971; James 1947; Lord et al. 1994; Payne and King 1970; Reed 1958; Smith 1975, 1986.

Humpbacked Flies or Scuttle Flies (Family Phoridae)

The humpbacked flies are a large family with more than 2,500 species worldwide. Nearly half of these species belong to the single genus *Megaselia*. Approximately 226 species of phorid flies are currently recognized in the United States, and a total of approximately 350 in North America.

Phorid flies are small, ranging in length from about 1.5 to 6 mm. They are easily recognized by their humpbacked appearance, which is especially noticeable when the flies are viewed in profile. Other taxonomic characters helpful in their identification are the flattened femora of the hind legs and the heavy veins found at the base of the anterior wing margin. Humpbacked flies may be black, brown, or yellow in color.

Adult phorid flies are found in a variety of situations. They are commonly associated with decaying plant matter and are often unwanted yet ubiquitous pests where live insect colonies are maintained. The adult insect will run in a very characteristic swift and erratic manner, which has earned them the common name of scuttle flies. The larvae typically develop in any decomposing organic matter, whether human, animal, or vegetative in origin. The puparia are easily recognized as they are dorsoventrally flattened, with a pair of horns or "breathing trumpets" emerging from the anterior end. Some species are predacious while others are parasitic. A few are commensal in the nests of ants and termites, while other larvae develop in fungi.

References: Arnett and Jacques 1981; Bland and Jacques 1978; Borror and White 1970; Borror et al. 1989; Castner et al. 1995; Hogue 1993; James and Harwood 1969; Peterson 1979, 1987.

Megaselia scalaris (Loew)
Common name: Humpbacked fly

This species is distributed throughout the world and its members are generally similar in appearance to the familiar fruit or vinegar flies (genus *Drosophila*). The thorax is yellow to yellowish brown, and on close inspection the abdomen is black above, with yellow to white sections both laterally and ventrally. The legs are yellow and the abdomen is yellowish with brown bands. The larvae of *M. scalaris* can develop in vegetation, car-

(a)

(b)

(c)

Figure 2.40 (a) This humpbacked or scuttle fly (*Megaselia scalaris*) will run in a very erratic manner, moving quickly for a short distance, then stopping briefly to change direction. (b) The small white larvae are often found feeding on the outer surfaces of the body, or developing in clothing that has become soiled from the process of decomposition. They are also serious pests, especially in laboratory insect colonies. (c) The small pupae, which possess distinctive respiratory horns, are sometimes recovered in enormous numbers from human remains recovered in indoor locations. Here, they are pictured on the sole of a shoe. (Photo courtesy of Dr. James L. Castner.)

Figure 2.41 *Conicera tibialis* can be common on human remains that have been buried. Exhumations have shown that this insect species can develop for many generations within a sealed casket. (Photo courtesy of Paul Beuk.)

rion, or excrement. This species has been recorded as causing cutaneous, intestinal, and ophthalmic myiasis.

References: Greenberg 1971; Hall 1948; Hall and Townsend 1977; James 1947; Smith 1986.

Conicera tibialis (Schmitz)
Common name: Coffin fly

This is a very small black species (1.5 to 2.5 mm) often found associated with buried remains. Eggs are laid on the soil surface, and the larvae of the coffin fly burrow down through holes and fissures in order to colonize bodies that have been buried. Research has shown that this and other species of phorid flies can reach remains that are buried from 30 to 100 cm deep. Material of this species recovered during exhumations has shown that apparently the coffin fly can develop and progress through many generations on buried human cadavers once they are colonized.

Large numbers of adults have been reported on the soil surface of gravesites. This behavior can be used to aid in the location of buried human remains. It should be noted, however, that adult flies tend to be associated with remains that have been buried for at least a period of months, and in some cases for years.

References: Colyer 1954a, 1954b, 1954c; Greenberg 1971; Hall 1948; Hall and Townsend 1977; James 1947; Payne et al. 1968; Smith 1986.

Moth Flies, Sand Flies, and Owl Midges (Family Psychodidae)

Moth flies are found throughout the world, with more than ninety species in the United States and Canada. They are very small as adults, usually no more than 3 to 4 mm in length. The body and wings are covered with scales (hairs) that give them a moth-like appearance. The wings come to a point at the tip and have straight, prominent longitudinal veins. The larvae develop in moist environments and may be up to 10 mm long.

Two groups (or subfamilies) within the family Psychodidae have medical importance. Members of the subfamily Phlebotominae are referred to as sand flies, and are characterized by nonaquatic larvae and adults that do not hold their wings roof-like over the body. They occur in the southern United States and in tropical areas of the world. Female sand flies suck blood and are responsible for vectoring the disease leishmanisasis, as well as several others.

Members of the subfamily Psychodinae are called moth flies or owl midges. The larvae are usually aquatic and typically develop in the slime that accumulates in sewers and drains. The adult flies tend to hold their wings roof-like over the body. Species in this subfamily do not take blood, but are often considered pests due to their sheer numbers. They are common in warm, damp places and are often seen indoors on the walls of restrooms or showers. Adults run about and fly very quickly in an erratic, haphazard, and very distinctive manner.

Psychodid flies seldom constitute more than a minor portion of the arthropod fauna that colonizes human remains. However, adult moth flies are occasionally attracted to carrion and are a commonplace insect found indoors and associated with man.

References: Arnett and Jacques 1981; Bland and Jacques 1978; Borror and White 1970; Borror et al. 1989; Castner et al. 1995; Hogue 1993; James and Harwood 1969; Peterson 1979; Quate and Vockeroth 1981.

Psychoda alternata (Say)
Common name: Moth fly, drain fly, sewage fly, trickling filter fly

This species is found distributed throughout North America. The adults of these small, moth-like flies are gray to brown in color and range from 2 to 4 mm in length. Their wings and body are densely covered with hair. Their wings are noticeably pointed and may have black spots, giving them a mottled appearance. When at rest, the wings are held spread and at a slight angle over the body.

Adults occur throughout the summer and prefer damp places. They are usually seen walking or crawling on walls or other vertical surfaces, and they fly only very short distances. Their characteristic flight is very jerky and erratic. The larvae are long, slender, and grayish white with a noticeable head and dark-colored "tail," which is their respiratory horn or breathing apparatus. The larvae develop in liquid or semiliquid decaying matter,

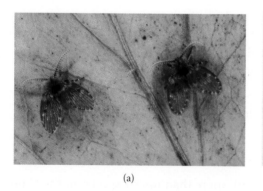

(a)

(b)

Figure 2.42 (a) Adult moth flies (*Psychoda alternata*) have wings that are covered with scales, which give them a moth-like appearance. (b) The long, slender larvae are often found in the slimy material coating drains. They are infrequently recovered from human remains, but may be found if the remains are in shaded moist habitats. (Photo courtesy of Dr. James L. Castner.)

including excrement. They are commonly found in foul water, sewage, and in the filter beds of sewage plants. The latter has resulted in their common name of the trickling filter fly. Any wet decomposing organic matter is suitable larval habitat, however, and they are particularly common in drain pipes and bathrooms.

The larvae of this fly will most likely not inhabit carrion itself, but may be found in the excrement occasionally associated with decomposing remains. Since they have been recovered from both floating and buried bodies, their occurrence on human remains should not be taken casually. Although this species is not considered of major forensic importance, its presence may contribute information when deriving postmortem interval estimations. This species has been reported as causing human myiasis.

References: Bohart and Gressitt 1951; Greenberg 1971; James 1947; Payne et al. 1968; Smith 1986.

Beetles (Order Coleoptera)

Coleoptera is the largest order, containing about a third of all known insects, which is about 300,000 species. About 30,000 of those species are found in North America. This family is of tremendous economic importance. Members of this family are characterized by having hard wing covers (called elytra) that cover and protect the membranous hind wings used for flight. Adult beetles possess chewing mouthparts, and most have the ability to fly. Their feeding habits vary greatly. They can be predacious, scavengers, or vegetarians, with a few being parasitic. The larvae are called grubs, and they vary widely in appearance (Figure 2.43). However, all grubs have a noticeable head and possess six legs. Beetles are known to attack plants, infest stored foods, act as important scavengers and decomposers, and some species serve as pollinators of plants.

Carrion Beetles (Family Silphidae)

The carrion beetles comprise a large family that is nearly worldwide in distribution and has more than 1,500 species. In North America there are approximately forty-six species of silphid beetles, which are widely distributed. Certain members of this family are also

Figure 2.43 Beetle larvae vary in size and shape depending on the family and species. The four larvae pictured here (from left to right) belong to the families Scarabaeidae, Silphidae, Staphylinidae, and Dermestidae. (Photo courtesy of Dr. James L. Castner.)

Figure 2.44 Carrion beetles, depending on species, may have diurnal or nocturnal activity. In some habitats, beetles may be more abundant than flies. Heavy predation on fly eggs by large numbers of beetles can delay colonization and possibly alter an entomological-based postmortem interval assessment that is not carefully based on the time of colonization determination. (Photo courtesy of Susan E. Ellis.)

known as sexton or burying beetles, derived from the fact that some species bury small carcasses they intend to use as their food source. Although the larvae are present and often abundant during the later stages of decomposition, the adults may arrive very early, often within the first 24 hours after death. During this early arrival the adults will feed on the eggs and newly hatched larvae of flies (Figure 2.44).

Silphid beetles are usually medium to large in size, typically ranging from 10 to 35 mm. Although adults vary greatly in size and shape, certain reliable physical characteristics can be used in their identification. The antennae are clubbed and either knob-like or broaden gradually. The wing covers (or elytra) that cover the back are often short and leave several abdominal segments exposed. The body also tends to be broader toward the posterior end rather than the anterior. The tarsi (foot) or terminal portion of each leg has five segments. The body is usually black, but marked with orange, yellow, or red patches of color.

Carrion beetle larvae also vary in size and shape, but are generally from 15 to 30 mm long. Most tend to be flattened, with some species (e.g., *Oiceoptoma inaequale*) almost trilobite-like in appearance (Figure 2.45). All larvae seem capable of mobility even when they remain in the same place throughout larval development.

The behavioral habits of silphid beetles are unusual among insects, and they have not been entirely observed. While some larvae develop in rotting vegetable material, others are predacious. However, most species are attracted to and feed on decaying animal carcasses. In the case of the sexton or burying beetles (genus *Nicrophorous*), adults bury small animal carcasses upon which they lay their eggs. In some species, a depression is made in the decaying flesh to house a group of developing larvae that the parents feed and protect. The burying of the food source may be a means of eliminating competition with other carrion-feeding insects, such as flies and other beetles. In both the burying beetle group and the carrion beetle group (genus *Silpha*) of the silphids, the larvae apparently feed on carrion while the adults consume primarily maggots.

References: Abbott 1937; Anderson and Peck 1985; Arnett and Jacques 1981; Arnett et al. 1980; Bland and Jacques 1978; Borror and White 1970; Borror et al. 1989; Castner et al.

Figure 2.45 The larvae of Silphidae are flattened and armored in appearance. Although they may be difficult for the untrained to identify as to species, they are easily recognized at the family level. (Photo courtesy of Dr. James L. Castner.)

1995; Dorsey 1940; Hogue 1993; Illingworth 1927; James and Harwood 1969; Payne and King 1970; Peterson 1979; Steele 1927; White 1985.

Heterosilpha ramosa (Say) *(= Silpha ramosa)*
Common name: Garden carrion beetle

The garden carrion beetle is found throughout the western United States from California east to New Mexico, and north to Nebraska and Montana. It also occurs in the southern and western half of Canada, as well as northern portions of Mexico. The adult ranges from 14 to 18 mm in length. The pronotum and elytra are a velvety black to dark brown. The pronotum is smooth and contrasts with the rough elytra, which have noticeable ridges and heavy punctuation. The elytra are short, exposing the last two abdominal segments. The larvae are dark brown to black and have a light brown stripe along the dorsal surface.

Garden carrion beetle adults are most active in spring and remain active throughout the summer months. They are found on fresh remains and throughout the later stages of decomposition. The eggs are typically laid in the soil around a carcass rather than directly on decomposing tissue. The adults and larvae feed on carrion and other insects, particularly maggots.

References: Anderson and Peck 1985; Brewer and Bacon 1975.

Figure 2.46 The adult of the garden carrion beetle (*Heterosilpha ramosa*) is flattened in appearance and has a dark body and legs. The elytra have a series of longitudinal ridges that contrast sharply with the smooth pronotum. (Photo courtesy of Dr. James L. Castner.)

(a)   (b)

(c)

Figure 2.47 (a) The Suriname carrion beetle (*Necrodes surinamensis*) is almost totally black, except for orange markings at the tip of the elytra. The elytra have pronounced longitudinal ridges. (b) The fast-moving larva is predacious and has a characteristic shape and armored appearance. (c) Larvae are often found in association with human remains, where they are predacious on maggots. (Photo courtesy of Dr. James L. Castner.)

Necrodes surinamensis (Fabricius)

Common name: Suriname carrion beetle, red-lined carrion beetle

This species is commonly found throughout the United States (except in the southwestern portions) and throughout central and southern Canada. It is distinguished by a shiny black pronotum, while the elytra are dull black. Each elytron has a reddish orange marking that runs laterally near the tip. However, the pattern on the wing covers can vary greatly, from being completely black to having two transverse red bands (or rows) of red spots near the tip. There are pronounced lengthwise ridges along the elytra, which are rounded at the tip and not truncate as in *Nicrophorus*. The elytra do not quite cover the entire abdomen and leave the tip exposed. The larvae are a dark reddish brown with a light brown dorsal stripe.

Adult Suriname carrion beetles are primarily nocturnal and emerge in early spring after overwintering in the adult stage. Both the adults and larvae are most common on large carcasses, such as bear, deer, and human. The adults can secrete an offensive odor as a mode of defense when disturbed.

References: Anderson 1982a; Anderson and Peck 1985; Dorsey 1940; Ratcliffe 1972.

Necrophilia americana (Linnaeus) *(= Silpha Americana)*

Common name: American carrion beetle

(a)   (b)

Figure 2.48 (a) The American carrion beetle (*Necrophila americana*) has noticeably rounded elytra and a large yellow pronotum with a black spot in the center. (b) The larva are quick moving and predacious on maggots and the larvae of other beetles. They also feed on the decomposing tissues, and are commonly found directly on the remains (in contrast to *Nicrophorus* larvae). (Photo courtesy of Dr. James L. Castner.)

Possibly the most common carrion beetle, the American carrion beetle is found east of the Rocky Mountains from eastern Texas north to Minnesota, and from Maine to Florida in the United States. They are also present in southeastern Canada. The adults are 12 to 22 mm in length. The adult body is oval in shape and is distinguished by a large yellow pronotum with a black center. The elytra are dull brownish black to black, with three raised ridges connected by smaller cross-ridges. In the northern extent of its range the elytra have a yellow tip, and in the southern portions the elytra are entirely black. The elytra are shortened, exposing the tip of the abdomen. The head, legs, antennae, and underside of the beetle are black. The larvae of this species are black and appear armored. A light brown dorsal stripe may be visible on some specimens.

Adult and larval American carrion beetles will feed on carrion, fly larvae, and the larvae of other beetles. They are found from spring to fall, and overwinter in the adult stage. The adults of this species are diurnal, active during the daylight hours.

References: Anderson and Peck 1985.

Nicrophorus americanus (Olivier)

Common name: American burying beetle

Historically, this species is most common in the eastern and central portions of the United States, from Florida to eastern Texas and north to Michigan. The former habitat of this species encompassed thirty-five states as well as southern portions of three Canadian provinces. Currently, the American burying beetle is found in only four states: Arkansas, Nebraska, Oklahoma, and Rhode Island. In 1988, this species was placed on the federal endangered species list. The American burying beetle is critically endangered and may become extinct. Scientists believe that the drastic population decline is due to habitat destruction.

This is the largest member of the genus *Nicrophorus*, with the adults ranging up to 30 to 35 mm. It is distinguished by the pronotum being a distinctive reddish orange, while the elytra are black with two orange bands.

References: Anderson 1982b; Anderson and Peck 1985.

Figure 2.49 The American burying beetle (*Nicrophorus americanus*) has distinctive red-orange and black coloring, including a large colored spot on and behind the head. This beetle has been placed on the endangered list because much of its habitat has been destroyed by development. (Photo courtesy of Dr. James L. Castner.)

Nicrophorus carolinus (Linnaeus)

Common name: Burying beetle

This burying beetle can be distinguished from other species in its genus by the extremely narrow pronotal margin. The pronotum appears slightly dull and punctate in the anterior half, and more smooth and shiny in the posterior portion. This beetle is typically 12 to 18 mm in length, and can be found in the Atlantic coastal states from Virginia to southern Georgia, throughout Florida, and along the gulf coastal states from southern Alabama to east and northern Texas. This species also ranges north of Texas into eastern Colorado and southeastern Wyoming. It has been recovered from isolated localities in Arizona. Although little is known about the biology and natural history of this species, it is restricted to open, sandy habitats in sparsely wooded areas. Although most beetles display the common orange and black color pattern found in other burying beetles, some individuals of this species are entirely black in populations of the desert southwest. Many carrion beetles carry large numbers of phoretic Gamasid mites from one carcass to another. These mites are not parasitic, but simply use the beetle as a mode of transportation. The beetles carry the mites from carcass to carcass, and it is believed that beetles benefit from this relationship because the mites will consume fly eggs, thus reducing the resource competition between maggots and beetle larvae. Additionally, the mites may also help to keep the beetle clean, and thus free of harmful bacteria.

References: Anderson and Peck 1985.

Nicrophorus investigator (Zetterstedt)

Common name: Banded burying beetle

Unlike *Nicrophorus carolinus*, the margin of the pronotum is very wide and pronounced. This is a relatively large beetle, reaching 22 mm in length, and has a variable color pattern on the elytra. This species is found primarily in the northwestern United States; however, individuals have been collected as far south as Arizona and New Mexico. They are also commonly collected along the U.S. and Canadian border in the east, and throughout Canada. *N. investigator* overwinters in the pupal stage, and the adults appear in early summer. The adults have been shown to be both diurnal and nocturnal, and therefore able to

(a)   (b)

(c)   (d)

Figure 2.50 (a) *Nicrophorus carolinus* has the common orange and black of other burying beetles, but with a very smooth and unsculptured pronotum. (b) Sclerotized plates and spines adorn the back of the slow-moving larva. These larvae are found in the soil underneath (or adjacent to) the decomposing remains. Rarely are they seen directly on the remains. (c) Lateral view of the exarate pupa showing the developing appendages. (d) Phoretic mites are common on carrion beetles, and are sometimes found on flies. (Photo courtesy of Susan Ellis.)

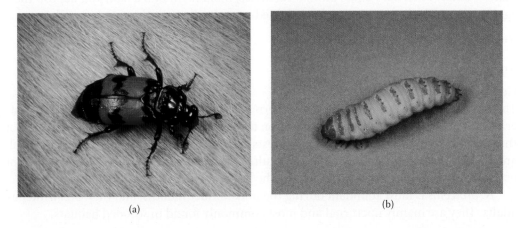

(a)   (b)

Figure 2.51 (a) The adult *Nicrophorus investigator* has large red-orange bands of color on the elytra and a smooth but sculptured pronotum. (b) The grub-like larva has large, sclerotized dorsal plates with spines on every body segment. (Photo courtesy of Dr. James L. Castner.)

Figure 2.52 The margined burying beetle (*Nicrophorus marginatus*) has the orange and black coloration typical of many other silphid species. However, this species has a distinctive band of orange hairs behind the head. (Photo courtesy of Dr. James L. Castner.)

seek carcasses during both light and dark hours. Variations in color and elytral patterns are extensive in this species, depending on geographic habitat.

References: Anderson and Peck 1985.

Nicrophorus marginatus (Fabricius)

Common name: Margined burying beetle, margined sexton beetle

This colorful silphid beetle is found in southern Canada and throughout the United States, except for southern Georgia and Florida. It is 15 to 22 mm in length, with shortened or truncated elytra that expose the last couple of abdominal segments. Their bodies are shining black, with two orange bands on the elytra. The pronotum is a shining black disc with a distinctive row of orange hairs behind the head.

The adults are mainly nocturnal and are most active during the summer months. The larvae and the adults are predacious on fly larvae. This species can be found at fresh remains as well as throughout the later stages of decay. *N. marginatus* overwinter as adults. Pairs of adults often bury small animal carcasses, mold them into a ball, and deposit their eggs on them. The larvae are cared for and fed periodically once they are hatched.

References: Anderson and Peck 1985.

Nicrophorus orbicollis (Say)

Common name: Sexton beetle

This sexton beetle is found in central-southern and southeastern Canada, as well as in eastern Texas north to the Dakotas. It is common throughout its range. The primary habitat of this species is the temperate hardwood forests of North America. The pronotum is black and slightly oblong in appearance on the adults. The elytra are also black with two rows of orange spots, and they are short, exposing the last few abdominal segments.

N. orbicollis is most abundant during the midsummer months, and they overwinter as adults. They are mainly nocturnal and most commonly found in wooded habitats.

References: Anderson and Peck 1985.

(a)   (b)

(c)   (d)

Figure 2.53 (a) The elytra of the sexton beetle (*Nicrophorus orbicollis*) are orange and black and sparsely covered with fine hairs. Two brown phoretic mites are visible on the specimen above. (b) The slow-moving larva has clearly visible sclerotized dorsal plates. (c) Larvae often develop clumped together feeding from a semisolid food ball formed by the parent beetles. (d) The developing eyes, mandibles, and legs are clearly visible on this exarate pupa. (Photo courtesy of Dr. James L. Castner.)

Nicrophorus tomentosus (Weber)

Common name: Gold-necked carrion beetle

Nicrophorus tomentosus is easily recognized by the dense covering of gold hairs on the pronotum, which gives this species its common name. This species is smaller than other species in its genus, generally only about 15 mm long. *N. tomentosus* is found throughout the eastern and central United States (primarily east of the Rocky Mountains), with the exception of Florida and southern Texas. It is also found in southeastern and south-central Canada. As with most other silphid beetles, they are active in the summer; however, unlike other species in its genus, the adults do not actually bury a carcass. Instead, they remove soil from under the carcass so that it sinks into the excavation. They then cover the freshly dug pit and carcass with leaf litter and surface debris. The larvae feed on the carcass and pupate in the adjacent soil. There is no known geographic variation in color or elytral pattern with this species, and it occurs in almost any geographic habitat throughout its range.

References: Anderson and Peck 1985; Shubeck 1976; Pirone 1974.

(a)   (b)

Figure 2.54 (a) The gold-necked carrion beetle (*Nicrophorus tomentosus*) has a pronotum that is completely covered with dense golden hair, a character unique to this species. (b) In older specimens, the distinct golden hair may be lost and not as apparent. (Photo courtesy of Susan Ellis, www.bugwood.org.)

Oiceoptoma inaequalis (Fabricius)

Common name: Carion beetle

This carrion beetle is found throughout the eastern United States from Texas to Florida and north to New York. It is also found in the extreme southeastern portions of Canada. Adults range from 13 to 15 mm in length. The pronotum and the elytra are black in color. The larvae vary from light brown to a dark reddish brown.

    O. inaequalis is most common in forested areas, where it arrives at carcasses before *Ocieoptoma noveboracense* and *N. americana*. The adults are diurnal and are most abundant in the spring through fall. They are very similar in appearance to *O. rugulosum*.

References: Anderson 1982a; Anderson and Peck 1985; Dorsey 1940; Reed 1958.

(a)   (b)

Figure 2.55 (a) The adult *Oiceoptoma inaequalis* is entirely black and distinctly oval in appearance. (b) The larva is flattened, armored, and trilobite-like in shape. Larvae such as this are fast moving, and can be found either directly on the remains or in the nearby soil. They rapidly disperse when the remains are disturbed, or when human activity is present. (Photo courtesy of PDNR.)

Figure 2.56 The carrion beetle *Oiceoptoma novaboracense* has dark, ridged elytra and a distinctive dark pronotum with a wide orange-red margin. This species is most common in forested habitats. (Photo courtesy of Susan Ellis.)

Oiceoptoma noveboracense (Forster)

Common name: Carrion beetle

This silphid beetle is found in the northeastern and central United States as far west as Wyoming and as far south as Texas and east to Georgia. It also occurs in southern Canada. The adults are 13 to 15 mm in length. This species is easily identified by the dull orange margins of the pronotum. The elytra vary from reddish brown to brownish black to black in color, with prominent lengthwise ridges. The larvae are light brown to a dark reddish brown.

This species is most abundant in the early spring, and it overwinters in the adult stage. The adults are diurnal and are most common in forested habitats.

References: Anderson and Peck 1985.

Oiceoptoma rugulosum (Portevin)

Common name: Carrion beetle

This carrion beetle is found in Florida, Louisiana, and Texas. It replaces *O. noveboracense* and *O. inaequale* in these localities. This is a dull black species with an oval body, and the

(a)

(b)

Figure 2.57 (a) *Oiceoptoma rugulosum* is an oval, black carrion beetle with smooth, ridged elytra. This species, found in the southern United States, can be found attracted to remains at any stage of decomposition where it feeds on fly larvae. (b) The fast-moving and predacious larval stage is flattened, armored, and distinctly colored. They are easily disturbed by human activity, and often hide in nearby soil or surface debris. (Photo courtesy of Dr. James L. Castner.)

Figure 2.58 The Lapland carrion beetle (*Thanatophilus lapponicus*) has dark, ridged elytra with rows of bumps. The body is sparsely covered with a yellowish pubescence that sometimes causes the beetle to appear gray. This species can be found during both the fresh and advanced stages of decay. (Photo courtesy of Dr. James L. Castner.)

elytra have three to four distinct longitudinal ridges. *O. rugulosum* can be found on human remains in most any stage of decomposition, where both the adults and larvae are predatory on fly larvae.

References: Anderson and Peck 1985.

Thanatophilus lapponicus (= Silpha lapponica) (Herbst)

Common name: Lapland carrion beetle, northern carrion beetle, common carrion beetle

The Lapland carrion beetle is sometimes also referred to as the northern or common carrion beetle. This species is Holarctic in distribution. In the eastern United States, it is found chiefly in the northernmost portion, but in the western United States, it ranges south to California, New Mexico, Arizona, and then north to Alaska. It is also distributed throughout Canada. It is most common in the northern latitudes of its range, at high altitudes. Adult beetles are 10 to 14 mm in length. The pronotum is black and has gray to gold pubescence. The elytra are brownish black and appear very bumpy. The body is gray to black, but it is covered with golden yellowish hair that sometimes appears gray. The body is slightly oblong in shape and appears very punctate when viewed from above.

This species prefers open habitats, where it is predominant in the summer. Lapland carrion beetles can be found at both fresh remains and during the advanced stages of decay. The adults and larvae are predacious on other insect larvae. This species is very cold tolerant and overwinters in the adult stage.

References: Anderson 1982a; Anderson and Peck 1985; Dorsey 1940.

Skin Beetles, Leather Beetles, Hide Beetles, Carpet Beetles, Larder Beetles (Family Dermestidae)

Members of this family have also been given the common names of hide beetles, carpet beetles, and larder beetles, based on their food preferences.

Dermestid beetles are worldwide in distribution with over 500 species, approximately 123 of which are found throughout North America. Dermestids are generally small beetles, ranging from 2 to 12 mm in length. They are rounded to oval in shape and covered with scales that may form distinctive and colorful patterns. The larvae range from 5 to 15 mm and are usually covered with tufts of long, dense hair.

This family represents one of the most economically important groups of beetles in the world. The carpet beetle species damage rugs, clothing, and furniture. Others, like the khapra beetle, infest grains and inflict serious losses on stored products. Still others, like the hide beetles, may ruin leather goods or destroy irreplaceable museum specimens, especially mounted insects. Almost all species are scavengers and feed on various types of dried animal tissue.

Skin beetles, especially in the genus *Dermestes*, can be of considerable forensic importance. In sufficient numbers, they have been reported as reducing a human body to a skeleton in only 24 days. Due to this uncanny ability, beetles have been employed for decades in the removal of flesh from the bones of museum specimens. The larvae are typically found on human corpses during the dry and skeletal stages of decomposition. They move away from light and will hide in any cavity or recess that is available. A close and detailed examination of remains may be required to collect the small, young larvae.

The adults are cannibalistic and will eat young larvae and puparia. For this reason, adult dermestid beetles should be put in separate containers from the immatures. These beetles can be found in indoor situations on bodies throughout the year, but species of the genus *Dermestes* are most active during the warmer months. The presence of dermestid beetles or their sawdust-like frass (fecal material) is often an indication that considerable time has elapsed since death. The mere presence of their frass is proof that dermestid beetles have fed extensively on the tissues. In some cases, where remains are mummified, a period of years may have passed and living dermestid adults and larvae may still be associated with the remains.

References: Arnett and Jacques 1981; Arnett et al. 1980; Bland and Jacques 1978; Borror and White 1970; Borror et al. 1989; Castner et al. 1995; Hogue 1993; James and Harwood 1969; Peterson 1979; Smith 1986; Voigt 1965; White 1985.

Dermestes ater (DeGeer)

Common name: Black larder beetle, incinerator beetle

The black larder beetle is cosmopolitan in distribution and appears similar to *D. maculatus*, but the elytra are not serrated. The elytra appear a dark to light brown in color and have scattered yellow hairs. This species is also distinctive in that the ventral pattern is yellowish and not white. The larvae can be easily distinguished from others by the two spines near the posterior end that extend backward and are not strongly curved. This beetle is a serious pest of dried fish, mushrooms, and cheese, and it is particularly attracted to the protein-rich tissues of decaying vertebrates. This beetle is also attracted to dried pet food and can sometimes be found associated with the waste materials burned in incinerators.

Dermestes caninus (Germar)

This species of dermestid beetle is distributed throughout the United States except for the Pacific Northwest. Like all dermestid beetles, they are attracted to stored food products and carrion of all types. This species has been found in the nests of predatory birds, apparently attracted by the prey remains. *D. caninus* and other members of this family overwinter in the adult stage and become active during the early summer months, when the females readily enter dwellings in search of egg-laying sites.

(a)   (b)

Figure 2.59 (a) Dorsal view of the black larder beetle (*Dermestes ater*) showing the scattered yellow hairs on the elytra. (b) As with all *Dermestes* larvae, they appear very bristly, and seem relatively unaware of human activity unless directly disturbed. (Photo courtesy of Dr. James L. Castner.)

(a)   (b)

Figure 2.60 (a) Dorsal view of the skin beetle (*Dermestes caninus*). Beetle species such as this are found on remains in both indoor and outdoor locations. However, dry remains seem to be preferred, and they are most commonly recovered in indoor habitats. (b) The light-colored markings on the underside produce patterns that are characteristic of individual species. (Photo courtesy of Dr. James L. Castner.)

Dermestes maculatus (DeGeer)

Common name: Hide or leather beetle

This species of skin beetle is worldwide in distribution. The adults are 5 to 10 mm in length and black to reddish brown on the dorsum or upper surface. They also have characteristic black and white markings on the ventral surface. The apex of the elytra are serrate or saw-toothed and end in a terminal spine where they come together. The larvae of *D. maculatus* are dark brown and have a broad light brown to yellow stripe extending lengthwise along the body. They are easily identified by the fact that the spines near the tip of the tail curve forward, toward the head of the larvae. This and other species of dermestid larvae are covered with tufts of long dark hairs.

Dermestid Frass (Peritrophic Membrane)

Beetles in the family Dermestidae produce a protective membrane that lines their gut, and surrounds the food meal. This lining protects the gut walls from abrasion during the

(a)   (b)

Figure 2.61 (a) The hairs or setae on the body of the skin beetle (*Dermestes maculatus*) produce distinct patterns that can aid in field identification. The serrated edge and small spine at the tip of the elytra are characteristic of this species. (b) *Dermestes maculatus* larvae prefer to feed on dry tissues and may remain associated with decomposing tissues for several months. (Photo courtesy of Dr. James L. Castner.)

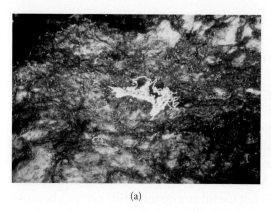

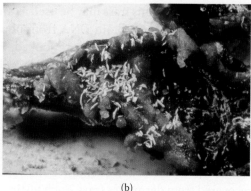

(a)   (b)

Figure 2.62 (a) The dried frass (peritrophic membrane) of a skin beetle. This substance is frequently found when dermestid beetles have fed on human tissues in indoor environments. The presence of such material, even in the absence of the beetles themselves, is valuable entomological evidence. (b) Freshly deposited skin beetle frass. (Photo courtesy of Dr. James L. Castner.)

digestive process. As the digested food is passed from the beetle's body, the fecal material is wrapped in the protective membrane, which may be passed in an unbroken chain. This frass material is light brown, stringy in appearance, and quite dry. It crumbles very easily when disturbed. Dermestes beetles generally prefer dry tissues on which to feed, and it is not uncommon to find these beetles associated with decomposed and mummified human remains found in indoor locations. Presence of the peritrophic membrane is generally indicative of an extended PMI. However, this frass has been found associated with remains from as little as 4 months to more than 10 years. Due to this variability, a qualified forensic entomologist should be consulted before a PMI determination based on the simple presence of peritrophic membrane is attempted. A trained forensic entomologist is likely to find additional entomological evidence to help support the PMI estimation. However, in itself, dermestes frass and the associated peritrophic membrane are valuable entomological evidence since they indicate the conditions that were likely present throughout the decomposition process.

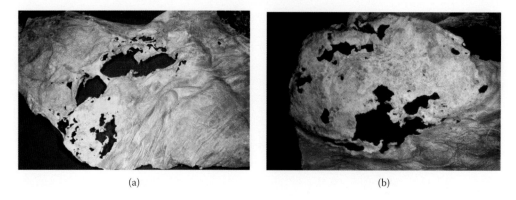

(a)   (b)

Figure 2.63 (a) Feeding damage by skin beetles to human skin. The feeding of dermestid larvae can produce postmortem artifacts that resemble possible sites of trauma. (b) Some of these feeding sites may resemble the small symmetrical entrance and exit holes made by feeding maggots, but most typically are identified by larger irregular holes with irregular edges. (Photo courtesy of Dr. James L. Castner.)

Additionally, this frass may be used as an alternative DNA or toxicology sample. While it may be possible to obtain DNA from bone samples or teeth, collection and analysis of the frass present should not be overlooked. Frequently, frass may yield more DNA because the beetles feed on the tissue while it is still relatively fresh, and their digestive process efficiently extracts moisture from the tissues, and dry partially digested tissue is passed protectively sealed in the peritrophic membrane. While it may be possible to obtain an alternate DNA sample, it may not always be possible to find a better substitute for toxicological analysis. Once the soft tissues of the body have completely decomposed, dermestes frass may be the only item available for toxicological testing. In such cases, the toxicology results will be useful as a qualitative analysis. Many prescription and illicit drugs have been recovered from dermestes frass. Such testing is not useful as a quantitative analysis because the rate of assimilation of the drug by the larvae is not known, nor can it be known on which tissues the larvae have fed. Due to the pharmacokinetics of drug deposition within the human body, different tissues and organ systems will absorb and eliminate drugs and their metabolites at different rates.

Dermestid Damage to Human Skin

In many cases that involve extensively decomposed or mummified human remains, skin tissue often remains intact. The remaining tissues often contain many holes, and they are sometimes mistaken for gunshot wounds. The feeding maggots usually produce symmetrically round holes that are of a uniform size. Dermestes adults and larvae often create the same symmetrical artifacts. However, they also produce holes that are irregular in size, as well as tears that resemble lacerations and abrasions. The edges of these artifacts are often irregular and jagged, not smooth, as with those produced by feeding maggots.

Rove Beetles (Family Staphylinidae)

The rove beetles are one of the largest families of insects in North America north of Mexico, with over 47,700 species. They are widely distributed and found in various habitats, where they feed on decomposing animal tissue, plant debris, and fungi. The adult beetles vary greatly in size, ranging from 1 to 25 mm. The characteristic shape of many species attracted to carrion, however, makes them easy to identify to the family level. Other members of this family do not have the typical shape, but they are not frequently found at carrion.

Typical rove or staphylinid beetle adults are slender, elongate, and have very short wing covers or elytra. The elytra typically appear square, and are approximately as long as they are wide. Although the membranous hind wings remain folded beneath and completely concealed (except during flight), six to seven abdominal segments are exposed. This makes the rove beetle appear to be divided into four sections. The head, thorax, and wing covers make up the first three sections and are approximately equal to each other in size. The fourth section is the exposed abdomen, which is roughly equal to all of the first three together.

Although earwigs (order Dermaptera) (Figure 1.23) are vastly fewer in number of species than are the rove beetles, they seem better known to the general public. Because earwigs are elongate insects with short elytra, people often mistake staphylinid adults for earwigs, although staphylinds do not have pincer-like cerci at the tip of the abdomen.

Staphylinid larvae are typically long, slender, pale in color, and may have a darker head. Many larvae and adults have mandibles that are long and curved, and which may cross over in front of the head. Larvae and adults are typically quick moving and predacious on smaller insects. Some, however, eat fungi or diatoms. Even among those that are predacious, some have specialized diets, whereas others are generalists. The species attracted to carrion feed on maggots and the larvae of other insects. The adults are strong flyers and often run about with the abdomen raised in the air, as if they were capable of stinging, although they are unable to do so.

References: Abbott 1937; Arnett 1961; Arnett and Jacques 1981; Arnett et al. 1980; Bland and Jacques 1978; Borror and White 1970; Borror et al. 1989; Castner et al. 1995; Hogue 1993; James and Harwood 1969; Mank 1923; Peterson 1979; Stehr 1991; Voris 1939; White 1985.

Creophilius maxillosus (Gravenhorst)

Common name: Hairy rove beetle

Adults range in length from 12 to 18 mm, but all are black bodied and covered with patches of pale yellow hairs. This species is found throughout the eastern United States. The body is very slender and is typical of rove beetles, appearing to be divided into four sections. Both the adults and larvae are predacious on maggots. The adults may be found on carcasses only hours after death, as well as during the advanced stages of decomposition. When threatened or disturbed, they often hold their abdomen curved upward and forward above the head, as if to sting. Although they are unable to sting, they can emit an offensive odor as a defense mechanism.

Platydracus comes (LeConte)

Common name: Brown rove beetle

The adults of this species range in length from 10 to 15 mm and are brown with black markings. The body is very slender and appears divided into four sections. Very little is known about this species, but it is commonly attracted to carrion and frequently recovered from bodies in the southeastern United States.

(a)

(b)   (c)

Figure 2.64 (a) Patches of pale yellow hairs adorn the otherwise black body of the hairy rove beetle (*Creophilus maxillosus*), making it the most easily recognizable of all rove beetles. (b) The distinctive, rapidly moving predatory larva has a pair of slender dorsal appendages at the tip of the abdomen and a single stout ventral appendage. (c) The nearly mature pupa with most anatomical features easily recognizable. (Photo courtesy of Dr. James L. Castner.)

Figure 2.65 The brown rove beetle (*Platydracus comes*) is one of many staphylinid species attracted to carrion. Its cryptic coloring and shy nature make it seldom seen or recovered at carrion. (Photo courtesy of Dr. James L. Castner.)

Figure 2.66 The spotted rove beetle (*Platydracus fossator*) has two reddish orange spots that contrast sharply with the black body, making it easily recognizable in the field. (Photo courtesy of Dr. James L. Castner.)

Platydracus fossator (Gravenhorst)

Common name: Spotted rove beetle

The adults of this rove beetle range from 12 to 18 mm in length. The body is dark blue (and may appear black, depending on lighting conditions) with two red spots on the elytra. Some specimens may have a faint band of orange hair near the tip of the abdomen. Although detailed information is lacking on the habits of this species, it is often encountered on carrion and frequently found in association with human remains in the southeastern United States.

Platydracus maculosus

Common name: Rove beetle

Adults are shaped like other *Platydracus* adults and are mottled dark brown with golden hairs. Little is known about the biology of this species, but it is frequently recovered on decomposing remains where it is a predator on fly larvae. The adults and larvae of this species should be collected as evidence whenever encountered at a crime scene.

Figure 2.67 *Platydracus maculosus* is a predator like other staphylinids attracted to carrion. The sharp, curving mandibles are used for grabbing its prey. (Photograph courtesy of Susan E. Ellis.)

Figure 2.68 *Platydracus tomentosus* is a rove beetle commonly found at carrion. It is easily disturbed and runs very rapidly, making it one of the least commonly collected beetles. (Photo courtesy of Dr. James L. Castner.)

Platydracus tomentosus (Gravenhorst)

Common name: Rove beetle

This rove beetle is 7 to 13 mm in length and dull brown in color. The body has the same general shape as all other *Platydracus* adults, and this species is frequently found at carrion.

Clown Beetles (Family Histeridae)

This is a large family of over 3,000 species, more than 500 of which are widely distributed throughout North America. Clown beetles are usually small, seldom getting beyond 10 mm in length. They are rounded, shiny beetles that are black or sometimes metallic green. The elytra are short and squared at their apex, exposing the last two abdominal segments. Therefore, the adults appear to be divided into three longitudinal sections, with the central section bearing a line down the middle. The antennae of clown beetles are both elbowed and clubbed.

Clown beetles are very common on carrion and excrement, as well as on fungi and decaying plant material. When on carcasses, they tend to stay concealed in the soil underneath during the daylight hours, becoming active at night. Both the larvae and adults are predacious and feed readily on maggots and fly puparia. They have also been observed feeding on the larvae of dermestid beetles. As with other predatory species that are collected from a crime scene to be maintained alive, clown beetles should be isolated into separate containers.

References: Arnett and Jacques 1981; Arnett et al. 1980; Bland and Jacques 1978; Borror and White 1970; Borror et al. 1989; Castner et al. 1995; Hogue 1993; James and Harwood 1969; Peterson 1979; White 1985.

Hister sp.

Common name: Clown beetle

Species in this genus are commonly 4 to 5 mm in length and are worldwide in distribution. Their body color is most frequently a shiny jet black, but in some species can be brown, red,

(a)

(b)

(c)

Figure 2.69 (a) Clown beetles (*Hister* spp.) have a distinctive shape, and multiple species are usually recovered from human remains. The time of arrival of these beetles is useful in establishing the time of colonization. (b) This view shows the strongly convex body shape of the clown beetles. (c) Clown beetle larvae (*Hister* spp.) have a heavily sclerotized head and mandibles, giving them a distinct appearance. However, because of their small size, they are frequently overlooked at the crime scene. (Photo courtesy of Dr. James L. Castner.)

or metallic green. The body shape is very convex in profile, and the elytra are short and cut square at the apex, exposing the last two abdominal segments.

Hister species are found on cadavers from bloat through the dry stages of decomposition. The adults and larvae are chiefly nocturnal, feeding on carrion, maggots, and other insect eggs and larvae. The adults fly very well and can run swiftly. When disturbed, they pull their legs and antennae tightly against their body and lie motionless, feigning death. Due to this feigning behavior they are often overlooked and not collected.

References: Arnett and Jacques 1981; Arnett et al. 1980; Bland and Jacques 1978; Borror and White 1970; Borror et al. 1989; Castner et al. 1995; Hogue 1993; Holdaway and Evans 1930; James and Harwood 1969; Nuorteva 1970; Peterson 1979; White 1985.

Saprinus pennsylvanicus

Common name: Clown beetle

The body of this clown beetle is oval, convex, and a shiny metallic green. This species, like other clown beetles, feigns death when disturbed by pulling the legs close to the body and lying motionless. *S. pennsylvanicus* feeds on fly eggs and maggots, and is often found underneath the body. This species is worldwide in distribution and is noted to occur from the fresh throughout the later stages of decomposition.

Figure 2.70 This clown beetle (*Saprinus pennsylvanicus*) has a metallic green coloring and surface sculpturing on the elytra. Although a different color, the body shape is the same as with other clown beetles. These beetles often occur in large numbers, and they may arrive at carrion as little as 24 hours after death. (Photo courtesy of Dr. James L. Castner.)

Checkered Beetles (Family Cleridae)

There are approximately 3,500 checkered beetle species worldwide, more than 500 of which occur in North America. The bodies of most species are covered with bristly hairs and are often brightly colored. Adults range from 3 to 12 mm in length. The head is often wider than the pronotum (neck area), which is narrower than the wing bases. This gives the appearance of a narrowing between the head and the point where the wings start. The antenna types found in this family are variable.

Both the larval and adult clerid beetles are predacious. Most species prey on the immature stages of various wood-boring beetles, while others feed on maggots. They are common visitors to decomposing animal matter in the later, drier stages of decomposition. Adult checkered beetles are also often frequently found on flowers.

References: Arnett and Jacques 1981; Arnett et al. 1980; Bland and Jacques 1978; Borror and White 1970; Borror et al. 1989; Castner et al. 1995; Hogue 1993; James and Harwood 1969; Peterson 1979; White 1985.

Necrobia rufipes (DeGeer)

Common name: Red-legged ham beetle, copra beetle

This species is worldwide in distribution and can be recovered throughout the United States. This is a small beetle ranging from 3 to 7 mm in size. It is very distinctive with a shining, metallic blue body and red legs. The larvae are brightly colored with purple specks on the body and a reddish brown head and tail.

The common name originates from the habit of this species of infecting cured meat products. The adults and larvae are also predacious on eggs and maggots, and will sometimes feed on carrion. For most of the United States, they are commonly found throughout the year on human corpses in outdoor habitats during the drier stages of decomposition. The adults may be found on carrion earlier in the sequence of decomposition, but colonization does not usually occur until much later. The adults are slow fliers and typically run instead of flying. However, their small size, relatively low numbers, and rapid movement can make them difficult to collect despite the fact that they do not often take flight.

References: Clark 1895; Payne and King 1970.

Hide Beetles (Family Trogidae)

The trogid hide beetles are a small family of which approximately fifty species are widely distributed throughout North America. This group is sometimes considered or treated as a subfamily (Troginae) of the extremely large and cosmopolitan family of beetles known as the scarabs (family Scarabaeidae).

Trogids range from 5 to 20 mm in length and are quite distinctive in appearance. They are usually oblong to oval in shape and are basically similar in form to scarabs such as the June beetle. However, trogids tend to be brown in color and have the back or dorsum of the body rough and covered with ridges.

Trogid hide beetles are often present on carrion and carcasses. As many as eight different species have been collected from a single set of animal remains. However, adults often become covered with mud and animal tissue that becomes encrusted on the body. This results in the beetle looking very much like a small piece of debris, which can be easily overlooked. Trogids are typically attracted to dry remains and are found during the advanced stages of decay.

References: Abbott 1937; Arnett and Jacques 1981; Arnett et al. 1980; Bland and Jacques 1978; Borror and White 1970; Borror et al. 1989; Castner et al. 1995; Hogue 1993; James and Harwood 1969; Payne and King 1970; Peterson 1979; Spector 1943; White 1985.

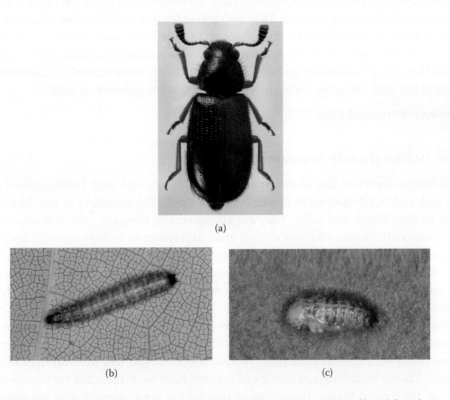

(a)

(b)   (c)

Figure 2.71 (a) Red-legged ham beetle (*Necrobia rufipes*) adults are small and fast, but easy to identify by their metallic blue body and red legs. (Photo courtesy of Dr. Mike Thomas.) (b) The larva is small, quick moving, and an aggressive predator on insect eggs and young maggots. (c) The pupa has an appearance similar to that of the adult form, but it is hard to detect due to its subtle coloring. (Photo courtesy of Dr. James L. Castner.)

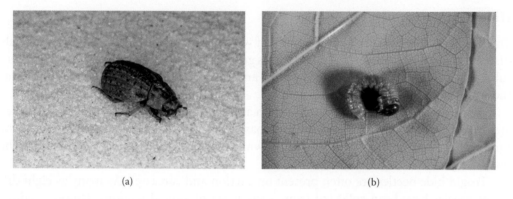

(a)   (b)

Figure 2.72 (a) The hide beetle (*Trox suberosus*) is often found covered with debris as an adult. (b) The larva is C shaped (like its close relative the scarabs) with a heavily sclerotized head. (Photo courtesy of Dr. James L. Castner.)

Trox suberosus (Fabricius) *(= Omorgus suberosus)*

Common name: Hide beetle

This species can be found throughout the United States and Canada. The dorsal surface of the body is very rough, convex, and light brown in color. When alive, they are often covered with dirt and debris, and thus are often overlooked. *T. suberosus* is one of the last in the succession of insects on decomposing remains. They are carrion feeders, and like most trogids occur on carcasses from advanced decomposition through the dry decay stages. When disturbed, they draw in their legs and lie motionless, resembling dirt or rubbish. This behavior increases the chances that they go unnoticed and uncollected by forensic investigators. *T. suberosus* over-winters in the adult stage, becoming most active during the summer months.

References: Payne and King 1970.

Scarab Beetles (Family Scarabaeidae)

Scarab beetles represent one of the largest beetle families with over 19,000 species world-wide, and with 1,400 species in North America alone. The members of this family vary greatly in size, shape, and color. They are also generally elongate, robust beetles that are usually convex in profile, and they exhibit great differences in biology, ecology, and behavior. Most beetles in this family feed on dung, carrion, or decomposing plant materials, while others feed on flowers, fruits, and foliage of living plants. Some are serious pests of ornamental plants and agricultural crops. A few feed on fungi, and some live in the nests and burrows of vertebrate animals.

Dung beetles, or tumblebugs, belong to the subfamily Scarabaeinae and are found associated with dung and carrion. The adults of those species most likely to be encountered in association with cadavers are dull-colored, rounded, and less than an inch in length (Figure 2.73). At least fourteen species of scarabs have been recorded on vertebrate carcasses in the United States. The larvae of this family are C shaped, and are commonly referred to as grubs (Figure 2.74). The larvae of dung beetles develop from an egg laid in a ball of dung that is collected by the adults. These larvae are generally white with a brown head. Occasionally, these larvae can be found buried in the soil underneath the body.

References: Woodruff 1973.

Figure 2.73 This adult scarab beetle illustrates the typical body shape and coloring of many species of carrion frequenting scarabs. (Photo courtesy of Dr. James L. Castner.)

Figure 2.74 A typical scarab larva known as a white grub (*Strategus antaeus*). Scarab larvae may be found in the soil surrounding the remains. They are not typically found crawling directly on the remains. (Photo courtesy of Dr. James L. Castner.)

Deltochilum gibosum gibosum (Fabricius)
Common name: Tumblebug

These are large and robust beetles that generally attain a length of 30 mm, making them one of the largest species of dung beetles in North America. The beetles are dull black in color, with pronounced ridges (striae) on the elytra. They occur in Alabama, Florida, Georgia, Kentucky, Louisiana, Mississippi, North Carolina, South Carolina, Tennessee, and Texas.

Figure 2.75 The tumblebug (*Deltochilum gibosum gibosum*) is one of the larger scarabs often attracted to carrion. It has unusually long hind legs, which it uses to roll the large balls of dung it will use to feed its young. (Photo courtesy of Dr. James L. Castner.)

Although they occur throughout the year in Florida, they are primarily nocturnal, and as a result, few behavioral observations have been recorded. The adults roll almost any decaying plant or animal substance into a tightly packed ball, which is then buried. The eggs are deposited within the rolled ball, and it serves as the larval food source. Although this species is attracted to many decomposing substances, it is particularly attracted to chicken. It is also commonly found on decomposing vertebrate carcasses and human dung.

References: Woodruff 1973.

Phanaeus vindex (MacLeay)

Common name: Dung beetle, rainbow scarab

This prominent beetle is widely distributed from Massachusetts to Florida (except in the Everglades and the Florida Keys), west to Texas, and north to South Dakota. The male has a rhinoceros-like horn, which is not found on the female. Both sexes have dark green elytra that are very iridescent. The head and pronotum are a reddish to coppery-gold color that has a metallic luster. Both sexes are very conspicuous on sunny days in open pastures and fields.

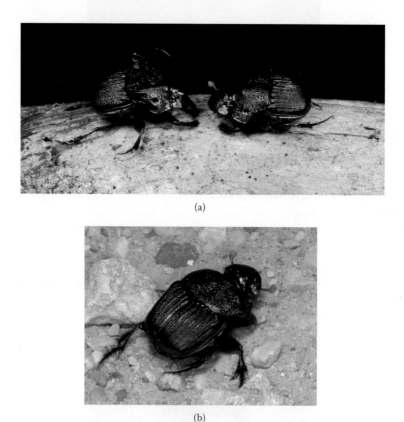

(a)

(b)

Figure 2.76 (a) The iridescent dung beetles (*Phanaeus vindex*) are more commonly found at dung than at carrion. However, they are attracted to the dung associated with decomposing remains, and may be attracted to any exposed gut contents. The male (left) has a large horn, which is not present in the female. (Photo courtesy of Dr. James L. Castner.) (b) The iridescent color of this beetle gives it the common name of the rainbow carrion beetle. (Photo courtesy of Jay Cossey, www.photographsfrom nature.com.)

This well-known and widely distributed beetle has a preference for dung, and in particular, human dung. *P. vindex* rolls dung into tightly packed balls that are stored in vertical burrows a few inches beneath the dung source. The beetle's eggs are then deposited in the dung ball.

References: Woodruff 1973.

Sap Beetles (Family Nitidulidae)

This is a large cosmopolitan family with more than 2,500 species worldwide. In North America, approximately 183 species are found. Sap beetles are considerably variable in appearance, most species being dark in color and from 4 to 12 mm long. All have distinctly clubbed antennae, with the club portion rounded and ball-like. Many species are oval in shape, but some are very similar to rove beetles (family Staphylinidae) in appearance. These species have short elytra that leave from one to three abdominal segments exposed. Rove beetles, however, do not have clubbed antennae.

Most sap beetles are attracted to rotting fruit and decaying or fermenting vegetable matter. Some are also associated with fungi or flowers. A few species are attracted to carrion and decomposing animal remains, and are usually found during the more advanced stages of decay. One such species is *Omosita colon*, a very small, dark brown sap beetle that is only 2 to 4 mm in length (Figure 2.77). Although commonly found at the same time as skin beetles (family Dermestidae), the sap beetles appear to prefer a moister environment. Little work has been done to establish the value of nitidulid beetles in forensic investigations.

References: Arnett and Jacques 1981; Arnett et al. 1980; Bland and Jacques 1978; Borror and White 1970; Borror et al. 1989; Castner et al. 1995; Hinton and Corbet 1975; Hogue 1993; James and Harwood 1969; Payne and King 1970; Peterson 1979; White 1985.

Figure 2.77 The tiny sap beetle (*Omosita colon*) has the tips of the elytra light in color, while the bases of the elytra are dark. These are some of the last insects attracted to remains. They can be found months to years after death, when very little detectable tissue can still be found. (Photo courtesy of Susan Ellis.)

Other Arthropods of Forensic Importance

Venomous Arthropods

Forensic investigators often consider insects and other arthropods solely for their importance and value in determining the time of death, forgetting that they may be the causal agents of death themselves. Venomous arthropods cause only a small number of deaths each year in the United States, yet still should be mentioned since the possibility of involvement in such cases exists. There are a number of circumstances where arthropod envenomization can lead to fatalities.

Spiders such as the brown recluse (*Loxosceles reclusa*) (Figure 2.78) and the widow group, including the black widow (*Latrodectus mactans*) (Figure 2.79), red widow (*Latrodectus bishopi*), and brown widow (*Latrodectus geometricus*) (Figure 2.80a and b), all have potent venom, which are introduced via fangs when the spiders bite. The bite of the brown recluse causes extreme destruction of flesh, often requiring the removal of significant amounts

Figure 2.78 Dorsal view of the brown recluse spider (*Loxosceles reclusa*) showing the characteristic fiddle mark in the head area. These spiders are commonly found in indoor habitats. (Photo courtesy of Dr. James L. Castner.)

Figure 2.79 Lateral view of the black widow (*Latrodectus mactans*) showing the red hourglass mark on the center of the abdomen. This species is typically found in outdoor habitats, and also may nest under homes, on porches, and along the eaves of houses. Although venomous, it is not aggressive. (Photo courtesy of Dr. James L. Castner.)

(a)   (b)

Figure 2.80 (a) The brown widow spider (*Latrodectus geometricus*). This species has a behavior and habitat similar to the black widow spider. (b) This distinct red hourglass marking is also present. (Photo courtesy of Dr. James L. Castner.)

Figure 2.81 Scorpions are venomous arthropods with a stinger at the tip of the tail. They can be found attracted to decomposing remains where they are predacious on other insects. (Photo courtesy of Dr. James L. Castner.)

of tissue. Widow bites are usually less severe, but more frequent due to their tendency to invade dwellings and take refuge in dark, enclosed areas. In the most common scenario, a person comes in contact with the spider without seeing it by sitting on it or disturbing it in some similar manner. Scorpions (Figure 2.81) inject venom through a stinger on the tail, often quickly penetrating the skin in several different spots. Fortunately, few scorpions in the United States are deadly. Centipedes (Figure 2.82) are found throughout most of the United States. Although many are capable of breaking the skin and inflicting a severe bite, the pain typically diminishes quickly and gradually disappears. The bite of a centipede has been compared in severity to that of a wasp sting. Only one fatality due to the bite of a centipede has been recorded.

Bees (Figure 2.83), wasps (Figure 2.84), hornets, and yellowjackets (Figure 2.85) all belong to the insect order Hymenoptera. They are social insects and live in colonies that can sometimes grow to proportions where sheer numbers are dangerous even though the amount of venom carried by an individual is small. In a worst-case scenario, the insects are both dangerous and aggressive, as seen with the Africanized or killer bees (*Apis mellifera scutellata*) that have recently invaded the United States. Even when social insects are not

Figure 2.82 The centipede (*Scolopendra viridula*) can bite and inject venom through a modified pair of legs. Centipedes are frequently attracted to decomposing remains where they are predacious on other insects. (Photo courtesy of Dr. James L. Castner.)

Figure 2.83 A small number of people each year die as a result of allergic reactions to the stings of the honeybee (*Apis mellifera*). Occasionally, honeybees can be found feeding from the fluids on decomposing remains. (Photo courtesy of Dr. James L. Castner.)

Figure 2.84 Paper wasps such as this species (*Polistes dorsalis*) often make open nests in doorjambs, which brings them in close contact with people. Fortunately, their colonies are usually small and typically produce little threat to most people. (Photo courtesy of Dr. James L. Castner.)

Figure 2.85 The yellowjacket wasp (*Vespula squamosa*) aggressively defends its nest against intruders and has a painful sting. The adults forage for food over long distances, and frequently come in contact with humans. Frequently, they are attracted to decomposing remains where they feed on the blow flies that are present. (Photo courtesy of Dr. James L. Castner.)

Figure 2.86 Some perennial yellowjacket nests like this one may contain over 200,000 individuals. Colonies of yellowjackets can grow to enormous sizes, and many nests become a serious danger to people that wander too close. (Photo courtesy of Dr. James L. Castner.)

naturally aggressive, most will vigorously defend their nest against any intruder. Problems occur when the nests are constructed in close proximity to human habitations and activities. The inadvertent disturbance of a nest can lead to dozens of stings for the unfortunate individual, or in some cases even death. In the semitropical areas of Florida, yellowjacket (*Vespula squamosa*) nests (Figure 2.86) have been removed that contained over 200,000 adult wasps.

It should be noted that certain individuals are allergic to arthropod venom in any amount. Such people, upon being stung or bitten, can experience anaphylactic shock within minutes. Swift and immediate treatment, usually via the administration of epinephrine, is necessary to save the life of those allergic to insect stings. In cases where death results, an examination of internal organs may be necessary to confirm the cause, as external signs of trauma can be minimal.

People allergic to bees and wasps (or merely believing themselves to be allergic) will sometimes act in an uncontrollable and irrational manner in such an insect's presence. Although comical under certain circumstances, all too often this occurs during the operation of a motor vehicle. A certain amount of traffic deaths every year result from the strenuous efforts and overreaction of drivers attempting to rid themselves of bees or wasps that have entered their car.

Crime scene technicians and death investigators often receive stings while attempting to process death scenes since wasps are attracted by the large numbers of flies on which they feed. Accidentally collecting or crushing a wasp during scene processing often results in stings.

Blood-Feeding Arthropods

It is worthy to mention that blood-feeding arthropods may be of immense value to a legal investigation. The gut content of these arthropods by contact with biological material from either the suspect or victim, and proper collection and preservation of these insects may allow for the analysis of the contents for development of a useful DNA profile. While the recovery of a DNA profile from either the victim or suspect may not be possible in all cases, crime laboratory analytical techniques have progressed sufficiently to make such testing worthwhile in almost any case that involves the recovery of blood-feeding arthropods during the investigation. Human DNA has been recovered from the gut contents of maggots, beetle larvae, ticks, mosquitoes, and lice.

Most maggots have a nearly transparent integument through which the internal structures can be easily viewed. The crop containing the undigested tissues of the food source is visible and distinct (Figure 2.87). Larval metabolism is extremely rapid, and they should

Figure 2.87 The crop of a maggot is clearly visible through the larval integument. The crop contains undigested tissue, and upon careful dissection and analysis, human DNA can be recovered from the larvae. This may provide information that will be useful in many ways for a diversity of legal investigations. (Photo courtesy of Dr. James L. Castner.)

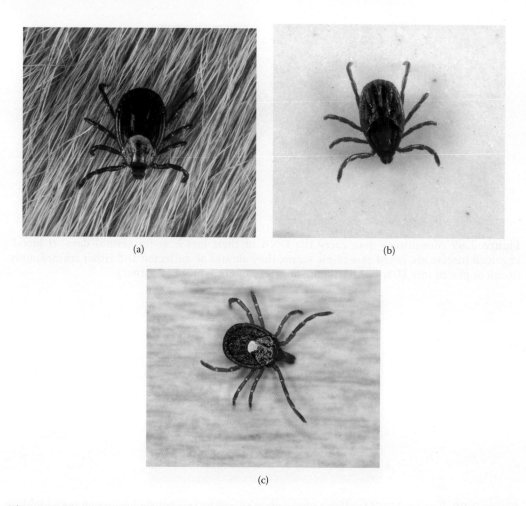

(a)   (b)

(c)

Figure 2.88 Ticks contain undigested blood consumed from the host. If the insect is frozen or otherwise preserved immediately, the DNA of the ingested blood can be recovered and profiled. (a) The American dog tick (*Dermacentor variabilis*). (b) The brown dog tick (*Rhipicephalus sanguineus*). (c) The lone star tick (*Amblyomma americanum*). These ticks are considered to be "hard" ticks due to the obvious scutellum, or shield, on the dorsum. They are common tick species, and are likely to be encountered in wooded environments. (Photo courtesy of Dr. James L. Castner.)

be preserved quickly by either freezing, or placement into 80% ETOH as soon as possible at the death scene. Beetle larvae have a slower metabolism, and human DNA can be successfully recovered from their gut, or from the frass, which may remain for many years in protected environments. Ticks also have a remarkably slow metabolism and contain comparatively large amounts of blood. The sedentary nature of some tick species and their ability to retain and slowly digest their blood meal make them an excellent source for the possible recovery of human DNA when found in association with crime scenes (Figure 2.88a–c). Mosquitoes are ubiquitous in most outdoor environments for a majority of the year. Only the females of each species are blood feeders, and if allowed to feed to completion, the female may not feed again for more than a week. Also, many species may not disperse far during this time. Such a relatively prolonged digestive period may allow the investigator the opportunity to recover these blood-engorged insects and submit them

Figure 2.89 Mosquitoes may carry the DNA of their last host for several days. If blood-engorged insects are noted at a crime scene, they should be collected and either immediately frozen or placed into 80% ETOH. (Photo courtesy of Dr. James L. Castner.)

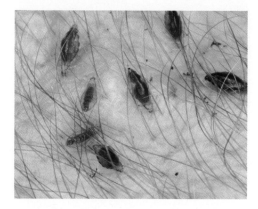

Figure 2.90 Lice are blood-feeding arthropods, and can be transmitted from one individual to another by direct contact, or contact with infested clothing or bedding materials. The transfer of pubic lice during the commission of a sex crime has resulted in the recovery of the DNA of the criminal offender. (Photo courtesy of Dr. James L. Castner.)

for molecular analysis (Figure 2.89). Recovery of human DNA from the gut of a mosquito is possible and, if successful, can lead to valuable information on suspect or victim linkages to the scene. Similarly, human body and pubic lice are blood feeders and may be collected as evidence during a legal investigation. Human DNA has been recovered from lice, and such evidence can be crucial in creating a suspect-victim association (Figure 2.90).

Scavengers

Many scavengers and predatory insects that do not specialize in feeding on carrion will utilize it opportunistically as a food source when it is available. For example, paper wasps and yellowjacket wasps (both family Vespidae) may sometimes be observed tearing off chunks of tissue and flying away with them. While the actual amount of flesh removed is usually minimal, the feeding activities of insects, as with much larger vertebrate scavengers and predators, alter the remains and may leave artifacts that are difficult to interpret. For example, ants will often feed on human skin and body tissues when the remains are

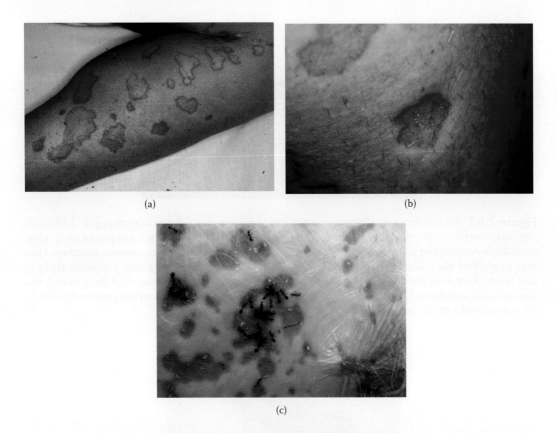

(a)

(b)

(c)

Figure 2.91 (a) Fire ant (*Solenopsis invicta*) feeding on human skin can leave artifacts similar in appearance to chemical burns. (b) Close-up of fire ant feeding damage on human remains. Damage such as this can be extensive and can occur in a very short period of time. (c) In some cases extensive damage has been produced less than 1 hour after death, but this is dependent on how closely the remains are located to the fire ant mound. (Photos courtesy of Dr. James L. Castner.)

left outside and exposed. In the southeastern United States, at least one species of fire ant (*Solenopsis invicta*) is very common and aggressive in its foraging habits. The feeding of these tiny, 3 to 4 mm long ants often leaves postmortem damage to tissue that appears to be premortem burns (Figure 2.91a–c). Ants also affect the fauna of primary colonizers on remains by feeding on fly eggs and maggots (Figure 2.92). In some instances the predation rate of ants on fly eggs may be so great that initial colonization may be delayed 2 to 3 days.

In indoor conditions, cockroaches may alter a body in a manner similar to that of fire ants in outdoor conditions. Cockroaches are scavengers on filth and refuse, but some species will feed on carrion when available, or even on living animals as well as the skin of living humans. Often babies that are left unattended in tenements and housing where unsanitary conditions prevail have been found with wounds indicating cockroach feeding. Cockroaches belong to the insect order Blattaria and are cosmopolitan in distribution. The species typically found indoors in the United States are the German cockroach (*Blatella germanica*) (Figure 2.93), the American cockroach (*Periplaneta americana*) (Figure 2.94), and the Australian cockroach (*Periplaneta australasiae*) (Figure 2.95).

Several other outdoor scavengers are frequently found in association with bodies and animal carcasses. Among the most common are pillbugs and sowbugs, sometimes called

Figure 2.92 An acrobat ant, *Crematogaster lineolata*, carries off a young maggot. Although ants are routinely observed at carrion feeding on fly eggs and larvae, they are not of great concern. However, in the southern United States, fire ants occur in such enormous numbers that they can affect the postmortem interval estimation. In some cases, ants may consume fly eggs and young larvae as soon as they are deposited. This may continue for up to 3 days before the flies are able to successfully colonize the body, after scavenging by the ants begins to diminish. (Photo courtesy of Dr. James L. Castner.)

Figure 2.93 A female German cockroach (*Blatella germanica*) with egg case. It prefers indoor environments where it can grow to enormous populations. This roach is attracted to human remains (especially in indoor locations), where it commonly produces postmortem damage to human skin that is often confused with antemortem trauma. (Photo courtesy of Dr. James L. Castner.)

Figure 2.94 The American cockroach (*Periplaneta americana*) is one of the most common roach species associated with man. This species has strong, hard mandibles that are capable of producing extensive damage to human skin that may resemble antemortem abrasions or burns. (Photo courtesy of Dr. James L. Castner.)

Figure 2.95 An adult (right) and nearly mature nymph (left) of the Australian cockroach (*Periplaneta australasiae*). This is another large roach species frequently responsible for post-mortem damage to human skin. Like the American cockroach, it is attracted to remains in outdoor and indoor locations. (Photo courtesy of Dr. James L. Castner.)

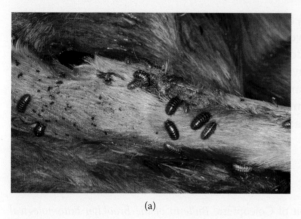

(a)   (b)

Figure 2.96 (a) Pillbugs (order Isopoda) are sometimes found associated with carrion, such as on the deer carcass pictured above. (b) These small armored terrestrial crustaceans are capable of rolling into a ball as a defense. (Photo courtesy of Dr. James L. Castner.)

isopods (Figure 2.96a and b). These small, gray, armored-looking creatures represent the only terrestrial crustaceans known. Some are capable of rolling up into a tight ball for protection, at which time they are about the size of a small pea. They are usually found in mulch, under rocks, or in damp earth, but sometimes occur in large numbers at carrion. Like other scavenging soil organisms, such as millipedes (class Diplopoda) (Figure 2.97), they are often in the protected area underneath the remains, where the body comes in contact with the soil.

Conclusion

This chapter has featured color photography of some of the most common species of insects of forensic importance in North America. Although this work is not inclusive of every insect species that an investigator might recover at a death scene, it concisely represents the general appearance of most insects of forensic importance. It is hoped that investigators will be able to improve their collection techniques by learning the general appearance of

Figure 2.97 Millipedes such as this one (*Narceus americanus*) are scavengers and harmless to man. (Photo courtesy of Dr. Jason H. Byrd.)

forensically important insects, and thereby have a higher and more successful recovery rate of those particular specimens. If the investigator becomes familiar with the appearance of the insects listed in this chapter, and follows the collection methodologies described in Chapter 3, he or she will be able to provide the forensic entomologist with entomological evidence that will yield the maximum amount of information possible. Additionally, procedures at the crime scene will be optimized, as the investigator will spend less time collecting transient and incidental insects and focus his or her collection efforts on the insects with the most evidentiary value.

References

Abbott, C. E. 1937. The Necrophilous habit of Coleoptera. *Bulletin of the Brooklyn Entomological Society* 32:202–4.

Abbott, C. E. 1937. The development and general biology of *Creophilus villosus* (Grav.) [= *C. maxillosus*]. *Journal of the New York Entomological Society* 46:49–52.

Aldrich, J. M. 1916. *Sarcophaga and allies in North America*. La Fayette, IN: Thomas Say Foundation.

Anderson, R. S. 1982a. Resource partitioning in the carrion beetle (Coleoptera: Silphidae) fauna of southern Ontario: Ecological and evolutionary considerations. *Canadian Journal of Zoology* 60:1314–25.

Anderson, R. S. 1982b. On the decreasing abundance of *Nicrophorus americanus* (Olivier) (Coleoptera: Silphidae) in eastern North America. *Coleopterist Bulletin* 36:362–65.

Anderson, R. S., and S. B. Peck. 1985. *The insects and arachnids of Canada*. Part 13. *The Carrion Beetles of Canada and Alaska*. Coleoptera: Silphidae and Agyrtidae. Research Branch Agriculture Canada. Publication 1778.

Archer, M. S., and M. A. Elgar. 2003. Effects of decomposition on carcass attendance in a guild of carrion-breeding flies. *J. Med. Vet. Entomol.* 17:263–71.

Arnett, R. H., Jr. 1961. *The beetles of the United States: A manual for identification*. Washington, DC: Catholic University of America Press.

Arnett, R. H., Jr., N. M. Downie, and H. E. Jacques. 1980. *How to know the beetles*. 2nd ed. Dubuque, IA: William C. Brown Company Publishers.

Arnett, R. H., Jr., and R. L. Jacques, Jr. 1981. *Guide to insects*. New York: Simon and Schuster.

Baumgartner, D. L. 1986. The hairy maggot blowfly, *Chrysomya rufifacies* (Macquart), confirmed in Arizona. *Journal of Entomological Science* 21:130–32.

Baumgartner, D. L. 1993. Review of *Chrysomya rufifacies* (Diptera: Calliphoridae). *Journal of Medical Entomology* 30:338–52.

Baumgartner, D. L., and B. Greenberg. 1984. The genus *Chrysomya* (Diptera: Calliphoridae) in the New World. *Journal of Medical Entomology* 21:105–13.

Bishopp, F. C. 1915. Flies which cause myiasis in man and animals—Some aspects of the problem. *Journal of Economic Entomology* 8:317–29.

Bland, R. G., and H. E. Jacques. 1978. *How to know the insects*. Dubuque, IA: William. C. Brown Company Publishers.

Bohart, G. E., and J. L. Gressitt. 1951. *Filth-inhabiting flies of Guam*. Bulletin of the Bernice P. Bishop Museum 204.

Borror, D. J., C. A. Triplehorn, and N. F. Johnson. 1989. *An introduction to the study of insects*. 6th ed. Philadelphia: Saunders College Publishing.

Borror, D. J., and R. E. White. 1970. *A field guide to insects—America north of Mexico*. Boston: Houghton Mifflin Company.

Brewer, J. W., and T. R. Bacon. 1975. Biology of the carrion beetle, *Silpha ramosa*. *Annals of the Entomological Society of America* 68:786–90.

Castner, J. L., J. H. Byrd, and J. F. Butler. 1995. *Forensic insect field identification cards*. Colorado Springs, CO: Forensic Sciences Foundation.

Chapman, R. K. 1944. An interesting occurrence of *Musca domestica* L. larvae in infant bedding. *The Canadian Entomologist* 76:230–32.

Clark, C. U. 1895. On the food habits of certain dung and carrion beetles. *Journal of the New York Entomological Society* 3:61.

Cockburn, A., R. A. Barraco, T. A. Reyman, and W. H. Peck. 1975. Autopsy of an Egyptian mummy. *Science* 187:115–1158.

Colyer, C. N. 1954a. The 'coffin fly' *Conicera tibialis* Schmitz (Diptera: Phoridae). *Journal of the Society for British Entomology* 4:203–6.

Colyer, C. N. 1954b. More about the 'coffin fly' *Conicera tibialis* Schmitz (Diptera: Phoridae). *The Entomologist* 87:130–32.

Colyer, C. N. 1954c. Further emergences of *Conicera tibialis* Schmitz, the 'coffin fly.' *The Entomologist* 87:234.

Cragg, J. B., and J. Hobart. 1995. A study of the field population of the blowflies *Lucilia caesar* and *L. sericata* (Meigen). *Annals of Applied Biology* 43:645–663.

Cragg, J. B. 1956. The olfactory behavior of *Lucilia* species (Diptera) under natural conditions. *Annals of Applied Biology* 44:467–77.

Cragg, J. B., and P. Cole. 1952. Diapause in *Lucilia sericata* (Diptera: Calliphoridae). *Journal of Experimental Biology* 29:600–4.

Cushing, E. C., and Hall, D. G. 1937. Some morphological differences between the screwworm fly *Cochliomyia americana* (C. & P.) and other closely allied or similar species in North America (Diptera: Calliphoridae). *Proceedings of the Entomological Society of Washington* 39(7).

Davis, W. T. 1928. *Lucilia* flies anticipating death. *Bulletin of the Brooklyn Entomological Society* 23:118.

Denno, R. F., and W. R. Cothran. 1975. Niche relationships of a guild of necrophagous flies. *Annals of the Entomological Society of America* 68:741–45.

Dorsey, C. K. 1940. A comparative study of the larvae of six species of *Silpha*. *Annals of the Entomological Society of America* 33:120–39.

Dunn, L. H. 1916. *Hermetia illucens* breeding in a human cadaver. *Entomological News* 27:59–61.

Erzinclioglu, Y. Z. 1985. Immature stages of British *Calliphora* and *Cynomya* with a re-evaluation of the taxonomic characters of larval Calliphoridae (Diptera). *Journal of Natural History* 19:69–96.

Fuller, M. E. 1934. *The insect inhabitants of carrion: A study in animal ecology*. Bulletin 82. Commonwealth of Australia, Council for Scientific and Industrial Research.

Gagne, R. J. 1981. *Chrysomya* sp. Old World blowflies (Diptera: Calliphoridae), recently established in the Americas. *Bulletin of the Entomological Society of America* 27:21–22.

Greenberg, B. 1971. *Flies and disease*. 2 vols. Princeton, NJ: Princeton University Press.

Greenberg, B. 1988. *Chrysomya megacephala* (F.) (Diptera: Calliphoridae) Collected in North America and notes on *Chrysomya* species present in the new world. *Journal of Medical Entomology* 25:199–200.

Hall, D. G. 1948. *The blowflies of North America.* Lafayette, IN: The Thomas Say Foundation.

Hall, R. D., and K. E. Doisy. 1993. Length of time after death: Effect on attraction and oviposition or larviposition of midsummer blowflies (Diptera: Calliphoridae) and flesh flies (Diptera: Sarcophagidae) of medicolegal importance in Missouri. *Annals of the Entomological Society of America* 86:589–93.

Hall, R. D., and L. H. Townsend, Jr. 1977. The blow flies of Virginia (Diptera: Calliphoridae). In *The Insects of Virginia* 11. Virginia Polytechnic Institute and State University Research Division Bulletin 123.

Hinton, H. E., and A. S. Corbet. 1975. *Common insect pests of stored products: A guide to their identification.* 5th ed. London: British Museum (Natural History).

Hogue, C. L. 1993. *Latin American insects and entomology.* Berkeley: University of California Press.

Holdaway, F. G., and A. C. Evans. 1930. Parasitism a stimulus to pupation: *Alysia manducator* in relation to the host *Lucilia sericata*. *Nature* 125:598.

Huckett, H. C., and J. R. Vockeroth. 1987. *Manual of Nearctic Diptera.* Vol. 2, monograph 28. Hull, Quebec: Canadian Government Publishing Centre, 1115–32.

Hudson, H. F. 1914. *Lucilia sericata* (Meigen) attacking a live calf. *The Canadian Entomologist* 46:416.

Illingworth, J. F. 1927. Insects attracted to carrion in southern California. *Proceedings of the Hawaii Entomological Society* 6:397–401.

James, M. T. 1947. *The flies that cause myiasis in man.* Miscellaneous Publication 631. U.S. Department of Agriculture, 1–175.

James, M. T. 1981. *Manual of Nearctic Diptera.* Vol. 1, monograph 27. Hull, Quebec: Canadian Government Publishing Centre, 497–512.

James, M. T., and R. F. Harwood. 1969. *Herm's medical entomology.* 6th ed. Toronto: The Macmillan Company.

Kamal, A. S. 1958. Comparative study of thirteen species of sarcosaprophagous Calliphoridae and Sarcophagidae (Diptera) 1. Bionomics. *Annals of the Entomological Society of America* 51:261–71.

Knipling, E. F. 1936. A comparative study of the first instar larvae of the genus *Sarcophaga*, with notes on the biology. *Journal of Parasitology* 22:417–54.

Liu, D., and B. Greenberg. 1989. Immature stages of some flies of forensic importance. *Annals of the Entomological Society of America* 82:80–93.

Lord, W. D., T. R. Adkins, and E. P. Catts. 1992. The use of *Synthesiomyia nudesita* (Van Der Wulp) (Diptera: Muscidae) and *Calliphora vicina* (Robineau-Desvoidy) (Diptera: Calliphoridae) to estimate the time of death of a body buried under a house. *Journal of Agricultural Entomology* 9:227–35.

Lord, W. D., M. L. Goff, T. R. Adkins, and N. H. Haskell. 1994. The black soldier fly *Hermetia Illucens* (Diptera: Stratiomyidae) as a potential measure of human postmortem interval: Observations and case histories. *Journal of Forensic Science* 39:215–22.

Mank, H. G. 1923. The biology of the Staphlinidae. *Annals Entomological Society of America* 16:220–37.

Marshall, S. A., and O. W. Richards. 1987. *Manual of Nearctic Diptera.* Vol. 2, monograph 28. Hull, Quebec: Canadian Government Publishing Centre, 993–1006.

Mégnin, P. 1894. *La Faune des Cadavers: Application de L'entomologie à la Médecine Légale.* Paris: Villars, Masson et Gauthier.

Melvin, R. 1934. The incubation period of eggs of certain muscoid flies at different constant temperatures. *Annals of the Entomological Society of America* 27:406–10.

McAlpine, J. F. 1987. *Manual of Nearctic Diptera.* Vol. 2, monograph 28. Hull, Quebec: Canadian Government Publishing Centre, 845–52.

Nuorteva, P. 1970. Histerid beetles as predators of blowflies (Diptera: Calliphoridae) in Finland. *Annales Zoologici Fennici* 7:195–98.

Nuorteva, P. 1974. Age determination of a blood stain in a decaying shirt by entomological means. *Forensic Science* 3:89–94.

Nuorteva, P. 1977. Sarcosaprophagous insects as forensic indicators. In *Forensic medicine, a study in trauma and environmental hazards.* Vol. II. *Physical trauma,* ed. C. G. Tedeschi, W. G. Eckert, and L. G. Tedeschi. Philadelphia: Saunders, 1072–95.

Nuorteva, P. 1987. Empty puparia of *Phormia terraenovae* R.-D. (Diptera: Calliphoridae) as forensic indicators. *Annales Entomologici Fennici* 33:53–56.

Oldroyd, H., and K. G. V. Smith. 1973. Eggs and larvae of flies. In *Insects and other arthropods of medical importance,* ed. K. G. V. Smith. London: British Museum (Natural History), 289–323.

Payne, J. A. 1965. A summer carrion study of the baby pig *Sus scrofa* Linnaeus. *Ecology* 46:592–602.

Payne, J. A., and E. W. King. 1970. Coleoptera associated with pig carrion. *Entomologist's Monthly Magazine* 105:224–32.

Payne, J. A., F. W. Mead, and E. W. King. 1968. Arthropod succession and decomposition of buried pigs. *Nature* 219:1180–81.

Peterson, A. 1979. *Larvae of insects.* Part II. Ann Arbor, MI: Lithographed by Edwards Brothers.

Peterson, B. V. 1987. *Manual of Nearctic Diptera.* Vol. 2, monograph 28. Hull, Quebec: Canadian Government Publishing Centre, 689–712.

Pirone, D. J. 1974. Ecology of Necrophilus and Carpophilus Coleoptera in a southern New York woodland (pehnology, aspection, trophic, and habitat preferences). PhD thesis, Fordham University, New York.

Prins, A. J. 1982. Morphological and biological notes on six South African blow-flies (Diptera: Calliphoridae) and their immature stages. *Annals of the South African Museum* 90:201–17.

Quate, L. W., and J. R. Vockeroth. 1981. *Manual of Nearctic Diptera.* Vol. 1, monograph 27. Hull, Quebec: Canadian Government Publishing Centre, 293–300.

Rabinovich, J. E. 1970. Vital statistics of *Synthesiomyia nudiseta* (Diptera: Muscidae). *Annals of the Entomological Society of America* 63:749–52.

Ratcliffe, B. C. 1972. The natural history of *Necrodes surinamensis. Transactions of the American Entomological Society* 98:359–410.

Ratcliffe, F. N. 1935. Observations on the sheep blowfly (*Lucilia sericata* Meigen) in Scotland. *Annals of Applied Biology* 22:742–53.

Reed, H. B. 1958. A study of dog carcass communities in Tennessee, with special references to the insects. *The American Midland Naturalist* 59:213–45.

Richard, R. D., and E. H. Ahrens. 1983. New distribution record for the recently introduced blow fly *Chrysomya rufifacies* (Macquart) in North America. *Southwestern Entomologist* 8:216–18.

Rodriguez, W. C., and W. M. Bass. 1983. Insect activity and its relationship to decay rates of human cadavers in east Tennessee. *Journal of Forensic Sciences* 28:423–32.

Shewell, G. E. 1987. *Manual of Nearctic Diptera.* Vol. 2, monograph 28. Hull, Quebec: Canadian Government Publishing Centre, 1133–46.

Shubeck, P. P. 1976. Carrion beetle responses to poikilothermic and homiothermic carrion. *Entomological News* 89:265–69.

Simmons, P. 1927. *The cheese skipper as a pest in cured meats.* Department Bulletin 1453. U.S. Department of Agriculture, 1–55.

Smit, B. 1931. A study of the sheep blowflies of South Africa. Report of the Director of Veterinary Service 17. Onderstepoort, 299.

Smith, K. G. V. 1956. On the Diptera associated with the stinkhorn (*Phallus impudicus* Pers.) with notes on other insects and invertebrates found on this fungus. *Proceedings of the Royal Entomological Society of London* 31:49–55.

Smith, K. G. V. 1975. The faunal succession of insects and other invertebrates on a dead fox. *Entomologist's Gazette* 26:277–87.

Smith, K. G. V. 1986. *A manual of forensic entomology.* Ithaca, NY: Cornell University.

Spector, W. 1943. Collecting beetles (*Trox*) with feather bait traps (Coleoptera: Scarabaeidae). *Entomological News* 54:224–29.

Steele, B. F. 1927. Notes on the feeding habits of carrion beetles. *Journal of the New York Entomological Society* 35:77–81.

Stehr, F. W. (Ed.). 1991. *Immature insects*. Vol. 2. Dubuque, IA: Kendall/Hunt.

Streyskal, G. C. 1987. *Manual of Nearctic Diptera*. Vol. 2, monograph 28. Hull, Quebec: Canadian Government Publishing Centre, 945–50.

Townsend, C. T. 1937. *Manual of myiology*. Vol. 5. pp. 65–176.

Tullis, K., and M. L. Goff. 1987. Arthropod succession in exposed carrion in a tropical rainforest on Oahu Island, Hawaii. *Journal of Medical Entomology* 24: 332–39.

Vockeroth, J. R. 1987. *Manual of Nearctic Diptera*. Vol. 2, monograph 28. Hull, Quebec: Canadian Government Publishing Centre, 1085–98.

Voigt, J. 1965. Specific postmortem changes produced by larder beetles. *Journal of Forensic Medicine* 12:76–80.

von Zuben, C. J., S. F. Dos Reis, J. B. do Val, W. A. Godoy, and O. B. Ribeiro. 1993. Dynamics of a mathematical model of *Chrysomya megacephala* (Diptera: Calliphoridae). *Journal of Medical Entomology* 30:443–48.

Voris, R. 1939. The immature stages of the genera *Ontholestes, Creophilus* and *Staphylinus. Annals of the Entomological Society of America* 32:288–300.

Wasti, W. S. 1972. A study of the carrion of the common fowl, *Gallus domseticus*, in relation to arthropod succession. *Journal of the Georgia Entomological Society* 7:221–29.

Wells, J. D., and B. Greenberg. 1992. Interaction between *Chrysomya rufifacies* and *Cochliomyia macellaria* (Diptera: Calliphoridae): The possible consequences of an invasion. *Bulletin of Entomological Research* 82:133–37.

White, R.E. 1985. *Beetles of North America*. Norwalk, CT: Easton Press.

Woodruff, R. E. 1973. *The scarab beetles of Florida*. Vol. 8. *Arthropods of Florida and neighboring land areas*. Gainesville, FL: Florida Department of Agriculture and Consumer Services, Division of Plant Industry.

Zumpt, F. 1952. Flies visiting human faeces and carcasses in Johannesburg, Transvaal. *South African Journal of Clinical Science* 3:92–106.

Zumpt, F. 1965. *Myiasis in man and animals in the old world*. London: Butterworths.

Collection of Entomological Evidence during Legal Investigations

3

JASON H. BYRD
WAYNE D. LORD
JOHN R. WALLACE
JEFFERY K. TOMBERLIN

Contents

Introduction

The general acceptance of arthropods as indicators of critical forensic parameters has steadily increased throughout the world over the last several years. Scientific analysis and the expert opinion of qualified forensic entomologists are now routinely solicited by law enforcement and legal professionals in both criminal and civil investigations. This widespread acceptance of forensic entomology within the criminal justice system has created a high demand for entomological services by the law enforcement community. Such casework makes it necessary for crime scene analysts and death investigation professionals to become increasingly involved in the documentation, collection, and shipment of insect evidence to qualified forensic entomologists.

Dr. Peter Pizzola once stated that forensic science does not begin at the crime laboratory. This principle certainly holds true for forensic entomology as a discipline within the forensic sciences. Forensic entomology starts at the death scene, or site of recovery. Many of the arthropods associated with the general fauna of decomposition are not on the remains, but distributed around the remains, or even inhabiting the soil under the remains. If entomological collections are made once the remains are removed from the site of recovery, there is an increased likelihood that some of the associated arthropod fauna will not be documented and collected. This may or may not have a negative impact on the ensuing analysis. However, it does result in less evidence retained, and thus standard practice should be to document and collect the entomological evidence while at the site of body recovery.

A core element to the utilization of arthropods as forensic indicators is the proper recognition, documentation, collection, and shipment of entomological evidence from human or animal remains. For maximum evidence yield, this task should be performed at the death scene or body recovery site. However, due to the fact that in some cases the major feeding areas of the arthropods are internal, some collections must be performed at the morgue, prior to or during the autopsy procedure. In these instances, the investigator should return to the scene of recovery and make a collection from the soil under and around where the remains were once located. Proper collection procedures dictate collections of live and preserved samples at the scene and autopsy be performed.

Proper entomological collections are critical because the accurate determination of specimens to the species level (using non-DNA methods) can be attained only if certain morphological characteristics are intact and on collected specimens. The live samples may aid in species identification by allowing the insects to develop to a more advanced life stage with more readily identifiable characters, whereas the preserved samples are essential to demonstrate the stage of insect development attained at the time of remains recovery. Preservation stops the arthropod development and prevents an error in developmental estimation from improperly documented temperatures during transport or storage prior to autopsy.

In addition to the actual collection of both preserved and live samples of arthropods, the investigator must thoroughly document the circumstances of the death, including all insect activity, create evidence labels for all collected entomological samples, and document environmental habitats at the scene as well as current weather data. The judicial system mandates the utilization of best practices and proceedings that ensure that proper custody is maintained. Thus, a proper chain of custody should be established for entomological evidence as soon as it is collected.

Entomological Collection Procedures at the Scene

There is a bewildering and seemingly endless array of environmental circumstances in which human remains are recovered. In the most broad perspective those habitats can be categorized into terrestrial (including subterranean), aquatic, and marine environments. Insects and their arthropod relatives have been recovered from human and animal remains in each of these habitats. When considering the unique aspect of artificial environments created by humans in areas such as houses, apartments, abandoned buildings, car trunks, various sealed containers, and landfills, the science of forensic entomology becomes quite complex. Interpreting arthropod activity in more natural environments such as forests, mountain regions, deserts, swamps, riverbanks, lakes, ponds, ditches, or agricultural fields may be more straightforward, but each environment presents its own set of unique circumstances that must be taken into account.

This chapter presents best-practice procedures for most habitats and geographic areas. These procedures may be altered to suit the particular requirements of individual recovery sites and to account for various departmental policies for evidence procedures. However, technical aspects of the collection (such as preservative fluids and temperature data collection) should not be altered. Previous works detail procedural instructions and may also be consulted (Lord and Burger 1983, Smith 1986, Catts and Haskell 1990, Wecht 1995, Haglund and Sorg 1997). However, collection procedures for entomological samples have changed over the years, and more recent works should be consulted to remain current on proper collection techniques (Byrd and Castner 2001, Amendt et al. 2007).

Entomological Collection at the Death Scene

It is unlikely that a qualified forensic entomologist will actually be present on site for the collection and documentation activities. Therefore, it is essential that the crime scene analyst or medicolegal death investigator become well educated on the various aspects of entomological documentation and collection procedures. Properly trained death scene personnel can make entomological collections just as adequate as those of a forensic entomologist. It is not essential to have a forensic entomologist present at the scene to collect the evidence before it can be analyzed. If a forensic entomologist is present at the scene, he or she should be one of the first investigators to approach the remains for an initial assessment and photography. Prior to the entomological evidence collection, it is desirable for the remains to be undisturbed, and that access to the remains is limited to few individuals. Human activity can impact the presence of both flying and crawling arthropods on and around the remains.

Of critical importance to the use of entomological evidence in a legal investigation is the creation of a detailed overview of the physical surroundings. This description includes written notes and detailed photographs of the scene and the surrounding environment. This documentation is part of the initial assessment that should be completed for each investigation that will utilize entomological evidence. If present at the scene, the forensic entomologist will work as part of the investigative team. As such, he or she must also be advised of routes of ingress and egress to the body, and be informed of what physical evidence is still *in situ*, what has been recovered, and what must not be disturbed. Since investigative agencies differ widely in their scene protocols, and since evidence is often commingled, the forensic entomologist may find it best to coordinate all collection and preservation activities related to the entomological evidence with the assistance of the primary crime scene investigator, medicolegal death investigator, medical examiner, or coroner.

The process of collecting entomological evidence may be somewhat intrusive, depending on the location of the insects and the extent of collection. In some circumstances, the collection will result in minor and unavoidable disturbance to the remains. Therefore, it is good practice to receive permission to enter the scene and collect from the remains from the medical examiner or coroner with jurisdiction over the scene. In all cases, the utmost caution, care, and coordination with crime scene personnel is required for the entomological collection. Prior to collection, the medical examiner or coroner should be advised of the intended collection areas, and the extent of the collection in those areas. In some situations, the best practice may be to make a surface collection from the body at the scene, collect from under the body and surrounding scene once the remains are removed, and then complete the collection during the autopsy process by collecting insects from within the tissues of the body during the autopsy under the guidance of the medical examiner or coroner. This three-step collection methodology will generally result in a more representative sample of the faunal population present on the remains.

Effective coordination prior to the entomological collections will minimize unwanted disturbance or alteration to the remains, and minimal interference with other physical evidence will occur. Information exchange is also a critical component to the entomological investigation. Crime scene personnel should thoroughly brief the forensic entomologist as to the circumstances surrounding the discovery of the body recovery site, and provide all available information regarding prior evidence collection attempts at the scene. Information on past or suspected drug use is also critical information to convey to the forensic entomologist. In a similar manner, the forensic entomologist should inform the crime scene personnel about the entomological collection procedures that need to take place and what type of information may be yielded from an entomological collection. In the majority of forensic cases involving entomological evidence, it is the crime scene analyst or medicolegal death investigator that will perform the entomological collection procedures (Figure 3.1). The entomological collection will take place in relation to the other physical evidence collections and will depend on the types of evidence present, the circumstances of the death, and the environmental factors at the scene. No rigid sequence of events should be set, but primary evidence should be collected first, as with any best management practice. The use of "General Guidelines for the Collection and Recording of Entomological Evidence" in the appendix of this chapter will greatly assist the investigator in following the proper sequence of events for the entomological collection.

Figure 3.1 For most jurisdictions, it is not practical to have a forensic entomologist respond to the scene for the purpose of collecting the entomological evidence. Therefore, for almost all cases, crime scene personnel properly trained in the collection and preservation of entomological evidence can effectively recover the insect samples. (Photo courtesy of Dr. Jason H. Byrd.)

Figure 3.2 Visual and written observations, as well as photography of the general scene and surrounding area, should begin at a distance. Crime scene technicians should do the preparations for entomological collection well away from the remains, so that the arthropods present on the remains will not be disturbed. (Photo courtesy of Dr. Jason H. Byrd.)

The Collection Process

Once the site has been assessed and a scene plan developed for the collection of evidence, the importance of the entomological evidence will be ranked with all other physical evidence. At the time for collecting the entomological evidence, slow movement on the initial approach to the remains is very important. Such an approach will minimize disturbance to the arthropod fauna, particularly the flying adult insects. Visual observations, photography, and written notations about the degree and position of insect infestation on the body should begin several feet from the body (Figure 3.2). At this stage, it is important to determine what arthropods are present, the location of the major areas of colonization (as evidenced by the presence of eggs, larvae, or pupae), and the location of any insect activity on the ground or substrate near the body. It is also good practice to record the distances from the body and compass direction to remote sites of insect activity, and other appropriate notations written. Written habitat documentation can be made on the "Forensic Entomology Data" form (see appendix), and the collected samples can be recorded on the "Entomological Sample Log Sheet" (see appendix).

Entomological documentation and collection at the scene can be broken down into several major steps (also see guidelines in appendix):

1. Photographic and written notation of general scene characteristics and habitat
2. Photographic and written notation of insect infestations on and around the remains
3. Collection of meteorological data
4. Collection of adult flies and beetles on and flying above the remains
5. Collection of eggs, larvae, and pupae on the remains
6. Collection of specimens from the surrounding area (up to 20 ft [6 m]) from the body
7. Collection of specimens from directly under and in close proximity to the remains (3 ft [1 m] or less) after the body has been removed
8. Collection of soil samples from under remains (head, chest, abdomen)
9. Notations of the ecological characteristics (soil, plant, water, etc.) at the recovery site

10. Collections of entomological samples during autopsy
11. Retrieval of historical climate/weather data from the nearest weather station

In many instances, the forensic entomologist can provide information valuable to the overall death investigation in the form of written notes, photographs, and physical evidence. Several instances exist in which the forensic entomologist, during the course of his or her investigation and analysis of the scene, has found hairs, fibers, teeth, and small bone fragments that were previously overlooked by other investigators. Such evidence has provided valuable information on suspects, victims, and trauma analysis, and has assisted medical examiners and coroners in determining cause and manner of death. Thus, a cooperating entomologist, often with the aid of high-powered microscopes and an eye for detail, becomes another forensic analyst in the general task of collecting any and all physical evidence associated with a legal investigation.

Crime scene investigators have a tremendous challenge and responsibility in their attempts to identify the circumstances of a crime, and to locate, document, and collect all evidence related to the crime. This task can be arduous since a wide diversity of evidence may be present, and many different and sometimes conflicting procedures may be required to recover and preserve such evidence. In many jurisdictions only one person is available to work the scene, which requires prioritization of the procedures that can be handled within the time constraints allowed. Given the pressures of the media, administrators, and the public for immediate answers regarding a death scene, important but obscure evidence may be passed over and not collected or noted, thus ending up under the microscope of the forensic entomologist.

The list of equipment and supplies outlined in the appendix of this chapter will facilitate proper documentation and collection of specimens from the body recovery site and during autopsy for delivery to a forensic entomologist in the proper physical condition necessary for analysis. The equipment listed may be purchased in assembled kits from a multitude of biological supply and scientific equipment companies, or each agency can compile its own kits to suit its specific needs (Figure 3.3).

Photographic and Written Notation of General Scene Characteristics and Habitat

The first step in the entomological examination of a death scene is a visual observation and written notations of the habitat of the surrounding scene (Figure 3.4). Habitat documentation is of great importance because it can assist the entomologist in determining the species of insect that should be present in that type of environment. If the arthropod species recovered are not what would be expected in that environment, it could be indicative of the remains being colonized elsewhere and then being relocated to the site of discovery. Habitat documentation should include completion of the "Forensic Entomology Data" form, and complete photographic documentation of the area (Figure 3.5). Photographs taken should include long-range photos toward the body from each of the four major compass points (north, south, east, west), as well as 360° views from the remains facing outward. In addition, if in an outdoor location, photographs should be taken directly over the remains. These types of photographs aid in showing the degree of exposure to direct sunlight, extent of tree canopy, and even atmosphere conditions at the time of scene processing.

Figure 3.3 It is essential to have assembled the proper equipment necessary to complete the entomological collection process before responding to a scene. Although preconfigured commercial collection kits are available, many agencies compose their own kit to suit their specific needs. (Photo courtesy of Dr. Jason H. Byrd.)

Figure 3.4 Specific written documentation of the scene and the surrounding area should be completed prior to making the collection. Notation of habitat type and surrounding environment is essential information for the entomological analysis. (Photo courtesy of Dr. Jason H. Byrd.)

Photographic and Written Notation of Insect Infestations on and around the Remains

Once the surrounding environment has been documented with written notations and photographs, the investigator should focus on documentation of the body at the scene. Written notations should focus on the amount and location of activity on or around the remains and, if possible, descriptions of what types of insects are present. These observations should be started at a distance, before anyone approaches the remains. During this observational stage, the investigator should always maintain some distance from the remains so as not to disturb the flying adult arthropods. In addition to distance, another method to minimize the disturbance to the insects and associated arthropods is for the

Figure 3.5 The written documentation of the scene should also include notations as to the general stage of decomposition, amount of insect activity, and position of the body. These notes are supplemented with detailed photography of the remains and insect activity. (Photo courtesy of Dr. Jason H. Byrd.)

investigator to make sure that his or her shadow does not fall across the remains. Always be aware of the position of the sun and artificial light sources because sudden shifts between light and dark, such as a shadow passing over the insect, may startle some species and prompt them to suddenly disperse. Many fly species have acute vision, and they often disperse when approached closely. They may not return until the human activity has subsided. Many arthropods have distinct feeding periods, and thus may not be present or visible on the remains for an extended period of time. Due to the fleeting nature of some insects and their distinct feeding periods, some species may be observed when personnel first arrive at the scene, but they may not be present when the collection process begins. In particular, adult beetles and most larvae located underneath the remains often bury themselves quickly once the body is removed. If these arthropods cannot be collected at the time they are first observed, then a photographic and written record of their presence will assist in the proper documentation of the scene.

It is good practice to make notations of the relative abundance of insects and arthropods present. A photographic documentation should be made to support the comments made in the written documentation. All photographs should include a size reference scale to assist the forensic entomologist in assessing type and age of arthropods present.

The scene also may be recorded with a video camera. While this method is excellent for recording overall elements of the scene and observing entomological information from a gross perspective, the fine details and close-up imagery required in entomological investigations usually cannot be viewed due to the lower resolution of video camera technology. Therefore, video should not be used as the sole source for recording the details of the body recovery site. Since a video camera can be used to record the relative positions of items at a death scene much as a human eye would depict the scene, videography is an extremely useful way to help document and preserve the scene for future analysis. For the documentation of entomological evidence, still photography provides the highest resolution of the insect specimens present, and a macro lens (105 mm) and flash should be routinely employed to record the details of entomological evidence.

Collection of Meteorological Data

After habitat documentations are completed and the areas of insect activity are noted and recorded, the investigator can begin collecting meteorological data. Proper documentation of the meteorological conditions at the scene is critical to the analysis of the collected arthropod specimens. Accurate climatological information is critical when estimating the time of colonization (TOC) or period of insect activity (PIA) by entomological means. The TOC or PIA can be equivalent to the minimum estimation of the postmortem interval (PMI). The time required for arthropods to undergo their life cycle development is determined largely by the temperatures and relative humidity in the particular environment to which they are exposed. Other climatological conditions (e.g., rainfall, full sun, snow cover, and fog) also may influence insect development rates, behavior, and carrion-feeding habits. Therefore, the forensic entomologist should develop a working knowledge of climatology and its influence on carrion insect ecology. Since the proper collection and interpretation of climatological data are essential when determining the TOC or PIA, having a working relationship established with a qualified climatologist can be of great value when seeking meteorological data from a specific area or region, or with the analysis and interpretation of collected weather information.

Documentation of climate data is a multistep process that consists of documenting the current weather conditions at the scene during the scene processing, and obtaining the climate record and historical weather data for the period when the person was last reliably seen alive (or the suspected time of death) and discovery of the remains. Additionally, the investigator may be required to monitor the meteorological conditions at the scene for a time period of several days after body removal.

Provided adequate food resources are available, temperature is the most important factor influencing insect growth and development. Arthropods are cold blooded (poikilotherms), and temperature drives the rate of enzymatic action for insect growth. Generally speaking, insects develop slower and age more slowly when temperatures are cooler. Conversely, they are able to develop, and thus age more quickly, under warmer temperatures. The larvae of flies, commonly called maggots, have the ability to regulate their environmental temperatures with the formation of an aggregation, which is termed the maggot mass. As with any other organism, upper and lower temperature thresholds exist that go beyond enzyme capabilities to produce expected and desired reactions within the insect. When these limits are exceeded, the effects can be lethal to the developing organism. The biological adaptation of a maggot mass helps ensure that these thresholds are not surpassed, and allows the larvae within the mass to develop at a more optimum temperature, as desired by the insects.

To estimate arthropod age, and subsequent determinations of the TOC or PIA, the investigator should record several temperature readings while processing the body recovery site. The suggested minimum temperature documentations are:

1. Ambient air temperature recorded by readings taken at 1 and 4 ft (0.3 and 1.3 m) heights in close proximity to the body (Figure 3.6a–c).
2. Ground surface temperatures obtained by placing a thermometer on the ground on top of the soil surface (Figure 3.7a and b).
3. Body surface temperatures obtained by placing a thermometer on the upper surface of the body (Figure 3.8a and b).

(a)

(b)

Figure 3.6 (a) Ambient air temperature should be recorded at a height of approximately 4 ft (1.3 m) above (or in close proximity to) the remains. This temperature reading will be used to correlate scene temperatures with National Weather Service observations. (b) All temperature readings should be taken with the thermometer in shade. Do not expose the thermometer to direct sunlight. (c) An ambient air temperature reading should also be recorded at a height of 1 ft (0.3 m) above (or in close proximity to) the remains. This temperature reading will allow the forensic entomologist to determine the occurrence and effect of microclimatic conditions at the scene. (Photos courtesy of Dr. Jason H. Byrd.)　　　　*Continued*

4. Underbody interface temperatures obtained by sliding the thermometer between the body and the ground surface (Figure 3.9).
5. Maggot mass temperatures obtained by inserting the thermometer into the center of the maggot mass (Figure 3.10a and b). Care should be taken not to damage the remains.
6. Soil temperatures taken immediately following body removal at a ground point that was under the remains prior to removal (Figure 3.11).

(*Note*: The direct rays of the sun should not be allowed to shine on the thermometer-sensing element. Radiant heat from the sun will cause readings far in excess of the true environmental temperatures. Always shade the thermometer from direct sunlight when taking temperature data).

(c)

Figure 3.6 *Continued.*

(a)

(b)

Figure 3.7 (a) Ground surface temperature should be recorded from the top surface of the soil. (b) The soil surface temperature should be taken at a location that does not contain any fluid exudates from the decomposing remains. (Photos courtesy of Dr. Jason H. Byrd.)

(a)

(b)

Figure 3.8 (a) Body surface temperature should be recorded directly from the upper surface of the body. In many cases, a temperature reading should be taken from both the skin surface (if exposed) and the upper surface of clothing or wrappings (if present). However, the remains should not be unwrapped at the scene. (b) The thermometer's probe should be touched lightly to the upper surface of the remains. Do not insert the probe into the remains. (Photos courtesy of Dr. Jason H. Byrd.)

Figure 3.9 A temperature reading should always be taken from the interface between the body and the substrate on which it rests (such as soil, vegetation, concrete, asphalt, or flooring materials). (Photo courtesy of Dr. Jason H. Byrd.)

(a)   (b)

Figure 3.10 If a centralized mass of larvae is present, the internal temperature of the mass should be recorded by simply inserting a thermometer probe into the center of the active mass (a). Often the internal mass temperature is over 100°F (b) Movement of the probe will disrupt the larval mass. It is important to keep the thermometer motionless once inserted into the mass during the temperature reading. (Photos courtesy of Dr. Jason H. Byrd.)

Figure 3.11 Immediately upon removal of the body, soil temperatures should be recorded from directly underneath the prior location of the remains. Additionally, it is important to record soil temperatures at a point 3 to 6 ft (1 to 3 m) from the body. These should be recorded from under any ground cover at a depth of 4 in. (10 cm) and 8 in. (20 cm). (Photo courtesy of Dr. Jason H. Byrd.)

It is good practice to also record soil temperatures taken from a second point 3 to 6 ft (1 to 2 m) from where the body lay. This temperature reading should be recorded from three levels: directly under any ground cover (grass, leaves, etc.), at a soil depth of 4 in. (10 cm), and at a soil depth of 8 in. (20 cm).

The investigator should also make a written documentation of the time duration the remains are exposed to direct sunlight, broken sunlight, and shade for the total daylight hours. This can be accomplished by observing the surrounding and overhead vegetation

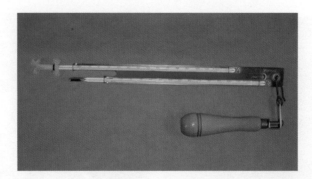

Figure 3.12 A sling type psychrometer (wet and dry bulb thermometer) used for determining relative humidity. The sling psychrometer is spun rapidly on its handle for approximately 1 minute before the reading is taken. Evaporative cooling of the moistened cotton wick of the wet bulb thermometer produces a lower than ambient temperature reading. The difference in temperatures is then utilized to determine the relative humidity. (Photo courtesy of Dr. Jason H. Byrd.)

and structures or the location of windows, and noting their compass direction relative to the position of the remains. If there is any question as to this relationship, observation of the site periodically throughout a sunny day will provide additional information. When direct sunlight is shining on the body, external temperatures and some internal temperatures close to the surface may be higher than when the remains are shaded.

Relative humidity can be obtained at the scene by using a sling (wet and dry bulb) (Figure 3.12) or battery-powered psychrometer. From these data, dew points and dew wettings of the remains can be estimated. Under certain conditions, the moisture attributed to dewfall can be of greater cooling and wetting influence than that resulting from rain or snowfall. Even though there has been no precipitation recorded, a body may have been coated with surface moisture daily. Air movement and cooling from dew evaporation (evaporative cooling) can reduce the temperature of the remains below ambient levels.

Collection of Specimens before Body Removal

Once the needed temperature recordings have been made, the investigator should begin the collection of flying insects above the body. This collection should be performed quickly after the completion of the temperature recording because many adult flying insects will disperse from the remains once human activity begins. Some will return within a few minutes after disturbance; however, some species may not return for an extended period of time.

The first step in the collection of flying adult insects is the aerial collection. This method is accomplished with the use of an aerial insect net and employing a back-and-forth or figure 8 sweeping motion over the body. Flying arthropods associated with carrion are strong, fast fliers; thus, netting of specimens requires some experience and practice. Appropriate netting techniques use several rapid, back-and-forth sweeping motions of the net (six to ten sweeps) with reversal of the opening of the net 180° on each pass (Figure 3.13a–c). On the last pass, the opened portion of the net is brought up to about chest level, with rotation of the opening 180° (Figure 3.14a and b). This movement causes the netting material to be folded over the top edge of the large net ring opening, thus trapping the arthropods in the net bag (Figure 3.15). Another technique that can be employed to collect flying arthropods is to hold the tail of the

(a)

(b)

(c)

Figure 3.13 Aerial netting of flying arthropods is one of the first entomological collection procedures to undertake. The insect net is swept rapidly back and forth above the body (a and b), with a rotation of the net opening 180° after each pass (c). (Photos courtesy of Dr. Jason H. Byrd.)

(a)

(b)

Figure 3.14 (a) After the last pass, the net is quickly brought up to chest height and the open hoop is rotated 90°. (b) This rotation is continued for another 90° until the tail of the net lies across the wire hoop. (Photos courtesy of Dr. Jason H. Byrd.)

net up and approach the arthropods from above with a swatting motion (Figure 3.16). The natural escape behavior of the insect will cause it to fly up and into the net. With either technique the arthropods can be easily confined in the end of the net (Figure 3.17a and b). Care to not disturb the remains should be taken when using a sweep net.

The end of the net, with arthropods inside, can then be placed into a wide-mouth killing jar, which is then capped (Figure 3.18a and b). The killing jar should contain either gypsum cement (plaster of paris) or a few cotton balls freshly soaked with ethyl acetate. This compound will kill the adult arthropods after a few minutes (2 to 5 minutes of exposure is usually adequate). Following immobilization, the arthropods can be transferred into vials containing 80% ethyl alcohol (ETOH) by placing a small funnel into the vial and carefully dumping the contents of the net into the funnel (Figure 3.19). If the investigator prefers to not use the kill jar technique, the flying adult arthropods can also be placed directly into

Figure 3.15 In final position, the wire hoop has simply been brought to chest height and rotated 180°. This effectively traps the flying arthropods within the net and minimizes the chance for escape. (Photo courtesy of Dr. Jason H. Byrd.)

Figure 3.16 An alternative technique to the sweep method involves approaching the arthropods from above while holding the tail of the net. A swatting motion is then used to collect arthropods in the net. This is an effective method because arthropods will have a tendency to fly up in order to escape. (Photo courtesy of Dr. Jason H. Byrd.)

Figure 3.17 (a) Grasping the net at its midpoint with one hand will confine the arthropods in the tail of the net. (b) The confined arthropods can be easily transferred into a kill jar or preservative vial. (Photos courtesy of Dr. Jason H. Byrd.)

alcohol by holding the end of the net up and reaching under the wire hoop and into the net with an alcohol vial and gently tapping the insect into the vial (Figure 3.20a and b).

In outdoor locations, after the aerial insect collection is complete, a similar method should be employed to collect insects from the surrounding vegetation. Many flies associated with carrion can be found resting on nearby vegetation. By sweeping the vegetation within 10 to 20 ft (4 to 6 m) from the remains, some of these adult arthropods may be collected (Figure 3.21). The aerial net can be used for this survey, but if the foliage is too thick or dense, or the plants are woody and stiff, the lightweight material of the aerial net will snag and tear. Instead, heavy-bagged sweep nets (such as those used by agricultural insect specialists for surveying arthropods on field crops) can be employed for collecting forensic arthropods resting on nearby foliage.

In order to conduct the "vegetation sweep," the aerial insect net is held extended outward and downward in front of the investigator, and the figure 8 sweeping motion used. This vegetation sweep is accomplished by sweeping the net over and along vegetation near

(a)

(b)

Figure 3.18 (a) With the arthropods confined, the tail of the net can be placed inside the kill jar and capped. (b) The net should remain inside of the kill jar for 3 to 5 minutes while the ethyl acetate takes effect. (Photos courtesy of Dr. Jason H. Byrd.)

Figure 3.19 Once the adult arthropods have been killed, they can be easily transferred to the alcohol vials by emptying the end of the net into a funnel. (Photo courtesy of Dr. Jason H. Byrd.)

(a)   (b)

Figure 3.20 (a) It is possible to place the adult arthropods directly into alcohol, thus avoiding the use of kill jars and funnels. This can be accomplished by holding the end of the net up, and reaching under the wire hoop with a vial of alcohol and gently tapping the insect into the vial. (b) The arthropods trapped in the net will have a natural tendency to walk upwards. The open vial filled with alcohol can be brought up from underneath, and the adult insect can be tapped into the vial. (Photos courtesy of Dr. Jason H. Byrd.)

Figure 3.21 A net sweep of the surrounding vegetation should also be conducted. The vegetation provides a resting place for arthropods that have been disturbed from the remains by the activity of the crime scene personnel. (Photo courtesy of Dr. Jason H. Byrd.)

the remains. Generally, a sweep of the vegetation 5 to 10 ft (1.5 to 3 m) from the remains and in a 360° circle around the remains is completed. All collected insects from this sweep can first go into the kill jar, and later be transferred into a single vial with an 80% ETOH solution, and marked with the companion dual-labeling system described below.

Specimens from the kill jar may be placed into a dry, clean vial for pinning. If stored in this dry condition, the arthropods must be processed in a few hours because excessive moisture on the arthropods (condensation arising from within the closed vial) can promote the growth of mold and fungus that will quickly damage or destroy the specimens. Aerial sweep-netting procedures should be repeated three to four times to ensure a representative sample of all of the flying arthropods present. Many arthropods have spurs or claws that can catch on the netting material and hinder their falling free. Care should be taken that all arthropods are shaken out of the net and into the vial. The bag portion of the net should be cleaned thoroughly between each use to prevent any cross-contamination from one scene to the next.

Sticky traps can be placed on site to collect flying adult arthropods. These traps consist of white or yellow cardboard coated with a sticky substance. There are specific traps available for collecting arthropods; however, these usually have to be hung to work. In most instances, these traps are not useful at a body recovery site. Sticky traps for catching rodents should be used instead. At the scene, the protective paper should be removed and the trap folded in half to form a tent. Clothes pins can be fastened to each corner to form legs for the trap, and then placed on the ground near the remains. Once all other evidence collections have been made and the remains removed from the scene, the clothes pins can be removed and the trap inverted and folded into a cylinder with the sticky side on the inside and fastened. The trap can be labeled and stored in a Ziploc bag in a freezer until it can be transferred to a forensic entomologist for examination.

All of the collected adult insects from the kill jar or net can be placed directly into the 80% ETOH solution. Vials containing the insects should be appropriately labeled. The labeling of evidence containers for entomological samples is conducted differently from that of most other items of physical evidence. One of the major differences is that the labels for entomological evidence are written in pencil, not ink! The 80% ETOH and many other solutions for the preservation of entomological evidence will remove the ink from the paper. Labels made with a graphite pencil will not be affected by solutions used to preserve specimens. Another difference is the utilization of a dual-labeling technique. This is called companion labeling, and the twin labels contain the same information, both written in pencil. The difference between the two labels is simply the paper on which they are written. The first label is written on plain paper, generally 1 × 3 in (2.5 × 7.5 cm). This label is placed directly inside of the vial with the insect samples immersed in 80% ETOH (Figure 3.22a and b). The second label is written on adhesive paper stock and affixed to the outside of the specimen container (Figure 3.23a). A vial with dual labels is completed for each area of insect colonization (Figure 3.23b). Precut address labels are generally used for this purpose.

Both labels contain the same information, consisting of the case number, agency, sample number, date, time (hour and minute), and initials of the collector, as a minimum of data. Double labeling is done to ensure that data will not be lost due to external labels being lost or damaged (Figure 3.24). Frequently the preservative chemicals leak or spill onto the outside of the vial, and the ink is smeared or washed away. The companion labeling technique is also helpful when working with specimens in the entomology laboratory. The internal label will be placed in the examination container that contains a portion of

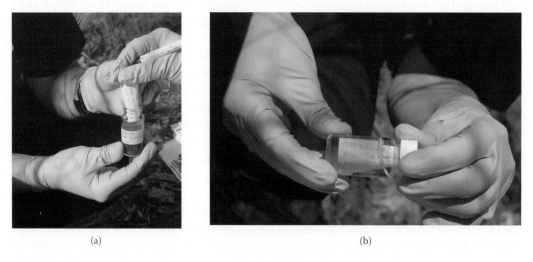

(a)   (b)

Figure 3.22 (a) After completion of the aerial netting of the adult arthropods, the collection of the preserved larval samples can begin. Containers for preserved samples have dual labels (inner and outer). (b) Both labels are written in pencil, with one label being placed directly into the preservative solution. (Photos courtesy of Dr. Jason H. Byrd.)

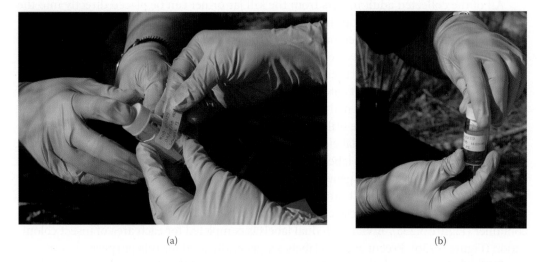

(a)   (b)

Figure 3.23 (a) The duplicate outer label written in pencil on adhesive label, and affixed to the outside of the container. (b) The collection areas should not be mixed, and each container should have two labels (interior and exterior) written in pencil. If ink is used, the alcohol will dissolve the print from the paper surface. (Photos courtesy of Dr. Jason H. Byrd.)

Figure 3.24 One container of preserved insects should be made for each area of collection. Samples from varying locations should not be mixed. The officer responsible for completion of the data labels should always include the case number, time of collection, date, geographic location, location of insect on the remains, and the name of the collector. Duplicate labels should be made on regular cotton bond paper (to be placed inside of the collection vial), and another printed on adhesive paper to be affixed to the outside of the collection container. (Photo courtesy of Dr. Jason H. Byrd.)

the specimens from the vial, while the remaining specimens in the collection vial are still properly labeled, thus reducing the chances for mixing similar-appearing samples.

Further documentation of the vial sample number and the data pertaining to where the sample was collected from the remains should then be recorded on the "Insect Specimen Disposition and Identification" and also noted on the "Entomological Sample Log Sheet" (see appendix).

Proper collection and preservation techniques, specimen labeling, and data recording are necessary for entomological data to be accurately evaluated by the forensic entomologist and legally accepted by the criminal justice system. The accurate recording of entomological data is essential for entry into the court record. On occasion, opposing attorneys have had opportunities to attack the expert forensic entomologist when well-meaning evidence technicians, untrained in entomological recovery techniques, have omitted steps in the sequence of collection and documentation. To avoid such criticism, it is important to complete a thorough collection with supporting documentation regardless of the circumstances of the scene.

Collection of Ground-Crawling Arthropods on and around the Body

Forceps or gloved fingers can be used to collect many ground-crawling adult arthropods. Preservation should be conducted in the same manner as with the adult flying arthropods. These arthropods may include some of the beetles (Coleoptera), ants, bees and wasps (Hymenoptera), true bugs (Hemiptera), springtails (Collembola), and newly emerged flies (Diptera). Some of these arthropods are quite fast, so it may be necessary to grab them with vigor and purpose, as they are likely to disperse into the soil or under other materials present.

Collection of Entomological Samples from the Body

Collection of specimens from the body can begin once the surrounding area has been processed. In accordance with general death scene protocol, the individual having overall authority of the scene must give permission to approach the body. Often this permission is granted by the medical examiner or coroner; however, such permissions should have been obtained prior to conducting the collections around and over the remains.

It is suggested that the entomological collection first be conducted prior to the removal of the body from the scene. When this is done, it is extremely important that nothing be moved or taken from the corpse except the arthropods that are on the surface and clearly visible. Limit the amount of disturbances to any portion of the clothing, body, the immediate area around the body, or the body itself. Postmortem artifacts inflicted inadvertently while collecting specimens may be misleading and can cause needless questions and speculation. If any type of postmortem trauma is inflicted on the remains during the collection of entomological samples, it is imperative that the trauma be photographically documented and reported to the medical examiner or coroner. A more thorough examination of the clothing and body for arthropods not collected at the body recovery site can be conducted at the time of the autopsy.

The collection directly from the remains will typically consist of eggs and a mixed-size sample of larvae. The average sample size is 50 to 100, and these should be preserved in one of the specified preservatives (KAA solution is preferred; see Chapter 4). A dime-sized sample of fly eggs will often consist of several hundred. Since larvae are also often found in abundance, it is not difficult to collect several hundred. Although fifty is a suggested minimum sample size, collections of several hundred larvae are not discouraged. It is important for the investigator to ensure that a representative sample of the eggs and larvae is collected, being careful to also collect the largest and smallest larvae present on the remains in order to document the range of size variation with the larvae.

Separate areas of insect colonization should be sampled and preserved separately, with temperature recordings made from each site of collection. The larvae can be collected with forceps, brush, gloved hand, or disposable spoon (Figure 3.25a–d). The collected eggs or larvae should be placed directly into a preservative solution. The preservation will cease larval development, thus preserving the exact stage of development noted while processing the body recovery scene. The suggested preservative fluid is KAA (see appendix), but several different methods exist. An excellent technique for preservation is to blanch the larvae in hot (nearly boiling) water for 60 to 120 s, and then place the blanched larvae into 80% ETOH. It is important to understand that with soft-bodied insect larvae, simple placement of the insect directly in 80% ETOH is not an adequate method of preservation. With soft-bodied insect larvae, it is the choice of the investigator to use either a preservative solution or blanching. All preserved samples from each area of colonization should be placed in separate vials, and each vial labeled with the companion label system written in pencil.

The second stage of the collection of insects on remains is the live collection. Live samples are collected because it may be difficult to conclusively identify the preserved larvae to species. It may be necessary to allow the insects to complete their development to the adult stage before the species identification may be obtained. In order to keep the collected arthropods alive during shipment to the forensic entomologist, a specialized shipping container must be constructed so that the larvae can have access to food. Using the reserved larval samples as a guide, a second equal-sized portion of the larvae should be placed alive into maggot-rearing containers. These containers can be prepared while on scene before the collection process begins. They consist of plastic containers with sealable, tight-fitting

(a)

(b)

(c)

(d)

Figure 3.25 Both live and preserved specimens should be collected from the remains. This companion sampling of living and preserved arthropods is standard practice in forensic entomology. Both samples should consist of arthropods from the same body areas, and should reflect a representative sample of the larvae found on the body. Samples from the body can be collected using forceps (a), brush (b), gloved hand (c), or spoon (d). (Photos courtesy of Dr. Jason H. Byrd.)

lids. A substrate such as soil or vermiculite is added to the bottom of the container to a depth of approximately 0.5 in. (1.25 cm) (Figure 3.26a and b). A larval-rearing pouch is constructed from aluminum foil and placed into the container, on top of the soil substrate. The larval-rearing pouch can be constructed from aluminum foil by folding a 6 × 7 in. (15 × 18 cm) piece of foil into thirds horizontally, and then again into thirds vertically, ending with a rectangular piece approximately 2 × 2.5 in. (5 × 7 cm). This rectangle is then unfolded and the corners crimped together, forming an open-topped three-dimensional rectangular pouch.

(a)

(b)

Figure 3.26 Preserved samples collected from the aerial and vegetation sweep, and those specimens collected from the body can be placed directly in a preservative solution. However, those specimens to be kept alive for shipment to the forensic entomologist require a different collection process. (a) The live insect container is made by adding approximately ½ to ¾ in. soil or vermiculite to a plastic container with a tight-fitting lid. (b) Vermiculite (or soil) will provide the larvae a substrate in which to burrow if they reach the wandering stage during shipment, and also helps to absorb any fluids that may leak from the maggot pouch. (Photos courtesy of Dr. Jason H. Byrd.)

A palm-sized (approximately 3 to 5 oz, or 90 to 150 g) piece of lean pork, or other rearing medium, is placed within the foil pouch. Larvae (50 to 100 individuals) can then be placed into the foil pouch containing the food substrate (Figure 3.27a and b). Once the larvae are placed on the food substrate, the top edges of the foil should then be tightly crimped together to reduce desiccation and to help prevent the larvae from becoming dislodged from their food source during shipment (Figure 3.28a and b). This pouch is placed in a vented pint-sized 16 oz (453 g) cardboard or plastic container with approximately 1.0 in. (2.5 cm) of medium-sized vermiculite or sand in the bottom (Figure 3.29a and b). In especially dry climates, a wet paper towel can be placed into the pouch in order to prevent desiccation of the food substrate and the larvae. If plastic sandwich bags are used, the top should be sealed, providing that adequate ventilation has been provided by placing small holes in the plastic with a pin.

(a)   (b)

Figure 3.27 (a) An aluminum foil pouch that contains a food substrate of beef liver or ground beef is placed in the live insect container. (b) A palm-sized piece of ground beef or liver will keep 50 to 100 insects alive for approximately 24 hours. (Photos courtesy of Dr. Jason H. Byrd.)

(a)   (b)

Figure 3.28 (a) After the larvae are added, the top edges of the pouch should be tightly crimped. This prevents the larvae from being separated from the food source during shipment, and also helps prevent desiccation. (b) The crimped pouch is then placed into the live insect container for shipment. (Photos courtesy of Dr. Jason H. Byrd.)

Eggs are treated in the same manner as larvae. However, live pupae can be placed directly into shipping containers with vermiculite or sand in the bottom (Figure 3.30). It is not necessary to provide pupae with the pouch and food source, as this stage does not feed. At the death scene, beetle larvae (generally recognized by having three pairs of legs) also may be collected. However, many beetle larvae are predacious on fly larvae. Therefore, live beetle larvae should not be placed in the same shipping container with live fly larvae. Extra fly larvae should be collected for use as a food source for the beetle larvae. These fly larvae will be consumed by the beetle larvae during shipment to the forensic entomologist. Portions of beef liver or other substrates utilized as a maggot food source may be used as an alternative food source for immature beetles.

Areas of the body where concentrated insect activity most likely will be encountered during the early stages of infestation are the nasal openings, ears, mouth, eyes, and sites of trauma (e.g., cuts, gunshot wounds, and blunt force injuries where the skin is broken). Skin creases of the neck may also contain egg masses and larvae, which may also be found along the hairline and close to the natural body openings or matted in bloody hair. Wounds may

(a)   (b)

Figure 3.29 (a) The container holding the live insect samples should have duplicate labels, as with all other collection vials. The plain paper label is placed inside the collection container with the live insects. (b) The adhesive label should be affixed to the outside of the container. (Photos courtesy of Dr. Jason H. Byrd.)

Figure 3.30 A representative sample of the pupae found at the scene should be collected and immediately placed into a preservative solution. A companion sample (shown here) should be placed into a collection cup containing approximately 1 in. of soil or vermiculite. Providing a food source is not necessary since the pupal stage does not feed. (Photo courtesy of Dr. Jason H. Byrd.)

have both egg masses and larvae associated with them, and exposed genital and anal areas may contain egg masses and larvae, especially if these areas were traumatized prior to or after death.

Collection of Entomological Samples away from the Body

Once samples have been collected from the body, the investigator should then focus on collecting the arthropods that have potentially completed feeding and dispersed from the body. Generally, these arthropods will be older than those found on the remains and thus are extremely valuable in a death investigation. A search to recover these dispersing larvae or resulting pupae should be conducted at all death scenes. During subsequent trial testimony it is difficult for the investigator to say that no dispersing larvae were present unless a search is conducted to detect their presence. Dispersal is a normal part of the development for many insect species. If the remains are recovered in an outdoor environment, many larval species will disperse and burrow under surface debris or into the top couple of inches (2.5 cm) of topsoil. Most of these dispersing arthropods can be found within a radius of 20 ft (6 m), but this distance varies greatly, depending on terrain and degree of soil compaction. It is typical that most dispersing arthropods will aggregate in areas of soft soil, in thick clumps of vegetation, and around the trunks of trees, as well as under rocks and fallen tree limbs. These arthropods are easily recovered by gently removing the surface debris with a trowel or hand shovel (Figure 3.31a and b). The topsoil should also be sifted in order to recover other insect species that may have burrowed deeper than under the surface debris (Figure 3.32). The soil should be sifted over progressively smaller screen sizes so that arthropods, and insect fragments, that passed through the larger screen can still be recovered.

These dispersing larvae should be collected in the same manner as all other entomological samples. A representative sample should be preserved, and a second nearly identical sample should be kept alive for rearing purposes, as described above. The distance from the body, and the compass direction of travel, should be documented. It is also good

(a)

(b)

Figure 3.31 (a) The surface debris in the immediate surrounding area should be removed to expose any dispersing larvae or pupae. (b) If no insect activity is found on the soil surface, small trowels should be used to excavate the first 1 to 2 in. of soil to expose any larvae or pupae that may have already burrowed into the soil. (Photos courtesy of Dr. Jason H. Byrd.)

Figure 3.32 At some stages, dispersing larvae or pupae may be difficult to detect visually. However, certain stages may be colorful and thus easy to see. (Photo courtesy of Dr. Jason H. Byrd.)

practice to photograph the migrating insects so that their exact stage of development can be adequately documented.

Collection of Specimens from Body Recovery Site after Body Removal

In cases where bodies are outdoors and heavily colonized, many arthropod larvae, pupae, and adults will remain on the ground after the body is removed. The procedures described above should be followed for each of the different arthropod stages seen after removal of the body. A number of specimens of each immature stage should be collected and preserved, while a second sample should be collected alive for rearing.

Litter samples (e.g., leaves, grass, bark, and humus) or any material on the ground surface close to or under the remains should be collected and labeled. Many carrion-feeding arthropods will hide in this material close to the body, and these materials should be thoroughly examined in a laboratory for additional faunal evidence. Collect handfuls of the litter down to the exposed soil, particularly litter in close proximity to the ground surface. This material can be placed into two-quart (1.8 L) cardboard or plastic containers.

Soil samples may yield arthropod specimens as well as provide samples for biochemical assay of decomposition fluids. Soil samples should be approximately 4 in. cubes or cylindrical cores (100 ml) of material from areas associated with different body regions (head, torso, and extremities). Soil samples of this size should be placed into paper evidence bags, or plastic evidence containers. Soil samples (approximately six total) should be taken from under, adjacent to, and up to 3 ft (1 m) from the body, noting the origin of each sample in reference to the position of the body (Figure 3.33a and b). These samples should be labeled according to the technique described for the insect vials.

Collection of Entomological Specimens during Autopsy

It is best for the investigator conducting the entomological evidence collection at the body recovery scene to also perform the collection prior to and during the autopsy. A collection at autopsy will often allow the full potential of forensic entomology to be realized. This collection is done in conjunction with the cooperation and permission of the forensic pathologist performing the autopsy, and it is helpful due to the ability to document and collect insect samples that are feeding deep within bodily cavities and tissues. The forensic

(a)   (b)

Figure 3.33 (a) Soil samples from under the remains should also be collected. A small sample should be taken from the area under the head, thorax, and pelvis. (b) Each sample should be placed within individual plastic bags, and labeled as with all other entomological evidence containers. (Photos courtesy of Dr. Jason H. Byrd.)

pathologist should be informed of the extent of the collection process and the location from which the samples will need to be collected. If the forensic pathologist prefers to make the collection, he or she should easily be able to gather the required entomological samples provided he or she follows the appropriate collection procedures. If a certain arthropod life stage is found during the autopsy, the investigator should return to the scene in an effort to recover and confirm the presence of the missing insect life stage.

Most likely the remains will be enclosed in a zippered vinyl body bag when it arrives at the morgue from the death scene, particularly if the remains are in a state of advanced decomposition. Once the body bag is opened, the inner surfaces of the bag should be examined thoroughly for the presence of arthropods that may have crawled away from the body due to changes in temperatures or physical disturbance. This evidence should not be overlooked; these arthropods should be collected and labeled using procedures described previously.

After the remains have been taken from the body bag and placed on the autopsy table, an external examination of them is conducted, and this provides an excellent opportunity for a second collection of arthropods from the exterior of the body to be made. Generally the medical examiner or coroner will readily allow the crime scene analyst to conduct the collection at this stage of the autopsy procedure. If the body is clothed, a complete and detailed examination of the clothing is essential and may yield a variety of stages and kinds of arthropods. Folds in the clothing where eggs, larvae, pupae, or adults may be sheltering should be gently opened and examined. Areas of clothing that are moist or contaminated with excrement are sites with a high probability of yielding entomological evidence. If the case is a homicide, suicide, or questionable death, bagging the hands of the deceased in paper bags taped to the wrists is often a standard practice. This exercise is done to preserve foreign trace evidence, such as skin fragments or hair adhering to the hands. Inspection of these hand bags after their removal at autopsy is necessary, as arthropods infesting hand wounds may crawl off the remains during transit. After the clothing has been examined and removed, the areas of the body where concentrations of arthropod activity are found should be noted and photographed with a macro lens to show the extent of the infested

Figure 3.34 Area of larval activity 5 days after the infliction of a postmortem stab wound on a pig carcass with and ice pick. (Photo courtesy of Dr. Jason H. Byrd.)

area and arthropod composition. As with the scene collection, representative samples from each major area of infestation should be performed. On fresh bodies, the face is the most likely area to have arthropod activity. Flies, in particular, will seek external openings (e.g., nostrils, mouth, and eyes) for deposition of their eggs or larvae. The genital or rectal areas, if exposed, will sometimes provide shelter and moisture attracting egg-laying (ovipositing) flies, especially if those areas have been traumatized or soiled with excretions. However, the urogenital-anal area is generally not a primary site for oviposition. It has been seen in research on human cadavers at the Anthropological Research Facility, Knoxville, Tennessee (and in case studies), that a delay exists of several hours to several days before colonization of the pelvic area and genital-anal openings if these areas have not been traumatized. If trauma has occurred, there appears to be an increased attraction to these areas, and they may be colonized simultaneously as other sites, and in some cases exclusive of the face, even though the face was exposed and available for colonization. Traumatized areas of the limbs and torso, where breaks in the skin occur, may also contain patchy areas of arthropod colonization (Figure 3.34). For example, arthropod infestations on the hands or forearms may suggest the victim had sustained defensive wounds. This scenario may not always be the case, and caution should be exercised when drawing this conclusion.

Very small arthropods such as fleas, ticks, mites, lice, or nits (lice eggs) may be present on both fresh remains and associated clothing. Many may be attempting to leave the cooling body, and therefore can be found in the clothing, or they may even be attracted to the investigator. Small ectoparasites may also be present on or within the tissues of the body itself. Thus, it is important to examine the hair close to the scalp for the presence of nits. The eye lashes or sebaceous gland areas of the face may harbor follicle mites, *Demodex folliculorum* var. *hominis* (Simon) (Acari: Demodicidae). A few lashes are typically plucked and examined microscopically if the presence of this mite is suspected. Estimation of a minimum time of colonization (TOC) range may be determined based on whether or not these arthropods are still alive.

Once the internal portion (e.g., surgical entry into the torso and skull) of the autopsy has begun, major sites of arthropod activity may include the skull, with natural body openings, hair and scalp, respiratory tract (including inner nasal passages), esophagus, ante- or perimortem wound sites, and anal-genital areas. Also, the chest cavity, and areas under desiccated skin, should be examined thoroughly. Arthropods collected from any of these body areas should be labeled, preserved, and their location noted. As with all entomological samples, the containers should be labeled so that the area of the body from which they were collected is clearly documented.

Many times remains are stored for some period in coolers or refrigeration units prior to autopsy. This period may range from several hours to several days. Therefore, notation should be made of both the total time the body was in the cooling chamber and the temperature of the chamber. Information on the duration of time required from transport from the scene to the morgue should also be noted. If possible, a recording thermometer should be attached to, or placed within, the body bag while in the transport vehicle. Also, the temperature of any maggot mass present should again be recorded when the body is removed from the cooler. There may be little or no effect of the lower temperatures on arthropod development if the maggot mass was well established prior to the body being placed in the cooler. Maggot mass temperatures are commonly between 80 and 100°F (27 to 37°C), even if the temperatures in the cooler are maintained between 30 to 40°F (–1 to 4°C). Bodies heavily infested with larvae should be autopsied as soon as possible. If even a weekend passes before the body is autopsied, the voraciously feeding larvae may consume valuable evidence. If necessary, the forensic entomologist can easily provide telephone advice and direction to those tasked with specimen collection and preservation at any stage of the death investigation process.

Retrieval of Historical Climate/Weather Data

When insect collecting is completed at the scene, weather data for the time period should be obtained. This time period should extend from 1 to 2 weeks prior to the rough estimate of when death occurred, or to the last reliable time the individual was seen alive. This time period should also include 3 to 5 days past the time the body was discovered. Weather data retrieval can be accomplished by contacting the nearest National Weather Service (NWS) station or other climate data-gathering agency. Locating the closest meteorological recording station to the site of body recovery is a necessary part of the investigation. Data from the closest recording station must be collected for use in the proper correlation of the historical meteorological data collected to the temperatures documented during the processing of the scene. Often the investigator can get assistance from the Office of State Climatology maintained by most states. These offices have records of other data collection sites and agencies and can easily determine which station is in closest proximity to the body recovery site. Some NWS stations (including state and local municipal facilities as well as most airport facilities) can give extensive weather data, including hourly temperatures, humidity, extent of cloud cover, precipitation, and wind speed and its direction. In addition, soil temperatures, water temperatures, river stages, tidal swings, soil moisture conditions, and evaporation rates may also be obtained from first-order NWS stations. All climate data are eventually sent to the National Oceanic and Atmospheric Administration (NOAA), where they are stored at the National Climatic Data Center in Ashville, North Carolina. The needed weather data can be obtained in the form of recorded material officially certified for court documentation. Smaller meteorological stations may have only daily maximum and minimum temperatures, and total precipitation. In many instances, such data may be all that are available for the entomologist to use in making the appropriate time interval evaluation.

Finally, it is important to conduct periodic temperature observations (three to four readings over a 24-hour period for 3 to 4 days) at the body recovery scene. In particular, these readings should be taken during the times of temperature maximums and minimums. The investigator can return to the scene to make these recordings with a traditional thermometer. However, inexpensive data loggers are now commonly available. These can be set to record temperatures at specific time intervals, eliminating the need to return to the scene multiple times. These observations are valuable for the correlation of the site microtemperatures with those of the closest NWS recording station data. Temperature differences can exist even within short spatial distances, and the possibility for this type of error must be considered.

For example, in a California case where a body was found next to a river, extensive periodic temperature data collections were conducted at the site. Temperature data were recovered over a period of days from the scene and compared to those documented at the airfield weather station. It was found that the site temperatures were approximately 11°F (6°C) lower than the same hourly temperatures recorded from the NWS station at a nearby airport. Once the calibration was made between these two sites, an additional 48 hours was added to the growth and development of the fly larvae due to the much cooler temperatures under the culvert at the waterline. By making periodic visits (including times after dark) to the site over several days, accurate correlation could be made between the NWS station and the body recovery site. When an NWS station is recording data on an hourly basis, an even greater degree of accuracy can be achieved by recording coincidental death scene readings.

It is recommended that these body recovery scene temperature comparisons be taken at a time when weather conditions are similar to those noted during the time that the body was at the recovery location. If a major frontal system passes shortly after recovery of the remains, it may be advisable to wait until temperatures more closely approximate the levels that are representative of the time the body was *in situ*. Over a period of 4 to 6 days, if four or five temperatures can be taken at the scene per day, an adequate number of data points can be generated for a linear regression statistical analysis of the site versus the NWS. To help accomplish this task, remote electronic temperature sensors may be calibrated and placed at the site for the required period of time. These sensors can record thousands of data points and recover the temperature every second, if desired. Once the climatological data are collected, appropriate analysis of the pertinent climatic information can be made.

A recent case from New York demonstrated that solar radiation played a limited part in altering the decomposition of a body hidden in a car trunk for several days. As would be expected, and as research from the Anthropological Research Facility in Knoxville, Tennessee, has demonstrated (W. Bass, personal communication), full sunlight on the car would likely have caused temperatures in the trunk of the car to be very high. However, upon examination of the cloud cover data for the region on the dates in question, it was found that only two days showed full sun, with two more showing scattered or broken clouds. The rest of the period was rainy and overcast; thus, only limited influence was attributed to solar radiation on the heat loading of the car trunk, where much more might have been expected.

Another evaluation rendered from climate records involved the estimation of duration of the covering of a body with snow. In this case from northern Ohio, the remains of a young girl were found during the second week of February in a rural agricultural field. It was suggested by the land owner, whose house was approximately ½ mile from the roadside ditch where the body was found amongst the tall weeds, that the remains could not have been there since late October or early November. It was theorized that family members, field workers, and hunters would have seen the body either during hunting season or during the process of harvesting crops. Once weather records had been studied, it was found that there had been a 10 in. (25.4 cm) snowfall during the second week of November and temperatures were never above freezing (32°F) until the later part of January. Another 5 in. (12 cm) of snow fell and covered the remains until the first week of February, when there was a considerable thaw. Thus, the girl's body had been covered with several inches of snow for over two months, conceivably concealing the remains from observation.

Collection of Specimens from Buried Remains

Although burial slows the decomposition process, it does not necessarily exclude arthropod colonization. Flies in the families Calliphoridae, Sarcophagidae, and Muscidae have been catalogued from remains covered with several centimeters of soil. Blow fly pupae recovered from 100-year-old Native American skeletal remains from South Dakota suggested the season of the year in which the remains were buried (Gilbert and Bass 1967). A giant bison skull exposed by erosion from a stream bank in Alaska was collected for study. The internal cavity of the skull contained two species of fly larvae and pupae (one species was identifiable), which were carbon dated back approximately 22,000 years (Catts, personal communication). Valuable entomological evidence can be gained from buried remains, often regardless of burial depth. It is essential to have a person experienced in the proper archaeological exhumation techniques and procedures present when unearthing remains. For a detailed discussion of this process, see Chapter 7.

Collection of entomological specimens from burial sites is nearly identical to the procedures described in the section on collection of specimens at the scene. Preserved and live collections are made from each area of colonization, and companion labeling is utilized throughout the collection process. The collection of soil samples for examination for arthropods, or their exoskeletons shed during the molting process should be undertaken by the forensic investigator as the excavation of the burial site proceeds. Soil from the ground surface down to the upper surface of the body should be sifted and examined, as well as soil from the side and bottom of the burial pit after body removal. Arthropod larvae, pupae, adults, or any arthropod fragments may be found in the soil from within or surrounding the gravesite. If life stages are seen on the remains, a separate collection of samples should be made with permission from the medical examiner or coroner.

Collection of Specimens from Enclosed Structures

Enclosed environments present several problems for the forensic investigator in the collection and evaluation of arthropod colonization. First, if the enclosed structure is tightly sealed (e.g., newer automobiles with the windows and doors closed; tightly sealed rooms; and newer, well-insulated houses), the chemicals emitted by the decomposing tissues used as attraction cues by arthropods do not dissipate as rapidly as those from bodies that are left with open exposure. In these investigations, the question arises as to how much time has elapsed before the odors finally emanate from the restrictive confines of the body enclosure? In some rare circumstances, the odors never permeate from the container. However, even in such cases, extensive visual searches should be conducted at the body recovery site to make certain no arthropod life stages are present.

Even if the attractant odor has emanated from the structure or container, arthropods may still be excluded from direct access to the decomposing remains by some type of mechanical barrier (i.e., screening or some type of synthetic cloth). Thus, there may be considerable numbers of flies or other carrion arthropods found outside, attempting to gain access to the remains. A concentration of blow flies outside a structure or container likely indicates that something inside the enclosure is dead. Flying arthropods should be collected by aerial netting, and ground-crawling arthropods should be hand collected. The duration of time that the remains have been in place may be indicated by certain assemblages of arthropods from a successional group (Megnin 1894). This niche formation, or succession, is due to the production of different chemical cues associated with the changes of advancing decomposition.

It also must be taken into account that the enclosed structure may not have internal temperatures comparable to those reported from the local NWS station. As explained in "Collection of Meteorological Data," temperatures in a particular outdoor habitat may be considerably different from those recorded at the NWS station, and independent temperature data should be collected from the scene to determine environmental conditions more accurately. This calculation is even more crucial when dealing with enclosed structures. A car parked on a black asphalt surface with the windows closed, even on a mild sunny day (75°F or 24°C), may exhibit temperatures 30 to 40°F (20°C+) higher inside the passenger compartment or trunk than the outside ambient temperature. This condition could accelerate larval development by several days, creating a considerable error in estimating the minimum period of insect activity (PIA). By following the procedures described above or recreating the circumstances of the enclosed structure and recording temperature data, an accurate correlation can be obtained between the NWS data and the enclosed environment in question.

When investigating a body recovery site inside a building, always check the thermostat setting controlling the heating or cooling system. If activated, relatively constant temperatures may have existed for the period in question. This information allows for a more accurate estimation of the rate of development of arthropods associated with the remains. Correlation of these data with on-site temperatures should be completed. Other problems with enclosed structures can occur. The first responder to a dwelling may open windows in the house in an attempt to dissipate decomposition volatiles. In this event, the actual temperatures to which the arthropods were subjected may be impossible to evaluate accurately until the building is closed again and the temperature allowed to stabilize.

Collection of specimens from enclosed structures is conducted more efficiently with some knowledge of the likely places to which arthropods may disperse. In houses where the remains are in advanced stages of decomposition (i.e., skeletal and mummified), it is possible that more than one generation of larvae has dispersed from the remains, developed to pupae, and emerged as adults. Inspection of the edges of the room where the walls and floor make contact as well as under carpeting or carpet pads may reveal postfeeding (dispersing) larvae or fly pupae. These larvae, that have reached the stage in which they no longer feed, may be found under carpeting, rugs, or any other covering that provides protection and seclusion for their transformation to the pupal stage. Fly pupae are small (approximately 9 mm), cylindrical objects that may be red, brown, or black in color. Crime scene investigators encountering them often ask why there were so many "rat droppings"

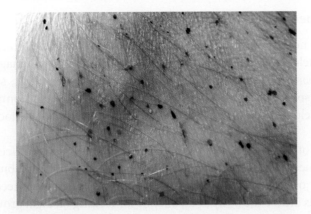

Figure 3.35 Fly specks. The circular drops which are light in color are spots of fly regurgitate. The darker comma shaped spots which are pointed on one end are the fecal deposits of adult flies. (Photo courtesy of Dr. Jason H. Byrd.)

in the room where the remains were found? These are most likely not rat droppings, but fly pupae. Close examination will show segmental lines (sutures) on the fly pupae, which are not found on rat droppings (Figure 2.18b).

The larvae may disperse to pupate into other rooms as well. Occasionally, larvae may travel up to 150 ft (50 m) in search of a suitable pupation site. So, larval dispersal in an enclosed environment can easily extend to the outer limits of most structures. Do not be surprised if larvae or pupae are located on structures, such as countertops and appliances as well. A basement or crawl space under the structure will be an additional location where arthropods might be collected if the remains are in advanced decomposition and fluids have seeped through flooring and into the space below. When dealing with remains in automobiles, one should look under the floor mats and carpeting, between the seats, and even under the upholstery in close proximity to the remains. Larvae also may disperse into the car trunk to pupate, so this area must also be inspected.

Because newly emerged adult flies and beetles will seek light and the outdoors, the backside of window blinds, shades, and curtains should be searched. Window sills and ledges on the inside of the structure may contain adult flies or other species that have completed development on the remains, emerged, and sought dispersal to their natural outdoor habitat. If the enclosed area is too hot, these arthropods may die.

Dark fly specks (fecal spots), and lighter-colored food regurgitation spots, are frequently deposited by flies on the surrounding ceilings, floors, and walls (Figure 3.35) or even directly on the remains. Take note of the density of this spotting, as it may give some indication of the relative size of the fly population attracted to the remains. In advanced decomposition, arthropod feces (frass) can accumulate in conspicuous amounts around the remains. Usually the cast skins of immature stages will be mixed in with the fecal material. In some cases, the dried feces still may be inside an extruded intestinal sheath, called the peritrophic membrane. A mass of dermestid beetle fecal material has the general appearance in color and form of pencil shavings or sawdust (Figure 2.51a and b).

The same collection, preservation, and labeling techniques described earlier should be followed when collecting from these enclosed environments. Dead and dry specimens must be handled very carefully because they are extremely fragile. They should be stored in preservative to allow for rehydration, or stored dry if toxicological analysis is anticipated.

Collection of Specimens from Aquatic Habitats

At times it may be necessary to collect entomological evidence from bodies found in water. Such sites may include shorelines, rivers, ponds, irrigation or drainage ditches, sewage ponds, sewers, open wells, or rain barrels. In each of these habitats, specific arthropod species may have special survival adaptations unique to these environments. Specialized arthropods may help the investigator identify a particular geographic location or a specific time of year.

The majority of aquatic species collected from remains do not feed directly on the decomposing tissues. However, some aquatic arthropods will use submerged or floating bodies for shelter or as a solid surface for attachment to facilitate feeding. Most often, aquatic arthropods utilize the body as an anchor for feeding on algal growth, or filter-feeding on small organisms, or for hiding from predators. A recent study of dead rats placed in an aquatic environment suggested that certain species of midges (Chironomidae), also known as bloodworms, would colonize as time increased (Keiper et al. 1997). These aquatic arthropods may hold the answer to determining the length of submersion time in certain areas of the country. Aquatic arthropods (e.g., crawfish, crabs, or shrimp) will feed extensively on human tissue (Smith 1986), producing postmortem artifacts.

Most aquatic arthropods spend a major portion of their life cycle in water, sometimes passing several years as a developing aquatic immature form before reaching the adult stage. In some cases, all life stages may be found in aquatic habitats, but in many species the growing and feeding stages are in water while the reproductive portion of the life cycle (winged flying adults) is of short duration and terrestrial. Therefore, death scene investigators usually will encounter only larvae, pupae, or other immature stages on submerged bodies. However, during warmer times of the year, newly emerged winged adults may be encountered on corpses found on the surface or along shorelines or riverbanks.

While common terrestrial arthropods (e.g., blow flies, flesh flies, clown beetles, carrion beetles) usually are not found on submerged bodies, floating bodies may contain many arthropods common to terrestrial scenes. These arthropods should be handled by using the same procedures described previously. If a corpse was colonized by terrestrial arthropod species prior to immersion, these arthropods may have drowned, which could indicate that the body was on the surface of the water for some time before it sank to the bottom. Additionally, live larvae collected from human remains recovered from below the water surface might suggest that the remains have not been submerged for very long.

Some police agencies have procedures detailing recovery of bodies from water habitats. The protocol usually calls for some type of shroud or sheet to be fitted beneath the corpse before movement (if the corpse is floating) or for the body to be placed immediately upon a shroud if partially in the water. The shroud helps retain evidence on the body that may otherwise flush off when disturbed. This procedure is essential for proper recovery of aquatic arthropod specimens on the corpse. Most arthropods living on aquatic substrates will quickly detach at the slightest disturbance. Therefore, if a sheet or large, fine-weave mesh net can be slipped under and around the corpse before it is disturbed, most of the arthropods using the body as shelter or as a food source will be captured in this shroud.

Proper collection and preservation of the aquatic arthropods is essential for accurate identification of the organisms. The specimens may be collected for preservation or for rearing to adults. There are several standard techniques available for rearing and preserving these fragile, soft-bodied arthropods (Merritt and Cummins 1996).

Live immature specimens should be collected and transported in water taken from the environment in which they are found. During transportation, the collecting jars (approximately 1 L) should be filled completely to reduce damage to the arthropods caused by excessive splashing. Also, the water must not be allowed to elevate in temperature because many of these immature arthropods are heat sensitive. Keeping the collecting jars shaded, covered with a wet cloth, or against chemical ice will help reduce excessive heating during transit. The collecting jars containing the specimens could be placed in a Styrofoam or similar type ice chest with ice or icepacks inside. Portions of naturally occurring substrate found in close proximity to the corpse should be collected and placed into separate collecting jars.

Collecting procedures for preserving aquatic arthropods are very similar to those described under general collecting procedures. The processing and labeling techniques also are the same. Data labels should always be placed in the vials immediately after the collection is made using the same label format as discussed previously.

Collection of flying adult arthropods that are in close proximity to the body should be accomplished using techniques previously described. Eggs, larvae, and pupae also can be picked off the body (substrate), by using forceps or fingers. If a net or shroud is not used in recovery, specimens should be collected at the body recovery site directly from the body before it is moved. If collection is delayed until the time of autopsy, many of these aquatic arthropods will have crawled off or dropped from the disturbed body and may not be located.

Often, immature stages of aquatic arthropods are difficult to see due to their size and their camouflaged appearance. In one case, aquatic midge larvae were collected from a corpse discovered in a river and mistaken for red carpet fibers (Hawley et al. 1989). The investigator must look very carefully and closely at the outer portions of the skin and clothing. A hand lens may be necessary for these observations. Larvae also may be found under the slime and algae that coat the skin or clothing. These larvae may not be visible until this slime covering is scraped off. Caddisfly (Trichoptera) larvae or pupae may be found on and in clothing from bodies found in fast-moving streams. Once the specimens are collected, they should be preserved in one of the solutions listed in Chapter 4.

There has been very little forensic utilization of entomological evidence in aquatic environments (Holzer 1936, Hawley et al. 1989, Haskell et al. 1989, Haskell et al. 1990, Merritt, personal communication). Individuals lacking entomological training do not recognize the varieties of arthropods living in these environments. Even with formal entomological training, many are difficult to identify to species, and there is little known of their behavior. This situation could be improved by involving entomologists knowledgeable of aquatic arthropods when bodies are recovered from aquatic habitats. The aquatic groups that may prove useful in yielding minimum PMI and post mortem submergence interval (PMSI) (Chapter 6) location information for death scene investigations include midges, caddisflies, mayflies (Ephemeroptera), and some other small flies. If these arthropods can be identified and tabulated, data relating to their developmental time intervals might be used for making a minimum TOC or PIA estimation.

Soft-bodied stages of aquatic arthropods should be preserved in fluids to prevent extreme distortion of the morphological characteristics resulting from drying. In contrast, several of the larger hard-bodied adult arthropods (dragonflies [Odontata], dobsonflies, and damselflies [Neuroptera]) can be killed by using the killing jar method, after which they may be pinned and labeled. Additional information on techniques of collecting and preserving the aquatic arthropods can be found in McCafferty (1981), Borror et al. (1989), Peterson (1967), Simpson and Bode (1980), and Merritt et al. (1984).

Processing Litter and Soil Samples

Insects and other arthropods that live in litter and humus can be sorted from the materials and collected by using a Berlese/Tullgren funnel (Figure 3.30). The funnel contains a wire screen platform to hold the litter sample in place, but allows insect and other arthropods to pass through the screen and down through the funnel. A small bottle or vial containing 75% ethanol at the bottom of the funnel cone collects the arthropods. Insects and other arthropods, such as mites, are driven to the bottom of the funnel by the heat from a low-wattage (10- to 25-watt) lightbulb or by fumes from a chemical repellant. Litter samples are placed in the funnel on top of the wire screen. The low-wattage light or cloth bag containing the chemical repellant is suspended a few centimeters over the sample material, and a vial or jar of alcohol is placed under the lower funnel opening. In 3 to 4 days arthropods present will be driven from the litter material and into the alcohol bottle. These specimens can then be processed and examined.

Soil samples can be examined by using different-sized meshes of metal screens. First, a large ¼ in. (6 mm) opening mesh screen should be used to separate large particles of soil and large arthropods. Sifting the soil onto a large, flat, white pan helps keep all of the evidence confined and allows the material to be spread out in a thin layer to expose the active arthropods. The white background makes the smaller, moving arthropods more visible. The arthropods should be preserved in the same manner as described earlier: adults in 75% ethanol and larvae in Kahle's or 75% ethanol. After these screenings are examined thoroughly, the finer particles are screened again using a 0.055 in. (1.4 mm) mesh screen. This will separate the large larvae, pupae, and medium-sized adults from the very tiny adults and larvae. A technique used for picking up very small arthropods and mites is to dip one point of a pair of forceps or a fine artist's paint brush into the collecting vial of preservative solution and touching the wet tool to the arthropod. The surface tension of the liquid will hold the arthropod as it is transferred to the collecting vial and the arthropod is "washed off" by the solution. Another method is to use a blow type or electric aspirator to suck up the arthropods into a collection chamber for placement into ethyl alcohol. Mites may be collected from the soil samples by using the Berlese/Tullgren funnel technique for the soil sifted through the 0.055 in. (1.4 mm) screen.

The entomologist should also keep in mind that as the leaf litter or soil samples are being processed, other physical evidence may be discovered. Bone fragments, hair, teeth, projectiles, or other physical evidence essential to the case can be contained in samples. This evidence should be handled in the manner prescribed for collecting and labeling any other physical evidence, and should be reported to the primary investigator or the forensic pathologist.

Conclusion

Within the past few years, forensic entomology has gained widespread acceptance within the forensic sciences. It is difficult for law enforcement agencies to have a forensic entomologist respond to and process the body recovery scene for entomological evidence. However, it is fortunate that a forensic entomologist appearing at the scene is not essential. Specialized training of crime scene analysts, medical-legal investigators, medical examiners, and coroners will enable these professionals to properly document and collect entomological evidence. Established scene protocols for the collection of entomological evidence will ensure that valuable evidence does not go unnoticed, and that the maximum information can be obtained from this unique item of physical evidence. Every jurisdiction involved in the death investigation process should establish a working relationship with a forensic entomologist long before an active case investigation occurs.

References

Amendt, J., C. P. Campobasso, E. Gaudry, C. Reiter, H. N. LeBlanc, and M. J. R. Hall. 2007. Best practice in forensic entomology—Standards and guidelines. *International Journal of Legal Medicine* 121:90–104.

Borror, D. J., C. A. Triplehorn, and N. F. Johnson. 1989. *An introduction to the study of insects.* 6th ed. Philadelphia: Sanders College Publishing.

Byrd, J. H., and J. L. Castner. 2001. *Forensic entomology: The utility of arthropods in legal investigations.* 1st ed. Boca Raton, FL: CRC Press LLC.

Catts, E. P., and N. H. Haskell. 1990. *Entomology and death: A procedural guide.* Clemson, SC: Joyce's Print Shop.

Gilbert, B. M., and W. M. Bass. 1967. Seasonal dating of burials from the presence of fly pupae. *American Antiquity* 32:534–35.

Haglund, W. D., and M. H. Sorg. 1997. *Forensic taphonomy.* Boca Raton, FL: CRC Press.

Hall, R. D., and N. H. Haskell. 1995. Forensic entomology: Applications in medicolegal investigations. In C. Wecht, Ed., *Forensic sciences.* New York: Matthew Bender & Company.

Haskell, N. H., D. G. McShaffery, D. A. Hawley, R. E. Williams, and J. E. Pless. 1989. Use of aquatic insects in determining submersion interval. *Journal of Forensic Sciences* 34:622–32.

Haskell, N. H., A. J. Tambasco, D. A. McShaffery, and J. E. Pless. 1990. Identification of Trichoptera (caddisfly) larvae from a body in a stream. Paper presented at American Academy of Forensic Science, 42nd Annual Meeting, Cincinnati, OH.

Hawley, D. A., N. H. Haskell, D. G. McShaffrey, R. E. Williams, and J. E. Pless. 1989. Identification of red "fiber": Chironomid larvae. *Journal of Forensic Sciences* 34:617–21.

Holzer, F. J. 1936. Zerstorung an Wasserleichen durch larven des Kocherfliege. *Zeitschrift fur die gesamte gerichliche Medizin* 31:223–28.

Keiper, J. B., E. G. Chapman, and B. A. Foote. 1997. Midge larvae (Diptera: Chironomidae) as indicators of postmortem submersion interval of carcasses in a woodland stream: A preliminary report. *Journal of Forensic Sciences* 42:1074–79.

Lord, W. D., and J. F. Burger. 1983. Collection and preservation of forensically important entomological materials. *Journal of Forensic Sciences* 28:936–44.

McCafferty, W. P. 1981. *Aquatic entomology.* Boston: Jones and Bartlett Publishers.

Megnin, J. P. 1894. *La Fauna des Cadavres: Application de la Entomologie a la Medecin Legale.* Paris: Encyclopedie Scientifique des Aide-Memoires, Masson et Gauthiers, Villars.

Merritt, R. W., and K. W. Cummins. 1996. *An introduction to the aquatic insects of North America.* Dubuque, IA: Kendall/Hunt Publishing.

Merritt, R. W., K. W. Cummins, and T. M. Burton. 1984. The role of aquatic insects in the processing and cycling of nutrients. In: *The ecology of aquatic insects in North America.* 3rd ed. R. W. Merritt and K. W. Cummins (Eds.). Dubuque, IA: Kendall/Hunt Publishing.

Peterson, A. 1967. *Larvae of insects: An introduction to Nearctic species.* 6th ed. Ann Arbor, MI: Edwards Brothers.

Simpson, K. W., and R. W. Bode. 1980. Common larvae of the Chironomidae (Diptera) from New York state streams and rivers, with particular reference to the fauna of artificial substrates. *Bulletin of the New York State Museum,* no. 439, 1–105.

Smith, K. G. V. 1986. *A manual of forensic entomology.* Ithaca, NY: Cornell University Press.

Wecht, C. 1995. *Forensic sciences.* New York: Matthew Bender & Company.

Appendix
General Guidelines for the Collection and Recording of Entomological Evidence

1. Complete written notations and photography of environment surrounding the body.
2. Make written notations on the number and kinds of arthropods observed.
3. Make written notations on the location of major insect infestations. (These infestations may include insect eggs, larval, pupal, or adult stages in combination or by themselves.)
4. Photograph each area of colonization on the body.
5. Note immature stages of particular arthropods observed. (These stages can include eggs, larvae, and pupae. Include notations on empty (eclosed) pupal cases, cast larval skins, fecal material (frass), and exit holes or feeding marks on the remains).
6. Complete "Forensic Entomology Data" form (see below).
7. Collect adult insects flying above body.
8. Collect insects on surrounding vegetation.
9. Collect larval insects on the body (live and preserved samples).
10. Note any insect predators such as carrion and rove beetles (e.g., silphids and staphylinids), ants and wasps (e.g., formicids and vespids), or insect parasites (e.g., ichneumonid and chalcid wasps).
11. Note insect activity within 10 to 20 ft (3 to 6m) of the body. Observe flying, resting, or crawling insect adults and larvae or pupae.
12. Search for and collect insects that have migrated away from the body (live and preserved samples).
13. Complete "Insect Specimen Disposition and Identification" record (see below).
14. Note the exact position of the body, including the compass direction of the main axis, position of the extremities, position of the head and face; parts in contact with substrate, and areas in sunlight and shade.
15. Note any unusual naturally occurring, man-made, or scavenger-caused alterations that could modify the decomposition of the remains (e.g., trauma or mutilation of the body, burning, covering or enclosing of the body, burial, movement, or dismemberment).

Collecting Materials and Equipment

1. Aerial or sweep insect nets (e.g., 15 or 18 in. diameter bags with 3 ft handles; 15 or 18 in. collapsible nets with variable length handles).
2. Collecting vials (1 to 2 dram size) with neoprene stoppers; screw cap collecting vials (4 dram size).
3. Wide-mouth (8 oz) pharmacy bottles with screw tops.
4. Light tension larval forceps; needlepoint watchmaker's forceps; medium or fine-point dissecting curved forceps; dental picks.
5. Camel hair brush (no. 2).
6. Plastic or heavy cardboard containers (16 to 32 oz size).
7. Plastic specimen cups, screw cap lid type (4 oz size).
8. Paper labels, heavy quality paper (for placement inside of collection containers).
9. Paper labels, adhesive backed (for placement on the exterior of collection containers).

10. Dark graphite pencil (no. 2) (for marking paper labels). *Important*: Use only pencil for labeling, since ink will run in alcohol-based fluids.
11. Hand trowel or 4 to 6 in. core sampling tool.
12. Thermometers, electronic or mercury.
13. Psychrometers, electronic or sling type (for measuring relative humidity).
14. Camera, 35 mm with macro lens and flash; dual slide/print film; video camera/recorder.
15. Paper towels and tissue paper. Can be used in rearing of live specimens (growing a particular insect stage to adult) or cleaning thermometer probes, forceps, or other equipment after collecting.
16. Solutions for preserving specimens (uses and formulations described in Chapter 4).
17. Disposable surgical or polyethylene gloves.
18. Eyedropper pipettes (for collecting very small arthropods in fluids).
19. Insect aspirator, battery powered or blow type (for collecting small crawling and flying arthropods).
20. Flashlight or other portable light source.
21. Measuring devices, rulers, grids, tapes, etc. (for photographic purposes or for obtaining measurements and distances of other evidence).
22. Berlese funnel (for extracting fauna from leaf litter, soil samples, or other material).
23. Shipping containers—Styrofoam boxes with lids, small rectangular cardboard boxes (5 × 10 × 12 cm; 2 × 4 × 5 in.), or cardboard and metal screw cap cylinders (7 × 15 cm; 3 × 6 in.); shock-absorbing type packing.
24. Chemical ice (not dry ice) (for cooling and maintaining specimens collected alive).
25. Log book (for recording location, scene data, date, etc.).
26. "Forensic Entomology Data" form.
27. "Insect Specimen Disposition and Identification" record.

Note: Not all of this equipment is essential for conducting an adequate or satisfactory entomological case evaluation. An insect net, forceps, vials with the proper preservation solutions, labels, and maggot shipping containers have been used to facilitate many death scene investigations. However, having the above-listed material available when going to a scene or autopsy will enhance the success of the insect recovery. Much of the listed equipment may be purchased from biological supply and scientific equipment companies.

Insect Label Format

Labels for the collected insect specimens must contain information as to date and time of collection, case number, location (state, county, and city), and sample number. This is the suggested minimum information to place on a collection label. Additional information can be included if space allows. Labels containing this information should be printed on both heavy bond paper and adhesive paper, since one label will be affixed to the outside of the sample container and one label will be placed inside of the container with the insect samples. A suggested label size is 1 × 3 in., which is the approximate size of a standard return address label. It is good practice to have labels preprinted. Preformatted label templates within word processing programs, and widely available precut adhesive paper labels, make preprinting an easy task. However, it must be remembered that a laser printer should be used (ink jet and bubble jet printers will not suffice). If the labels are printed by hand, they must be printed with pencil; do not use ink! The alcohol used to preserve insect specimens will dissolve the ink and remove printed letters from the paper surface.

Sample Label

Date/Time:
Case #:
Location:
Sample #:

To conserve space, the text of the label can be printed in an abbreviated format:

Date:
Cs #:
Loc:
Sa #:

Forensic Entomology Data

Date: _____   Case number: _____

County/state: _____   Agency: _____

Decedent name: _____

Age: _____   Sex: _____   Race: _____

Last seen alive: _____   Date and time found: _____

Date reported missing: _____   Time removed from scene: _____

Recovery site description:

_____

_____

_____

Description of Death Scene Habitat:

Rural: Forest_____   Field_____   Pasture_____   Brush_____

    Roadside_____   Barren area_____   Closed building_____

    Open building_____

Other: _____

Urban/suburban: Closed building_____   Open building_____

    Vacant lot_____   Pavement_____   Trash container_____

Other: _____

Aquatic habitat: Pond_____   Lake_____   Creek_____

    Small river_____   Large river_____   Irrigation canal_____

    Ditch_____   Gulf_____   Swampy area_____

    Drainage ditch_____   Salt water_____   Fresh water_____

    Brackish water_____

Other: _____

Exposure: Open air_____   Burial/depth_____

Clothing entire_____   Partial_____   Nude_____

Portion of body clothed_____

Description of clothing_____

Type of debris on body_____

Stage of decomposition: Fresh_____   Bloat_____   Active decay_____

    Advanced decay_____   Skeletonization_____   Saponification_____

    Mummification_____   Dismemberment_____

Other: _____

Evidence of scavengers: _____

Possible traumatic injury sites (comment or draw below):

_____

_____

_____

_____

Scene temperatures:

Ambient_____ Ambient (1')_____   Body surface_____

Ground surface_____   Underbody interface_____

Maggot mass_____   Water temp, if aquatic_____

Enclosed structure_____   AC/heat—on/off_____

Ceiling fan—on/off_____   Soil temperature—1"_____   2"_____

Number of preserved samples_____   Number of live samples_____

Note: Record all temperatures periodically each day at the site for 3–5 days after body recovery, or place a data logger at the site for temperature recording.

Entomological Sample Log Sheet

Case number:	Agency:	Date:

Number of Samples

Preserved:	Live:

Weather Data

Sun	☐ Full	☐ Partly	☐ None		
Clouds	☐ Completely	☐ Mostly	☐ Partly	☐ Scattered	☐ None
Rain	Current rainfall:	☐ Heavy ☐ Light	☐ None	Approx. 24 h total:	
Wind	Direction:	Approx. speed:		Gusts:	
Snow	Current snowfall:	☐ Heavy ☐ Light	☐ None	Approx. 24 h total:	

Sample Information

Sample 1:	Date:	Time:	Method: ☐ Aerial; ☐ Hand
Location on Body:		Type: ☐ Maggots; ☐ Adult flies; ☐ Puparia Beetles	
		☐ Preserved	☐ Live for rearing
Sample 2:	Date:	Time:	Method: ☐ Aerial; ☐ Hand
Location on body:		Type: ☐ Maggots; ☐ Adult flies; ☐ Puparia Beetles	
		☐ Preserved	☐ Live for rearing
Sample 3:	Date:	Time:	Method: ☐ Aerial; ☐ Hand
Location on body:		Type: ☐ Maggots; ☐ Adult flies; ☐ Puparia Beetles	
		☐ Preserved	☐ Live for rearing
Sample 4:	Date:	Time:	Method: ☐ Aerial; ☐ Hand
Location on body:		Type: ☐ Maggots; ☐ Adult flies; ☐ Puparia Beetles	
		☐ Preserved	☐ Live for rearing
Sample 5:	Date:	Time:	Method: ☐ Aerial; ☐ Hand
Location on body:		Type: ☐ Maggots; ☐ Adult flies; ☐ Puparia Beetles	
		☐ Preserved	☐ Live for rearing
Sample 6:	Date:	Time:	Method: ☐ Aerial; ☐ Hand
Location on body:		Type: ☐ Maggots; ☐ Adult flies; ☐ Puparia Beetles	
		☐ Preserved	☐ Live for rearing

Insect Specimen Disposition and Identification

Case #:				Date:		
Agency:						
Person completing form:						
Date Obtained	Sample Number	Location on Body	Specimen Length (Range)	Number of Specimens	Stage/ Instar	Family/ Species

Identifier: _____

Indoor vs. Outdoor Temperature Calibration Checklist

Outdoor	Inside Structure
Approximately 1.5 meters above remains Sun: _____ Shade: _____ *12" aboveground* Sun: _____ Shade: _____ *Body surface* Sun: _____ Shade: _____	*Approximately 1.5 meters above remains* Sun: _____ Shade: _____ *12" above floor* Sun: _____ Shade: _____ *Body surface* Sun: _____ Shade: _____

Recording thermometer type: _____

Structure type: _____

Wall construction: _____

Door type and position: _____

Window type and position: _____

Location of body relative to doors or windows: _____
_____

Completed by: _____ Date: _____

Laboratory Rearing of Forensic Insects

4

JASON H. BYRD

JEFFERY K. TOMBERLIN

Contents

Introduction

If the concept of "companion sampling" is followed at the death scene, the investigator will collect both preserved and live samples. For most cases, the preserved samples will be utilized for the determination of the time of colonization (TOC) or the period of insect activity (PIA), with the live samples being held in the forensic entomology laboratory to allow the insects to complete their development to the adult stage. The laboratory rearing of the live collection is conducted because some species of fly and beetle larvae are not distinctive in the larval stage. The adult form may be necessary to conduct definitive species identification since that is the insect life stage bearing the most distinct morphological characteristics. While this is not true for all species, it is difficult for the forensic investigator to know which species can be readily identified in their larval stage, so the routine practice of live samples should be employed for scene processing. In all cases, the species determination should be conducted by a qualified entomologist.

Another purpose for laboratory rearing is to allow for a more refined estimation of the TOC or PIA, since the rearing of subsequent life stages provides a better estimation of the amount of development the insects had undergone at the time of collection. Proper documentation of onset and duration for all subsequent life stages may allow the forensic entomologist to more accurately determine the time at which the eggs or larvae were deposited on the body. This information can result in a more accurate determination of the TOC. Additionally, the laboratory rearing of forensically important insects can provide the forensic entomologist with an invaluable database of insect development under controlled environmental conditions, and such data can be accurately applied to future cases to assist in the further refinement of the TOC or PIA determination.

Depending on the insect species involved, laboratory rearing can prove to be an extremely difficult task, and some species do not survive to adulthood under artificial conditions. Fortunately, the growth and development needs of most insects of forensic importance are easily met in captivity. The major component of a typical rearing setup in the forensic laboratory is the environmental chamber with programmable temperature, humidity, and lighting regimes (Figure 4.1). Such units are commercially available and relatively inexpensive when compared to the cost of other equipment commonly utilized within a forensic laboratory. Elaborate "walk-in" chambers are commercially available that are more costly and less space efficient (Figure 4.2a and b). However, for some research applications, they are more versatile and effective than traditional environmental chambers. However, if such environmental chambers are not available, effective laboratory

Figure 4.1 Typical reach-in type environmental chamber that controls light, temperature, and humidity for large numbers of insects. Such chambers are used for casework or research on insect growth and development. (Photo courtesy of Dr. Jason H. Byrd.)

(a)   (b)

Figure 4.2 (a) The walk-in type rearing facility is utilized for rearing large numbers of insects. Environmental controls in rooms such as this allow for the control of temperature, humidity, and lighting, as with a smaller chamber. (b) Rack systems provide for the efficient use of floor space to rear the maximum number of insects. Individual insect colonies are maintained in small portable containers that are easily washed and disinfected. (Photos courtesy of Dr. Jason H. Byrd.)

rearing can still be conducted in a room in which temperature is kept between 27 and 30°C and 80 and 90% relative humidity (Byrd 1995). These environmental conditions are tolerated by most insects of forensic importance, but specific conditions may have to be created for particular species. A fume hood should serve as room exhaust, and be large enough to house several 10-gallon (38 L) or 20-gallon (76 L) "long" aquariums as depicted in the section on "Rearing Containers." Insect rearing can be conducted within the fume hood to minimize odor; however, this practice is generally not recommended due to the high evaporation rates caused by the rapid airflow that occurs under a fume hood.

Aquariums such as those previously described are recommended for use in laboratory rearing operations since their size allows for the normal larval dispersal of most species to occur, and their lightweight construction allows them to be handled with little effort. Aquariums are also recommended because the vertical glass sides are easily cleaned, and when coated with a liquid or powder Teflon® such as Fluron® PTFE, the glass becomes an effective barrier to crawling insects that may attempt to escape. Aquariums can also be fitted with screen lids that will help to contain any flying insects that inadvertently hatch from laboratory-rearing projects. Most aquarium and lid combinations are inexpensive and found at most pet supply dealers. While aquariums are best suited for rearing the larval stages of flies, as well as immature beetles, they are not effective containers for adult flies. Adult flies should be housed within a screened enclosure, which is described in detail later in this chapter. Naturally, the number of aquariums and cages needed depends solely on the laboratory caseload and the number of species in each case that will need to be reared simultaneously.

When conducting laboratory rearing of insects from forensic casework, it is obvious that samples from differing cases should not be mixed. But it may not be obvious that

Figure 4.3 Although much more difficult to maintain in captivity than most fly species, adult beetles such as *Necrodes surinamensis* can be reared in the laboratory provided the proper environmental conditions are maintained. (Photo courtesy of Dr. Jason H. Byrd.)

Figure 4.4 With some beetle species, such as *Nicrophorus orbicolis*, the male and female will remain as a pair and care for the young. This parental feeding is not common among insects, and because of this behavior, laboratory rearing can be difficult. (Photo courtesy of Dr. Jason H. Byrd.)

different species from the same case should also not be reared within the same containers. Likewise, adult stages should generally not be mixed with larval stages. This standard operating procedure is particularly true with beetles collected from the body recovery scene. Large adult Silphidae (Coleoptera) species such as *Necrodes surinamensis* (Figure 4.3) and *Nicrophorus orbicollis* (Figure 4.4), as well as larval and adult Staphylinidae (Figure 4.5a and b), are best kept isolated or in low numbers in large containers that provide adequate hiding places. Many fly larvae are both predatory and cannibalistic. As a result of mixing species, or overcrowding of individuals, you could lose valuable and irreplaceable entomological evidence due to predation and cannibalism. Thus, it is best to keep like stages of the same species together, and use as many rearing containers as necessary to accommodate the requirements of the caseload.

Insect Eggs

When collecting entomological evidence at a body recovery site, the forensic investigator is likely to encounter insect eggs. Due to the successional nature of insect activity, insect eggs may be found on remains that are many weeks old. The presence of eggs does not necessarily mean that the remains are fresh. The number of eggs will vary greatly, but those

(a)   (b)

Figure 4.5 (a) Adults of the family Staphylinidae can be abundant at decomposing remains; however, little is known about the development of the larvae (B) due to the difficulty in species identification at the larval stage. (Photos courtesy of Dr. Jason H. Byrd.)

most commonly encountered are fly eggs. These eggs are deposited in clusters, or masses, that accumulate around the eyes, nose, mouth, and ears, as well as at sites of trauma, and possibly the genital/anal area. Unfortunately, many forensic investigators have discounted them as "sawdust," or similar materials of inconsequence, and have promptly brushed them away in their search for evidence. Fly eggs are often clustered together, and individual eggs are difficult to detect unless the cluster is inspected closely (Figure 4.6a and b). Although

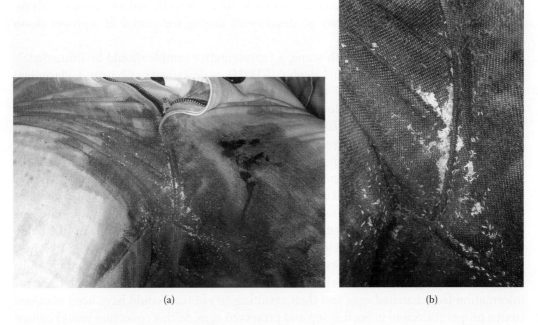

(a)   (b)

Figure 4.6 (a) Due to their small size, fly eggs often go unnoticed by the forensic investigator. In many cases, they are misidentified by the investigator as sawdust. The eggs are usually deposited around or in the natural orifices of the head, at sites of trauma, or in the genital/anal area. However, when excrement soiled clothing is present, the flies often deposit their eggs on the outer surface of clothing, as seen here. (b) Detail of fly eggs and newly emerged larvae on the outer surface of the clothing seen in (a). Note the length of the eggs and larvae in comparison to the length of the stitch in the denim jeans. (Photos courtesy of Dr. Jason H. Byrd.)

Figure 4.7 Slight magnification shows the distinct shape of the fly egg. It is good practice for forensic investigators to carry a small hand lens. Magnifications of 10× to 15× will reveal the true nature of many items of trace evidence often found at crime scenes. (Photo courtesy of Dr. Jason H. Byrd.)

individual fly eggs are visible to the unaided eye, a hand lens or a low-power dissection microscope will readily discern their distinct oblong shape (Figure 4.7). Collectively and singularly, the eggs have a creamy-white to yellowish appearance, are dry to the touch, and will readily "flake" away from the substrate on which they were deposited. Because of these characteristics, it is easy to understand why they are frequently overlooked, or mistaken as sawdust. However, general crime scene protocol should dictate that when nonidentifiable substances are found at a crime scene, they should always be collected for further analysis. Therefore, insect eggs should never go uncollected during the course of a proper death scene investigation.

If fly eggs are noted at a death scene, a representative sample should be immediately placed into an 80% solution of ethyl alcohol (ETOH), as described in Chapter 3. Placement of the collected eggs in alcohol is usually suitable for long-term archival storage of such specimens. It is not a prerequisite to first place insect eggs in a fixative solution such as KAA, or boil them, as is done with the larvae of flies. However, if these practices are completed, it will ensure proper specimen preservation. The preservation process will be described in detail later in this chapter. It is important for the forensic investigator to remember that the egg stage is very short in duration, and if eggs are not immediately preserved, hatch may occur only a short time later. If all collected eggs hatch, leaving none preserved, a valuable marker for the time interval becomes permanently altered. If the time and duration of egg hatch are not adequately documented, this could result in a less precise TOC or PIA being derived from the entomological evidence than what would otherwise be possible. The short duration of this stage makes it an extremely valuable life stage for the forensic entomologist to recover and analyze. It would be difficult for the forensic entomologist to obtain the same information from hatched eggs and their resulting larvae than could have been obtained from a proper collection of both living and preserved eggs. Such a collection would consist of a representative sample being immediately preserved, and another separate but equally representative sample being kept alive for rearing purposes. This concept of companion sampling should be followed for all insect life stages encountered at a death scene.

Second to the importance of proper collection techniques is the immediate shipping of live samples to a forensic entomologist for analysis and rearing purposes. In most cases, live eggs must be allowed to continue development in order to make a definitive species identification. Unless scanning electron microscopy or DNA analysis is utilized, it is not always possible to determine species identification utilizing only the egg stage. Even with the species that have a distinctive egg stage, meticulous preparation of the eggs for scanning electron microscopy is a difficult, time-consuming, and expensive task. Therefore, the egg stage is probably the most important stage in which the practice of companion sampling should be employed. Fortunately, it is probably the easiest insect life stage to work with since the eggs need little care, except to be protected from extreme temperatures. Thus, they are easily kept alive for rearing purposes and shipment to the forensic entomologist.

Larval Rearing

Larvae of many fly species of forensic importance can be easily reared on several different substrates. Various cuts of beef, chicken, and pork are commonly used as well as artificial diets (Byrd 1998). Artificial diets are popular in mass rearing operations because of their low odor and increased sanitation. However, they are not recommended for use when rearing insects to support a forensic investigation because the main component of many artificial diets is dry cat food (Mandeville 1988). The growth and development durations that would occur on human and animal tissues is not always similar to that produced by an artificial diet. Since the type of larval rearing media can alter insect growth rates, insects whose growth data will be utilized in legal investigations should be reared only on animal tissue.

Pork is the preferred rearing substrate for most forensic entomology rearing. It is inexpensive, commonly available, and physiologically similar to human tissue. Insect growth on pork does not differ significantly from insect development on human tissue. Pork is commonly selected because it does not produce excess liquid during the decomposition process, and it does not undergo rapid desiccation. Beef liver also works well and has been in common use; however, it has the undesirable property of drying very quickly on the surface. During the advanced stages of decomposition, liver liquefies too rapidly, and developing larvae may drown in the liquid environment of the rearing container. Successful development will also occur on chicken, but during decomposition it produces more odor than pork, and some larval species may develop differently than on human tissue.

Rearing Containers

Fly larvae that are part of the live insect collection made during the course of a death investigation should be placed into a rearing environment as soon as possible following removal from the death scene. Refer to Chapter 3 for proper collection procedures at the scene. Once at the laboratory, the insects should be immediately placed into plastic, foil-lined cardboard, or wax paper containers that have been filled with approximately 1/3 lb (150 g) beef or pork, and approximately ½ in. (1.25 cm) of soil or vermiculite to absorb excess moisture. Use as many containers as necessary to rear all of the living samples collected (see Figure 4.2b). If needed, collected insects can be temporarily held (24 to 48 hours) in a refrigerator at 3 to 6°C (Haskell 1990).

For most laboratory rearing purposes, each larval container (Figure 4.8) should be placed in the center of an aquarium floor that has been covered with either sand or vermiculite to a depth of ¾ in. (2 cm) (Figure 4.9) to facilitate larval dispersal. Each container

Figure 4.8 A typical larval rearing (or collection) container should be a short-sided plastic cup (16–32 oz size) with a tight-fitting and ventilated lid. Vermiculite, kept in a resealable sandwich bag until needed, provides a convenient way to provide a burrowing substrate for the larvae once collected. Containers such as this are inexpensive, can be assembled quickly, and kept in storage until needed. (Photo courtesy of Dr. Jason H. Byrd.)

will adequately rear approximately 75 to 150 larvae, and experience demonstrates that a 16 oz (0.5 L) plastic container (a pint-sized yogurt cup) with short sides works well. The short-sided container serves to keep the larvae associated with their food source, but it will not restrict larval dispersal (Byrd 1998), as the insects will need to crawl away from the food source and out of the food container. A lid is generally not necessary, or recommended, as long as an adequate and continuous food supply is maintained. The larvae will always remain associated with their food source until the onset of dispersal.

If a lid is used, it should be well ventilated with small holes or slits, and the lid must be removed at the onset of larval dispersal. Equipping the rearing container with a tight-fitting lid could alter the larvae's normal dispersal behavior, or restrict the airflow too much,

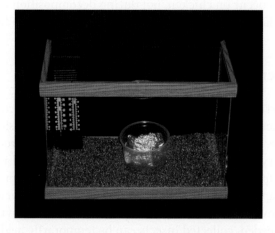

Figure 4.9 Larval-rearing containers such as that shown in Figure 4.8 should be placed into a larger container, such as a small aquarium, to more adequately allow for the larvae to undergo their natural migratory stage. Providing this extra space is critical to the successful development of some species, or when large numbers of larvae are within the rearing container. (Photo courtesy of Dr. Jason H. Byrd.)

Figure 4.10 *Synthesiomyia nudiseta* pupae on the outer surface of clothing. Pupation in areas such as this is common with this species since the larvae typically do not migrate far away from the remains, as in other species. (Photo courtesy of Dr. Jason H. Byrd.)

causing an increase in ammonia that is lethal to the developing larvae. High ammonia levels can accumulate to a toxic level in less than 6 hours, killing the larvae within the rearing container. Interfering with the normal dispersal behavior of the insect larvae reared as part of a TOC or PIA determination can be detrimental to a forensic investigation, as this may cause a delay in pupation and alter the time sequence of all subsequent life stages.

If the larvae are confined at the time of dispersion onset, subsequent pupation and emergence can be delayed or even prevented. This delay can be detrimental to a forensic investigation because some insect species will die instead if they are confined in too small of an enclosure. Providing for the natural dispersal behavior is critical in artificial rearing environments. A few species, such as *Chrysomya rufifacies* (the hairy maggot blow fly) and *Synthesiomyia nudiseta*, will successfully pupate in a confined environment. Due to this behavior, it is always necessary to give any clothing found with the victim a thorough examination, as the pupae of some species may be attached to the inner surfaces of the material, in folds of clothing (Figure 4.10), or along contact points between the victim and ground. Since other species, such as *Cochliomyia macellaria* (the secondary screwworm fly), will typically die before pupation if trapped in a confined location, any large numbers of dead or inactive larvae should be noted and collected. A complete search of the surrounding area should also be made to recover any larvae dispersing, or to collect puparia that have formed after dispersal. Since there is typically more than one species inhabiting decomposing remains that prefer different pupation sites, it is always necessary to search all items in contact with the body as well as the surrounding area. A detailed discussion of such a search procedure is provided in Chapter 3.

As a result of the alteration in larval development due to improper rearing protocols, the TOC or PIA determination may not be valid, and should be used in legal proceedings with great caution. To eliminate this potential source of error, the use of an open container for larval rearing will allow for larval dispersion to occur naturally and without delay, thus reducing the possibility of improper rearing and alteration of larval development.

Monitoring Larval Growth and Archival Preservation

Once the larvae are placed on the food source in the plastic, foil-lined cardboard, or waxed-paper-rearing containers, they should begin to feed immediately. If they do not, it could

Figure 4.11 Dispersing larvae of *Musca domestica*. Larvae of this species may pupate directly on the remains or on clothing associated with the remains in indoor environments. However, in outdoor locations, the larvae usually migrate away from the remains before forming the pupal stage. (Photo courtesy of Dr. Jason H. Byrd.)

be because they have already entered the prepupal dispersal stage (Figure 4.11). Additional food substrate should be added as necessary, and a few larvae should be collected daily and immediately preserved in order to document the various stages of development. For those collected and preserved, detailed records should be kept as to stage of development, larval behavior, and time and manner of preservation. The number of larvae that should be preserved during rearing is strictly dependent on population size of the collected larvae. It is more important to have a few individuals reared to adulthood, and less preserved in the immature stage, than to have several dozen preserved immature forms and no adults.

The proper preservation of larvae is an essential part of the collection and analysis process, and it should not be overlooked. Fly larvae can be preserved utilizing a variety of methods. Many of these methods have equally good results, but it is imperative to inform the forensic entomologist of the method of preservation utilized in each case because each method produces various alterations of the larval body, which will affect the time of colonization estimate.

Larval preservation is a two-step process. In order for the larvae to be properly preserved, they must first be fixed within 24 to 48 hours after larval death. Larvae must be fixed because ethyl and isopropyl alcohol will not fully penetrate the larval cuticle. This inability of the preservative to penetrate the cuticle allows the bacterial fauna present within the insect gut to decompose the larva even though it is submerged in alcohol. Decomposition of the larval body will alter its size and color, eventually causing the loss of all internal tissue and subsequent loss of the sample. Improper preservation will quickly make the specimens unsuitable for use in a TOC or PIA estimation. Proper preservation requires fixing the larvae, which can be accomplished by either placing the larvae in a fixative solution such as KAA (see appendix) or blanching the larvae in nearly boiling water. If a fixative solution is used, the larvae should not remain within that solution for over 12 hours. Once removed from the fixative solution, placement in 80% ETOH will make the specimens suitable for archiving. Leaving the larvae in the fixative for too long will result in the specimen becoming hard and brittle, making analysis difficult. Due to this reason, and the difficulty sometimes encountered by some law enforcement agencies when attempting to purchase and keep chemical solutions on hand, the most recommended method for fixing larvae is by boiling them in water. Since it is difficult to have boiling water at a crime scene, it is recommended that the larvae be placed into 80% ETOH for on-scene preservation, and then removed from the alcohol and blanched as soon as possible (or within 24 hours after initial preservation) once back in the laboratory. After blanching, the specimens should be promptly removed from the water and placed into 80% ETOH alcohol. It should be noted that the act of blanching larval specimens will fully extend the larval body, which could produce a specimen approximately 10% longer than it would be while living (Tantawi and Greenberg 1993). This extension would result in an inaccurate TOC or PIA estimation unless the forensic entomologist was advised of the preservation method and compensated for such known changes. Therefore, the forensic entomologist must always be completely advised of each step in the specimen preservation process. It is the duty of the individual making the collection to keep detailed notes as to the time and method of preservation.

Mass Rearing of Insects Related to Forensic Investigations

To rear larger numbers of larvae in situations where it is not necessary to keep samples separate (i.e., mass rearing operations), a setup similar to that shown in Figure 4.2b can be utilized. This process simply uses multiple food substrate containers, as described earlier, which keeps the larvae/food substrate ratio low and serves to keep larval masses small and manageable. It also helps reduce the odor produced when rearing large numbers of larvae in the laboratory. However, the most important aspect of utilizing multiple small food containers is the ability to keep the larvae/food ratio low. If there are a high number of larvae per gram of food substrate, the metabolic heat generated by the larvae will rise above the ambient temperature of the environmental chamber and shorten the developmental duration of the larvae (Goodbrod and Goff 1990; Marchenko 1988). Such an event is particularly undesirable since the TOC derived from their development will also be shorter than it should be unless maggot mass heat is taken into account. Thus, it is a good laboratory practice to keep a temperature probe inserted into the larval food source to monitor temperatures and account for any unexpected temperature rises (Figure 4.12a–d). If any temperature increases occur, the event must be taken into account (or otherwise compensated for) when making the TOC or PIA estimation, and included in the final report on the thermal history of the larvae.

Larval Dispersal in the Laboratory

When the feeding stage has completed, postfeeding larvae will attempt to leave their food source and exit their rearing container. This is known as larval dispersal, and these larvae are known as prepupal, dispersing, or wandering. Dispersing fly larvae will pose a problem since they can easily climb out of their food substrate container, up the glass sides of the aquarium, and then become widely dispersed among the equipment of a laboratory. One method commonly used to deter the larvae from crawling up the sides of the container is to cover the aquarium bottom with about 2 cm of builder's sand, or vermiculite. They will readily crawl up the sides of their rearing container and drop onto the vermiculite-, sand-, or leaf litter-covered aquarium floor. Most fly larvae have a natural tendency to bury themselves in soil in order to complete pupation, and a loose substrate in the bottom of the

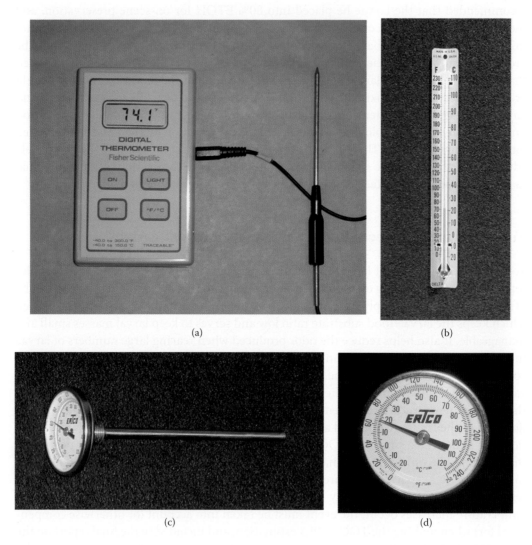

(a)   (b)

(c)   (d)

Figure 4.12 (a) A digital thermometer, with a multipurpose temperature probe, is ideal for recording the thermal history of entomological evidence. Such instruments are waterproof and have lighted LCDs for easy reading in low-light conditions. The probe can be inserted into soil, fluids, and directly into larval masses for proper documentation of scene and laboratory rearing temperatures. (b) The least expensive and most widely available thermometer is the familiar mercury type; however, they no longer contain mercury. With reasonable care, these thermometers will provide adequate service. (c) A "dial face" or "stem type" thermometer is inexpensive, very durable, and reasonably accurate. Thermometers such as this are ideal for rough service situations, and can be kept as equipment in crime scene units and field vehicles for long periods of time. (d) The face or dial of stem type thermometers can be purchased in any size. Thermometers with a dial size of 1 to 1.5 in. (2.5 to 3.8 cm) are generally too small for the temperature to be read easily. However, many models of stem type thermometers have a dial face of 2 in. (5 cm), making the dial more clearly visible under field conditions. These models are available with increments of either 1° or 2° Fahrenheit or Celsius (shown here). (Photos courtesy of Dr. Jason H. Byrd.)

Figure 4.13 Demonstration of the color progression of the pupal stage. It is important to photograph this life stage so that the color can be documented. The color can help establish the age of the insect. It is good practice to include a photo-gray ruler so that length can be documented, and color can be calibrated. (Photo courtesy of Dr. Jason H. Byrd.)

rearing aquarium will avoid this problem. This substrate will also act to absorb the excess fluid that will accumulate upon dispersal. Once the larvae locate a suitable place for burrowing and concealment, they usually will not continue to wander unless they are crowded and cannot readily disperse. Allowing the insects to undergo their natural behavior will not create any unnatural delays in the time of pupation, which could alter TOC or PIA estimations calculated from arthropod development.

Once on the substrate covering the aquarium floor, the larvae will behave in a sluggish manner and their body length will start to decrease, becoming entirely immobile within 24 to 72 hours after the start of dispersal. When fully shrunk and immobile, the larvae will begin to darken from the normal creamy-white color typical of the larval skin to a light red color, then changing to darker red, light brown, and finally to a dark brown that appears almost black (Figure 4.13). Once dispersal has been completed, allow a couple of days for the larvae to complete pupation and for the pupae to harden before working with the insects. After this time period has elapsed, it is safe to gently sift the pupae from the sand or vermiculite substrate. Samples of the dispersing larvae should also be preserved on a daily basis until pupation. A written record of the dates of preservation should be kept and added to the case file.

Once the pupae have been sifted from the burrowing substrate, they should be placed into an emergence container (Figure 4.14) and monitored closely for adults. The emergence container should be checked at frequent intervals, and any emerged adults should be removed, counted, and placed into rearing cages and a small sample preserved. This procedure should be followed repeatedly until no adults are noted to emerge. This practice will provide both the duration and rate of emergence, which is often essential and invaluable data in proper TOC and PIA estimations from entomological evidence.

Adult Emergence

At the onset of emergence, the adult inflates a balloon-like organ, called the ptilinum, on its head. The inflation of this fluid-filled organ causes the end of the puparium to split, and the adult fly pulls itself out with its legs (Figure 4.15a). When newly emerged adults appear within the emergence container, they will look spider-like instead of fly-like (Figure 4.15b).

Figure 4.14 A commercially available rearing container that can be used for rearing terrestrial flying and some aquatic flying insects. The container is fitted with a cone that directs newly emerged adults into the vented top portion, where they are held until the container is emptied. (Photo courtesy of Dr. Jason H. Byrd.)

(a)   (b)   (c)

Figure 4.15 (a) An emerging *Cochliomyia macellaria* adult. Once the adult has split open the end of the puparium, it simply pulls itself out with its legs. (b) The newly emerged adult looks spider-like, as it does not have its adult color or fully formed wings. They run about rapidly in search of shelter while they harden their exoskeleton and, as a result, are often mistaken by forensic investigators as spiders. (c) The emerging adult seen in Figure 4.10A has finally attained its final adult color. This process usually requires several hours, but can vary widely, depending on temperature. (Photos courtesy of Dr. Jason H. Byrd.)

The body will be gray in color and the small wings will not yet be fully extended. Shortly after emergence is complete, the adult will have expanded its wings, but not yet attained the familiar shades of metallic green and blue commonly associated with blow flies. The adult fly usually gains the characteristic adult appearance within 8 to 24 hours of emergence under typical laboratory conditions (Figure 4.15c). Once these characteristics appear, the adults can be moved from the emergence container into the fly cages after a representative sample has been preserved. Until the time they attain their "normal" adult appearance, they are extremely soft bodied and easily damaged. Care must be taken to minimize contact with them before they are able to harden their exoskeleton.

Figure 4.16 Properly pinned adult insect for archival preservation in a forensic entomology reference collection. Adult insects collected from forensic investigations can be pinned and placed into museum type storage compartments. This properly preserves the adult insect, and provides all necessary reference information on cotton bond paper affixed to the pin. However, if pinned improperly, critical identification characteristics can be damaged. (Photo courtesy of Dr. Jason H. Byrd.)

Some of the laboratory-reared adult insects should be euthanized and put on pins or card points and properly labeled for further study (Figure 4.16). If the proper pins or points are not available, or the investigator is unfamiliar with pinning techniques, placement of the adult insects directly into an 80% ETOH solution will adequately preserve most adult insects. Specimens can also be placed in Ziploc bags, labeled with pertinent information, and stored in a freezer; however, care should be taken to not crush the specimens while in storage. Additionally, specimens removed from the freezer should be transferred to the custody of the entomologist within 24 hours in order for them to be properly preserved.

It is good practice to preserve the adult form of most insects of forensic importance, utilizing any of these methods. However, the forensic investigator should first consult with the cooperating forensic entomologist to determine method of preference. It should be noted that placing the adult insects in alcohol may alter their color, or it may mat together the small bristles and hairs on the fly body, making taxonomic identification more difficult. All described methods serve to accomplish the task of preserving the insect, and it is the sole discretion of the forensic entomologist as to which method he or she prefers. In all cases, the date and time of pupation and adult emergence should be noted and entered into a written laboratory record, as well as a label placed in the alcohol vial or container with the preserved insects. Proper museum (or archival) preservation requires the adult insects to be "pinned" or "pointed," and such mounting should be left to individuals properly trained in this procedure. Improper mounting can permanently distort or destroy key identification features, and possibly render the specimen useless for the species identification that is potentially necessary for legal investigations. However, DNA analysis might be possible with damaged specimens.

Figure 4.17 In laboratory-rearing colonies, an open cup of water can cause many flies to drown, eventually possibly decimating the colony population. A small specimen container, with a dental wick inserted through a hole in the lid, works well as a constant source of freshwater for the fly colony. (Photo courtesy of Dr. Jason H. Byrd.)

Adult Insect Colonies in the Laboratory

If necessary, adult flies can be maintained in colony quite easily, and a standard size fly cage can adequately contain about 250 adult flies. Adult flies can be kept in a screened enclosure of any desired size, and commercially available fly cages that are approximately 14 in.³ (35 cm³) are a good choice between functionality and space-saving measures. Most commercially manufactured fly cages are designed with a collapsible aluminum frame for storage and easy cleaning. One side is fitted with an elastic "sock," so colony maintenance and the addition of food, water, and meat for egg laying can be accomplished without opening the cage and allowing flies to escape. Flies contained within the colony should be provided with a 50:50 mixture of table sugar and powdered milk. Additionally, a supply of freshwater should be available at all times. A fly-drinking station can be fashioned from a plastic specimen cup with a hole bored into the lid to allow for a cotton wick (Figure 4.17). A tray with a solid bottom should be placed under the fly cage to help contain the debris that will accumulate from the activity of the fly colony, and to aid in maintaining laboratory sanitation.

Rearing Aquatic Insects

With some forensic casework, it may be necessary to rear aquatic insects that have been collected as part of a legal investigation. Species of aquatic insects are commonly found in man-made and natural freshwater and brackish habitats. These include lakes, rivers, sewers, ditches, and artificial containers. These insect species generally do not feed directly upon the decomposing remains as do terrestrial insects, but they will utilize the carcass as a substrate on which to attach themselves as they filter the water column for food. Or, they may use the body as a substrate on which to attach a small tube made from debris in the water that they use for shelter. It is the insects and their protective cases attached to human remains and inanimate objects associated with crime scenes that allow investigators to determine both the geographic location of body deposition and the seasonality encompassing the TOC, PIA, or minimum period of submersion.

Unfortunately, a number of aquatic insects require a habitat of flowing water, and thus a rather elaborate rearing system must be devised if they are to be kept alive outside of their natural environment (see Merritt and Wallace, Chapter 7). If they are deprived of these conditions, they can die in a period of minutes or hours. A further complication is that many aquatic insect larvae will leave the body as soon as it is disturbed. For these reasons, it is usually a good practice to make detailed notes, complete all photography, and preserve aquatic specimens while at the body recovery scene.

For some species found in still water habitats, laboratory rearing is relatively easy, and a small aquarium or similar container will suffice (see Merritt and Wallace, Chapter 7). The collected insects should be kept in a sample of the water found at the scene. If necessary, tap water can be used, but only after it has been allowed to stand at least 24 hours so common chlorine additives can dissipate. The collection container should be filled completely to reduce damage to the insects from excessive jostling and wave action during transport. Additionally, if collected during the summer, the containers holding these insects should be stored in the shade or on ice to prevent the water temperature from rising too high and killing the insects. The rearing container utilized in the laboratory should be aerated by an air diffuser connected to a simple air pump, such as those commonly used for freshwater fish aquariums. Sticks, rocks, or other vegetation protruding above water level should be provided within the container so that emerging adult insects can have a refuge out of the water. Feeding aquatic insects can be best accomplished by submerging aquatic plant material found at the scene into the rearing container, and supplementing that with commercial flake fish food. As with terrestrial larvae, periodic collection and preservation should be conducted, and a detailed account of the time, date and method of preservation should be kept to provide the forensic entomologist with a detailed life history of the insects.

A conical screen "tent" apparatus with a removable clear plastic receptacle can be constructed over the top of the rearing container, so that emerging adults will become trapped after they leave the water and take flight. Emerged adult aquatic insects can be preserved by placement directly in 80% ETOH. Aquatic insect-rearing containers are commercially available and are the same as the container earlier described as an adult emergence container for terrestrial flies. This container will work well for both terrestrial and aquatic insects and is an indispensable part of the rearing laboratory equipment.

Unique Species Requirements

The rearing of fly species such as *Musca domestica* (the house fly) and *Hydrotaea anescens* (the bronze dump fly) is conducted differently from that of the blow flies, as house flies and dump flies often do not feed directly on the tissues of the carcass. These species most commonly feed on exposed gut contents, fecal matter, or clothing and material that have become fetid with urine or feces. In the past, laboratory rearing of these species was conducted by using a mixture of horse and hog manure, despite its disagreeable nature. Fortunately, Richardson (1932) developed the first artificial rearing medium, which consisted of 3.75 lb (1.7 kg) of wheat bran mixed with 1.75 lb (0.8 kg) of alfalfa meal. To this was added a mixture of 5 L water (5,000 cc); 3 L yeast suspension (3,000 cc), which consisted of 1 lb (0.5 kg) baker's yeast and 2 L of water (kept refrigerated until used as needed); and 25 g malt sugar (25 cc) (Galtsoff et al. 1937). These ingredients are stirred thoroughly and added to rearing jars or tin cake pans. It is not necessary to mix such a large quantity of rearing media. The recipe can be proportionally reduced to suitable size. Larvae can be added to this mixture as soon as it is complete. There is no developmental difference between muscoid flies reared on natural versus artificial substrates like the one described above, so it is a suitable alternative to the natural larval food source. Thus, artificial rearing media is used extensively when dealing with muscoid flies due to decreased odor and increased sanitation.

Rearing Beetles in the Laboratory

Skin Beetles (Family: Dermestidae)

The rearing of various beetle species likely to be encountered during a forensic investigation will require a few modifications of the fly-rearing process. For the various skin beetle species that will be frequently collected in death investigations, a food substrate of beef, chicken, or pork will suffice. However, this food source must be relatively dry before being placed into the rearing chamber, and kept dry throughout the rearing process. *Dermestes* spp. (skin beetles) can be easily bred on almost any kind of dried animal matter (Abbott 1937), and beetles in the family Trogidae (hide beetles) have habits that allow rearing methods similar to those utilized for the various species of *Dermestes*. The larvae of the dermestid beetles do not prefer moist or wet conditions, as do many fly larvae. They are negatively phototrophic (they move away from light), and often seek shelter and conceal themselves within or under decomposing remains. The beef or pork utilized for rearing dermestids should be whole and not ground, and it is usually not a problem to keep whole beef or pork dry during decomposition in the rearing environment. Chicken, if used, can be kept dry by cutting it in small strips. Although the food substrate and container should be kept dry at all times, a moist cotton pad covering a shallow petri dish of water should be available to both the adult and larval dermestids at all times.

As with the fly larva-rearing protocol, a burrowing substrate should be provided to allow for dispersal and concealment of the prepupal larvae. If it is not provided, the rearing chamber should be checked daily, and the newly formed pupae should be removed and placed into a separate container. If they are not removed, the feeding larvae may feed on the pupae. Cannibalism is not uncommon, especially if the food source becomes depleted. Beetles in the family Dermestidae, unlike any other forensically important insect, can undergo a molting process to reduce their body size if food resources become scarce. Therefore, food must be available in excess, and any periods of food depletion must be accounted for when using the development of dermestids as a basis for TOC or PIA estimations.

Carrion Beetles (Family: Silphidae)

The insects of forensic importance in the family Silphidae are large beetles (typically 10 to 35 mm in length), frequently observed associated with decomposing organic material (Anderson and Peck 1985). Their most common habitat is that of the carcass, and some

beetles in this insect family even possess the uncanny behavior of interring small vertebrate carcasses. It is this behavior that has given some species the common name of carrion beetles, sexton beetles, or burying beetles. Most of the silphids are necrophagous as
adults, and of the necrophagous species, the larvae feed exclusively on decaying animal
flesh. However, the adults may feed on the carrion as well as the other insects present,
especially the fly larvae that they readily find on carcasses.

The two subfamilies within the Silphidae, Silphinae and Nicrophorinae, have radically different methods in which to utilize the carrion resource and avoid competition.
Pukowski (1933) first described the life history of the beetles in the genus *Nicrophorus*.
A large amount of data has been collected about their development and behavior. When
the *Nicrophorus* adults locate a carcass, they first crawl over the outer surface to assess its
size and suitability. If the carcass size is small enough (i.e., a small rodent carcass), they
will either bury it where it was found or attempt to move it to a different site. Milne and
Milne (1976) describe in detail the complex behavior the beetle adults employ when moving a carcass. Once the carcass has been buried, the female then excavates a tunnel leading
away from the carcass, where she will lay approximately thirty eggs. When the eggs hatch,
the larvae crawl toward the carcass and into a hole chewed by the female. Once inside the
carcass, the female regurgitates and feeds the larvae a liquid diet for at least 6 hours before
the larvae begin to feed on their own. This parental feeding behavior can last through the
first couple of months, but it is not required, as the larvae will continue to develop even if
the female is removed.

The larvae undergo three stages, or instars, which are completed in about 22 to 26
days (Brewer and Bacon 1975). Once the larval stages are complete, the larvae leave the
carcass and crawl into the surrounding soil to pupate. Therefore, it is important to check
the surrounding soil at the crime scene for larvae and pupae of the Silphidae, as well as
for dispersing fly larvae. The pupal stage typically lasts 15 days, depending on the species and ambient temperature. Usually *Nicrophorus* spp. do not compete with Silphinae
spp., which commonly utilize larger carcasses (i.e., deer, bear, and human) as their food
resource. However, large numbers of *Nicrophorus* adults may be seen on large carcasses,
including human cadavers, where they consume both the carrion and the abundant fly
larvae (Wilson and Knollenberg 1984).

The life cycle of the Silphinae is not as complex, and unlike *Nicrophorus* that arrive
early when the fly larvae are present, the Silphinae avoid competition for food with the fly
larvae by waiting until larval dispersal has occurred before they colonize a carcass. There
are no known parental-larval interactions within this subfamily, and development proceeds much more slowly than in *Nicrophorus*. Silphinae adults mate after locating a suitable carcass, and the females oviposit (lay eggs) in the surrounding soil. Larvae hatch from
the eggs 2 to 7 days after being laid and move to the carcass to feed. On the carcass they
undergo three larval instars, and pupation takes place in the soil away from the carcass 10
to 30 days after hatch. The pupa has a duration of 14 to 21 days (Anderson and Peck 1985).
Again, sifting the soil surrounding the body is a critical step in ensuring that all life stages
have been recovered. This procedure is particularly important since the larvae and pupae
found in the surrounding soil are typically older than those found directly associated with
the carcass, which is critical information when determining the TOC or PIA based on
entomological evidence.

Both adult and larval forms of the Silphidae are quite common and easily collected.
Laboratory rearing of the Silphidae is also relatively simple. Beetle larvae should be placed

into an aquarium (Figure 4.9) containing about 5 cm of moist soil, and a few leaves or other detritus should be added on the soil surface to serve as coverage. A piece of carrion (beef, chicken, or pork) equivalent in size to a mouse carcass should be added as well as a shallow dish containing moist cotton to serve as a water supply. Once settled, the larvae will then move to the food substrate and burrow into the soil. They will remain in the soil, directly underneath their food source, and feed from below for about 2 weeks. If adults are present, they will bury the food source so that it is not visible from above. The successful rearing of the beetles in the family Silphidae is largely dependent on seasonality and species, so expect to meet with varying degrees of success.

The practice of feeding fly larvae and dead adult flies to adult beetles in the family Silphidae, in addition to vegetative matter and decomposing meat, can improve rearing success in the laboratory. It should also be noted that with extremely young larvae of the genus *Nicrophorus*, parental feeding might still be required. It is difficult to collect both the larvae and the parental adults, unless the parental adult is observed in direct association with the larvae. However, the extremely young larvae that still require parental feeding are so small and inconspicuous that they are not likely to be recovered unless by a trained entomologist. Therefore, the parental feeding that occurs with the *Nicrophorus* is not a likely hindrance to laboratory rearing for forensic purposes. Beetles in the families Cleridae, Histeridae, and Nitidulidae can be reared in much the same manner as the Silphidae.

Rove Beetles (Family: Staphylinidae)

The life histories and habits of beetles in the family Staphylinidae are little known, primarily because few are of economic importance. They are also not brightly colored enough to be of notice to the casual observer. However, some are large and conspicuous, and a few species are of economic importance as predators of insect pest species, and some life history information has been documented (Mank 1923; Voris 1939). As a result, only a select few species are useful in TOC or PIA estimations, and both adults and larvae should be collected whenever possible.

Staphylinid larvae can be expected to inhabit a decomposing substrate (animal, vegetable, or human) from about 9 days to over 2 months. While associated with carrion, they are mainly predators of other insects, and their larvae, that are attracted to the decomposing remains. Unlike most insects of forensic importance, the amount of decomposition has little effect on their abundance. However, moisture is an extremely important factor, as they seem to prefer drier remains. Staphylinid larvae typically will not colonize a decomposing substrate that is extremely moist. Like many other beetle species, they prefer to feed on fly larvae, and thus should never be placed in the same container as fly larvae when collected. The collector should be advised that both adult and larval forms of these beetles remain well hidden and motionless within the cracks and crevices of a carcass and are sometimes overlooked. Adults seem to be found much more commonly than larvae. When they are observed, they are usually walking, but they are also strong fliers. When the adults are threatened or disturbed, they often turn their abdomen up and hold it over their head in a "stinging" posture. They have no stinging apparatus, and thus this is merely a hollow threat. However, they are able to emit a very disagreeable odor when alarmed that cannot be easily washed away.

Both the staphylinid larvae and adults are hardy in captivity and are easily kept in small containers with about 1 in. (2.5 cm) of soil. They should be fed a diet of vegetable material, decomposing meat, and provided with a shallow dish of powdered milk. Live fly larvae and dead adults can also be added to their diet to improve rearing success. In their natural environment these beetles seem to prefer fly larvae and adults to decaying meat. Moistened cotton should also be provided at all times as a source of water. The beetles should be kept in a dark rearing environment, as they feed much more readily under cover of darkness. The egg-to-adult duration can be expected to be about 6 weeks, but will vary according to temperature (Mank 1923). The larvae are also known to be cannibalistic, and thus placing one larva per small container for rearing purposes may be the best method to utilize, preventing the loss and destruction of entomological evidence.

Photographing Evidence

Taking macro images of specimens collected, preserved, or to be reared to the adult stage can be done. Doing so preserves some information, such as actual size during collection, that might be lost due to preservation technique. Accordingly, images can serve as evidence in cases where specimens are lost in the mail or destroyed. However, when taking such pictures, it is important to always include a scale for comparison. Scales can be rulers or coins of known denomination. Photographs should also contain, or be labeled, with pertinent information, such as case number, date, and investigator taking the picture.

Rehydration Methods

Some fly larval specimens that are preserved might dehydrate due to being stored in improper containers. Such containers do not have appropriate lids to prevent the preservative from evaporating. Consequently, resulting specimens will dehydrate and assume a raisin appearance. Identifying specimens in this condition is near impossible; however, methods for rehydration are available. It is suggested that such specimens be placed in 5% TSP (Alfa Aesar, A. Johnson Matthey Company, Lancashire, United Kingdom) for approximately 15 hours prior to transfer to 80% ETOH for 3 days before examination (M. Sanford, personal communication). The TSP solution will break down the lipid layer covering the larvae, allowing for the ETOH to penetrate and reinflate the dehydrated larvae. Larval lengths recorded after rehydration should not be interpreted as actual length of specimens at the time of collection. Solely larval instar and species identification should be recorded for these specimens.

Conclusions

The rearing of entomological evidence is a task not often undertaken by the crime scene technician, law enforcement agency, or medical examiner's office during the course of a death investigation. Typically, the laboratory rearing of forensic insects is left to the forensic entomologist. However, in an effort to best preserve forensic evidence, some law enforcement agencies and medical examiner's offices are starting to rear collected specimens on their own in addition to the living sample collected for shipment to their cooperating forensic entomologist. Therefore, if the original shipment is lost or the larvae perish in transit, the investigating agency has living samples to serve as backup. Also, rearing operations at some agencies are undertaken simply due to the keen personal interest of some investigators in entomological evidence. These specimens are often incorporated into display or teaching collections at the respective offices. It is expected that more investigators will take it upon themselves to rear entomological evidence, at least temporarily, until the evidence can be forwarded to a professional forensic entomologist. This practice will help ensure the use of entomological evidence in criminal investigations, and help support the efforts of the forensic entomologist.

References

Abbott, C. E. 1937. The necrophilous habit in coleoptera. *Bulletin of the Brooklyn Entomological Society* 32:202–4.

Anderson, R. S., and S. B. Peck. 1985. *The insects and arachnids of Canada*. Part 13. *The carrion beetles of Canada and Alaska*. Publication 1778. Research Branch, Agriculture Canada.

Brewer, J. W., and T. R. Bacon. 1975. Biology of the carrion beetle *Silpha ramosa* Say. *Annals of the Entomological Society of America* 68:768–90.

Byrd, J. H. 1995. The effect of temperature on flies of forensic importance. Thesis, University of Florida, Gainesville.

Byrd, J. H. 1998. Temperature dependent development and computer modeling of insect growth: Its application to forensic entomology. Dissertation, University of Florida, Gainesville.

Galtsoff, P. S., F. E. Lutz, P. S. Welch, and J. G. Needham. 1937. *Culture methods for invertebrate animals*. Ithaca, NY: Comstock Publishing Company.

Goodbrod, J. R., and M. L. Goff. 1990. Effects of larval population density on rates of development and interactions between two species of *Chrysomya* (Diptera: Calliphoridae) in laboratory culture. *Journal of Medical Entomology* 27:338–43.

Haskell, N. H. 1990. Procedures in the entomology laboratory. In *Entomology and death: A procedural guide*, ed. E. P. Catts and N. H. Haskell. Clemson, SC: Joyce's Print Shop, pp. 111–122.

Mandeville, D. J. 1988. Rearing *Phaenicia sericata* (Diptera: Calliphoridae) on dry cat food with CSMA. *Journal of Medical Entomology* 25:197–98.

Mank, H. 1923. The biology of the Staphylinidae. *Annals of the Entomological Society of America* 16:220–37.

Marchenko, M. I. 1988. The use of temperature parameters of fly growth in medicolegal practice. General trends. Paper presented at Proceedings of the International Conference on Medical and Veterinary Dipterology, Ceske Budejovice.

Milne, L. J., and M. J. Milne. 1976. The social behavior of burying beetles. *Scientific American* 235:84–89.

Pukowski, E. 1933. Ecological investigation of *Necrophorus*. *Zeitschrift fur Morphologie und Oekologie der Tiere* 27:518–86.

Richardson, H. H. 1932. An efficient medium for rearing house flies throughout the year. *Science* 76:350.

Tantawi, T. I., and B. Greenberg. 1993. The effect of killing and preservative solutions on estimates of maggot age in forensic cases. *Journal of Forensic Sciences* 38:702–7.

Voris, R. 1939. Immature stages of the genera *Ontholestes*, *Creophilus* and *Staphylinus*. Staphylinidae (Coleoptera). *Annals of the Entomological Society of America* 32:288–303.

Wilson, D. S., and W. G. Knollenberg. 1984. Food discrimination and ovarian development in burying beetles (Coleoptera: Silphidae: *Nicrophorus*). *Annals of the Entomological Society of America* 77:165–70.

Appendix

Preservation Solutions for Terrestrial Insects

A. **Ethyl alcohol (ethanol or ETOH):** This is commonly known as ethanol or ETOH. The ethanol is best suited for entomological purposes at 75–80%, and can be used to kill and preserve adult specimens and for preserving larvae after fixing in either Kahle's solution or KAA.

Ethanol is usually purchased in bulk at 95% concentration. An 80% solution can be produced by adding 15 parts distilled water to 80 parts of the 95% ethanol.

B. **KAA (KAAD):**

95% ethanol	80–100 ml
Glacial acetic acid	20 ml
Kerosene	10 ml

This solution should be used only for killing larval specimens. The specimens should not remain in this solution for over 12 hours, as they become brittle and unsuitable for examination. Specimens in this solution should be transferred into a 75–80% solution of ethanol as soon as they have been killed.

C. **Kahle's solution:**

95% ethanol	30 ml
Formaldehyde	12 ml
Glacial acetic acid	4 ml
Water	60 ml

This is another popular solution that should be kept available in the forensic laboratory. This solution can be used for killing and preserving adult insects, and for the preservation of larval specimens.

D. **XAA:** This solution can be used in place of KAA (or KAAD), but it is not commonly used.

Isopropyl alcohol	60 ml
Xylene	40 ml
Glacial acetic acid	50 ml

Preservation Solutions for Aquatic Insects

A. **Carnoy fluid:**

Chloroform (30%)	30 ml
Ethyl alcohol (95%)	60 ml
Glacial acetic acid	10 ml

Suitable as a killing agent and preservative for most soft-body aquatic insects. However, it is not commonly used due to the danger and restricted use of chloroform.

B. **Ethyl alcohol** (90–95%): Most popular solution for the preservation of most eggs, larvae, and pupae of aquatic insects. The adult form of most aquatic insects can be preserved in only 75–80% ethanol.

C. **Kahle's solution:** See comments for Kahle's solution above.
D. **Pampel's solution:**

Formalin	10 ml
Ethyl alcohol (95%)	30 ml
Glacial acetic acid	7 ml
Water	53 ml

Factors That Influence Insect Succession on Carrion

5

GAIL S. ANDERSON

Contents

Introduction

Insects are attracted to a body immediately after death, and they colonize in a predictable manner. A corpse, whether human or animal, is a large food resource for a great many creatures, and supports a large and rapidly changing ecosystem as it decomposes. The body progresses through a recognized sequence of decompositional stages, from fresh to skeletal, over time. During this decomposition, it goes through dramatic physical, biological, and chemical changes (Coe and Curran 1980; Henssge et al. 1995; Van den Oever 1976). Each of these stages of decomposition is attractive to a different group of sarcosaprophagous arthropods, primarily insects. Some are attracted directly by the corpse, which is used as food or an oviposition medium, whereas other species are attracted by the large aggregation of other insects they use as a food resource.

When the sequence of insects colonizing carrion is known for a given area and set of circumstances, an analysis of the arthropod fauna on a carcass can be used to determine the time of death. This procedure can provide accurate and precise methods for estimating elapsed time since death, and is used in many homicide investigations worldwide.

However, the sequence of insects that colonize the remains, the species involved, and the time of arrival and tenure are impacted by a variety of parameters.

When remains are found weeks, months, or more after death, insect evidence is often the only method available to determine reliably the time of death. Insects colonize in a predictable sequence, with some species being attracted to the remains very shortly after death; others are attracted during the active decay stage, and still others are attracted to the dry skin and bones. Insects continue to colonize a body until it no longer provides a food source. Each group of insects that colonizes the remains changes the resource, making it undesirable for its own species but desirable for others. This is known as the facilitation model of succession (Connell and Slatyer 1977; Hobischak et al. 2006; Payne 1965).

When the early-arriving insects migrate from the remains they invariably leave evidence of their presence behind, such as cast larval skins, empty pupal cases, and even peritrophic membrane. Meanwhile, the remains themselves have changed and entered a stage of decomposition that is attractive to other, later colonizers. Therefore, when remains are found, the forensic entomologist will study not only the insects that are present on the remains at the time of discovery, but the evidence left behind by earlier colonizers. He or she will also note the species that are absent, but normally expected to be present, in the colonization sequence. From this information an accurate time of death can be established. However, insect succession on a corpse is impacted by many factors, including geographical region, exposure, season, and habitat.

Attraction to the Remains

Insects are attracted to the body immediately after death, frequently within minutes (Anderson and VanLaerhoven 1996; Dillon 1997; Dillon and Anderson 1995; Erzinclioglu 1983; Nuorteva 1977; Smith 1986). Blow flies are usually the first colonizers and can be attracted over great distances by odor. In South Africa, marked flies of the genus *Chrysomya* were caught in baited traps up to 63.5 km away from the point of release (Braack 1981). Vision, color, and the presence of other conspecifics on the remains also play a role (Hall et al. 1995). Oviposition is elicited primarily by the presence of ammonia-rich compounds, as well as moisture, pheromones, and tactile stimuli (Ashworth and Wall 1994). The remains also appear to be more attractive once one female has begun to lay eggs, with many females immediately laying large numbers of eggs in one area (Anderson, unpublished data; Barton Browne et al. 1969). This may be an evolutionary strategy to minimize desiccation and predation during the egg stage, which eventually results in large numbers of maggots on the body. Large maggot masses can break down a body faster than individual invertebrate scavengers, and also generate heat that can protect the insects against adverse temperatures. This mass egg-laying behavior was found to be partially mediated by pheromones in Australian fly species (Barton Browne et al. 1969), although experiments on British species indicated that the odor produced by larval activity was neither attractive nor repellent to gravid females (Erzinclioglu 1996). However, if this behavior is pheromone induced, it is possible that the pheromone is deposited on the egg chorion itself and dissipates after the insect ecloses (Erzinclioglu 1996).

The odor emanating from a corpse changes as the body decomposes, becoming more attractive to some species and less attractive to others as time progresses. Although blow flies arrive very shortly after death, they are no longer attracted when the remains have passed a certain stage of decomposition, or become mummified or dry (Nuorteva 1977), although some species, such as *Calliphora vicina* Robineau-Desvoidy, preferred decomposed remains to fresh when given a choice (Erzinclioglu 1996).

In England, *C. vicina* and *Lucilia caesar* (L.) appeared on rodent remains within hours or minutes of death. *L. illustris* (Meigen) was not attracted to corpses in woodland until 76 h after death, and to corpses in open grassland until 48 h after death (Lane 1975). This seems strange, as *L. illustris* is one of the main flies involved in blow fly strike in sheep in Britain, colonizing living animals (Wall et al. 1992) and is found to cause myiasis in man (Bauch et al. 1984; Greenberg 1984; Khan and Khan 1987). It is also used in maggot debridement therapy (Sherman and Pechter 1988). Since this species is clearly attracted to living animals, it would be expected that it would also be more attracted to a fresh carcass than one several days after death. However, this delay was not seen in British Columbia, where it was found on pig carrion in open pasture within minutes of death, with eggs being laid within an hour (Anderson and VanLaerhoven 1996).

Irrespective of carcass size, *Phormia regina* (Meigen) is often reported to arrive later on remains than other blow flies, being attracted a day or two after death (Denno and Cothran 1976; Goddard and Lago 1985; Lord and Burger 1984). In Missouri, experiments showed that although a few adult *P. regina* were collected in the first 24 h, many more were collected on carcasses 24 to 28 h old, and significantly more again on carcasses 48 to 72 h old (Hall and Doisy 1993). In British Columbia, however, *P. regina* adults were collected from the remains immediately after death, although no eggs were laid until 2 days postmortem (Anderson and VanLaerhoven 1996). This indicates that in some cases adults are attracted to the remains immediately, perhaps to obtain the protein meal required for ovary and testes development (Erzinclioglu 1996). Apparently the remains were not considered an attractive oviposition media until a few days after death in these areas. However, this may be geographically variable, as in other studies, early colonization and oviposition by *P. regina* has been observed in Ontario, Canada (Rosati and VanLaerhoven 2006). *Cochliomyia macellaria* (F.) is also reported to be a later arrival and has been attracted to remains 18 to 48 h after death (Hall and Doisy 1993).

The size of the carcass also seems to effect its attractiveness to blow flies (Erzinclioglu 1996). For instance, *Calliphora vomitoria* (L.) has been reported to prefer larger carcasses (Davies 1990; Nuorteva 1959), while *Lucilia richardsi* Collin prefers to colonize small rodent carcasses (Nuorteva 1959).

Blow flies are diurnal species and usually rest at night. Therefore, eggs are not usually laid at night, and a body deposited at night may not attract flies until the following day. One study in an urban area found that some blow fly species oviposited in low numbers on rat carcasses placed near sodium vapor lamps (Greenberg 1990), but this is rare, and nocturnal oviposition has not been observed in large-scale studies in other areas (Anderson, unpublished data; Haskell et al. 1997). In a recent study in Texas, flies did not oviposit between 2100 to 0600 h Central Daylight Savings Time on a variety of baits, including fresh and aged beef liver, freshly killed rodents, freshly thawed pigs, or on pigs aged for 2 days, except in one case, in which a small number of eggs were laid on one bloated pig carcass in the first 20 min of the timeframe, but not subsequently (Baldridge et al. 2006). The sites tested were in a variety of urban and rural areas, and some were lit and others were not. The single recording of oviposition at the beginning of the experimental period occurred at a lit, rural site. The authors observed that, with this single exception, fly interest in the baits ceased less than an hour after sunset, and no activity was then noticed until after 0600 h the next day (Baldridge et al. 2006).

Although blow flies rarely lay eggs at night, they will often lay eggs in dark areas during daytime. These areas include under wrappings, inside closets, in dark basements, in containers, and in chimneys (Anderson, unpublished data; Erzinclioglu 1996). In fact, it has sometimes appeared that turning off the light in a lab situation can induce egg laying (Anderson, unpublished data). Therefore, it does not seem to be the darkness that inhibits oviposition, but rather the insect's diurnal rhythm. Although nocturnal oviposition is not normally observed in Hawai'i, it has occurred in rare situations when a victim has been dumped close to resting blow flies (Goff 1998). In Canada, two young bear cubs were shot and disemboweled for their gall bladders late one evening at a large garbage dump. Despite the season (summer) and the fact that large numbers of blow flies were in the vicinity (due to the presence of garbage and other recent bear kills), no eggs were laid on the remains until the following morning (Anderson 1999a). This difference may be geographical in nature or reflect differences in nocturnal temperatures between Hawai'i and Canada. Fly oviposition is strongly influenced by temperature (Erzinclioglu 1996) and generally does not occur below 10°C, unless the substrate itself is warmer than the ambient temperature, for instance, if it is in direct sunlight (Erzinclioglu 1996).

Geographical Differences in Succession

Insect colonization of carrion is dependent on many factors, but one of the most important is the geographic region or biogeoclimatic zone in which the remains are found. The biogeoclimatic zone defines the habitat, vegetation, soil type, and meteorological conditions of the area. This obviously has a major impact on the types and species of insects present, as well as their seasonal availability. It also affects the decomposition of the remains, which in turn impacts the insects that colonize them. Many families of carrion insects are relatively ubiquitous, but the individual species involved in decomposition vary from region to region. Decomposition itself is also quite different in various biogeoclimatic zones (MacGregor 1999a, 1999b).

The species involved in the sequential colonization of the remains and their times of arrival will vary from region to region. Invariably, certain groups will colonize first, such as blow flies (Family: Calliphoridae) and flesh flies (Family: Sarcophagidae), but the species involved will vary. In tropical regions such as Hawai'i, the first colonizers were found to be *Lucilia* (= *Phaenicia*) *cuprina* (Wiedemann), *Chrysomya megacephala* (F.), and *Chrysomya rufifacies* (Macquart) in the family Calliphoridae, and *Bercaea* (= *Sarcophaga*) *haemorrhoidalis* (Fallén), *Parasarcophaga ruficornis* (F.), *Sarcophaga occidua* (F.), and *Helicoba morionella* (Aldrich) in the family Sarcophagidae (Early and Goff 1986). In contrast, the first colonizers in a Tennessee study were *Lucilia* (= *Phaenicia*) *coeruleiviridis* (Macquart) and *Phormia regina* (both calliphorids) (Reed 1958), while in South Carolina the first colonizer was *Cochliomyia macellaria* (F.) (a calliphorid) (Payne 1965). However, any given state or province may contain several biogeoclimatic zones, and colonization can vary between zones. In the Coastal Western Hemlock region of British Columbia, frequent early colonizers include *L. illustris* and *P. regina* as well as *C. vomitoria* (Anderson and VanLaerhoven 1996; Dillon 1997). Further north, *Protophormia terraenovae* Robineau-Desvoidy was the more common species in both casework (Anderson 1995) and experimental studies (Dillon 1997; Dillon and Anderson 1996a).

Times of colonization of insect species and groups also vary greatly with geographical region. In many areas, dermestid beetles (Coleoptera: Dermestidae) are considered to be very late colonizers, frequently arriving when only skin and bones remain, sometimes months after death (Easton and Smith 1970; Fuller 1934; Payne and King 1970; Reed 1958; Rodriguez and Bass 1983; Smith 1986). In Hawai'i, however, some adult dermestid beetles were collected as early as 3 to 10 days after death (Early and Goff 1986; Hewadikaram and Goff 1991), although larvae were not collected until later (Hewadikaram and Goff 1991). In Coastal British Columbia, dermestid larvae were first collected from pig carrion in exposed pasture 21 days after death, during the early stages of advanced decay. The majority was collected more than 43 days after death (Anderson and VanLaerhoven 1996). In the northern region (Sub-boreal Spruce biogeoclimatic zone) and the interior region of British Columbia (Interior Douglas Fir biogeoclimatic zone), dermestid adults were first found on pig carcasses as early as the bloat stage, and larvae were found by the decay stage (Dillon 1997; Dillon and Anderson 1996a). During the summer in the interior region, dermestid larvae were found as early as the bloat stage, but this was rare (Dillon 1997; Dillon and Anderson 1996a).

The Piophilidae (skipper flies) are another family with species often found at very different postmortem intervals, depending on the region. In early reports from Europe, piophilid larvae were considered to be late-stage insects attracted to remains 3 to 6 months after death, arriving after Dermestidae (Leclercq 1969; Smith 1986). This was confirmed in Illinois when the majority of piophilid flies were collected during the decay stage, although no larvae were reported (Johnson 1975). Isolated reports such as that by Smith indicated that piophilid larvae could be collected as early as 30 days after death (Smith 1975), but these incidents were still considered anomalous. More recently, larvae have been collected from human cases in British Columbia as early as 26 days after death (Anderson 1995) and from pig carcasses in open pastureland 29 days after death (Anderson and VanLaerhoven 1996). Further large-scale studies confirmed these results (Dillon 1997; Dillon and Anderson 1995, 1996). In Hawai'i, piophilid flies have been collected from carcasses 33 to 36 days after death (Goff and Flynn 1991).

Even when a species is found to be present in many different regions, it is possible that there may be within-species differences. Byrne and colleagues (Byrne et al. 1995) noted biochemical differences among geographic populations of *P. regina*. This could be of value in determining whether remains have been moved from the original death site to a secondary site after death (Byrne et al. 1995).

Geographical region obviously has a major effect on arrival times of different species of insects. This means that data generated in one region or biogeoclimatic zone should not be used to determine time of death in a different region. Databases should be developed for every biogeoclimatic zone in which insects are being used to estimate time of colonization (TOC). Data are presently available for certain regions in North America, including Hawai'i (Goff et al. 1986; Goff and Flynn 1991; Tullis and Goff 1987), South Carolina (Payne 1965; Payne and Crossley 1966; Payne and King 1969, 1970, 1972; Payne et al. 1968a, b; Payne and Mason 1971), Virginia and West Virginia (Joy et al. 2002, 2006; Tabor et al. 2004, 2005), Louisiana (Watson and Carlton 2003, 2005), and Canada.

In British Columbia, areas covered include the Coastal Western Hemlock biogeoclimatic zone along the coast of British Columbia, including Vancouver island (Anderson and VanLaerhoven 1996; Dillon 1997; Dillon and Anderson 1995; Hobischak 1997; MacDonell and Anderson 1997; VanLaerhoven 1997; VanLaerhoven and Anderson 1996), the Sub-boreal Spruce biogeoclimatic zone, the northern region of British Columbia (Dillon 1997; Dillon and Anderson 1996a; VanLaerhoven 1997; VanLaerhoven and Anderson 1996), and the Interior Douglas Fir biogeoclimatic zone in the interior of British Columbia (Dillon 1997; Dillon and Anderson 1996a).

Across Canada, this author is working with colleagues and graduate students to develop databases for several provinces. Data on pig carcasses in several seasons, regions, habitats, and in some cases, over several years, are now complete for Alberta (Anderson et al. 2002; Hobischak et al. 2006), Saskatchewan (Sharanowski 2004), and Manitoba (Gill 2005a, 2005b). Work in Nova Scotia has examined succession on pig carcasses in urban and rural areas (LeBlanc 1998; Simpson 1999) and is ongoing. Extensive work is presently being conducted in several areas of Ontario (Rosati and VanLaerhoven 2006).

Although these databases should cover most of the major biogeoclimatic zones in Canada, most biogeoclimatic zones cover vast areas, and there are many variations within these zones. Even over short distances, there can be a great deal of difference in microclimate that can have an effect on decomposition and colonization (Cornelison 1999; Rosati and VanLaerhoven 2006). This means that further studies within each biogeoclimatic zone will also be warranted.

Studies are now being conducted in many areas of the world, for instance, Brazil (Carvalho et al. 2000, 2004; Carvalho and Linhares 2001; de Souza and Linhares 1997), Argentina (Centeno et al. 2002; Oliva 2001), Colombia (Martinez et al. 2006; Perez et al. 2005; Wolffe et al. 2001), India (Bharti and Singh 2003), Spain (Arnaldos et al. 2001, 2005; Martinez-Sanchez et al. 2000), France (Bourel et al. 1999), South Africa (Kelly 2006; Kelly et al. 2005a, 2005b, 2008, 2009), and Australia (Archer 2004; Archer and Elgar 2003a, 2003b).

In all such studies, the investigators have used repeated sampling of a single or several carcasses in order to develop baseline data. This may have included extensive or limited sampling as well as other such manipulations, such as lifting the carcass to assess biomass loss. It is always a concern whether such manipulations impact the insect colonization. In those studies using control carcasses, no such impact was observed (e.g., Anderson and VanLaerhoven 1996); however, a recent study designed specifically to access the impact of investigator disturbance on succession and faunal community composition also found that neither daily sampling nor weighing impacted the carrion community significantly (De Jong and Hoback 2006).

Effects of Season

Season has a major impact on weather, and the flora and fauna of a region. Thus, the faunal colonization of a body is also impacted. Many blow fly species vary in abundance, depending on season. For instance, in Mississippi *Lucilia* (= *Phaenicia*) *coeruleiviridis* and *Cochliomyia macellaria* were dominant in the warmer summer months, from April to September, whereas *Calliphora livida* Hall and *Cynomya* (= *Cynomyopsis*) *cadaverina* (Robineau-Desvoidy) dominated in the winter months from October to March, with *P. regina* being found throughout the year (Goddard and Lago 1985). In Maryland, *P. coeruleiviridis* and *P. regina* were found in both spring and summer, whereas *C. vicina*, *C. livida*, and *L. illustris* were found only in spring, and *L. sericata* was found only in summer (Introna et al. 1991).

In an exposed rural area in northeastern England, *C. vicina* dominated mouse carcasses in spring, but was replaced by *Lucilia caesar*, *L. illustris*, *L. silvarum* Meigen, *L. sericata*, and *L. richardsi* in summer. However, in a more urban garden, *C. vicina* dominated throughout the insect season (Davies 1999). The authors suggest that these differences may relate to carcass size and species availability, as *Lucilia* species are commonly found on sheep carcasses as well as being common in sheep strike or myiasis (Davies 1999).

In a study of 117 human deaths in Japan, distinct seasonal differences were seen in blow fly colonization, with only one species, *Calliphora vicina*, being found all year (Shroeder et al. 2003). In a study in Finland, considerable seasonal fluctuation and regional differences were seen between blow fly species (Nuorteva 1959). Work in Australia has also shown a dramatic effect of season and sun exposure on pig carcass decomposition (MacGregor 1999a, 1999b), and several species are considered seasonally distinctive in some areas (Archer and Elgar 2003b).

In South Africa, distinct seasonal differences in blow fly composition were noted, although Coleoptera diversity remained fairly constant (Kelly et al. in press; Kelly et al. 2008; Kelly 2006). In summer and fall, *Chrysomya marginalis* (Wiedemann) and *C. albiceps* (Wiedemann) were most prominent on pig carcasses, with *C. chloropyga* (Wiedemann) as well as *C. albiceps* found most commonly in spring. In these seasons, the same species were collected as adults as well as raised from maggot masses (Kelly 2006). Interestingly, in winter, the most common species collected were *Sarcophaga cruentata* Meigen, *C. chloropyga*, *Calliphora vomitoria*, and various *Lucilia* species, although adult trapping did not reflect the species reared from maggot masses (Kelly 2006).

In Alberta, no difference in decomposition stages or rate of decomposition was seen between pig carcasses placed in direct sunlight in comparison with those in shade, although species abundance was impacted (Hobischak et al. 2006). This was particularly true of *Protophormia terraenovae* as well as several Coleoptera species (Hobischak et al. 2006).

Whether a species is found in a specific season is strongly impacted by geography. For instance, *Phormia regina* is sometimes considered a cool weather species, but in British Columbia it was collected in spring, summer, and fall despite high summer temperatures (Anderson and VanLaerhoven 1996; Dillon 1997; Dillon and Anderson 1995, 1996a). However, blow fly colonization also appears to be a function of altitude more than season in some regions (Dillon 1997; Dillon and Anderson 1995).

In addition to the blow flies, many other carrion insects are impacted by season and have specific peaks of activity. For instance, *Necrophilus hydrophiloides* Guéurin-Méneville (Coleoptera: Agyrtidae) was collected 128 days after death on pig carcasses placed out during the summer in British Columbia (Anderson and VanLaerhoven 1996), but this elapsed time since death was reached in mid-October, which is the beginning of the species' period of activity, with peak activity reported between November and May (Anderson and Peck 1985). Later work in this region showed that *N. hydrophiloides* was characteristic of temperatures below 10°C (Dillon 1997). Therefore, time of colonization for some species may relate less to time since death, and more to season. Colonization of remains in water by aquatic organisms is also influenced by season and corresponds to the life cycle of the organism (Hobischak 1997).

The seasonality, or relative abundance, of certain insects and the potentially differing times of colonization of the remains in different seasons are important for several reasons. First, it means that carrion studies should be performed throughout the year in order to develop a valid database for an area. Second, it means that insects may be valuable in determining season of death. This may be useful when remains are discovered several years after death, although insects will probably be of little use in determining a precise time of death. Such seasonal dating has been used to determine time of death even in prehistoric studies. In Belgium, the presence of fossilized puparia from *Protophormia terraenovae* found with mammals from the Pleistocene era were used to determine season of death (Germonpre and LeClercq 1994), and in Canada, puparia from a variety of species, including *Protophormia terraenovae*, *Phormia regina*, and *Cynomya* (= *Cynomyopsis*) *cadaverina*, were used to indicate season of interment (Teskey and Turnbull 1979).

In more modern forensic investigations, a case involving skeletonized remains was discovered in a shallow grave more than 10 years after the disappearance of the victim. The total lack of entomological evidence indicated to the investigators that death had occurred in winter (Rodriquez et al. 1993). This supported the police investigation that indicated a winter death and refuted the defendant's claim that an associate had killed the victim in summer (Rodriquez et al. 1993). However, care must be taken when interpreting such evidence, as the time it takes for the natural decomposition of insect material (such as empty pupal cases and beetle exuviae) will be affected by many factors, such as amount of exposure, level of moisture, soil pH, etc. During the excavation of Arikara burials in South Dakota, empty pupal cases of blow flies and flesh flies were found in graves known to be 130 to 160 years old. This indicated that death had occurred between late March and mid-October, when such flies are active in that area (Gilbert and Bass 1967). The excellent preservation was attributed to the low annual rainfall in the region, and to generally dry conditions (Gilbert and Bass 1967).

Conversely, in a case from the Lower Mainland Region of British Columbia (a much wetter area), remains found after only 27 years had very few associated empty blow fly puparia. In this case, the decedent went for a walk in a park area of a temperate rainforest and disappeared during December. The decedent was found shallowly buried under the large root mass of a fallen tree. A large quantity of pill vials, previously containing therapeutic drugs, was found associated with the remains. It was believed that the decedent had wandered into the forest after taking the drugs, had become confused, and had taken refuge under the upturned roots of a fallen tree. No foul play was detected. The decedent had died, and over many years the body had slowly become buried by the soil falling from the roots. The remains were skeletonized but had not been heavily scavenged, indicating that insects had probably removed the flesh. As the decedent went missing in December, and probably laid exposed for months or years before natural burial, the remains would probably have been colonized by blow flies in early spring and most of the flesh was probably removed before the body was gradually buried. Larvae would have entered the surrounding soil or the clothing, pupated and emerged, leaving behind the empty pupal cases. Therefore, it would be expected that large numbers of such puparia would have been found at the scene. However, despite the fact that the scene was very carefully searched and the soil in and around the natural grave was screened, only four or five such puparia were found, and these were highly deteriorated. It is probable that the high moisture level in the soil contributed to the destruction of the insect evidence. Therefore, a lack of insect evidence in an old grave can only be interpreted to indicate a winter death if the conditions would normally be considered suitable for the preservation of puparia, or other insect artifacts.

Effects of Sun Exposure

The placement of the corpse has an effect on the decomposition and faunal colonization of the remains. The most obvious effect is that of sunlight and heat. Bodies found in direct sunlight will be warmer, heating up more rapidly and decomposing faster. They will lose biomass more rapidly than bodies in shade and progress through the decompositional stages faster (Dillon 1997; Dillon and Anderson 1995, 1996a; Reed 1958; Shean et al. 1993). Work in Australia has shown dramatic differences between decomposition rates of pig carcasses in sun versus shade (MacGregor 1999a, 1999b). Vertebrate scavengers also impact the decomposition and were found less frequently in sunny habitats (Dillon 1997; Dillon and Anderson 1995), therefore having less effect on remains in such locations.

In a comparison of pig carcasses placed in sun and shade in West Virginia, sunlit carcasses decomposed faster than shaded carcasses, with maggot mass temperatures being positively correlated with ambient temperatures in sunlit carcasses but not in shaded carcasses (Joy et al. 2006).

Blow flies exhibit habitat preferences within their regional distribution (Erzinclioglu 1996), but these may vary by region. In Britain, *C. vicina* and *L. illustris* were found most commonly in open conditions, and *C. vomitoria* and *L. ampullacea* Villeneuve seemed to require dense cover. *L. caesar* appeared to be an intermediate species, preferentially being found in scrub areas with sparse trees (MacLeod and Donnelly 1957). A later study in Britain revealed that *C. vicina* was found in both sun and shade (Lane 1975), and in France, *L. caesar* was found to prefer shady habitats (Holdaway 1930). In British Columbia, *L. illustris*, traditionally considered to be found only in direct sunlight (Smith 1986), was found on pig carcasses in both open pasture (Anderson and VanLaerhoven 1996) and dense forest (Dillon 1997; Dillon and Anderson 1995, 1996a). However, most of the work that refers to *L. illustris*, and other flies in the tribe Luciliini colonizing remains only in direct sunlight, originates from studies performed in northern Europe (Smith 1986). Therefore, it seems probable that habitat preferences may vary between Europe and Canada, explaining the behavioral differences observed. This is supported by observations in Germany (Haskell, personal communication) in which only flies in the tribe Calliphorini were attracted to bait when conditions were overcast, but flies in the tribe Luciliini were attracted as soon as the sun shone directly on the bait. This was in contrast to studies in the United States (Haskell et al. 1997).

Calliphora vomitoria is traditionally considered to be primarily a shade species (Smith 1986). In British Columbia, it was usually found on pig carcasses in dense forest, although during the fall it was collected from carcasses in direct sunlight in open regions of the forest (Dillon 1997; Dillon and Anderson 1995, 1996a). It was not collected in open pasture (Anderson and VanLaerhoven 1996). In the state of Washington, *Lucilia illustris* and *C. vomitoria* were collected from a carcass in direct sun and from a carcass in shade. However, more *C. vomitoria* were collected in the shade, and more *L. illustris* were collected in the sun (Shean et al. 1993). *Phormia regina* was collected in both scenarios (Shean et al. 1993). The arrival times of several species of beetles in the families Carabidae, Staphylinidae, Silphidae, and Histeridae varied with exposure, and members of some families such as the Dermestidae, Cleridae, and Nitidulidae were only found on the exposed pig in Washington State (Shean et al. 1993). However, only one carcass was examined in each habitat, so it is difficult to determine whether these are actual trends for this area. In Canada, where a large number of carcasses were studied in a similar habitat, some beetle species were regularly found on both sun-exposed and shaded carcasses. Others varied in their preferences or times of arrival, although this was also impacted by season and geographic region (Dillon 1997; Dillon and Anderson 1995, 1996a).

In direct sunlight during summer in British Columbia, a study using clothed and unclothed pig carcasses showed that the unclothed carcasses were heavily inundated by blow fly eggs, and they also mummified rapidly due to high temperatures. This resulted in the mass migration of undersized second- and third-instar calliphorid larvae in search of other food sources (Dillon 1997; Dillon and Anderson 1995). Such depletion of resources was not observed on shade carcasses (Dillon 1997; Dillon and Anderson 1995), or on clothed carcasses in either sun or shade (Dillon and Anderson 1996a). In Tennessee, Reed (1958) noted that insect populations were smaller at carcasses in pasture areas than in wooded, shaded areas, but this was not supported in other studies (Dillon 1997; Dillon and Anderson 1995, 1996a; Goddard and Lago 1985).

Urban versus Rural Scenarios

Some insect species are found in both urban and rural areas, yet others are very specific to one or the other, which indicates resource partitioning. The early-colonizing blow flies include rural, urban, and ubiquitous species. This can be useful in forensic analyses, as certain species of blow flies found on remains may be used to indicate that the remains have been moved from an urban to a rural environment or vice versa (Catts and Haskell 1990; Erzinclioglu 1989). However, caution must be exercised since only some species are found exclusively in one or the other habitats, while many species can be collected in both. In an analysis of casework from British Columbia, *Protophormia terraenovae* and *Calliphora vomitoria* were found almost exclusively in rural areas, whereas species such as *Lucilia sericata* were found exclusively in urban areas (Anderson 1995). Others, such as *Phormia regina*, *Eucalliphora latifrons* (Hough), and *Calliphora terraenovae* Macquart, were collected in both habitats (Anderson 1995). Pig carrion studied in forested regions of British Columbia attracted only rural and ubiquitous species such as *P. regina*, *P. terraenovae*, *L. illustris*, and *C. vomitoria*. It did not attract more urban species such as *C. vicina* and *L. sericata* (Dillon and Anderson 1995, 1996a, 1997). Pig carrion in open fields in rural southwestern British Columbia attracted only *P. regina* and *L. illustris* (Anderson and VanLaerhoven 1996). Species such as *C. vicina* and *L. sericata* are commonly considered urban species (Reiter 1984), but have been collected in rural regions (Anderson 1995; Haskell et al. 1997). Therefore, caution must be used in determining whether remains have been moved based on insect evidence.

Rural blow flies survive on natural animal carrion, whereas urban blow flies are primarily associated with human refuse, in the form of discarded food. Therefore, a further complication arises as many so-called rural areas are close to human habitation, with their associated human garbage, which encourages urban fly colonization, and may increase the chances of accidental transport of urban species to rural environments by humans. Parks in urban areas also provide settings for rural insects to be found close to human habitation. At Simon Fraser University in British Columbia, both rural and urban species have been collected (Anderson 2000). The university can be considered a large urban area due to its high human population and presence of human garbage, but is situated on a mountain surrounded by parkland, which in turn is surrounded by a large city. In New Zealand, *Lucilia* (= *Phaenicia*) *sericata* was commonly collected in urban parks and gardens, but was not collected in native bush remnants in urban regions (Dymock and Forgie 1993). Similarly, *Lucilia cuprina* (Wiedemann) was only rarely collected in garbage despite being a common rural species (Dymock and Forgie 1993). *Lucilia* (= *Phaenicia*) *eximia* Robineau-Desvoidy, *L. sericata*, *L. cuprina*, *Chrysomya putoria* (Wiedemann), *C. albiceps*, *C. megacephala*, *Musca domestica* L.,

Ophyra sp., *Fannia* sp., and several species of Sarcophagidae were associated with urban garbage in Goias, Brazil, although whether these species were considered to be solely urban was not reported (Ferreira and Lacerda 1993).

A small study in an urban backyard in Vienna, Austria, resulted in colonization by forty-two arthropod species (Grassberger and Frank 2004). The two carcasses were dominated by *Calliphora vomitoria* and *Chrysomya albiceps*. *C. vomitoria* is often considered to be a rural species, but outnumbered all other blow fly species in this urban study. As well, *C. albiceps* is considered to be a tropical and semitropical species, but was common in this more northern European study (Grassberger and Frank 2004). This again highlights the need for geographically specific studies.

In Britain, a series of habitats was sampled that ranged from the urban center of London through the suburban outskirts of the city to the rural region southwest of London (Hwang and Turner 2005). The authors sampled 127 times and caught over 3,000 Diptera from 20 families, with three-quarters being in the family Calliphoridae. Based on their collections, the authors identified three separate habitat types that were typified by different species: a rural woodland habitat, which the authors found to be dominated by *Calliphora vomitoria*, *Phaonia subventa* (Harris), *Neuroctena anilis* (Fallen), and *Tephrochlamys flavipes* (Zetterstedt); a rural grassland habitat typified by *Lucilia caesar*; and an urban habitat typified by *C. vicina*, *L. illustris*, and *L. sericata* (Hwang and Turner 2005).

Bodies Found Inside Buildings

The public and police alike often believe that insects will not colonize remains inside a building. However, insects will colonize remains indoors as easily as outdoors. The succession will be limited by the species that can and will enter a dwelling, and on how well the dwelling is sealed.

Blow flies are strong fliers that can follow an odor plume over a long distance (Erzinclioglu 1996) and can easily enter buildings. In British Columbia, an analysis of cases over a 5-year period showed that *Lucilia* (= *Phaenicia*) *sericata* and *Phormia regina* were commonly collected from victims found inside houses, while *Calliphora vicina* and *C. terraenovae* were sometimes collected indoors (Anderson 1995). *Protophormia terraenovae*, *Eucalliphora latifrons*, and *C. vomitoria* were not collected indoors (Anderson 1995). Other insect species such as *Piophilidae* sp. (Piophilidae), *Hydrotaea* sp. (Muscidae), *Thanatophilus lapponicus* (Herbst) (Coleoptera: Silphidae), and *Necrophilus hydrophiloides* (Coleoptera: Agyrtidae) were collected from human cases indoors, but not exclusively (Anderson 1995).

Insects will be found even in high-rise apartments. In Germany, numerous carrion-frequenting flies were collected inside a multilevel apartment building, including *Fannia cannicularis* (L.) (the most numerous), *L.* (= *Phaenicia*) *sericata*, *Sarcophaga carnaria* L., *C. vicina*, *Muscina stabulans* (L.), and *F. manicata* (Meigen) (Schumann 1990). In Gdansk, Poland, *F. cannicularis*, *M. stabulans*, and *L. sericata* were found in an eleven-story apartment building (Piatkowski 1991). In Germany, a body in a warm, dry apartment was almost completely skeletonized in less than 5 months by dermestid beetles (*Dermestes maculatus* DeGeer, Dermestidae) (Schroeder et al. 2002).

In Canada, *C. vicina* has been collected on the eighteenth floor of an apartment complex (Anderson, unpublished data). In all these cases, no remains were present to lure insects, so they were presumably attracted by the presence of normal household garbage. In Germany, the presence of *Parasarcophaga argyrostoma* (Robineau-Desvoidy) (Diptera: Sarcophagidae) is considered to be an indicator that remains have been outside at some time, as it is considered to be an exclusively outdoor species (Benecke 1998).

In comparing insects collected from thirty-five cases of decomposing remains in Hawai'i, in both indoor and outdoor situations, Goff (1991) found a greater variety of fly larvae associated with indoor deaths, and a greater variety of beetles associated with outdoor deaths. Some species were restricted to remains discovered indoors, while some were restricted to outdoor deaths. Certain taxa were considered to be sufficiently restricted to one environment in this region (i.e., Hawai'i) to serve as indicator species (Goff 1991). Therefore, in some cases, it may be possible to determine whether remains have decomposed *in situ*, or have been moved either from an indoor to an outdoor scenario or vice versa. In a case in England, remains were concluded to have decomposed completely inside a house due to the presence of *Leptocera caenosa* Rondani (Diptera: Sphaeroceridae), a species that the author reported to be common in human habitations but rarely collected out of doors (Erzinclioglu 1985). Care should be taken with such conclusions, however, as other members of this genus have been found in large numbers on remains buried outdoors in rural areas in both experimental studies (VanLaerhoven 1997; VanLaerhoven and Anderson 1996, 1999) and in actual cases (Anderson unpublished data). It should be noted that Smith (1975) collected this species from a dead fox in an outdoor environment. As well, a recent study of twenty-two legally interred and exhumed cadavers in northern France regularly reported the presence of *L. caenosa* (Bourel et al. 1999). Moreover, species that may be considered to be indicators of an indoor scenario in one region may not necessarily be restricted to an indoor scenario in another region. For example, *Stomoxys calcitrans* (L.) (Diptera: Muscidae) was considered to be sufficiently restricted to indoor situations in the Hawaiian Islands to serve as an indicator species (Goff 1991). However, this species is a major livestock pest frequently recovered in and around stables, and commonly found outdoors in other geographic regions (Kettle 1990).

Most data on bodies found indoors are naturally anecdotal, as researching such a scenario poses obvious difficulties. However, this author was recently presented with an excellent opportunity to compare insect colonization and decomposition between pig carcasses outside in a suburban area and pig carcasses inside a residential house (Anderson 2005a).

Three freshly killed pig carcasses were placed inside three different rooms in a house in a residential area of Edmonton, Alberta, and a further three freshly killed pigs were placed outside in a grassed area close to the residence (Figure 5.1a and b) (Anderson 2005a). The pigs were killed by pin gun and were not exsanguinated. The average weight was 42 kg. Each carcass was clothed identically in panties and shirt, and the experiment was begun in late May. The indoor carcasses were placed in the living room, bedroom, and bathroom, respectively. Two windows in the house were open but screened, with a small hole in the mesh of one screen. Carcasses were examined and sampled regularly for 6 weeks. Dataloggers placed in the house and outdoor scene recorded temperature every hour. Outside temperatures predictably fluctuated much more than inside, but overall, the temperatures were very similar, with an indoor mean temperature for the duration of the experiment of 17.8°C and outside of 16.4°C (Anderson 2005a).

Outside, flies were attracted to the remains within minutes of placement, and eggs were laid immediately. Indoors, no flies or eggs were found on the remains until 5 days

(a)

(b)

Figure 5.1 Single-story residential house site (a) and outdoor site (b) for comparison of indoor and outdoor decomposition. (Photos courtesy of Dr. Gail Anderson.)

postmortem, but by 7 days, large numbers of flies had penetrated the house. Decomposition proceeded more rapidly outdoors. Figure 5.2a–d shows a comparison of decomposition indoors and outdoors. Blow fly diversity was similar in both habitats, with *Lucilia sericata*, *Cynomya cadaverina*, *Calliphora vicina*, *Phormia regina*, and *Protophormia terraenovae* being found both indoors and outside. Only *Eucalliphora latifrons* was found solely outside. Indoors, the only colonizers were Calliphoridae, whereas outdoors, other insect families colonies such as Phoridae (Diptera) and Staphylinidae, Silphidae, Histeridae, and

(a)

(b)

Figure 5.2 Comparison of decomposition rate and insect colonization of carcasses inside a residence and outside. (a) Indoor carcass 17 days postmortem. (b) Outside carcass 17 days postmortem. (c) Indoor carcass 32 days postmortem. (d) Outside carcass 32 days postmortem. (Photos courtesy of Dr. Gail Anderson.)

Continued

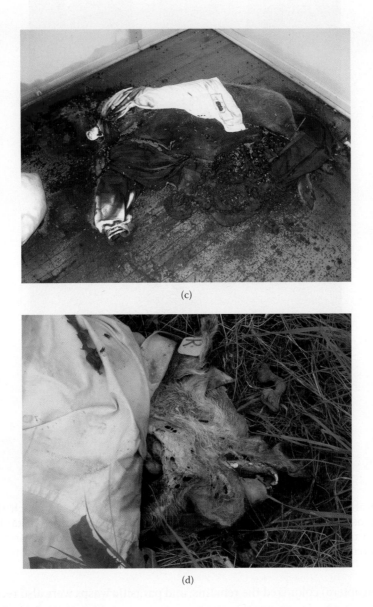

(c)

(d)

Figure 5.2 *Continued.*

(a)

(b)

Figure 5.3 Calliphoridae pupae and puparia in basement of house in which three pig carcasses decomposed. All carcasses were on main floor, but migrating third-instar larvae moved throughout all rooms of the house and into the ducts to the basement (a). In the basement, larvae attempted to get underneath loose pieces of cement (b). (Photos courtesy of Dr. Gail Anderson.)

Cleridae (Coleoptera) colonized the remains, and parasitic wasps were also recovered from the Diptera pupae (Anderson 2005a).

Very large numbers of puparia were found throughout the house. Some, such as those of *P. terraenovae*, were found close to the remains. *P. terraenovae* is a species that is commonly reported to pupate on or close to the remains. Others were found many meters away throughout the rooms on the main floor and in the basement (Figure 5.3a and b), suggesting they had migrated through heating ducts. One carcass was placed in the bathtub. Migrating larvae attempted to leave the confines of the tub but were unable to climb the smooth surfaces. Eventually, they pupated in the tub, almost burying the carcass in pupae and puparia (Figure 5.4). As the outdoor carcasses were colonized faster and decomposed

Figure 5.4 Pig carcass in bathtub 42 days postmortem. Migrating third-instar Calliphoridae of several species were unable to climb the smooth walls of the tub and pupated on the body, almost covering it. (Photo courtesy of Dr. Gail Anderson.)

more rapidly, it is probable that similar numbers of puparia were present around the outdoor carcasses. However, due to the presence of soil and grass, they were very hard to find. The numbers found inside the house indicate the vast numbers of puparia that are probably present but are missed at an outdoor scene, and suggest possible search patterns (Anderson 2005a).

Effects of Burial

Disposal of the remains is often of paramount concern to a killer, but a human body is a surprisingly difficult object of which to dispose, and a commonly chosen method is burial. Bodies buried feloniously, however, are rarely deeply buried, as burying a full-sized human body at the traditional 2 m or 6 ft depth requires a great deal of work and time. The more time a criminal spends with the victim, the greater the chance that evidence will be transferred, and also that the killer will be found with the remains. Therefore, a hasty, shallow grave is usually all that is dug.

Buried remains are still colonized by insects, but burial influences the time required for insects to reach the remains, the sequence of colonization, the species involved, and the rate of decomposition (Payne et al. 1968a; Rodriguez and Bass 1985; Rodriguez 1997; Smith 1986; VanLaerhoven 1997; VanLaerhoven and Anderson 1996, 1999). The above are also affected by geographical region, soil type, whether the remains are disturbed after death, and the elapsed time since death before burial (Rodriguez 1997; VanLaerhoven 1997; VanLaerhoven and Anderson 1996, 1999), as well as the depth of burial (Mann et al. 1990; Rodriguez and Bass 1985).

Many studies have looked at insect colonization of buried bodies, although much of the early work centered on human exhumations, rather than empirical studies (e.g., Gilbert and Bass 1967; Motter 1898; Schmitz 1928; Stafford 1971). Research on buried baby pig carcasses was performed in South Carolina (Payne et al. 1968a), and results indicated that several insect species were confined to buried pigs, and that decomposition was greatly slowed by burial. However, in this study, the carcasses were placed in small coffins, which may mimic a legal burial somewhat, but is unlikely to be representative of an illicit burial. In Tennessee, human burial experiments were performed using six cadavers buried at different times of the year in an effort to study decomposition and insect colonization (Rodriguez and Bass 1985). This work is noteworthy, as it was the first experimental work of its kind to utilize human cadavers. However, due to the limitations imposed by using human cadavers, relatively few were studied and no replication was possible. In addition, the cadavers were often received in varying conditions.

An extensive burial project using pig carcasses was conducted in British Columbia in 1995–1996 (VanLaerhoven 1997; VanLaerhoven and Anderson 1996, 1999). In these studies, carcasses were buried during the summer in the forested areas of two biogeoclimatic zones. The Coastal Western Hemlock Zone and the Sub-boreal Spruce Zone (characteristic of the Vancouver and Coastal region, and the Cariboo regions, respectively) were selected. A large number of carcasses were buried, with three carcasses exhumed at various time intervals over a 16-month period. Studies were also performed on carcasses buried 48 h after death, and on carcasses that had been disturbed after burial (VanLaerhoven 1997). The results indicated that, in these areas, muscid flies were much more common on buried remains than on aboveground carcasses. Although some calliphorid flies did colonize the remains, maggot masses did not form, and carcass temperatures remained very similar to ambient soil temperature (VanLaerhoven and Anderson 1999). This is in contrast to results obtained from work in Tennessee on human cadavers in which temperature increases were recorded in the cadavers (Rodriguez and Bass 1985). The difference may be due to geographic region and the resultant soil type, as soil type has a major influence on insect colonization (VanLaerhoven 1997; VanLaerhoven and Anderson 1996).

Although a predictable sequence of insect colonization was clearly identifiable, it was quite different from that of exposed carcasses used in the control group of this and other studies performed in these regions (Anderson and VanLaerhoven 1996; Dillon 1997; Dillon and Anderson 1995, 1996a). Some species were much more commonly found on buried bodies, and others more commonly on exposed bodies. Also, species that colonized both exposed and buried bodies frequently colonized them at different elapsed times since death (VanLaerhoven 1997; VanLaerhoven and Anderson 1996, 1999). Most studies noted that decomposition was greatly slowed by even shallow burial (Payne et al. 1968a; Rodriguez and Bass 1985; Rodriguez 1997; VanLaerhoven and Anderson 1999). Figure 5.5a–f represents a comparison of decomposition rates over time in pig carcasses aboveground and those shallowly buried (≈30 cm of soil above carcass) for the Coastal Western Hemlock biogeoclimatic zone of British Columbia. VanLaerhoven and Anderson (1999) provide a detailed description of the research.

Insect colonization of remains was also delayed by burial by 2 or more weeks in some cases (VanLaerhoven 1997; VanLaerhoven and Anderson 1996, 1999). However, when carcasses were left exposed for 48 h prior to burial (a common homicide scenario), the carcasses were colonized by calliphorid flies. These remains were presumably being colonized prior to burial, as maggots were present on the carcasses exhumed 2 weeks postmortem after 48 h exposure, but not on carcasses that had been buried immediately after death (VanLaerhoven 1997; VanLaerhoven and Anderson 1996).

Depth of burial also has an impact on decomposition and insect colonization (Mann et al. 1990; Rodriguez and Bass 1985), but has not been extensively studied. Some reports have indicated that carrion beetles will not orient to carrion more than a few centimeters

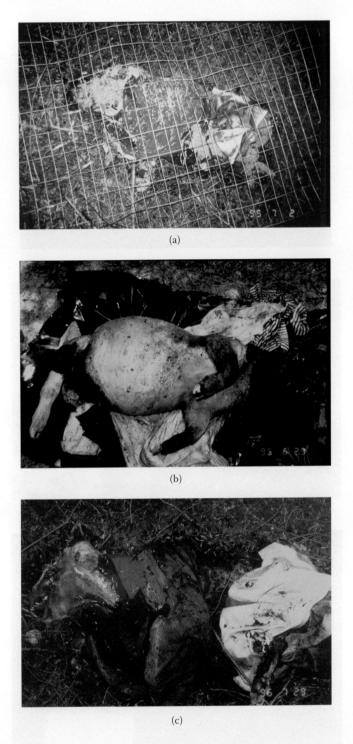

(a)

(b)

(c)

Figure 5.5 Clothed pig carcasses decomposing aboveground compared with those shallowly buried. (a) Two weeks postmortem aboveground. (b) Two weeks postmortem, shallow burial (~30 cm soil above body). (c) Six weeks postmortem aboveground. (d) Six weeks postmortem, shallow burial. (e) Three months postmortem aboveground. (f) Three months postmortem, shallow burial. (Photos courtesy of Dr. Sherah VanLaerhoven. Used with permission.)

Continued

(d)

(e)

(f)

Figure 5.5 *Continued.*

below the surface (Shubeck 1985; Shubeck and Blank 1982). However, both adult and larval carrion beetles were found on carcasses buried approximately 30 cm deep in soil in British Columbia (VanLaerhoven 1997; VanLaerhoven and Anderson 1996, 1999).

Season of interment will also impact insect colonization. In Britain, experiments were conducted to simulate a homicide case in which a victim was strangled and placed in a shallow grave in December (Turner and Wiltshire 1999). Pig carcasses were interred in similar graves close to the original site, in December. No insects colonized the remains until scavengers exposed the carcasses in April, at which point they were colonized by *Calliphora vomitoria*. Low soil temperatures had preserved the carcasses so that they were still attractive to blow fly colonization 4 months postmortem (Turner and Wiltshire 1999).

Geographic region, habitat, and season all play a major role in insect succession on exposed carrion. Further investigation is required to study the effects of these parameters on buried carrion.

Most cases of burial refer to illicit burial, in which the victim is usually buried hurriedly and shallowly in an effort to rapidly dispose of the corpse. However, even in traditional, legal burials, certain species of insects have been reported to colonize a body buried deeply and housed in a coffin. Such species have undergone several generations on the remains (Colyer 1954a,b; Stafford 1971). *Conicera tibialis* Schmitz (Diptera: Phoridae) was the most common species reported to inhabit coffins, and this species is thought to burrow down to the corpse as an adult in order to lay eggs closer to the corpse (Colyer 1954a,b). Adult flies have been observed to burrow upwards through the soil after adult eclosion (Colyer 1954a,b). These reports are more than 50 years old, and so date from a time when it might appear that older-style coffins and the lack of a vault would make it easier for such insects to penetrate. However, a recent study from Lille, in northern France, suggests otherwise (Bourel et al. 2004). Twenty-two exhumations were performed on legal interments from 1996 to 2002. Burial custom varied, but all cadavers were clothed and placed in a wooden coffin at a depth of 60 cm. Some were also placed in vaults, and 30% had received postmortem treatment with formalin. Death had occurred between 2 and 29 months prior to exhumation (Bourel et al. 2004). All cadavers had been colonized by insects. The most common species collected were *Ophyra* (=*Hydrotaea*) *capensis* (Wiedmann), *Conicera tibialis*, and *Leptocera caenosa*. Surprisingly, the authors noted no difference between cadavers that had received postmortem treatment and those that had not. As well, although not specifically identified, some of the cadavers were interred in coffins that were then placed inside vaults, and yet all cadavers were colonized (Bourel et al. 2004), showing that vaults do not prevent colonization.

Bodies in Water

Bodies are frequently found in water, whether as a result of a recreational death or an illegal disposal after homicide. When remains are found in water, faunal succession will be very different from that seen on land. This will be impacted by many factors, including the body of water (e.g., lake, stream, ditch, ocean), temperature of water, season, presence/absence and type of clothing, scavenging, and biogeoclimatic zone (Hobischak 1997; MacDonell and Anderson 1997). In some cases, when the remains are only partially submerged, both terrestrial and aquatic fauna may colonize.

A study in the Coastal Western Hemlock region of the Lower Mainland of British Columbia, Canada, compared decomposition and arthropod colonization in lentic and lotic systems (Hobischak 1997; Hobischak and Anderson 2002; MacDonell and Anderson 1997), and later compared these data with human death cases (Hobischak and Anderson 1999; Petrik et al. 2004). Four freshly killed, partially clothed, 23 kg pig carcasses were placed in a series of small ponds, and a similar four carcasses were placed in running streams. The carcasses were examined and sampled regularly over a 1-year period. The carcasses were protected from scavenging by large carnivores by being placed in metal cages with bars approximately 10 cm apart, which still allowed small scavengers such as mink to access the carcasses. The cages allowed the carcasses to sink or float naturally, but prevented them from floating away. The carcasses were observed to go through the usual decompositional stages seen on land, including fresh, bloat, active decay, post decay, and remains stages, although post decay was represented by adipocere formation and the remains stage was completely submerged. In the pond habitat, most of the carcasses did not reach the remains stage, and one remained in the decay stage (Hobischak 1997; Hobischak and Anderson 2002; MacDonell and Anderson 1997).

Both terrestrial and aquatic species colonized the remains, depending on the level of flotation. Some species were carrion dependent, but many were only associated with the carrion or simply using it as a convenient substrate. Once adipocere tissue formed, it resulted in a hard shell around the carcass and prevented further decomposition and colonization, and also resulted in extended flotation (Hobischak 1997; Hobischak and Anderson 2002; MacDonell and Anderson 1997). Blow fly larvae were found in both lotic and lentic systems, although no maggot masses formed. Staphylinidae were also found in both habitats. Chironomidae larvae were found on the carcasses in the streams but not ponds, and large numbers of Trichoptera larvae were found on the pond carcasses, as well as Gerridae and Leptodiridae. In both habitats, clothing provided shelter and reduced scavenging (Hobischak 1997; Hobischak and Anderson 2002; MacDonell and Anderson 1997).

In comparisons with human coroners' cases, the pig carcasses went through similar decompositional changes, including marbling, loosening of hair, skin flaking, nail (hoof) shedding, adipocere formation, and exhibited similar scavenging patterns (Hobischak and Anderson 1999). One of the greatest problems noted was the lack of standardization of descriptions of human decomposition in bodies recovered from water, and a complete lack of entomological evidence either recognized or recovered (Hobischak and Anderson 1999).

In an effort to improve this, a study was undertaken to document human decomposition in bodies recovered from a variety of freshwater environments (Petrik et al. 2004). The human cases consisted of drowning deaths in which the bodies had been recovered by the Canadian Amphibious Search Team (CAST). All recoveries were recorded by video and still images. Decompositional states, water characteristics, anthropophagy, and faunal colonization were recorded. In all cases it was found that time of death had been determined by date last seen alive rather than by decompositional or faunal characteristics (Petrik et al. 2004). This determination may be valid in an accidental drowning, but is unlikely to be accurate in a homicide.

In further studies, the effect of submergence on pig carcasses in the marine environment was investigated in the Howe Sound Region, near Vancouver, British Columbia (Anderson and Hobischak 2002, 2004). Twelve freshly killed pig carcasses were placed at two depths: shallow, 7.6 m (25 ft); and deep, 15.2 m (50 ft). The research area was off the coast close to a small island. Each carcass was tied with rope and attached to a heavy anchor made of concrete blocks. The rope allowed the carcass to float or sink freely, but moored it close to its placement site. The carcasses were separated by a minimum of 150 m. Six carcasses were placed in spring (May) and six in fall (October), three at each depth. The carcasses were observed, videotaped, and sampled by divers as often as possible, although sampling times were restricted by diver and boat availability (Anderson and Hobischak 2002, 2004).

All the carcasses floated when first placed, but rapidly sank to the substrate in a few hours. Recognizable decompositional stages of fresh, bloat, active decay, and remains were seen. In some cases, gases remained in the stomach and held the body up as if bloated for far longer than the actual bloat stage. A large variety of arthropods were attracted to the remains, some to feed, others for shelter, and many to feed on those already at the carcass. Bacteria and algae buildup on the carcasses also was attractive. Large fauna, such as the sunflower seastar, *Pycnopodia helinthodes* (Brandt), completely covered the remains at some stages and left artifacts that could be mistaken for wounds (Anderson and Hobischak 2002, 2004). Significantly more fauna were found on carcasses at 15 m than at 7 m, and some differences in species were noted. Season had a strong influence on both diversity and quantity of fauna colonizing the carcasses, with considerably greater numbers of species found in the spring trials (Anderson and Hobischak 2002, 2004).

Although no deliberate attempt was made to place carcasses on different substrates, the sea floor in this area is extremely heterogeneous, so when carcasses sank after bloating, some fell onto rock and others onto sand. This had a tremendous influence on decomposition and fauna, as those that fell on rock were mostly colonized by species that swim, whereas those that fell on sand were also colonized by those that live in or on sand, and thus were skeletonized much faster (Anderson and Hobischak 2002, 2004).

Some carcasses remained floating a meter or two above the substrate; the height they rose to was constrained by the rope. This was caused by gases remaining in areas such as the stomach and keeping the carcass up beyond the bloat stage. This considerably slowed decomposition, as it only allowed access to fauna that swam. On land, wound areas and natural orifices are normally colonized first, but although wounds were attractive, non-wound areas were also rapidly colonized (Anderson and Hobischak 2002, 2004).

One of the major limitations of this work was accessing the carcasses, as this required divers and boats, which were not always available. Therefore, this author is presently conducting ongoing work with the Victoria Experimental Network under the Sea (VENUS) (Anderson 2008). VENUS is a large undersea laboratory in the Saanich Inlet near Vancouver Island, British Columbia. It involves a network of cameras, instrument nodes, and sensors on the ocean floor that allows researchers real-time remote access to their experiments on the sea bed (www.venus.uvic.ca). VENUS includes a power and communications node connected to an instrument platform and a camera platform. The Venus Instrument Platform (VIP) houses various instruments, such as an oxygen optode, measuring temperature and oxygen levels; the Falmouth and Seabird CTDs, measuring conductivity, pressure, salinity, and temperature at 1 and 60 s intervals, respectively; a transmissometer, which measures light transmission (which gives clarity of water); and a gas tension device, which measures dissolved gas pressure. The VIP and camera are all connected via fiber optics to the node that supplies power.

The Remote Operated Platform for Ocean Science (ROPOS), a remote operated submarine (Figure 5.6), was used to place the camera and tripod at a depth of 94 m on a silty substrate. The digital camera is controlled by remote computer and records both still and video images, and so can be controlled and accessed via the Internet, anywhere in the world (Figure 5.7). A single 26 kg weighted pig carcass was attached to a transponder and deployed from a Canadian Coast Guard vessel in August. Once it had sunk to the

Figure 5.6 The Remote Operated Platform for Ocean Science (ROPOS), used to place equipment, the camera tripod, and the carcass in the Saanich Inlet, British Columbia, for underwater study. (Photo courtesy of Dr. Gail Anderson.)

Figure 5.7 Camera, housing, and lighting array used to observe underwater carcass in Saanich Inlet in VENUS project. Camera and tripod deployed at a depth of 94 m in Saanich Inlet by ROPOS (Figure 5.6). (Photo courtesy of Dr. Gail Anderson.)

seabed it was located via the transponder by ROPOS and positioned at the camera site. Once deployed and positioned, the only access to the carcass was remotely through the camera and chemical sensors. Although technically the VENUS equipment would allow camera recording 24 h a day, this is not feasible, as the ocean bed at 94 m is completely dark, so a large array of lighting was required to take images. Therefore, the lights and cameras were only turned on for brief periods of time every day. On average, the carcass was observed two to four times a day. The carcass was immediately attractive to a variety of crustacea. Figure 5.8a–f shows the carcass from day 1, one day after submergence, to day 18. The primary site of arthropod feeding was at the rear end of the pig. Although there

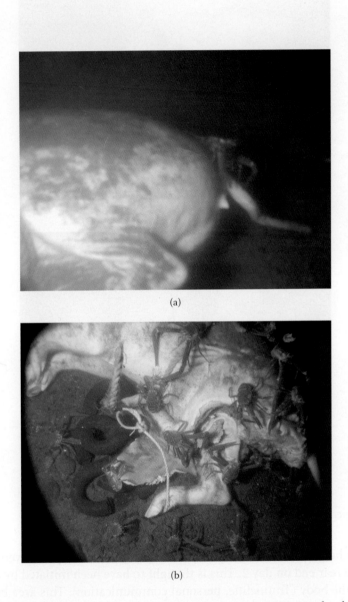

(a)

(b)

Figure 5.8 Decomposition and arthropod feeding on a pig carcass at a depth of 94 m in the Saanich Inlet, British Columbia. (a) Day 1, 24 h after placement, squat lobster at tail. (b) Day 2, considerable tissue removal at rear end of pig, probably initiated by shark bite. All activity concentrated in this region. (c) Day 5, further tissue removal, parts of viscera removed. (d) Day 9, spinal column picked clean and most of viscera removed. (e) Day 14, most of hind end skeletonized. (f) Day 18, hind end skeletonized and disarticulated. Note how carcass has been moved. (VENUS Project, University of Victoria. Photos used with permission.)   *Continued*

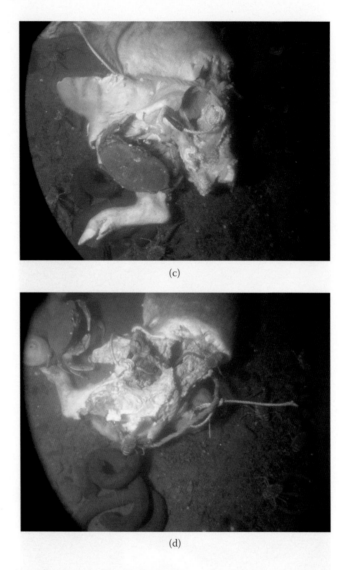

(c)

(d)

Figure 5.8 *Continued.*

was interest in both the head and rear end of the pig at first, rapid scavenging suddenly occurred at the rear end on day 2. This is thought to have been initiated by a shark bite in this region of the body (Tunnicliffe, personal communication). This area became rapidly consumed and released the carcass from the ropes tethering it under the camera. By day 22 the upper part of the pig that had not been consumed was dragged out of camera range. This may have been due to crustacean action alone, as video footage illustrated that crabs and small lobsters could rock and move the carcass. However, seals in the area were seen to push and "play" with the carcass and dogfish sharks and, although not usually interested in human or pig flesh as food (Tunnicliffe, personal communication), may have pushed or bitten at it and helped to move it. The carcass's position moved dramatically day to day, sometimes completely rotating from head to hoof. No clear stages of decomposition were seen, and the carcass did not enter bloat, although this is to be expected at such a great depth due to pressure and temperature (Teather 1994). At the last sighting of the carcass,

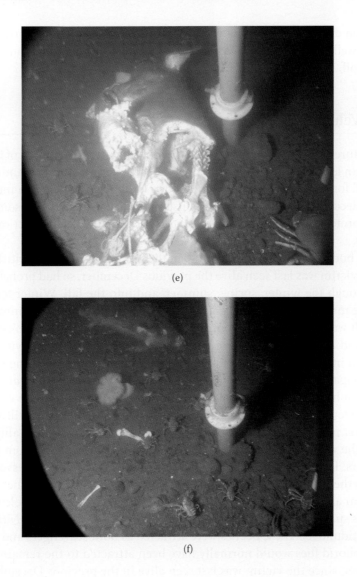

(e)

(f)

Figure 5.8 *Continued.*

tissue in the head and shoulders was still mostly intact, whereas the rest of the carcass was mostly skeletonized and eviscerated. Tissue had been removed over the snout and in the orbit region. This work is still under analysis and is ongoing. The drawback of this work is that only one carcass can be observed at any one time. However, the obvious advantage of real-time underwater observation at such frequent intervals makes this an invaluable tool in examining underwater decomposition (Anderson 2008).

In an intertidal study in Hawai'i, pig carcasses were placed in two intertidal regions and decomposition was compared between these carcasses and adjacent carcasses in terrestrial habitats (Davis and Goff 2000). At both sites, the carcasses in the intertidal regions decomposed more slowly than the terrestrial carcasses, with most of the biomass loss and decomposition being attributed to wave action and bacteria rather than faunal activity. At one site, no permanent insect colonization was seen, and at the second, Diptera larvae did remove tissue until the carcass was no longer available above the waterline.

Decomposition did progress through the usual stages. The authors reported the presence of marine as well as terrestrial scavengers, but these had little impact on decomposition (Davis and Goff 2000).

Bodies in Vehicles

Due to the nature of the crime, homicide victims are sometimes disposed of in unorthodox places. This can lead to a restricted or changed succession pattern. Cars and other vehicles are often used for suicide, or for the disposal of a body. They provide an interesting environment for decomposition, as the vehicle itself may act as a barrier to some species, but will act as a protectant from rain and predators, and also have an effect on temperature and humidity.

A human body was found in a car trunk in the Lower Mainland of British Columbia in October. The victim was last seen alive the previous December, so had probably been in the car trunk through the winter, spring, summer, and into the fall. Winter temperatures in this region are mild, rarely going much below 0°C. The remains were mummified and relatively intact. It was apparent that a number of different species had colonized the remains over time. Dead adult blow flies were found in the car interior, along with a quantity of empty puparia in the car trunk and carpeting, as well as in the clothing of the cadaver. Calliphorid puparia were found along with the larvae of *Lucilia sericata* and *Phormia regina*. When a body is exposed, the largest numbers of insects are the early colonizers, primarily the blow flies. The species that are considered later colonizers will also arrive, but usually in lower numbers than the blow flies. In this case, although the Calliphoridae were represented, the vast majority of insects and insect remains collected from the body were later colonizers. Larval sarcophagid (*Liopygia argyrostoma* Robineau-Desvoidy) flies were collected, together with puparia and larvae of two species of the skipper flies, *Stearibia nigriceps* (Meigen) and *Piophila casei* (L.). Also collected were species of Coleoptera, including *Dermestes* sp. (Dermestidae), *Necrobia ruficollis* (F.) (Cleridae), and *Nitidula carnaria* (Schaller) (Nitidulidae). These beetles were attracted as both scavengers and predators.

The calliphorid flies would normally have been attracted to the remains shortly after death. However, since the victim was last seen alive in the previous December and disappeared mysteriously at that time, it is probable that death took place in the winter. Therefore, it is probable that insects did not colonize the remains until the following spring (probably March), when temperatures in this region become consistently warm enough for insect colonization. If the body had been outside in this region from December until spring, it would not have mummified, but the dry interior of the car trunk allowed mummification to occur. This would have made the remains less attractive to blow flies, but still attractive to later colonizers.

The fact that a large range of species, including both Diptera and Coleoptera, were found on the remains indicates that the car itself did not provide much of an obstacle to the entrance of the insects. In fact, there are many entrances into most vehicles, including drainage holes in the trunk, and rusted-out areas in older vehicles. This vehicle remained parked on a city street for many months, until the landlord insisted that it be moved in late summer. The landlord objected to the smell of the "stuff" that was appearing under the drainage holes in the trunk, and to the large numbers of flies that appeared to be attracted. The defendant, who had borrowed the vehicle from a friend, apparently sprayed it with

Figure 5.9 Hatchback car parked into bushes on city street. Body was in trunk under a blanket. (Photograph courtesy of Vancouver City Police, Homicide Squad. Used with permission.)

a commercially available fly spray, and then drove the car around the corner, parking it within meters of its original place. This apparently satisfied the landlord. Some time later, however, a police officer recognized the odor emanating from the vehicle and the killer was arrested (still living in the same apartment). He was convicted of murder.

It might be supposed that a vehicle would prove somewhat of a barrier to insects, and that the reduced number of calliphorid flies collected in the above case may have been due to a reluctance to enter enclosed spaces. However, in a more recent case under very similar circumstances, it was shown that large numbers of blow flies could easily enter a vehicle. Human remains were discovered in the summer in the trunk of a small hatchback car parked on a city street in British Columbia (Figure 5.9). A blanket loosely covered the remains. The car was older with rusted areas, and one front window was open about 3 cm. Blow flies had colonized the remains, and larval and pupal *Phormia regina* and *Protophormia terraenovae* were collected along with a few empty puparia of both species (Figure 5.10). In addition, a few larval *Calliphora vomitoria* were collected inside the vehicle. This was a rare case in which *P. terraenovae* and *C. vomitoria* were found in an urban environment, possibly indicating that the vehicle had been moved. The numbers of blow flies on the remains were considerably less than those normally expected on remains exposed outside. This indicates that the vehicle did provide somewhat of a barrier, although a number of adults had obviously entered the vehicle for oviposition. Some pupae even showed evidence of parasitism, which indicated that pupal parasites had also been attracted into the vehicle.

Under the drainage holes in the trunk, a quantity of body fluids had leached out, and a secondary colonization had occurred. Attracted by the fluids were *P. regina*, *P. terraenovae*, and *C. vomitoria*. Also collected at this secondary site were *Eucalliphora latifrons*, and *Hydrotaea* sp. (Muscidae). Beetles had not yet entered the vehicle, but were also attracted to this drainage area. These included *Omosita colon* (L.) (Nitidulidae) and *Aleochara curtula* Goeze (Staphylinidae) as well as tenebrionid larvae.

The fact that the remains were inside an almost sealed vehicle in summer would have a strong impact on the temperatures to which the remains were exposed, which in turn would have an impact on the insect's development. Therefore, after the crime scene vehicle had been removed, a vehicle of similar design and color was placed in the same spot for 10 days. A SmartReader 1® datalogger was placed in the trunk under a blanket, and recorded

Figure 5.10 Calliphoridae pupae and puparia were found on the clothing and in the trunk of the car depicted in Figure 5.9. (Photograph courtesy of Vancouver City Police, Homicide Squad. Used with permission.)

the vehicle temperature every half-hour. The temperature data showed that the daytime temperatures inside the car were much higher than the ambient temperatures; however, nighttime temperatures both inside the car and out were very similar (Figure 5.11). Certified weather records from the nearest weather station were collected from Environment Canada, and these data were compared using a regression analysis with those from the datalogger for the same time period (i.e., the 10 days after the discovery of the remains). A regression analysis is a mathematical way of using two known sets of data (such as the known temperatures in the car recorded by the datalogger, and the known temperatures for the same time period from a weather station), and determining the relationship between the two. If there is a good correlation between the two sets of data (traditionally R^2 above 0.5, or 50%),

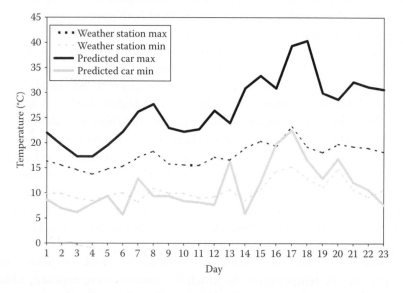

Figure 5.11 Minimum and maximum temperatures recorded inside vehicle over a 9-day period compared with temperatures recorded at Environment Canada weather station for same time period. (Figure courtesy of Dr. Gail Anderson.)

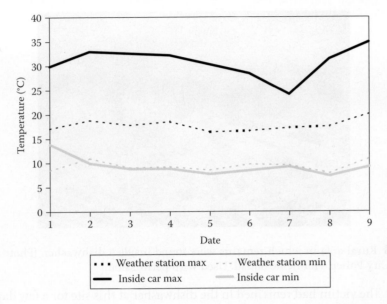

Figure 5.12 Actual temperatures recorded at Environment Canada weather station, and car temperatures predicted with regression analysis. (Figure courtesy of Dr. Gail Anderson.)

then the relationship can be used to predict the temperature at the death scene for any given day, using the known temperature at the weather station for that day. Utilization of the method allows the scene temperature to be predicted. The regression analysis showed a good correlation between the two sets of data, with an R^2 of 0.74. The regression equation was then used to predict the temperature inside the vehicle (Figure 5.12). The entomological evidence indicated that oviposition had begun 15 to 17 days before discovery. However, since the victim was inside the car, it is probable that oviposition was delayed (possibly for a few days) until the odor from the remains attracted insects from outside. This case remains unsolved at this time, and, thus, the period of delay involved cannot be verified as yet.

Bodies in Enclosed Spaces

A human body was found inside a domestic dishwasher that had been dumped down the side of a cliff in a rural area of British Columbia (Figure 5.13). The dishwasher had obviously been rolled over the edge and was found suspended in trees 60 to 100 cm above the ground. The remains were wrapped in a sleeping bag. A large number of maggots were present, together with several adult carrion beetles. Several maggot masses were present in the body, and the internal temperature of the remains was 34°C. The dishwasher was closed tightly, but the front handle was missing, resulting in a small hole approximately 1 cm in diameter, which had allowed entrance of the adult insects. This was the only entrance found. Large quantities of decompositional fluid were present in the bottom of the dishwasher, indicating that the base area was sealed. The blow flies all belonged to the same species, *Phormia regina*, and the oldest were in the third instar when discovered. Two species of beetles, *Creophilus maxillosus* (L.) (Staphylinidae) and *Nicrophorus defodiens* Mannerheim (Silphidae), had also located the remains and entered via the small hole. It was later determined that the victim had been killed inside a residence and placed into the

Figure 5.13 Rural area in which remains were found inside a dishwasher. (Photo courtesy of Vancouver City Police, Homicide Squad. Used with permission.)

dishwasher. The victim had remained in the dishwasher at this site for a few days, and was then moved to a storage area. After several days the remains, still inside the dishwasher, were moved to the rural site, where they were dumped. Timing of colonization indicated that it was at this time that the insects entered the dishwasher.

Hanged Bodies

Hanging, as a result of either suicide or accident (or more rarely, homicide), is not an uncommon form of death. If the body is suspended above the ground, it could present a unique environment for insect colonization. Hanging does affect the many factors in the decomposition process. Goff and Lord (1994) noted in Hawai'i that hanging altered the insect colonization by excluding soil-dwelling taxa, thus changing the drying pattern of the body, and consequently limiting the activities of fly species. This reduced the numbers of insects collected and influenced which species colonized the remains as well as their times of colonization (Goff and Lord 1994).

Later work in Hawai'i comparing a hanged pig carcass with a control carcass in contact with the ground showed that decomposition was retarded in the hanging pig in comparison with that on the ground (Shalaby et al. 2000). As well, although internal carcass temperature was raised as normal by maggot activity in the carcass on the ground, no such elevation was seen in the hanging carcass. The hanging carcass also supported a lesser diversity of arthropod species and lower overall numbers. The primary site of insect activity in the hanging carcass was found to be in the soil beneath the carcass in the drip zone (Shalaby et al. 2000).

Burnt Remains

Remains may often be burned either perimortem or postmortem. Little research has been published on the effects of burning on insect succession, but a study from Hawai'i has shown that burned remains are colonized in a manner different from unburned remains

(Avila and Goff 1998). In this study, the remains were burned to give a Crow-Glassman Scale (CGS) level 2 burn. The CGS is divided into five levels, depicting increasing destruction to the body relative to burn injury (Glassman and Crow 1996). At level 2, the remains were charred with cracked skin (Avila and Goff 1998). The arthropod fauna that colonized the burned and unburned carcasses were basically the same, but appeared slightly earlier on the burned carcasses (Avila and Goff 1998), presumably due to the openings caused by the cracked skin. The burnt carcasses attracted much more fly oviposition than the unburnt carcasses, showing that burnt carcasses are still extremely attractive to calliphorid flies (Avila and Goff 1998). Other work had previously suggested that oviposition was deterred by burning (Catts and Goff 1992), but this no doubt depends on the level of burning and amount of incineration. In two cases in Italy, insect evidence (primarily larval calliphorid and sarcophagid flies) could still be used to determine time since death, despite the fact that both victims were burned (Introna et al. 1998). Current studies being conducted on pig carcasses at the University of Alberta support the contention that burned bodies become more attractive to blow flies as the skin splits simulate wounds, resulting in large numbers of eggs being laid in these regions. This in turn results in rapid biomass loss (Samborski, personal communication).

Killers often try to dispose of a victim by burning the body, but are unaware of the tremendously high temperatures and the time required to completely incinerate a human body. Even in the extreme heat of a professional crematorium, recognizable pieces of human remains are still present (Kennedy 1996; Murray and Rose 1993). The level and amount of colonization of burned remains by insects will no doubt be strongly influenced by the amount of flesh remaining, with more complete incineration reducing insect colonization. This author has participated in several cases of extreme burning in which no or very few insects colonized the remains.

After a homicide, a killer often tries to dispose of the evidence. This may occur immediately after the crime, but often the killer flees the scene, only to return later to see what is left. People are often surprised that, despite decomposition, much of the evidence remains for a long time. In such cases, the killer may attempt to dispose of the body some time after death. In an extension of experiments on carcass decomposition conducted inside a house (Figures 5.1 to 5.4) (Anderson 2005a), this author carried out a series of experiments in order to determine whether insect evidence could be recovered and analyzed after an arson fire (Anderson 2005b). Three freshly killed pig carcasses were placed in a house, in the living room, bathroom, and bedroom, and allowed to decompose naturally for 42 days. At this time, all three were in advanced decomposition and had been colonized by five species of blow fly: *Lucilia sericata*, *Cynomya cadaverina*, *Calliphora vicina*, *Phormia regina*, and *Protophormia terraenovae*. The insects had progressed through the entire life cycle, as evidenced by empty puparia and teneral adults (Figure 5.14). Also present were live pupae, nonfeeding third-instar larvae, and some feeding third-instar larvae. Pupae and puparia were found on and around the carcasses and throughout the house (Figure 5.3). One carcass, in the bathtub, was more than 10 cm deep in puparia and pupae (Figure 5.4) (Anderson 2005b).

The first fire was started using an accelerant poured directly onto a blanket covering one of the carcasses. It was allowed to burn until smoke was seen coming from the house. It was then assumed that a "concerned neighbor" would see the smoke and call emergency services. The area in which the fires were set was serviced by two fire stations, and a dispatch and response time of 3½ min would be expected from either station. Therefore, 3½ min after the

Figure 5.14 Pig carcass after decomposing in a residential house for 42 days, prior to being burned in an arson experiment. (Photo courtesy of Dr. Gail Anderson.)

first smoke was seen, firefighters were dispatched and began extinguishing the fire. After the fire, the room in which the carcass had lain was examined. The room itself was destroyed and the roof was open. Gas temperatures measured inside the room during the fire rose to over 600°C (uncorrected for radiation) (Dale et al. 2004). Both windows were removed and all walls were burned down to the wooden studs (Figure 5.15). The entire room was covered in burnt insulating material (sawdust). However, once this was cleared using normal firefighter procedures, the pig carcass was found to be remarkably intact (Figure 5.16). All the insect evidence from the carcass itself and close to the carcass was recoverable and identifiable (Anderson 2005b). Entomological evidence a distance from the remains was destroyed, but that on, under, and within approximately 10 cm of the carcass was intact and undamaged, although heat killed. The carcass rapidly became attractive to adult Calliphoridae, which entered easily as the windows had now been removed.

Figure 5.15 Inside a room in which a pig carcass had decomposed for 42 days, after a simulated arson. (Photo courtesy of Dr. Gail Anderson.)

Figure 5.16 Pig carcass and attendant insect evidence after fire is put out. Majority of insect evidence still intact. (Photo courtesy of Dr. Gail Anderson.)

The second fire was set in the living room, on furniture placed close to another carcass. It was set using combustible materials but no accelerant. Again, a 3½ min response time was used. Temperatures reached similar peaks as before (Dale et al. 2004) and the room was destroyed. Once the debris was cleared, however, the carcass was once again undamaged, and all stages of insect evidence were not only recoverable and identifiable, but those from under the carcass were still alive and were later reared to adulthood (Anderson 2005b). Species close to the body were recovered intact; however, those farther from the body were destroyed, which could result in a biased sample being recovered, as some species are more prone to pupate close to the carcass.

It could be argued that in both of these fires, firefighters responded so fast that the fire was extinguished before it could destroy the evidence. Although the response times used were appropriate for the area, this assumes that the fire either took place in daylight, when someone is likely to notice, or took place in a suburban area, with people around to take notice. Many arsons, however, take place at night or in remote locations, where fire can quickly take hold and burn a property to the ground. Therefore, in a final fire, accelerants were sprayed throughout the house (gasoline and diesel fuel), lit, and the fire allowed to burn until it burned itself out. Peak temperatures reached 900 to 1,000°C inside the house (Dale et al. 2004), and the house burned to the ground in 30 min (Figure 5.17). The fire was allowed to burn unchecked. It ran out of combustible fuel after about 3 h, leaving only the concrete basement, chimney, metal objects such as the bathtub, and rubble still present. The following day, when it was safe to enter, the three sites of the pig carcasses were examined. The first site examined was that of the pig that had decomposed in the bathtub. Although the carcass was completely incinerated, large quantities of identifiable puparia and pupae were found clumped under the bathtub (Figure 5.18). They clearly indicated that a carcass had been present, and that it had died prior to the fire. Estimation of elapsed time since death was still possible. Excavation of the basement area, into which the remains of the other two carcasses had fallen, revealed the presence of pupae and larvae, as well as hair and tissue (Anderson 2005b).

The experiments showed that even after complete immolation, insect evidence may still be recoverable and should be considered after a fire. It may be possible to determine

Figure 5.17 House fire in which fire was not extinguished but allowed to burn house to the ground. Insect evidence still recoverable from all three carcasses. (Photo courtesy of Dr. Gail Anderson.)

Figure 5.18 Recognizable Calliphoridae pupae and puparia found after a house fire. (Photo courtesy of Dr. Gail Anderson.)

time of death, and if recovered, would at least indicate that a victim had not succumbed to the fire (Anderson 2005b).

Wrapped Remains

Remains, whether whole or dismembered, are frequently found wrapped in some material. This may be an effort to disguise the remains, to facilitate handling, or to prevent bleeding onto a carpet or vehicle. The type and extent of the wrapping may affect the insect colonization pattern of the remains.

Figure 5.19 Large numbers of calliphorid puparia (empty pupal cases) on carpeting used to wrap body. (Photo courtesy of Royal Canadian Mounted Police. Used with permission.)

This author received a case in which a male victim was found halfway down a scree (loose rock) slope in British Columbia in spring. The remains were completely wrapped in an old carpet, but the carpeting was not sealed at the head or feet. The carpeting had not prevented insects from colonizing the remains, but resulted in the majority of prepupal insects remaining within the carpet to pupate. The remains were skeletonized, primarily due to insect activity, and large numbers of empty calliphorid puparia were collected (Figure 5.19). Later-colonizing flies (such as Fanniidae) had also been present, as evidenced by their empty puparia. Therefore, these species were not dissuaded from the remains by the presence of the wrapping. This is probably to be expected, as these species are collected in much larger numbers on buried remains than on exposed remains in British Columbia (VanLaerhoven 1997; VanLaerhoven and Anderson 1996, 1999), indicating a preference for protected, moister areas. Adult staphylinid beetles were also collected, but no other Coleoptera were found in this case. The entire remains were skeletonized, except for one foot located inside a boot, which had saponified to adipocere tissue. A few muscid flies (*Hydrotaea* sp.) were also collected from this area. In this case, it would appear that wrapping the body did not impede insect colonization, but possibly provided protection from predators and the elements. More secure wrapping may delay insect colonization, although this author has not yet seen a case where it was entirely prevented.

In Hawai'i, a young female victim was found heavily wrapped in blankets in a rural, outdoor habitat. In order to determine the possible delay of insect colonization due to the wrapping, Goff simulated the case experimentally by wrapping a freshly killed pig carcass in a similar manner and observing how long it took for insects to begin to colonize the remains (Goff 1992). Insects were first seen on the pig carcass 2.5 days after death, indicating a probable delay in colonization of 2.5 days in the human case (Goff 1992). This later proved to be the case.

The previous case was outdoors, but a similar scenario was seen indoors in British Columbia, where a man was found in a basement completely wrapped in a large cloth bag originally designed for sports equipment (Anderson 1999b). The basement was also used as a hydroponics facility for marijuana, and thus had controlled and known temperatures. *Phormia regina* had heavily colonized the body, with all stages, from early instars to empty puparia, being collected from the remains and the surrounding area. The insects indicated that oviposition had begun between 13.5 and 16 days prior to discovery. The

Figure 5.20 Garbage bag containing human thigh. (Photo courtesy of Vancouver City Police, Homicide Squad. Used with permission.)

victim had last been seen alive 18.5 days prior to discovery, indicating a delay of 2.5 to 5 days before oviposition. Both the wrapping of the remains and the fact that the body was indoors would have compounded this delay. Although only blow flies were found, later succession insects might have eventually colonized had the remains been present for a longer period of time.

In the following case, the plastic wrapping was complete and secure. A human thigh was found tightly sealed in a plastic garbage bag in British Columbia during the late summer (Figure 5.20). The remains were suspended in a bush when found. The bag was a commercially available plastic garbage bag with drawstrings that had been pulled closed. No tears or cuts were visible on the bag. No adult flies were collected, and there appeared to be no possible entrance for adult insects. However, a number of third-instar *P. regina* and *Lucilia illustris* were collected from the proximal portion of the leg, inside the bag (Figure 5.21). It is probable that the adult females detected the presence of the body part in the bag, and laid their eggs on the knot of the drawstrings. When the first-instar larvae eclosed, they would have oriented to the carcass and would have been small enough to crawl around the knot and to the body part. Therefore, although it is probable that the presence of the knotted bag delayed insect colonization, it was not prevented.

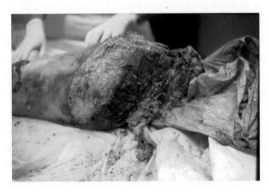

Figure 5.21 Human thigh, with third-instar *Phormia regina* and *Lucilia illustris* (Diptera: Calliphoridae) colonizing the distal, cut end. (Photo taken by Sgt. Allen Boyd, Vancouver Police Department, Homicide Squad. Used with permission.)

In a case from northern France, plastic bags wrapping an entire corpse were interpreted as having prevented insects from colonizing remains (Bourel et al. 1995). Flies in the families Calliphoridae, Piophilidae, Fanniidae, Phoridae, and Sphaeroceridae had colonized the remains, among other insect species. This was interpreted to indicate that the remains had been exposed for a period of time (long enough for blow fly colonization) before being wrapped. It was thought that this wrapping excluded second- and third-wave colonizers, then months later, tears developed in the plastic and fourth- and fifth-wave colonizers arrived. An elapsed time since death of 11 to 12 months was predicted from these data (Bourel et al. 1995). However, work in British Columbia has indicated that many of these so-called fourth- and fifth-wave insects actually colonize much earlier than was previously thought (Anderson and VanLaerhoven 1996; Dillon 1997; Dillon and Anderson 1995, 1996a; VanLaerhoven 1997; VanLaerhoven and Anderson 1996, 1999). It is thus possible that the remains had been colonized earlier, by the insects entering through knots or overlapping seams.

In South Africa, Kelly performed a series of experiments comparing pig carcasses wrapped in a sheet with unwrapped pig carcasses (Kelly 2006; Kelly et al. 2008, 2009). Carcasses were either naked, clothed but not wrapped, wrapped but not clothed, or clothed and wrapped. Trials were conducted in all four seasons. In all seasons except winter, dipteran oviposition was not delayed by wrapping or the presence of clothing (Kelly 2006; Kelly et al. 2008, 2009). In winter, a delay of 4 days was noted on the wrapped carcasses. The wrapping did influence decomposition rate, and in particular, the rate of drying. Wrapped carcasses did not enter the typical dry remains stage, but rather, stayed moist, although there was no change in the insect succession of the carcasses. In all seasons, except winter, the distribution of the maggot masses was impacted by the sheet, as wrapping appeared to facilitate maggot mobility (Kelly 2006; Kelly et al. 2008, 2009). Very high maggot mortality was noted on the wrapped carcasses in summer and fall, and this was thought to be due to extremely high temperatures in the wrapped carcasses as well as the presence of metabolic gases trapped under the sheet (Kelly 2006; Kelly et al. 2008, 2009).

Other Factors That May Affect Succession

Scavenging

Scavengers other than insects are also attracted to remains, and can remove large quantities of flesh and even clothing (Dillon 1997; Dillon and Anderson 1996a). This can have a major effect on the decomposition rate and consequent insect colonization. In British Columbia, vertebrate scavenging (primarily from small rodents) occurred most commonly in shaded areas. Despite the cooler temperatures, pig carcasses situated at shaded sites were scavenged and lost mass at the same rate as those in direct sun. This equal mass loss rate was attributed to scavenging (Dillon 1997; Dillon and Anderson 1995, 1996a).

Carrion in aquatic habitats was also scavenged more often in shade than in direct sun (Hobischak 1997; MacDonell and Anderson 1997) (Figure 5.22). Scavenging was also more common in winter than summer (Dillon 1997; Dillon and Anderson 1995, 1996a). Vertebrate scavenging increased the rate of decomposition, and in some cases eliminated decompositional stages in pig, bear, and cougar carcasses (Dillon and Anderson 1996b). This resulted in less tissue available for later colonizers, and consequently reduced the

Figure 5.22 Mink-scavenging pig carcasses in a creek in a forest in the Coastal Western Hemlock biogeoclimatic zone. (Photo courtesy of N. R. Hobischak. Used with permission.)

numbers of species and individual insects colonizing the remains over time (Dillon 1997; Dillon and Anderson 1995, 1996a,b, 1997). Vertebrate scavenging appeared to be virtually eliminated in shaded areas in Coastal British Columbia, and greatly reduced in northern British Columbia by shallow burial, although it did occur on rare occasions (VanLaerhoven 1997; VanLaerhoven and Anderson 1996, 1999).

Scavenging, in addition to affecting decomposition and insect colonization, may also produce postmortem artifacts that may be initially mistaken for wounds or mutilation (Anderson, unpublished data; Dillon 1997; Dillon and Anderson 1995, 1996a; Patel 1994). Conversely, wounds originally mistaken as rodent damage may actually have other causes (Patel 1995).

Scavengers, acting as opportunistic predators of insects, are common on remains. Although they may remove substantial numbers of colonizers (particularly blow fly larvae), they usually have little impact overall. However, some insect scavengers (due to their voraciousness and numbers) can have a substantial impact on arthropod colonization of remains. One such example is the fire ants (Hymenoptera: Formicidae: *Solenopsis* spp.), which may remove significant numbers of blow fly eggs and larvae (Early and Goff 1986; Greenberg 1991; Hayes 1994; Stoker et al. 1995; Wells and Greenberg 1994). Other species of ants also have been shown to have an impact (Cornaby 1974; Lord and Burger 1984), but most are present throughout decomposition as scavengers with little effect on overall decomposition rates (Anderson and VanLaerhoven 1996; Payne and Mason 1971). In British Columbia, a pig carcass accidentally placed close to an ant nest was scavenged almost clean of blow fly eggs for 2 days, until numbers of blow fly eggs were so large as to overwhelm the ant presence. After a few days, colonization proceeded as normal (Dillon and Anderson 1995).

Presence or Absence of Clothing

Human victims are frequently clothed. The clothing may be complete or partial. In forensic cases in British Columbia, the majority of victims were completely or partially clothed. Alternatively, the victim may be naked but wrapped in a variety of materials, such as carpet, blankets, sleeping bag, towels, etc., which act in a manner similar to that of clothing.

Clothing can be expected to have an effect on insect succession on a corpse, as it affects the temperature and humidity of the remains, the amount of shade, and protection the body provides, etc. In British Columbia, clothed pig carcasses were found to have increased numbers and diversity of successional insects (Dillon 1997; Dillon and Anderson 1995). Most early-instar larvae require liquid protein for survival (Smith 1986). As the clothing becomes saturated with decompositional fluids, it provides more sites for oviposition than a naked corpse, resulting in larger larval masses, and hence faster decomposition (Dillon 1997; Dillon and Anderson 1995, 1996a). In naked remains, whether human or animal, the skin is often left behind after maggot colonization as it dries, and quickly becomes unattractive to Calliphoridae larvae. It also acts as a protectant for the larvae, and frequently, breathing holes, looking like an area of puncture marks, are seen in the skin. However, when clothing is present, the clothing acts as a protectant, keeping the skin moist and often resulting in the skin being consumed (Kelly 2006; Kelly et al. 2008, 2009; Dillon 1997; Dillon and Anderson 1995, 1996a).

The clothing can also provide additional shelter for blow flies and their predators, increasing the number of Coleoptera on the remains and making the remains more attractive for species that prefer wetter environments (Dillon 1997; Dillon and Anderson 1995, 1996a). Conversely, clothing may make the remains less attractive to insects preferring dried environments. However, in a series of experiments in South Africa, clothing did not prevent the remains from drying out completely (Kelly 2006; Kelly et al. 2005a, 2008, 2009).

In aquatic environments, clothing provides shelter and extra attachment sites for aquatic fauna (Hobischak 1997; MacDonell and Anderson 1997). However, the effect of clothing depends on whether the body is completely or partially submerged. When the body is exposed above the waterline, clothing protected maggots from predation, whereas below the waterline, organisms such as crayfish and caddisflies (Trichoptera) fed on unclothed regions preferentially (Hobischak 1997; MacDonell and Anderson 1997). Clothes permeated with lubricants, paint, or combustibles may double the time for initial colonization and have been shown to retard decomposition (Marchenko 1980, cited in Greenberg 1991).

Pig carcasses are frequently used to mimic human remains and are considered to be an excellent model for human decomposition (Catts and Goff 1992). Most research has concentrated on naked pig carcasses (Anderson and VanLaerhoven 1996; Hewadikaram and Goff 1991; Payne 1965; Payne and Crossley 1966; Shean et al. 1993), although some work has included clothing (Dillon 1997; Dillon and Anderson 1995, 1996a; Hobischak 1997; Komar and Beattie 1998; MacDonell and Anderson 1997; VanLaerhoven 1997; VanLaerhoven and Anderson 1996, 1999). Therefore, clothing can have a considerable impact on the decomposition, and it is important that future studies include this aspect of colonization as a consideration.

Insects have also been shown to move and tear clothing in a manner that may mislead investigators into assuming that a sexual assault has taken place (Kelly 2006; Komar and Beattie 1998). Maggot masses have been able to move clothing from underneath a body, despite the overlying carcass weight. In carcasses clothed in skirts, the underwear and pantyhose were moved down to the distal hindlimbs, while the skirt was pushed up (Komar and Beattie 1998). Natural decompositional changes such as bloat tore clothing as well (Komar and Beattie 1998). Heavy clothing may deter carnivore scavenging (Haglund 1997), but carnivores can also cause clothing disarray. Such postmortem artifacts are usually easy to differentiate from that caused by insects (Komar and Beattie 1998).

In South Africa, Kelly found that clothing did not have an impact on Calliphoridae oviposition, decompositional rate, or insect succession (Kelly 2006; Kelly et al. 2005a,

2008, 2009). However, it did impact the distribution of maggot masses in spring, summer, and fall, as it appeared to facilitate maggot movement.

Pigs are chosen as good human models for many reasons, including the fact that they are relatively hairless. However, carcasses with a coat of fur have also been frequently used to generate carrion data (Bornemissza 1957; Braack 1981; Denno and Cothran 1976; Dillon 1997; Dillon and Anderson 1996b, 1997; Early and Goff 1986; Easton 1966; Jiron and Cartin 1981; Putman 1978; Reed 1958; Smith 1975). In British Columbia, the diversity of insects and times of colonization was found to be similar in bear and cougar carcasses to that seen in clothed pig carcasses, indicating that the fur acted in a similar manner to that of the clothing (Dillon 1997; Dillon and Anderson 1996b, 1997).

Conclusions

The predictable sequence of insect succession on a body has long been recognized as an excellent method to estimate the time since death. However, there are many diverse parameters that can affect the timing and species composition of the carrion fauna. It is vitally important to be aware of all the factors that can impact insect colonization of remains, and to take them into account when analyzing a death. In particular, further research is needed to develop more geographical databases of insect succession on carrion in a variety of habitats and scenarios in all regions in which forensic entomology is being used.

Acknowledgments

I thank my past and present graduate students, research assistants, and collaborators for their tremendous work on the Canadian database. I particularly thank Sharon Abrams, Anne Reusse, Janice Lacapra (RCMP Forensic Lab, Edmonton), Crystal Samborski (University of Alberta), John Amerongen and Ed Rostalski (Edmonton Fire Department), for their support in the indoor/outdoor and arson research. I am extremely grateful to the Canadian Amphibious Search Team, the Canadian Coast Guard, the RCMP, the Vancouver Aquarium Marine Science Centre, Dr. Verena Tunnicliffe and the entire VENUS team, for their support in the marine research. I also thank the Canadian Police Research Centre, the Information and Identification Services Directorate, and the Training Directorate of the Royal Canadian Mounted Police for their ongoing support of forensic entomology in Canada.

References

Anderson, G. S. 1995. The use of insects in death investigations: An analysis of forensic entomology cases in British Columbia over a five year period. *Canadian Society of Forensic Sciences Journal* 28:277–92.

Anderson, G. S. 1999a. Wildlife forensic entomology: Determining time of death in two illegally killed black bear cubs, a case report. *Journal of Forensic Science* 44:856–859.

Anderson, G. S. 1999b. Forensic entomology: The use of insects in death investigations. In *Case Studies in Forensic Anthropology*, ed. S. Fairgreave. Toronto: Charles C. Thomas. pp. 303–325.

Anderson, G. S. 2000. Minimum and maximum developmental rates of some forensically significant Calliphoridae (Diptera). *Journal of Forensic Science* 45:824–32.

Anderson, G. S. 2005a. Differences in insect colonization and decomposition rates between carcasses placed inside and outside a suburban house in Canada. Paper presented at European Association of Forensic Entomology Annual Meeting, Lausanne, Switzerland.

Anderson, G. S. 2005b. Effects of arson on forensic entomology evidence. *Canadian Society of Forensic Sciences Journal* 38:49–67.

Anderson, G. S. 2008. Investigation into the effects of oceanic submergence on carrion decomposition and faunal colonization using a baited camera. Part I. Technical Report TR-10-2008, Canadian Police Research Centre, Ottawa, Ontario.

Anderson, G. S. Unpublished data. School of Criminology, Simon Fraser University, Burnaby, British Columbia.

Anderson, G. S., and N. R. Hobischak. 2002. Determination of time of death for humans discovered in saltwater using aquatic organism succession and decomposition rates. Technical Report TR-09-2002, Canadian Police Research Centre, Ottawa, Ontario.

Anderson, G. S., and N. R. Hobischak. 2004. Decomposition of carrion in the marine environment in British Columbia, Canada. *International Journal of Legal Medicine* 118:206–9.

Anderson, G. S., N. R. Hobischak, O. Beattie, and C. Samborski. 2002. Insect succession on carrion in the Edmonton, Alberta region of Canada. Technical report TR-04-2002, Canadian Police Research Centre, Ottawa, Ontario.

Anderson, G. S., and S. L. VanLaerhoven. 1996. Initial studies on insect succession on carrion in southwestern British Columbia. *Journal of Forensic Sciences* 41:617–25.

Anderson, R. S., and S. B. Peck. 1985. *The carrion beetles of Canada and Alaska.* Ottawa, Ontario: Biosystematics Research Institute, Research Branch, Agriculture Canada.

Archer, M. S. 2004. Rainfall and temperature effects on the decomposition rate of exposed neonatal remains. *Science and Justice* 44:35–41.

Archer, M. S., and M. A. Elgar. 2003a. Effects of decomposition on carcass attendance in a guild of carrion-breeding flies. *Medical and Veterinary Entomology* 17:263–71.

Archer, M. S., and M. A. Elgar. 2003b. Yearly activity patterns in southern Victoria (Australia) of seasonally active carrion insects. *Forensic Science International* 132:173–76.

Arnaldos, M. I., M. D. Garcia, E. Romera, J. J. Presa, and A. Luna. 2005. Estimation of postmortem interval in real cases based on experimentally obtained entomological evidence. *Forensic Science International* 149:57–65.

Arnaldos, M. I., E. Romera, M. D. Garcia, and A. Luna. 2001. An initial study on the succession of sarcosaprophagous Diptera (Insecta) on carrion in the southeastern Iberian peninsula. *International Journal of Legal Medicine* 114:156–62.

Ashworth, J. R., and R. Wall. 1994. Responses of the sheep blowflies *Lucilia sericata* and *L. cuprina* to odour and the development of semiochemical baits. *Medical Veterinary Entomology* 8:303–9.

Avila, F. W., and M. L. Goff. 1998. Arthropod succession patterns onto burnt carrion in two contrasting habitats in the Hawaiian Islands. *Journal of Forensic Sciences* 43:581–86.

Baldridge, R. S., S. G. Wallace, and R. Kirkpatrick. 2006. Investigation of nocturnal oviposition by necrophilous flies in central Texas. *Journal of Forensic Sciences* 51:125–26.

Barton Browne, L., R. J. Bartell, and H. H. Shorey. 1969. Pheromone-mediated behaviour leading to group oviposition in the blowfly, *Lucilia cuprina. Journal of Insect Physiology* 15:1003–14.

Bauch, R., K. Ziesenhenn, and C. Groskoppf. 1984. *Lucilia sericata* myiasis (Diptera: Calliphoridae) on a gangrene of the foot. *Angewandte Parasitologie* 25:167–69.

Benecke, M. 1998. Six forensic entomology cases: Description and commentary. *Journal of Forensic Sciences* 43:797–805.

Bharti, M., and D. Singh. 2003. Insect faunal succession on decaying rabbit carcasses in Punjab, India. *Journal of Forensic Sciences* 48:1133–43.

Bornemissza, G. F. 1957. An analysis of arthropod succession in carrion and the effect of its decomposition on the soil fauna. *Australian Journal of Zoology* 5:1–12.

Bourel, B., L. Martin-Bouyer, V. Hedouin, J. C. Cailliez, D. Derout, and D. Gosset. 1999. Necrophilous insect succession on rabbit carrion in sand dune habitats in northern France. *Journal of Medical Entomology* 36:420–25.

Bourel, B., L. Martin-Boyer, V. Hedouin, E. Revuelta, and E. Gosset. 1995. Estimation of post mortem interval by arthropod study: The case of a corpse wrapped in a plastic bag. Paper presented at 47th Annual Meeting of the American Academy of Forensic Sciences, Seattle.

Bourel, B., G. Tournel, V. Hedouin, and D. Gosset. 2004. Entomofauna of buried bodies in northern France. *International Journal of Legal Medicine* 118:215–20.

Braack, L. E. O. 1981. Visitation patterns of principal species of the insect complex at carcasses in the Kruger National Park. *Koedoe* 24:33–49.

Byrne, A. L., M. A. Camann, T. L. Cyr, E. P. Catts, and K. E. Espelie. 1995. Forensic implications of biochemical differences among geographic populations of the black blow fly, *Phormia regina* (Meigen). *Journal of Forensic Sciences* 40:372–77.

Carvalho, L., P. Thyssen, A. Linhares, and F. Palhares. 2000. A checklist of arthropods associated with pig carrion and human corpses in Southeastern Brazil. *Memórias do Instituto Oswaldo Cruz* (Rio de Janeiro) 95:135–38.

Carvalho, L. M. L., and A. X. Linhares. 2001. Seasonality of insect succession and pig carcass decomposition in a natural forest area in southeastern Brazil. *Journal of Forensic Sciences* 46:604–8.

Carvalho, L. M. L., P. J. Thyssen, M. L. Goff, and A. X. Linhares. 2004. Observation on the succession patterns of necrophagous insects on a pig carcass in an urban area of southeastern Brazil. *Journal of Forensic Medicine and Toxicology* 5:33–39.

Catts, E. P., and M. L. Goff. 1992. Forensic entomology in criminal investigations. *Annual Review of Entomology* 37:253–72.

Catts, E. P., and N. H. Haskell. 1990. *Entomology and death—A procedural guide*. Clemson, SC: Joyce's Print Shop.

Centeno, N., M. Maldonado, and A. Oliva. 2002. Seasonal patterns of arthropods occurring on sheltered and unsheltered pig carcasses in Buenos Aires Province (Argentina). *Forensic Science International* 126:63–70.

Coe, J. I., and W. J. Curran. 1980. Definition and time of death. *Modern legal psychiatry and forensic science*, ed. W. J. Curran, A. L. McGarry, and C. S. Petty. Philadelphia: F.A. Davis Co., 141–69.

Colyer, C. N. 1954a. The 'coffin fly' *Conicera tibialis* Schmitz (Diptera: Phoridae). *Journal of British Entomology* 4:203–6.

Colyer, C. N. 1954b. More about the 'coffin fly' (Diptera: Phoridae), *Conicera tibialis* Schmitz. *The Entomologist* 87:130–32.

Connell, J. H., and R. O. Slatyer. 1977. Mechanisms of succession in natural communities and their role in community stability and organization. *The American Naturalist* 3:1119–41.

Cornaby, B. W. 1974. Carrion reduction by animals in contrasting environments. *Biotropica* 6:51–63.

Cornelison, J. B. 1999. Microenvironmental effects on the decomposition of pig carrion (*Sus scrofa* L.) and carrion arthropod communities in southeastern Idaho. Paper presented at 51st American Academy of Forensic Sciences Annual Meeting, Orlando, FL.

Dale, J. D., M. Y. Ackerman, D. A. Torvi, T. G. Threlfall, and P. A. Thorpe. 2004. Interior temperature and heat flux measurements during a house burn. Paper presented at Combustion Institute, Canadian Section, Spring Technical Meeting, Kingston, Ontario.

Davies, L. 1990. Species composition and larval habitats of blowfly (Calliphoridae) populations in upland areas in England and Wales. *Medical and Veterinary Entomology* 4:61–68.

Davies, L. 1999. Seasonal and spatial changes in blowfly production from small and large carcasses at Durham in lowland northeast England. *Medical and Veterinary Entomology* 13:245–51.

Davis, J. B., and M. L. Goff. 2000. Decomposition patterns in terrestrial and intertidal habitats on Oahu Island and Coconut Island, Hawaii. *Journal of Forensic Sciences* 45:836–42.

De Jong, G. D., and W. W. Hoback. 2006. Effect of investigator disturbance in experimental forensic entomology: Succession and community composition. *Medical and Veterinary Entomology* 20:248–58.

Denno, R. F., and W. R. Cothran. 1976. Competitive interaction and ecological strategies of sarcophagid and calliphorid flies inhabiting rabbit carrion. *Annals of the Entomological Society of America* 69:109–13.

de Souza, A. M., and A. X. Linhares. 1997. Diptera and Coleoptera of potential forensic importance in southeastern Brazil: Relative abundance and seasonality. *Medical and Veterinary Entomology* 11:8–12.

Dillon, L. C. 1997. Insect succession on carrion in three biogeoclimatic zones in British Columbia. MSc thesis, Department of Biological Sciences, Simon Fraser University, Burnaby, British Columbia.

Dillon, L. C., and G. S. Anderson. 1995. Forensic entomology: The use of insects in death investigations to determine elapsed time since death. Technical Report TR-05-95, Canadian Police Research Centre, Ottawa, Ontario.

Dillon, L. C., and G. S. Anderson. 1996a. Forensic entomology: A database for insect succession on carrion in northern and interior B.C. Technical Report TR-04-96, Canadian Police Research Centre, Ottawa, Ontario.

Dillon, L. C., and G. S. Anderson. 1996b. The use of insects to determine time of death of illegally killed wildlife. Technical report, World Wildlife Fund, Toronto, Ontario.

Dillon, L. C., and G. S. Anderson. 1997. Forensic entomology—Use of insects towards illegally killed wildlife. Technical report, World Wildlife Fund, Toronto, Ontario.

Dymock, J. J., and S. A. Forgie. 1993. Habitat preferences and carcass colonization by sheep blowflies in the northern North Island of New Zealand. *Medical and Veterinary Entomology* 7:155–60.

Early, M., and M. L. Goff. 1986. Arthropod succession patterns in exposed carrion on the island of O'ahu, Hawai'i. *Journal of Medical Entomology* 23:520–31.

Easton, A. M. 1966. The Coleoptera of a dead fox (*Vulpes vulpes* L.); including two species new to Britain. *Entomologists Monthly Magazine* 102:205–10.

Easton, A. M., and K. G. V. Smith. 1970. The entomology of the cadaver. *Medicine, Science and the Law* 10:208–15.

Erzinclioglu, Y. Z. 1983. The application of entomology to forensic medicine. *Medicine, Science and the Law* 23:57–63.

Erzinclioglu, Y. Z. 1985. The entomological investigation of a concealed corpse. *Medicine, Science and the Law* 25:228–30.

Erzinclioglu, Z. 1989. Entomology and the forensic scientist: How insects can solve crimes. *Journal of Biology and Education* 23:300–2.

Erzinclioglu, Z. 1996. *Blowflies.* Slough, England: Richmond Publishing Co. Ltd.

Ferreira, M. J. D. M., and P. V. D. Lacerda. 1993. Synanthropic muscoids associated with the urban garbage in Goiania, Goias. *Revista Brasileira de Zoologia* 10:185–95.

Fuller, M. E. 1934. The insect inhabitants of carrion: A study in animal ecology. *CSIRO Bulletin* 82:5–62.

Germonpre, M., and M. LeClercq. 1994. Pupae of *Protophormia terraenovae* associated with Pleistocene mammals in the Flemish Valley (Belgium). *Bulletin de l'Institut Royal des Sciences Naturelles de Belgique Sciences de la Terre* 64:265–68.

Gilbert, B. M., and W. M. Bass. 1967. Seasonal dating of burials from the presence of fly pupae. *American Antiquity* 32:534–35.

Gill, G. J. 2005a. Carrion ecology in forested regions of Manitoba and determination of time since death. MSc thesis, Department of Entomology, University of Manitoba, Winnipeg, Manitoba.

Gill, G. J. 2005b. Decomposition and arthropod succession on above ground pig carrion in rural Manitoba. Technical Report TR-06-2005, Canadian Police Research Centre, Ottawa, Ontario.

Glassman, D. M., and R. M. Crow. 1996. Standardization model for describing the extent of burn injury to human remains. *Journal of Forensic Sciences* 41:152–54.

Goddard, J., and P. K. Lago. 1985. Notes on blowfly (Diptera: Calliphoridae) succession on carrion in northern Mississippi. *Journal of Entomological Sciences* 20:312–17.

Goff, M. L. 1991. Comparison of insect species associated with decomposing remains recovered inside dwellings and outdoors on the island of Oahu, Hawaii. *Journal of Forensic Sciences* 36:748–53.

Goff, M. L. 1992. Problems in estimation of postmortem interval resulting from wrapping of the corpse: A case study from Hawaii. *Journal of Agricultural Entomology* 9:237–43.

Goff, M. L. 1998. Personal communication. University of Hawaii, Manoa, HI.

Goff, M. L., M. Early, C. B. Odom, and K. Tullis. 1986. A preliminary checklist of arthropods associated with exposed carrion in the Hawaiian Islands. *Proceedings of the Hawaiian Entomological Society* 26:53–57.

Goff, M. L., and M. M. Flynn. 1991. Determination of postmortem interval by arthropod succession: A case study from the Hawaiian Islands. *Journal of Forensic Sciences* 36:607–14.

Goff, M. L., and W. D. Lord. 1994. Hanging out at the sixteenth hole: Problems in estimation of postmortem interval using entomological techniques in cases of death by hanging. Paper presented at 46th American Academy of Forensic Sciences Annual Meeting, San Antonio, TX.

Grassberger, M., and C. Frank. 2004. Initial study of arthropod succession on pig carrion in a central European urban habitat. *Journal of Medical Entomology* 41:511–23.

Greenberg, B. 1984. Two cases of human myiasis caused by *Phaenicia sericata* (Diptera: Calliphoridae) in Chicago area hospitals. *Journal of Medical Entomology* 21:615.

Greenberg, B. 1990. Nocturnal oviposition behavior of blow flies (Diptera: Calliphoridae). *Journal of Medical Entomology* 27:807–10.

Greenberg, B. 1991. Flies as forensic indicators. *Journal of Medical Entomology* 28:565–77.

Haglund, W. D. 1997. Dogs and coyotes: Postmortem involvement with human remains. *Forensic taphonomy. The postmortem fate of human remains*, ed. W. D. Haglund and M. H. Sorg. Boca Raton, FL: CRC Press.

Hall, M. J. R., R. Farkas, F. Kelemen, M. J. Hosier, and J. M. El-Khoga. 1995. Orientation of agents of wound myiasis to hosts and artificial stimuli in Hungary. *Medical and Veterinary Entomology* 9:77–84.

Hall, R. D., and K. E. Doisy. 1993. Length of time after death: Effect on attraction and oviposition or larviposition of midsummer blow flies (Diptera: Calliphoridae) and flesh flies (Diptera: Sarcophagidae) of medicolegal importance in Missouri. *Annals of the Entomological Society of America* 86:589–93.

Haskell, N. H., R. D. Hall, V. J. Cervenka, and M. A. Clark. 1997. On the body: Insects' life stage presence and their postmortem artifacts. *Forensic taphonomy. The postmortem fate of human remains*, ed. W. D. Haglund and M. H. Sorg. Boca Raton, FL: CRC, 415–48.

Hayes, J. 1994. Fire ants—Forensic implications and potential impact upon determination of postmortem interval. Paper presented at 47th American Academy of Forensic Sciences Annual Meeting, San Antonio, TX.

Henssge, C., B. Madea, B. Knight, L. Nokes, and T. Krompecher. 1995. *The estimation of the time since death in the early postmortem interval*. London, England: Arnold.

Hewadikaram, K. A., and M. L. Goff. 1991. Effect of carcass size on rate of decomposition and arthropod succession patterns. *American Journal of Forensic Medicine and Pathology* 12:235–40.

Hobischak, N. R. 1997. Freshwater invertebrate succession and decompositional studies on carrion in British Columbia. MPM thesis, Department of Biological Sciences, Simon Fraser University, Burnaby, British Columbia.

Hobischak, N. R., and G. S. Anderson. 1999. Freshwater-related death investigations in British Columbia in 1995–1996. A review of coroners cases. *Canadian Society of Forensic Sciences Journal* 32:97–106.

Hobischak, N. R., and G. S. Anderson. 2002. Time of submergence using aquatic invertebrate succession and decompositional changes. *Journal of Forensic Science* 47:142–51.

Hobischak, N. R., S. L. VanLaerhoven, and G. S. Anderson. 2006. Successional patterns of diversity in insect fauna on carrion in sun and shade in the Boreal Forest Region of Canada, near Edmonton, Alberta. *Canadian Entomologist* 138:376–83.

Holdaway, F. G. 1930. Field populations and natural control of *Lucilia sericata*. *Nature* 126:648–49.

Hwang, C., and B. D. Turner. 2005. Spatial and temporal variability of necrophagous Diptera from urban to rural areas. *Medical and Veterinary Entomology* 19:379–91.

Introna, F. J., C. P. Campobasso, and A. Di-Fazio. 1998. Three case studies in forensic entomology from southern Italy. *Journal of Forensic Sciences* 43:210–14.

Introna, F. J., T. W. Suman, and J. E. Smialek. 1991. Sarcosaprophagous fly activity in Maryland. *Journal of Forensic Sciences* 36:238–43.

Jiron, L. F., and V. M. Cartin. 1981. Insect succession in the decomposition of a mammal in Costa Rica. *New York Entomological Society* 89:158–65.

Johnson, M. D. 1975. Seasonal and microseral variations in the insect populations on carrion. *America Midland Naturalist* 93:79–90.

Joy, J. E., M. L. Herrell, and P. C. Rogers. 2002. Larval fly activity on sunlit versus shaded raccoon carrion in southwestern West Virginia with special reference to the black blowfly (Calliphoridae). *Journal of Medical Entomology* 39:392–97.

Joy, J. E., N. L. Liette, and H. L. Harrah. 2006. Carrion fly (Diptera: Calliphoridae) larval colonization of sunlit and shaded pig carcasses in West Virginia, USA. *Forensic Science International* 39(2):392–97.

Kelly, J. 2006. The influence of clothing, wrapping, and physical trauma on carcass decomposition and arthropod succession in central South Africa. PhD thesis, Department of Entomology and Zoology, University of the Orange Free State, Blomfontein, South Africa.

Kelly, J., T. C. van der Linde, G. S. Anderson, and C. J. De Beer. 2005a. Forensic entomology: Factors that could change postmortem interval estimations. Paper presented at Entomological Society of Southern Africa Annual Meeting, Grahamstown, South Africa.

Kelly, J. A., T. C. van der Linde, G. S. Anderson, and C. J. De Beer. 2005b. The influence of knife wounds on insect succession on carcasses during summer and winter in central South Africa. Paper presented at Entomological Society of Southern Africa Annual Meeting, Grahamstown, South Africa.

Kelly, J. A., T. C. van der Linde, and G. S. Anderson. 2008. The influence of clothing and wrapping on carcass decomposition and arthropod succession: A winter study in central South Africa. *Canadian Society of Forensic Science Journal* 41(3):135.

Kelly, J. A., T. C. van der Linde, and G. S. Anderson. 2009. The influence of clothing and wrapping on carcass decomposition and arthropod succession during the warmer seasons in central South Africa. *Journal of Forensic Science*. In press.

Kennedy, K. A. R. 1996. The wrong urn: Commingling of cremains in mortuary practices. *Journal of Forensic Sciences* 41:689–92.

Kettle, D. S. 1990. *Medical and veterinary entomology*. Wallingford, UK: C A B International.

Khan, M. A. J., and R. J. Khan. 1987. Hematoma of scalp in a baby caused by the common green bottle—*Lucilia sericata* (Meigen) (Diptera: Calliphoridae) in Karachi, Pakistan. *Japanese Journal of Sanitary Zoology* 38:103–5.

Komar, D., and O. Beattie. 1998. Postmortem insect activity may mimic perimortem sexual assault clothing patterns. *Journal of Forensic Sciences* 43:792–96.

Lane, R. P. 1975. An investigation into blowfly (Diptera: Calliphoridae) succession on corpses. *Journal of Natural History* 9:581–88.

LeBlanc, H. N. 1998. A study of carrion insects in urban and rural sites in the Halifax, Nova Scotia region. Honor's thesis, Department of Biology, St. Mary's University, Halifax.

Leclercq, M. 1969. *Entomological parasitology: Entomology and legal medicine*. Oxford: Pergamon Press.

Lord, W. D., and J. F. Burger. 1984. Arthropods associated with Herring Gulls (*Larus argentatus*) and great black-backed gulls (*Larus marinus*) carrion on islands in the gulf of Maine. *Environmental Entomology* 13:1261–68.

MacDonell, N. R., and G. S. Anderson. 1997. Aquatic forensics: Determination of time since submergence using aquatic invertebrates. Technical Report TR-01-97, Canadian Police Research Centre, Ottawa, Ontario.

MacGregor, D. M. 1999a. Decomposition of pig carrion in southeast Queensland, Australia, during summer. Paper presented at 51st American Academy of Forensic Sciences Annual Meeting, Orlando, FL.

MacGregor, D. M. 1999b. Decomposition of pig carrion in southeast Queensland, Australia, during winter. Paper presented at 51st American Academy of Forensic Sciences Annual Meeting, Orlando, FL.

MacLeod, J., and J. Donnelly. 1957. Some ecological relationships of natural populations of calliphorine blowflies. *Journal of Animal Ecology* 26:135–70.

Mann, R. W., W. M. Bass, and L. Meadows. 1990. Time since death and decomposition of the human body: Variables and observations in case and experimental field studies. *Journal of Forensic Sciences* 35:103–11.

Marchenko, M. I. 1980. Impact of clothing and its soiliness on cadaver decomposition rate by insects (in Russian). In: *Current Problems of Medico-legal Expertise* USSR: Alma-Ata, 51–53.

Martinez, E., P. Duque, and M. Wolff. 2006. Succession pattern of carrion-feeding insects in Paramo, Colombia. *Forensic Science International* 166:182–89.

Martinez-Sanchez, A., S. Rojo, and M. A. Marcos-Garcia. 2000. Annual and spatial activity of dung flies and carrion in a Mediterranean holm-oak pasture ecosystem. *Medical Veterinary Entomology* 14:56–63.

Motter, M. G. 1898. A contribution to the study of the fauna of the grave—A study of 150 disinterrments, with some additional observations. *Journal of the New York Entomological Society* 6:201–31.

Murray, K. A., and J. C. Rose. 1993. The analysis of cremains: A case study involving the inappropriate disposal of mortuary remains. *Journal of Forensic Sciences* 38:98–103.

Nuorteva, P. 1959. Studies on the significance of flies in the transmission of poliomyelitis. *Annales Entomologia Fennici* 25:137–62.

Nuorteva, P. 1977. Sarcosaprophagous insects as forensic indicators. In *Forensic medicine: A study in trauma and environmental hazards*, ed. C. G. Tedeschi, W. G. Eckert, and L. G. Tedeschi. Vol. II. Philadelphia: W.B. Saunders Co., 1072–95.

Oliva, A. 2001. Insects of forensic significance in Argentina. *Forensic Science International* 120:145–54.

Patel, F. 1994. Artifact in forensic medicine: Postmortem rodent activity. *Journal of Forensic Sciences* 39:257–60.

Patel, F. 1995. Artifact in forensic medicine: Pseudo-rodent activity. *Journal of Forensic Sciences* 40:706–7.

Payne, J. A. 1965. A summer carrion study of the baby pig *Sus scrofa* Linnaeus. *Ecology* 46:592–602.

Payne, J. A., and D. A. J. Crossley. 1966. Animal species associated with pig carrion. Oak Ridge, TN: Oak Ridge National Laboratory.

Payne, J. A., and E. W. King. 1969. Lepidoptera associated with pig carrion. *Journal of the Lepidopterists Society* 23:191–95.

Payne, J. A., and E. W. King. 1970. Coleoptera associated with pig carrion. *Entomologist's Monthly Magazine* 105:224–32.

Payne, J. A., and E. W. King. 1972. Insect succession and decomposition of pig carcasses in water. *Journal of the Georgia Entomological Society* 73:153–62.

Payne, J. A., E. W. King, and G. Beinhart. 1968a. Arthropod succession and decomposition of buried pigs. *Nature* 219:1180–81.

Payne, J. A., and W. R. M. Mason. 1971. Hymenoptera associated with pig carrion. *Proceedings of the Entomological Society of Washington* 73:132–41.

Payne, J. A., F. W. Mead, and E. W. King. 1968b. Hemiptera associated with pig carrion. *Annals of the Entomological Society of America* 61:565–67.

Perez, S. P., P. Duque, and M. Wolff. 2005. Successional behavior and occurrence matrix of carrion-associated arthropods in the urban area of Medellin, Colombia. *Journal of Forensic Science* 50:448–54.

Petrik, M. S., N. R. Hobischak, and G. S. Anderson. 2004. Examination of factors surrounding freshwater decomposition in death investigations: A review of body recoveries and coroner cases in British Columbia. *Canadian Society Forensic Science Journal* 37:9–17.

Piatkowski, S. 1991. [Synanthropic flies in an 11-story apartment house in Gdansk] [English abstract]. *Wiad Parazytologie* 37:115–17.

Putman, R. J. 1978. The role of carrion-frequenting arthropods in the decay process. *Ecological Entomology* 3:133–39.

Reed, H. B. 1958. A study of dog carcass communities in Tennessee with special reference to the insects. *America Midland Naturalist* 59:213–45.

Reiter, C. 1984. Zum Wachstumsverhalten der Maden der blauen Schmeissfliege *Calliphora vicina*. *Zeitschrift für Rechtsmedizin* 91:295–308.

Rodriguez, W. C. 1997. Decomposition of buried and submerged bodies. *Forensic taphonomy. The postmortem fate of human remains*, ed. W. D. Haglund and M. H. Sorg. Boca Raton, FL: CRC Press, 459–67.

Rodriguez, W. C., and W. M. Bass. 1983. Insect activity and its relationship to decay rates of human cadavers in East Tennessee. *Journal of Forensic Sciences* 28:423–32.

Rodriguez, W. C., and W. M. Bass. 1985. Decomposition of buried bodies and methods that may aid in their location. *Journal of Forensic Sciences* 30:836–52.

Rodriquez, W. C. I., R. I. Sundick, and W. D. Lord. 1993. Forensic importance of entomological evidence in the determination of the seasonality of death in cases involving skeletonized remains. Paper presented at 45th American Academy of Forensic Sciences Annual Meeting, Boston.

Rosati, J., and S. L. VanLaerhoven. 2006. The trials and tribulations of using successional waves of insects for post-mortem interval estimation. Forensic identification: Common heritages, shared future? Paper presented at Joint Meeting of the Canadian Identification Society and the Canadian Society of Forensic Sciences, Windsor, ON.

Samborski, C. Personal communication. University of Alberta, Department of Anthropology, Edmonton, Alberta.

Schmitz, H. 1928. Occurrence of phorid flies in human corpses buried in coffins. *Natuurhistorisch Maandblad* 17:150.

Schroeder, H., H. Klotzbach, L. Oesterhelweg, and K. Puschel. 2002. Larder beetles (Coleoptera, Dermestidae) as an accelerating factor for decomposition of a human corpse. *Forensic Science International* 127:231–36.

Schumann, H. 1990. [The occurrence of Diptera in living quarters] [English abstract]. *Angewandte Parasitologie* 31:131–41.

Shalaby, O. A., L. M. L. de Carvalho, and M. L. Goff. 2000. Comparison of patterns of decomposition in a hanging carcass and a carcass in contact with soil in a xerophytic habitat on the island of Oahu, Hawaii. *Journal of Forensic Science* 45:1267–73.

Sharanowski, B. 2004. Decomposition and insect ecology of carrion in Saskatchewan. Department of Archaeology, University of Saskatchewan, Saskatoon.

Shean, B. S., L. Messinger, and M. Papworth. 1993. Observations of differential decomposition on sun exposed vs. shaded pig carrion in coastal Washington state. *Journal of Forensic Sciences* 38:938–49.

Sherman, R. A., and E. A. Pechter. 1988. Maggot therapy: A review of the therapeutic applications of fly larvae in human medicine, especially for treating osteomyelitis. *Medical Veterinary Entomology* 2:225–30.

Shroeder, H., H. Klotzbach, and K. Puschel. 2003. Insects' colonization of human corpses in warm and cold season. *Legal Medicine* (Tokyo) 5(Suppl. 1):S372–74.

Shubeck, P. P. 1985. Orientation of carrion beetles to carrion buried under shallow layers of sand (Coleoptera: Silphidae). *Entomological News* 96:163–66.

Shubeck, P. P., and D. L. Blank. 1982. Carrion beetle attraction to buried fetal pig carrion (Coleoptera: Silphidae). *Coleopterists Bulletin* 36:240–45.

Simpson, G. 1999. Carrion insects from Nova Scotia. Honor's thesis, Department of Biology, St. Mary's University Halifax, Nova Scotia.

Smith, K. G. V. 1975. The faunal succession of insects and other invertebrates on a dead fox. *Entomological Gazette* 26:277.

Smith, K. G. V. 1986. *A manual of forensic entomology.* London: Trustees of the British Museum (Natural History).

Stafford, F. 1971. Insects of a medieval burial. *Science and Anthropology* 7:6–10.

Stoker, R. L., W. E. Grant, and S. B. Vinson. 1995. *Solenopsis invicta* (Hymenoptera: Formicidae) effect on invertebrate decomposers of carrion in central Texas. *Environmental Entomology* 24:817–22.

Tabor, K. L., C. C. Brewster, and R. D. Fell. 2004. Analysis of the successional patterns of insects on carrion in southwest Virginia. *Journal of Medical Entomology* 41:785–95.

Tabor, K. L., R. D. Fell, and C. C. Brewster. 2005. Insect fauna visiting carrion in southwest Virginia. *Forensic Science International* 150:73–80.

Teather, R. G. 1994. *Encyclopedia of underwater investigations.* Flagstaff, AZ: Best Publishing Company.

Teskey, H. H., and C. Turnbull. 1979. Diptera puparia from pre-historic graves. *Canadian Entomology* 111:527–28.

Tullis, K., and M. L. Goff. 1987. Arthropod succession in exposed carrion in a tropical rainforest on O'ahu Island, Hawai'i. *Journal of Medical Entomology* 24:332–39.

Tunnicliffe, V. 2006. Personal communication. Project director, Canada Research Chair in Deep Oceans, Professor in Biology and in Earth and Ocean Sciences, University of Victoria, Victoria, British Columbia.

Turner, B., and P. Wiltshire. 1999. Experimental validation of forensic evidence; a study of the decomposition of buried pigs in a heavy clay soil. *Forensic Science International* 101:113–22.

Van den Oever, R. 1976. A review of the literature as to the present possibilities and limitations in estimating the time of death. *Medicine, Science and the Law* 16:269–76.

VanLaerhoven, S. L. 1997. Successional biodiversity in insect species on buried carrion in the Vancouver and Cariboo regions of British Columbia. MPM thesis, Department of Biological Sciences, Simon Fraser University, Burnaby, British Columbia.

VanLaerhoven, S. L., and G. S. Anderson. 1996. *Forensic entomology. Determining time of death in buried homicide victims using insect succession.* Technical Report TR 02-96, Canadian Police Research Centre, Ottawa, ON.

VanLaerhoven, S. L., and G. S. Anderson. 1999. Insect succession on buried carrion in two biogeoclimatic zones of British Columbia. *Journal of Forensic Sciences* 44:31–41.

Wall, R., N. French, and K. Morgan. 1992. Blowfly species composition in sheep myiasis in Britain. *Medical and Veterinary Entomology* 6:177–78.

Watson, E. J., and C. E. Carlton. 2003. Spring succession of necrophilous insects on wildlife carcasses in Louisiana. *Journal of Medical Entomology* 40:338–47.

Watson, E. J., and C. E. Carlton. 2005. Insect succession and decomposition of wildlife carcasses during fall and winter in Louisiana. *Journal of Medical Entomology* 42:193–203.

Wells, J. D., and B. Greenberg. 1994. Effect of the red imported fire ant (Hymenoptera: Formicidae) and carcass type on the daily occurrence of postfeeding carrion-fly larvae (Diptera: Calliphoridae, Sarcophagidae). *Journal of Medical Entomology* 31:171–74.

Wolff, M., A. Uribe, A. Ortiz, and P. Duque. 2001. A preliminary study of forensic entomology in Medellin, Colombia. *Forensic Science International* 120:53–59.

Insect Succession in a Natural Environment

6

K. LANE TABOR KREITLOW

Contents

Introduction

Insects typically are the first organisms to discover a body after death. They are capable of arriving at and colonizing within minutes of the victim's final breath (Anderson and VanLaerhoven 1996, Nuorteva 1977, Erzinclioglu 1983, Catts 1990, Catts and Goff 1992, Smeeton et al. 1984, Smith 1986, Dillon 1997, Wells and Lamotte 2001). As such, it is possible to infer a portion of the postmortem interval (PMI) from the age of the fly larvae that arose from the first eggs deposited on the body. The entomological basis for the postmortem interval estimation may also be referred to as the period of insect activity (PIA) or the time of colonization (TOC). When weather conditions are optimal and there are no extenuating circumstances preventing or delaying oviposition, the minimum PMI (or PIA/TOC) is obtained with the assumption that flies did, indeed, deposit eggs very shortly after death. Age is derived by comparing the degree of development of maggots collected from a corpse with maggots of the same species reared under constant conditions (Kamal 1958, Nuorteva 1977, Smith 1986, Catts 1990). Development data are available for a number of forensic flies under a number of different temperature/relative humidity regimes (Kamal 1958, Ash and Greenberg 1975, Byrd and Butler 1996, 1997, Greenberg 1991, Grassberger and Reiter 2001, 2002, Anderson 2000b, Byrd and Allen 2001). Likewise, historical weather data at or near the location of the body can be obtained from the hundreds of weather stations placed throughout the United States (http://lwf.ncdc.noaa.gov). Insect development as an indicator for portions of the postmortem interval is most relevant when the victim has been dead only a short period of time, before the first wave of flies has left the corpse.

Oftentimes, the body remains undiscovered until much later in the decomposition process, when the early-arriving flies have already departed. In such situations, the main focus on insects as indicators of the postmortem interval shifts from insect development to insect succession patterns, although the two methods are complementary (Wells et al. 1999). The successive nature of faunal visitation on the corpse is somewhat predictable (Payne 1965, Rodriguez and Bass 1983, Anderson 2001). Thus, a comparison of the fauna collected at the time of discovery (corpse fauna) with the fauna collected in controlled studies in similar environments (baseline fauna) can provide the forensic entomologist with the means with which to estimate the PMI, often when little or no other evidence is available (Schoenly et al. 1996). In addition, information on the diversity of species, relative abundance of taxa, life stages present, and the number of individuals in each life stage can be obtained from succession studies (Keh 1985).

Faunal Succession Defined

Payne (1965) describes faunal visitation on carrion as a process of ecological succession. Ecological succession refers to the orderly and predictable changes in structure or composition of an ecological community through time. Ecological succession is influenced by a variety of factors, including inter- and intraspecies dynamics, environmental conditions, random occurrences, etc. The corpse and all of its inhabitants, along with the environment in which it exists, is itself an ecological community. The attractiveness of the corpse to specific fauna changes with the progression of decay; thus, the composition of fauna utilizing the corpse changes with time as well. For example, early-arriving insects, such as flies in the families Calliphoridae and Sarcophagidae, are attracted to the corpse during the fresh stage of decay, which is characterized by the presence of soft tissue. Such flies deposit eggs or larvae in moist areas of the corpse, which subsequently utilize the soft tissue as the sole food source throughout larval development. In turn, later stages of decomposition characterized by the absence of soft tissue render the corpse inappropriate for the same early-arriving flies. Thus, the composition of the corpse community changes in a predictable way. As with most biological organisms, specific resources/nutrients are necessary for the proper development of carrion arthropods. Ergo, the type and composition of fauna found on a corpse are indicative of its stage of decomposition (Mégnin 1894, Reed 1958, Payne 1965, Early and Goff 1986). Conversely, faunal assemblages at a given stage of decay are somewhat predictable.

Figure 6.1 Many visitors to carrion exploit the insects feeding directly on the carcass. Insect predators are often seen feeding on maggots and other carrion feeders. (Photo courtesy of Dr. K. Lane Tabor Kreitlow.)

Postmortem Changes in the Animal Model

The concept of ecological succession is fundamental to the field of ecology, which is simply defined as the study of the interrelationships of an organism and its environment, but it is quite complex because it includes not only abiotic factors (e.g., matter and energy), but also biotic factors (e.g., genes, cells, tissues, organs, organisms, populations, communities). An organism's environment may further be defined by physical factors such as temperature, humidity, and light, and biological factors such as members of the same species, food sources, natural enemies, and competitors. With regard to forensic entomology, the ecology of the insect plays a crucial role in behavior and activity because the rate of decomposition can be greatly affected by environmental and biological factors. A decomposing corpse is a dynamic ecosystem, and these identifiable phases attract different fauna as decomposition progresses (Smith 1986, Goff 1993).

During decomposition, human remains undergo physical and chemical changes that are often described by distinct stages, although the process is continuous (Smith 1986, Bornemissza 1957, Early and Goff 1986, Goff 1993). Of the different groups of arthropods that have been categorized based on their attraction to the corpse at different stages of decomposition, four basic arthropod-corpse relationships have generally been accepted: necrophagous fauna, which feed and breed on carrion; predators and parasites of necrophagous fauna (Figure 6.1); omnivores, which feed on both the corpse and other colonizers; and other species that exploit the corpse as opportunists. Like the stages of decomposition, the attractiveness of decomposing remains to a particular insect is ephemeral (Smith 1986, Goff 1993).

The fate of the pig carcass is remarkably similar to that of human cadavers in its process of decay and attractiveness to insects. Pig carcasses can therefore be used to illustrate the postmortem changes that human corpses undergo when allowed to decompose.

Fresh Stage

The fresh stage of decay lasts from the moment of death until the first signs of bloating. While no outward signs of decomposition may be apparent, internal bacteria begin to digest the body's internal organs, producing odor cues that attract the first insects. During this stage the first colonizers arrive, typically blow flies (Calliphoridae) and flesh flies (Sarcophagidae), which can be collected with a standard sweep net (Figure 6.2).

Figure 6.2 When a carcass is in the fresh stage of decomposition, there may be no outwardly visible signs of decay, but a great deal of bacterial decomposition is occurring internally. During the fresh stage, collecting specimens primarily involves catching flying insects on or around the carcass. (Photo courtesy of Dr. Ksenia Onufrieva.)

Figure 6.3 A carcass in the bloated stage of decay may appear inflated and balloon-like as pressure from the gases produced by decomposition builds up in the body cavity. (Photo courtesy of Dr. K. Lane Tabor Kreitlow.)

Bloated Stage

The bloated stage marks the beginning of putrefaction. Metabolic processes result in the production of gases by anaerobic bacteria, which may cause the abdomen to distend. Later the corpse may appear balloon-like (Figure 6.3). During this stage, the soil beneath the corpse tends to become alkaline, affecting the normal soil fauna. Calliphorid and sarcophagid numbers usually peak during the bloated stage.

Decay Stage

The onset of decay is considered to begin when gasses escape and the remains become deflated. Dipteran larvae forming large maggot masses are predominant during this stage (Figure 6.4). In addition, large numbers of predaceous coleopterans begin to arrive. By the

Figure 6.4 Large maggot masses form from fly larvae during the decay stage of decomposition. These masses produce an incredible amount of metabolic heat, which may be detected by holding your hand a few inches above the mass. Many predaceous beetles appear to feed off the plentiful maggots. (Photo courtesy of Dr. K. Lane Tabor Kreitlow.)

Figure 6.5 At the point of postdecay, the soft tissue has been removed, leaving mostly skin, cartilage, and bones. Many beetles have very strong mandibles that enable them to utilize the tough remains. (Photo courtesy of Dr. K. Lane Tabor Kreitlow.)

end of the decay stage, most of the flesh has been removed from the corpse and most of the Calliphoridae and Sarcophagidae have departed the remains.

Postdecay/Dry Stage

The postdecay/dry stage of decomposition is characterized by the reduction of the corpse to skin, cartilage, and bones (Figure 6.5). At this stage, various beetles that feed on the dry remains are predominant.

Skeletal (Remains) Stage

By this final stage of decomposition, the remains consist of only hair and bones (Figure 6.6). Most of the previous taxa will have disappeared, leaving mainly mites as the useful indicators of the PMI during this stage (Early and Goff 1986, Goff 1993).

Figure 6.6 By the skeletal stage, the remains consist only of hair and bones, with mainly mites as the arthropod fauna. (Photo courtesy of Dr. K. Lane Tabor Kreitlow.)

The Faunal Succession Study

Data obtained from faunal succession studies can be applied to cases where a corpse is discovered in a similar area as where the studies were conducted. When a person dies unattended, insects can arrive and begin colonizing the corpse remarkably soon after death (Nuorteva 1977, Erzinclioglu 1983, Catts 1990, Catts and Goff 1992, Smith 1986, personal observation). In addition to the age of the earliest colonizers, correctly identifying the taxa and patterns of colonization is essential to deducing accurate information about time since death from the entomological evidence. Likewise, it is crucial to use data from a very similar area to where a corpse is discovered because there can be great variations in fauna and patterns of activity across geographic locales. A review of forensic succession studies is presented at the end of this chapter, but there is a continuous need to add to the database of information available for many areas around the globe. The typical faunal succession study is therefore discussed below.

The Utility of the Animal Model in Succession Studies

The major objective of insect succession studies is to study the colonization patterns of insects in a particular area, often with the goal of obtaining data that can be applied to forensic investigations involving insects in the event that an unassisted death occurs in a similar environment. Although there have been a few studies using human cadavers to conduct anthropological/entomological research (Rodriguez and Bass 1983, 1985), in general, the use of human cadavers for field research is restrictive, and for the most part unavailable, for political and logistical reasons. Thus, most insect succession studies have been performed on animal models of various sorts, including guinea pigs (Borzemissza 1957), chickens (Hall and Doisy 1993), rabbits (Bourel et al. 1999), bears, deer, and alligators (Watson and Carlton 2005), lizards and toads (Cornaby 1974), dogs (Reed 1958), domestic cats (Early and Goff 1986), rats (Moura et al. 1997, Tomberlin and Adler 1998), and most commonly, the domestic pig (Payne 1965, Anderson and VanLaerhoven 1996, Avila and Goff 1998, Tabor et al. 2005, Carvalho et al. 2000, Hewardikaram and Goff 1991, and many others). The domestic pig is an ideal model for humans because it is relatively hairless compared with most other mammals, has similar skin type and physiology, is readily available, and can grow to be the size of an average adult human (Catts 1990, Catts and Goff 1992, Anderson 2001). Pig carcasses can be easily modified to replicate the physical condition of the victim in death cases where a reenactment might be warranted. The pattern of arthropod succession has not been shown to alter with carcass size (Hewadikaram and Goff 1991), although a larger carcass will be able to support a greater number of individuals.

Sampling in Faunal Succession Studies

In a typical faunal succession study with forensic applications, a fresh carcass is placed at a field site under some type of cage to ensure protection from scavengers. The enclosure should be solid enough to withstand disturbance from large vertebrates such as dogs and coyotes, but still allow insects to freely colonize the carcass. The cage should be securely staked to the ground by remaining open at the bottom to allow direct contact with the ground. Figure 6.7 illustrates a typical cage that can be used in these types of studies. These cages were constructed out of 2.54 cm steel-welded tubing enclosed with 1.27 cm mesh hardware cloth. The number of replicates used is often dictated by the availability of resources, but a large amount of data can be collected from only one or two carcasses.

Samples can be collected through sweep-netting flying insects above and around the carcass, by removing specimens directly from the carcass, and by pitfall traps. A typical pitfall trap is a hollowed-out area adjacent to the carcass in which a receptacle (such as a plastic cup) is filled with soapy water. One or two drops of liquid dish soap to a cup-sized container filled with water are sufficient. The trap is then camouflaged with leaves or other ground debris so that crawling insects fall into the trap and become trapped in the soapy water. Several pitfall traps placed around a carcass can yield an enormous number of individuals. Pitfall traps also provide passive trapping of nocturnal insects when it is not possible to actively sample at night.

Figure 6.7 Cages used to house carcasses in faunal succession studies must be sturdy enough to prevent vertebrate scavenging. This cage has a trapdoor that provides easy access to the carcass without having to remove the rebar stakes securing it to the ground. (Photo courtesy of Dr. K. Lane Tabor Kreitlow.)

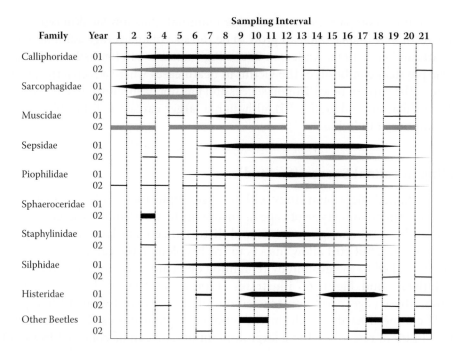

Figure 6.8 Occurrence and relative abundance (indicated by the thickness of bands) of families of insects found on pig carcasses in spring 2001 and 2002 in southwest Virginia. Sampling interval is 1 day. (Figure courtesy of Tabor et al. 2005.)

Data Collected from Faunal Succession Studies

Faunal succession studies can provide a huge amount of information about local carrion arthropods. Minimally, a checklist of the species found in a given geographic location can be identified. These data can be valuable in cases where a body has been moved and the fauna collected from the corpse does not match local species. Other approaches to data collection in succession studies may include taxonomic diversity, species richness, and relative abundance (Figure 6.8). Inter- and intraspecies dynamics can affect the relative abundance of taxa found on a corpse. Larvae of the hair maggot blow fly, *Chrysomya rufifaces* Macquart, are predaceous on maggots of other species; thus, their presence can alter the relative abundance of species arriving at the same time. Because of this predatory nature, along with their shorter life cycle than other larvae, the postmortem interval from entomological data can be altered by the presence of this species. Prior knowledge of the taxonomic diversity and relative abundance for an area can help preempt erroneous interpretation of the PMI where *C. rufifaces* is common. The presence of fire ants near a corpse can also affect the relative abundance of forensic taxa or delay colonization by feeding disproportionately on a given species or life stage (Castner and Byrd 2000). The red imported fire ant, *Solenopsis invicta*, an invasive species that has colonized the southern United States, is very common, and its presence near a corpse should be noted.

Identifying Specimens

Particularly fundamental to the successful application of entomological data to PMI estimation is the correct identification of the fauna collected from the corpse and in controlled studies. Entomological data can either be identified by the investigator himself or herself (in the case of controlled studies performed by an entomologist) or sent to an entomologist with expertise in taxonomy of forensic arthropods. Many forensic insects can be site-identified to family, but the more specifically an organism is taxonomically classified, the greater the expertise necessary. Likewise, knowledge of genera and species and their activity patterns provides a greater amount of information to be used in forensic applications. Using the black blow fly, *Phormia regina*, as an example, the levels of taxonomic classification are as follows:

Kingdom: Animalia
Phylum: Arthropoda
Class: Hexapoda (Insecta)
Order: Diptera
Family: Calliphoridae
Genus: *Phormia*
Species: *Phormia regina*

In accordance with Castner and Byrd (2000), the purpose of the information in this section is not to encourage forensic investigators to make a positive species identification, but rather to illustrate the methods and resources available for identifying entomological evidence in forensic applications.

The dichotomous key is a tool used across a variety of disciplines to identify a given taxon (taxonomic category) by using a series of paired statements (alternatives). Through a sequence of alternatives, the decision at each juncture dictates which direction toward the final taxonomic identification to head. The decisions are based on a given character (or set of characters) from which the taxonomic identification can be deduced. Typically, nonsubjective characters such as measurements or the presence or absence of structures is used. Many dichotomous keys specific to carrion arthropods are available, but they require a great amount of entomological expertise (Greenberg and Singh 1995, Smith 1986, Greenberg and Szyska 1984, Hall and Townsend 1977, Liu and Greenberg 1989, Anderson and Peck 1985, Ratcliffe 1996, Knipling 1939). For familiarization with the appearance and habits of common forensic species, insect identification cards provide an excellent resource for nonexperts that can be used in the field (Castner and Byrd 1998).

Estimating the PMI from Succession Patterns

The postmortem interval is derived from the timeline of arthropod assemblages described by controlled studies. Schoenly et al. (1996) present a method of organizing data from forensic succession studies that allows the PMI to be deduced from a comparison of the assemblage of taxa collected from the corpse at the time of discovery (corpse fauna), with the assemblage of taxa collected systematically in controlled studies along a timeline throughout decomposition (baseline fauna). Data are placed into an occurrence matrix, which describes the presence or absence of a taxon at a specific point within the timeline. When a corpse is discovered, sampling at the time of discovery yields an assemblage of invertebrates that ideally will match an assemblage at a certain point in time from the baseline data, from which the time since death can be deduced.

Ideally, the systematic arrival and departure of insects to a decomposing corpse provide a timeline of expected visitation, which can be used to predict the PMI from baseline data. At the family level of taxonomic classification, faunal assemblages are generally predictable across geographic locations (Bornemissza 1957, Reed 1958, Payne 1965, Carvalho et al. 2000, Anderson and VanLaerhoven 1996, Early and Goff 1986, Tabor et al. 2005, Archer and Elgar 2003). However, at more discrete levels of classification (e.g., genus or species) there is greater variability in faunal succession patterns with respect to geographic location (Payne 1965, Carvalho et al. 2000, Anderson and VanLaerhoven 1996, Early and Goff 1986, Tabor et al. 2005, Archer and Elgar 2003, Sukontason et al. 2003); thus, it is important to apply relevant baseline data when estimating portions of the postmortem interval from succession patterns.

For successful application of succession data to PMI, the occurrence and patterns of visitation for forensically important taxa in the specific geographic location (or a very similar area) in which decomposing remains are discovered must be understood (Anderson 2000a). Identifying the forensic arthropods for a given biogeographic zone is fundamental to entomological applications to the postmortem interval. Many faunal succession studies result in a checklist of carrion arthropods specific to the geographic location in which the study was conducted (Arnaldos et al. 2004, Carvalho et al. 2000, Anderson and VanLaerhoven 1996, Early and Goff 1986, Tabor et al. 2005, Archer and Elgar 2003, Sukontason et al. 2003, Braack 1986, Moura et al. 1997, and many others). In addition to knowing *which* taxa occur in a given area, the pattern of visitation of these taxa is important because behavioral differences within species are possible depending on geographic location (Anderson 2000a). Arthropod colonization patterns of a corpse are descriptive of a given area because there is such specificity.

Likewise, seasonal activity of forensic insects is important to consider, particularly in temperate environments where many insects exhibit seasonally specific behavior. Thus, patterns of visitation can be expected to be season specific as well (Reed 1958, Arnaldos et al. 2004, Archer and Elgar 2003, Tabor et al. 2005, Tomberlin and Adler 1998). As such, seasonal differences in activity of forensic insects do not preempt their utility in forensic investigations, provided they are taken into consideration when choosing the succession data on which to base colonization interval estimations.

Determining Season or Year from Succession Patterns

The utility of arthropod succession patterns to postmortem interval does not cease with the arrival of the skeletal stage of decomposition. A combination of alkalinity caused by ammonia excretions from the maggot mass and purged body fluids causes the pH of the soil beneath the corpse to change, in turn affecting the composition of soil fauna for many months or years following death (Bornemissza 1957, Goff 1991, Anderson and VanLaerhoven 1996). An analysis of the changes in soil faunal composition can therefore provide investigators with clues many months or even years following death.

Any information regarding postmortem interval can be valuable in death investigations, even if the interval is wide. The presence or absence of certain insects or insect artifacts can be an indication of the season in which death occurred (Archer and Elgar 2003). In cases where bodies are discovered months or years after death, remnants from long-departed seasonally active insects often remain persistent in the environment and can provide clues as to the season in which death occurred. The hard exoskeleton characteristic of insects enables remnants such as body parts and pupal cases to withstand degradation, so they may remain in the vicinity of the crime scene months or many years following death (Gilbert and Bass 1967, Teskey and Turnbull 1979, Nuorteva 1987, Archer and Elgar 2003). However, the absence of remnants should not be construed as the absence of a given species, as remnants are easily overlooked and subject to scavenger removal (Archer and Elgar 2003). Seasonal variation in activity of carrion arthropods has been observed in many studies (Reed 1958, Anderson and VanLaeurhoven 1996, Arnaldos et al. 2004, Archer and Elgar 2003, Gruner et al. 2007, Tabor et al. 2005, Tomberlin and Adler 1998).

Limitations of Using Succession Patterns to Estimate Postmortem Interval

Patterns of arthropod succession can provide valuable clues to an investigation, particularly when used in conjunction with development of the earliest colonizers of the corpse. The basis of using succession patterns to estimate PMI rests on the consistency of colonization patterns in a given area. Insects are cold blooded (poikilothermic), so activity and behavior are highly dependent on the ambient temperature. In mild climates, insects are generally active year-round, so succession patterns remain fairly consistent. In temperate areas, which most of the United States is considered, many insects undergo behavioral and physiological changes concurrently with changes in season. For example, the black blow fly, *Phormia regina*, undergoes diapause in many areas, which is a hormonally controlled period of reduced activity characterized by cessation of growth and reduction of metabolic activity (Stoffola et al. 1974). It is therefore possible to see a decrease in the population of certain insects in the fall, even when temperatures and climactic conditions are similar to what they were in the spring when activity for the same species was much higher (Tabor et al. 2005). Thus, for greatest accuracy, estimation of postmortem interval from succession data requires information of local carrion fauna and their colonization patterns at different times of the year (Anderson 2001, Wells and Lamotte 2001).

A Global Review of Succession Studies

Faunal succession studies on carrion insects provide information on the insects that occur in a particular geographical region that can sometimes be used to help estimate the PMI. Dozens of decomposition studies on the succession patterns of arthropods (based on arthropod visitation) have been conducted throughout the world in climactically different areas, many of which resulted in a checklist of the forensic arthropods that occur in the region in which the studies were conducted. The following review presents succession studies in the natural environment of a variety of geographic locations, with the intention of assisting investigators in locating data that might be useful in cases that occur in similar environments. While this is by no means an exhaustive list of all of the published succession studies that have been conducted worldwide, it provides a composite of what data are available in different regions of the world.

Africa, Asia, Australia, and New Zealand

Forensic entomology has been used for decades in southern Africa, but expressly forensic entomological research began in southern Africa only within the past few decades. As such, much of the data are difficult to obtain or unpublished. Williams and Villet (2006) present an excellent review of the history and status of forensic entomology research in southern Africa. Braack (1986) provides an extensive list of arthropods associated with carcasses in northern Kruger National Park in South Africa. A total of 227 species in 36 families were collected from impala carcasses placed in field sites at seasonal intervals. Abundance, food relations, and associations with carrion habitat are also described. No additional species were recovered from naturally occurring vertebrate carcasses (Braack 1986).

Sukontason et al. (2003) describe a survey of forensically relevant flies in six mountainous sites in Chiang Mai Province (17–21°N and 98–99°E) of northern Thailand. The summer studies were the first of their kind in mountainous areas of the province. The objective is to establish baseline data that can be useful in many cases of corpses found in various stages of decay each year in Thailand (Sukontason et al. 2003). Another preliminary study of insect succession on pig carrion in Asia was conducted in Phitsanulok, northern Thailand, located at 16°N 101°E (Vitta et al. 2007). The region is about 44 m above sea level and has average temperatures of 26.4, 30.5, and 27.4°C in the cold, hot, and rainy seasons, respectively, with an annual rainfall of approximately 1,400.8 mm. The authors identify five stages of decomposition and present species identification for each stage (Vitta et al. 2007).

Bornemissza (1957) characterized the succession of organisms on guinea pig carcasses in the Mediterranean-type environment of Western Australia. Five stages of decomposition were identified and were found to correlate with the fauna that were present. In addition to a diagram of succession patterns, the study found that carrion decomposition affected the physical properties of the soil and its arthropod population beneath it. The succession and rate of development of carrion blow flies were investigated in southern Queensland, Australia (O'Flynn 1983). The studies were carried out on sheep, pig, and dog carcasses on a farm on the outskirts of Brisbane and in grazing country in the Thallon district. Blow fly succession patterns are described for winter and summer for both field sites.

Lang et al. (2006) studied blow fly succession on brushtail possum in a sheep field in Tasmania (42°48'S, 147°25'E, 10–20 m above sea level) The species and number of flies colonizing the carcasses for up to 17 days are presented. Yearly succession patterns in a forest setting of southern Australia were described from a 2-year study performed on neonatal piglets (Archer and Elgar 2003). The profiles and activity patterns of ten seasonally active taxa, chosen for their liklihood to leave remnants (e.g., pupal cases, exoskeletons) that can indicate possible months in which death occurred, were described. Smeeton et al. (1984) performed the first published succession study conducted in New Zealand using human cadavers. Fifty corpses were used in the study and allowed to decompose for a few weeks or less; therefore, the species identified in the study were early arrivers.

Other studies focusing on insect succession patterns or community dynamics on carcasses include Ellison (1990), Braack (1981, 1987), Louw and van der Linde (1993), van Wyk et al. (1993), van der Linde and Hugo (1997), and Kolver and van der Linde (1999) (in reference only) in southern Africa, Bharti and Singh (2003) in India, Tantawi et al. (1996) in Egypt, and Chin et al. (2007) in Malaysia.

Central and South America

Decomposition of lizard and toad carcasses was studied in a tropical dry and a tropical wet forest in Costa Rica (Cornaby 1974). Over 170 species representing 49 families were identified. Differences in temporal succession patterns and species compositions were observed between the study sites. A survey of carrion insects was conducted in Curitiba, State of Paraná in Brazil, with the objective of providing a preliminary database of medicolegal importance in southern Brazil (Moura et al. 1997). Insects were collected from ~250 kg rat carcasses placed in an urban environment and in a forest. In addition to a checklist of the most common arthropods associated with carcasses in the region, the authors also provide a chart of the seasonal variation of necrophagous species in both environments, as well as the distribution of third-instar larvae of insect species throughout the phases of decomposition, which they describe as fresh, bloat, decaying, adipocere-like, and dry.

Observations of insect succession on decomposing a dog carcass were made during the dry season in a secondary forest in Costa Rica at an altitude of 1,200 m (Jirón and Cartín 1981). Patterns of decomposition in the tropical area of the study were the same as described for temperate zones, although the insect fauna and ecological complexity were different. The appearance of insect populations through the stages of decomposition and the principal insect consumers are presented.

Carvalho et al. (2000) provides a checklist of carrion arthropods collected from pig carcasses placed under natural conditions in an urban forest in Campinas São Paulo, Brazil. Sampling was conducted until the entire carcass was consumed. Other data included in the checklist were observed and collected from human corpses at the Institute of Legal Medicine, which is located in the same city. While the checklist does not give a temporal description of taxon occurrence, the authors do indicate whether individuals are post-mortem interval or area indicators. The frequency of Coleoptera and Diptera of forensic importance is also presented.

An investigation of the Coleoptera that inhabit carrion was conducted in Brazil during a 1-year study (Mise et al. 2007). One hundred and twelve species of twenty-six families, twelve of which were considered of forensic potential, were collected from daily sampling on a 15 kg pig carcass. The authors present an occurrence matrix of the families and species collected during the study throughout the decay process. Martinez et al. (2007) describe insect succession in the high-altitude plains of Colombia at 3,035 m above sea level, where annual temperatures fluctuate between 4.5 and 21.4°C and the average relative humidity is over 80%. A table of faunal data is presented for all five recognized stages of decomposition: fresh, bloated, active, advanced, and decay.

An ongoing study of insect succession on pig carrion was conducted in Medellín, Colombia, in which the insect sequence was characterized for over 207 days (Wolff et al. 2001). In addition to a comprehensive checklist of the arthropods observed throughout the study, the authors present the total percentage of families attracted to the carcasses at various stages of decay, from which the relative densities can be interpreted. Five decomposition stages and their characteristic fauna were observed (Wolff et al. 2001). Patterns of arthropod visitation on pig carcasses were examined in Buenos Aires Province, Argentina, in all seasons of the year (Centeno et al. 2002). The authors describe the site as humid warm-temperate, with warm summers where temperatures approach 40°C when dry, and may be as low as 18°C after the common occurrence of sudden showers, mildly cold winters, and an annual rainfall of 900 to 950 mm. A checklist of carrion arthropods is provided, as well as figures illustrating the succession patterns of important forensic insect families for spring, summer, winter, and fall.

The seasonality and relative abundance of dipteran and coleopteran of forensic significance were examined in southeastern Brazil (De Souza and Linhares 1997). The climate of the region is seasonal, ranging from dry and cool in the winter to warm and wet in the summer. The seasonal frequency and stages of ovarian development of the six species of Calliphoridae are presented. Also included is the relative abundance of adult and immature Coleoptera collected throughout the year.

Europe

A comprehensive study of the arthropods that colonize carrion was conducted in the southeastern Iberian Peninsula (Arnaldos et al. 2004). A thorough checklist of data is presented for all four seasons during four described stages of decomposition: fresh, decomposition, advanced decomposition, and skeletonization. In addition, relative abundance and differences in seasonal succession patterns are described.

Grassberger and Frank (2004) conducted arthropod succession studies in an urban backyard in the city of Vienna, Austria. The geographic position of the city ranges from longitude 16°11'03"to 16°34'43"east and from latitude 48°07'06" to 48°19'23" north. The elevation of the study site was 175 m above sea level. Arthropods were collected from clothed pig carcasses from May to November. The authors present a checklist of arthropods collected during the study, as well as a diagram of the chronology and relative abundance of insects visiting carrion for 60 days.

Lane (1975) investigated blow fly succession on mice carrion in Berkshire, England, in July. The relative abundance of local blow fly species is presented for different habitats. Results from the study indicate that blow fly populations are affected primarily by intra- and interspecific competition.

Bourel et al. (1999) examined insect succession on rabbit carrion in sand dune habitats in northern France during the spring seasons. A checklist of 33 arthropod species representing 7 orders and 25 families is presented. A chronology of insect appearance and relative abundance is given for up to 110 days of decomposition.

North America

Numerous studies of carrion succession have been conducted in North America. A comprehensive year-round study was conducted in Tennessee on dog carcasses (Reed 1958). The forty-three carcasses used in the study were distributed at different times of the year in wooded and nonwooded areas. Arthropods were classified according to the stage of decomposition during which they were found most frequently. Seasonal variations in activity are noted for many taxa. In another early study, Payne (1965) presents faunal succession patterns on pig carcasses in a hardwood-pine community in South Carolina. Six stages of decomposition are recognized, and each stage was found to be colonized by a characteristic group of arthropods. A total of 422 insect species representing 11 orders, 107 families, and 283 genera were identified in this study (Payne 1965).

Early and Goff (1986) described arthropod succession patterns in Hawaii using domestic cat carcasses. Stages of decomposition similar to those described by Bornemissza (1957) were recognized. A thorough checklist of arthropod taxa collected from carrion is presented. The pattern of arthropod succession observed in this study was found to be consistent with other studies in tropical and temperate areas (Bornemissza 1957, Reed 1958, Payne 1965).

A comprehensive examination of carrion arthropods was conducted in the summer in southwestern British Columbia (Anderson and VanLaerhoven 1996). The study site is in a rural farming area of the Coastal Western Hemlock biogeoclimatic zone. A thorough checklist of arthropods collected from pig carcasses is presented. In a separate table, arthropods collected from soil samples over time highlight the drastic changes in composition of soil arthropods caused by changes in soil pH from decomposition (Anderson and VanLaerhoven 1996).

Rodriguez and Bass (1983) collected succession data of the few studies that employed human cadavers in the study. Daily observations were made throughout the year. The data showed that there is a direct correlation between the rate of decay and the rate of succession of fauna found in association with the remains. Four distinct stages of decay were recognized.

An expansive list of carrion blow flies in Colorado was obtained from specimens collected from road-killed carcasses and specimens submitted by crime scene technicians (De Jong 1994). The elevation of Colorado ranges from 1,066 m to over 4,300 m. Thirty-three species in seventeen genera are reported, with notes on the carrion associations and forensic importance for particular species (De Jong 1994).

Tabor et al. (2005) present a checklist of insects collected from pig carrion in the spring and summer in the mountainous region of southwest Virginia. The study site is ~608 m above sea level and has average temperatures of 15, 20.7, and 15.5 °C in the spring, summer, and fall, respectively, with an average annual rainfall of 102 cm. In a similar article, the data were analyzed using the Jaccard similarity coefficient (Tabor et al. 2004). Results indicated that the succession patterns seem to be typical for the seasonal periods, and thus provide data on baseline fauna for estimating postmortem interval.

Conclusions

Variations in factors such as season, temperature, elevation, longitude, latitude, humidity, and sun exposure can have an effect on the colonization pattern of corpse insects (Smith 1986, Shean et al. 1993, Erzinclioglu 1996, Tantawi et al. 1996). As such, deviations from the physical factors of the geographic region from which baseline data are obtained and applied to casework can produce erroneous interpretations in PMI estimations (Anderson 2001, Rodriguez and Bass 1983). Thus, one would be remiss to assume that the cache of published data obtained from faunal succession studies on carcasses were sufficient in every scenario. The pronounced geographic variability at the genus and species level of taxonomic classification elucidates the need for continuous research in the area of insect succession studies across a variety of geographic locations. Faunal succession studies using animal models provide the means by which to add to the database of information available for potential use in estimating the the time since colonization.

References

Anderson, G. S. 2000a. Insect succession on carrion and its relationship to determining time of death. In *Forensic entomology: The utility of arthropods in legal investigations*, ed. J. H. Byrd and J. L. Castner. Boca Raton, FL: CRC Press, LLC, 143–75.

Anderson, G. S. 2000b. Minimum and maximum development rates of some forensically important calliphoridae (Diptera). *Journal of Forensic Sciences* 45:824–32.

Anderson, G. S. 2001. Forensic entomology in British Columbia: A brief history. *Journal of the Entomological Society of British Columbia* 98:127–35.

Anderson, G. S., and VanLaerhoven, S. L. 1996. Initial studies on insect succession on carrion in southwestern British Columbia. *Journal of Forensic Sciences* 41:617–25.

Anderson, R. S., and Peck, S. B. 1985. *The insects and arachnids of Canada*. Part 13. *The carrion beetles of Canada and Alaska*. Ottawa, Ontario: Biosystematics Research Institute.

Archer, M. S., and Elgar, M. A. 2003. Yearly activity patterns in southern Victoria (Australia) of seasonally active carrion insects. *Forensic Science International* 132:173–76.

Arnaldos, M. I., Romera, E., Presa, J. J., Luna, A., and Garcia, M. D. 2004. Studies on seasonal arthropod succession on carrion in the southeastern Iberian Peninsula. *International Journal of Legal Medicine* 118:197–205.

Ash, N., and Greenberg, B. 1975. Developmental temperature responses of sibling species *Phaenicia-sericata* and *Phaenicia-pallescens*. *Annals of the Entomological Society of America* 68:197–200.

Avila, F. W., and Goff, M. L. 1998. Arthropod succession patterns onto burnt carrion in two contrasting habitats in the Hawaiian Islands. *Journal of Forensic Sciences* 43:581–86.

Bharti, M., and Singh, D. 2003. Insect faunal succession on decaying rabbit carcasses in Punjab, India. *Journal of Forensic Sciences* 48:1133–43.

Bornemissza, G. F. 1957. An analysis of arthropod succession in carrion and the effect of its decomposition on the soil fauna. *Australian Journal of Zoology* 5:1–12.

Bourel, B., Martin-Bouyer, L., Hedouin, V., Cailliez, J., Derout, D., and Gosset, D. 1999. Necrophilous insect succession on rabbit carrion in sand dune habitats in northern France. *Journal of Medical Entomology* 36:420–25.

Braack, L. E. O. 1981. Visitation patterns of principal species of the insect-complex at carcasses in the Kruger National Park. *Koedoe* 24:33–49.

Braack, L. E. O. 1986. Arthropods associated with carcasses in northern Krueger National Park. *Wildlife Research Natuurnavorsing* 16:91–98.

Braack, L. E. O. 1987. Community dynamics of carrion-attendant arthropods in tropical African woodland. *Oecologia* 72:402–9.

Byrd, J. H., and Allen, J. C. 2001. The development of the black blow fly, *Phormia regina* (Meigen). *Forensic Science International* 120:79–88.

Byrd, J. H., and Butler, J. F. 1996. Effects of temperature on *Cochliomyia macellaria* (Diptera: Calliphoridae) development. *Journal of Medical Entomology* 33:901–5.

Byrd, J. H., and Butler, J. F. 1997. Effects of temperature on *Chrysomya rufifacies* (Diptera: Calliphoridae) development. *Journal of Medical Entomology* 34:353–58.

Carvalho, L. M. L., Thyssen, P. J., Linhares, A. X., et al. 2000. A checklist of arthropods associated with pig carrion and human corpses in southeastern Brazil. *Memorias do Instituto Oswaldo Cruz* 95:135–38.

Castner, J. L., and Byrd, J. H. 1998. *Forensic insect identification cards*. Gainesville, FL: Feline Press.

Castner, J. L., and Byrd, J. H. 2000. Insects of forensic importance. In *Forensic entomology: The utility of insects in legal investigations*. Boca Raton, FL: CRC Press, LLC, 43–79.

Catts, E. P. 1990. Analyzing entomological data. In *Entomology and death: A procedural guide*, ed. E. P. Catts and N. H. Haskell. Clemson, SC: Joyce's Print Shop, 124–37.

Catts, E. P., and Goff, M. L. 1992. Forensic entomology in criminal investigations. *Annual Review of Entomology* 37:253–72.

Centeno, N., Maldonado, M., and Oliva, A. 2002. Seasonal patterns of arthropods occurring on sheltered and unsheltered pig carcasses in Buenos Aires Province (Argentina). *Forensic Science International* 126:63–70.

Chin, H. C., Marwi, M. A., Salleh, A. F. M., Jeffery, J., and Omar, B. 2007. A preliminary study of insect succession on a pig carcass in a palm oil plantation in Malaysia. *Tropical Biomedicine* 24:23–27.

Cornaby, B. W. 1974. Carrion reduction by animals in contrasting tropical habitats. *Biotropica* 6:51–63.

De Jong, G. D. 1994. An annotated checklist of the Calliphoridae (Diptera) of Colorado, with notes on carrion assocations and forensic importance. *Journal of the Kansas Entomological Society* 67:378–85.

De Souza, A. M., and Linhares, A. X. 1997. Diptera and Coleoptera of potential forensic importance in southeastern Brazil: Relative abundance and seasonality. *Medical and Veterinary Entomology* 11:8–12.

Dillon, L. C. 1997. Insect succession on carrion in three biogeographic zones in British Columbia. MSc thesis, Department of Biological Sciences, Simon Fraser University, Burnaby, B.C.

Early, M., and Goff, M. L. 1986. Arthropod succession patterns in exposed carrion on the island of O'ahu, Hawaiian Islands, USA. *Journal of Medical Entomology* 23:520–31.

Ellison, G. T. H. 1990. The effects of scavenger mutilation on insect succession on impala carcasses in southern Africa. *Journal of Zoology* 220:679–88.

Erzinclioglu, Z. 1983. Application of entomology to forensic medicine. *Medicine, Science and the Law* 23:57–63.

Erzinclioglu, Y. Z. 1996. *Blowflies*. Naturalists' Handbook 23. Slough, Great Britain: Richmond Publishing Co. Ltd.

Gilbert, B. M., and Bass, W. M. 1967. Seasonal dating of burials from the presence of fly pupae. *American Antiquity* 34:534–35.

Goff, M. L. 1991. Comparison of insect species associated with decomposing remains recovered inside dwellings and outdoors on the island of Oahu, Hawaii. *Journal of Forensic Sciences* 36:748–53.

Goff, M. L. 1993. Estimation of postmortem interval using arthropod development and succession patterns. *Forensic Science Review* 5:81–94.

Grassberger, M., and Frank, C. 2004. Initial study of arthropod succession on pig carrion in a central European urban habitat. *Journal of Medical Entomology* 41:511–23.

Grassberger, M., and Reiter, C. 2001. Effect of temperature on *Lucilia sericata* (Diptera: Calliphoridae) development with special reference to the isomegalen- and isomorphen- diagram. *Forensic Science International* 120:32–36.

Grassberger, M., and Reiter, C. 2002. Effect of temperature on development of the forensically important blow fly *Protophormia terraenovae* (Robineau-Desvoidy) (Diptera: Calliphoridae). *Forensic Science International* 128:177–82.

Greenberg, B. 1991. Flies as forensic indicators. *Journal of Medical Entomology* 28:565–77.

Greenberg, B., and Singh, D. 1995. Species identification of Calliphorid (Diptera) eggs. *Journal of Medical Entomology* 32: 21–26.

Greenberg, B., and Szyska, M. L. 1984. Immature stages and biology of fifteen species of Peruvian calliphoridae (Diptera). *Annals of the Entomological Society of America* 77:488–517.

Gruner, S. V., Slone, D. H., and Capinera, J. L. 2007. Forensically important Calliphoridae (Diptera) associated with pig carrion in rural north-central Florida. *Journal of Medical Entomology* 44:509–15.

Hall, R. D., and Doisy, K. E. 1993. Length of time after death: Effect on attraction and oviposition or larviposition of midsummer blow flies (Diptera: Calliphoridae) and flesh flies (Diptera: Sarcophagidae) of medicolegal importance in Missouri. *Annals of the Entomological Society of America* 86:589–93.

Hall, R. D., and Townsend, Jr., K. E. 1977. The blow flies of Virginia (Diptera: Calliphoridae). In *The insects of Virginia*, No. 11. Virginia Polytechnic Institute and State University Research Division Bulletin 123, Blacksburg, VA.

Hewardikaram, K. A., and Goff, M. L. 1991. Effect of carcass size on rate of decomposition and arthropod succession patterns. *Journal of Forensic Medicine and Pathology* 12:235–40.

Jirón, L. F., and Cartín, V. M. 1981. Insect succession in the decomposition of a mammal in Costa Rica. *Journal of the New York Entomological Society* 89:158–65.

Kamal, A. S. 1958. Comparative study of thirteen species of sarcosaprophagous Calliphoridae and Sarcophagidae (Diptera). I. Bionomics. *Annals of the Entomological Society of America* 51:261–71.

Keh, B. 1985. Scope and applications of forensic entomology. *Annual Review of Entomology* 30:137–54.

Knipling, E. F. 1939. Key for blowfly larvae concerned in wound and cutaneous myiasis. *Annals of the Entomological Society of America* 32:376–83.

Kolver, J. H., and van der Linde, T. C. 1999. A comparison of the decomposition and insect succession of a hanging carcass versus a carcass lying on the ground. Paper presented at Proceedings of 12th Entomological Congress of the Entomological Society of Southern Africa, Potchefstroom, July 12–15.

Lane, R. P. 1975. An investigation into blowfly (Diptera: Calliphoridae) succession on corpses. *Journal of Natural History* 9:581–88.

Lane, M. D., G. R. Allen, B. J. Horton. 2006. Blow fly succession from possum (*Trichosurus vulpecula*) carrion in a sheep-farming zone. *Medical and Veterinary Entomology.* 20(4):445–52.

Liu, D., and Greenberg, B. 1989. Immature stages of some flies of forensic importance. *Annals of the Entomological Society of America* 82:80–93.

Louw, S. v.d. M., and van der Linde, T. C. 1993. Insects frequenting decomposing corpses in central South Africa. *African Entomology* 1:265–69.

Martinez, E., Duque, P., and Wolff, M. 2007. Succession pattern of carrion-feeding insects in Paramo, Colombia. *Forensic Science International* 166:182–89.

Mégnin, P. 1894. *La Faune des Cadavres.* Encyclopédie Scientifique des Aide-Memoire. Paris: G. Masson, Gauthier-Villars et Fils.

Mise, K. M., deAlmelda, L. M., and Moura, M. O. 2007. A study of the Coleopteran (Insecta) fauna that inhabit *Sus scrofa* L. carcass in Curitiba, Parana. *Revista Brasileira de entomologia* 51:358–68.

Moura, M. O., deCarvalho, C. J. B., and Monteiro, E. L. A. 1997. A preliminary analysis of insects of medico-legal importance in Curitiba, State of Parana. *Memorias do Instituto Oswaldo Cruz* 92:269–74.

National Satellite and Information Services. http://lwf.ncdc.noaa.gov.

Nuorteva, P. 1977. Sarcosaprophagous insects as forensic indicators. In *Forensic medicine: A study of trauma and environmental hazards.* Vol. 11. Philadelphia: W. B. Saunders Co., 1072–95.

Nuorteva, P. 1987. Empty puparia of Phormia terranovae R.-D. (Diptera, Calliphoridae) as forensic indicators. *Annales Entomologici Fennici* 53:53–56.

O'Flynn, M. A. 1983. The succession and rate of development of blowflies in carrion in southern Queensland and the application of these data to forensic entomology. *Journal of the Australian Entomological Society* 22:137–48.

Payne, J. A. 1965. A summer carrion study of the baby pig *Sus scrofa* Linnaeus. *Ecology* 46:592–602.

Ratcliffe, B. C. 1996. *The carrion beetles (Coleoptera: Silphidae) of Nebraska.* Bulletin of the University of Nebraska State Museum, Lincoln.

Reed, H. B. 1958. A study of dog carcass communities in Tennessee, with special reference to the insects. *The American Midland Naturalist* 59:213–45.

Rodriguez, W. C., and Bass, W. M. 1983. Insect activity and its relationship to decay rates of human cadavers in East Tennessee. *Journal of Forensic Sciences* 28:423–32.

Rodriguez, W. W., and Bass, W. M. 1985. Decomposition of buried bodies and methods that may aid in their location. *Journal of Forensic Sciences* 30:836–52.

Schoenly, K., M. L. Goff, J. D. Wells, and W. D. Lord. 1996. Quantifying statistical uncertainty in succession-based entomological estimates of the postmortem interval in death scene investigations: A simulation study. *American Engomologist* 42(2):106–12.

Shean, B. S., Messinger, L., and Papworth, M. 1993. Observations of differential decomposition on sun exposed v shaded carrion in coastal Washington-state. *Journal of Forensic Sciences* 38:938–49.

Smeeton, W. M., Koelmeyer, T. D., Holloway, B. A., and Singh, P. 1984. Insects associated with exposed human corpses in Auckland, New Zealand. *Medicine, Science and the Law* 24:167–74.

Smith, K. G. V. 1986. *A manual of forensic entomology.* Ithaca, NY: Cornell University Press.

Stoffola, J. G., Greenber, S., and Calabres, E. 1974. Facultative imaginal diapause in black blowfly, Phormia-regina. *Annals of the Entomological Society of America* 67:518–19.

Sukontason, K., Sukontason, K. L., Piangjai, S., Tippanun, J., Lertthamnongtham, S., Vogtsberger, R. C., and Olson, J. K. 2003. Survey of forensically-relevant fly species in Chiang Mai, northern Thailand. *Journal of Vector Ecology* 28:135–58.

Tabor, K. L., Brewster, C. C., and Fell, R. D. 2004. Analysis of the successional patterns of insects on carrion in southwest Virginia. *Journal of Medical Entomology* 41:785–95.

Tabor, K. L., Fell, R. D., and Brewster, C. C. 2005. Insect fauna visiting carrion in Southwest Virginia. *Forensic Science International* 150:73–80.

Tantawi, T. I., Wells, J. D, Greenberg, B., and El-kasy, E.M. 1996. Fly larvae (Diptera: Calliphoridae, Sarcophagidae, Muscidae) succession in rabbit carrion: Variation observed in carcasses exposed at the same time and the same place. *Journal of Egyptian German Society of Zoology* 25:195–208.

Teskey, H. J., and Turnbull, C. 1979. Diptera puparia from pre-historic graves. *Canadian Entomologist* 111:527–28.

Tomberlin, J. K., and Adler, P. H. 1998. Seasonal colonization and decomposition of rat carrion in water and on land in an open field in South Carolina. *Journal of Medical Entomology* 35:704–9.

van der Linde, T. C., and Hugo, L. 1997. The influence of freezing and burning on insect succession on dog carcasses. Paper presented at Proceedings of Joint Congress of the Entomological Society of Southern Africa (11th Congress) and the African Association of Insect Scientists (12th Congress), Stellenbosch, June 30–July 4.

van Wyk, J. M. C., Louw, S., Kriel, A., and van der Linde, T. C. 1993. Insects associated with covered carcasses in the central Orange Free State. Paper presented at Proceedings of 9th Entomological Congress of the Entomological Society of Southern Africa, Johannesburg, June 28–July 1.

Vitta, A., Pumidonming, W., Tangchaisuriya, U., et al. 2007. A preliminary study on insects associated with pig (*Sus scrofa*) carcasses in Phitsanulok, northern Thailand. *Tropical Biomedicine* 24:1–5.

Watson, E. J., and Carlton, C. E. 2005. Insect succession and decomposition of wildlife carcasses during fall and winter in Louisana. *Journal of Medical Entomology* 42:193–203.

Wells, J. D., Byrd, J. H., and Tantawi, T. I. 1999. Key to third-instar Chrysomyinae (Diptera: Calliphoridae) from carrion in the continental United States. *Journal of Medical Entomology* 36:638–41.

Wells, J. D., and LaMotte, L. R. 2001. Estimating the postmortem interval. In: *Forensic entomology: The utility of arthropods in legal investigations.* ed. J. H. Byrd and J. L. Castner. Boca Raton, FL: CRC Press, LLC. pp. 263–85.

Williams, K. A., and Villet, M. H. 2006. A history of southern African research relevant to forensic entomology. *South African Journal of Science* 102:59–65.

Wolff, M., Uribe, A., Ortiz, A., et al. 2001. A preliminary study of forensic entomology in Medellín, Colombia. *Forensic Science International* 120:53–59.

The Role of Aquatic Insects in Forensic Investigations

7

RICHARD W. MERRITT
JOHN R. WALLACE

Contents

Introduction

The interest to study aquatic insects has exhibited exponential growth from early roots in limnology and sport-fishery types of inquiry almost five decades ago (Merritt et al. 2008a). The areas of interest that aquatic ecologists have focused on in these studies include the major areas of ecological inquiry, e.g., predator-prey interactions, physiological and trophic dynamics, competition, population dynamics, pollution, disturbance, parasitism, and basic management applications of this research (Cairns and Pratt 1993; Rosenberg and Resh 1996; Merritt et al. 2008a). However, new avenues of research and application of aquatic ecological principles and models have arisen from the wealth of benthological knowledge applicable to medicolegal inquiries into suspicious human deaths (Keiper and Casamatta 2001). Historically, different organisms, such as plants, pollen, fungi, mammals, and insects, have proven useful in forming evidentiary linkages among suspects, victims, and property with specific locations, particularly at outdoor crime scenes (Hall 1990; Lane et al. 1990; Lord 1990; Smith 1986; Easton and Smith 1970).

In a court of law, medicocriminal forensic entomology is defined as the application of the study of insects and other arthropods to violent crimes such as murder, suicide, rape, as well as physical abuse. Lord (1990) stated that human corpses, whether they have been produced naturally or as the result of foul play, are processed by insect decomposers in the same manner as any other carrion. Although considerable amounts of important research and reviews in the field of forensic entomology were produced before 1990 (e.g., Smith 1986; Keh 1985; Meek et al. 1983; Rodriguez and Bass 1983; Nuorteva 1977), a large body of literature has emerged on arthropod colonization of terrestrial carrion (including humans) in the past 10 years. This has proved to be a great utility for forensic science (see reviews by Benecke 2004; Merritt et al. 2000; Haskell et al. 1997; Catts and Goff 1992; Catts and Haskell 1990). In spite of this tremendous body of research, the role of freshwater and marine fauna in forensic investigations in the past received very little attention (Hobischak 1997; Nawrocki et al. 1997; Vance et al. 1995; Catts and Goff 1992; Haskell et al. 1989). Lately, however, the potential use of aquatic organisms such as macroinvertebrates and algae in both freshwater and marine environments for the estimation of a postmortem submersion interval has received increased attention (Zimmerman and Wallace 2008; Wallace et al. 2008; Horton et al. 2006; Haefner et al. 2004; Hobishak and Anderson 2002; Keiper and Casamatta 2001; Casamatta and Verb 2000).

Despite this increase in interest, a review of the literature found that over 80% of studies pertained to terrestrial organisms, while less than 20% pertained to aquatic organisms and their role, if any, in death scene investigations (Hobischak and Anderson 1999; Goff 1993; Catts and Goff 1992; Smith 1986). To date, most empirical evidence examining insect colonization in aquatic systems has concentrated on blow flies (Calliphoridae), and a few other terrestrial species that colonize a corpse after it bloats and rises to the surface. Those species that are restricted to aquatic ecosystems for survival in one or more life stages have been largely ignored (Hobischak and Anderson 1999; Tomberlin and Adler 1998; Mann et al. 1990; Goff and Odom 1987; Smith 1986; Nuorteva et al. 1974; Payne and King 1972). In fact, submersion in freshwater environments can alter the terrestrial faunal succession on carrion or corpses, and subsequently alter the process of decomposition (Rodriguez 1997; Haskell et al. 1989; Smith 1986; Payne and King 1972).

With possibly few exceptions, evolutionarily speaking, there are no truly sarcophagous aquatic insects that have evolved functionally to feed on carrion alone (Chaloner et al. 2002;

Haskell et al. 1989). This is in contrast to the common terrestrial indicator species (e.g., blow flies and flesh flies) that can often provide a predictable timeframe of succession, assisting in determining the range for the postmortem interval. Frequently, dipteran larvae are the principal insect component within the terrestrial arthropod assemblage inhabiting vertebrate carrion, and therefore, forensic entomologists tend to focus on this group (Catts and Goff 1992; Greenberg 1991; Haskell et al. 1997). Because oviposition by terrestrial flies does not occur on a completely submerged carcass or corpse, it has been suggested by several investigators that the potential use of algae, silt, or sediment deposition, or the presence of aquatic macroinvertebrates and the casings or structures housing them, may be of significance (Horton et al. 2006; Haefner et al. 2004; Keiper and Casamatta 2001; Casamatta and Verb 2000; Keiper et al. 1997; Nawrocki et al. 1997; Siver et al. 1994; Haskell et al. 1989).

As noted by Sorg et al. (1997) and modified here, human remains found in freshwater or marine ecosystems serve several ecological functions. Depending on the nature of the resting site (e.g., pool or riffle, depth and substrate), degree of decomposition, and the composition of the local fauna, human remains in water may: (1) become immediate or eventual sources of food for a wide variety of invertebrates and fish, (2) provide a sheltered microhabitat for small nonscavenging species, (3) draw in a variety of secondary predatory species attracted to the original scavengers, (4) provide substrate upon which primary producers, e.g., algae and other periphyton, can colonize and grow, and (5) at advanced stages of decomposition, serve as substrata for invertebrate grazers attracted to bacterial or algal biofilms on bones, skin, or clothing, or to the bones themselves as a source of calcium and other minerals.

Because corpses are often found in aquatic environments (e.g., Hobischak and Anderson 1999; Haglund et al. 1990), it is important that forensic scientists and police visiting a crime scene have an increased knowledge of the aquatic organisms that could potentially colonize humans and nonhuman models. They should also be aware of the environmental factors affecting their distributions. The objectives of this chapter are to first discuss the different stages of carrion decomposition in aquatic environments, and then concentrate on the independent variables of importance (e.g., factors relative to the environment or to the corpse itself) that affect the postmortem submerged interval. The second part will characterize the aquatic insect community involved in decomposition and discuss their functional roles relevant to their importance in forensic science investigations. The third part will document and discuss several case histories involving aquatic insects. Throughout this review, we will concentrate on freshwater habitats and, to a lesser extent, marine environments.

Decomposition in Aquatic Ecosystems

Freshwater Ecosystems

The concept of phases or stages of decomposition in terrestrial habitats was developed in the nineteenth century with Megnin's (1894) description of eight series of changes and the arthropod assemblage associated with each stage. Presently, the five stages of decomposition recognized for corpses found in terrestrial settings include the following: fresh, bloat, active decay, advanced decay, and the dry/remains stage (see Anderson and VanLaerhoven 1996; Tullis and Goff 1987; Smith 1986; Payne 1965). Decomposition of a body submerged in an aquatic environment occurs at a rate roughly half that of decomposition in air, primarily due to cooler temperatures and the inhibition of terrestrial insect activity (Knight

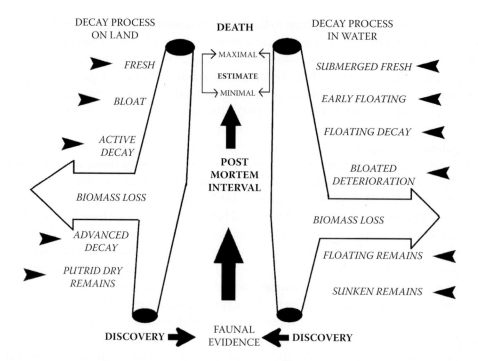

Figure 7.1 A comparison of the stages of decomposition on land and in the water. (Modified from Catts and Haskell 1990.)

1997; Rodriguez 1997). Based on studies using immersed pigs (*Sus scrofa*) from June to November, Payne and King (1972) revised the stages of decomposition to accommodate corpses found in aquatic habitats and further divided the process into six stages: submerged fresh, early floating, floating decay, bloated deterioration, floating remains, and sunken remains (Figure 7.1). As with most other studies on carrion and corpses in aquatic and semiaquatic environments, these are primarily characterized or defined by the presence or absence of terrestrial insects, without any mention of aquatic insect colonization.

Since Payne and King's (1972) research on pig decomposition in water, more recent studies on the decompositional stages in aquatic habitats have been conducted. They have provided information that includes a reduction in the number of stages, the duration of each decompositional stage, the condition of the animal carcass, and the presence or absence of certain aquatic insects on various animals, including pig, salmon, and rat carrion in freshwater ecosystems of North America (Haefner et al. 2004; Hobischak and Anderson 2002; Tomberlin and Adler 1998; Hobischak 1997; Keiper et al. 1997; Minakawa 1997; Vance et al. 1995).

The characterization of the stages of decomposition in aqueous environments has been reduced to five stages (Haefner et al. 2004; Hobischak and Anderson 2002) from six (Payne and King 1972) based on the physical characteristics and the difficulty in distinguishing between two stages of decay (Table 7.1). In Payne and King's 1972 treatise, one pig carcass was used, and it was permitted to float to the surface, at which time terrestrial insects were allowed to colonize the carcass. This difference altered the decomposition rate of the carcass compared to later studies, in which carcasses were maintained submerged the entire duration of the decomposition process. The five stages are discussed in depth here and are a compilation of observations from several studies.

Table 7.1   Observed Stages of Decomposition with Corresponding Physical Descriptions from Submerged Pigs in Two Pennsylvania Streams

Stage of Decomposition	Physical Description of Carcass
1. Submerged fresh	Fresh; no outward signs of decomposition; still sunken; stage ends when body floats to surface
2. Early floating	Bloated; floating on surface of water; cage indentations on carcass as it presses against the top of the cage; algal growth evident
3. Early floating decay	Minor decay becoming apparent; sloughing of flesh; loss of muscle mass or thinning of hind limbs; eyes and soft tissues becoming disarticulated; head and legs remain intact; identity of carcass as being that of a pig still evident
4. Advanced floating decay	Major deterioration visible; ribs and skull exposed; breaks in and loss of bones, including skull; leg bones gone; carcass identity becoming indistinguishable as a result of major appendage and skull loss; stage ends as remains sink
5. Sunken remains	Remains sunken to bottom of cage; any skin takes on a soup-like consistency; stage ended arbitrarily, with mostly small pieces of bones remaining

Source: Modified from Haefner et al. 2004.

First Stage: Submerged Fresh

The submerged fresh stage for pig carcasses in a stream is best characterized as the period of time between when the carcass is initially submersed and when it begins to bloat and rise to the surface (Figure 7.2a). Depending on the geographic location of a running or standing water habitat, the microhabitat within this water body, and the time of year, a pig carcass may not begin to bloat and rise to the water surface for 2 to 13 days in the spring in mid-latitudes and 11 to 13 days in more northern latitudes (Haefner et al. 2004; Hobischak and Anderson 2002; Hobischak 1997). Truly aquatic insects in their immature stages, e.g., hydropsychid caddisflies (Trichoptera: Hydropsychidae), chironomid midges (Diptera: Chironomidae), and heptageniid mayflies (Ephemeroptera: Heptageniidae), were observed on carcasses by several investigators during this stage in running and standing water habitats (Haefner 2005; Hobischak and Anderson 2002; Hobischak 1997; Keiper et al. 1997; Schultenover and Wallace, unpublished data; Vance et al. 1995), while adult hydrophilid beetles (Coleoptera: Hydrophilidae) also were collected on pigs in standing water (Vance et al. 1995).

Second Stage: Early Floating

As gases produced from anaerobic bacterial respiration in the abdomen increase, either the pig carcass floats to the surface and pushes the abdomen above the water surface or, in the case of a totally submerged pig in a cage, bloating will force the carcass against the roof of the cage and cause an indentation from the wire to form on the carcass (Haefner et al. 2004; Schultenover and Wallace, unpublished data) (Figure 7.2b). Pig, rat, fish, or human carcasses projecting above the water surface attract terrestrial insect species, e.g., blow flies (Calliphoridae) and related families (Muscidae, Sarcophagidae), which lay eggs or larvae on exposed areas of the carcass. Carrion and rove beetles (Silphidae and Staphylinidae) may feed on larval blow flies or animal flesh. Yellowjacket and bald-faced hornets (Vespidae) generally prey on the adult and larval blow flies (Tomberlin and Adler 1998; Goff and Odom 1987; Nuorteva et al. 1974; Payne and King 1972). However,

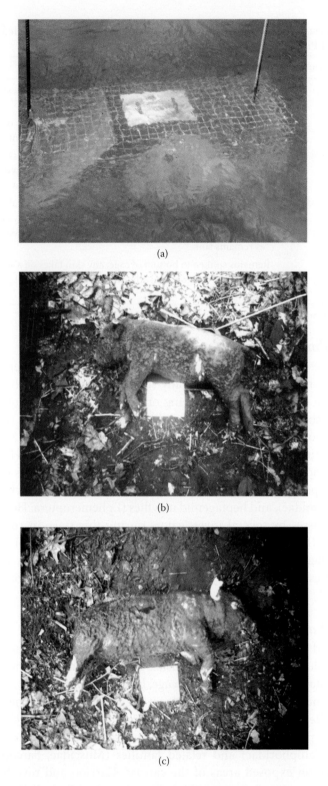

(a)

(b)

(c)

Figure 7.2 (a–f) Different stages of pig decomposition (days after submergence) in a riffle area of a small southern Indiana stream. Pigs were removed from stream and placed on bank for photographs. (Photos by J. R. Wallace.)   *Continued*

(d)

(e)

(f)

Figure 7.2 *Continued.*

for totally submerged carcasses, aquatic invertebrates such as hydropsychid caddisflies, chironomid midges, aquatic isopods, and heptageniid mayflies were found inhabiting the carcass during this stage (Haefner 2005; Hobischak and Anderson 2002; Hobischak 1997; Schultenover and Wallace, unpublished data; Haskell et al. 1989, Hawley et al. 1989). It also was noted that the decay odor was quite evident and pronounced during this stage, tissues turned from pinkish hues to shades of blue-green, yellow fluids and gases oozed from the anus, and nails begin to slough off the hooves of pigs (Haefner et al. 2004; Schultenover and Wallace, unpublished data). In addition, wire indentations remained on the dorsal side of each pig in both riffles and pools. Algal or periphyton growth increased significantly more on pigs than on artificial substrates that were placed as controls in riffle and pool habitats (Figure 7.2b). This type of primary production may in fact provide food upon which mayfly larvae (Ephemeroptera) can graze.

The time of year when such studies are initiated plays an important role in determining the duration of each stage of decomposition. For example, Schultenover and Wallace (unpublished data) found that in the midwestern United States during spring, the early floating stage persisted for approximately 8 days in a riffle habitat and 6 days in a pool habitat. Conversely, in a cooler climate, Hobischak and Anderson (1997, 2002) observed the bloat stage lasted from 23 to 37 days in pond and stream habitats in coastal British Columbia; however, this time period may have encompassed part of the floating decay stage described below.

Third Stage: Floating Decay

Intense feeding activity by calliphorid maggots has been observed on pig and rat carcasses floating above the water surface, creating many openings in the exposed skin (Tomberlin and Adler 1998). Silphid, staphylinid, and histerid beetles were observed in high abundance during this time, searching for prey and copulating (Payne and King 1972). Several investigators have reported that aquatic macroinvertebrate colonization of pig carcasses that were totally submerged varied both temporally and spatially between riffle and pool microhabitats (Haefner 2005; Hobischak and Anderson 2002; Schultenover and Wallace, unpublished data). For example, Schultenover and Wallace (unpublished data) observed perlodid stonefly (Plecoptera: Perlodidae) larvae preying on both black fly larvae (Diptera: Simuliidae) attached to the carcass and chironomid midge larvae crawling on and in the carcass. Crayfish feeding activity during this stage increased on pigs in both riffle and pool habitats. Depending on the geographic region in North America from which decomposition studies have been conducted, the duration of this stage has varied from 8 (Indiana) to 331 (British Columbia) days in the pool or pond habitats and approximately 24 (Indiana) to 161 (British Columbia) days in the riffle or stream habitats. Although wire indentations were still present, skin sloughed off the hind limbs of the pig in a riffle habitat, and both carcasses assumed a blackened color (Figure 7.2c). Gas released from the pig in a pool habitat was great enough to cause the abdomen to collapse even though it remained floating.

Fourth Stage: Advanced Floating Decay

During this stage, with carcasses floating at the surface of the water, most of the exposed tissues of pig or rat carcasses floating or projecting above the water surface disappear due to the continual feeding activities of blow fly maggots (Figure 7.2d) (Payne and King 1972). As maggots mature, they will migrate away from the carcass. However, pig or rat carcasses floating at the tops of cages that were totally submerged showed significant sloughing of flesh and disarticulation of phalangial and limb bones (Haefner et al. 2004; Keiper et al. 1997; Schultenover and Wallace, unpublished data; Haglund and Reay 1993) (Figure 7.2d). Mainly chironomid and black fly larvae colonized pigs submerged completely underwater. The duration of the bloated deterioration stage varied between pool and riffle habitats from an average of 12 (Indiana) to 171 (British Columbia) days in pools or pond habitats and 8 (Indiana) to 161 (British Columbia) days, respectively, and among seasons (Hobischak and Anderson 2002; Tomberlin and Adler 1998; Hobischak 1997; Schultenover and Wallace, unpublished data).

Mostly chironomid midge larvae and some black fly larvae were recorded on pig or rat carcasses in both riffle and pool habitats during this stage (Haefner 2005; Hobischak and Anderson 2002; Keiper et al. 1997; Schultenover and Wallace, unpublished data). Some vertebrate predators, e.g., sunfish (Centrarchidae), dace (Cyprinidae), and sculpin (Cottidae), were observed feeding on carcass flesh or on aquatic macroinvertebrates found on the carcass (J. Wallace, personal observation). In a case history reported later in this chapter, Merritt found that amphibians and fish were feeding on maggots that were exiting a victim's body in a pond during this stage. Hobischak (1997) found that carcasses also might experience scavenging from vertebrate species, such as mink (*Mustela vison*). Singh and Greenberg (1994) suggested that the duration of submergence of corpses in forensic investigations may be estimated from the pupae of some aquatic insects that colonize, develop, and are later found adhering to or entangled in decaying flesh, hair, or clothes of a corpse even after complete submergence. For example, in a case described later in this chapter, Wallace reports that aestivating or pupating caddisfly larvae attached to human remains can be used to roughly estimate a time interval during which the body was submerged long enough for this type of aquatic insect to attach, but cautions this is season dependent, as pupating caddisfly larvae will not attach their cases until typically the early to mid-summer months for most species of case-building caddisflies.

Fifth Stage: Sunken Remains

The duration of this stage is quite variable, but is primarily characterized with only bones and bits of skin remaining on the substrate. Payne and King (1972) noted that decomposition is completed by bacteria and fungi during this stage, while Haefner et al. (2004) as well as Schultenover and Wallace (unpublished data) noted that skull segments had become disarticulated, the remaining flesh resembled a soupy texture, and the decay odor was negligible (Figure 7.2e). In addition to fish scavengers, benthic aquatic fauna such as crayfish (*Orconectes propinqus*) may be found within the carcass remains, as well as chironomid and mayfly (*Ephemerellidae*) larvae, annelids, snails, leeches, and amphipods (Haefner 2005; Hobischak and Anderson 2002; Hobischak 1997; Nawrocki et al. 1997; Schultenover and Wallace, unpublished data; Vance et al. 1995).

Marine Ecosystems

Sorg et al. (1997) provided an excellent review of forensic taphonomy in marine contexts and noted that little research has been done on the process of decomposition of human remains, mainly because marine death assemblages including human remains are rare, and the recovery of such is even more rare. In addition, human remains are often found long after the primary scavenging sequences have occurred and the participants departed, making it difficult to associate decomposition with a particular scavenger. Until recently, those studies that have focused on human remains in marine waters have done so as a by-product of forensic casework (e.g., Boyle et al. 1997; Ebbesmeyer and Haglund 1994; Haglund and Reay 1993; Giertsen and Morild 1989; Donoghue and Minnigerode 1977).

Whereas specific insects and their allies are generally the dominant scavengers of terrestrial and freshwater environments, crustaceans, fish, gastropod mollusks, and echinoderms are the dominant scavengers in marine environments (Sorg et al. 1997). Möttönen and Nuutila (1977) reported cases where crustaceans destroyed all the soft tissues of a body, including parenchymal organs, and inflicted crater-like lesions of varying size.

Lord and Burger (1984) observed that the process of decomposition and faunal succession by terrestrial arthropods of stranded marine harbor seals on islands was generally similar and predictable to that reported for sheep carcasses in Australia (Fuller 1934), dogs in Tennessee (Reed 1958), rodents in Great Britain (Putman 1978), and pigs in South Carolina (Payne 1965). They also noted that typical marine intertidal and wrack decomposers, such as crabs or amphipods, had a negligible impact on seal decay, as did the activities of scavenging seabirds. The absence of several beetle taxa (e.g., Silphidae) was of further interest in that carcass size and the harsh maritime conditions characteristic of rocky supratidal zones appeared to be important factors precluding these insects from visiting the carcasses. However, specimens of the beetle family Silphidae were encountered on gull and rodent carcasses located in more protected island habitats (Lord and Burger 1984). Sorg et al. (1997) concluded in his review that only a few published studies (Sorg et al. 1995; Skinner et al. 1988) have actually described the decomposition of submerged remains in marine environments, and more documentation of the condition of human remains and the associated marine organisms is necessary to build sufficient case series from which marine models of postmortem change can be constructed.

Recently, Anderson and Hobischak (2004) provided one of the most complete descriptions of each stage of decomposition for pig carcasses in a marine environment. They found that sediment type significantly influenced rate of decomposition. In addition, they observed that one of the primary consumers of pig carrion in deep marine habitats was mollusks such as the wrinkled amphissa (*Strongylocentrotus droebachiensis*) and the sunflower sea star (*Pycnopodia helianthoides*). Moreover, they concluded that many marine animals fed directly on the skin of the pig carcasses, and that the marks created from feeding could be misidentified as an antimortem wound (Anderson and Hobischak 2004).

Estimating a Postmortem Submergence Interval

A portion of the postmortem history, and circumstances surrounding the discovery of the morbid remains of a corpse in a terrestrial setting, has often been provided by the arthropod community inhabiting the body (Keiper et al. 1997; Catts and Haskell 1990). A range for the postmortem interval, i.e., the time interval between the earliest insect colonization to corpse discovery for a body, can be estimated by an investigation of the composition and age of this invertebrate assemblage (Catts and Haskell 1990; Smith 1986). For those corpses found totally submerged in aquatic environments, such as rivers, streams, ponds, or lakes in which oviposition and larval development of terrestrial sarcophagous insects are prevented, determination of the postmortem submersion interval (PMSI) has proven more problematic. Ecological context is very important because even within one species, variations in behavior may occur in slightly different habitats or different seasons.

Although few indicators of time since submergence for corpses found in aquatic ecosystems are comparable in precision to the insect indicators used in terrestrial cases, there are observations that can be useful in suggesting or ruling out an approximate PMSI. In particular, the time intervals needed for certain growth phases of marine plants or animals that attach themselves to the remains can be used to estimate a minimum PMSI (Sorg et al. 1997). Sorg et al. (1997) recommended that research on marine cases be focused on sessile forms of fauna because with these organisms one can be sure that the animal or plant is truly associated with the remains. Further, in aquatic as well as in terrestrial habitats, the stages of decomposition (which embody the postmortem interval) can be affected by several environmental factors. These factors include, but are not limited to, temperature, water current regime, the aquatic organisms present, and other factors related to the corpse itself (e.g., presence or absence of clothing, and body habitus, i.e., submerged or floating).

Environmental Factors

The major variables that affect the rate of decomposition of human remains found in aquatic environments are summarized in Table 7.2. Physical-chemical parameters of water, such as temperature, current regime, oxygen content, and those associated with the corpse itself (e.g., clothing and trauma), not only play a part in decomposition of human corpses, but also influence the dominant pathway of decomposition. This decomposition is mediated through biological mechanisms, such as microbial and macroinvertebrate interactions.

Physical and Chemical Parameters

Temperature is probably the foremost environmental factor that influences the rate of corpse decomposition via the timing of insect oviposition or colonization, larval insect growth, and survivorship, and hence is the most important factor in the estimation of the PMI in terrestrial ecosystems (Greenberg 1991). Because temperature plays such an important role in forensic investigations in these ecosystems, it is reasonable to expect that temperature has the same effect in aquatic systems. As with terrestrial insects, aquatic insects respond to the summation of thermal units (i.e., degree-days) as well as absolute

Table 7.2   Mean Temperature and Total Degree-Days per Stage of Decomposition in Riffle and Pool Habitats in a Pennsylvania Stream

Stage of Decomposition	Temperature (°C)		Total Degree-Days	
	Riffle	Pool	Riffle	Pool
	Mean (SE)	Mean (SE)		
1. Submerged fresh	9.6 (0.60)	6.3 (0.57)	76.8	151.2
2. Early floating	3.9 (0.25)	2.0 (0.18)	118.1	88.6
3. Early floating decay	1.6 (0.15)	3.5 (0.25)	70.0	152.3
4. Advanced floating decay	4.1 (0.33)	6.7 (0.25)	64.9	242.0
5. Sunken remains	4.2 (0.52)	10.7 (1.05)	34.0	139.2
Total		363.8		773.3

Source: Modified from Haefner et al. 2004.

ambient temperatures. For example, riffle and pool temperatures in a Pennsylvania stream had a significant impact on total degree-days between the two microhabitats (Table 7.2) (Haefner et al. 2004). Seasonal temperature fluctuations characterize most natural lotic (running water) and lentic (standing water) freshwater environments (Ward 1992; Ward and Stanford 1982). However, streams and rivers differ from lakes and ponds in that: (1) even though the annual temperature range is less in lotic waters, the seasonal rate of temperature change is often greater than in lakes and may affect surface and subsurface current patterns, particularly in large or deep rivers (Nawrocki et al. 1997); (2) typically diurnal temperature changes are often greater in streams than in the nonlittoral areas of lakes; and (3) thermal stratification is very rare in natural streams and, if present, persists for only short periods relative to that in lakes (Brittain 1976). In a specific case report of a corpse discovered in a swamp in Hawaii, Goff et al. (1988) found that oceanic island temperatures tended to remain relatively stable on both day/night and annual bases. This stability, unlike diurnal and seasonal fluctuations encountered in temperate continental areas, minimizes temperature-related variations in rates of arthropod development (Goff et al. 1988).

In general, cold water temperatures, whether in freshwater or marine ecosystems, retard the processes of decomposition, especially microbial breakdown. However, warm temperatures due to seasonal fluctuations in temperate climates or more stable environments, as in tropical seas, accelerate decomposition (Sorg et al. 1997). This, in turn, influences trophic dynamics of aquatic invertebrates directly through its effects on feeding rates, and indirectly through the food base available to these animals (Ward and Stanford 1982). Higher temperatures increase larval growth rates of aquatic macroinvertebrates by altering the quantity (e.g., density or productivity of attached algae) and quality (microbial populations, such as bacteria and fungi) of organic matter associated with a corpse (Cummins and Klug 1979). Direct effects on aquatic macroinvertebrate life histories can result by temperature influencing the feeding rates, digestion, and respiration, as well as food conversion efficiencies, enzymatic kinetics, and endocrine processes (Sweeney and Vannote 1981; Vannote and Sweeney 1980). The dynamic nature of environmental temperatures, and their effects on invertebrate life histories, must be incorporated into forensic studies to further our knowledge of the role of aquatic organisms in the decomposition of human and other animal remains.

The physics of corpse movement in water, both horizontally (downstream) and vertically (to depths) have been summarized by Dilen (1984) and modeled by Ebbesmeyer and Haglund (1994). Dilen (1984) described four stages of motion of the body: sinking to the bottom, motion along the bottom, ascent to the surface, and drift at the surface. Dispersal of human remains in the sea is much more similar to that in lotic environments than in freshwater lentic or terrestrial sites. The principal agents of distribution in marine ecosystems are tidal, current, and wave action, as well as motile scavengers (Sorg et al. 1997).

Many factors can affect postmortem condition in aquatic environments, and can interact in predictable and unpredictable ways. For example, soon after death a body may be clad in heavy clothing, but after deposition in a physically harsh aquatic habitat (e.g., riffles or rapids in streams and rivers, or intertidal zones of oceans), the force of the current or wave action could move the body between riffle and pools or remove the clothing in the immediate postmortem interval (Hobischak 1997; Keiper et al. 1997; Sorg et al. 1997). Likewise, if a corpse is bleeding in the ocean, large marine scavengers, such as sharks, may be attracted and cause severe alteration of the remains (Sorg et al. 1997).

Small-scale variations in temperature may occur over short distances in streams and rivers, and therefore the specific area where the body settles out within a stream reach may influence the rate of decomposition. For example, slow-flowing pools may attain higher temperatures in summer than adjacent rapids, and therefore any knowledge of body movement over time in an aquatic habitat may provide important information when determining the PMSI.

Oxygen level in the water also is an important factor to consider when determining the PMSI. Oxygen solubility in water is negatively correlated with water temperature, and oxygen levels vary with current speed and turbulence. That is, small, fast-flowing, unpolluted streams are usually saturated with oxygen, whereas polluted streams, stream pools, small ponds, and stagnant bays (especially those with a high organic load of dead leaves or high sediment loads) can have relatively low levels of oxygen.

Certain groups of aquatic invertebrates have specific respiratory and behavioral adaptations to deal with low oxygen conditions in polluted streams or highly eutrophic lentic habitats, allowing them to survive in such habitats (Eriksen et al. 1996). However, many insect species are not able to deal with this respiratory demand. Many insects show a clear preference for cold waters (e.g., Plecoptera or stoneflies), possibly owing to the effects of temperature on oxygen availability as opposed to just the effects of temperature per se (Giller and Malmqvist 1998; Ward 1992). Therefore, depending on the oxygen concentration in a specific aquatic habitat where a corpse is found, the faunal community colonizing a body may be quite different (Hobischak 1997). As a general rule, very low oxygen environments favor lower species diversity. However, higher numbers of individuals of those species that are able to tolerate low oxygen conditions and outcompete pollution-intolerant groups are usually found (Hynes 1960). In streams, ponds, or swamps that are highly polluted or anaerobic part of the time, there may be few, if any, organisms present that would be available to colonize a corpse, therefore resulting in a slower rate of decomposition. In contrast, an unpolluted stream generally has a higher diversity of organisms with fewer numbers of individuals per species (Hynes 1960). These water quality differences in oxygen concentrations may therefore affect the diversity and abundance of invertebrates colonizing a corpse in a specific habitat, and the subsequent rate of decomposition over time.

Biological Mechanisms

The breakdown of dead organic matter such as a corpse is primarily a biological process involving three types of organisms: large- and small-particle detritivores (commonly referred to as macroinvertebrates), fungi, and bacteria. Although the critical role of microorganisms such as bacteria and fungi in stream ecosystems has been clearly established (Suberkropp and Klug 1980, 1976; Kaushik and Hynes 1971), little detailed work has been conducted on the specific microbial assemblages and the succession of different microorganisms associated with decomposing organic matter. Bärlocher (1982) reported that typically four to eight species of aquatic fungi dominate throughout the decomposition of leaves, but apparently no particular succession occurs on a single leaf. Fungi initially appear to dominate microbial assemblages in leaves as long as the tissue is more or less intact, while bacteria tend to increase when leaves become partially broken down and dominate the terminal processing stage (Baldy et al. 1995; Bengtsson 1992; Suberkropp and Klug 1976). If this is the case, a clear distinction between fungal and bacterial activity and diversity could be important in determining the postmortem submerged interval of a corpse colonized by these microorganisms in running waters. We are not aware of any published studies on the role of stream or pond microbes in the decomposition of carrion or human corpses. However, Siver et al. (1994) reported the successful employment of the use of diatoms and planktonic algal communities in a pond to link three subjects to a freshwater crime scene in southern New England.

In aquatic systems, decomposition processes involve the dissipation of energy stored in organic matter (Allan 1995). The major sources of energy in most stream ecosystems are: (1) terrestrial inputs of organic matter in the form of leaf litter (commonly referred to as allocthonous material) (Cummins 1974; Fisher and Likens 1973), and (2) instream primary production (termed autochthonous material) brought about by photosynthesizing organisms such as algae (usually diatoms) and mosses (Lamberti and Moore 1984). Nutrient recycling of this decomposed organic matter has long been recognized as an important function of aquatic ecosystems (Merritt et al. 1984).

Decomposing remains, such as salmon carcasses, may begin to influence nutrient dynamics as soon as they enter an aquatic ecosystem through the excretion of nitrogenous compounds resulting from protein catabolism (Kline et al. 1997; Schuldt and Hershey 1995; Mathisen et al. 1988). Studies have shown that the decomposition of salmon carcasses in Alaskan streams increased algal biomass and primary production (Chaloner and Wipfli 2002; Chaloner et al. 2002; Wipfli et al. 1998; Kline et al. 1990, 1997). Schultenover and Wallace (unpublished data) found that the introduction of pig carcasses to riffle and pool microhabitats in an Indiana stream also resulted in a significant increase of algal growth on these decomposing carcasses in both microhabitats over a 30-day period (Figure 7.2a–f), as compared to artificial substrata (i.e., ceramic tiles), which showed a lesser increase in algal growth. Keiper et al. (1997) also observed algal colonization of rat carcasses in riffle and pool areas of an Ohio woodland stream. Because algae is often a dominant component of primary production in streams, it is of major importance to organisms such as aquatic insects that utilize living plant material as a food source for growth and reproduction (Lamberti and Moore 1984). Brusven and Scoggan (1969) observed algal-feeding caddisfly larvae feeding on dead squawfish after a fish kill, and concluded that these larvae contributed directly and indirectly to the removal of the fish from the stream. Minakawa (1997) found twenty-four insect taxa associated with salmon carcasses in Pacific Northwest streams, with ten caddisfly genera and two stonefly genera directly feeding on salmon flesh. He observed that one stonefly genus and three caddisfly genera associated with the carcasses had fungi in their mouths, suggesting that these insects might be feeding on the fungi and other microbes growing on the salmon carcasses. Kline et al. (1997) noted that macroinvertebrates invaded mouth and gill areas of salmon in fast currents and abraded body parts (tails, fins, and nose) in slower currents. They appeared to feed on gill membranes in the oral cavity, but chose exposed muscle in the abraded areas, or fungal patches on the exposed parts of salmon. These findings may be important to forensic scientists in explaining how and why aquatic insects begin to colonize and utilize a human corpse in freshwater ecosystems, and which groups one would expect to find on a corpse after a given time interval. Schultenover and Wallace (unpublished data) have shown that the algal species present and abundance of algal growth on a corpse (in this case a pig) in a given stream system may be helpful in determining the PMSI in forensic investigations.

In addition to microbes (primarily bacteria, fungi, periphyton, or attached algae), the other basic food resource categories include:

1. CPOM, coarse particulate organic matter (particles greater than 1 mm in size): Represented by litter accumulations consisting of leaves, needles, bark, twigs, and other plant parts, large woody debris, macrophytes (including macroalgae), and rooted and floating vascular plants. By definition, this also would include carrion of wild or domestic animals and human corpses.
2. FPOM, fine particulate organic matter (particles less than 1 mm and greater than 0.5 μm in size): Generally composed of unattached living or detrital material, including that created through the physical and biological reduction of CPOM.
3. Prey (all invertebrates captured by predators) (Merritt and Cummins 2006).

These nutritional resource categories form the basis of a functional feeding group classification system designed to identify macroinvertebrates involved in the processing of organic matter in aquatic ecosystems, and assist with evaluating the role of these animals in death scene investigations.

Macroinvertebrate Functional Feeding Groups

Aquatic insects have been a major focus of many ecological studies in aquatic habitats. Due to their ubiquitous distribution, relative abundance, ease of collection, and large size (observable with the unaided eye) (Merritt et al. 2008a), they lend themselves well to be the organism of choice in the study of death scene investigations. As pointed out earlier, unlike terrestrial habitats, the primary problem in aquatic environments is that there are no purely sarcophagous aquatic insects to compare with the common terrestrial indicator species, such as blow flies (Calliphoridae) (Haskell et al. 1989). There does not appear to be a clear, predictable successional pattern on submerged carrion by different species of aquatic insects, yet Hobischak (1997) suggested a predictable succession of invertebrates colonizing pig carcasses (exposed and submerged) in the aquatic habitats she studied in British Columbia. However, she noted environmental conditions and organism habitat preferences influenced this pattern, and indicated that discretion should be used when evaluating succession for use of determining time of submergence or death.

Terrestrial indicator species of insects are ecologically characterized according to their trophic relationship with carrion; e.g., necrophagous taxa feed mainly on the corpse itself; omnivores feed on either the corpse or associated fauna; and parasites and predators parasitize or feed on the fauna associated with the corpse. The aquatic functional feeding group (FFG) method is based on the association between a limited set of feeding adaptations found in freshwater invertebrates and their basic nutritional resource categories elaborated on later in this section. The same morphobehavioral mechanisms can result in the ingestion of a wide range of food items, the intake of which constitutes herbivory, detritivory, or carnivory (Merritt and Cummins 2006). Although food type intake would be expected to change from season to season, habitat to habitat, and with growth stage, limitations in food acquisition mechanisms have been shaped over evolutionary time, and these are relatively more fixed (Cummins and Merritt 1996).

Corpses initially enter the aquatic system as coarse particulate organic matter (CPOM), much like that of leaf litter. Some of this material will enter the fine particulate organic matter (FPOM) and dissolved organic matter (DOM) pools following leaching, physical abrasion, and biotic processing, and some will enter the inorganic nutrient pool as a result of microbial activity (e.g., Cummins 1974). Shredders (e.g., aquatic isopods and amphipods [Crustacea], some caddisflies [Trichoptera: Limnephilidae], and crane flies [Diptera: Tipulidae]) are dependent on large pieces of organic matter (CPOM), such as leaves, needles, wood, and other plant parts derived primarily from the riparian zone. Collectors (e.g., net-spinning caddisflies [Hydropsychidae], black flies [Diptera: Simuliidae], and midges [Diptera: Chironomidae]) utilize small particles of organic matter (FPOM), either by filtering from the passing water or by gathering from deposits in the sediments on the stream bottom. Scrapers (e.g., some caddisflies [Trichoptera: Glossosomatidae] and mayflies [Ephemeroptera: Heptageniidae]) are adapted to remove algae or periphyton attached to rock or log surfaces. Predators (e.g., caddisflies [Rhyacophilidae], stoneflies [Plecoptera], and beetle larvae [Coleoptera]) are adapted to capture prey organisms through behavioral mechanisms or specialized body parts. The development of an aquatic insect community on an introduced substrate, such as a corpse, is highly dependent on available colonizers in addition to other factors, such as seasonality, temperature, current speed, and depth, as discussed previously (Haskell et al. 1989).

The FFG approach can be used to track the decomposition of carrion or a human corpse over time, and it can document the changes in functional feeding groups based on changes in the nutritional resource categories available to the different taxa. The presence and abundance of the various FFGs and their percent composition is a direct reflection of the availability of the required food resources and the condition of the related environmental parameters. Recent empirical evidence by Schultenover and Wallace (unpublished data), presented below, demonstrates the utility of this approach to track some type of predictable successional change in stream macroinvertebrate FFGs on, in, or around a corpse as decomposition proceeds over time.

Figure 7.3a shows that the greatest percentage of FFGs on pigs in riffle and pool habitats were collector-gatherers (Col-gath.) and collector-filterers (Col-filt.) during the freshly submerged and early floating stages of decomposition. The percentage of the FFG collector-gatherers was higher in the pool (46%) versus riffle (25%) habitat, where chironomid midge larvae made up the major portion of the Col-gath., and net-spinning caddisflies (Hydropsychidae) made up the Col-filt. Although present, shredders such as crayfish and crane fly (Tipulidae) larvae, as well as scrapers (Ephemeroptera: Heptageniidae) and predators, were not very abundant in either habitat. The diversity of aquatic macroinvertebrates was greatest in these early stages of decomposition as compared to the remaining stages.

During the floating decay and bloated deterioration stages, the Col-gath. FFG was the most abundant on pigs in both habitats compared to all other feeding groups (Figure 7.3b). In fact, midge larvae were the only macroinvertebrates found on the pig carcass in the pool habitat. This may have been due, in part, to the sediment accumulation on the pig providing an ideal habitat in which these insects could construct feeding burrows or tubes. The pigs in the riffle habitat were colonized by a more diverse assemblage of insects, including some scraper mayflies and net-spinning caddisflies (Figure 7.3b).

As pigs in both stream habitats reached the floating remains–sunken remains stages of decomposition (Figure 7.3c), shredders represented by amphipods and limnephilid caddisflies, and Col-gath. midges became the dominant groups in the pool habitat. However, in the riffle habitat, Col-gath. chironomid midges were the most abundant FFG group represented (79%). During these two stages, predator species such as hellgrammites

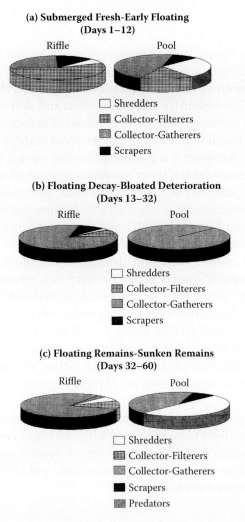

(a) Submerged Fresh-Early Floating (Days 1–12)

Riffle   Pool

☐ Shredders
▦ Collector-Filterers
▨ Collector-Gatherers
■ Scrapers

(b) Floating Decay-Bloated Deterioration (Days 13–32)

Riffle   Pool

☐ Shredders
▦ Collector-Filterers
▨ Collector-Gatherers
■ Scrapers

(c) Floating Remains-Sunken Remains (Days 32–60)

Riffle   Pool

☐ Shredders
▦ Collector-Filterers
▨ Collector-Gatherers
■ Scrapers
▨ Predators

Figure 7.3 (a–c) Percent composition and comparative changes in macroinvertebrate functional feeding groups (FFGs) colonizing pigs at different stages of decomposition in riffle and pool habitats in a southern Indiana stream.

(Megaloptera: Corydalidae) and fish were observed feeding on or around the pig in the riffle habitat.

Aquatic Insects of Importance to Forensic Science

Although only 3% of all species of insects have aquatic or semiaquatic stages (Daly 1996), in some freshwater biotopes insects may comprise over 95% of the total individuals or species of macroinvertebrates (Ward 1992). Identification of aquatic macroinvertebrates is the first step toward a basic understanding of the role these organisms play in death scene investigations. It is not the purpose of this chapter to provide an all-inclusive identification key to aquatic insects; this can be found in other major reference works (e.g., Merritt et al. 2008a, Thorp and Covich 1991). Rather, we have chosen to include a simplified, illustrated, dichotomous key to functional feeding groups of lotic macroinvertebrates modified from Cummins and Wilzbach (1985) and Merritt and Cummins (2006) (see appendix).

The key emphasizes higher-level taxonomic separations that permit reliable categorization of functional feeding groups. It is organized in two levels of resolution. The first level can be done in the field with a minimum of taxonomic skill, and the second level should be done in the laboratory. The amount of taxonomic effort and skill required and the need for the use of a microscope increases with the second level of resolution. While this key is primarily aimed at insects found in streams and rivers, most of these groups also are found in standing water habitats. Although it is possible to key out aquatic insects to major families using this key, we recommend that if lower resolution is required, or if the specimens are to be used as evidence in court, they be taken to an aquatic entomologist for verification. For a discussion of marine invertebrates associated with human decomposition, see Anderson and Hobischak (2004) and Sorg et al. (1997), and references contained therein.

The evolution of a vast array of morphological, physiological, and behavioral adaptations in aquatic insects enables these organisms to inhabit virtually all bodies of water. Truly aquatic or semiaquatic insects occur in every conceivable aquatic habitat where a human body could be found. This includes such specialized habitats as hot and cold springs, intertidal pools, temporary and aestival ponds, water-filled tree holes, intermittent streams, saline lakes, marine intertidal zones, as well as in less harsh running and standing water habitats (Ward 1992). In fact, rarely are conditions in natural or even polluted waters so extreme as to totally exclude insects. The virtual absence of insects in the open sea suggests that other marine organisms are of more importance to forensic investigators in this habitat.

Approximately eight of the thirteen orders of insects containing species with aquatic or semiaquatic stages are likely to be associated with carrion or corpses in aquatic habitats. The following brief synopsis describes those orders of aquatic and semiaquatic macroinvertebrates, their functional role in aquatic ecosystems, and their relative importance in previous or ongoing forensic case histories.

Mayflies (Order Ephemeroptera)

Except for a very few species that venture into brackish areas, mayflies occur exclusively in freshwater. The ephemeral nature of the adult stage, generally 2 to 3 days' duration or less, accounts for the Latin name of the order Ephemeroptera, or "short-lived" (Ward 1992). Mayflies are morphologically and behaviorally diverse, and the larvae have been grouped into four life forms: (1) swimming, (2) creeping and climbing, (3) flattened and streamlined, and (4) burrowing (Ward 1992) (see appendix, keys 5 and 6). Both swimming (Baetidae) and flattened and streamlined (Heptageniidae) mayflies have been observed feeding on or near pig carcasses (Haefner 2005; Hobischak 1997; Schultenover and Wallace, unpublished data). Heptageniid larvae and the larvae of other families (e.g., Ameletidae) remove attached algae and periphyton from substrata such as rocks, logs, or in some instances, corpses. Collector-gatherer mayfly larvae, such as in the Baetidae or Ephemerellidae, obtain their food (largely fine particulate organic matter [FPOM]), by simply gathering it from wherever they can find it, such as under rocks, in deposition zones, or on the surface of stones or other substances. FPOM accumulates in many places on the streambed, wherever the current slackens enough to permit it to settle from the water column and accumulate in these deposition zones. Some scraper mayfly families Ameletidae and Heptageniidae, and

collector-gatherer families Baetidae and Leptophlebiidae are associated with salmon carcasses in Pacific Northwest streams (Wipfli et al. 1998; Minakawa 1997). Vance et al. (1995)
collected baetid and caenid (Caenidae) mayflies off pig carcasses submerged for only 2 days
in a small lake. Interestingly, Schultenover and Wallace (unpublished data) found heptageniid mayflies grazing on periphyton or algae that had accumulated over time on pigs
in riffle and pool habitats. They provided baseline data that indicated periphyton growth
increases significantly over time on pig carcasses compared to growth on control tiles in
the same aquatic habitat. This enriched feeding substrate may attract increased numbers of
scraper insect species, such as heptageniid mayflies, as has been demonstrated in salmon
carcass-enriched streams elsewhere (e.g., Wipfli et al. 1998). These trends are preliminary
and require further testing; however, such a strong correlation between periphyton growth
on corpses and increased abundance of a specific species utilizing this resource may provide some important entomological evidence useful in future death scene investigations.

Stoneflies (Order Plecoptera)

Stoneflies are associated primarily with clean, cool, running waters, although a number of
species are adapted to life in large oligotrophic, alpine lakes. Stonefly larvae have specific
water temperature (cold stenotherms), substrate type (usually cobbles, boulder surfaces),
and stream size (small to medium) requirements in their distribution and succession
along the course of streams and rivers (Stewart and Harper 1996). While eggs may hatch
soon after deposition, the larvae grow slowly and reach maturity nearly a year later. Major
growth occurs during the winter, and most species reach the adult stage early in the year
(Wallace 1996). Some adult stoneflies are the earliest species of aquatic insects to emerge
during the spring when snow still covers the ground.

Most species of stoneflies are clingers or sprawlers and can be divided into two functional feeding groups, shredders or predators (see appendix, key 5). Families of both shredders (Taeniopterygidae, Nemouridae, Pteronarcyidae) and predators (Perlodidae, Perlidae)
have been observed colonizing or feeding on submerged salmon carcasses or pig carrion in
streams, with larvae of Pteronarcyidae feeding on salmon flesh in the laboratory (Hobischak
1997; Minakawa 1997; Schuldt and Hershey 1995; Schultenover and Wallace, unpublished
data). Based on the restricted habitat requirements of stoneflies and the predatory nature of
many species, the presence of larvae on a body may indicate that it was probably deposited
and remained in a riffle zone of a stream or river. Thus, it may have been in the stream for
some time to allow for the colonization of other insects, which may serve as prey for the
stoneflies associated with the corpse at that time.

Caddisflies (Order Trichoptera)

Caddisflies occur on all continents except Antarctica, and are found in freshwaters (e.g.,
streams, rivers, ponds, lakes), brackish waters, and occasionally in marine intertidal areas.
The high ecological diversity of caddisflies has been attributed to their ability to produce
silk. The larvae use silk to construct fixed retreats or nets that trap and collect food particles in the current, or to build their portable cases (Wiggins 1996) (see appendix, keys 2
and 3).

Different families of caddisflies use different organic materials (e.g., twigs, leaves,
and other plant parts) or inorganic matter (sand grains, pebbles, stones) and sometimes

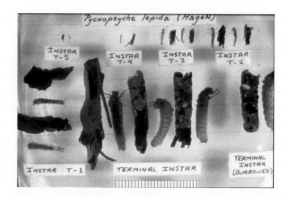

Figure 7.4 Larvae and cases of first through fifth instar of *Pycnopsyche lepida* showing changes in the type of case building materials with increasing size and age of larva. Instar T-5 = first instar; instar T-4 = second instar; and so forth to terminal instar = fifth instar; terminal instar (burrowed) = prepupa. (Photo by R. W. Merritt and K. W. Cummins.)

a combination of each to construct their cases (see appendix, key 2). As a caddisfly larva grows, it either adds on to its present case or changes to another type of case to accommodate its increase in size, depending on the availability of case-building materials. Often different stages of the same species use different materials to make their cases. For instance, some early larval stages of two caddisfly species belonging to the family Limnephilidae, *Pycnopsyche lepida* (Hagen) and *Pycnopsyche luculenta* (Betten), construct cases composed of leaf disks surgically cut from tree leaves that have accumulated in the streams (Figure 7.4). As these larvae grow and mature, and as the available leaf material significantly declines over time in these streams, *P. lepida* changes its case type to include more mineral deposits and sand grains glued together with silk, and also moves to faster-moving water (Cummins 1964) (Figure 7.4). Conversely, *P. luculenta* switches from the same type of leaf disk case to a case lined with twig portions as ballast, and moves to slower water, such as that found along stream margins (Wallace et al. 1992). Therefore, knowing the case material used by the larva in a particular stage of its development, one could identify the season of the year, or even a specific month, that a body may have entered the water if a caddisfly larva and its associated case were found on the victim's remains.

Benecke (2004) reported on a homicide from the 1950s involving a larval caddisfly case (most likely *Limnophilus flavicornis*) that contained fibers of the red socks that were worn by the deceased victim. However, the fibers were only found at the very top and the very bottom of the case, indicating that the larval caddisfly had, for the most part, already constructed the case and then finished it at the corpse (fibers on top), and after that attached it to the red sock (fibers on bottom). Because the attachment procedure takes a few days, it was estimated that the body had been in the water for at least 1 week.

Caddisfly cases found on a corpse also may indicate whether the corpse had been moved from a specific habitat within a lotic environment (e.g., riffle or pool), or transported by the current to a different location downstream. Many caddisflies also have known geographic ranges and phenologies (i.e., the timing of specific biological events within their life histories). Based on the species collected on the corpse, one could determine both spatial and temporal relationships (such as the likelihood of the corpse being transported long distances, or the time of year a corpse may have been placed in a given aquatic system).

Shredder species of caddisflies (Limnephilidae: *Pycnopsyche* sp.) and filtering-collector species (Hydropsychidae: *Hydropsyche* spp.) have been observed on pig carcasses in stream habitats in Indiana (Schultenover and Wallace, unpublished data) (Figure 7.5a). Brusven and Scoggan (1969) observed caddisfly larvae of the family Limnephilidae feeding on dead squawfish in a river in northern Idaho, and Hobischak (1997) found the same family associated with submerged pig carrion in ponds in British Columbia. Minakawa (1997) and Kline et al. (1997) observed larvae of several genera of Limnephilidae feeding on dead salmon flesh in the field and in the laboratory (Figure 7.5b), with some individuals actually penetrating the salmon skin (Figure 7.5c). The latter authors found over 1,000 caddisfly larvae (Linnephilidae: *Ecclisomyia*) on one fish head in an Alaskan stream. Limnephilid

(a)

(b)   (c)

Figure 7.5 (a) Coho salmon carcass in stream with caddisfly larvae (Limnephilidae: *Ecclisomyia* sp.) attached to skin. (b) Coho salmon carcass showing feeding marks in skin made by caddisfly larvae (Limnephilidae: *Ecclisomyia* sp.). (Photographs by Jason Walter and Brian Fransen.) (c) Larva of net-spinning caddisfly (Trichoptera: Hydropsychidae) on skin of submerged pig in southern Indiana stream. (Photo by J. R. Wallace.)

caddisfly larvae were observed moving to salmon carcasses and feeding for up to 15 minutes before leaving, whereas chironomid midges colonized the carcass fungal mat for longer periods (Kline et al. 1997). Benecke (2004) and Haskell et al. (1989) have discussed the utility of caddisfly casings as forensic entomological evidence, and Nawrocki et al. (1997) described a case where several caddisfly cases made of tiny stones (possibly Limnephilidae) were cemented to the floor of the nasal cavity of a corpse found in the water.

True Flies (Order: Diptera)

Over half of all known species of aquatic insects are dipterans, with midges (family Chironomidae) constituting the largest family of freshwater insects (Ward 1992). Many dipteran families that have aquatic representatives are largely composed of species that inhabit water as larvae (Ward 1992). Dipterans are found in virtually every conceivable aquatic environment, and may in fact be the only insects in freshwater habitats with extreme environmental conditions. These include hot springs, petroleum pools, and saline lakes. This order contains the most successful insect colonizers of the marine intertidal environment.

The chironomid midges are of particular interest to forensic investigators because of their overall diversity and presence in nearly every aquatic habitat. They often account for over 50% of the total macroinvertebrate species diversity, occurring at densities in excess of $50,000/m^2$ (Coffman and Ferrington 1996; Armitage et al. 1995; Berg and Hellenthal 1991). As an example, in a small woodland stream in eastern North America, 143 different chironomid species were recognized (Coffman 1973). The midges also are the most diverse family of Diptera in their selection of aquatic habitats, occupying every conceivable water body, from the smallest streams to the largest rivers, and from large lakes to very small pools. They occupy both fresh and salt water environments as well as clean and polluted waters of varying depths (Pinder 1995). The red color of some midges (Figure 7.11), leading to the name bloodworms, is caused by the respiratory pigment hemoglobin, which enables the larvae to recover rapidly from anaerobic periods in low oxygen environments (i.e., sediments) (Eriksen et al. 1996).

Chironomid larvae are known to feed on a great variety of organic materials representing all functional feeding groups. These include (Coffman and Ferrington 1996; Berg 1995):

1. Coarse detrital particles (leaf shredders)
2. Medium detrital particles deposited in or on sediments (gatherers and scrapers)
3. Fine detrital particles in suspension, transport, or deposited (collector-filterers and -gatherers, and scrapers)
4. Algae (benthic, planktonic, or in transport) (scrapers, collector-gatherers, and some filterers)
5. Vascular plants (miners)
6. Fungal spores and hyphae (gatherers)
7. Other animals (predators)

The ability of chironomid larvae to spin silk has great adaptive importance and plays a major role in filtering collectors and gatherers (Wallace and Merritt 1980). It allows them to construct a silken tube that houses the larva, and spin a conical catchnet across the

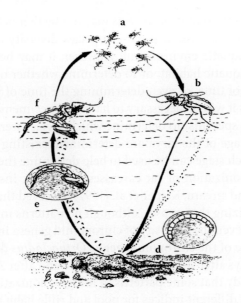

Figure 7.6 Generalized diagrammatic life cycle of a chironomid midge (Diptera: Chironomidae) colonizing a human corpse. For explanation, see text. (Used with permission from Haskell et al. 1989.)

lumen of the tube (appendix, key 3). For those inhabiting soft substrates in lotic and lentic habitats, a U- or J-shaped burrow is constructed in which they live. Because of their diverse food habits and mode of existence, it is not surprising that most studies dealing with submerged carrion of any type have noted the presence of chironomid midges (Wipfli et al. 1998; Hobischak 1997; Keiper et al. 1997; Minakawa 1997; Vance et al. 1995; Haskell et al. 1989, Hawley et al. 1989). They are often found to be the most abundant insect on corpses, and are usually represented by several different genera and species. Also, because egg masses and young larvae of chironomids are known to drift in streams and rivers and disperse in ponds and lakes (Pinder 1995), they are one of the first colonizers to arrive on a corpse.

Midges have four distinct life stages: egg, larva, pupa, and adult (Figure 7.6a–f). The duration of the larval stage, with four instars, may last from less than 2 weeks to several years, depending on species and environmental conditions. However, chironomids in temperate climates usually have from one to two generations per year and, in general, exhibit shorter life cycles in warm, nutrient-enriched waters (Coffman and Ferrington 1996). As noted above, early instars may be planktonic and in the water column, while later instars are usually benthic and on substrates or in the sediments (Figure 7.6d). Most identification keys to the Chironomidae are written for mature larvae. The pupal stage (Figure 7.6e) begins with the separation of the larva from the underlying pupal integument. After ecdysis, or shedding of the old skin, the pupa usually remains hidden in debris until it swims to the surface, where adult emergence takes place (Figure 7.6a and f). Chironomid adults usually live a few days, although some species survive for several weeks. Adults generally do not feed, and this stage performs the functions of reproduction and dispersal. Mating takes place in aerial swarms (Figure 7.6a), on the water surface in skating swarms, or on solid substates. Females may broadcast the eggs at the water surface, where they sink to the bottom sediments (Figure 7.6c), or more frequently, deposit gelatinous egg masses on the open water or on emergent vegetation (Coffman and Ferrington 1996).

Family-level identification of these insects may not be that helpful in determining the PMSI in an investigation because of the high species diversity and numbers of midges one could find in any aquatic environment. However, it may help to associate a corpse with a specific type of aquatic habitat, or to determine whether or not a body was totally submerged for a period of time, or even determining the time of year when the body was deposited. In most cases it will be necessary to have the specimens sent to an aquatic entomologist or chironomid specialist to determine which species were found on the remains. Once the species and stage of the larva are determined, existing information on the life cycle and duration of each stage can be used to help determine the PMSI.

In studying the colonization of rat carcasses by aquatic insects in riffles and pool areas of a small woodland stream, Keiper et al. (1997) observed that midge larvae were the dominant insects colonizing the corpses, although no patterns in numbers of larvae over time were evident. However, the diversity of chironomid genera increased after 29 days in the riffle area, and larvae of the chironomid genus *Orthocladius* did not begin to colonize the carcasses until 13 days after submersion in the riffle and after 20 days in the pool. They concluded from this study that some patterns in midge colonization were detectable over time, but suggested that different indices for pool and riffle habitats need to be developed when determining the PMSI on corpses based on midge colonization rates. Schultenover and Wallace (unpublished data) found that aquatic insect diversity on submerged pigs generally decreased over time; however, midges were the dominant organism found in or on pig carcasses. Wipfli et al. (1998) observed Chironomidae burrowing into salmon carcasses in the field, and which were abundant within the salmon tissue after 30 days. The presence of midges in all of these studies suggests that the carcasses were submerged for some time.

Studies of midges occurring in some Iceland and Canadian lakes (Lindegaard 1992; Harper and Cloutier 1986) have shown that specific species complexes are associated with different depth zones of the lake, such as the surf zone (lake margin), littoral (0 to 10 m), sublittoral (10 to 20 m), and profundal (>20 m). The highest number of individuals and species were found in the shallowest sites, and the lowest number of both occurred in the deepest sites. If this information was known about a lake in which a corpse containing chironomids was found, then it might be possible to determine where in the lake a body was originally deposited before it floated to the surface, and if it was possibly transported by wind or wave action. Also, there may be published life history information on some of the species present that could assist with determining a PMSI.

Midges and black flies (Simuliidae) have both been evaluated in death scene investigations during the past decade and are discussed in more detail in the "Case Histories" section.

Collecting and Rearing Aquatic Insects from Corpses

Collecting and preserving insects from aquatic environments is somewhat different than from terrestrial sites because one is mostly dealing with larvae rather than adults. All larvae and adults collected from aquatic habitats should be placed in vials, labeled, and preserved in 70 to 80% ethanol (ethyl or isopropyl), as soon as they are carefully removed from substrates, clothing, or the corpse with forceps. It is helpful to have waders or hip boots and shoulder-length rubber gloves (trapper gloves) to gain access to the body in cold water and remove insects before moving the body to land. A large magnifying glass or hand lens in the field is helpful to spot insects on the habitus. Labeling instructions for insect specimens are detailed in Catts and Haskell (1990) and in this volume (Chapter 3). It also is important to keep larval caddisflies with different cases in separate vials, as they tend to abandon their case when placed in alcohol, and sometimes the case is a key character in identifying the specific taxa of Trichoptera. Although there are other solutions that can be used to preserve aquatic insects (e.g., Carnoys, Hood's, Kahle's [see Catts and Haskell 1990]), ethanol is the preferred medium (Merritt et al. 2007b). If one is collecting stream-bottom samples containing detritus and insects, it is preferable to use 95% ethanol to account for dilution from water in the samples (Merritt et al. 2007b). A white enamel pan is helpful to provide contrast when sorting through silt and detritus for insect specimens. It is very important that specimens are *never* placed in water as a preservative because they will decay rapidly (within 1 to 2 hours) and not be identifiable. In most instances, one will not see many adult insects flying around the habitat or corpse that are associated with the larvae in aquatic environments.

In some instances, it may be necessary to transport live larval insects from the field to the laboratory so that they can be reared to adults for specific identification. One of the most common problems encountered in rearing aquatic insects involves mortality during transport due to inadequate oxygen supply or temperature control. Agitation during transport will maintain oxygen levels, but may damage delicate specimens. Alternative methods include transporting the animals in damp moss, burlap, or paper towels with a small amount of cold water and using small "bait bucket" aerators that can be purchased at most pet stores. To maintain cool temperatures, thermal containers or ice coolers should be used (Merritt et al. 2007b).

Laboratory rearings can be maintained at field temperatures using an immersible refrigeration unit or by recirculating water through a cooling reservoir. If laboratory temperatures do not match those in the field, mortality can be reduced by allowing temperatures to equilibrate slowly. To maintain water quality in the laboratory, tap water should be dechlorinated and distilled, or spring water added to replace evaporation loss. Algal and detrital food supplies are often best maintained by periodic replenishment from the field (Merritt et al. 2008b). For a further discussion of rearing techniques, see Merritt et al. (2008b) and Chapter 4 in this volume.

Case Histories Involving Aquatic Insects

Case 1

In late June 1989, a pair of recreational scuba divers were exploring the waters of the Muskegon River in western Michigan when they discovered a car lying upside down on the bottom of the river, with the dead body of a woman inside (Figure 7.7a). The car was found in a 15-foot-deep hole, so it was not observable from the riverbank or the bridge above. Police hauled the car from the river and traced it to the woman's husband. Medical examiners found contusions on the dead woman's head, which did not appear to be caused by the accident. Based on other evidence, the husband became the prime suspect. He initially claimed he had argued with his wife in late September of the previous year, and that she had driven away upset into a foggy night. He further stated that he had not seen her since that night. However, the car in which his wife was found was in relatively good condition, suggesting that it had been pushed into the river and had not gone into the water as the result of a crash.

Position of car with
body in the
Muskegon River

Black flies on windshield

(a)

(b)

Figure 7.7 (a) Drawing showing position of car with body inside, found by recreational divers near bridge abutments in the Muskegon River, Michigan. Inset shows larvae and pupal cocoons of black flies on windshield. (b) Recovered car was found in relatively good condition. Detectives noticed aquatic insects attached to windshield, fenders, and door panels. (c) Close-up photograph of car door panel with arrow pointing to attached black fly pupae. (Photos by Det./ Sgt. Richard Miller, Michigan State Police.)

Continued

Even though the police thought the husband was lying about the circumstances of the case, the cold water had preserved the woman's body, making it difficult to determine the time of death. Interestingly, when the car was brought up from the river, the detectives noticed that there were aquatic insects attached to the windshield, fenders, and door panels (Figure 7.7b and c). Specimens were collected from these substrates and sent to one of us (Merritt) in the Department of Entomology at Michigan State University for identification. Three different insect taxa were identified from the car: (1) caddisfly cases, (2) chironomid midges, and (3) black fly pupal cocoons and some larvae. It was difficult to establish a time

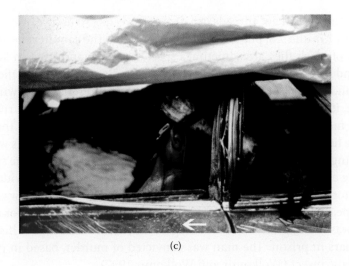

(c)

Figure 7.7 *Continued.*

that the car had gone in the river using the caddisflies or midges; however, the black fly cocoons provided evidence that was significant to establishing the PMSI.

Adult female black flies, in this case belonging to the genus *Prosimulium*, lay eggs in the late spring or early summer. These settle into the sediments and undergo obligate diapause (an arrested state of development) until the following fall or early winter (late November to January) (Figure 7.8; eggs). The eggs hatch into larvae, and these attach to a specific substrate in the stream (e.g., rocks or vegetation, in this case a car) where they filter very small particles from the current with fan-like structures on their head (Figure 7.8; larva). Larval growth is slow at the low water temperatures prevailing in the river during January and February. Following snowmelt, growth increases and pupation on the substrate occurs in late March or April (Figure 7.8; pupa).

Emergence follows shortly after (early to mid-May), and adults are present for 1 to 2 months, during which time they mate, take a suitable blood meal (if required), lay eggs, die, and the life cycle begins again (Figure 7.8; adult). In this particular case, numerous pupal

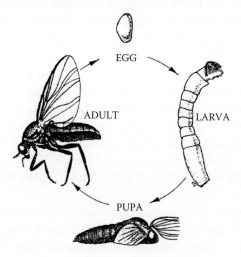

Figure 7.8 Generalized life cycle of a black fly (Diptera: Simuliidae) showing all stages.

cocoons (consisting of "silken" threads in a dense sleeve-like structure of specific shape and texture) (Figures 7.7a and 7.8; pupa) remained attached to the car after the adults emerged. Pupal cocoons of black flies are sometimes species specific and used in keys to identify specimens found in lotic habitats (Merritt et al. 2008b). Based on the specific identification of the cocoon and known life cycle of the black fly species present on the windshield of the car in late June, it was determined that the car had to have gone into the river long before June 1989 and most likely the previous fall (October/November 1988). The suspect later claimed that he had spoken with friends who had been in contact with his wife during late winter and spring of 1989. This would not have been possible based on the species of black fly found on the car. In April 1990, based on autopsy, insect, and other evidence presented by the prosecutor and expert witnesses, it was proven that the husband was lying and his wife had disappeared the previous fall and the car had been in the river for approximately 9 months. He was found guilty of second-degree murder in the death of his wife and sentenced to 20 years in prison. The man was convicted of murder, based in part on the life cycle of an aquatic insect (Wolkomir and Wolkomir 1992).

Case 2

On July 5, 1997, the dead body of a partially submerged 19-year-old female was discovered by turtle hunters in a small lake in western Michigan, 3 days before she was supposed to testify against a man who had sexually assaulted her (Figure 7.9). Unconfirmed reports placed her alive on the 5th or 6th of June 1997. The woman's eyes and mouth were shut with duct tape, and the body had been weighted down with two cinder blocks and bound with chains before she was thrown into the lake alive. The body was submerged within the lake prior to rising to the surface due to gaseous buildup of putrefaction. Insect larvae (fly maggots belonging to the family Calliphoridae) were collected by the police on July 7 from the head and face exposed to the air when the body was pulled from the water. No puparia were observed on the body by the detective in charge. Specimens were sent to one of us (Merritt) at Michigan State University for identification.

An analysis of the fly larvae showed that they were mature third larval instars of *Phormia regina* (Meigen), the black blow fly, a common blow fly in the Midwest during the summer months. This species is a late arriver at carrion, and at temperatures in the 80s and 90s, *P. regina* will generally have a 24- to 48-hour delay in oviposition (Hall and Doisy 1993). However, this situation was somewhat different and more difficult to interpret because the body had been underwater for some time and surfaced in a different physiological state than would have occurred if the victim had just been killed minutes earlier and never exposed to water. Based on the degree-days required for larval development of the oldest and largest specimens, the floating remains of the body would most likely have been exposed to colonization by this species between July 1 and 2. This PMSI date turned out to be extremely close to when police suspected the victim had surfaced, and was in line with the estimation by the forensic pathologist of how long the body had been submerged in the lake.

The time required for the body to resurface from being submerged is primarily dependent on water temperature and typically decreases with depth (Rodriguez 1997). With fairly warm water temperatures, as was the case for this shallow lake in July, and the observed presence of algal blooms, a body would be expected to surface within a few days to a week. Other factors that affect the rate of decomposition in water include bacterial content and

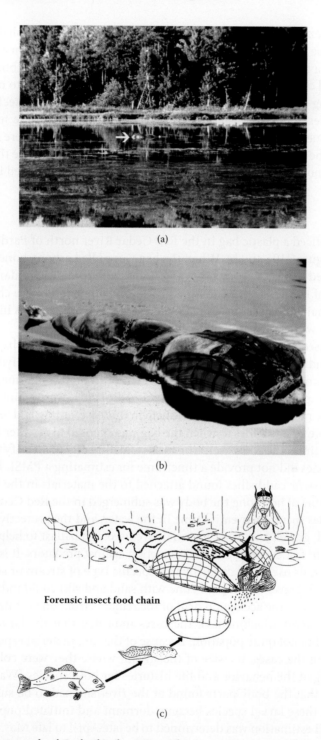

Figure 7.9 (a) Photograph of Oxford Lake, MI, with arrow pointing to a human corpse floating on the surface. (b) Close-up photograph of the body floating on the lake surface shows chains and cinder blocks used to drown the victim. Maggots of blow flies colonized the head region after the body floated to the surface. (c) Forensic insect food chain in a lake environment showing a case where blow fly maggots colonizing a body exited the body into the water. Tadpoles were observed trying to feed on maggots and then being eaten by fish. Fish also were observed feeding on maggots. (Drawing by Ethan Nedeau.)

salinity (Rodriguez 1997). A corpse submerged in a highly eutrophic shallow lake such as the one in this case will decompose much more rapidly than a corpse in a relatively cooler, deeper lake with a lower bacterial count. Of course, the weight of the body, as well as the cinder blocks and chains used to keep it submerged, also were factors to consider.

An interesting aspect to this case was the observation by the detective at the crime scene that tadpoles and small fish, probably bass and blue gills, were feasting upon the decomposing tissue of the body, as well as the dipteran larvae that were exiting the body and falling into the water. From an ecological standpoint, this could be the first published account, to our knowledge, of an aquatic forensic food chain as depicted in Figure 7.9.

Case 3

An individual noticed a plastic bag in the Red Cedar River north of Pardee, Michigan, on June 13, 2005 (Figure 7.10a and b). When the bag was pulled ashore, bones and remaining flesh were observed inside. A decoy bag was found as well containing a dark-colored plastic bag inside. Part of the decoy bag contained rocks and a portion of the chest containing a bone, rib bones, hair, and other fragments. A closer look of the remains indicated the presence of a forearm with an intricately designed tattoo.

At autopsy, insect specimens collected from the decoy and plastic garbage bags contained a muscid larva, and several caddisfly larvae belonging to two families, Hydropsychidae (net spinners) and Limnephilidae (case-makers). A total of twenty-one limnephilid caddisflies belonging to two species, *Pycnopsyche guttifer* and *P. lepida*, were counted and identified (Figure 7.10c). The muscid larva was of little help in this case, as it would only take the earliest oviposition date by adult flies to when the bag was exposed to air after being submerged in water for some time. The complex life histories and difficulty to identify the hydropsychid caddisflies to species did not provide a timeframe for estimating a PMSI. However, the two species of *Pycnopsyche* caddisflies found attached to the materials in the bag were used to provide a timeframe of how long the body was submerged in the Red Cedar River. The larvae of these species spend their entire time in the water, and the protective cases they construct out of sand, gravel, and other materials are used as a ballast to help weight them for bottom dwelling in fast currents and for protection against predators. It is possible that the type of case materials used by a caddisfly in a specific type of stream or stream reach (e.g., in a riffle zone with gravel versus pool zone with sand and silt) could indicate the planned movement of a body or the accidental drift of the remains from one habitat to another.

In this case, larval attachment of the cases indicated that the larvae had entered a quiescent period but not quite pupation. Because of the case material type, size of the mineral pieces used in the cases, the size of the stream where they were collected, as well as an understanding of the behavior and life histories for both species (Wallace et al. 1992), it was estimated that the body parts found at the river site had to be submersed prior to the period when these larval species become dormant and initiated pupation (Table 7.3). Therefore, a PMSI estimation was determined to be late April to late May. It is possible that the remains entered the water prior to this time interval, but the period defined as the colonization period by these larval caddisflies would have been late April to late May 2005.

In this particular case, the larvae of the caddisflies found on the victim's clothing were instrumental in estimating a portion of the PMSI. They were used to confirm that the body had been in a river for a given period of time and probably not moved or transported there from another type of aquatic habitat. We strongly feel, as discussed earlier and illustrated

(a)

(b)

(c)

Figure 7.10 (a, b) Decoy bag found in the Red Cedar River, north of Pardee, Michigan, containing bones, flesh, and other fragments. (Photos by Det. Billy Mitchell, Ingham County Sheriff's Office, Mason, Michigan.) (c) Pycnopsyche caddisfly larvae with cases. (Photo by R. W. Merritt.)

Table 7.3 Life Cycles of *P. lepida* and *P. guttifer* as Established by Howard (1975) on Gull Creek, Kalamazoo, Michigan

```
                                    Month
              J  -  A  -  S  -  O  -  N  -  D  -  J  -  F  -  M  -  A  -  M  -  J
P. lepida      -- V --- pupae ----
                      -- Emerge
                      --- Adults ------
                      ---- Eggs ---------
                         ------------ I ----
                            ------------ II ----
                         ------------ III ----
                            ------------ IV ----
                                     ------------------- Burrowed, V—dormant --

                                    Month
              J  -  A  -  S  -  O  -  N  -  D  -  J  -  F  -  M  -  A  -  M  -  J
P. guttifer    -- V --- pupae ----
                      -- Emerge
                      --- Adults ------
                      ---- Eggs ---------
                         ------------ I ----
                            ------------ II ----
                         ------------ III ----
                            ------------ IV ----
                                     ------------------- V ---------- dormant --
```

in this case study, that a knowledge of the life history features of caddisflies could be very important in certain aquatic forensic investigations.

Case 4

A missing person report was made on March 26, 1993, for a 16-year-old white female who was last seen at her place of employment in LaPorte, Indiana. Her empty car with the hood up and the keys in the ignition was found the following night a short distance from the above site. On April 27, fishermen discovered the partially clothed body of the missing female in a large pond near where the car was discovered. Her body was floating face down in less than 6 feet of water with some tree branches lying across part of her remains. An autopsy revealed that the cause of death was the result of strangulation. The body was in a bad state of decomposition, and it was determined that it had been in the water an estimated time of 30 days.

When the body was removed from the water, numerous aquatic invertebrates were collected from the victim's body and clothing. These included species from the following taxa: amphipods (*Hyalella azteca*), snails (*Physella, Planorbula, Promenetus*), clams (*Sphaerium*), water mites (*Lebertia*), dragonflies (Gomphidae; *Gomphus*), damselflies (Coenagrionidae; *Ischnura*), chironomids (*Phaenopsectra*), pigmy backswimmers (Pleidae), crane flies (Tipulidae; *Erioptera*), and biting midges (Ceratopogonidae; *Bezzia*).

The above information collected from the remains did not provide evidence that could be used to establish a PMSI in this investigation. However, after reviewing the habitat, mode of existence, and trophic information on the invertebrate taxa collected, a few conclusions could be drawn. First, the habitat in a pond or lake where most of these invertebrates occur and where the victim was found is termed the littoral zone. This is the shallow shoreline zone where sunlight penetrates to the bottom and is sufficient to support rooted plant growth. Most of the taxa collected from the corpse are climbers on rooted vegetation growing in this habitat or burrowers in sediments, primarily silt (Merritt and Cummins 2006). Also, several taxa found belong to the scraper functional feeding group and feed on periphyton or attached algae. This would indicate that the victim had been in the lake long enough for these plants to colonize the surface of her body or clothing and provide a food resource for these invertebrates to utilize (J. Wallace, personal observation). This kind of ecological information would have been more useful if it was not known when the victim originally disappeared, as the presence of different invertebrate life stages are seasonal and substrate colonization rates of the different taxa vary over time. Also, based on the invertebrates found on the body at that depth and location, one could possibly determine whether there was movement of the body from one site to another within the pond (e.g., deep to shallow zone), possibly due to bloating, wind, or wave action.

Case 5

This case history actually describes two separate murders reported by Hawley et al. (1989) involving the same forensic insect evidence. The first case was a 21-year-old male found dead floating in a stream in an urban area. The cause of death was impalement per rectum by a 3-foot length of pipe. A preliminary examination of the body showed the presence of a few 3 to 5 mm coiled translucent red strands or "fibers" on the clothing and body, thought to be carpet fibers. These fibers were initially treated as such and placed in a dry plastic petri dish and sealed for later analysis. When analyzed, the fibers turned out to be larvae of chironomid midges, or blood worms, as they are called due to the presence of hemoglobin (Figure 7.11).

The second case occurred a year later and involved an unidentified adult female found floating in a farm drainage ditch. She had been killed by blows to the face with aspiration of blood (Hawley et al. 1989). Examination of the partially clothed body again disclosed similar translucent red "fibers," but this time a forensic entomologist was available to recognize these as insect larvae belonging to the family Chironomidae (see appendix, key 4). Although it was stated by Hawley et al. (1989) that the stage of development of these larvae may be useful in estimating the duration of submersion if life cycle data were available, no

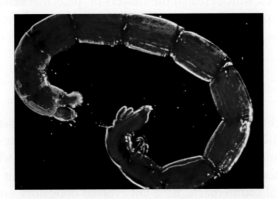

Figure 7.11 Larva of a blood worm, or chironomid midge (*Chironomus* sp.), showing red color due to the presence of the respiratory pigment, hemoglobin. (Photograph by R. W. Merritt.)

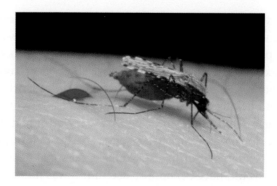

Figure 7.12 Photograph of engorged adult female *Anopheles stephensi* completing a blood meal on a human. (Photograph by R. W. Merritt.)

information was provided as to whether or not these larvae were important in these cases. We feel, as others do, that chironomid midges may prove to be very useful in forensic investigations (Haskell et al. 1989; Hawley et al. 1989; Keiper et al. 1997).

Case 6

In most instances, it is the immature stages of aquatic insects, particularly Diptera, that are associated with a corpse in a stream or a pond, rather than the terrestrial adults. However, one case involving the latter occurred in the northwestern United States on the lower slopes of a mountain range, and proved to be interesting and unique. The story was passed on to Haskell et al. (1997) by the late Dr. Paul Catts, an outstanding forensic entomologist at Washington State University. The body of a dead female clad in a swimsuit was found to have petechial hemorrhages arranged in clustered patterns on exposed portions of her thighs. She was found in a wet swampy habitat that was heavily forested and a major mosquito breeding area. The hemorrhages were inconsistent with the circumstances surrounding the death, and the pathologist had no explanation. A forensic entomologist who was participating in the investigation determined that these "hemorrhages" were actually multiple probing marks on her exposed skin caused by a mosquito's mouthparts (proboscis), as the insect was trying to take a blood meal on the still warm, but dead body (Figure 7.12). Unable to bring up blood in the proboscis due to lack of blood pressure, the female mosquito probed more sites, which created the clusters of "petechial hemorrhages." When the assailant was finally apprehended, he confessed to having placed the body in the swampy habitat shortly after he had killed her (Haskell et al. 1997).

Conclusions

At the beginning of this chapter we showed the great discrepancy between the number of forensic studies published dealing with terrestrial insects and those dealing with aquatic insects. This discrepancy has occurred for several reasons, some of which have been made evident throughout the text. First, we and others have pointed out that aquatic insects, unlike their terrestrial counterparts, have not evolved obligatory sarcophagous habits to feed on carrion in their environment. Crayfish, a crustacean, but found in the same habitats, have probably come the closest to this functional feeding mode. Studies on salmon decomposition in Pacific Northwest streams, where carcasses have been a natural part of the landscape for eons of time, indicate that some caddisflies and possibly other groups of aquatic insects have become facultative scavengers and utilize this food resource during different times of the year. However, most insect decomposers in lotic and lentic systems feed on leaf litter and similar forms of organic matter that enters the stream or lake on a seasonal basis. This is in contrast to marine systems where a wide variety of scavenging fish and invertebrates other than insects occur.

Second, to date we have not discovered a successional insect model to follow in aquatic systems, as exists with many blow fly species in terrestrial systems. This is partly because many aquatic insects are natural drifters and will drift onto a substrate (carcass) for attachment only, and not necessarily to feed. Therefore, when an aquatic insect is collected from the victim or his or her clothes, it is difficult to interpret whether or not the insect in question was actually colonizing the corpse or arrived there by accidentally drifting onto another substrate in its movement downstream with the current. Even if it did colonize the corpse and was identified, it is much more difficult to trace it back to a specific instar or developmental stage knowing that some aquatic insect orders have twenty to forty larval instars. This also is confounded by the fact that developmental times for the majority of species have not been worked out in detail enough to establish a PMSI.

We also noticed in recent studies that the growth rates of algae on submerged carrion over time, in conjunction with the utilization of this food resource by certain aquatic insects (i.e., scrapers), may be of significance in helping to determine the PMSI. In fact, our preliminary results indicated that algal growth on an aquatic substrate such as a pig or even a human corpse occurs in a successional manner similar to the successional sequence we observe with terrestrial insect colonization of human corpses. In our discussion of aquatic insects of forensic importance and in the case histories given, we have provided examples of different families, based on their functional feeding groups, that could be or have been of assistance in determining the PMSI. Specifically, these include larvae of mayflies (Ephemeroptera), stoneflies (Plecoptera), caddisflies (Trichoptera), midges (Chironomidae), black flies (Simuliidae), and in a unique case, adult mosquitoes (Culicidae). We know that others exist and will be found to play a significant role, but there first needs to be much more experimental research conducted on the aquatic insect colonization of submerged carrion or corpses in lotic and lentic systems, and the elucidation of life history features on those specific species involved.

Another problem that has hampered studies in freshwater and marine aquatic systems has been the wide variety of confounding environmental and biological factors acting on the corpse and influencing the PMSI. These were discussed in some detail and include current and wave action, water temperature, oxygen concentration, salinity, depth, nature of the substrate, and other parameters that may influence water quality and the diversity of aquatic insects or other arthropods colonizing the corpse. These, in addition to body factors such as whether or not it is floating or submerged, or clothed, make the determination of a PMI or PMSI a difficult task.

Lastly, the collection and rearing of aquatic insects from submerged corpses by untrained personnel has not been emphasized in the literature, and is a more difficult procedure to accomplish than in terrestrial situations. This statement may have support from a recent study by Hobischak and Anderson (1999), who carefully examined water-related deaths reported by British Columbia's Coroners Service over a 2-year period. After looking at reports from forty-seven cases, they found entomological evidence from only three. Of these, all the insect species reported were terrestrial; none were aquatic, and no cases reported using insects in determining the PMI. As they alluded to in their study, this was probably due to ignorance by the investigator(s) as to the potential importance of the aquatic fauna present on a corpse, lack of recognition of these kinds of insects, poor collecting techniques, or a combination of the above.

If a corpse is found in a stream or other shallow body of water, an investigator should come prepared to the scene with waders, alcohol vials, forceps, and gloves to collect specimens off the body before it is removed from the water. This procedure is probably rarely done and is more difficult than collecting maggots with forceps off a terrestrial corpse and placing them in an alcohol vial. If one wishes to rear immature aquatic insects back in the laboratory to obtain adults for specific identifications, he or she generally will have to carry out the procedures described earlier, which are much more labor-intensive than simply placing the maggots collected from a terrestrial corpse on a piece of fresh liver in a container with sand and waiting for a week for adult emergence. The different techniques used for collecting and rearing aquatic insects have surely prevented investigators from involving these kinds of insects more in crime scene investigations.

In summary, our chapter has reviewed the literature and presented the stages of decomposition for different types of carrion (mainly pigs) in freshwater environments and the associated aquatic insect fauna. The ecological approach we have chosen to determine which aquatic insect groups may be important in corpse decomposition is based on the functional feeding group classification. This general classification system distinguishes invertebrate taxa within aquatic ecosystems according to the different morphological-behavioral adaptations used to harvest nutritional resources. These feeding mechanisms determine the food resources that are processed: shredders feed on CPOM, collectors on FPOM, scrapers on periphyton or algae, and predators on prey.

A simplified pictorial dichotomous key is provided to assist investigators and researchers with the identification of these different functional groups that may be found associated with or feeding on corpses in aquatic environments. As we have discussed above, determining the postmortem submersion interval (PMSI) in aquatic studies has been, and will continue to be, problematic due to several factors, but it is critical to determining the time between death and corpse discovery in these environments. Hopefully, we have identified those groups that may be important and ones that need further research emphasis. Collecting and rearing techniques for aquatic insects have been reviewed, and the logistical problems have been discussed, as compared to those used in terrestrial environments. Finally, we have provided several case histories where aquatic insects have been used to either determine the PMSI or associate a body with a particular type of aquatic habitat or location within the habitat. It is abundantly clear that we still have a long way to go in determining the PMSI using aquatic insects, and it will be some time before we can approach the level of sophistication and accuracy that we have achieved with terrestrial insects. We hope that this chapter has provided a start in that direction.

Acknowledgments

The authors thank the following individuals for their contributions: master's students James Haefner (Millersville University) and Nikki Hobischak (Simon Frazier University, Vancouver, British Columbia, Canada) for their contributions through their theses;

Dr. Joyce Deyong, Forensic Pathologist, Sparrow Hospital in Lansing, Michigan, Dr. Eric Benbow, Ryan Kimbarauskas, and Mollie McIntosh for their assistance in case 3; Sara Schultenover for use of her data on pig decomposition in streams; Neal Haskell for helpful comments on various cases; and Ethan Nedeau for his drawing of the forensic food chain. We also acknowledge Det. Billy Mitchell, Ingham County Sheriff's Office, Mason, Michigan, for allowing us permission to use his photographs in the chapter, and Dr. Richard Glenn (Millersville University), who commented on the manuscript. A portion of this research was supported in part by a grant from the USDA-CSREES National Research Initiative Competitive Grants Program (Ecosystem Science Program). Research conducted by James Haefner was supported by Millersville University Faculty Grants Program.

References

Allan, J. D. 1995. *Stream ecology—Structure and function of running waters.* London: Chapman & Hall.

Anderson, G. S., and N. R. Hobischak. 2004. Decomposition of carrion in the marine environment in British Columbia, Canada. *International Journal of Legal Medicine* 118:206–9.

Anderson, G. S., and S. L. VanLaerhoven. 1996. Initial studies on insect succession on carrion in southwestern British Columbia. *Journal of Forensic Sciences* 41:617–25.

Armitage, P., P. S. Cranston, and L. C. V. Pinder, Eds. 1995. *The Chrionomidae: The biology and ecology of non-biting midges.* London: Chapman & Hall.

Baldy, V., M. O. Gessner, and E. Chauvet. 1995. Bacteria, fungi, and the breakdown of leaf litter in a large river. *Oikos* 79:93–102.

Bärlocher, F. 1982. The contribution of fungal enzymes to the digestion of leaves by *Gammarus fossarum* Koch (Amphipoda). *Oecologia* 52:1–4.

Benecke, M. 2004. Arthropods and corpses. In *Forensic pathology reviews*, ed. M. Tsokos. Vol. II. Totowa, NJ: Humana Press, 207–40.

Bengtsson, G. 1992. Interactions between fungi, bacteria and beech leaves in a stream microcosm. *Oecologia* 89:542–49.

Berg, M. B. 1995. Larval food and feeding behaviour. In *The Chironomidae: Biology and ecology of non-biting midges*, ed. P. D. Armitage, P. S. Cranston, and L. C. V. Pinder. London: Chapman & Hall, 136–67.

Berg, M. B., and R. A. Hellenthal. 1991. Secondary production of Chironomidae (Diptera: Chrionomidae) in a north temperate stream. *Freshwater Biology* 25:497–505.

Boyle, S., A. Galloway, and R. T. Mason. 1997. Human aquatic taphonomy in the Monterey Bay Area. In *Forensic taphonomy: The postmortem fate of human remains*, ed. W. D. Haglund and M. H. Sorg. Boca Raton, FL, 605–14.

Brittain, J. E. 1976. The temperature of two Welsh lakes and its effect on the distribution of two freshwater insects. *Hydrobiologia* 48:37–49.

Brusven, M. A., and A. C. Scoggan. 1969. Sarcophagous habits of Trichoptera larvae on dead fish. *Entomological News* 80:103–5.

Cairns, J., Jr., and J. R. Pratt. 1993. A history of biological monitoring using benthic macroinvertebrates. In *Freshwater biomonitoring and benthic macroinvertebrates*, ed. D. M. Rosenberg and V. H. Resh. New York: Chapman & Hall, 10–27.

Casamatta, D. A., and R. G. Verb. 2000. Algal colonization of submerged carcasses in a mid-order woodland stream. *Journal of Forensic Science* 45:1280–85.

Catts, E. P., and M. L. Goff. 1992. Forensic entomology in criminal investigations. *Annual Review of Entomology* 37:253–72.

Catts, E. P., and N. H. Haskell. 1990. *Entomology and death: A procedural guide.* Clemson, SC: Joyce's Print Shop.

Chaloner, D. T., and M. S. Wipfli. 2002. Influence of decomposing Pacific salmon carcasses on macroinvertebrate growth and standing stock in sourtheastern Alaska streams. *Journal of the North American Benthological Society* 21:430–42.

Chaloner, D. T., M. S. Wipfli, and J. P. Caouette. 2002. Mass loss and macroinvertebrate colonization of Pacific salmon carcasses in southeastern Alaskan streams. *Freshwater Biology* 47:263–74.

Coffman, W. P. 1973. Energy flow in a woodland stream ecosystem. II. The taxonomic composition and phenology of the Chironomidae as determined by the collection of pupal exuviae. *Archiv für Hydrobiologie* 71:281–322.

Coffman, W. P., and L. C. Ferrington. 1996. Chironomidae. In *An introduction to the aquatic insects of North America*, ed. R. W. Merritt and K. W. Cummins. 3rd ed. Dubuque, IA: Kendall/Hunt Publishing Co, 635–754.

Cummins, K. W. 1964. Factors limiting the microdistribution of the caddisflies *Pycnopsyche lepida* (Hagen) and the *Pycnopsyche guttifer* (Walker) in a Michigan stream (Trichoptera: Limnephilidae). *Ecological Monographs* 34:271–95.

Cummins, K. W. 1974. Structure and function of stream ecosystems. *Bioscience* 24:631–41.

Cummins, K. W., and M. J. Klug. 1979. Feeding ecology of stream invertebrates. *Annual Review of Ecology and Systematics* 10:147–72.

Cummins, K. W., and R. W. Merritt. 1996. Ecology and distribution of aquatic insects. In *An introduction to the aquatic insects of North America*, ed. R. W. Merritt and K. W. Cummins. 3rd ed. Dubuque, IA: Kendall/Hunt Publishing Co., 74–86.

Cummins, K. W., and M. A. Wilzbach. 1985. *Field procedures for analysis of functional feeding groups of stream macroinvertebrates*. Contribution 1611, Appalachian Environmental Laboratory, University of Maryland, Frostburg.

Daly, H. V. 1996. General classification and key to the orders of aquatic and semiaquatic insects. In *An introduction to the aquatic insects of North America*, ed. R. W. Merritt and K. W. Cummins. 3rd ed. Dubuque, IA: Kendall/Hunt Publishing Co., 108–12.

Dilen, D. R. 1984. The motion of floating and submerged objects in the Chattahoochee River, Atlanta, GA. *Journal of Forensic Sciences* 29:1027–37.

Donoghue, E. R., and G. C. Minnigerode. 1977. Human body buoyancy: A study of 98 men. *Journal of Forensic Sciences* 22:573–79.

Easton, A. M., and K. G. V. Smith. 1970. The entomology of the cadaver. *Medicine, Science and the Law* 10:208–15.

Ebbesmeyer, C. C., and W. D. Haglund. 1994. Drift trajectories of a floating human body simulated in a hydraulic model of Puget Sound. *Journal of Forensic Sciences* 39:231–40.

Eriksen, C. H., V. H. Resh, and G. A. Lamberti. 1996. Aquatic insect respiration. In *An introduction to the aquatic insects of North America*, ed. R. W. Merritt and K. W. Cummins. 3rd ed. Dubuque, IA: Kendall/Hunt Publishing Co., 29–40.

Fisher, S. G., and G. E. Likens. 1973. Energy flow in Bear Brook, New Hampshire: An integrative approach to stream ecosystem metabolism. *Ecological Monographs* 43:421–39.

Fuller, M. E. 1934. The insect inhabitants of carrion: A study in animal ecology. *Bulletin Council of Science and Industry Research in Australia* 82:1–62.

Giertsen, J. C., and I. Morild. 1989. Seafaring bodies. *American Journal of Forensic Medicine and Pathology* 10:25–27.

Giller, P. S., and B. Malmqvist. 1998. *The biology of streams and rivers*. New York: Oxford University Press.

Goff, M. L. 1993 Estimation of postmortem interval using arthropod development and successional patterns. *Forensic Science Review* 5:81–94.

Goff, M. L., and C. B. Odom. 1987. Forensic entomology in the Hawaiian Islands: Three case studies. *American Journal of Forensic Medicine and Pathology* 8:45–50.

Goff, M. L., A. I. Omori, and K. Gunatilake. 1988. Estimation of postmortem interval by arthropod succession. *American Journal of Forensic Medicine and Pathology* 9:220–25.

Greenberg, B. 1991. Flies as forensic indicators. *Journal of Medical Entomology* 28:565–77.

Haefner, J. N. 2005. Forensic studies in aquatic ecosystems. Master's thesis, Millersville University, Millersville, PA.

Haefner, J. N., J. R. Wallace, and R. W. Merritt. 2004. Pig decomposition in lotic aquatic systems: The potential use of algal growth in establishing a postmortem submersion interval (PMSI). *Journal of Forensic Science* 49:330–36.

Haglund, W. D., M. A. Reichert, D. G. Reay, and T. Donald. 1990. Recovery of decomposed and skeletal human remains in the "Green River Murder" investigation. *American Journal of Forensic Medicine and Pathology* 11:35–43.

Haglund, W. D., and D. T. Reay. 1993. Problems of recovering partial human remains at different times and locations: Concerns for death investigators. *Journal of Forensic Sciences* 38:69–80.

Hall, R. D. 1990. Medicocriminal entomology. In *Entomology and death: A procedural guide*, ed. E. P. Catts and N. H. Haskell. Clemson, SC: Joyce's Print Shop, 1–8.

Hall, R. D., and K. E. Doisy. 1993. Length of time after death: Effect on attraction and oviposition or larviposition of midsummer blow flies (Diptera: Calliphoridae) and flesh flies (Diptera: Sarcophagidae) of medicolegal importance in Missouri. *Annals of the Entomological Society of America* 86:589–93.

Harper, P. P., and L. Cloutier. 1986. Spatial structure of the insect community of a small dimictic lake in the Laurentians (Quebec). *Internationale Revue der Gesamten Hydrobiologie* 71:655–85.

Haskell, N. H., R. D. Hall, V. J. Cervenka, and M. A. Clark. 1997. On the body: Insects' life stage presence, their postmortem artifacts. In *Forensic taphonomy: The postmortem fate of human remains*, ed. W. D. Haglund and M. H. Sorg. Boca Raton, FL: CRC Press, 415–48.

Haskell, N. H., D. G. McShaffrey, D. A. Hawley, R. E. Williams, and J. E. Pless. 1989. Use of aquatic insects in determining submersion interval. *Journal of Forensic Sciences* 34:622–32.

Hawley, D. A., N. H. Haskell, D. G. McShaffrey, R. E. Williams, and J. E. Pless. 1989. Identification of a red "fiber": Chironomid larvae. *Journal of Forensic Sciences* 34:617–62.

Hobischak, N. R. 1997. Freshwater invertebrate succession and decompositional studies on carrion in British Columbia. Master's thesis in pest management, Simon Fraser University, Burnaby, British Columbia, Canada.

Hobischak, N. R., and G. S. Anderson. 1999. Freshwater-related death investigations in British Columbia in 1995–1996. A review of coroners cases. *Canadian Society of Forensic Sciences* 32:97–106.

Hobischak, N. R., and G. S. Anderson. 2002. Time of submergence using aquatic invertebrate succession and decompositional changes. *Journal of Forensic Science* 47:142–51.

Horton, B. P., S. Boreham, and C. Hillier. 2006. The development and application of a diatom-based quantitative reconstruction technique in forensic science. *Journal of Forensic Science* 51:643–50.

Howard, F. O. 1975. Natural history and ecology of *Pycnopsyche lepida, P. guttifer,* and *P. scabripennis* (Trichoptera: limnephilidae) in a woodland stream. Ph.D. thesis, Michigan State University, East Lansing, MI.

Hynes, H. B. N. 1960. *The biology of polluted waters.* Liverpool, UK: Liverpool University Press.

Kaushik, N. K., and H. B. N. Hynes. 1971. The fate of the dead leaves that fall into streams. *Archives fur Hydrobiologie* 68:465–515.

Keh, B. 1985. Scope and applications of forensic entomology. *Annual Review of Entomology* 30:137–54.

Keiper, J. B., and D. A. Casamatta. 2001. Benthic organisms as forensic indicators. *Journal of the North American Benthological Society* 20:311–24.

Keiper, J. B., E. G. Chapman, and B. A. Foote. 1997. Midge larvae (Diptera: Chironomidae) as indicators of postmortem submersion interval of carcasses in a woodland stream: A preliminary report. *Journal of Forensic Sciences* 42:1074–79.

Kline, T. C., J. J. Goering, O. A. Mathison, and P. H. Poe. 1990. Recycling of elements transported upstream by runs of Pacific salmon. 1. N and C evidence in Sashin Creek, southeastern Alaska. *Canadian Journal of Fisheries and Aquatic Sciences.* 47:136–44.

Kline, T. C., J. J. Goering, and R. J. Piorkowski. 1997. The effect of salmon carcasses on Alaskan freshwaters. In *Freshwaters of Alaska: Ecological synthesis*, ed. A. M. Milner and M. W. Oswood. New York: Springer-Verlag, 179–204.

Knight, B. 1997. *Simpson's forensic medicine.* 11th ed. London: Arnold.

Lamberti, G. A., and J. W. Moore. 1984. Aquatic insects as primary consumers. In *The ecology of aquatic insects*, ed. V. H. Resh and D. M. Rosenberg. New York: Praeger Scientific, 164–95.

Lane, M. A., L. C. Anderson, T. M. Barkley, J. H. Bock, E. M. Gifford, D. W. Hall, D. O. Norris, T. L. Rost, and W. L. Stern. 1990. Forensic botany. Plants, perpetrators, pests, poisons, and pot. *Bioscience* 40:34–39.

Lindegaard, C. 1992. Zoobenthos ecology of Thingvallavatn: Vertical distribution, abundance, population dynamics and production. *Oikos* 64:257–304.

Lord, W. D. 1990. Case histories of the use of insects in investigations. In *Entomology and death: A procedural guide*, ed. E. P. Catts and N. H. Haskell. Clemson, SC: Joyce's Print Shop, 9–37.

Lord, W. D., and J. F. Burger. 1984. Arthropods associated with harbor seal (*Phoca vitulina*) carcasses stranded on islands along the New England coast. *International Journal of Entomology* 26:282–85.

Mann, R. W., W. M. Bass, and L. Meadows. 1990. Time since death and decomposition of the human body: Variables and observations in case and experimental field studies. *Journal of Forensic Sciences* 35:103–11.

Mathisen, O. A., P. L. Parker, J. J. Goering, T. C. Kline, P. H. Poe, and R. S. Scalan. 1988. Recycling of marine elements transported into freshwater by anadromous salmon. *Verhandlungen der Internationalen Vereinigung fur Theoretische und Angewandte Limnologie* 23:2249–58.

Meek, C. L., M. D. Andis, and C. S. Andrews. 1983. Role of the entomologist in forensic pathology, including a selected bibliography. *Bibliographies of Entomolgical Society of America* 1:1–10.

Megnin, J. P. 1894. *La Fauna des Cadavres: Application de la Entomologie a la Mededin Legale.* Encyclopedie Scientifique des Aide-Memoires. Paris: Masson et Gauthiers, Viullars.

Merritt, R. W., and K. W. Cummins. 2006. Trophic relationships of macroinvertebrates. In *Methods in stream ecology*, ed. F. R. Hauer and G. A. Lamberti. 2nd ed. London: Elsevier Inc., 585–609.

Merritt, R. W., K. W. Cummins, and M. B. Berg, Eds. 2008a. *An introduction to the aquatic insects of North America.* 4th ed. Dubuque, IA: Kendall/Hunt Publishing.

Merritt, R. W., K. W. Cummins, and T. M. Burton. 1984. The role of aquatic insects in the processing and cycling of nutrients. In *The ecology of aquatic insects*, ed. V. H. Resh and D. M. Rosenberg. New York: Praeger Scientific, 134–63.

Merritt, R. W., K. W. Cummins, V. H. Resh, and D. P. Batzer. 2008b. Sampling aquatic insects: Collection devices, statistical considerations and rearing procedures. In *An Introduction to the aquatic insects of North America*, ed. R. W. Merritt, K. W. Cummins, and M. B. Berg. 4th ed. Dubuque, IA: Kendall/Hunt Publishing, 15–37.

Merritt, R. W., M. J. Higgins, and J. R. Wallace. 2000. Entomology. In *Encyclopedia of forensic sciences*, ed. J. A. Siegel, P. J. Saukko, and G. C. Knupfer. New York: Academic Press, 699–705.

Minakawa, N. 1997. The dynamics of aquatic insect communities associated with salmon spawning. PhD dissertation, University of Washington, Seattle.

Möttönen, M., and M. Nuutila. 1977. Post mortem injury caused by domestic animals, crustaceans, and fish. In *Forensic medicine: A study in trauma and environmental hazards*, ed. C. G. Tedeschi, W. G. Eckert, and L. G. Tedischi. Philadelphia: W. B. Saunders, 1096–98.

Nawrocki, S. P., J. E. Pless, D. A. Hawley, and S. A. Wagner. 1997. Fluvial transport of human crania. In *Forensic taphonomy: The postmortem fate of human remains*, ed. W. D. Haglund and M. H. Sorg. Boca Raton, FL: CRC Press, 529–52.

Nuorteva, P. 1977. Sarcosaprophagous insects as forensic indictators. In *Forensic medicine: A study in trauma and environmental hazards*, ed. C. G. Tedeschi, W. G. Eckert, and L. G. Tedischi. Philadelphia: W. B. Saunders, 1096–98.

Nuorteva, P., H. Schumann, M. Isokoski, and K. Laiho. 1974. Studies on the possibilities of using blowflies (Diptera: Calliphoridae) as medicolegal indicators in Finland. *Annales Entomologici Fennici* 40:70–74.

Payne, J. A. 1965. A summer carrion study of the baby pig, *Sus scrofa*. *Ecology* 46:592–602.

Payne, J. A., and E. W. King. 1972. Insect succession and decomposition of pig carcasses in water. *Journal of the Georgia Entomological Society* 7:153–62.

Pinder, L. C. V. 1995. The habitats of chironomid larvae. In *The chironomidae: Biology and ecology of non-biting midges*, ed. P. D. Armitage, P. S. Cranston, and L. C. V. Pinder. London: Chapman & Hall, 107–35.

Putman, R. J. 1978. The role of carrion-frequenting arthropods in the decay process. *Ecological Entomology* 3:133–39.

Reed, H. B., Jr. 1958. A study of dog carcass communities in Tennessee, with special reference to the insects. *American Midland Naturalist* 59:213–45.

Rodriguez, W. C., III. 1997. Decomposition of buried and submerged bodies. In *Forensic taphonomy: The postmortem fate of human remains*, ed. W. D. Haglund and M. H. Sorg. Boca Raton, FL: CRC Press, 459–81.

Rodriguez, W. C., and W. M. Bass. 1983. Insect activity and its relationship to decay rates of human cadavers in east Tennessee. *Journal of Forensic Sciences* 28:423–32.

Rosenberg, D. M., and V. H. Resh. 1996. Use of aquatic insects in biomonitoring. In *An introduction to the aquatic insects of North America*, ed. R. W. Merritt and K. W. Cummins. 3rd ed. Dubuque, IA: Kendall/Hunt Publishing Co., 87–97.

Schuldt, J. A., and A. E. Hershey. 1995. Effect of salmon carcass decomposition on Lake Superior tributary streams. *Journal of the North American Benthological Society* 14:259–68.

Singh, D., and B. Greenberg. 1994. Survival after submergence in the pupae of five species of blow flies (Diptera: Calliphoridae). *Journal of Medical Entomology* 31:757–59.

Siver, P. A.,W. D. Lord, and D. J. McCarthy. 1994. Forensic limnology: The use of freshwater algal community ecology to link suspects to an aquatic crime scene in southern New England. *Journal of Forensic Sciences* 39:847–53.

Skinner, M. F., J. Duffy, and D. B. Symes. 1988. Repeat identification of skeletonized human remains: A case study. *Journal of the Canadian Society of Forensic Science* 21:138–41.

Smith, K. G. V. 1986. *A manual of forensic entomology*. London: British Museum (Natural History).

Sorg, M. H., J. H. Dearborn, E. I. Monahan, H. F. Ryan, K. G. Sweeney, and E. David. 1997. Forensic taphonomy in marine contexts. In *Forensic taphonomy: The postmortem fate of human remains*, ed. W. D. Haglund and M. H. Sorg. Boca Raton, FL: CRC Press, 567–604.

Sorg, M. H., J. H. Dearborn, K. G. Sweeney, H. F. Ryan, and W. C. Rodriguez. 1995. Marine taphonomy of a case submerged for 32 years. *Proceedings of the American Academy of Forensic Sciences* 1:156–57.

Stewart, K. W., and P. P. Harper. 1996. Plecoptera. In *An introduction to the aquatic insects of North America*, ed. R. W. Merritt and K. W. Cummins. 3rd ed. Dubuque, IA: Kendall/Hunt Publishing Co., 217–66.

Suberkropp, K., and M. J. Klug. 1976. Fungi and bacteria associated with leaves during processing in a woodland stream. *Ecology* 57:707–19.

Suberkropp, K., and M. J. Klug. 1980. The maceration of deciduous leaf litter by aquatic hyphomycetes. *Canadian Journal of Botany* 58:1025–31.

Sweeney, B. W., and R. L. Vannote. 1981. *Ephemerella* mayflies of White Clay Creek: Bioenergetic and ecological relationships among six coexisting species. *Ecology* 67:1396–410.

Thorp, J. H., and A. P. Covich. 1991. *Ecology and classification of North American freshwater invertebrates*. San Diego: Academic Press.

Tomberlin, J. K., and P. H. Adler. 1998. Seasonal colonization and decomposition of rat carrion in water and on land in an open field in South Carolina. *Journal of Medical Entomology* 35:704–9.

Tullis, K., and M. L. Goff. 1987. Arthropod succession in exposed carrion in a tropical rainforest in O'ahu Island, Hawaii. *Journal of Medical Entomology* 24:332–39.

Vance, G. M., J. K. VanDyk, and W. A. Rowley. 1995. A device for sampling aquatic insects associated with carrion in water. *Journal of Forensic Sciences* 40:479–82.

Vannote, R. L., and B. W. Sweeney. 1980. Geographic analysis of thermal equilibria: A conceptual model for evaluating the effect of natural and modified thermal regimes on aquatic insect communities. *American Naturalist* 115:667–95.

Wallace, J. B. 1996. Habitat, life history, and behavioral adaptations of aquatic insects. In *An intro-duction to the aquatic insects of North America*, ed. R. W. Merritt and K. W. Cummins. 3rd ed. Dubuque, IA: Kendall/Hunt Publishing Co., 41–73.

Wallace, J. B., and R. W. Merritt. 1980. Filter-feeding ecology of aquatic insects. *Annual Review of Entomology* 25:103–32.

Wallace, J. R., F. D. Howard, H. E. Hays, and K. W. Cummins. 1992. The growth and natural his-tory of the caddisfly *Pycnopsyche luculenta* (Trichoptera: Limnephilidae). *Journal of Freshwater Ecology* 7:399–405.

Wallace, J. R., R. W. Merritt, R. Kimbarauskas, M. E. Benbow, M. McIntosh. 2008. Caddisfly cases assist homicide case: Determining a postmortem submersion interval (PMSI) using aquatic insects. *Journal of Forensic Science.* 53(1):1-3.

Ward, J. V. 1992. *Aquatic insect ecology.* 1. Biology and habitat. New York: J. Wiley & Sons, Inc.

Ward, J. V., and J. A. Stanford. 1982. Thermal responses in the evolutionary ecology of aquatic insects. *Annual Review of Entomology* 27:97–117.

Wiggins, G. B. 1996. Trichoptera families. In *An introduction to the aquatic insects of North America*, ed. R. W. Merritt and K. W. Cummins. 3rd ed. Dubuque, IA: Kendall/Hunt Publishing Co., 309–49.

Wipfli, M. S., J. Hudson, and J. Caouette. 1998. Influence of salmon carcasses on stream productivity: Response of biofilm and benthic macroinvertebrates in southeastern Alaska, U.S.A. *Canadian Journal of Fisheries and Aquatic Sciences* 55:1503–11.

Wolkomir, R., and J. Wolkomir. 1992. When scientists become sleuths. *National Wildlife* 30:8–15.

Zimmerman, K., and J. R. Wallace. 2008. Estimating a postmortem submersion interval using algal diversity on mammalian carcasses in brackish marshes. *Journal of Forensic Sciences.* 53(4):935-41.

Appendix

KEY TO FUNCTIONAL FEEDING GROUPS

⊢—⊢ ⊢————————⊣ Indicates size or range of sizes

1. ANIMALS IN HARD SHELL (Phylum Mollusca)

(a) LIMPETS (Class Gastropoda)

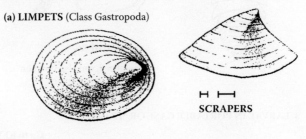

SCRAPERS

(b) SNAILS (Class Gastropoda)

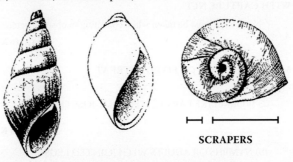

SCRAPERS

**Snails are generalized (facultative) feeders
and can also function as Shredders.**

(c) CLAMS OR MUSSELS (Class Pelecypoda)

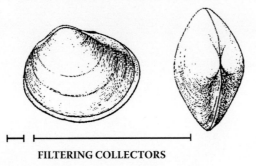

FILTERING COLLECTORS

2. SOW BUG OR SHRIMP-LIKE AMIMALS

 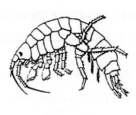

SHREDDERS

Generalized, can also function
as Gathering Collectors.

3. LARVAE IN PORTABLE CASE OR "HOME"

Go to KEY 2

**4. LARVAE IN FIXED RETREAT,
WITH CAPTURE NET**

Note: Care must be taken when collecting to observe nets.

Go to KEY 3

5. WITHOUT CASE OR FIXED RETREAT

(a) WORM-LIKE LARVAE WITHOUT JOINTED LEGS

Go to KEY 4

(b) NYMPHS OR ADULTS WITH JOINTED LEGS

Go to KEY 5

6. DOES NOT FIT KEY 5 EXACTLY.   **GO TO KEY 6.**

KEY 2

FIRST LEVEL OF RESOLUTION

LARVAE IN PORTABLE CASE
Caddisfies (Order Trichoptera)

CASES ORGANIC
Leaf, stick, needle, bark

CASES MINERAL
Sand, fine, gravel

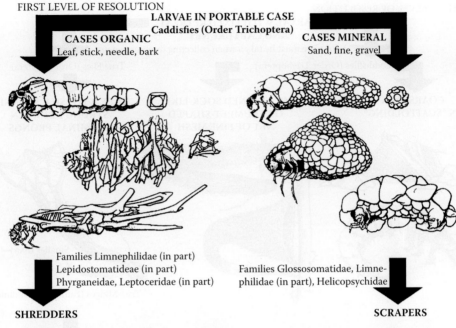

Families Limnephilidae (in part)
Lepidostomatideae (in part)
Phyrganeidae, Leptoceridae (in part)

Families Glossosomatidae, Limnephilidae (in part), Helicopsychidae

SHREDDERS

SCRAPERS

SECOND LEVEL OF RESOLUTION considers a few fairly common caddisflies that would be
misclassified above on the basis of case compositon alone.

CASES ORGANIC

CASES MINERAL

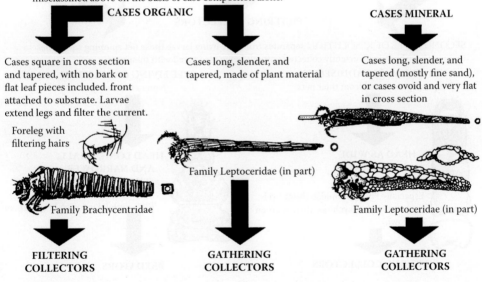

Cases square in cross section
and tapered, with no bark or
flat leaf pieces included. front
attached to substrate. Larvae
extend legs and filter the current.

Foreleg with
filtering hairs

Cases long, slender, and
tapered, made of plant material

Cases long, slender, and
tapered (mostly fine sand),
or cases ovoid and very flat
in cross section

Family Leptoceridae (in part)

Family Brachycentridae

Family Leptoceridae (in part)

FILTERING
COLLECTORS

GATHERING
COLLECTORS

GATHERING
COLLECTORS

KEY 3

FIRST LEVEL OF RESOLUTION

**LARVAE WITH FIXED RETREAT
AND CAPTURE NET**

Note: Care must be taken when collecting to observe nets.

Caddisflies (Order Trichoptera)   True Flies (Order Diptera)

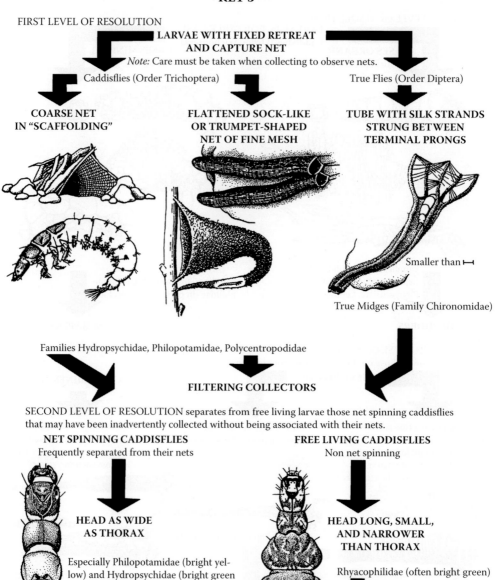

| **COARSE NET IN "SCAFFOLDING"** | **FLATTENED SOCK-LIKE OR TRUMPET-SHAPED NET OF FINE MESH** | **TUBE WITH SILK STRANDS STRUNG BETWEEN TERMINAL PRONGS** |

Smaller than ⊢⊣

True Midges (Family Chironomidae)

Families Hydropsychidae, Philopotamidae, Polycentropodidae

FILTERING COLLECTORS

SECOND LEVEL OF RESOLUTION separates from free living larvae those net spinning caddisflies that may have been inadvertently collected without being associated with their nets.

NET SPINNING CADDISFLIES
Frequently separated from their nets

FREE LIVING CADDISFLIES
Non net spinning

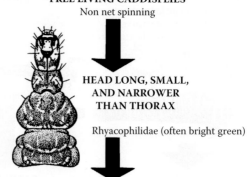

**HEAD AS WIDE
AS THORAX**

**HEAD LONG, SMALL,
AND NARROWER
THAN THORAX**

Especially Philopotamidae (bright yellow) and Hydropsychidae (bright green or brown)

Rhyacophilidae (often bright green)

FILTERING COLLECTORS   **PREDATORS**

KEY 4

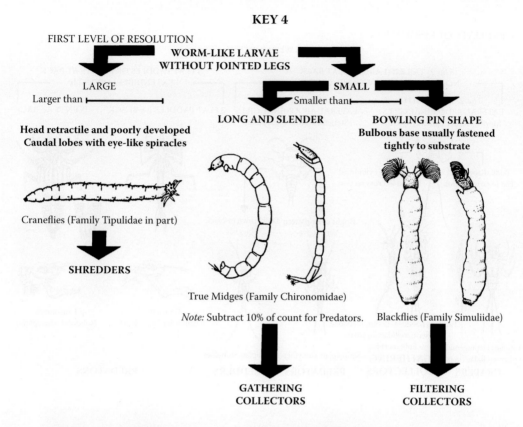

FIRST LEVEL OF RESOLUTION

WORM-LIKE LARVAE
WITHOUT JOINTED LEGS

LARGE

Larger than |————————————|

SMALL

Smaller than |————|

LONG AND SLENDER

BOWLING PIN SHAPE
Bulbous base usually fastened
tightly to substrate

Head retractile and poorly developed
Caudal lobes with eye-like spiracles

Craneflies (Family Tipulidae in part)

SHREDDERS

True Midges (Family Chironomidae)

Note: Subtract 10% of count for Predators.

Blackflies (Family Simuliidae)

GATHERING
COLLECTORS

FILTERING
COLLECTORS

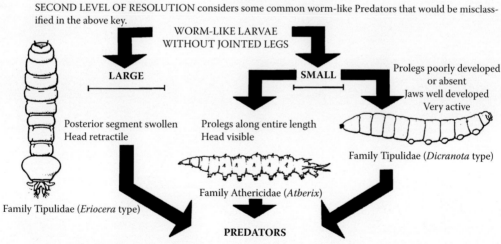

SECOND LEVEL OF RESOLUTION considers some common worm-like Predators that would be misclassified in the above key.

WORM-LIKE LARVAE
WITHOUT JOINTED LEGS

LARGE
|————————————|

SMALL
|————|

Prolegs poorly developed
or absent
Jaws well developed
Very active

Posterior segment swollen
Head retractile

Prolegs along entire length
Head visible

Family Tipulidae (*Dicranota* type)

Family Athericidae (*Atherix*)

Family Tipulidae (*Eriocera* type)

PREDATORS

KEY 5

FIRST LEVEL OF RESOLUTION

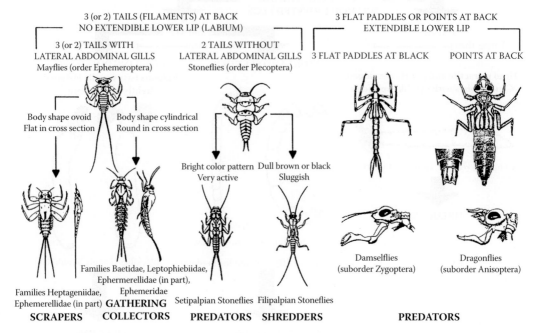

NYMPHS WITH JOINTED LEGS

3 (or 2) TAILS (FILAMENTS) AT BACK
NO EXTENDIBLE LOWER LIP (LABIUM)

3 FLAT PADDLES OR POINTS AT BACK
EXTENDIBLE LOWER LIP

3 (or 2) TAILS WITH
LATERAL ABDOMINAL GILLS
Mayflies (order Ephemeroptera)

2 TAILS WITHOUT
LATERAL ABDOMINAL GILLS
Stoneflies (order Plecoptera)

3 FLAT PADDLES AT BLACK

POINTS AT BACK

Body shape ovoid
Flat in cross section

Body shape cylindrical
Round in cross section

Bright color pattern
Very active

Dull brown or black
Sluggish

Damselflies
(suborder Zygoptera)

Dragonflies
(suborder Anisoptera)

Families Baetidae, Leptophlebiidae,
Ephermerellidae (in part),
Ephemeridae

Families Heptageniidae,
Ephemerellidae (in part) **GATHERING**
SCRAPERS   **COLLECTORS**

Setipalpian Stoneflies   Filipalpian Stoneflies

PREDATORS   **SHREDDERS**   **PREDATORS**

KEY 6

SECOND LEVEL OF RESOLUTION considers some fairty common insects that do not fit in the above key or would be misclassified on the basis of body shape alone.

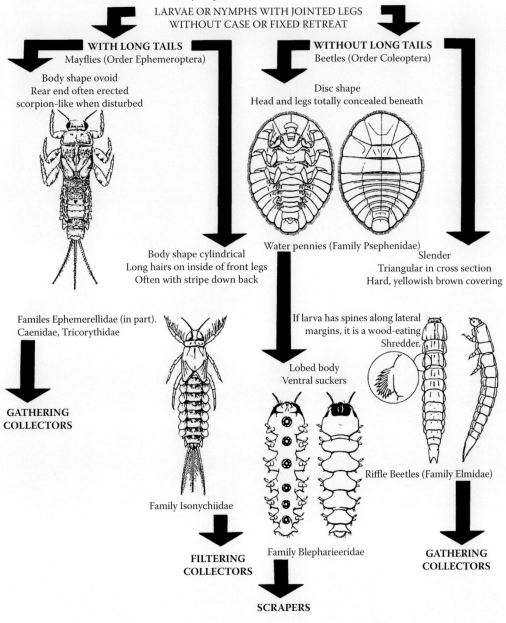

LARVAE OR NYMPHS WITH JOINTED LEGS
WITHOUT CASE OR FIXED RETREAT

WITH LONG TAILS
Mayflies (Order Ephemeroptera)

WITHOUT LONG TAILS
Beetles (Order Coleoptera)

Body shape ovoid
Rear end often erected
scorpion-like when disturbed

Disc shape
Head and legs totally concealed beneath

Body shape cylindrical
Long hairs on inside of front legs
Often with stripe down back

Water pennies (Family Psephenidae)

Slender
Triangular in cross section
Hard, yellowish brown covering

Familes Ephemerellidae (in part).
Caenidae, Tricorythidae

If larva has spines along lateral
margins, it is a wood-eating
Shredder.

Lobed body
Ventral suckers

**GATHERING
COLLECTORS**

Family Isonychiidae

Riffle Beetles (Family Elmidae)

Family Blepharieeridae

**FILTERING
COLLECTORS**

**GATHERING
COLLECTORS**

SCRAPERS

Recovery of Anthropological, Botanical, and Entomological Evidence from Buried Bodies and Surface Scatter

8

HEATHER A. WALSH-HANEY

ALISON GALLOWAY

JASON H. BYRD

Contents

Introduction

Excavation of buried remains or the recovery of bodies from the surface requires adherence to collection strategies, attention to detail in documentation and recording of data, and understanding of the nature of remains and the taphonomic processes they have encountered during the postmortem interval. Once a crime scene is disturbed, it can never be fully reconstructed. Skeletonized or decomposing remains present particular challenges to recovery due to the effects of segmentation, scattering, and camouflaging that may hinder identification of all the elements.

It is essential that forensic investigators be meticulous in recording the features of the scene and all stages of the recovery process (Dirkmaat and Adovasio 1997). This attention to detail also minimizes the possibility of contamination and maximizes the recovery of the skeletal material and items of physical evidence, including entomological and botanical, which may be overlooked or destroyed in a hasty collection. The primary aim is to use these data to discern the events of the death and disposal of the body and, if necessary, prosecute the perpetrators, so all work must be done with an eye toward presentation in the courtroom.

Most cases in which the remains are skeletonized or badly decomposed will eventually be sent to a forensic anthropologist. Forensic anthropological analysis is a multi-stage process based on the close examination of the osseous material in the laboratory or morgue. Importantly, accurate laboratory analysis requires that the forensic anthropologist know as much as possible concerning scene taphonomy, e.g., the substrate, weather, plant, insect, and animal (including human) variables that have an effect on decomposition rates, bone diagenesis, scattering, and bone trauma. In cases where the forensic anthropologist does not participate in the scene recovery, law enforcement or medical examiner crime scene investigators must provide scene and field specimen pictures, which should include the following:

- Photograph the interior crime scene with 360° photos while standing adjacent to or on the subdatum and in an overlapping series using a normal lens.
- Photograph the overall scene from the permanent datum using a wide-angle lens.
- Photograph the condition of the evidence before vegetation is removed and recovery has started.
- With a scale in place, take one photo at a medium distance that shows the evidence relative to other materials and a second photo that fills the frame.

Along with photos of the crime scene, the forensic anthropologist should be provided with law enforcement and medical examiner case numbers, GPS coordinates for the scene and permanent datum, and a brief narrative regarding how and when the remains or scene was discovered. Lastly, the investigating agency that has called upon the forensic anthropologist to serve as a consultant must clearly state the types of analyses expected, such as a report of osteological analysis for identification, trauma, or time interval since death.

Once the forensic anthropologist accepts custody of remains and has received the ancillary documents mentioned above, the laboratory analysis begins. First, the anthropologist provides a biological profile of the decedent (i.e., age, sex, ancestry, and stature) through the collection of anthroposcopic and metric data. Aging the skeletal remains may include analysis of epiphyseal (e.g., growth plate) fusion and dental exfoliation and eruption in subadults, and histomorphology or comparison of joint degeneration with standard exemplars for adults and elderly persons (Scherer and Black 2000, France 1998). Sex is often determined through size and robusticity differences between males and females (e.g., sexual dimorphic characteristics) noted on the skull, pelvis, or long bones (Walsh-Haney et al. 1999). Ancestry tends to be determined through visual assessment of the cranial mid-face bone morphology and metric assessment of the crainial vault using FORDISC 3.0 (Gill and Rhine 2004, Ousley et al. 2006). Forensic stature is determined through measurement, particularly the femur and tibia, which are input into the appropriate FORDSIC 3.0 stature equations, which are sex and ancestry specific (Jantz et al. 2006).

Second, the forensic anthropologist analyzes the trauma observed on the bones, states whether this occurred in the antemortem (during life), perimortem (at or around the time of death), or postmortem (after death) period, and discusses the mechanisms of injury that could have produced these defects. Trauma analysis may include but is not limited to the following:

- Microscopic analysis of histological sections for skeletal pathology assessment
- Radiographic examination of bones with gunshot trauma in order to document bullet trajectory and bullet wipe
- Stereoscopic analysis of fracture margins in order to record striations associated with sharp force trauma wounds and particular sharp implement types (e.g., serrated vs. flat blades)
- Casting of tool marks in bone left by blunt or sharp-edged implements for later comparison with confiscated weapons
- Reconstruction of fragmentary remains to help trace fracture line propagation to the related impact site, which in turn helps to determine the minimum numbers of blows that may have been inflicted
- Measurement of punctures, furrows, and incised grooves left by scavengers in order to match the puncture size to the type of animal, which may lead scene investigators to animal burrows containing missing skeletal elements

Third, the anthropologist estimates the interval since the death of the individual and discusses the probable sequence of events in the postmortem period (body movement due to scavenging, water action, etc.). The biological profile is rarely of particular interest in court, although it is critical for beginning the search to identify the remains. Court testimony usually focuses on the issues of trauma and disposal, for how and when the death occurred are crucial points that may make or break the case for the prosecution or the defense.

Anthropologists can play a critical role in the recovery of evidence, especially when local law enforcement agencies lack investigators with experience and training in working with human skeletons, or when the agency has come to appreciate the level of expertise that the anthropologist can bring to the scene recovery. The anthropologist is often better equipped to locate and map the scene of initial decomposition, to eliminate nonhuman bones, to identify carnivore or rodent activity on the bones, and to predict the direction of travel for missing elements. He or she is able to recognize skeletal elements amidst an assortment of ground cover that often closely matches the bones in coloration and shape, ensuring maximum recovery. For example, collecting bones from a heavily forested area is often complicated, as the bones have taken on staining from the leaf cover and are difficult to identify, especially since ribs resemble twigs and vertebrae may be concealed under the fallen leaves. As a case in point, forensic anthropologists were an important part of the recovery of human remains from the World Trade Center attacks. In particular, one author was stationed at the Staten Island Landfill from September through October 2001 as part of the federal government's Disaster Mortuary Operations Response Team (DMORT). The forensic anthropologists working the landfill during the September–October deployment cycle were specifically tasked with differentiating human from nonhuman bone as well as other organic or inorganic materials. As one can imagine, the identification of the fragmentary human remains was difficult for crime scene investigators, Federal Bureau of Investigation (FBI) special agents, and firefighters who searched the debris field because they lacked extensive training in the identification of fragmentary human remains that were covered in dirt, vegetative matter, and other debris. Therefore, the forensic anthropologists were stationed in a tent, where the human remains were identified by element, assigned a specimen number, photographed, and stored as evidence to be turned over to the medical examiner. Indeed, nearly 75% of the material gathered by the landfill searchers was determined to be nonhuman by forensic anthropologists. As such, the forensic anthropologist helped to reduce time that would have been otherwise wasted in the processing of nonhuman materials.

Since the anthropologist arrives with this knowledge, and often with the equipment for recovery (Appendix I), it tends to be the anthropologist's responsibility to organize and conduct the recovery of the remains themselves. Often it is the anthropologist, entomologist, or botanist's knowledge of evidence collection and mapping within their respective discipline that provides them with the skills needed to collect evidence for their scientific colleagues in other forensic disciplines. For this reason, no one should attempt to do scene recovery until he or she has either extensive work on prior supervised scenes or has a strong background in recovery and collection techniques, such as archaeology, and the ability to quickly adapt these for the forensic setting. Equipment pertaining to scene recovery should be preassembled, cleaned to prevent cross-contamination from other crime scenes, and kept up to date with modern technology.

Scene Recovery in the Caseload

Scene recoveries constitute only a small portion of the overall caseload, approximately 27% among board-certified forensic anthropologists (P. Willey, 1998, personal communication). The majority of cases are submitted directly to the forensic anthropology laboratory, in some cases via electronic images, which enables rapid identification of those items not of forensic interest, such as nonhuman bones. Indeed, about a third of our cases are nonhuman elements. As a case in point, one of the authors received a case from a Florida medical examiner investigating a potential crime scene involving fragmentary skeletal remains that were found alongside Interstate 75. The medical examiner investigator called the author and requested speedy resolution of the case because the highway was shut down to all northbound traffic. The investigator's initial impressions were that the case was nonhuman and probably canine. Law enforcement officers on the scene believed the remains were human and possibly those of an infant. The medical examiner investigator photographed the cache of bones *in situ* and then focused upon key osseous elements, including the vertebrae and long bones. The vertebrae were opisthocoelus and an unfused metapodial was present. It was determined that the remains were not human and probably deer (*Odocoileus virginianus*) post haste.

Many human bones come in as isolated specimens, such as an old anatomical skull found among the trash at the side of the road, a femur washed up on the beach, or a pelvis uncovered by hikers who can no longer recall where it was found. Isolated cases may be the result of intentional postmortem dismemberment by perpetrators endeavoring to prevent a positive identification of the victim. However, in many instances the isolated specimen, though human, may not be of forensic significance, e.g., in Florida, if the decedent has been dead for more than 75 years, thereby becoming categorized as historic or prehistoric remains that are the responsibility of the state archaeologist. Importantly, when the context is not known and skeletal morphology bears no indication of prehistoric or historic markers, radiocarbon dating may be requisite. Radiocarbon dating of teeth or bone can provide a date of birth reference point, as the radiocarbon signatures of individuals born before and after 1950's atom bomb testing differ (Ubelaker et al. 2006).

Often human remains are recovered intact in relatively early stages of decomposition by the medical examiner/coroner's personnel. In some instances, the remains are fresh and the anthropologist is only asked to render an opinion on the skeletal trauma. At other times, no skeletal evidence is available for analysis because the remains have been interred or cremated. Therefore, the forensic anthropology assessment is based on photographs and the reports of other experts.

Irregularities in the collection of evidence can undermine an otherwise apparently solid case. As a consequence, law enforcement personnel are turning to the forensic sciences to assist in this initial stage of the investigation. Educational programs aimed at exposing more law enforcement agencies to what anthropologists, entomologists, and botanists can offer also increases the likelihood that these forensic professionals will be called for scene assistance. Although many jurisdictions take pride in their own units, the mobility of officers between locations, interagency training programs, and the demand for more sophisticated analyses even in rural areas have increased the spirit of cooperation.

The Nature of Anthropological/Archeological, Botanical, and Entomological Evidence

For those cases in which the anthropologist is called to the scene, his or her responsibility is primarily for the recovery of skeletal evidence. This includes the bones of the body, the various epiphyseal portions for subadults, and the dentition and ossified cartilaginous material. During scene processing, however, the anthropologist may be responsible for the initial recovery and documentation of all other physical evidence associated with the remains, such as the soft tissue, expended bullets, shotgun casings and wads, ligatures, clothing, jewelry, medications, hair, and personal items. Insects associated with decomposition, those in direct contact with the remains, and those in the immediate surrounding environment also may be collected by the forensic anthropologist or other trained crime scene analysts or medicolegal death investigators. At the very least, these individuals must ensure that such collections are properly made if a forensic entomologist is not available. The responsibility of collecting botanical samples such as roots that have grown within the body, and broken branches may also fall to the forensic anthropologist, forensic entomologist, crime scene analyst, or medicolegal death investigator in the absence of the forensic botanist. Less tangible things, such as the degree of leaf cover, the shape and depth of the burial, digging instrument impressions, sunlight exposure, drainage, vegetation changes, or root growth through the disposal site, must also be recorded and documented by the trained forensic investigator.

These tangible and intangible items all have a role in the interpretation of the circumstances of the death and disposal of the remains. Some will pertain to the identification of the decedent if lifestyle or occupation had affected the bony tissue, and others may help determine the individual's health prior to the time of death. As an example of the latter, one author had a case involving a juvenile victim who was habitually beaten with a baton around the joint complexes of the shoulder, hip, and knee during the last months of his life. The bones of the victim were marked by Salter-Harris fractures (epiphyseal fractures) in various stages of healing and areas of hypertrophic ossification nodules on the long bone shafts. Also of interest was that the last time the victim was seen alive, he was of normal height and weight for his age. Months later he was discovered (hours before his death) with edema so severe his legs appeared elephantine. He was also approximately 30 pounds under his normal weight for height. While the soft tissue manifestation of malnutrition and beatings were clearly evidence, the forensic anthropologist was able to determine that multiple beatings had taken place over a period of months based upon the various stages of healing bone fractures.

Other tangible and intangible lines of evidence pertain to the circumstances of death: Was the victim killed? What was the mechanism of death? Did the victim arrive at the disposal site under his or her own power? Was the decedent clothed at the time? Were personal items, such as a watch and jewelry, on the victim? Some evidentiary lines pertain to the exact injuries produced around the time of death, and others relate to the events of the postmortem interval: How long was it between the deposition of the body and its discovery? Were the remains moved? Was there predation on the remains? If questions such as these arise during the trial process, all of these points must be investigated, and the interpretations made from them must be shown to have strong evidentiary support.

Two of the authors participated in a case that involved primary, secondary, and tertiary crime scenes that were linked based upon associated physical evidence mentioned above, and particular evidentiary attention was paid to the multiple locations the victim, or parts of the victim, were transported to by the perpetrator. In particular, the victim had been shot in the front seat of a borrowed automobile and then stored in the trunk for a period of days. Fearing discovery of the body, the perpetrator removed the clothed victim from the trunk and carried it to the back of wooded property owned by a relative and hastily left it on the ground surface. Days later, the victim's remains were dragged to a large body of water on the property and submerged with branches from adjacent trees. The careful recovery of entomological and anthropological evidence from all three scenes was requisite in linking the victim and perpetrator to the various scenes. Careful examination of the entomological evidence illustrated how long the victim's body had remained at each of the scenes before the body was relocated to the next site. Based on the entomological evidence, it could be shown that the body was first contained in the automobile, then removed and located under nearby trees, and then moved once again and deposited into the swamp.

Finding the Location

Scene recoveries can begin slowly, building from rumors of a body, or quickly, by the accidental discovery of remains. Because these events are rarely predictable, it is important to have the procedures well established, equipment and forms at the ready, supplies on hand, and personnel contact methods in place.

Once the scene location is suspected, one of the first responsibilities is determining who is in charge of the search and possible recovery. A single law enforcement agent should be in charge of the entire investigation (Morse et al. 1976, Stoutamire 1983, Haglund et al. 1990). Initially, this person decides what areas will be searched, the methods to be utilized, and the sequence in which the search will be conducted. He or she also decides who will be involved, where the crime scene boundaries are, where the press will be permitted, and when to call off work in an unproductive search. Once remains are located, this person determines the entry routes to the scene to be taken by personnel, what should be collected, and to whom it should be delivered. Obviously, since scene recovery usually involves a number of people and often a number of agencies, these decisions usually are made after discussion and consultation. For the forensic scientist coming into this arena, it is important to "scope out" the organization and internal politics of the scene to avoid disturbing the chain of command (Killam 1990). All affiliated personnel who will be overseeing the forensic anthropologist, entomologist, or botanist's work or assisting in the actual recovery should be recorded (name, agency affiliation, and contact information). The exchange of business cards often facilitates this process, but it is the information contained on the crime scene log that will be used as the legal documentation of those present at the scene. When bringing students to crime scenes, be sure to have a worksheet prepared in advance that includes their names, student status, and contact information, and be ready to turn this information over to the crime scene manager. It is important to note that as more universities offer upper-division programs that stress the importance of "education through service," law enforcement agencies, states attorneys, and medical examiner/coroner offices find student participation commonplace.

What Are the Methods Used to Locate Human Remains?

Human remains are often discovered casually, such as by hikers, hunters, campers, etc., who happen upon them in the course of their normal activities (Hochrein 1998). In some cases, homeowners will discover remains in undisclosed storage areas, yards, or basements. In most cases of this type, authorities are almost immediately notified of the grisly find. Unfortunately, in some instances, people have collected recognizable portions of the remains, such as the cranium or pelvis, and kept them for variable periods of time before realizing their significance, and identification of the original location may be problematic.

Another scenario for the initial discovery of a human body is from information provided by an involved person. This informant may be the perpetrator or someone who has inside knowledge of a crime or inappropriate disposal of a human body. Often sufficient time has elapsed so that their memory of the exact location is weakened. At the time of the crime, they may have reached the location in such a state of anxiety that judgment of distances is wildly inaccurate. For example, Rhine (1999) recalls one case in which the informant insisted that they had driven 10 miles down a dirt road, when in actuality the remains were only a mile or so off the main highway.

Such searches often demand considerable patience as well as an understanding of the thought processes involved in depositing a dead body, or in the wanderings of an individual whose death was suicidal, natural, or accidental. When moving a live or dead individual, transportation becomes a critical issue. A dead body is heavy and unwieldy; consequently, most body disposal sites will be located near where a vehicle can be easily driven. Camouflaging the remains is often desirable so bodies will be placed in thicker brush or ground cover. Live victims brought to the scene also usually require some vehicular transportation to get to the vicinity; however, they may be forced to move farther from roads before being killed (Duncan 1983). A possible exception to this is with children, whose lighter weight makes them easier to transport. One way in which some of these difficulties are circumvented is when disposal is done within the structure in which the person was killed. In some respects, those individuals whose disappearance is due to nonhomicidal circumstances may be more difficult to predict. Their bodies may be found in extremely remote areas or in the odd nooks and crannies of even the most urban of places.

Military situations, human rights violations such as the mass killings associated with genocide, and disasters such as explosions or earthquakes require more reliance on interviews with the local population. During a war, the craters left by plane crashes often replicate those left by other ordnance, and some may have since been converted to irrigation and fish ponds. Much of the plane may have been recycled into housing or farming instruments over the course of time. People also may be reluctant to discuss the collection of material from the crash or the fate of the occupants of the plane. Diplomacy and patience are required (Finnegan 1995). The killings occurring during genocide may remove many of the relatives who would normally monitor gravesites, either by killing entire families or by forcing remaining relatives to flee the area. In disasters, the usual landmarks by which people locate themselves are often obliterated, making identification of dead bodies difficult if the place of death is the primary means of generating an associated name. More critical from the perspective of the recovery team is the large amount of debitage that occurs with disaster and the health hazards with which recovery teams must deal.

Once the general locale has been identified, the anthropologist, entomologist, or botanist may be able to assist in the determination of the exact site. There are scene indicators in each of these respective disciplines that probably would not be noted by the casual or untrained observer. A general walk-through or drive-by is useful to determine the terrain, most probable extent of the scatter, and the best approach (Duncan 1983). This should be done by as few people as possible, and the pathway taken should be noted in the scene recovery records.

The location techniques differ due to the nature of the remains and the substrate. Underwater recovery will not be discussed here, as these activities require specialized training (Johnson and Steuer 1983). Terrestrial scene recoveries can be roughly divided into those involving surface remains and those in which burial occurs. In surface scatters, the terrain may indicate how extensive the distribution will be. For example, in an urban area, a body found between two fences may be confined within a narrow area, while in a rural area, where there are large numbers of coyotes, bears, or feral dogs, the remains may be distributed over areas ½ to 1 mile from the initial scene. Remains in or along a creek or river may have been transported much greater distances. In one instance, a body could be traced back to a river disposal after it was recovered in ocean waters (Ebbesmeyer and Haglund 1994). Buried remains tend to be confined to the grave, although body position varies and burials are often disturbed by scavengers or those making the discovery.

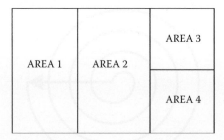

Figure 8.1 Example of a sector search plan. The area to be searched is divided into sectors, which are designated by letter or number. (Adapted from *Death Investigator's Handbook*. Used with permission of Louis N. Eliopulos.)

It is customary to establish a series of perimeters at the scene with varying search intensities (Haglund et al. 1990). The enclosed areas are restricted to those investigators who have assigned tasks relevant to the case. The inner perimeter is typically an area 50 to 100 feet in diameter within which the remains are concentrated and should include the perpetrator's entry and exit routes. Within this inner area is a second perimeter within which the body is concentrated. There may be a third inner area in which the search is most intense, usually shoulder-to-shoulder spacing while crawling. The outer perimeter is an area that is larger than the crime scene so as to exclude press, potential witnesses, and sightseers. Inner and outer scene perimeters are roped off and secured. Boundaries of these areas can be effectively marked by flagging tape. Peripheral searches in the outermost areas are usually less intense and may only be used when substantial portions of the evidence are known to be missing. Uniformed and armed law enforcement personnel are well trained in keeping onlookers outside of these perimeters and maintaining scene integrity. Smoking, eating, or drinking is prohibited at or near the established crime scene perimeters. Doing so will contaminate the scene and may jeopardize the integrity of the case.

Locating surface remains will require a systematic search. When searching a small constricted area, the sector or zone search is recommended (Eliopulos 1993). Relatively few personnel are necessary for this type of search (Figure 8.1). Also, requiring few personnel is a spiral pattern that begins at the main concentration of body parts (the center) and continues in widening circles moving away from the victim (Figure 8.2). Straight-line searches are used in very large scenes and when large numbers of people are available to search. This search technique requires that the searchers stand side by side and travel across the scene in a straight line (Figure 8.3). When few personnel are available but a large area must be searched, a line search can be used (Figure 8.4). Grid searches are usually rectangular areas (Figure 8.5) that cover the entirety of the general area. The scene is divided into multiple grid units and a search is made of each grid unit. A second search follows that is perpendicular to the first search, but within the same grid units.

Whatever the search technique, obvious accommodation is made for surface features (cliffs, trees, bodies of water, etc.) without ignoring the possibility that bodies, body parts, or other evidence may be located in poorly accessible areas. Intensity of the search varies by the visibility. When the ground cover is sparse, greater distance can be maintained between the searchers. If ground cover is thick, closer spacing is used and the searchers may crawl while hand raking the ground cover to expose the underlying surface and examine between the leaves.

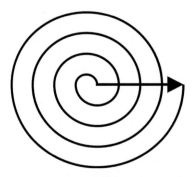

Figure 8.2 Spiral searches begin at the central location of the remains and move progressively outward. (Adapted from *Death Investigator's Handbook*. Used with permission of Louis N. Eliopulos.)

Figure 8.3 Searchers working at approximately arm's length can slowly work through leafy ground cover such as this outside of the innermost perimeters. As search intensity increases, distance between searchers decreases and they will need to work more slowly and closer to the ground. (Photo courtesty of Dr. Jason H. Byrd.)

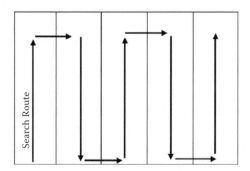

Figure 8.4 Line searches are conducted with searchers lining up and moving parallel to each other along the search route. (Adapted from *Death Investigator's Handbook*. Used with permission of Louis N. Eliopulos.)

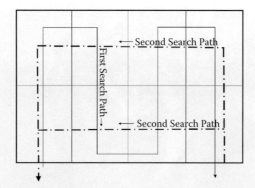

Figure 8.5 Grid searches begin with a strip search but follow this with a second search pathway that is perpendicular to the initial path. (Adapted from *Death Investigator's Handbook*. Used with permission of Louis N. Eliopulos.)

Each searcher marks anything that is suspicious, usually using colored flags on thin metal strands. In addition to the obvious items, such as bones, clothing, or weapons, other items, such as animal scat, which may contain bone fragments, trash items, and unidentified stains, should also be flagged. Searchers should be briefed as to what constitutes an item of interest prior to the onset of the search, especially on less obvious items, such as small bones, insects and their remains, as well as vegetation damage or changes, and other signs of soil disturbance.

Color-coded flags may be used to delineate the type of material flagged. For example, orange flags may indicate possible skeletal material, while red flags indicate associated physical evidence like a bullet jacket. Once all the material is flagged, the limits of the crime scene can be redefined and better delineated. If there is excessive vegetation at the scene, as much as possible is removed without disturbing the remains and associated physical evidence. However, this should only be done after the forensic botanist (or someone specifically trained to collect botanical evidence) has made all relevant collections and cleared the scene. Before cutting with shears or saw, ensure that all botanical evidence has been collected! All vegetation, except trees, is cut to within 1/2 to 3/4 inch of the surface, unless it will result in movement of evidence. This provides an unobstructed view of the entire scene to facilitate mapping. All vegetation that has been cut should be carefully placed in a suitable receptacle, removed from the cleared area, and examined to determine if the cuttings contain any further physical evidence. Upon completion of these operations, use a metal detector within and outside the grid (Figure 8.6). All positive recordings should be flagged.

Since it is difficult to acquire enough personnel to conduct a search only with anthropologically trained persons, it is useful for several trained people to be intermingled with the searchers. It is a good practice for the anthropologist, entomologist, botanist, investigator, and evidence technician to follow behind inspecting each flagged item, without moving it from its original location. This can allow for elimination of spurious evidence, such as large numbers of nonhuman bones, piles of animal feces without human bony inclusions, incidental or transient insect species associated with dung or decomposing vegetation, or naturally occurring botanical phenomena (oddly shaped sticks, decomposing wood stains, etc.). The remaining items are numbered in sequence, usually using standing plastic markers. It is not unusual to find that the entire search has been for naught in that the "body" consists of the disarticulated nonhuman remains of someone's or something's dinner.

Figure 8.6 A metal detector is useful for locating jewelry, metal items on clothing, as well as bullets and casings at the scene. (Photo courtesy of Dr. Jason H. Byrd.)

Frequently the bulk of the human remains are located in one general area with various items transported beyond that region. As noted above, a much larger peripheral area may be systematically searched, but frequently similar results can be obtained by accommodating the search to the terrain. Game trails are obvious areas to follow, as dogs, coyotes, and wolves tend to bring food away from the scene to dens or eating areas. Areas in which dens may be located should be searched, and also open areas are often favored as feeding areas. Nests of rodents should also be examined. Rodents, in particular, are frequent thieves of small bones that are used in the building of their elaborate nests. Birds may be attracted by shiny, glittering objects and store these in nearby nests. They may also frequently incorporate hair into their nests (Pickering and Bachman 1997). Carrion-eating bird species, like turkey vultures (*Cathartes aura*), may also damage the remains by removing soft tissues and bones. Additionally, most gulls within the family Laridae are opportunistic and will feed upon decomposing remains. Indeed, one author has observed gulls feeding upon human remains. Finally, gravity may play a role with remains being moved downslope. Crania are notorious for "rolling away" and may be found in streambeds well below the rest of the body (Behrensmeyer 1975).

It is often the case that a domestic dog out for its daily run returns with a portion of human remains. Increasingly, investigators exploit this ability to locate remains by using cadaver dogs in scene investigation (Figure 8.7). These highly trained animals may be introduced in the initial search to locate the remains. Since even the general scene location is not known, cadaver dogs can be used to track along the most likely areas, such as paths of egress or entrance. Dogs may also be used following a walk-through search when smaller bones may still be missing. They are an excellent alternative when larger peripheral searches are needed. They are often useful in locating the initial area of decomposition, which may be marked by drainage of body fluids and a hair mass, but often little else. This may be well masked by subsequent leaf fall or debris accumulation and effectively hidden from a walk-through search.

Cadaver dogs are usually trained with decomposing human flesh and bone, although some "pseudoscents" are available (Tolhurst 1991). Various breeds are represented, and training must begin relatively early in the dog's life to achieve a well-trained dog. Trainers tend to be very cautious in the use of their animals, usually avoiding areas where they can be exposed to hazardous materials such as landfills, and preferring to work with them

Figure 8.7 A cadaver dog, such as the one shown here, is useful in searching large areas for scattered remains.

in the early morning when the animals are fresh and when the scents are much more keenly sensed. If the remains are covered or buried, dogs require more stringent conditions. Temperatures should be between 40 and 60°F, humidity should be 20% or higher, the ground should be moist, and windspeed should be at least 5 mph (France et al. 1992).

Buried remains typically occur as the result of human activity, such as legally sanctioned burial or the illegal burial of a crime victim, or by natural deposition (e.g., covered by landslide). Buried remains usually are more confined and less obvious. One should expect that a clandestine grave will be irregular in shape and shallow. A hastily dug grave used to hide a crime victim is not going to be the classic grave (i.e., rectangular in shape and 6 feet deep). Initially, evidence of animal and insect activity may give clues as to the location of the remains. Animals are often drawn by the smell and will dig down into a grave, even exposing and removing portions of the body. Also, the body may have been dragged out of the grave. Flies may also be attracted to completely covered bodies with shallow internment. Aggregations of flies may be seen landing on the disturbed ground, or often clusters of small flies may be seen hovering above the gravesite. These flies are attracted to the decomposing odors emanating from fissures in the soil, and may often be found months after the remains are buried.

Burial location may be discerned by observation of differences in vegetation, soil compaction or slumping, and disturbed soil (Figure 8.8). As the decomposition process continues, soft tissue will be lost and the resultant voids will be filled by overlying soils (Duncan 1983). Rains also help dissolve the loose clods of dirt and fill the areas between the dirt clumps. The result is that the initial mound of dirt will subside, leaving a slight depression. Cracks may appear as the disturbed soil pulls away from the undisturbed soil outside the grave. Water may accumulate in this area and allow for more moisture-loving vegetation to develop.

During the first year after burial, vegetation may be different in areas over the grave (Duncan 1983, France et al. 1992, Owsley 1995). This difference is due to the disturbance of the soil and the ability of the seeds to take advantage of the cleared and loosened soil. In most cases, seeds are available in the soil at the site, which have likely been deposited during the previous years. Additionally, seeds from plants currently fruiting can be blown by the wind or otherwise transported to the site. In some cases, no seeds are available due to placement of the soil originally located on or near the surface too deeply in the grave pit to allow for seed germination. Seasonality and exclusion of light from the soil surface are also factors that may cause the soil surface to remain bare of vegetation. The number of

Figure 8.8 A burial depression of a shallow grave approximately 8 months after the remains were placed is seen as a dip in the surface and, in this case, by a scarcity of vegetation due to destruction of the original ground cover. (Photo courtesty of Dr. Jason H. Byrd.)

plants inhabiting a disturbed area such as a gravesite can vary from sparse to dense, but it is usually recognizable as being different from the surrounding vegetation. The inhabitants usually seen are those recognizable as weeds by botanists and gardeners.

Following the arrival of these plants, the growth may occur faster and the existing vegetation may appear greener as the less compacted soil allows moisture to accumulate and penetrate further. Seasonal differences in the vegetation growth over the gravesite were noted in one study (France et al. 1992). In dry areas, vegetation growth over the gravesite was delayed so that the burial area persisted virtually unchanged until after the rains returned. In contrast, in moister times, the soils broke down rapidly, masking pit edges. The soils became more fine-grained and compacted. France and associates showed that the actual nature of the initial vegetation regrowth depends largely on the seed sources in the immediate area and the microtopography. After about 3 years, it is the disturbance itself, rather than the presence of any decaying remains, that controls the vegetation differences. These differences are seen to persist for many years after the burial.

The area around the grave also is trampled during digging, so plants will be flattened, branches broken, and leaf cover disturbed. Soils removed from a grave are usually piled adjacent to the hole and mix with the preexisting materials. Rarely can the soil be completely returned to the grave, and loose dirt (e.g., the fill or overburden) will be spread over a wide area. Burials in areas that are cultivated are more difficult to detect (Duncan 1983). Typically depressions will only last for several months. Where the soil is fine-grained and loose, the depressed area may be slightly more visible. Where the dirt tends to be broken into large clumps, all traces are quickly obliterated. Flood irrigation may highlight the excavation since the more deeply disturbed soil can catch the water, cause additional subsidence, and remain damp longer. Fortunately (or unfortunately), shallow burials in cultivated areas are often discovered in the course of plowing and disking, superimposing massive postmortem trauma over any perimortem injuries.

While Duncan (1983) indicated that burials in sandy soils tend to be shallow because grave walls are weak, two of our authors have excavated clandestine graves in sandy soils that were in excess of 5 ft. (1.5 m) in depth. Excavation is also limited by the water table at or near sandy beaches. Cadaver dogs may be particularly helpful as the porous sands allow decomposition gases to escape.

Aerial photography can be utilized in locating graves (France et al. 1992). Changes in vegetation growth, soil marks due to the excavation, settlement, and subsidence of the ground are seen best in low-angle sunlight, which emphasize surface texture. Therefore, early morning and evening photography is best when the long shadows highlight even minor discrepancies in topography. Areas where there is new luxuriant growth can be more prominent from the air and may indicate disturbance of the soils.

Probes and penetrometers may be useful in locating graves. These are thin metal rods that are attached to a crossbar (Owsley 1995). Diameter is usually about ½ inch, and they can be extended to 4 to 5 feet. The probe is pushed into the ground, with the crossbar being used to continue the intrusion. In most cases, even if the topsoil is relatively loose, resistance will noticeably increase at 1 to 1½ feet below the surface. When the area has been disturbed, however, as in the burial of a body, the probe can easily be inserted to its maximum length. However, the opposite is true in sandy soils (Schultz et al. 2006). Additions of sensors can increase utility by monitoring changes in gases, soil pH, and subsurface temperature. The operator should be familiar with use of the probe and should "take the feel" of the area in a location away from any suspected burial before beginning to probe. Systematic probes over a large area can be useful in locating single or multiple graves, and positive hits should be mapped. Probes may also help release odors that can be detected by cadaver dogs and flies of forensic significance. Indeed, the probe operator may also be able to detect the odor of decomposition by smelling the end of the probe.

Ground-penetrating radar (GPR) may be used in the location of burial sites (Miller 1996). GPR provides both vertical and horizontal information on subsurface disturbances. The equipment is able to detect soils that have been disturbed, since these are less dense and have a different mixture of soils with different moisture contents and electrical properties. Larger objects are easier to identify, and the radar will not penetrate metal, so this is readily apparent.

GPR is useful for broad, flat areas with little vegetation, since the mobile unit rides on a sled that must be pulled over the surface of the area. Large grids may be established, and then the unit drawn back and forth over the entire area. However, it has some limitations. GPR requires a machine operator familiar with the general nature of the local soils. Rocky and rough areas are difficult to analyze, and soils with many natural inclusions may be difficult to interpret. Since the GPR cannot absolutely identify human remains, it will not prevent fruitless searches. Recently, one author was involved in a search in which about 10 to 15 feet of overlying roadway was removed by backhoe in search of a buried body reported by an informant and in which GPR confirmed some disturbance. After considerable effort by the law enforcement agency, the resulting finds consisted of one artiodactyl bone, one dog bone, and one turkey bone.

Scene Constraints and Integrity

When the location has been established, whether the anthropologist was involved from the start or has only been called as human remains are located, the proper authorities must be notified. Sometimes the local law enforcement agency will call in a forensic anthropologist, entomologist, or botanist but neglect to inform the coroner/medical examiner that human remains have been located. This leads to later conflicts when the authority of the coroner's/medical examiner's office over the human remains is effectively sidestepped. The situation may be confusing in some jurisdictions where both a coroner and a medical examiner system operate. It is important to be aware of the state of affairs in terms of responsibilities prior to beginning work.

Obviously, the law enforcement investigators should remain along with their evidence technicians throughout the recovery phase. It is also important to have uniformed personnel from the law enforcement agency with the recovery team, usually placed at the entry points at the outer perimeter. At times in high-tension situations or when the alleged perpetrator(s) may still be at large and potentially threatened by the recovery, having armed personnel with the recovery team is also important.

Recovery Constraints

At the outset it is useful to survey the constraints under which the recovery process will be conducted (Skinner and Lazenby 1983). In some cases these will be financial, where the requesting agency must support the anthropology team's time and expenses. Fees for recovery services vary considerably among forensic anthropologists, with some charging by the hour, some charging a flat fee regardless of the time, and others donating their efforts to community service. Out-of-pocket expenses should be covered in any event by the requesting agency.

Unlike archaeological excavations, where weeks to months may be allocated for a "dig," a forensic scene recovery normally is given less time despite its more critical nature. There is a fine balance of trying to work quickly while providing the utmost in documentation and recovery. At the beginning of the work, provide the agency in charge with an overall estimate of how long it will take to complete the work, barring unforeseen difficulties. This will allow it a chance to arrange for other contingencies, such as meals, hotel accommodations or campsites, toilets, security for the scene during nonworking hours, lights for night work, etc. Avoid attempting to work nonstop on a scene, as all workers must be alert and attentive to detail, which is impossible to maintain without rest, food, and breaks.

The terrain and climatic conditions may also provide a number of constraints (Skinner and Lazenby 1983). In northern climes, heavy snow cover or permafrost may prevent recovery efforts until there are more favorable conditions. In other cases, the need for additional safety equipment or specially trained personnel may necessitate delaying the recovery process. However urgent the recovery may appear, it is not worth risking the lives of the investigators. The presence of hazardous material (Galloway and Snodgrass 1998) and unexploded ordnance are additional concerns, particularly in military or paramilitary situations.

Because it is important that the forensic anthropologist derive his or her interpretations directly from the skeletal evidence, it is advisable to avoid any information about the scene that may influence this determination. For this reason, it is best to limit conversation about the possibilities of who the body may be or how long the person has been missing. This is often difficult, as the remains may be linked to a high-profile missing person, local law enforcement may already suspect the identity, or evidence found at the scene (purse, wallet, credit card) may point to a specific individual. If the agencies are aware of the anthropologist's preference to remain "in the dark" about these suspicions, they usually will respect this if assured they will get a field assessment of the biological profile as soon as possible.

Controlling Access, Paths of Travel, and Scene Photography

Once the crime or body disposal scene is identified, it is important to minimize damage due to trampling. The perpetrator's route may contain evidence of his or her passage, such as footprints, threads of clothing, hair or discarded trash (i.e., cigarette butts, soda cans, etc.), and broken foliage. Repeated passage would obliterate or dislodge these forms of evidence. For this purpose, it is customary to establish a line of entry, one route that is clearly marked by which all personnel enter and leave the scene. Brightly colored flagging tape is usually used to mark this route. In some cases, it is helpful to bring in long planks to use as walkways (Haglund et al. 1990). This may not be the most accessible route since the easiest way has often been the one used by those who placed the body in the area in the first place.

Written logs should be maintained for all personnel entering and leaving the scene. Similar logs are kept for all evidence found, recording how and when it is transported from the scene. This allows for a more accurate reconstruction of who was present when specific items were recovered, who could potentially be witnesses to the condition in which these items were recovered, and what happened to them during the recovery process. These logs are the first of a series of steps in the chain of custody that should be started during the scene recovery. The 24-hour clock should be used to prevent confusion as to when specific events occurred. All information, including written notes, may be subject to subpoena if the case goes to trial.

Photographs should be made of the area that assist in explaining what happened, as well as how it happened, and when it may have occurred. While the scene is in as close to pristine state as possible, photographs are taken of it from 360°, incorporating major landmarks (e.g., road signs, bridges, buildings, etc.). If the scene must be cleared of vegetation for investigation to proceed, continue to photograph the scene changes and evidence as it is collected. Pay attention to the vegetation surrounding the subject of investigation (remains or physical evidence) to support the forensic botany analysis. Macro photography of the vegetation in immediate proximity to (or in contact with) the remains or associated physical evidence should also be conducted.

A number of different formats are currently available with which such images can be recorded. Ideally, a dedicated camera and photographer should be available to document all fieldwork. Also helpful are camera lenses of 35, 50, and 100 mm, and a macro or macro zoom lens. Digital cameras provide for the rapid viewing of scene images, which can be readily transferred to a computer, flash drive, or compact disc. The advantage of this format is that the investigator can ensure that the photograph has proper exposure and composition before he or she departs the scene or manipulates the evidence. Fears concerning the evidentiary value of digital images in court have been quashed, as two 1995 court cases taken through *Kelly-Frye* hearings, *State of Washing v. Eric Hayden* and *State of California v. Phillip Lee Jackson* (1995), have found in favor of the use of digital images in court (Staggs 2003).

Video cameras also are increasingly used to document the sequence of events in the recovery process. These may also be used for training films and for other general audience purposes once the case is adjudicated, although the image quality of most handheld video cameras is relatively poor, particularly in the large format of a projection screen. Videos do allow the recovery process to be documented as it is being conducted, and they are particularly useful in complex cases, where the documentation of object relationships is critical.

The press often appears at the scene as quickly as the forensic anthropologist, entomologist, or botanist, if not much sooner. In the authors' experience, law enforcement agencies are becoming much more efficient in controlling the access to crime scenes and feeding information to the press via a spokesperson. In most cases, they will prefer that any information pertaining to a case be released via this method but, in some circumstances, may wish that one of the forensic scientists provide some airtime. This approach is most often used when a body is unidentified after a period of time. Discussions about what will be said should occur behind the scenes, and these specialists should consult with the agency before agreeing to do any interviews.

At the crime scene, media coverage can be extensive. Camera crews are able to provide close-up shots from great distances, such as from homes overlooking the scene, or from hovering helicopters. Similarly, microphones can pick up conversations well beyond normal hearing. Since actual coverage of the human remains can be excruciating for the relatives as well as damaging for the case, it is best to conceal the forensic work at the scene, where possible, with screens or covers, and avoid overly loud conversations. The extent of coverage may become a factor in dictating the speed of recovery, as added time can increase the press determination for more graphic footage.

Establishing the Location

As obvious as it sounds, one of the most important tasks early in the scene investigation is to determine your geographic location. When the written report is provided to other forensic investigators, they should be able to use it to return to the scene. This does not mean providing directions such as "Take Highway 1 north to milepost 26.3, turn west...," but rather: "The scene is located approximately 500 m due west of milepost 26.3 on Highway 1." The use of recognizable markers such as street addresses, road names or numbers, mileposts, and natural or architectural landmarks is helpful. An overview map placing the scene within this wider context can be useful (Figure 8.9). Computer map indexes are available to assist in this project.

Global Positioning Systems (GPS) allow for the rapid generation of a coordinate system to use for general map reference. Handheld or computer-linked systems are now available at a relatively low cost, which enables the field researcher to locate the scene onto a topographic map or a computer atlas. Unfortunately, such systems do not always function as desired due to the availability of an adequate number of satellites. Dense tree canopies may also prevent accurate GPS readings (Listi et al. 2007). GPS systems are, however, ideal when working in remote areas, as often the law enforcement investigators can locate the site only on the ground, which has limited use in producing the written report. However, it must also be noted that due to the "white noise" purposely introduced into the GPS units produced for public use, the signal will drift, causing the recorded GPS position to vary by approximately 50 yards. Therefore, GPS units should be used in setting the datum, but are not suitable for mapping the relative location of each piece of evidence. These items should be mapped in by running a tape measure from the datum to each item, in keeping with standard archaeological protocols (Listi et al. 2007).

Once the general location is determined, a prominent feature at the scene is often used to provide a link between mapping scales. For example, a house may be identified on the overview map by address, then the northeast corner of the house used to show the location of the remains (Figure 8.10) and, as importantly, the datum or data points, the point or points from which the scene is mapped. It may be helpful to think of these layers of maps as providing a means to drive to the scene (overview map), then the route to walk from where you parked to the actual scene (linkage map).

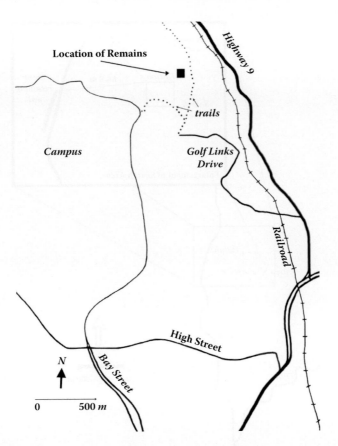

Figure 8.9 The overview map shows the location of the crime scene in relation to easily recognizable landmarks such as highways, towns, and named roads.

In order to map a site in greater detail, a datum must be established. All other locations at the scene are tied to this point. The datum must be something that has more permanence than a simple flag or other location marker, and it must be clear of any excavation area. Buildings, large boulders, trees, drainage structures, or communications towers and poles are often used. Structures such as the corner of a building or an electrical pole are excellent because these fixtures are usually documented in local government records. If there is no suitable item, a stake may be driven into the ground or the surface painted to denote a spot. A metal stake buried a few inches in the ground can mark the spot and can be easily found later with a metal detector. The indicated spot is then linked to other identifiable features through use of a tape measure and compass or by taping the distance from two other identifiable features (triangulation). Since often the measurements are taken over uneven terrain, the tape measures should be maintained at a horizontal level, with a plumb bob used to mark the point to which measurements are taken (Figure 8.11a–c). The datum should be established so that at least the bulk of the site can be covered from a single datum point. The use of multiple data points may be required in large scenes. Because the datum also serves as a depth indicator, it should be on a higher level than the scene, and a recorded point on the datum used as the depth reference. The distance from this point to the ground is recorded, and subtracted from subsequent depths to produce a profile. The datum should be photographed with a north arrow visible, and the direction of the photo indicated in the log or otherwise recorded.

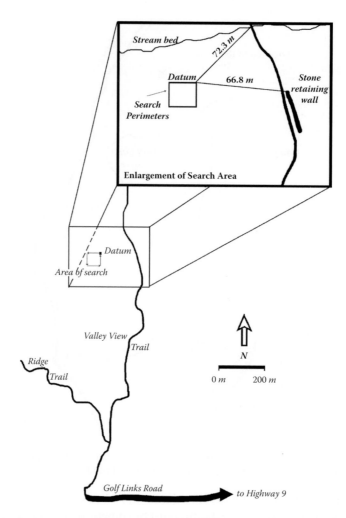

Figure 8.10 The linkage map shows more tightly defined areas and indicates the datum in relation to identifiable permanent landmarks in the local area.

If all permanent fixtures are far from the site, a subdatum can be established between the remains and the nearest permanent fixture. The exact location of the subdatum should be documented the same as for the datum. Documentation should include exact distance and compass direction from your datum in a manner typically referred to as configuring a compass rose (Figure 8.12). Often long distances are not drawn to scale but are marked by discontinuities in the lines between points, with the actual distance noted to the side.

Once the datum has been established, the excavation grid is created. However, a grid may be forgone at a simple site where the remains are in a confined area or in a simple grave. In such instances, a baseline may be used. The grid should include all of the bones and associated physical evidence. Usually a 3 × 4 m square unit is sufficient. If scattering is severe, a larger grid or several smaller grids separated from each other may be necessary. Establish a meridian by laying a north–south line if possible, and then lay a baseline (i.e., the east–west reference line). These lines should create the outermost borders of the grid (Figure 8.13a and b). To ensure the grid is symmetrical, make sure that the corners form right triangles. This can be accomplished by using a large drafting triangle, a 3-4-5

(a)

(b)

(c)

Figure 8.11 When measuring over longer distances, a tape is kept as close to horizontal as possible, with a plumb bob being used to sight the artifact or skeletal element being mapped. Where the plumb bob crosses the tape will indicate the distance. (Photos courtesy of Dr. Jason H. Byrd.)

triangle, or the basic formula for the length of the hypotenuse of a right triangle ($X = \sqrt{a^2 + b^2}$, where a and b are the lengths of the sides of the triangle). A wooden or metal stake is placed at each corner and connected by rope, and the corners are mapped in relative to the established datum. The elevation should also be measured for all four corners in order to establish ground level of your site in relation to your datum. Then, using more rope and stakes, establish transects within the grid and parallel to your meridian and baseline. These crossbars of the grid should be established at distances that best suit each particular site; however, 1 m units are commonly used. Identify grid unit coordinates by using letters in the E-W direction and numbers in the N-S direction (or vice versa). This will provide quick identification of a grid unit. Once the grid is set up and mapping begins, it is prudent to designate one individual as the scribe who plots the data on graph paper (in millimeters) while points are systematically called out by the individual tasked with measuring the location of each piece of evidence.

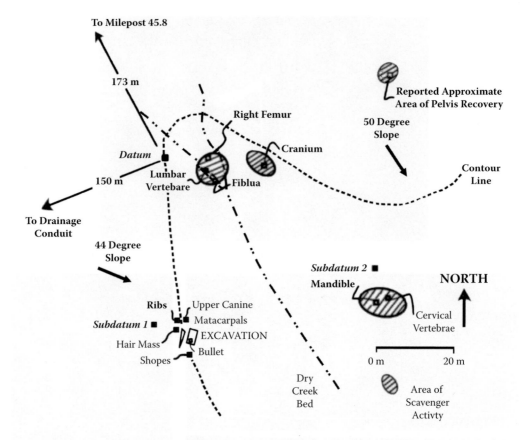

Figure 8.12 Map of scattered remains with subdatum, and discontinuities marking long distances.

Photograph the implementation of the grid at each stage, beginning with the first axes, through its completion and excavation. Take general shots across the area with datum included and indicate on the photo log the direction from which you were photographing (Appendix II).

Depth measurements should be taken across the grid. This allows a reconstruction of the topography of the site, such as the presence of creeks, cliffs, caves, or gullies. These features may help explain the movement of some items, which would not be understandable if the area was depicted as if it were a flat surface. Sometimes irregular topography does not easily permit accurate depth measurement, so access to more formal mapping equipment, such as a transit or total station, is necessary.

Forensic Excavation and Recovery Techniques

The recovery team should aim for completeness in recording so that the original layout of the scene can be reconstructed as well as how the actual recovery process was approached. Once the recovery is completed, this information cannot be recovered or recreated, so it is essential that it be done correctly from the start (Skinner and Lazenby 1983). Evidence often goes beyond the scope of being physically recoverable. Important information may lie in the position of the body, its location with regard to other physical features, the relative

Figure 8.13 (a) Search and excavation grids usually are laid out along a north–south axis, with each square being identified by a number and a letter. All items found within one grid are labeled with this designation. (Adapted from *Death Investigator's Handbook*. Used with permission of Louis N. Eliopulos.) (b) Excavation grid established over clandestine gravesite. All items recovered from inside and outside the grave pit are referenced to the grid in which they were recovered. (Photo courtesy of Dr. Jason H. Byrd.)

intensity of sunlight, etc. Evidence may even be less obvious. As Skinner and Lazenby point out, scorching of nearby trees can support claims that a body was burnt in a location, and absence of such scorching would argue against such events. To simply collect the physical remains of the victim without recording the context could substantially damage the case.

Before mapping begins, determine the spread of the remains and the scale of the drawing. Designate a notetaker, mapper, and photographer, and make sure they are familiar with the equipment and recording forms. Include a scale on the map from the start, and familiarize all personnel with that scale. Multiple drawings at different scales may be necessary in order to accommodate all the details. The positioning of the body in relation to magnetic north should be recorded (e.g., the head points SE, the feet point NW). Then placement of the head, hip, knees, feet, shoulders, elbows, wrists, and hands should be carefully mapped. Ensure that the remains and all physical evidence are photographed *in situ* and within the grid. When each item is mapped, the following information should be recorded on data sheets (Appendix III):

- Specimen number
- Description of item
- Grid unit in which the item was found
- Distance from the datum
- Compass direction in degrees (in relation to the datum)
- Elevation/depth

Voice-activated tape recorders are often useful to provide more complete detail than can be easily done on paper where dirt, wind, and biohazards interfere with moving between excavation and paper recording.

Photographs should be made frequently during the process of excavation. Inclusion of a photo board with the case number and date is helpful for establishing the case identity later. Cameras that include a date stamp also allow for some validation of the association. The victim and associated physical evidence must be photographed from many different directions. In general, when taking wide-angle shots of the subject, try to include a previously noted landmark for scale, or some other object of known size and dimension. If possible, include a metric scale on any macro shot. If this is impractical, take a larger photo with a scale to encompass the area later taken in extreme close-up. Additional specimen photography will most likely be necessary at the morgue or laboratory.

Surface Remains Recovery

In recovering surface-scattered remains, each item must be numbered, photographed from above with a scale and north arrow in the photograph, and mapped onto the scene map. One useful item that assists in mapping is a large 360° scale that fits over the datum point. Using a compass, orient this to north, then the entire scatter of remains can be plotted by running a tape measure from the datum to the item to be recorded. If this is not available, triangulation from two permanent points is used to place the location of the remains on the scene map.

Notes should record the position of all skeletal items, orientation, and degree of articulation. The loose leaf cover and topsoil may contain disarticulated bones, jewelry, hair ornaments, ties, insect fragments, botanical evidence, or other evidence that could easily be lost if this layer is simply discarded. Such small items will often be covered over and may "self-bury," in that the lighter soils will be washed over them and the evidentiary material will sink into the ground. Use of a metal detector is extremely helpful in searching such scattered scenes. A version that can distinguish between ferrous and nonferrous material is the most practical, as often bodies are deposited in areas already extensively used as trash sites. This type of machine will ignore the odd bits of barbed wire, rusted auto parts, or wrought iron, but will allow location of bullets and jewelry that could otherwise be easily lost.

Once photographs and notes are completed, the remains can be removed. Relatively complete remains are usually placed in body bags, with a body board or plank placed underneath to provide support and minimize shifting of the remains. Labels should include case number, agency name, date and time of collection, name of the collector, and a description of contents. Once the remains are lifted, it is often helpful to excavate several inches below the remains to recover any small bones, teeth, and physical evidence that have fallen from the body (Lipskin and Field 1983, Wolf 1986). This loosened soil should be screened.

Because remains are often scattered over a wide area and there is always the possibility of multiple victims present, running notes on the inventory are helpful. Charts are useful for a quick reading of which bones are still to be recovered. These charts also allow the anthropologist to record which portions of a bone are recovered, and may prevent some confusion, such as when two right proximal tibial fragments are found that are actually portions of one bone, rather than of two individuals.

Buried Remains Recovery

In the recovery of buried bodies, it is helpful to include an archaeologist with your recovery team (Morse et al. 1976), or to have significant excavation experience yourself. Excavations may require some additional personnel to assist with the multiple tasks of digging, removing soil from the site, and screening, as well as mapping and photography. However, usually only about three or four people are needed to work a single grave, and excessive personnel becomes unwieldy. It is also important to be aware of local conditions. Excavation techniques that work well in desert areas may be inappropriate in waterlogged areas, where lowering the water level becomes critical.

Once the prospective burial site is located, the area should be carefully photographed. Depict the overall area prior to and after the removal of soil and vegetation. As exhumation begins, show any sign of possible tool use by the grave digger during the original interment. Be sure to depict the sunken soil and soil color changes. Depth markers (meter or yard stick) and a north arrow should be included in each photo. Notes should be made with regards to vegetation, condition of the soil, contours of the presumed pit, odor and insect activity, and signs of disturbance, such as animal excavations. As scavengers may partially disinter and scatter the remains, the recovery should include a surface search, followed by excavation of the grave (Lipskin and Field 1983).

Grids outlined with tightly drawn string tied to large nails or surveyor pins are established to enclose the entire visible pit (this can be expanded if needed). It is best to position the nails or surveyor pins so they are not directly on the edge of your plot, where they can be undercut and prone to dislodging (Figure 8.14a and b). Again, photographs should be taken after the vegetation is removed. Loose leaf cover over the grave should be carefully examined. This layer also may conceal footprints or other evidence that may be useful for the investigation.

Often the outline of the grave can be estimated at this point. The edge may be highlighted by lightly tracing the line with the point of a trowel. In some cases, use of a short probe can be helpful in determining the edge, working from the outer nondisturbed soils in toward the grave. This also can be used as a reference for modifying the placement of grids and the excavation approach. It should be remembered that graves are often irregular in shape, having been dug in haste, with poor equipment, and in awkward circumstances, such as under low-hanging branches (Wolf 1986).

At this point the map of the excavation usually consists of the placement of a grid on the linkage map and a separate map of the grid. This second map should be at a large

(a)

(b)

Figure 8.14 (a) Large stakes or nails are used to form the string grid over the area to be intensively searched or excavated. Offsetting the nails prevents undercutting during excavation or clearing of topsoil. String should be set close to the ground to prevent tripping. (b) The squareness of the grid can be checked easily utilizing a builder's square. (Photos courtesy of Dr. Jason H. Byrd.)

enough scale that legible notes can be made as to the identification of items and depths. As each item is found, map its position with the tape measure, using the two grid sides to measure *into* the grid with the tape measure (held horizontally and parallel to the contiguous line from which the measurements are being taken) (Figure 8.15). Indicate the grave outline on the map and any exposed items or bones. Depth measurements should also be taken of the pit and the surrounding area. This allows the grave and eventually the remains to be shown in profile. To obtain depth measurements, a line level is attached to the string from the datum. The string is held taut over the item or area to be measured, and the tape measure is used to obtain the distance between the string and the surface (Figure 8.16a and b). Be sure to note how far above the ground the string attaches to the datum.

(a)   (b)

Figure 8.15 (a and b) Artifacts such as pieces of jewelry, bones, or other evidence are mapped by taking readings perpendicular to two of the grid lines. These are then drawn to scale on the site map. (Photos courtesy of Dr. Jason H. Byrd.)

(a)   (b)

Figure 8.16 (a) The bone in the burial pit is being measured for depth, baseline, and meridian simultaneously. (b) Detail illustrating the baseline, meridian, and depth measurements intersecting to capture the medical aspect of the tibia in the burial pit. (Photos courtesy of Dr. Jason H. Byrd.)

Because of the danger of contamination, loose soil is then removed from within the grid. First, any soil that appears to have been displaced by either animal predation or human discovery should be taken and screened. Then loose soil from around the presumed site is cleared and screened. Finally, the loose soil within the grave outline is swept up and screened.

Photograph frequently during the excavation process. Photos should be identified as to the case and orientation and recorded in the photograph log. Autofocus cameras are often useful for overhead shots since they can simply be held above the head and the photograph taken without requiring the operator to view the image. Some shots should also be taken without any identification, so they can be used for public presentations after the case is adjudicated.

At this point excavation may begin, but the approach often depends on the anticipated depth of the burial. The aim is to preserve as much of the pit outline as possible without endangering the remains or associated evidence. In shallow graves, it is possible for excavators to work from the surrounding areas by leaning into the pit. As the depth increases, however, leaning into the pit will endanger the sides, and there is the danger that either the sides or excavator, or both, will fall into the grave, possibly damaging the remains. In these cases, parallel pits are usually advisable so that the excavators can work at a safe and comfortable level with relation to the remains and complete full exposure of the body prior to removal of any portions (Figure 8.17). While this will eventually destroy one side wall, the outline of this wall should be recorded as the excavation moves downward. The side pit also facilitates the removal of the remains by providing space in which a support plank may be placed, onto which the body is moved.

Excavation should proceed in levels of approximately 10 cm within the grave. Hand trowels are used in a "scraping" manner to cut off broad thin layers of dirt. Frequent brushing exposes features. Brushing should always move in one direction, not back and forth as in paint strokes, as soil changes will be obscured or evidence dislodged. If possible, it may be useful to excavate only half of the grave (lengthwise). This exposes a profile of the feature, showing the pit length and any disturbances within the soils overlying the body. Often, however, the pit is relatively small, and adherence to archaeological standards is of less practical use. Screening, of course, should be used for all the soils because small bones and entomological and botanical evidence (i.e., pupal cases, grass, leaves, and broken twigs interred while filling the grave) can be found at all grave depths. In particular, adult insects emerging from the puparium will burrow up through the soil in order to reach the surface.

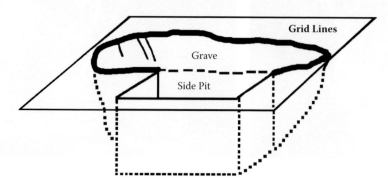

Figure 8.17 In excavations that exceed the excavator's ability to work safely from the edge, a side trench is dug. While sacrificing the integrity of one wall, this prevents edge damage from leaning into the gravesite, the danger of falls into the pit, and facilitates removal of the remains once excavation is completed.

Since many insects die in their attempt to reach the surface, it is not unusual to find adult insects distributed throughout the soil between the body and the ground surface. Be careful to place the screening area well away from any potential excavations. For instance, if you find that the body is in an unusual position, the grid may have to be expanded, necessitating moving the backdirt you had already accumulated.

The edges of the pit can be "felt" with the trowel or digging implement. Because this outside soil has not been disturbed, it is more densely compacted, making a noticeable difference in the difficulty of digging (Pickering and Bachman 1997). Experience in excavation teaches this skill. If the anthropologist lacks any archaeological training, it may be useful to take an archaeological field class if only to acquire the skill of feeling the differences in the soil. Careful excavation along with light brushing can, in soils with relatively high clay content, reveal indications of the instruments initially used to dig the grave (Hochrein 1998). The curved profile of a shovel, the narrow band of a pick, or the repeated grooves of a gardener's trowel can all be retained in the pit outline.

Once a bone is exposed, the body can be followed using careful excavation techniques and being sure to move from the known to the unknown. Specifically, once a bone or other piece of evidence is found, it should be exposed *in situ* by following the item until it reaches another, and so on. Many texts advocate switching from trowels to wooden or plastic implements. Where practical, this should be followed, but in many areas, reliance on wooden implements would be an exercise in futility—you could not move earth. Use of smaller tools (dental picks and probes) and frequent brushing of the surface are essential.

All remains must be mapped in three dimensions (Figure 8.18). This means not only that measurements are taken from the sides of the grid, but also that depth measurements are taken from the datum. This information permits a scene profile to be drawn. As described above, bring the string with the line level taut across the vertical tape measure. For large objects, such as the skull, depth measurements are taken on the top and bottom at a minimum. Indeed, a skull can easily have twelve data points (each having two longitudinal measurements and one depth measurement) recorded, so that its orientation can be accurately portrayed on the map. It is often easiest to map in most of the bone's peaks and troughs as well as anatomical points of interest. For example, the lateral side of a cranium that lies on the bottom of the burial pit may have the following points mapped: glabella, nasion, and the bottom of the nasal aperture, anterior-most aspect of the maxilla, foramen magnum, lambda, and so on. The mapper/scribe can then connect the dots on the graph paper in a way that somewhat resembles the lateral view of the cranium.

As much as possible, it is best to expose the entire scene for photographs prior to the removal of any part of the body. In fact, hasty removal of the remains can be one of the primary errors in scene recovery; they can be safely lifted only when their relationship to the other features of the scene are fully realized (Pickering and Bachman 1997). Photograph all physical evidence detailing clothing color, condition, size, and brand name. Macro photos of injuries and indications of restraints must be taken.

After the body is exposed, conduct a final check to make sure everything that is present at the scene is recorded on the map. Working systematically from one end of the scene to the other with two people, one calling out each item and the other checking the map(s), ensures that there will be no errors in recording. Remember that not only location but also depth may be needed.

In many cases the remains will not be completely skeletonized. This is particularly true in cases where the body has been wrapped in plastic or some type of container,

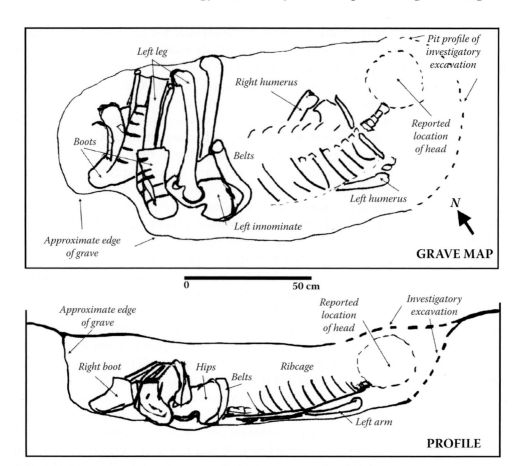

Figure 8.18 The scene is mapped from directly overhead and a profile view is also drawn that shows the excavation along the midline. Such maps become extremely important when presenting the case to a jury and require accurate in-field recording of all items. Map and profile relate to Figure 8.19.

or when there are multiple, stacked victims. Often the intent of the perpetrator was to accelerate decomposition, but instead he or she inadvertently dramatically decreased the rate of decay. The remaining flesh may be sufficiently intact even years afterward for at least a partial autopsy. In such cases, the odor usually is a warning to proceed carefully with the excavation and avoid any intrusive digging. Gentle scraping of the surface accompanied by frequent passes with the brush will expose the covering. Such remains can be lifted directly into a body bag if it is possible to provide sufficient support underneath the wrapping so that the body is not supported by the original container. Lifting by the wrapping could superimpose additional damage that may interfere with interpretation of any other indicators of tearing or stretching from the time of disposal.

When the remains are completely skeletonized, there are two approaches. In both cases it is important to completely expose the remains prior to removal of the body. In many cases, once exposed, photographed, mapped, and recorded, individual bones can be removed. Also in many cases, portions of the body will cover other body portions. For example, a body may be in the grave in a supine position, but with the hands tied behind the back (Figure 8.19). In this case, separate photographs of the hand positions

Figure 8.19 In this excavation, an individual is recovered whose head was removed during the initial search by police personnel. The remains lay supine, with the hands tied behind the back. Expended bullets were located in the thoracic cavity. (Photo courtesy of Dr. Alison Galloway.)

will be required. A second method, which also has advocates, is to completely pedestal the remains. This means that the body is left on a "stand" or pedestal of dirt. The pedestal is then undercut and the body, still encased in dirt, is slid onto a wooden support board for transport to the laboratory, where the excavation process is finished. This method is particularly helpful when the bones are in very fragile condition or when staining and other less tangible features can be recorded more carefully than in field conditions.

Soil samples from the body, particularly the abdominal and chest areas, may be helpful. On bodies buried close to the surface, large quantities of pupal cases are often found within or on the body and associated clothing, bedding, or other material, as the maggots were unable to migrate before pupation. Loose soil at the bottom of the pit must also be checked for evidence.

Once the body is removed, a profile of the grave should be completed by taking depth measurements of the floor of the pit. Using the midline of the overhead map and the same scale for depth, mark the depth to which the grave was originally dug.

Collection Techniques

If material is to be transported directly back to the laboratory for analysis, then one must establish the chain of custody, recognizing that the material is coming from the supervision of the primary investigator to the anthropologist, entomologist, or botanist. This documentation is essential for showing the court that the material has been safely and securely protected from the time of the excavation until it is presented in court. Since it is not unusual for the local agency to fail to have such a form, it is helpful to have a supply (Appendix IV). This form includes places for the case number(s), agency, items, the person's name and signature releasing the material, and the person's name and signature obtaining custody. Each transfer should be dated and timed, with both agencies retaining a copy for their records. When the material is returned to the agency, the next transfer should be noted on the same forms. This is true also for any transfers of evidence away from the laboratory, such as samples taken and sent for DNA sampling, dental arcades sent to the forensic odontologist, or prosthetic devices sent to an orthopedic consultant or manufacturer.

Anthropological Evidence

Bone does not preserve well in plastic bags. The moisture present in the bones along with the nutrients present in the soil and bone provide a perfect medium for molds and mildew. Left unchecked, this will lead to the breaking apart of the bone so that even basic morphology can be lost. Instead, bones should be packed in clean paper bags labeled with the case number, agency, item number keyed to the site map, date, contents, and the initials of the person responsible for collection. Identification of the element by the anthropologist also should be included. Since bags may differ in size and shape, it is useful to have a moving box into which to pack the bags of bones for transport to the morgue. Make sure that heavy bones (i.e., pelvis or femur) are not put on top of smaller, delicate bones. For bones that are less sturdy, bubble-pack or foam-pack can be wrapped around them. Cotton batting is difficult to extract from the bones and leaves excessive fibers, and therefore is less useful.

Human remains that are only partially skeletonized, with soft tissue and body fluids present, should be packed in heavy plastic airtight containers or evidence bags, labeled, and then placed inside "hazardous biological specimen" bags or body bags. If the soft tissues are still intact on the hands, place the hands in paper bags prior to moving the body. Watertight "hard-sided" equipment bags function as airtight (and odor-tight) containers for the transfer of this type of decomposing material. All the same, these remains should be immediately transferred to the morgue or laboratory for processing.

Bones collected from a single area such as a burial can be collected together, depending on the quality of the bone. For scattered remains, however, each area should be bagged separately and labeled as to its location within the overall scene (Howard et al. 1988).

Do not clean the bones in the field (Haglund et al. 1990). Since there may be trace evidence retained with the bones, cleaning may dissociate it and raise questions as to whether the evidence was from the body or surrounding area. For example, a bullet retained within the cranium could be lost, or questions raised as to the possibility that it was only an inadvertent association actually due to hunting or target shooting.

If material is dry, or will dry during transportation, then special care must be given to the cranium (Bass 1987). The drying soils will create a clump of dirt that forms a "bomb" within the cranial vault. Left intact, this will bounce around within the vault and can break the bones apart. The cranium should be examined for presence of dirt and carefully packed to minimize jostling during transport. Obviously, the dirt should not be removed at the scene, but packing material can be placed into the vault to control the movement of the clod. Handling the skull also requires care. No skull, in any circumstances, should be held by fingers in the orbit or nasal aperture, and forensic cases are no exception (Wolf 1986). The fragile bones in these areas are easily broken and almost impossible to reconstruct. Because the excavated skull may be severely fragmented by peri- or postmortem trauma, it also should not be lifted by the foramen magnum. Instead, it should be cradled in both hands until it can be packed for transport. Care also must be taken not to lose the teeth, especially the anterior ones.

Some more friable material requires special handling. With authorization, burned bones can be consolidated by using diluted glue, which is dribbled onto the bone/teeth using a pipette. This works well for preserving dentition during transportation to morgue and allows gentle handling. Only those areas critical to analysis should be treated and carefully recorded. These substances may interfere with chemical, serological, genetic, or other laboratory results if treated material is unknowingly submitted (Wolf 1986).

Botanical Evidence

Botanical samples from the scene can include macrobotanical samples as well as pollen. The larger samples include roots that have grown through the remains, ends of limbs that have been cut or broken to be used as weapons or coverings, or leaves that have fallen over the body. Analysis of new plant growth and dead plants at the scene may be used to establish the postmortem interval (Lipskin and Field 1983). In addition, because many plants have parts that are easily detached (i.e., seeds, spines, hooks, or hairs) and can inadvertently attach themselves to the clothing or hair of the victim or perpetrator, combined with the fact that plants are habitat specific, botanical evidence can be used to place the victim or perpetrator at a particular location (Lane et al. 1990, Hall 1988, 1997). Pollen samples are also particularly useful in establishing associations among victim, suspect, and crime scene.

The macrobotanical evidence should be documented and collected before the scene is processed for anthropological and entomological evidence. The habitat around and including the scene should be assessed. Broken branches, crushed plants, overturned planters, and places where the suspect(s) may have walked through or by a hedge or other dense vegetation should be photographed (with macro and normal lenses). Samples from the major trees and shrubs from around the scene are collected to help the botanist reconstruct the scene's habitat. Preferably, at least one 12-inch-long branch (with its leaves and fruits) from each woody plant and examples of all herbaceous (nonwoody) plants should be collected. Also, collect a handful of leaf litter and other vegetative detritus. After the samples for habitat reconstruction are collected, then proceed to collection of samples from around, on, and under the body. Collect all vegetative material on the victim and other objects associated with the crime. Because identification of species may be crucial for interpreting the information, it is best to also collect samples of branches with leaves from the same tree as the specimen taken into evidence or from surrounding vegetation if the plant of origin is not immediately evident. Searchers should take precautions against handling poison oak and poison ivy.

When collecting the herbaceous samples, include the plant's roots and be sure to shake out most of the dirt clumps. Plant material in body fluids or other muck should be put into plastic containers and be kept cool, but not frozen. All other samples are to be put into paper bags, or they can be wrapped in newspaper and allowed to fully dry in order to prevent decay from molds and mildew. The specimens must lie flat. On the bag or newspaper write the case number, sample number, date and time of collection, and name of collector in indelible ink. In addition, the following relevant data should be recorded on ancillary sheets (Appendix V):

- Where the sample was taken in relation to the established datum and GPS coordinates
- Where on the victim or object the sample was taken (i.e., adjacent to the left elbow)
- The type (as in grass, tree, or shrub) and size of plant from which the sample was taken (i.e., an approximately 25-foot tree)

- The plant's color (i.e., leaves are green on top and white with red veins underneath, and purple fruit is present)
- The relative frequency of occurrence for the plant within the scene (i.e., frequent or infrequent—percentages if possible)

Most importantly, a forensic botanist must examine the collected material and photographs immediately. It is prudent to maintain an on-call list of forensic experts with whom your agency wishes to work.

Entomological Evidence

The collection of entomological evidence should occur after the botanical evidence has been collected. The proper collection of entomological evidence is dealt with extensively in Chapter 3. What follows, however, is a brief discussion concerning the alteration of normal insect activity on buried human or animal remains. It has long been observed that buried remains decompose at a much slower rate than those placed aboveground. The fact that the act of burial slows decomposition is due to a number of factors. These factors include the exclusion of some bacterial organisms and vertebrate scavengers, and the cooler temperatures that decrease with soil depth. However, the most influential factor that slows the decomposition of buried remains is the exclusion (either partial or total) of insect activity and the alteration of normal insect succession (Lundt 1964, Payne et al. 1968). Mégnin (1887) and Motter (1898) were the first to formally report the alteration of normal insect succession of buried remains. Schmitz (1928) and Leclercq (1975) supported and enhanced earlier works by showing that buried remains have their own unique insect fauna that varies with the habitat and type of burial.

With buried remains, the insect fauna is generally limited in direct proportion to the amount of soil or debris covering the body. However, Rodriguez and Bass (1983) noted that as little as 1 inch (2.5 cm) of soil would exclude the majority of blow fly species. This is due in part to the fact that many adult female blow flies require physical contact with the remains before oviposition will commence. Many chemical receptors, including the sense of taste, are located on small hairs on various parts of the exoskeleton. The blow flies are stimulated to oviposit by the free moisture on the surface of the remains, as well as by oviposition pheromones released from other nearby female flies when a suitable oviposition substrate is found. Exclusion of physical contact, even by a thin layer of soil, may be enough to inhibit or delay oviposition due to lack of the proper chemical clues necessary to prompt the behavior. Colonization of a buried body by insects requiring direct access to the body is an indication that the remains may have been exposed to the environment for a period of time. Such a fact can be crucial in reconstructing the sequence of events leading to the burial of the corpse.

Flesh flies (Sarcophagidae) and various relatives of the ubiquitous house fly (Muscidae) are not as readily deterred by thin layers of soil. An evolutionary divergence from the blow flies has led to flies in the family Sarcophagidae not requiring tarsal contact with the remains, and their unusual ability to deposit active larvae instead of an egg. The first-instar larva hatches from the egg while it is inside of the female's body. Thus, it can crawl and burrow into the decomposing tissue immediately upon being deposited by the female. Flesh flies that are in flight have also been observed to directly deposit larvae onto carcasses, or on the soil surface above buried remains. These newly deposited larvae can rapidly burrow through several inches of soil to reach the remains. Colonization of buried remains by these flies is more common during the later stages of decay, when bloat may produce fissures in the soil surface and allow closer access by the adult flies. In most cases, it is generally assumed that bodies buried 1 foot (0.3 m) or less will be colonized by one or more species of dipterous larvae (Rodriguez and Bass 1985). The adults of smaller flies, such as those of the Phoridae (coffin flies), will burrow into the soil to oviposit directly on the remains. This adult burrowing behavior is also seen in several beetle families, such as the rove beetles (Staphylinidae) and burying beetles (Silphidae), all of which have been recovered from remains buried from 10 to 100 cm deep.

Some insect species (such as *Conicera tibialis*) that successfully colonize decomposing remains underground may undergo several generations and continue to live for extended periods of time on the buried remains. However, the typical scenario is that the larvae that were deposited on the soil surface and burrowed down to the remains will continue to develop and eventually pupate underground (either on the remains or in cavities in the soil created by the decomposing tissue). These species can also undergo adult emergence while still confined underground. The newly emerged adults will then burrow upward through the soil and reach the surface. However, if buried too deeply, many die while attempting to reach the soil surface. Therefore, it is not uncommon to find dead adult insects at all levels in the soil directly above the remains. The soil excavated from the defined burial site (from surface to grave floor) should be thoroughly sifted and examined by an entomologist for the recovery of entomological evidence. In addition to finding adult insects in underground locations, larvae, pupae, pupal cases, and insect body fragments can be expected. All stages of the insect's life cycle should be actively sought during the excavation process. If living insects are recovered from the burial site, live and preserved collections should be made, as described in Chapter 3.

It is impossible to overstate the simple fact that photographs of the scene are needed to augment the insect samples. The forensic entomologist is trained to observe small fragmentary remains of insects and their castings that most individuals on the forensic science team will overlook. Therefore, it is beneficial to take numerous photographs of the colonized body in reference to its surroundings, photographs depicting the location of the insect colonization on the remains, and overall body shots that depict the body's state of decomposition. Always include macro photographs of the insects that have colonized the body.

Field Assessment

A brief assessment of the biological profile, any obvious defects, and the length of the postmortem interval is appreciated at the scene. It is best not to be too restrictive on the ranges, and temper any interpretation with the precaution that this is a preliminary assessment and may be expanded during closer examination. Such information, however, allows law enforcement to begin amassing the information that may determine the positive identification.

In cases that were well publicized, it is not uncommon for an impromptu memorial to be established by the public as the sympathetic and the curious visit the crime scene. This frequently leads to a rash of submittals in the subsequent weeks, as every bone or oddly shaped rock or stick is brought to the sheriff's office by concerned citizens convinced they have found part of the victim's body. To avoid lengthy trips to confirm or deny these items, transmission of pictures by e-mail has proven quite effective and efficient.

Preservation Techniques

Anthropological specimens are best preserved in dry, well-aired conditions. Bones will need to be cleaned in the laboratory in most cases. This process can be accomplished by allowing dermestid beetles to consume remaining desiccated tissue, cooking in detergent-treated water, defleshing of significant amounts of soft tissue, or a light brushing under running water. Cleaning occurs only after there has been a close examination of the remains by the anthropologist, and of course, after the pathologist has completed his or her analysis and the remains have been inspected for trace evidence. If identification is in question, untreated portions of long bones and teeth should be retained for DNA analysis.

During the processing and analysis, and until the remains are returned to the coroner/ medical examiner, all human remains are to be treated as evidence. This requires that they only be handled in secure conditions. Only individuals directly associated with the investigation should be given access to the remains, and they should never be used for demonstrations or teaching purposes. When the bones are no longer being actively examined, they should be securely locked in storage cabinets or safes. Storage in archival-quality acid-free boxes and paper bags is preferable to plastic containers, which frequently encourage the formation of mold on bones. Containers should be clearly labeled with the case number and contents.

Summary

The resolution of civil and criminal trials depends upon the continuity and protection of evidence, whether skeletal, entomological, or botanical. In the best of circumstances, an experienced evidence response team, such as those of the Federal Bureau of Invetigation, will work together to systematically search and recover the evidence, including the human remains. However, it is best to be prepared for the worst situation, and the forensic scientist should be ready to handle a larger share of the investigatory roles at the death scene. Preparation in terms of equipment and data forms and in terms of excavation skill, as well as knowledge of evidence collection procedures and the rationale behind recovery techniques, is essential. To include the forensic anthropologist early in the scene investigation and recovery operations will help to ensure maximum recovery and preservation of both the tangible and intangible evidence (Galloway et al. 1990). Together we can strive to obtain the best recovery of evidence that may be the victim's last chance for justice to be brought forth.

References

Bass, W. M. 1987. *Human osteology: A laboratory and field manual.* 3rd ed. Columbia, MO: Missouri Archaeological Society.

Behrensmeyer, A. K. 1975. The taphonomy and paleoecology of plio-pleistocene vertebrate assemblages east of Lake Rudolf, Kenya. *Bulletin of the Museum of Comparative Zoology*, Vol. 126, No. 10.

Dirkmaat, D. C., and J. M. Adovasio. 1997. The role of archaeology in the recovery and interpretation of human remains from an outdoor forensic setting. In W. D. Haglund and M. H. Sorg, Eds., *Forensic taphonomy: The postmortem fate of human remains.* Boca Raton, FL: CRC Press, 39–74.

Duncan, J. 1983. Search techniques. In D. Morse, J. Duncan, and J. Stoutamire, Eds., *Handbook of forensic archaeology and anthropology.* Tallahassee, FL: J. Duncan, 4–19.

Ebbesmeyer, C. C., and W. D. Haglund. 1994. Drift trajectories of a floating human body simulated in a hydraulic model of Puget Sound. *Journal of Forensic Sciences* 39:231–40.

Eliopulos, L. N. 1993. *The death investigators handbook: A field guide to crime scene processing, forensic evaluations, and investigative techniques.* Boulder, CO: Paladin Press.

Finnegan, M. 1995. Killed in action—Body not recovered: Forensic anthropology and archaeology applied to the recovery of U.S. military remains in Vietnam. Paper presented at the Annual Meeting of the American Academy of Forensic Sciences, Seattle.

France, D. 1998. Observational and metric analysis of sex in the skeleton. In: K. Reichs, Ed., *Forensic osteology: Advances in the identification of human skeletal remains.* Springfield, IL: Charles C. Thomas Press, pp163-86.

France, D. L., T. J. Griffin, J. G. Swanburg, J. W. Lindemann, G. C. Davenport, V. Trammell, C. T. Armburst, B. Kondratieff, A. Nelson, K. Castellano, and R. Hopkins. 1992. A multidisciplinary approach to the detection of clandestine graves. *Journal of Forensic Sciences* 37:1445–58.

Galloway, A., W. H. Birkby, T. Kahana, and L. Fulginiti. 1990. Physical anthropology and the law: Legal responsibilities of forensic anthropologists. *Yearbook of Physical Anthropology* 33:39–57.

Galloway, A., and J. J. Snodgrass. 1998. Biological and chemical hazards of forensic skeletal material. *Journal of Forensic Sciences* 43:940–48.

Gill, G., and S. Rhine. 2004. *Skeletal attribution of race.* Maxwell Museum of Anthropology. University of New Mexico Press. Albuquerque, NM.

Haglund, W. D., D. G. Reichert, and D. T. Reay. 1990. Recovery of decomposed and skeletal human remains in the "Green River Murder" investigation: Implications for medical examiner/coroner and police. *American Journal of Forensic Medicine and Pathology* 11:35–41.

Hall, D. W. 1988. The contributions of the forensic botanist in crime scene investigations. *The Prosecutor: Journal of the National District Attorneys Association* 22:35-38.

Hall, D. W. 1997. Forensic botany. In W. D. Haglund and M. H. Sorg, Eds., *Forensic taphonomy: The postmortem fate of human remains.* Boca Raton, FL: CRC Press, 353–62.

Hochrein, M. H. 1998. An autopsy of the grave: The preservation of forensic geotaphonomic evidence. In *Proceedings: American Academy of Forensic Sciences 50th Anniversary Meeting, San Francisco.* Vol. IV, p. 206.

Howard, J. D., D. T. Reay, W. D. Haglund, and C. L. Fligner. 1988. Processing of skeletal remains: A medical examiner's perspective. *American Journal of Forensic Medicine and Pathology* 9:258–64.

Johnston, R. E., and P. Steuer. 1983. Underwater crime scene investigation. In *Handbook of forensic archaeology and anthropology*, ed. D. Morse, J. Duncan, and J. Stoutamire. Tallahassee, FL: J. Duncan, 48–75.

Killam, E. W. 1990. *The detection of human remains.* Springfield, IL: Charles C. Thomas.

Lane, M. A., L. C. Anderson, T. M. Barkley, J. H. Bock, E. M. Gifford, D. W. Hall, D. O. Norris, T. L. Rost, and W. L. Stern. 1990. Forensic botany. *BioScience: American Institute of Biological Science* 40:34–39.

Leclercq, M. 1975. Entomologie et Médecine légale. Etude des Insectes et Acariens Nécrophages pour Déterminer la Date de la Mort. *Spectrum* 17:1–7.

Lipskin, B. A., and K. S. Field. 1983. *Death investigation and examination: Medicolegal guidelines/ checklists.* Colorado Springs, CO: Forensic Sciences Foundation.

Listi, G., M. Manhein, and M. Leitner. 2007. Use of global positioning system in field recovery of scattered human remains. *Journal of Forensic Sciences* 52:11–15.

Lundt, H. 1964. Ecological observations about the invasion of insects into carcasses buried in soil. *Pedobiologia* 4:158–80.

Mégnin, P. 1887. La Faune des Tombeaux. *Compte Rendu Hebdomadaire des Séances del' Académie des Sciences* 105:948–51.

Miller, P. S. 1996. Disturbances in the soil: Finding buried bodies and other evidence using ground penetrating radar. *Journal of Forensic Sciences* 41:648–52.

Morse, D., D. Crusoe, and H. G. Smith. 1976. Forensic archaeology. *Journal of Forensic Sciences* 21:323–32.

Motter, M. G. 1898. A contribution to the study of the fauna of the grave. A study of one hundred and fifty disinterments, with some additional experimental observations. *Journal of the New York Entomological Society* 6:201–301.

Ousley, S. R. Jantz, and P. Moore-Jansen. 2005. FORDISC 3.0 manual. Knoxville, TN: University of Tennessee Press.

Owsley, D. W. 1995. Techniques for locating burials, with emphasis on the probe. *Journal of Forensic Sciences* 40:735–40.

Payne, J. A., E. W. King, and G. Beinhart. 1968. Arthropod succession and decomposition of buried pigs. *Nature* 219:1180–81.

Pickering, R. B., and D. C. Bachman. 1997. *The use of forensic anthropology.* Boca Raton, FL: CRC Press.

Rhine, S. 1999. *Bone voyage: A journey in forensic anthropology.* Albuquerque: University of New Mexico Press.

Rodriguez, W. C., and W. M. Bass. 1983. Insect activity and its relationship to decay rates of human cadavers in east Tennessee. *Journal of Forensic Sciences* 28:423–32.

Schmitz, H. 1928. Occurrence of phorid flies in human corpses buried in coffins. *Natuurhistorisch Maandblad* 17:150.

Schultz, J., M. Collins, and A. Falsetti. 2006. Sequential monitoring of burials containing large pig cadavers using ground penetrating radar. *Journal of Forensic Sciences* 51:607–15.

Skinner, M. S., and R. A. Lazenby. 1983. *Found! Human remains: A field manual for the recovery of the recent human skeleton.* Burnaby, British Columbia: Archaeology Press, Simon Fraser University.

Staggs, S. 2003. The admissibility of digital photographs in court. www.crime-scene-investigator.net/admissibilityofdigital.html

Stoutamire, J. 1983. Excavation and recovery. In *Handbook of Forensic Archaeology and Anthropology,* ed. D. Morse, J. Duncan, and J. Stoutamire. Tallahassee, FL: J. Duncan, 20–47.

Tolhurst, W. 1991. *The police textbook for dog handlers.* Sanborn, NY: Sharp Printing.

Ubelaker, D., B. Buchlotz, and J. Stewart. 2006. Analysis of artificial radiocarbon in different skeletal and dental tissue types to evaluate date of death. *Journal of Forensic Sciences* 51:484–88.

Walsh-Haney, H., C. Katzmarzyk, and A. Falsetti. 1999. Identification of human skeletal remains: Was he a he or was he a she? In: S. Fairgrieve, Ed. *Forensic osteological analysis: a book of case studies.* Springfield, IL: Charles C. Thomas Press, 17–35.

Willey, P. 1998. Personal communication—report on 1997 annual update forms for ABFA.

Wolf, D. J. 1986. Forensic anthropology scene investigations. *Forensic Osteology,* ed. K. J. Reichs. Springfield, IL: C.C. Thomas, 3–23.

Appendix I: Anthropology, Entomology, and Botany Equipment Lists

Recommended tool kits for use during field recovery are items listed below. Add or delete items depending upon agency budget, preference, and time allowances.

Anthropology

Search Equipment

- Probes
- Flags
- Walkie talkies
- Rope
- Leather gloves
- Knee pads

Excavation Equipment

- Pointed and square shovels
- Mason's trowels (4.5 or 5 inch)
- Pruning shears
- Saw
- Metal stakes
- Probes (2)
- Evidence flags (various colors)
- Dustpan (2 or 3)
- Small camper's shovel
- $^1/_8$- and ¼-inch screens
- 5-gallon buckets (3 or 4)
- Visual markers
- Tarps (for excavated soil)
- Metal detector
- Wheelbarrow
- Flashlight(s)
- Paintbrushes (½ to 4 inches wide)
- Rake (2–4)
- Dental picks
- Wooden picks
- Scene tape
- String, twine, rope, or surveyor's tape
- Brightly colored spray paint (red is suggested because it contrasts well with grass and shrubbery)
- Thin wooden sticks with one end painted to mark positive metal detector readings

Supplies for Transportation

- Paper evidence bags (various sizes)
- Plastic evidence bags (various sizes)
- Cardboard boxes
- Foam wrap/bubble wrap (acid-free)

- Plastic containers with lids
- Integrity tape
- Indelible markers
- Biohazard waste disposal bags
- Body bags

Mapping Equipment

- Compass
- 50 m tape measure
- 10 m tape (2)
- Two-way line level, transit, or farmer's level with stadia rod
- GPS equipment
- Plumb bob
- Graph paper
- Protractor
- Stakes
- String (heavy gauge)
- Pencils
- Erasers
- Pencil sharpener
- Metric straightedge
- Flagging pins
- Flagging tape
- Long nails (6 to 8 inches)
- Nails for stakes
- Drafting compass

Documentation and Field Analysis

- Notepaper
- Transfer of custody forms
- State map
- Anthropometer
- Inventory form
- Sliding calipers
- Spreading calipers
- Osteometric board
- Stature chart
- Soil cards
- Photography log
- Indelible pen (2 or 3)
- Indelible marker (2)
- Pencils
- Pencil sharpener
- Clipboard
- Hand lens
- FORDISC 3.0 data bank forms

Photographic Equipment

- 35 mm camera (2)
- Digital camera
- Flash unit
- Slide film
- Color print film
- Normal lens
- Extra camera and flash batteries
- Photo board or slate board with chalk
- North arrow
- Metric scales
- Video camera
- 50 mm macro lens
- Ladder

Personal Equipment

- Disposable gloves
- Antimicrobial cleansing soap
- Insect repellent
- Sunscreen
- Ivy block
- Drinking water
- Cell phone
- First aid kit
- Tarps for protection from the elements
- Toilet paper
- Transportation vehicle
- Identifying clothing
- Boots

Entomology

- Collapsible insect net and 12-inch extension handle
- Mason's trowel
- "Whirlpack" specimen bags, 6 × 9 inches
- Fine-point forceps, curved tip
- Fine-point forceps, straight tip
- Soft, featherweight forceps
- Glass vials with polyseal caps
- Surgical gloves
- Dual-scale thermometer, 6 inches
- Self-adhesive labels
- Nonadhesive labels
- Preservation and collection chemicals
- Paper towels
- Disposable gloves

- ¼-inch screen
- Styrofoam shipping containers
- Death scene form
- Aluminum foil
- Vermiculite
- Liquid-approved container with 70–80% ethyl alcohol
- Distilled water
- Indelible marker
- Pencil
- Pencil sharpener
- Eraser

Botany

- Pruning shears (2 pair)
- Sharpening stone
- Trowel (one per collector)
- Pointed shovel
- Indelible marker (2)
- Paper bags (various sizes)
- Plastic containers with lids
- Cooler
- Ice packs
- Surgical gloves
- Gardening gloves (leather)
- Ivy block

Appendix II: Photography Log

Case number: _____   Date: _____

ME number: _____   Time: _____

Photographer: _____   Recorder: _____   Page ___ of ___

Film Roll No.	Item No.	Description	Location	Compass Direction (degrees)

Appendix III: Evidence Log

Case no.: _____   Date: _____

ME no.: _____   Time: _____

Recorder: _____   Location: _____

Collector: _____   Page ____ of ____

Item No.	Item Description	Distance from Datum (meters)	Compass Direction (degrees)	Grid Unit (if applicable)	Elevation/ Depth (if applicable)

Appendix IV: Transfer of Evidence

Case no.: _____

ME no.: _____

The following items:

Item No.	Description

Transferred from	Time/Date	Transferred to	Time/Date

Appendix V: Botanical Evidence Collection Sheet

Recorder: _____   Case no.: _____

Agency: _____   Date: _____ Time: _____ Page___ of ___

Item No.	Tree/Grass/ Shrub/Other	Color Description	Distance from Datum (meters)	Compass Direction (degrees)	Location Relative to Body or Other Physical Evidence	Relative Frequency

Estimating the Postmortem Interval

9

JEFFREY D. WELLS
LYNN R. LAMOTTE

Contents

Introduction

The time elapsed since death, or postmortem interval (PMI), is a matter of crucial importance in investigations of homicides and other untimely deaths. Such information can help to identify both the criminal and the victim by eliminating suspects and connecting the deceased with individuals reported missing for the same amount of time (Catts 1990, Geberth 1996). Even when the cause of death is natural, the time of death can have important implications for legal matters such as inheritance and insurance (Henssge et al. 1995). Crucial information, such as when the deceased was last seen alive, may indicate a maximum possible PMI, but this can occur only if he or she has been identified. Often, however, it is the condition of the body itself that must tell us when death occurred (Henssge et al. 1995).

Postmortem changes in a body depend upon many factors (Micozzi 1991), and the PMI can be a remarkably difficult thing to determine (Bass 1984). Obviously, any physical or biological change that is a function of time since death provides a potentially useful clue in this matter. Initially, the predictable physical and chemical consequences of death are usually the most reliable PMI indicators (Henssge et al. 1995). But, as the time since death increases, the above methods become less useful, and more accurate results are often obtained using ecological information. A decomposing body can dramatically alter both the behavior and composition of species at a site. A cadaver attracts a variety of vertebrate and invertebrate scavengers (Putman 1983), while the products of decay can produce changes in the underlying soil flora and fauna (Bornemissza 1957).

It has long been observed that insects associated with vertebrate carrion display PMI-dependent processes (Hall 1990). One of these processes is the development of insect species whose larvae consume dead tissue. Flies in the families Calliphoridae, Sarcophagidae, and Muscidae are the most noticeable because of size, number, and ubiquity. These are frequently succeeded on carrion by beetles in the families Silphidae and Dermestidae (Smith 1986). For convenience, in this chapter we will use terms appropriate for the most common forensic situation. Typically, this is a case in which the evidence consists of *insects* collected at a *crime scene* from the body of what is suspected to be a *victim* of a homicide. The discussion, however, applies equally well if noninsect arthropods are collected, or if no crime is thought to have been committed.

The estimated age of an immature insect that has fed on a body provides a minimum PMI because, with very rare exception (see below), adult females do not deposit their offspring on a live host. In its most conservative form, such an approach does not estimate the maximum PMI, because an unknown period of time may elapse between death and the deposition of eggs or larvae. Depending on the insect species and conditions at the scene, the degree of development can indicate a PMI from less than 1 day to more than 1 month (Smith 1986).

Another PMI-dependent process is the succession of arthropod species found on and within a body (Schoenly and Reid 1987). In contrast to larval development, a succession model includes information about the time elapsed between death and the appearance of a particular arthropod species and stage. It can therefore be used to estimate both the minimum and maximum PMI (Schoenly et al. 1992). According to our definition, the simplest succession model is used when an investigator estimates both the age of a larva and the time interval between death and that individual insect's arrival at the body. Succession data have been used to very accurately calculate a PMI as large as 52 days (Schoenly et al. 1996), and could be applied to a much greater interval.

Reference Data

Faced with the need to estimate a portion of the PMI from entomological data, the investigator must select a model of insect development or succession. This can involve a comparison to other death investigations, or experimental data may be generated *a posteriori* to match a case (Goff 1992, Introna et al. 1989), but most often previously existing experimental data are consulted. Larval growth rates are usually studied using small containers in the laboratory. Larval growth in such an artificial setting suitably mimics growth in the field as long as certain other conditions match those of the scene (see below). Developmental data have been gathered for a large and constantly growing list of species and conditions (e.g., Byrd and Butler 1998, Goodbrod and Goff 1990, Greenberg 1991, Greenberg and Wells 1998, Introna et al. 1989, Nishida 1984, Reiter 1984, and many others).

Carrion succession is a classical subject in ecology (Fuller 1934, Payne 1965, Reed 1958), and much of the recent work has had a forensic objective. While some experiments have involved human corpses (Rodriguez and Bass 1983, 1985), these are extremely difficult to obtain legally, and most published information on insects involved in human decay comes from case studies (Lord 1990, Smith 1986). As useful as these may be, they are usually just a "snapshot" of the process. Therefore, researchers usually employ nonhuman carcasses (such as pigs) in order to include replication and repeated sampling in a succession study (e.g., Anderson and VanLaerhoven 1996, Hewadikaram and Goff 1991, Tantawi et al. 1996, and many others).

Factors That Influence Carrion Insect Development and Succession

In order to choose an appropriate model of growth or succession, conditions at the crime scene and the manner in which the specimens were handled must be determined. The closer the match between conditions at the scene and those used to generate reference data, the less margin of error in estimating the entomological-based portion of the PMI. If possible, the entomologist should visit the site, consult the reports of other investigators, and obtain the most reliable weather records (Haskell and Williams 1990).

Many biotic and abiotic factors are known to influence carrion insect growth and activity. Determining these factors and their effects has been the most active area of research in forensic entomology. The following factors are of particular importance. This is, however, not meant to be a comprehensive list, and investigators should carefully consult the primary literature and consider all biological information about any species used for analysis.

Individual Species Characteristics

Accurate identification of samples is usually the first priority in a forensic analysis of the entomological evidence (see Chapters 2 and 19). The most important implication for PMI estimation is that carrion insect species differ in terms of growth rate, arrival time, and position within the order of succession. However, other relevant physiological or natural history factors can only be considered following proper identification. For example, although it is generally known that Sarcophagidae deposit only live larvae (Shewell 1987), certain common Calliphoridae show a limited version of this behavior. Erzinclioglu (1990) described the retention of a single fertilized egg by a gravid *Calliphora vicina* Rodineau-Desvoidy. This egg can hatch immediately after oviposition, and the resulting larva will be "precociously" developed compared to its siblings. Such precocious eggs have been found to be common among blue bottle (Calliphorini) species, but not other calliphorid taxa in northern California (J. D. Wells and J. King, unpublished).

Weather and Seasonality

Temperature has a profound effect on insect metabolic and development rate (Andrewartha and Birch 1954, Chapman 1982). Generally, within a certain range of temperatures, development is accelerated as temperature is increased, but this does not hold true at temperature extremes that may prove lethal to the insect. Both air temperature and exposure to sunlight will affect corpse temperature, and these are important considerations that require thorough documentation in all death investigations involving entomological evidence.

Some species enter larval or pupal diapause (arrested development) in response to seasonal cues (Denlinger and Zdarek 1994), and this can greatly increase the time spent in that particular life stage, even if temperatures are relatively warm. However, some closely related species may display quite different diapause behavior (Ash and Greenberg 1975), thus illustrating the need for comprehensive knowledge of insect life history for PMI estimation.

Presence of a Maggot Mass

Carrion fly larvae can have a metabolic and feeding rate greater than most immature insects (Hanski 1976, Levot et al. 1979). A large number of larvae in the same location can generate a temperature substantially higher than ambient, causing what is commonly termed the maggot mass effect (Cianci and Sheldon 1990, Goodbrod and Goff 1990, Greenberg 1991, Turner and Howard 1992). Because larval growth is inhibited as lethal high temperatures are approached, the difference between ambient and maggot mass temperatures is likely to be greatest during cold weather (e.g., Deonier 1940). However, this is a complicated phenomenon, since the oldest larvae on the body may have developed prior to any elevation in temperature. In all cases, the temperature and size range of larvae in the mass should be recorded at the scene before the body is disturbed.

Food Type

Some carrion fly larvae can develop on a range of food types. The most extreme example is the phorid *Megaselia scalaris* (Loew), which can eat live or dead invertebrates, as well as a variety of other organic materials, and has even been found feeding on paint (Disney 1994). *Lucilia sericata* Meigen grew more slowly in a vegetable medium than when larvae were fed meat (Povolny and Rozsypal 1968). Therefore, it seems likely that an extreme change in diet would affect development rate in other species as well. Occasionally, a forensic entomologist may be asked to age larvae from something other than a corpse. For example, a vendor may be liable if larvae in contaminated food are shown to have been present at the time of sale, and larvae from feces in a diaper may suggest a case of abuse or neglect (Goff et al. 1991). It is important to recognize that development rates observed for a meat medium may not be appropriate for such cases.

Drugs and Other Toxins

Chemicals in or on the victim, such as might accompany drug overdose or suicide, may have a variety of effects on carrion insects (Goff and Lord 1994) (see Chapter 12). Insect-mediated decay processes can be accelerated or decelerated depending on the substance and concentration. The presence of such substances may be indicated by a toxicological analysis of victim or insect tissue, or by other evidence, such as a container at the crime scene.

Geographic Region

Little information is available on possible regional variation in forensic entomological phenomena. Although the development of some common species has been studied by laboratories in widely separated locations (e.g., see discussion of *Chrysomya megacephala* [Fabricius] in Wells and Kurahashi 1994), differences in the methods used and in presentation of the data make it difficult to interpret and compare the results. In one study designed to test the possibility of regional variation, Cyr (1993) found no statistically significant difference in the duration of developmental stages of *Phormia regina* (Meigen) from the states of Washington, Indiana, Texas, and Louisiana.

In contrast, a number of observations suggest that carrion fly behavior can indeed vary according to region. For example, throughout most of its distribution *Chrysomya rufifacies* (Macquart) invades carrion or live vertebrates only after other larvae are present (Bohart and Gressitt 1951, Wells and Greenberg 1994, Zumpt 1965), but in particular sites it has been observed to lay eggs on fresh carrion (O'Flynn and Moorehouse 1979) and on uninfested newborn calves (Shishido and Hardy 1969).

Whether any regional differences, if they exist, have a genetic basis or simply reflect a behavioral response to different environmental conditions is unknown. Although it is possible for a phenotypic gradient to develop from natural selection (one well-studied example is the diapause response) (Kurahashi and Ohtaki 1989), any regional genetic variation is unlikely to be the result of isolation. The larger carrion flies are extremely mobile, and within the limits set by preferences for general types of habitat, they are commonly observed to disperse several kilometers in one day (Baumgartner and Greenberg 1984, Norris 1965). *Cochliomyia macellaria* (Fabricius) typically survives the winter only as far north as southern Texas and Florida, yet it spreads throughout almost the entire contiguous United States and into Canada each year (Hall 1948). Such behavior should promote extensive gene flow between locations, and the limited number of molecular genetic studies done so far suggest that widespread carrion-feeding species encounter no barriers to gene flow over distances of thousands of kilometers (Stevens and Wall 1997, Taylor et al. 1996).

Preservation Method

Most plots of larval growth (growth curves) describe the change in body length with age. Typically, larvae are killed and preserved in some fluid prior to measurement. The type of fluid and the method of processing influence the length of the preserved larva (Tantawi and Greenberg 1993). These differences must be taken into account if the insect evidence is preserved using methods other than those of the reference study.

Did Insect Colonization Precede Death?

There are exceptions to the rule that carrion insects are only attracted to a dead body. Infestation of a live vertebrate by fly larvae is called myiasis (James 1947, Zumpt 1965). Although this situation is extremely rare, if a person was already infested when death occurred, then insects present at that time could cause one to overestimate the PMI. The larvae of many carrion fly species have occasionally been found feeding in necrotic wounds, where they can actually promote healing (Sherman and Pechter 1988). Investigators should be alert to contributing factors, which include a medical history of a persistent wound, such as might result from surgery, cancer, or diabetes, and physical or mental incapacitation (Greenberg 1984, Hall et al. 1986, Seaquist et al. 1983).

Deviations from the common decay pattern, in which the head is the first site of infestation, *may* indicate antemortem trauma and insect activity. As mentioned above, larvae may initially feed on feces within clothes. This would be suggested if the deceased wore diapers, or if larvae in the anal region were the oldest present. However, we stress that the typical head-down pattern of decomposition documented on humans provides only a suggestion of normal decay. Postmortem insect feeding can begin anywhere other than the head, such as when insects are attracted to a wound, and it is also important to understand that antemortem activity can also occur in the head. Some very close relatives of carrion flies, the screwworms *Chrysomya bezziana* Villeneuve and *Cochliomyia hominivorax* (Coquerel), will only deposit eggs on a live host. A nonspecialist may find it difficult to distinguish the immature stages of these species from those commonly used as forensic evidence (Spradbery 1991). Humans may be killed by an infestation that started in the nasal sinuses (Spradbery 1994), and this is a situation that could be misinterpreted as the normal pattern of decay if discovered after death, although we are not aware of this having happened. In most cases, the proper identification of specimens will prevent such a mistake.

Estimating the Postmortem Interval

In almost all cases, a sample of the collected insects are killed and preserved prior to being used to estimate the entomological-based portion of the PMI, while others within the sample or similar sample may be reared to the adult stage for identification. The moment of preservation is the point in time from which one calculates backward to the time of death, and it is of critical importance to document this time for the evidentiary record. The circumstances of every death investigation are unique, and there is no single best algorithm for PMI estimation. Both the conditions at the scene and the quality of the data available to the entomologist vary widely.

It almost goes without saying that one must make certain assumptions in order to reach any estimation of PMI. These may reflect some obvious lack of information, such as whether there was a maggot mass, or exactly how the remains were handled between the time of discovery and the time insect samples were collected. Frequently, collection of the insect samples occurs during the autopsy, which may take place many hours after the discovery of the body. However, even if the circumstances of the case appear to be easily reconstructed, one must usually assume that information such as that contained in other investigators' reports is correct. An opposing attorney may challenge these assumptions, however reasonable, and the forensic entomologist may be required under cross-examination to describe how the PMI estimate would be different if different assumptions were made (see Chapter 14). It is best to think of these things in advance so that an adequate defense may be prepared.

As a general approach, the analysis should begin by reaching those conclusions that require the least amount of knowledge of the scene. Perhaps the most conservative conclusion is an absolute lower limit on specimen age. Each insect species has a range of conditions under which development rate is highest. If known, this optimum rate provides a lower limit on age; i.e., a larva of a given size or instar cannot be younger than the time it takes to reach that state under optimal conditions.

Estimating PMI from Degree of Development

If the reference model for species development is a growth curve, then the best estimate of age for a larva is the value corresponding to its size on the curve. That is, a horizontal line from the value of the length or weight of the larva will intersect the growth curve directly above its age. This calculation is likely to be most precise if the intersection occurs where the growth curve is most steep, because a small change in size results in only a small change in the estimation of age (Figure 9.1). Maggot growth curves, particularly those showing weight as a function of age, may form an S shape, with slow growth during the first two larval intars and a slow decrease in size between the cessation of feeding by the third instar and the onset of pupariation. Within these flat regions of the curve other information is likely to be as useful as size for estimating age.

Larvae of the same age hatch and molt in relative synchrony (Davies and Ratcliffe 1994, Kamal 1958, Wells and Kurahashi 1994, Wells and LaMotte 1995), and these qualitative changes in form are easily recognized. The age of an individual insect from the crime scene is likely to fall within the range of age for which the same instar has been observed in the model. Other information, such as the presence of a pharate larva (the next growth stage that is visible through the cuticle immediately prior to molting), will narrow the range. Unfortunately, the postfeeding stage can last as long as the rest of larval development, and lacks such abrupt changes in appearance. Postfeeding larvae undergo an emptying

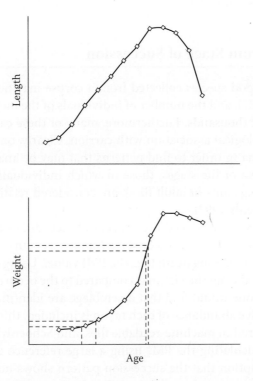

Figure 9.1 Hypothetical maggot growth curves showing the shapes often observed for the change in length or weight between egg hatching and pupariation. Dashed lines illustrate the fact that a prediction of age based on size is likely to be most precise where the curve is most steep, because at that point an error in measurement results in a relatively small change in the estimate of age.

of the crop, and the physical length of the crop provides some information about the time since feeding ceased (Greenberg 1991). However, pupariation, metamorphosis, and adult emergence produce dramatic morphological changes that may be used to estimate age (Greenberg 1991).

Although many development studies employed constant temperatures, thermal conditions at the scene will almost certainly have fluctuated to some extent. Data from growth at constant temperatures may be used for such situations by dividing the time period under consideration into short intervals (e.g., 12 hours [Williams 1984]), and then applying a model of development that is closest to the mean temperature during each period. Williams (1984) recommended a 10% correction factor for this method to accommodate the fact that the *Lucilia illustris* (Meigen) grew more slowly under fluctuating temperatures than was predicted by growth at constant temperatures (Hanski 1976). It appears, however, that this is not a universal effect, and development at fluctuating temperatures can be accelerated, retarded, or the same as development at a constant temperature with the same mean value (Davies and Ratcliffe 1994). Some development studies have included temperature fluctuations designed to mimic a typical diurnal cycle (Byrd and Butler 1996, 1997, 1998, Davies and Ratcliffe 1994).

Another approach has been to model development in terms of accumulated degree-hours or degree-days, a process known as thermal summation (Wigglesworth 1972). This method is described in greater detail in Chapter 10, and is the method used to create the computer model of insect growth.

Estimating PMI from Stage of Succession

The number of arthropod species collected from a corpse may number in the hundreds (Schoenly and Reid 1987), and the number of individuals of the most common species can easily be in the tens of thousands. Furthermore, many of these can be incidental species with no particular ecological association with carrion. Clearly, one must focus on only a subset of the total fauna in order to find patterns that may be analyzed for forensic purposes. Reoccurring taxa or life stages, those in which individuals may visit and leave a corpse several times (e.g., ants or adult flies), are considered relatively poor indicators of time since death (Schoenly 1992).

Schoenly et al. (1992) introduced the concept of recording succession as an "occurrence matrix." In this system, a species or stage is noted as being either present or absent at a given point in time following death (i.e., the PMI value). Using this system, the assemblage of species collected from the victim is compared to the occurrence matrix, and those PMI values for which one would find that assemblage are identified. Although no information about the relative abundance of each taxon is included, this system has the advantage of being easily stored in machine-readable files, and Schoenly et al. (1992) developed software capable of calculating the PMI using a large reference data set (provided that one accepts the assumption that the succession pattern shows no random variation for a given set of conditions—see the following). Furthermore, adoption of a standardized record-keeping method such as an occurrence matrix would prevent the difficulties we have found in comparing the results of succession papers that present their results in narrative form (see the following).

What If Conditions from the Scene Don't Match Any Experiment?

Conditions at the crime scene sometimes do not closely resemble those of any experiment on development or succession. In such a case, one must use qualitative judgement as to whether ecological processes proceeded at a faster or slower rate than for a chosen model. Investigators should develop their own baseline data for the species and conditions likely to be encountered at their location.

Dealing with Inherent Uncertainty in Growth and Succession

As with any natural phenomenon, there is random variation in forensic entomological phenomena. The common practice of using only larvae that are relatively big for their age to construct a growth curve suggests to us that most forensic entomologists realize that even under the same conditions not all larvae grow at the same rate. However, only a few laboratory studies have reported the complete range in size found among larvae of equal age (Wells and Kurahasi 1994, Wells and LaMotte 1995). These studies have shown that the larval size distribution is skewed toward the low end of the scale (Figure 9.2). In other words, there are a few extreme "runts," and this is particularly noticeable with third instars. Similar measurements of wild populations are almost nonexistent, but the single sample

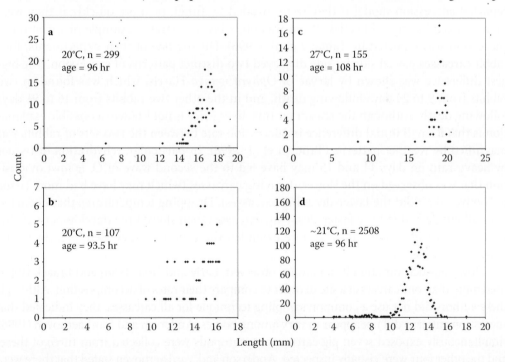

Figure 9.2 Size distribution for calliphorid larvae of equal age. (a) *Calliphora nigribarbis*, Tokyo, Japan. (b) *Aldrichina grahami*, Tokyo, Japan. (c) *Chrysomya megacephala*, Bangalore, India. (d) *Chrysomya megacephala*, Okinawa, Japan [84]. (a–c) were produced in the laboratory using the methods of [84]. Larvae for (d) were produced by exposing a rat carcass to oviposition by wild flies for a 2-hour period. The carcass was then held at a shaded outdoor location, and the larvae were preserved by flooding the carcass with ethanol.

we have taken fits this same pattern (Figure 9.2d). It may be that the stunted individuals are moribund (Davies and Ratcliffe 1994), and the fact that they have a relatively small cephalopharyngeal skeleton (Wells, unpublished) indicates that they are already under-sized at the time of the final larval molt.

Similarly, little is known about random variation in succession rate. Perhaps this is not surprising given the need for replicate bodies in order to measure such variation, and the effort that succession studies can entail. The replication of experimental conditions quickly leads to a huge amount of work, and the intensive sampling and analysis of even a couple of large carcasses can produce enough data for a master's thesis (Hewadikaram and Goff 1991). Although some ecological studies of carrion involved a large number of carcasses, the authors did not describe their methods in enough detail to know what conditions might have been replicated (Abell et al. 1982, Fuller 1934, Johnson 1975). Other authors who clearly used more than one carcass under the same conditions and sampled in the same manner presented their results as a composite "typical" succession pattern, with no information about random variation (Bornemissza 1957, Cornaby 1974, Hall and Doisy 1993, Nabaglo 1973, Reed 1958).

Despite some difficulties in interpreting the older papers, there are strong indications that dramatic variation in succession patterns can occur. It is interesting that Reed (1958) excluded some carcasses from analysis because they were "atypical" (his quotation marks), while Nabaglo (1973) mentioned that some carcasses decomposed more slowly than the reported succession model if they were invaded by fungi, or more quickly if they were monopolized by silphid beetles. What was probably an extreme example of variation in succession was reported by Tantawi et al. (1996). During one of their experiments, four rabbit carcasses placed side by side displayed two distinct patterns of succession. The biggest difference was shown by larvae of *Ophyra ignava* Harris, which was found in two rabbits from 9 to 23 days following death, and in the other two rabbits from 16 to 91 days following death. Although the reason for this divergence is not known, a possible explanation is that a small initial difference in succession rate between the two sets of rabbits was magnified by weather patterns (Tantawi et al. 1998). In particular, rehydration of tissues by heavy rain on days 14 and 15 may have led to the second wave of *O. ignava* oviposition that was observed on the slower-decaying carcasses (which may have had more tissue to "revive" than did the faster-decaying carcasses). Dropping temperatures that occurred after all larvae had left the faster-decaying carcasses then slowed the development of this second cohort of larvae, and therefore extended the time that they were present in the remaining carcasses.

However, such variation is not always observed. Early and Goff (1986) and Braack (1987) used more than one carcass at a site in order to compare their rates of decomposition. Although these authors did not use a common sampling technique for all carcasses, they indicated that succession patterns did not appear to vary among bodies. Anderson and VanLaerhoven (1996) simultaneously exposed seven pig carcasses. Arthropods were collected from three of these, and the other four were visually inspected. Anderson and VanLaerhoven stated that there were no differences among the pigs in the daily changes in gross morphology (e.g., bloating), and that the timing of colonization by individual arthropod species varied by no more than one day.

The implications of such variation for the conclusions reached by forensic entomologists must be addressed. More specifically, we believe that the field of forensic entomology will benefit from the development of the statistical reasoning that is so common in other areas of science. Statistical methods are essential for establishing the precision of a PMI estimate and for evaluating any conflict of opinion among experts (that is, different PMI estimates may not, in fact, be significantly different). Furthermore, we believe that they will help to direct forensic entomology research, because different techniques for PMI estimation may more easily be evaluated according to the precision of the conclusion. In the following sections, the efforts made by Wells and LaMotte to develop statistical approaches to PMI estimation based on insect evidence will be described.

Statistical Considerations and Methods

The technical problems involved in estimating PMI and constructing a confidence interval on it are different when dealing with development data or succession data. Within life stages, development occurs smoothly and continuously with age, while the set of species present is categorical and changes discretely with time. These differences correspond to distinctly different sets of statistical procedures, those dealing with quantitative or continuous measurements and those dealing with categorical measurements.

In spite of these fundamental differences, the underlying rationale is the same in both settings. The factual basis is similar: in both, experimental material is measured at several points in time. Denote this reference data symbolically as $Y(t_1)$, $Y(t_2)$, ..., $Y(t_k)$, where $Y(t_i)$ denotes the reference data gathered at time t_i. Information (call it y) is collected at the crime scene, such as weight of a maggot or a checklist of species present or absent. The datum y is compared to the reference data $Y(t_i)$ to assess whether y and $Y(t_i)$ are consonant. This assessment is in terms of the question "What is the probability that results as different as y and $Y(t_i)$, or more so, would occur if experimental material for both were measured at time t_i?" Times for which this probability (called a p-value) is not small (a 5% level of significance often is used as the cutoff between small and not small) are tenable in light of the data, while times for which this probability is small are untenable. Under appropriate assumptions, the set of times found tenable in this way constitutes a confidence set on the time t from which y resulted.

Gaps between times for which reference data are available may be wide. Despite the lack of information in such gaps, it may be reasonable to assume that development or succession characteristics change smoothly within this time interval, so that some form of interpolation can be justified. This will always require models and assumptions, the validity of which may or may not be possible to assess. For the sake of this discussion, though, denote this interpolated information for time t by $Y(t)$; in general, $Y(t)$ depends on all the reference data, but it may interpolate only information from the two times that bracket t.

Now the statistical method for constructing an interval estimate on t may be seen in general. For each t in some range of possibilities, assess the statistical significance of the difference between y and $Y(t)$. Those values of t for which y and $Y(t)$ are significantly different are rejected as being untenable in light of the data; those for which the difference is not significant form a confidence set on t. This process is described separately for development data and succession data.

An Interval Estimate of Age from Development Data

We assume that, among the population of subjects of age t, the distribution of the development characteristic y (such as weight or length) is approximately normally distributed with mean μ_t and variance σ_t^2. The reference data comprise measurements of y on independent samples of n_i subjects at times t_i, $i = 1, \ldots, k$. Denote the sample means by Y_1, DOTSLOW, Y_k and the sample variances by s_1^2, dotslow, s_k^2. From this data set, estimates μ_t of μ_t and σ_t^2 of σ_t^2 can be calculated for each t. We shall assume that $\hat{\sigma}_t$ is a linear combination of the sample means $Y_1, \ldots, Y_k$:

$$\mu_t = \hat{\sigma}_{i=1}^k a_{ti} Y_i$$

Then the estimate of the variance of μ_t is

$$\mathrm{Var}(\mu_t) = \sigma_{i=1}^k \frac{a_{ti}^2}{n_i} \sigma_{t_i}^2 = \sigma_{i=1}^k \frac{a_{ti}^2}{n_i} s_i^2$$

We assume that σ_t^2 is a linear combination of the sample variances:

$$\sigma_t^2 = \hat{\sigma}_{i=1}^k b_{ti} s_i^2$$

Given y from a mystery specimen, the test of the hypothesis that y came from a population with mean μ_t and variance σ_t^2 is based on the difference $y - \mu_t$ and

$$\mathrm{Var}(y - \mu_t) = \sigma_t^2 + \mathrm{Var}(\mu_t) = \hat{\sigma}_{i=1}^k \left(b_{ti} + \frac{a_{ti}^2}{n_i} \right) s_i^2$$

The test statistic is

$$T = \frac{y - \mu_t}{\sqrt{\sigma_t^2 + \mathrm{Var}(\mu_t)}}$$

Unless the population variances are equal (an unrealistic assumption in this setting), T does not follow a Student's t distribution. Satterthwaite's approximation (Kotz et al. 1988) can be used, in which probabilities for T are approximated from a Student's t distribution with degrees of freedom computed as

$$df_t = \frac{(\hat{\sigma}_{i=1}^k c_{ti} s_i^2)^2}{\hat{\sigma}_{i=1}^k \dfrac{(c_{ti} s_i^2)^2}{df_i}}$$

where $c_{ti} = b_{ti} + a_{ti}^2/n_i$ and $df_i = n_i - 1$, $I = 1, \ldots, k$. The hypothesis is rejected if the p-value computed from T is less than the chosen level of significance (say 5%); otherwise, the proposition is tenable in light of the data. By a result in Lehmann (1986), the set of times t for which this proposition is tenable forms a 95% confidence set on the age of the mystery specimen.

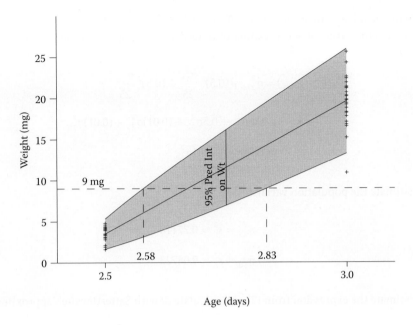

Figure 9.3 Simulated dry weights (mg) of *Cochliomyia macellaria* larvae by age (days), twenty-five each at 2.5 and 3.0 days, plotted as +. The region swept by 95% prediction intervals on weight at each age is shaded. Means and variances are interpolated linearly. The ages 2.58 and 2.83 days, at which the horizontal line at the weight 9 mg of the mystery specimen intersects the curves of upper and lower prediction limits, are the endpoints of an approximate 95% confidence interval on the age of the mystery specimen.

Example

Data for this example were simulated to correspond to the data in Wells and LaMotte (1995). They represent dry weights (mg) of larvae of *Cochliomyia macellaria*, twenty-five each collected at ages 2.5 and 3.0 days, under controlled laboratory conditions. Figure 9.3 shows scatterplots of the two sets of weights, plotted as +. A linear interpolation model is assumed for the means μ_t (corresponding to the middle line in Figure 9.3 joining the two sample means) and variances σ_t^2 of weights for ages between 2.5 and 3.0 days. It is assumed further that weights at time t are normally distributed with mean μ_t and variance σ_t^2. Based on the data and these assumptions, a 95% prediction interval on weight y at time t can be constructed. This interval is shown in Figure 9.3 for $t = 2.75$ days. The shaded area is the area swept by all such prediction intervals for t between 2.5 and 3.0 days.

Summary statistics for the simulated reference weights of *Cochliomyia macellaria* larvae are listed in the following table:

t_i	n_i	Y_I	s_i^2	df_i
2.5	25	3.4880	0.77110	24
3.0	25	19.6160	9.08723	24

Suppose the weight of a mystery specimen of *Cochliomyia macellaria* is 9 mg. Is it reasonable to think that such a larva might be 2.75 days old? Denote by μ the mean weight in the population from which the mystery specimen came. The t-statistic for testing $H_0 \, (\mu - \mu_{2.75} = 0)$ is constructed as follows:

1. Estimate $\mu - \mu_{2.75}$ by $y - \mu_{2.75} = 9 - (0.5Y_{2.5} + 0.5Y_{3.0}) = -2.552$.
2. Express the variance of this estimator in terms of the population variances:

$$\text{Var}(y - \mu_{2.75}) = \sigma^2_{2.75} + (0.5)^2 \frac{\sigma^2_{2.5}}{25} + (0.5)^2 \frac{\sigma^2_{3.0}}{25}$$

$$= 0.5\sigma^2_{2.5} + 0.5\sigma^2_{3.0} + (0.01)\sigma^2_{2.5} + (0.01)\sigma^2_{3.0}$$

$$= 0.51\sigma^2_{2.5} + 0.51\sigma^2_{3.0}$$

3. Estimate the population variances:

$$\sigma^2_{2.5} = s^2_1 = 0.77110$$

$$\sigma^2_{3.0} = s^2_2 = 9.08723$$

4. Estimate the expression from (2) and calculate df with Satterthwaite's approximation:

$$\text{Var}(y - \mu_{2.75}) = 0.51\sigma^2_{2.5} + 0.51\sigma^2_{3.0} = 5.02770$$

$$df_{2.75} = \frac{(0.51s^2_1 + 0.51s^2_2)^2}{\dfrac{(0.51s^2_1)^2}{df_1} + \dfrac{(0.51s^2_2)}{df_2}} = 28.04$$

The test statistic is

$$T = \frac{-2.552}{\sqrt{5.02770}} = -1.138$$

from which a two-tailed p-value from Student's t distribution with 28.04 degrees of freedom is 0.26. H_0 is not rejected at the 5% level of significance, and so $t = 2.75$ days is a tenable age for a larva weighing 9 mg. This procedure can be repeated for other times t, replacing the coefficients of $Y_{2.5}$ and $Y_{3.0}$ by $1 - r_t = 3.0 - t/3.0 - 2.5$ and r_t, respectively, and following those substitutions through the rest of the formulas. Tedious by hand, these calculations are easy to program on a statistical computing package. The least and greatest values of t for which this null hypothesis is not rejected are 2.58 and 2.83 days, respectively. Thus, given the mystery specimen weighing 9 mg and the reference data, a 95% confidence interval on the age of the mystery specimen is the range from 2.58 to 2.83 days, a span of 6 hours.

Comments

The inverse prediction approach we have described depends on the assumption that y is (at least approximately) normally distributed, and it depends on the assumed form of the relation between the population mean μ_t and time t, and between the population variance σ^2_t and t.

The closer together the time points are for which reference data are available, the less important will be the assumed relations for μ_t and σ^2_t, and linear interpolation should be sufficient. However, if μ_t and σ^2_t change rapidly and nonlinearly with t, then some other sort of interpolation should be undertaken. It is appealing at first to consider fitting growth-curve models, or to use smoothing or nonparametric regression techniques here. However, each such approach involves its own set of assumptions and arbitrary modeling choices. We feel that the approximations we have used are generally sufficient, and they are easy to understand. In a case where linear interpolation is troublesome, one might use three-point quadratic interpolation. Otherwise, if it is practicable, reference data could be gathered at one or more intermediate time points.

If the distribution of y at time t is not reasonably bell shaped, so that an assumption of normality is unreasonable, it may be possible to transform y so that its distribution is more nearly normal. There is an extensive literature on classes of such transformations (see, for example, Atkinson 1987).

Other approaches are possible that do not involve distributional assumptions. As an example, prediction bounds might be approximated by upper and lower percentiles (2.5%, for example) of the empirical relative frequency distributions, interpolated linearly between reference data time points. We have experimented with this approach, and its results were not strikingly different from results of the approach we described in the preceding section.

The approach we have described can be extended to use multivariate development data, such as weight and length, to construct a prediction interval on age (Oman and Wax 1984).

Estimating Age from Succession Data

The statistical questions involved in estimating PMI from successional data can be described in terms of an example. Consider a hypothetical field experiment in which $n = 10$ carcasses were exposed for 7 days. Each day, the presence or absence of each of two species, A and B, was recorded for each carcass. The occurrence data are shown in the following table. Denote the four possible species combinations as follows: A and not B = AB'; A and B = AB; B and not A = A'B; and not A and not B = A'B'. On day 3, for example, nine carcasses had AB, one had A'B', and none had either AB' or A'B.

			Day				
Species	1	2	3	4	5	6	7
AB'	9	1	0	0	0	0	0
AB	1	8	9	5	0	0	1
A'B	0	1	0	4	10	4	0
A'B'	0	0	1	1	0	6	9

In a population of subjects in which each subject falls into exactly one of c disjoint and exhaustive categories, if n subjects are sampled independently and at random, then the joint probability distribution of the frequencies with which the n sampled subjects fall into the c categories is a multinomial distribution. Denote the set of observed frequencies by $f = (f_1, \ldots, f_c)$; each f_i is a nonnegative integer, and the sum of the c f_i values is n. Over all possible samples, the expected value of f/n is $\pi = (\pi_1, \ldots, \pi_c)$, where π_i is the proportion of the subjects in the population that are in category I.

On day 4, the observed frequencies of the four species combinations were 0 for AB', 5 for AB, 4 for A'B, and 1 for A'B'; denote this set of frequencies by $f_4 = (0, 5, 4, 1)$. Suppose a mystery carcass is found, and on it species A is found and species B is not found, so it has species combination AB′; denote its set of frequencies by $y_* = (1, 0, 0, 0)$. The comparison between the reference data f and the mystery datum y_* can be represented as a contingency table, shown below.

Category	Reference Data	Mystery	All
1	0	1	1
2	5	0	5
3	4	0	4
4	1	0	1
All	10	1	11

The question of whether the PMI of the mystery carcass could be 4 days becomes the question of whether it is reasonable to think that y_* and f_4 could have resulted from random samples from the same population.

Denote the set of species combination proportions in the day 4 population by π_4, and in the population from which y_* came as π_*; we want to know whether $\pi_4 = \pi_*$. Standard methods for testing this hypothesis in the contingency table lead to a chi-squared statistic of 11, with a p-value of 0.012 (from the chi-squared distribution with three degrees of freedom) and a p-value for Fisher's exact test (FET) of 0.182. These tests are easy to compute, even by hand, but unfortunately they appear to lead to contrary conclusions. With a sample size of one, it is unreasonable to expect the chi-squared approximation to be good. The FET p-value, although it is exact, is a conditional probability, fixing the row and column totals, and it tends to be conservative when compared to corresponding unconditional probabilities (McDonald 1977). LaMotte and Wells (2000) present exact, unconditional p-values for this setting, one by "unconditioning" the Fisher exact test, and the other based on a likelihood ratio (LR) test. They are more difficult to compute than the FET p-value or the chi-squared approximation. Our intent here is to describe the nature of these tests, but without going into all details of the calculations (see LaMotte and Wells [2000] for those).

The set of frequencies $f_4 = (0, 5, 4, 1)$ is one possible outcome from a sample of 10 subjects. Any other set of frequencies that sum to 10 is also a possible outcome. There are 286 distinct such outcomes. Given the probabilities π_4, a multinomial probability can be computed for each outcome. For the mystery subject, there are four possible outcomes; given the category probabilities π_* in the population from which the mystery subject came, the probability of each outcome y_* is just the probability π_{*j} for the category into which the mystery subject falls. Assuming that the eleven subjects are sampled independently, the probability of the joint outcome f_4 and y_* is the product of their respective probabilities. Calculating the probability of any particular outcome, or any set of outcomes, is straightforward, and that probability depends on π_4 and π_*.

We want a p-value to assess the proposition H_0: $\pi_4 = \pi_*$. Given the outcome (f_4, y_*), suppose that we can list all the other outcomes that are as unfavorable to H_0 as, or more so than, this outcome is. Call this set of outcomes, including the realized outcome (f_4, y_*), the *extreme set* for (f_4, y_*), and denote it by $C(f_4, y_*)$. The probability of this set of outcomes, computed as if H_0 is true, is a p-value, but it depends on the particular value of the set π of category probabilities. A conservative approach is to present the greatest value this probability could be over all possible values of π_4. Finding the maximum value of the probability of $C(f_4, y_*)$ is where the computational difficulties begin. This is a multidimensional, constrained, nonlinear optimization problem. We have found an efficient way to perform this computation when extreme sets are defined in terms of likelihood ratios. For extreme sets corresponding to the FET, though, we have not found any such solutions, so we have had to rely on general-purpose computational algorithms, and they can be prohibitively slow for practical applications.

Different definitions of outcomes as unfavorable to H_0 as the observed outcome lead to different p-values. "Unfavorable" is defined in terms of an ordering, usually by a real-valued function that assigns a value to each possible outcome. The orderings corresponding to the likelihood ratio statistic and the chi-squared statistic are equivalent in the current setting, while the ordering corresponding to the FET conditional p-value is different.

Figure 9.4 shows four p-values for the hypothetical data given above, for each day separately. The "true" p-value is included for comparison. It is calculated from the known probabilities from which the data were simulated, and it is defined, for mystery category l, as the sum of the category probabilities π_i that are less than or equal to π_l. The LR p-value is the probability of outcomes with a likelihood ratio statistic greater than or equal to the likelihood ratio statistic of the observed outcome. FET(C) is the two-sided Fisher exact test p-value. FET(U) applies the FET ordering to all outcomes, not just those with the observed

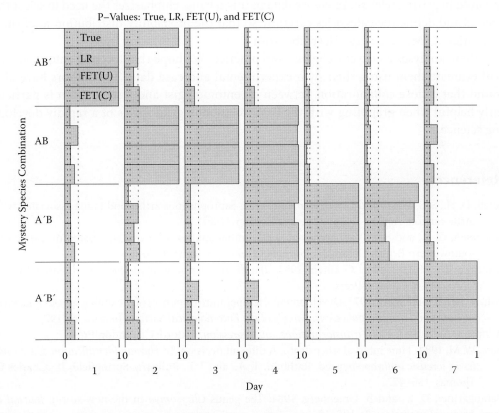

Figure 9.4 p-values for each possible species combination from a mystery specimen, by PMI (days), computed from the occurrence data. Vertical dotted lines are drawn at 5, 10, and 25%. In each of the four panels for each day, the top bar depicts the "true" p-value; the second bar (LR) depicts the likelihood ratio p-value; the third (FET(U)), the unconditional Fisher's exact test p-value; and the bottom (FET(C)), the conditional Fisher's exact test p-value.

marginal totals. Vertical dotted lines show 5, 10, and 25%. Looking at the top row, corresponding to a mystery species combination AB′, the LR, FET(U), and FET(C) p-values are large for day 1, moderate for day 2, and fairly small for days 3 to 7. The FET(C) p-value is generally greater than the LR p-value and the FET(U) p-value. Comparisons over the contents of the figure reflect our experience: the FET(C) p-value is consistently greater than the FET(U) p-value; hence, the FET(C) p-value is unnecessarily conservative. The FET(U) p-value is a little more refined than the LR p-value for outcomes that have somewhat small p-values. However, the as yet unsolved computational difficulties of the FET(U) p-value make it unattractive for now. The FET(C) p-value is so simple to calculate that it can easily be done by hand. The LR p-value can be calculated easily, but it requires a short computer program to perform the optimization.

Comments

Other than a basic independent-sampling model, no assumptions are required to justify the p-values described in the last section. The number of categories c required to describe the species combinations for m species is 2^m. An alternative to keeping a category for every possible species combination is to group them into a more manageable number of categories. That the number of categories may be large, coupled with the expense and difficulty involved in getting relevant reference data in this setting, emphasizes the need to select the species and define the categories carefully so that the probability distribution π_t of the c categories at PMI t changes discernibly with t.

These analyses have yet to be put into practice. We hope that forensic entomologists will evaluate them using their own experimental and case data. The authors have also found that a close collaboration between an entomologist and a statistician is particularly helpful when grappling with the basic methodological issues of a rapidly developing science.

References

Abell, D. H., S. S. Wasti, and G. C. Harmann. 1982. Saprophagous arthropod fauna associated with turtle carrion. *Applied Entomology and Zoology* 17:301.

Anderson, G. S., and S. L. VanLaerhoven. 1996. Initial studies on insect succession on carrion in southwestern British Columbia. *Journal of Forensic Sciences* 41:617.

Andrewartha, H. G., and L. C. Birch. 1954. *The distribution and abundance of animals.* Chicago: University of Chicago Press.

Ash, N., and B. Greenberg. 1975. Developmental temperature responses of the sibling species *Phaenicia sericata* and *Haenicia pallescens*. *Annals of the Entomological Society America.* 68:197.

Atkinson, A. C. 1987. *Plots, transformations, and regression.* Oxford: Clarendon Press.

Bass, W. M. 1984. Time interval since death. A difficult decision. In *Human identification: Case studies in forensic anthropology*, ed. Rathbun, T. A., and J. E. Buikstra. Springfield, IL: Charles C. Thomas, 136–47.

Baumgartner, D. L., and B. Greenberg. 1984. The genus *Chrysomya* in the new world. *Journal of Medical Entomology* 21:105–13.

Bohart, G. E., and J. L. Gressitt. 1951. Filth-inhabiting flies of Guam. *Bulletin of the Bernice P. Bishop Museum* 204:1–152.

Bornemissza, G. F. 1957. An analysis of arthropod succession in carrion and the effect of its decomposition on the soil fauna. *Australian Journal of Zoology* 5:1–12.

Braack, L. E. O. 1987. Community dynamics of carrion-attendant arthropods in tropical African woodland. *Oecologia* 72:402–9.

Byrd, J. H., and J. F. Butler. 1996. Effects of temperature on *Cochliomyia macellaria* (Diptera: Calliphoridae) development. *Journal of Medical Entomology* 33:901–5.

Byrd, J. H. and J. F. Butler. 1997. Effects of temperature on *Chrysomya rufifacies* (Diptera: Calliphoridae) development. *Journal of Medical Entomology* 34:353–58.

Byrd, J. H., and J. F. Butler. 1998. Effects of temperature on *Sarcophaga haemorrhoidalis* (Diptera: Sarcophagidae) development. *Journal of Medical Entomology* 35:694–98.

Catts, E. P. 1990. Analyzing entomological data. In *Entomology and death: A procedural guide*, ed. Catts, E. P., and N. H. Haskell. Clemson, SC: Joyce's Print Shop, 124–35.

Chapman, R. F. 1982. *The insects. Structure and function*. Cambridge, MA: Harvard University Press.

Cianci, T. J., and J. K. Sheldon. 1990. Endothermic generation by blow fly larvae *Phormia regina* developing in pig carcasses. *Bulletin of the Society for Vector Ecology* 15:33.

Cornaby, B. W. 1974. Carrion reduction by animals in contrasting habitats. *Biotropica* 6:51–63.

Cyr, T. L. 1993. Forensic implications of biological differences among geographic races of *Phormia regina* (Meigen) (Diptera: Calliphoridae). Unpublished MS thesis, Pullman, WA: Washington State University.

Davies, L., and G. G. Ratcliffe. 1994. Development rates of some pre-adult stages in blowflies with reference to low temperatures. *Medical and Veterinary Entomology* 8:245–54.

Denlinger, D. L., and J. Zdarek. 1994. Metamorphosis behavior of flies. *Annual Review of Entomology* 39:243–66.

Deonier, C. C. 1940. Carcass temperatures and their relation to winter blowfly activity in the Southwest. *Journal of Economic Entomology* 33:166–70.

Disney, R. H. L. 1994. *Scuttle flies: The Phoridae*. London: Chapman & Hall.

Early, M., and Goff, M. L. 1986. Arthropod succession patterns in exposed carrion on the island of O'ahu, Hawaiian islands. *Journal of Medical Entomology* 25:520–31.

Erzinclioglu, Y. Z. 1990. On the interpretation of maggot evidence in forensic cases. *Medicine, Science and Law* 30:65.

Fuller, M. E. 1934. *The insect inhabitants of carrion: A study in animal ecology*. Bulletin 82. Commonwealth of Australia, Council for Scientific and Industrial Research.

Geberth, V. J. 1996. *Practical homicide investigation*. Boca Raton, FL: CRC Press.

Goff, M. L. 1992. Problems in estimating the postmortem interval resulting from wrapping of the corpse: A case study from Hawaii. *Journal of Agricultural Entomology* 9:237–43.

Goff, M. L., S. Carbonneau, and W. Sullivan. 1991. Presence of fecal material in diapers as a potential source of error in estimation of the postmortem interval using arthropod developmental rates. *Journal of Forensic Sciences* 36:1603–6.

Goff, M. L., and W. D. Lord. 1994. Entomotoxicology. A new area for forensic investigation. *American Journal of Forensic Medicine and Pathology* 15:51–57.

Goodbrod, J. R., and M. L. Goff. 1990. Effects of larval population density on rates of development and interactions between two species of *Chrysomya* (Diptera: Calliphoridae) in laboratory culture. *Journal of Medical Entomology* 27:338–43.

Greenberg, B. 1984. Two cases of human myiasis caused by *Phaenicia sericata* (Diptera: Calliphoridae) in Chicago area hospitals. *Journal of Medical Entomology* 21:615.

Greenberg, B. 1991. Flies as forensic indicators. *Journal of Medical Entomology* 28:565–77.

Greenberg, B., and J. D. Wells. 1998. Forensic use of *Megaselia abdita* and *Megaselia scalaris* (Phoridae: Diptera): Case studies, development rate and egg structure. *Journal of Medical Entomology* 35:205–9.

Hall, D. G. 1948. *The blowflies of North America*. La Fayette, IN: Thomas Say Foundation.

Hall, R. D. 1990. Medicocriminal entomology. In *Entomology and death: A procedural guide*, Catts, E. P., and N. H. Haskell. Clemson, SC: Joyce's Print Shop, 1–6.

Hall, R. D., P. C. Anderson, and D. P. Clark. 1986. A case of human myiasis caused by *Phormia regina* (Diptera: Calliphoridae) in Missouri, USA. *Journal of Medical Entomology* 19:578–79.

Hall, R. D., and K. E. Doisy. 1993. Length of time after death: Effect on attraction and oviposition or larviposition of midsummer blow flies (Diptera: Calliphoridae) and flesh flies (Diptera: Sarcophagidae) of medicolegal importance in Missouri. *Annals of the Entomological Society of America* 86:589–93.

Hanski, I. 1976. Assimilation by *Lucilia illustris* (Diptera) larvae in constant and changing temperatures. *Oikos* 27:288–99.

Haskell, N. H., and R. E. Williams. 1990. Collection of entomological evidence at the death scene. In *Entomology and death: A procedural guide*, ed. Catts, E. P., and N. H. Haskell. Clemson, SC: Joyce's Print Shop, 82–96.

Henssge, C., B. Madea, B. Knight, L. Nokes, and T. Krompecher. 1995. *The estimation of the time since death in the early postmortem period*. London: Arnold.

Hewadikaram, K. A., and M. L. Goff. 1991. Effect of carcass size on rate of decomposition and arthropod succession patterns. *American Journal of Forensic Medicine and Pathology* 12:235–40.

Introna, F., B. M. Altamura, A. Dell'Erba, and V. Dattoli. 1989. Time since death definition by experimental reproduction of *Lucilia sericata* cycles in growth cabinet. *Journal of Forensic Sciences* 34:478–80.

James, M. T. 1947. *The flies that cause myiasis in man*. USDA Miscellaneous Publication 631.

Johnson, M. D. 1975. Seasonal and microseral variations in the insect populations on carrion. *American Midland Naturalist* 93:79–80.

Kamal, A. S. 1958. Comparative study of thirteen species of sarcosaprophagous Calliphoridae and Sarcophagidae (Diptera). 1. Bionomics. *Annals of the Entomology Society of America* 51:261–70.

Kotz, S., N. L. Johnson, and C. B. Read. 1988. *Encyclopedia of statistical sciences*. Vol. 8. New York: John Wiley & Sons.

Kurahashi, H., and T. Ohtaki. 1989. Geographic variation in the incidence of pupal diapause in Asian and Oceanian species of the flesh fly *Boettchersca* (Diptera: Sarcophagidae). *Physiological Entomology* 14:291–98.

LaMotte, L. R., and J. D. Wells. 2000. P-values for postmortem intervals from arthropod succession data. *Journal of Agricultural, Biological and Environmental Statistics*. 5(1):37–47.

Lehmann, E. L. 1986. *Testing statistical hypotheses*. 2nd ed. New York: John Wiley & Sons.

Levot, G. W., K. R. Brown, and E. Shipp. 1979. Larval growth of some calliphorid and sarcophagid Diptera. *Bulletin of Entomological Research* 69:469–75.

Lord, W. D. 1990. Case histories of the use of insects in investigations. In *Entomology and death: A procedural guide*, ed. E. P. Catts and N. H. Haskell. Clemson, SC: Joyce's Print Shop, 9–37.

McDonald, L. L., B. M. Davis, and G. A. Milliken. 1977. A nonrandomized unconditional test for comparing two proportions in 2×2 contingency tables. *Technometrics* 19:145.

Micozzi, M. S. 1991. *Postmortem change in human and animal remains: A systematic approach*. Springfield, IL: Charles C. Thomas.

Nabaglo, L. 1973. Participation of invertebrates in decomposition of rodent carcasses in forest ecosystems. *Ekologia Polska* 21:251–70.

Nishida, K. 1984. Experimental studies on the estimation of postmortem intervals by means of larvae infesting human cadavers. *Japanese Journal of Forensic Medicine* 38:24–41.

Norris, K. R. 1965. The bionomics of blowflies. *Annual Review of Entomology* 10:47–48.

O'Flynn, M. A., and D. E. Moorehouse. 1979. Species of *Chrysomya* as primary flies in carrion. *Journal of the Australian Entomological Society* 18:31–32.

Oman, S. D., and Y. Wax. 1984. Estimation of fetal age by ultrasound measurements: An example of multivariate calibration. *Biometrics* 40:947–60.

Payne, J. A. 1965. A summer study of the pig, *Sus scrofa* Linnaeus. *Ecology* 46:592602.

Povolny, D., and J. Rozsypal. 1968. Toward the autecology of *Lucilia sericata* (Meigen, 1826) (Dipt., Call.) and the origin of its synanthropy. *Acta Scentiarum Naturalium Academiae Scientiarum Bohemoslovacae Brno* 8:1–32.

Putman, R. J. 1983. Carrion and dung: The decomposition of animal wastes. In *Studies in biology*. London: Institute of Biology.

Reed, H. B. 1958. A study of dog carcasses in Tennessee, with special reference to the insects. *American Midland Naturalist* 59:213–45.

Reiter, C. 1984. Zum Wachstumsverhalten der Maden der blauen Schmeissfliege *Calliphora vicina*. *Zeitschrift für Rechtsmedizen* 91:295–308.

Rodriguez, W. C., and W. M. Bass. 1983. Insect activity and its relationship to decay rates of human cadavers in east Tennessee. *Journal of Forensic Sciences* 28:423–32.

Rodriguez, W. C., and W. M. Bass. 1985. Decomposition of buried bodies and methods that aid in their location. *Journal of Forensic Sciences* 30:836–52.

Schoenly, K. 1992. A statistical analysis of successional patterns in carrion-arthropod assemblages: Implications for forensic entomology and the determination of the postmortem interval. *Journal of Forensic Sciences* 37:1489–513.

Schoenly, K., M. L. Goff, and M. Early. 1992. A BASIC algorithm for calculating the postmortem interval from arthropod successional data. *Journal of Forensic Sciences* 37:808–23.

Schoenly, K., M. L. Goff, J. D. Wells, and W. Lord. 1996. Quantifying statistical uncertainty in succession-based entomological estimates of the postmortem interval in death scene investigations: A simulation study. *American Entomologist* 42:106–12.

Schoenly, K., and W. Reid. 1987. Dynamics of heterotrophic succession in carrion arthropod assemblages: Discrete series or a continuum of change? *Oecologia* 73:192–202.

Seaquist, E. R., T. R. Henry, E. Cheong, and A. Theologides. 1983. *Phormia regina* myiasis in a malignant wound. *Minnesota Medicine* 66:409–10.

Sherman, R. A., and E. A. Pechter. 1988. Maggot therapy: A review of the therapeutic applications of fly larvae in human medicine, especially for treating osteomyelitis. *Medical and Veterinary Entomology* 2:225–30.

Shewell, G. E. 1987. Sarcophagidae. In *Manual of Nearctic diptera*, ed. J. F. McAlpine. Vol. 2. Ottawa: Agriculture Canada, 1159–86.

Shishido, W. H., and D. E. Hardy. 1969. Myiasis of new-born calves in Hawaii. *Procedures of the Hawaiian Entomological Society* 20:435–38.

Smith, K. G. V. 1986. *A manual of forensic entomology*. Ithaca, NY: Cornell University Press.

Spradbery, J. P. 1991. *A manual for the diagnosis of screw-worm fly*. Canberra: AGPS Press.

Spradbery, J. P. 1994. Screw-worm fly: A tale of two species. *Agricultural Zoology Reviews* 6:1.

Stevens, J., and R. Wall. 1997. The evolution of ectoparasitism in the genus *Lucilia* (Diptera: Calliphoridae). *International Journal for Parasitology* 27:51–59.

Tantawi, T. I., E. M. el-Kady, B. Greenberg, and H. A. el-Ghaffar. 1996. Arthropod succession on exposed rabbit carrion in Alexandria, Egypt. *Journal of Medical Entomology* 33:566–80.

Tantawi, T. I., and B. Greenberg. 1993. The effect of killing and preservative solutions on estimates of maggot age in forensic cases. *Journal of Forensic Sciences* 38:702–7.

Tantawi, T. I., J. D. Wells, B. Greenberg, and E. M. el-Kady. 1998. Fly larvae (Diptera: Calliphoridae, Sarcophagidae, Muscidae) succession in rabbit carrion: Variation observed in carcasses exposed at the same time and in the same place. *Journal of the Egyptian German Society of Zoology* 25:195–208.

Taylor, D. B., A. L. Szalanski, and R. D. Peterson. 1996. Mitochondrial DNA variation in screwworm. *Medical and Veterinary Entomology* 10:161–69.

Turner, B., and T. Howard. 1992. Metabolic heat generation in dipteran larval aggregations: A consideration for forensic entomology. *Medical and Veterinary Entomology* 6:179–81.

Wells, J. D., and B. Greenberg. 1994. Resource use by an introduced and native carrion flies. *Oecologia* 99:181–87.

Wells, J. D., and H. Kurahashi. 1994. *Chrysomya megacephala* (Fabricius) (Diptera: Calliphoridae) development: Rate, variation and the implications for forensic entomology. *Japanese Journal of Sanitary Zoology* 45:303–9.

Wells, J. D., and L. R. LaMotte. 1995. Estimating maggot age from weight using inverse prediction. *Journal of Forensic Sciences* 40:585–90.

Wigglesworth, V. B. 1972. *The principles of insect physiology*. London: Chapman & Hall.

Williams, H. 1984. A model for the aging of fly larvae in forensic entomology. *Forensic Science International* 25:191–99.
Zumpt, F. 1965. *Myiasis in man and animals in the old world.* London: Butterworths.

Insect Development and Forensic Entomology

10

LEON G. HIGLEY
NEAL H. HASKELL

Contents

Introduction

When using insects as indicators of postmortem interval, two major approaches are possible. The first is to use the presence or absence of a species as an indicator of time of death, based on understandings of insect successional patterns. The second is to consider the degree of development for insects found on the cadaver. These approaches are complementary, although measuring insect development requires that immature insects are still present on the body. This latter approach depends on backtracking from the observed degree of development to the time of oviposition, to convert insect development into a time estimate. The time of death then can be estimated by adding the interval between death and oviposition for the species in question (recognizing that various factors can influence initial oviposition). Measuring insect development is a powerful method for providing estimates of postmortem interval, but there are many crucial considerations and potential limitations in making such estimates. In this chapter we will explore these details of estimating insect development, and explore their implications for forensic entomology. Some general reviews on temperature and development include Wagner et al. (1984), Higley et al. (1986), and Higley and Peterson (1996), and Catts and Goff (1992) includes a discussion of age determination of maggots by temperature for forensic entomology.

Temperature and Insect Development

The key observation regarding insect development is that the time of development depends on temperature. This point was first recognized in the early 1700s by the French scientist Reaumur (1735), but methods for using this understanding to describe or predict insect development mostly date from the 1950s (Arnold 1959, 1960) through to the present. Of course, the growth and development of all organisms is temperature dependent, but for organisms with constant body temperature, we can speak of development merely in terms of time without explicitly considering temperature. However, organisms in the size range of insects cannot maintain constant body temperature; these are what are called poikilo-thermic animals. Larger organisms can maintain a constant (or at least more constant) temperature through metabolic heat, but this is not possible, even for the largest insects. Also, for some organisms, plants, for example, other factors (such as photoperiod) may be almost as important as temperature in determining the sequence of developmental events.

All growth depends on temperature because the biochemical reactions that are the ultimate basis for growth are themselves temperature dependent. At this level, heat of reactions, enzyme thermal properties, and membrane permeabilities all contribute to the temperature requirements for growth. In a very influential paper, Sharpe and DeMichele (1977) proposed that temperature limits on growth were a consequence of rate-limiting reactions controlled by corresponding rate-limiting enzymes. From theoretical arguments, practical models for determining insect development rates based on the notion of rate-limiting enzymes have been developed and used for many insect species (Wagner et al. 1984). Unfortunately, the idea that development is only a function of rate-limiting enzymes ignores the potential importance of other factors, such as membrane permeability, and experimental evidence does not support many assumptions that follow from this model (Hilbert and Logan 1983, Lamb et al. 1984, Higley and Peterson 1994). Consequently, there is no single biologically based model to account for how temperature alters growth rates.

In the absence of a definitive theoretical understanding of the relationship between temperature and growth, empirical understandings of these relationships must be used. The general relationship between development rate and temperatures is described by the temperature development curve indicated in Figure 10.1 (Higley and Peterson 1994). The relationship is curvilinear at low and high temperatures and linear in between. The lowest temperature at which development can proceed is called the developmental minimum (or minimum threshold), and the highest temperature is called the developmental maximum (or maximum threshold). Threshold temperatures can be difficult to estimate (as we will discuss shortly), and the upper threshold typically occurs near the upper lethal tempera-ture for a species. The temperature development curve exists for all insect species, although the specifics of the curve will vary by species. Additionally, within species variation occurs, particularly at the extremes of the temperature range of a species. Why the low-temper-ature portion of the development curve is curvilinear has been the topic of some spec-ulation. Although some have argued that this part of the curve merely reflects genetic variation among insects, in an elegant experiment using pea aphid (*Acyrthosiphon pisum*) clones, Lamb (1992) demonstrated that the curvilinear relationship between temperature and development rate is a reflection of the underlying physiology of development and not genetic variation among individuals.

Describing the temperature development curve for an insect species is difficult, in that it requires substantial replication of development with many individual insects at

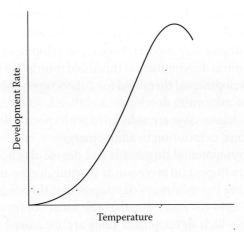

Figure 10.1 The thermal development curve—the generalized relationship between temperature and rate of development.

multiple temperatures. Also, obtaining reliable development rates at low and high temperatures is particularly challenging, because these often are close to or at lethal temperatures for the species.

Measuring Insect Development

Broadly speaking, there are three approaches for measuring insect development: physiological, curvilinear, and linear (degree-day). However, the physiological approach (based on the Sharp and DeMichele model) seems to have no more theoretical validity than the curvilinear approach, so it might be more accurate to speak of only two approaches: curvilinear and linear. Essentially, the curvilinear approach seeks to describe the temperature development curve and determine insect development from this. The linear approach approximates most of the development curve as a line, with cutoffs at high and low temperature. Both methods are essentially regression approaches.

Because the temperature development curve is undoubtedly curvilinear, it seems logical that a curvilinear approach should be better than a linear approach. However, this does not necessarily seem to be the case. First, there is no consensus around a single curvilinear equation to use to describe the temperature development curve (see Wagner et al. 1984 for a review of various approaches). Second, determining a curvilinear model for the temperature development curve requires a great deal of research and is relatively complicated to use in practice. Finally (and perhaps most importantly), various field studies have shown either no improvement of curvilinear models over linear models or better predictions from linear models for several insect species (e.g., Hochberg et al. 1986, Stinner et al. 1988, McClain et al. 1990). Perhaps the failure of curvilinear models to outperform linear models in some instances is related to accumulation of errors in the estimates of parameters needed to drive such models. In any event, on a practical level, linear models often seem to perform adequately (and occasionally better than their curvilinear counterparts.)

Linear models of development assume a constant increase in development rate with increasing temperature. They further assume no development occurs below the minimum developmental threshold, and a constant development rate occurs at or above the maximum threshold (in reality, the reverse may be true, although these are probably at a species' lethal limits, so the issue of development rate becomes moot). Linear models are most commonly called degree-day models, because development is regarded as the temperature above the minimum developmental threshold multiplied by time. Thus, 5 degrees above the minimum developmental threshold for 2 days represents 10 degree-days (5 × 2), just as 1 degree above the minimum developmental threshold for 10 days does (1 × 10). An accumulated number of degree-days are associated with specific developmental events like egg hatch, stage transitions, oviposition to adult emergence, etc.

Determining the developmental thresholds and degree-day accumulations for specific developmental events is an important prerequisite to using degree-days. Various approaches are available for estimating the minimum developmental threshold (see Higley et al. 1986 or Higley and Peterson 1994 for more details). The most commonly used method is the x-intercept approach, in which development rates are measured in the low-temperature range and results are fit in a linear regression (Arnold 1959). The linear regression can then be extrapolated to the x-axis, where the development rate is zero. As Arnold (1959) emphasized, the base temperature determined by the x-intercept has no biological meaning—it is a mathematical consequence of approximating the development rate through linear regression. Consequently, the base temperature used for calculating degree-days might be above or below the actual biological base temperature. Using the biological minimum (the empirically determined temperature at which development no longer occurs) rather than the x-intercept minimum violates the mathematical premise of the degree-day approach and leads to errors in developmental estimates.

As previously discussed, the conventional approach for relating temperature to insect development is to regress the development rate (1/days to reach a stage, the y-value) versus temperature (the x-value). Unfortunately, this method distorts the underlying variance structure of the data (it minimizes variation at the low-temperature end and maximizes variation at the high-temperature end), which is important because estimates of variability in the regression can be wrong. Ikemoto and Takai (2000) addressed this limitation by changing the nature of regression, to avoid skewing data variance. They advocate regressing time of development multiplied by temperature (degree-days or degree-hours) versus time of development. In this regression, the slope equals the developmental minimum, and the x-intercept equals the accumulated degree-days required to complete a stage. Although less commonly used than other procedures, the Ikemoto and Takai method is valuable for identifying outliers.

Once a developmental minimum and degree-day accumulations for life history events are determined, all that remains is to calculate degree-days with actual temperature data. It must be remembered that it is critical to use a minimum developmental threshold temperature even if temperatures are above this threshold. Calculations without an appropriate threshold will overestimate degree-day accumulations. When daily temperatures are below the developmental minimum, it is important that degree-day accumulations are set to zero (otherwise, you would have the absurd situation of calculating negative degree-days, implying an insect is growing younger).

This degree-day (or degree-hour) process involves estimating the area under a daily temperature curve that is above the minimum developmental threshold (Figure 10.2). Continuous, or even hourly, temperature records would provide the most accurate degree-day determination, but these are rarely available in practice (particularly for forensic determinations). However, hourly temperatures may be available from class I weather

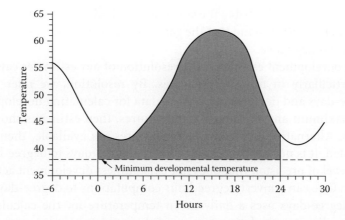

Figure 10.2 Degree-days, a measure of physiological time, indicated as the area under a daily temperature curve and above a minimum developmental temperature for a species.

service stations, and can be useful if the station is not too far from the crime scene. Higley et al. (1986) describe the various estimation methods, compare the results for different locations in the United States, and offer recommendations on appropriate methodology. Briefly, the standard technique is called the rectangle method, because daily degree-days are measured as a box with a length of 1 day and a height of the daily average temperature minus the minimum developmental threshold (Figure 10.3). As illustrated in Figure 10.3, in principle the area of the daily temperature curve above the mean daily temperature corresponds to those areas below the mean daily temperature and above the daily temperature curve. Consequently, it is possible to calculate the area under the curve as a rectangle, with sides 1 day × mean daily temperature minus developmental minimum temperature (as shown in Figure 10.3). Arnold (1960) discusses this method in detail, and points out limitations in estimating the mean daily temperature by averaging the daily high and low temperature, including possible seasonal bias. Other approaches for measuring the area under the curve include estimating the daily temperature curve on a half-day basis with

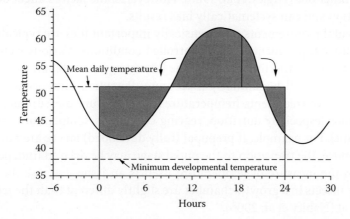

Figure 10.3 Rectangle calculation method for determining degree-days. The area calculated is that of a rectangle 1 day long and the mean daily temperature less minimum developmental temperature wide. As indicated in the figure, this method works because the area of the rectangle not under the daily temperature curve is equal to that portion of the curve above the rectangle (see text or [Arnold 1960] for additional explanation).

triangles, or as a sine wave. Probably the sine wave method is the most accurate approach, although Higley et al. (1986) found that the gain in accuracy of the sine wave over the rectangle approach was negligible.

In making development estimates, the resolution of our estimates can be an important issue, particularly in forensic estimates. By resolution, we refer to differences between degree-days and degree-hours. If the data for calculating development are limited to daily maximum and minimum temperatures, then estimates should be limited to degree-days. Alternatively, if hourly temperatures are available, then degree-hours may be calculated. It is not appropriate to convert degree-days to degree-hours, because this implies a level of precision that does not exist, unless development actually is determined by hour. One can convert degree-hour computations to degree-days, because the definition of degree-days uses a daily mean temperature for the calculation. A similar argument applies in calculating fractional degree-days. Under most instances, the thermal accumulation from midnight to noon will not equal the thermal accumulation from noon to midnight (typically noon to midnight is warmer). Also, the movement of frontal systems over a location can dramatically alter thermal accumulations, because the daily pattern of temperature change is radically altered. However, in the absence of radical temperature change, it may be possible to calculate degree-days by half days, by using the daily high and low coupled with the previous day's low (see Higley et al. 1986 for more information). When fractional degree-day accumulations are significant for determining insect development, it is essential to have data on weather patterns (frontal systems), precipitation, and most importantly, temperatures at intervals more frequent than once or twice a day.

Limits to Estimates of Insect Development

A great many factors can influence estimates of insect development. Frequently, errors will occur in both directions (retarding or accelerating development), so the net effect is to cancel each other out (Higley et al. 1986). However, some factors affect development in only one direction and can systematically bias results.

Developmental requirements for forensically important flies are typically determined through laboratory experiments under controlled conditions. Various factors may influence results from these experiments, and if the laboratory estimates are not accurate, obviously PMI estimates will be similarly inaccurate. Potentially important considerations include replication of treatments (temperatures), temperature measurement, and experimental conditions (especially nutrition, rearing conditions, temperature cycling, and the light:dark regime). For example, if prepupal (fully developed) larvae are not provided sufficient area to move, the transition to the pupal stage may be delayed (independent of temperature). Even the measure of temperature itself is an issue, because the temperatures experienced by insects in a growth chamber are slightly different than the temperature the chamber is set at (Nabity et al. 2006).

Perhaps the most obvious issue is that insects actually experience fluctuating temperatures; however, most developmental thresholds and accumulations are determined under constant temperature. This area has received much study and debate. On a practical level, the influence of fluctuating temperatures seems to be more important for some species than others. On a theoretical level, the issues of fluctuating temperature are more predictable.

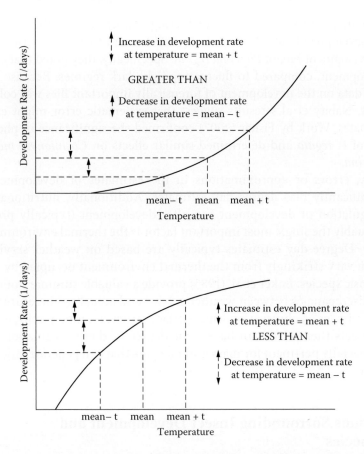

Figure 10.4 The rate summation effect. Low temperature fluctuations (t) above and below a given mean produce a development rate greater than that predicted by the mean temperature, because fluctuations above the mean increase development rates proportionally more than fluctuations below the mean reduce development rates. The reverse is true for fluctuations at high temperatures.

For simple mathematical reasons, fluctuations in the lower-temperature portion of the development curve will allow development to proceed more rapidly than predicted from a constant temperature, whereas fluctuations in the upper-temperature portion of the development curve will allow development to proceed more slowly than predicted from a constant temperature. This phenomenon is called the rate summation effect, and it is a reflection of the fact that contributions to development around a mean in the curvilinear portions of the temperature development curve are not equal (see Figure 10.4 for a graphical explanation of the rate summation effect). Specifically, in the lower (concave) portion of the temperature development curve, the differences in development at an interval above and below a mean temperature are not equal (greater development occurs above the mean). The reverse is true for the high-temperature (convex) portion of the curve. Tanigoshi et al. (1976) and Worner (1992) provide more comprehensive descriptions of this principle. Worner points out that it is not clear if there are effects on development under fluctuating temperature beyond the rate summation effect. The implications of fluctuating temperatures on development of forensic fly species are offered in Catts and Goff (1992) and Greenberg (1991).

Another emerging issue related to fluctuating temperatures is the influence of photoperiod on development rate. Nabity et al. (2007) found that photoperiod significantly altered the development rate of *Phormia regina*. Specifically, they noted that constant light delayed development, compared to fluctuating light:dark regimes. Because much of the experimental data on the development of forensically important flies was collected under constant light, Nabity et al.'s results suggest that systematic error might exist in some of these estimates. Work by Fisher (2007) confirmed the significance of photoperiod on development of *P. regina* and determined similar effects on *Cochlionyia macellaria* and *Calliphora vicina*.

Obviously, errors or approximations in the estimates of developmental thresholds can significantly bias degree-day estimates. Additionally, nutritional status and hormonal regulation of development can alter development (typically prolonging it). However, arguably the single most important factor is the thermal environment in which insects occur. Degree-day estimates typically are based on weather service data, but these data can vary strikingly from the thermal environment occupied by insects, particularly forensic species. Baker et al. (1985) provide a valuable summary of many factors resulting in discrepancies between weather service and actual temperatures in the field, including such issues as observation time, surface structure, topography, and urbanization. Also, insects themselves can have some degree of thermoregulation (May 1979), and this is especially pertinent for masses of maggots that can generate large amounts of metabolic heat.

Considerations Surrounding Insect Development and Forensic Species

Almost all work surrounding insect development has focused on agricultural or medically important insect species. In these instances, the question typically involves estimating time of potential occurrence of a pest species. In contrast, for ecological research and forensics, development is typically important in backtracking to determine time of occurrence of a specific event (oviposition).

Agricultural versus Forensic Species

Are there differences between agricultural and forensic species regarding development? Obviously all insects are poikilotherms, so there is no a priori reason to think these types of insects should differ. Moreover, agricultural and forensic designations are arbitrary human designations, not indications of biological difference per se. A more telling question is to consider if herbivorous species and saprophagous or necrophagous species are different regarding thermal development. Given that some agricultural pests are also saprophagous (for example, the anthomyiid fly, *Delia platura*), direct evidence indicates that the same principles apply to both groups. The important observation is that differences in the thermal environment, rather than differences intrinsic to species' ecological roles, are likely to have the greatest impact on development.

A more significant issue is the differences in uses and expectations for insect development in a forensic setting, compared to an agricultural application. Development estimates in agriculture are typically associated with making predictions about occurrence of an injurious insect population (Higley and Peterson 1994). This prediction is for a widely distributed population or populations of an insect species over an equally wide geographic range. In contrast, forensic uses of insect development are related to a very small population in a very specific site. Ideally, we want forensic estimates to be as accurate and precise as possible—certainly much more than is necessary for agricultural predictions.

Developmental Parameters for Forensic Species

Ironically, current insect development information is much more precise for agricultural pests than for forensic species. There is a critical need for thorough determinations of developmental thresholds and stage-specific degree-day accumulations for all major forensic species. Currently, many, even most, of our estimates of developmental thresholds and stage-specific accumulations for forensic species are based on only a few studies with relatively limited data sets. Work by Kamal (1958) on developmental thresholds for thirteen carrion-feeding flies has been a standard reference in forensic entomology, although some arithmetic errors in that publication have been noted (Catts and Goff 1992). Subsequent examinations include work by Nishida (1982), Reiter (1984), Vinogradova and Marchenko (1984), Nishida et al. (1986), Introna et al. (1989), Greenberg (1991), Byrd and Butler (1996), Anderson (2000), Grassberger and Reiter (2001, 2002a, 2002b), Clarkson et al. (2004), Huntington (2005), and Nabity et al. (2006, 2007).

In Tables 10.1 to 10.3, we used data from various sources to develop estimates of accumulated degree-days for three key forensically important species: *Calliphora vicina*, *Lucilia sericata*, and *Phormia regina*. Two estimates of the accumulated degree-days (ADD) are calculated: (1) ADD from variables in the inverse development versus temperature linear regression and (2) ADD from the mean of all calculated degree-days with parameters from the inverse development versus temperature regression. If assumptions for use of degree-day models are met, the ADDs determined with the same regression equations should be approximately equal.

Values in Tables 10.1 to 10.3 provide some practical basis for making development estimates, particularly in estimating potential minimum development times. Nevertheless, there is a continuing need to refine and improve these values. Additionally, studies characterizing variation in these parameters between geographically distinct populations of the same species would be of great value. Although it may be true that "most workers favor the use of locally generated development-rate data" (Catts and Goff 1992), ultimately, the forensic entomology community must move to consensus regarding temperature development parameters for forensic insect species. If significant geographic variation in developmental parameters exists, this variation needs to be quantified. Similarly, clear expressions of variability in these parameters are essential to provide meaningful indications of variation in estimates of insect development.

Ultimately, the degree-day approach to estimating development of forensic insects is likely to be supplanted by the use of curvilinear developmental models. Currently, insufficient data are available to allow the calculation of such models, but as these data become available, many of the problems associated with the degree-day approach (such as rate summation and temperature limits of linear approximation) may be avoided.

Table 10.1   Developmental Estimates for *Calliphora vicina* Calculated from Literature Data

Parameter	ADD 1/d vs. T (x-intercept = Dev min; 1/slope = ADD)				
	Egg	Larva	Pupa	Egg-pupae	Egg-adult
Data source: Reiter 1984					
Dev min by regression parameters				−4.51	
ADD by regression parameters				102.9	
ADD by mean of individual ADD				49.20	
SE of mean				53.71	
n				2	
Valid temperature range for ADD				18.5–35	
Data source: Greenburg 1991					
Dev min by regression parameters	2.2	−6.81	−3.752	−5.883	−7.71
ADD by regression parameters	13.3	245.2	413.9	256.1	616.3
ADD by mean of individual ADD	7.8	119.2	205.1	76.5	304.3
SE of mean	5.55	126.01	208.83	61.31	312.01
n	2	2	5	2	5
Valid temperature range for ADD	19–22	19–22	19–22	19–22	19–22
Data source: Anderson 2000					
Dev min by regression parameters	39.81	−16.960	−35.750	4.624	−32.900
ADD by regression parameters	−1.0	32.3	446.6	148.5	470.5
ADD by mean of individual ADD	−15.02	301.14	552.10	148.62	995.38
SE of mean	1.58	2.02	114.25	4.78	112.63
n	3	3	3	3	3
Valid temperature range for ADD	20.6–23.3	20.6–23.3	20.6–23.3	20.6–23.3	20.6–23.3
Data source: Huntington 2005					
Dev min by regression parameters	0	−10.4	−0.2829	−9.424	−15.27
ADD by regression parameters	2.0	182.5	290.0	195.4	685.5
ADD by mean of individual ADD	12.8	187.8	290.9	200.7	687.1
SE of mean	0.46	12.06	4.79	12.30	13.14
n	9	8	9	8	7
Valid temperature range for ADD	20.4–29	16.8–27.3	16.8–29	16.8–27.3	20.4–27.3

Note: Accumulated degree-days (ADD) to complete a stage are reported in centigrade degree-days, developmental minima are reported in °C, and ADD estimates are only valid across the indicated temperature range.

Table 10.2 Developmental Estimates for *Luclilia sericata* Calculated from Literature Data

Parameter	ADD 1/d vs. T (x-intercept = Dev min; 1/slope = ADD)				
	Egg	Larva	Parameter	Egg	Larva
Data source: Ash and Greenberg 1975					
Dev min by regression parameters	10.22	−7.263	15.43	−5.974	13.25
ADD by regression parameters	10.4	421.6	91.4	429.9	254.3
ADD by mean of individual ADD	10.5	516.9	91.4	518.6	266.8
SE of mean	0.20	137.42	0.01	132.99	26.79
n	3	3	2	3	3
Valid temperature range for ADD	19–35	19–35	19–27	19–35	19–35
Data source: Greenburg 1991					
Dev min by regression parameters					−6.7090
ADD by regression parameters					423.4
ADD by mean of individual ADD					423.8
SE of mean					5.63
n					7
Valid temperature range for ADD					19–29
Data source: Anderson 2000					
Dev min by regression parameters	−0.07694	7.647	16.13	4.479	13.55
ADD by regression parameters	239	133	−35	352	118
ADD by mean of individual ADD	241	133	−26	354	119
SE of mean	14.16	0.86	13.27	12.64	2.36
n	3	3	3	3	3
Valid temperature range for ADD	15.8–23.3	15.8–23.3	15.8–23.3	15.8–23.3	15.8–23.3
Data source: Grassberg and Reiter 2001					
Dev min by regression parameters	10.9	0.560	5.071	6.321	8.356
ADD by regression parameters	8.1	172.2	108.7	136.8	213.2
ADD by mean of individual ADD	8.1	172.9	110.0	137.0	228.0
SE of mean	0.07	3.46	5.77	1.70	14.46
n	7	8	5	6	7
Valid temperature range for ADD	20–34	19–34	20–28	19–28	17–28

Note: Accumulated degree-days (ADD) to complete a stage are reported in centigrade degree-days, developmental minima are reported in °C, and ADD estimates are only valid across the indicated temperature range.

Table 10.3   Developmental Estimates for *Phormia regina* Calculated from Literature Data

Parameter	Egg	Larva	Parameter	Egg	Larva
			ADD 1/d vs. T (x-intercept = Dev min; 1/slope = ADD)		
Data source: Greenberg 1991					
Dev min by regression parameters			−6.82		−6.82
ADD by regression parameters			416.4		416.4
ADD by mean of individual ADD			417.1		417.1
SE of mean			6.082		6.082
n			8		8
Valid temperature range for ADD			19–35	19–35	19–35
Data source: Anderson 2000					
Dev min by regression parameters	9.2	9.399	7.633	9.378	8.755
ADD by regression parameters	12.4	111.1	97.2	123.5	219.2
ADD by mean of individual ADD	12.4	111.1	97.2	123.5	219.2
SE of mean	1.78E-15	2.88E-03	1.09E-03	5.68E-03	1.14E-02
n	2	2	2	2	2
Valid temperature range for ADD	16.1–23	16.1–23	16.1–23	16.1–23	16.1–23
Data source: Byrd and Allen 2001					
Dev min by regression parameters	−25.98	−40.91	13.17	−39.35	−0.05648
ADD by regression parameters	40.5	549.5	58.8	587.5	360.5
ADD by mean of individual ADD	40.4	550.2	58.8	588.3	360.6
SE of mean	0.15	14.09	3.46	13.96	1.77
n	3	3	3	3	3
Valid temperature range for ADD	20–30	20–30	20–30	20–30	20–30
Data source: Nabity et al. 2006					
Dev min by regression parameters		5.895	5.987	5.895	5.831
ADD by regression parameters		162.2	108.1	162.2	274.2
ADD by mean of individual ADD		165.5	110.0	165.5	277.0
SE of mean		2.93	2.03	2.93	3.53
n		60	60	60	60
Valid temperature range for ADD		19.5–32.4	19.5–32.4	19.5–32.4	19.5–32.4

Note: Accumulated degree-days (ADD) to complete a stage are reported in centigrade degree-days, developmental minima are reported in °C, and ADD estimates are only valid across the indicated temperature range.

Development and Maggot Mass Temperatures

A related issue of particular importance to forensic uses of insect development information is the maggot mass temperature. The metabolic heat generated by a mass of maggots on dead tissue can be sufficient to raise the temperature of the mass well above ambient temperatures. Maggot mass temperatures may routinely occur over 20°C above ambient (Deonier 1940, Catts and Goff 1992, Haskell 1989, personal observation). At moderate ambient temperatures (about 25°C) maggot mass temperatures can exceed 45°C, and in cool or cold ambient conditions (even below freezing), maggot mass temperatures may be sufficient to allow continued development. This also is true when a heavily infested body is placed into a morgue cooler over a weekend. Maggot growth and development may not be arrested by the cool temperatures, and feeding can continue with additional destruction of tissues. Even with cooler temperatures in the morgue cooler (at approximately 38°F), temperatures have been recorded as high as 90°F within the maggot mass on a body that had been in the cooler for over 48 hours (Haskell, personal observation). Obviously, it is essential to account for maggot mass temperature in determining insect development. Principally, this is an issue for later-stage maggots, rather than eggs or earlier stages that do not generate substantial metabolic heat. Consequently, ambient temperatures may be used to determine development for earlier stages and maggot mass temperatures for later instars.

How can we account for maggot mass temperature? The maggot mass temperature will be a function of the size of the maggot mass, stage of development, location on corpse (which affects the relationship of mass temperature to ambient temperature), and ambient temperatures (Catts and Goff 1992). A common approach is to measure the maggot mass temperature and use this as a constant in determining development. This temperature can be well over 100°F and may be at or near the lethal high temperature for a species. Obviously, other influences are occurring because maggots remain alive in the mass at such a high temperature. A key principle here is that maggot mass temperatures are not uniform. From the center of the mass to the outside edge, a temperature drop of 6°C or more is possible (personal observation). Because maggots in the mass move throughout the mass for feeding, breathing, and thermoregulation, maggots will be exposed to this range of temperatures. Also, because mass temperatures often are near the upper temperature limit for a species, it is likely that using a mean mass temperature may overestimate development (based on rate summation effects in the upper portion of the development curve). A reasonable alternative to raw maggot mass temperatures would be to use a temperature that would provide the fastest growth of maggots (around 90 to 95°F for most species in the United States). This temperature would provide the minimum time for growth and development. Another more quantitative approach is to base an estimate of development on an average maggot mass. In determining the appropriate temperature to use, it is helpful to have maggot mass temperatures from the outside edge, as well as the center, of the mass. Improving our estimates of insect development in maggot masses is another important research issue in forensic entomology.

Developmental Delays and Confounding Factors

At the beginning of this section we highlighted the common need for backtracking from a given insect development stage to the time of oviposition. Such an approach provides a method for using insect development to help estimate the postmortem interval. Beyond the details and limitations in making development estimates, there are additional considerations that can influence those estimates.

To provide the most accurate estimate of insect development times, it is essential that we have an accurate understanding of the developmental history of the insect. Much of this relates to the thermal environment of the insect (as we have discussed), but other factors also are important. Often estimates are based on maggots collected from the corpse and maintained under constant (known) environmental conditions until pupation. One potential source of error is in the interval between collection and maintenance under controlled conditions. Handling maggots does not delay development, but chilling at 3°C does arrest development without increasing mortality (Johl and Anderson 1996), provided the maggots are in low numbers and maggot mass heating is not present. Unless maggots are maintained in conditions to explicitly stop development, a precise record of the thermal environment of insect from collection to deposition in a controlled rearing environment is essential for accurate development estimates.

Factors affecting the condition of a corpse may also affect insect development. Anything that might alter oviposition (such as coverings on the corpse, deposition of a body at night, etc.) will obviously influence developmental estimates, because initial oviposition may be delayed. Similarly, physical factors that constrain larvae migration, such as when the corpse is tightly wrapped, also may delay pupation, and therefore development (Wells and Kurahash 1994, Byrd and Butler 1996). Physical features, such as black cloth or plastic on or over remains, can increase temperatures, which will influence development estimates. More subtly, the presence of some antemortem drugs has been reported to alter insect development rates (Catts and Goff 1992). It may be possible to account for the potential influence of these various factors in making development estimates if these factors are recognized, which highlights the importance of having comprehensive information about conditions in and around remains where insects are collected.

Temperature Data

Although we have emphasized a number of potential problems and limitations in the use of temperatures for determining insect development, perhaps the key limitation is using the appropriate temperature data to determine development. In reviewing degree-day calculation methods, Higley et al. (1986) concluded that the temperature data used in degree-day calculations was likely the most significant single source of error. They also pointed out that variation in degree-day estimates would be most significant when ambient temperatures spanned the developmental threshold. Both of these conclusions seem particularly relevant in determining development of forensic insect species.

In practice, obtaining temperature data for determining insect development is a crucial need. Given that most situations will require use of data from the nearest recording site, direct comparisons are vital for daily maximum and minimum temperatures from the crime scene with recording station temperatures to allow for possible adjustments in estimates. Assuming weather conditions are not highly variable, temperature comparisons between the crime scene and weather service station should be conducted for 3 to 5 days (Haskell and Williams 2008). If such direct comparisons cannot be made, details on as many features of the insect site (topography, vegetation, covering, clothing color) as possible are vital to allow for informed adjustment of weather station data.

Describing Variation in Estimates

Another important consideration in forensic estimates is the need to provide courts with statements of variation in the estimate. This has not been done in agricultural estimates, because in making predictions it is possible to move forward the prediction to accommodate variation (Higley and Peterson 1994). In contrast, forensic estimates need to be as accurate as possible and have a clear, objective statement of variation in the estimate. Although no approach currently is available, we are exploring the use of Monte Carlo simulations in making such estimates of variation, and it may be other approaches can also be used as other investigators consider this issue. Despite the many sources of variation possible in making insect development estimates, undoubtedly many errors cancel out. This is another arena in which standard methods, agreed upon by the forensic entomology community, are clearly needed.

Conclusions

Insect thermal development is a powerful tool of forensic entomology, and additional research will improve its usefulness. Given the complexity of insect development and the many factors that influence development, the determination of insect development is necessarily an estimation process. With more research, we can expect the variation in our estimates to decrease. But thermal development estimates are unlikely to ever provide precise estimates of postmortem interval, except for those situations where we have very accurate temperature data. Instead, thermal development estimates are most useful in establishing a range for the postmortem interval, helping to indicate when death might or might not have occurred. Also, thermal development estimates are a valuable line of independent evidence when used in conjunction with other data, such as successional patterns or direct age grading of insects. Consequently, as our understandings and methods improve, it seems certain that estimates of insect thermal development will become increasingly useful in forensic entomology.

References

Anderson, G. S. 2000. Minimum and maximum developmental rates of some forensically important Calliphoridae (Diptera). *J. Forensic Sci.* 45:824–32.

Arnold, C. Y. 1959. The determination and significance of the base temperature in a linear heat unit system. *Proc. Am. Soc. Hortic. Sci.* 74:430–45.

Arnold, C. Y. 1960. Maximum-minimum temperatures as a basis for computing heat units. *Proc. Am. Soc. Hortic. Sci.* 76:682–92.

Ash, N., and Greenberg, B. 1975. Developmental temperature responses of the sibling species *Phaenicia sericata* and *Phaenicia pallescens*. *Ann. Entomol. Soc. Am.* 68:197–200.

Baker, D. G., Kuehnast, E. L., and Zandlo, J. A. 1985. Climate of Minnesota. Part XV. Normal temperatures (1951–80) and their application. University of Minnesota Agricultural Experiment Station, AD-SB-2777.

Byrd, J. H., and Allen, J. C. 2001. The development of the black blow fly, *Phormia regina* (Meigen). *Forensic Sci. Int.* 120:79–88.

Byrd, J. H., and Butler, J. F. 1996. Effects of temperature on *Cochliomyia macellaria* (Diptera: Calliphoridae) development. *J. Med. Entomol.* 33:901–5.

Catts, E. P., and Goff, M. L. 1992. Forensic entomology in criminal investigations. *Annu. Rev. Entomol.* 37:253–72.

Clarkson, C. A., Hobischak, N. R., and Anderson, G. S. 2004. A comparison of the development rate of Protophormia terraenovae (Robineau–Desvoidy) raised under constant and fluctuating temperature regimes. *Can. Soc. Forensic Sci.* 37:95–101.

Deonier, C. C. 1940. Carcass temperatures and their relation to winter blowfly populations and activity in the southwest. *J. Econ. Entomol.* 33:166–70.

Fisher, M. 2007. The effects of photoperiod on development rates of three species of forensically-important blow flies. M.S. thesis, University of Nebraska-Lincoln, Lincoln, NE.

Grassberger, M., and Reiter, C. 2001. Effect of temperature on *Lucilia sericata* (Diptera: Calliphoridae) development with special reference to the isomegalen—and isomorphen—diagram. *Forensic Sci. Int.* 120:32–36.

Grassberger, M., and C. Reiter. 2002a. Effect of temperature on development of *Liopygia* (= *Sarcophaga*) *argyrstoma* (Robineau–Desvoidy) (Diptera: Sarcophagidae) and its forensic implications. *J. Forensic Sci.* 47: 1–5.

Grassberger, M., and Reiter, C. 2002b. Effect of temperature on development of the forensically important Holarctic blow fly *Protophormia terraenovae* (Robineau-Desvoidy) (Diptera: Calliphoridae). *Forensic Sci. Int.* 128:177–82.

Greenberg, B. 1991. Flies as forensic indicators. *J. Med. Entomol.* 28:565–77.

Haskell, N. H. 1989. Calliphoridae of pig carrion in northwest Indiana: A seasonal comparative study. MS thesis, Purdue University, West Lafayette, IN.

Haskell, N. H., and Williams, R. E. 2008. Collection of entomological evidence at the death scene. In N. H. Haskell and R. E. Williams, Eds., *Entomology and death: A procedural guide*, 2nd ed. Clemson, SC: East Park Printing.

Higley, L. G., Pedigo, L. P., and Ostlie, K. R. 1986. DEGDAY: A program for calculating degree days, and assumptions behind the degree day approach. *Environ. Entomol.* 15:999–1016.

Higley, L. G., and Peterson, R. K. D. 1996. Initiating sampling programs. In L. P. Pedigo and G. D. Buntin, Eds., *Handbook of sampling methods for arthropods in agriculture*. Boca Raton, FL: CRC Press, 119–36.

Hilbert, D. W., and Logan, J. A. 1983. Empirical model of nymphal development for the migratory grasshopper, *Melanoplus sanguinipes* (Orthoptera: Acrididae). *Environ. Entomol.* 12:1–5.

Hochberg, M. E., Pickering, J., and Getz, W. M. 1986. Evaluation of phenology models using field data: Case study for the pea aphid, *Acyrthosiphon pisum*, and the blue alfalfa aphid, *Acyrthosiphon kondoi* (Homoptera: Aphididae). *Environ. Entomol.* 15:227–31.

Huntington, T. E. 2005. Temperature-dependent development of blow flies of forensic importance and the effects on the estimation of the postmortem interval. MS thesis, University of Nebraska, Lincoln.

Ikemoto, T., and Takai, K. 2000. A new linearized formula for the law of total effective temperature and the evaluation of line fitting methods with both variables subject to error. *Environ. Entomol.* 29:671–82.

Introna, F., Jr., Altamura, B. M., Dell'Erba, A., and Dattoli, V. 1989. Time since death definition by experimental reproduction of *Lucilia sericata* cycles in growth cabinet. *J. Forensic Sci.* 34:478–80.

Johl, H. K., and Anderson, G. S. 1996. Effects of refrigeration on development of the blow fly, *Calliphora vicina* (Diptera: Calliphoridae) and their relationship to time of death. *J. Entomol. Soc. Brit. Columbia* 93:93–98.

Kamal, A. S. 1958. Comparative study of thirteen species of sarcosaprophagous Calliphoridae and Sarcophagidae (Diptera). I. Bionomics. *Ann. Entomol. Soc. Am.* 51:261–70.

Lamb, R. J. 1992. Developmental rate of *Acyrthosiphon pisum* (Homoptera: Aphididae) at low temperatures: Implications for estimating rate parameters for insects. *Environ. Entomol.* 21:10–19.

Lamb, R. J., Gerber, G. H., and Atkinson, G. F. 1984. Comparison of development rate curves applied to egg hatching data of *Entomoscelis americana* Brown (Coleoptera: Chrysomelidae). *Environ. Entomol.* 13:868–72.

May, M. L. 1979. Insect thermoregulation. *Annu. Rev. Entomol.* 24:313–49.

McClain, D. C., Rock, G. C., and Stinner, R. E. 1990. San Jose scale (Homoptera: Diaspididae): Simulation of seasonal phenology in North Carolina orchards. *Environ. Entomol.* 19:916–25.

Nabity, P. D., Higley, L. G., and Heng-Moss, T. M. 2006. Effects of temperature on development of *Phormia regina* (Diptera: Calliphoridae) and use of development data in determining time intervals in forensic entomology. *J. Med. Entomol.* 43:1276–86.

Nabity, P. D., Higley, L. G., and Heng-Moss, T. M. 2007. Light-induced variability in the development of the forensically important blow fly, *Phormia regina* (Meigen) (Diptera: Calliphoridae). *J. Med. Entomol.* 44: 351–58.

Nishida, K. 1982. Experimental studies on the estimation of post mortem intervals by means of flies infesting human cadavers. *Jpn. J. Leg. Med.* 38:24–41.

Nishida, K., Shionaga, S., and Kano, R. 1986. Growth tables of fly larvae for the estimation of post mortem intervals. *Ochanomizu Med. J.* 34:9–24.

Reaumur, R. A. F. de. 1735. Observation du thermometre, faites a Paris pendant l'annee 1735, com- parees avec cells qui ont ete faites sous la ligne a Isle de France, a Alger et en quelque-unes de nos isles de l'Amerique. *Mem. Acad. Sci.* 545, Paris.

Reiter, C. 1984. Zum wachstumsverhalten der maden der blauen schmeißfliege *Calliphora vicina*. *Zeitschrift Rechtsmed.* 91:295–308.

Sharpe, J. H., and DeMichele, D. W. 1977. Reaction kinetics of poikilotherm development. *J. Theor. Biol.* 64:649–70.

Stinner, R. E., Rock, G. C., and Bacheler, J. E. 1988. Tufted apple budmoth (Lepidotera: Tortricidae): Simulation of postdiapause development and prediction of spring adult emergence in North Carolina. *Environ. Entomol.* 17:271–74.

Tanigoshi, L. K., Hoyt, S. C., Browne, R. W., and Logan, J. A. 1976. Influence of temperature on population increase of *Tetranychus mcdanieli* (Acarina: Tetranychidae). *Ann. Entomol. Soc. Am.* 69:712–16.

Vinogradova, E. B., and Marchenko, M. I. 1984. The use of temperature parameters of fly growth in the medicolegal practice. *Sud. Med. Ekspert.* 27:16–19.

Wagner, T. L., Wu, H., Sharpe, P. J. H., Schoolfield, R. M., and Coulson, R. N. 1984. Modeling insect development rates: A literature review and application of a biophysical model. *Ann. Entomol. Soc Am.* 77:208–25.

Wells, J. D., and Kurahashi, H. 1994. *Chrysomya megachephala* (Fabricius) (Diptera: Calliphoridae) development: Rate, variation and the implications for forensic entomology. *Jpn. J. Sanit. Zool.* 45:303–9.

Worner, S. P. 1992. Performance of phenological models under variable temperature regimes: Consequences of the Kaufmann or rate summation effect. *Environ. Entomol.* 21:689–99.

The Soil Environment and Forensic Entomology

11

SHARI L. FORBES
IAN DADOUR

Contents

Introduction

Cadaver decomposition involves complex chemical processes that are readily influenced by surrounding conditions. A body decomposing in an outdoor environment will be affected by ecological conditions such as climate, vegetation, and the soil environment. The soil environment is particularly important, as the process of decomposition can be altered considerably by soil type, pH, moisture, and oxygen content. The type of soil environment will also affect the invertebrate fauna associated with decomposition. As a result, the rate of decomposition will vary for a body decomposing on the soil surface compared to a body decomposing in a burial environment. It is therefore important to consider the effect of the soil environment on entomological evidence collected as part of a forensic investigation. This review will discuss the chemical processes of decomposition and preservation, as well as the effect of the soil environment, and subsequently the insect fauna, on these processes. A better understanding of these processes is imperative in the application of forensic entomology.

Human Decomposition

Chemical Processes of Decomposition

Decomposition of soft tissue is an inherently complex process that involves the degradation of the biological macromolecules that comprise the human body (Vass et al., 2002). The process of decomposition commences soon after death and undergoes numerous phases before reaching a final stage. The phases of decomposition are intrinsically dependent on the surrounding ecological conditions and, as a result, may vary considerably from one environment to another (Galloway, 1997). Decomposition follows the early and late postmortem changes, which include settling of the blood due to gravitational forces (livor mortis), acclimatization of the body to ambient temperature (algor mortis), and postmortem stiffening of the muscles (rigor mortis). It refers to the entire process, from the onset of the chemical degradation of soft tissue through to skeletonization (if indeed it occurs), and may be further subdivided into autolysis and putrefaction (Clark et al., 1997).

Autolysis is governed by hydrolytic enzymes (lipases, proteases, and amylases) that are released into the cytoplasm and digest the lipids, proteins, and carbohydrates. Self-digestion of the cell membrane causes these molecules to be released, thus providing nutrient-rich fluids that are utilized by microorganisms present in the body (Vass et al., 2002). The onset of autolysis results from cell deprivation of oxygen, which leads to an increase in pH and the accumulation of waste material. Autolysis will occur at different rates in different cell types and organs (Dix and Graham, 2000) and will generally commence sooner in cells containing a higher number of hydrolytic enzymes (e.g., liver, pancreas, stomach) (Gill-King, 1997). Like all stages of decomposition, the process of autolysis is temperature dependent. A cold environment such as a refrigerator will considerably retard the process. Conversely, high ambient temperature or high body temperature as a result of antemortem fever or physical exertion will accelerate autolysis (Clark et al., 1997). Initially, autolysis can only be seen microscopically; however, within approximately 48 hours after death, physical changes may be apparent on the body. Observable characteristics commonly associated with autolysis include skin slippage, postmortem bullae (fluid-filled blisters), and marbling as a result of intravascular hemolysis.

As the process of autolysis nears completion, putrefaction will commence and is characterized by the breakdown of the proteins, lipids, and carbohydrates that were released during autolytic degradation (Clark et al., 1997). The soft tissue disintegrates through the actions of microorganisms (bacteria, fungi, and protozoa) that invade the body immediately after death (Vass et al., 2002). The rapid growth of bacterial inhabitants promotes the formation of distinctive decomposition gases and odorous organic compounds. Accumulation of these gases in the soft tissue and gastrointestinal tract produces bloating of the cadaver, a common characteristic of the putrefactive stage (Gill-King, 1997). Additional characteristics of this stage include purging of fluids from orifices, green discoloration due to the formation of sulfhemoglobin, and observable degeneration of tissue (Figure 11.1).

Putrefaction is also referred to as wet decomposition or active decay because it is during this stage that the major chemical reactions occur to disintegrate soft tissue into a liquefied mass (Dix and Graham, 2000). Proteolysis of the muscle protein will yield amino acids, which in turn will decompose to form biogenic amines and volatile fatty acids. Carbohydrate fermentation results in the production of gases, organic acids, and alcohols

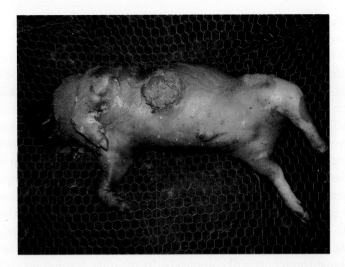

Figure 11.1 Pig (*Sus domesticus*) carcass showing discoloration, purging of fluids, and degeneration of the soft tissue following placement on the soil surface in a southern Ontario environment during summer 2007. (Photo courtesy of Dr. Shari Forbes.)

(Gill-King, 1997). Lipid degradation will produce glycerol, aldehydes, ketones, and both short- and long-chain fatty acids. Release of the cadaveric material is commonly observed as a sheath or island of liquefied tissue surrounding the cadaver (Figure 11.2). This sheath has been recently referred to as a "cadaver decomposition island," and represents a flux of nutrient material as a result of the decomposing remains (Carter et al., 2006b). As with autolysis, the process of putrefaction is significantly affected by surrounding environmental conditions, and the timeframe for putrefaction to occur may vary considerably.

The later stage of decomposition that follows putrefactive degradation is often referred to as advanced decay or dry decomposition. Advanced decay is characterized by the loss of soft tissue with only the skin, cartilage, and bone remaining (VanLaerhoven and Anderson, 1999).

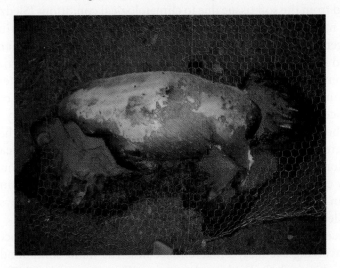

Figure 11.2 Pig (*Sus domesticus*) carcass showing the release of decomposition fluids and the formation of a cadaver decomposition island (CDI) following placement on the soil surface in a southern Ontario environment during summer 2007. (Photo courtesy of Dr. Shari Forbes.)

Due to the significant loss of moisture and nutrients from the body, and the inability for bacteria to proliferate, decomposition is minimal at this stage (Carter et al., 2006b). Desiccation of the tissue will be apparent, and very little activity will be present on the remains.

Advanced decay leads to skeletonization, the final stage of decomposition, which is characterized by desiccated material and exposed skeletal remains (Galloway, 1997). The cranium is often the first to skeletonize, followed by exposed extremities, including the arms and lower limbs (Roksandic, 2002). Eventually all soft tissue will become desiccated or completely lost to the environment, thus exposing the skeleton and allowing disarticulation to commence (Clark et al., 1997). The time taken for skeletonization to occur will be entirely dependent on the microenvironment. Remains that have decomposed in an outdoor environment will show bleaching within 2 to 6 months and exfoliation after approximately 12 to 18 months (Galloway, 1997). The environment will act on the exposed bone and disintegrate it over time through weathering, physical breaking, and dissolution by water. Carnivore activity may also assist the breakdown of the bone material. Eventually the remaining bone fragments will completely disintegrate, leaving no visible evidence of the decomposition process.

Chemical Processes of Preservation

Decomposition is the predominant process that occurs in human bodies following death. However, modifications to this process may occur based on the surrounding ecological conditions and their variable impact on the degradation stages. Modifications of decomposition that are based on the environment result in the natural preservation of the body. The preservation may be partial or complete, depending on whether the entire body, or only portions of it, is exposed to the environment (Sledzik and Micozzi, 1997). The chemical processes of preservation most often encountered in forensic investigations are mummification and adipocere formation.

Mummification is the product of desiccation, a process that occurs when moisture is rapidly removed from the soft tissue due to climatological conditions (Sledzik and Micozzi, 1997). Initially, moisture is lost from those regions of the body that contain minimal fluid, such as the fingers and toes. The skin becomes desiccated, turning dark and leathery, and can form a hard layer over the underlying tissue (Galloway, 1997). Internal organs, if not affected by decomposition, will shrink and desiccate, although they are often the last tissues to become mummified (Dix and Graham, 2000). Mummification is most often observed in the scalp and face, followed by the chest, back, and extremities (Galloway, 1997). Extreme environments, both hot and cold, can facilitate natural mummification; however, hot, dry environments are more commonly associated with the process of desiccation due to the rapid evaporation of moisture from the body. The time required for mummification to occur will vary considerably with different environmental climates. Mummification has been reported in as little as 8 days in a dry, heated apartment and 10 days in an open-air environment (Sledzik and Micozzi, 1997).

As with decomposition, mummification of skin and other tissue occurs via chemical pathways involving the body's macromolecules. The main compound classes typical in postmortem skin include lipids and proteins (Bereuter et al., 1997). Once the process of mummification commences, the proteins and lipids are released within the tissue. Desiccation will cause changes in the protein content along with possible changes in the secondary protein structure (Gniadecka et al., 1999). Disintegration of the epidermis during mummification will lead to a loss of proteins and a predominance of triacylglycerols as a result of lipid degradation (Bereuter et al., 1997). Triacylglycerols will eventually decompose to their constituent fatty acids, which will predominate the mummified tissue (Mayer et al., 1997). The fatty acids present in mummified tissue are similar to those found in adipoceratous tissue but vary in ratio and composition.

The formation of adipocere results from the degradation of lipids in soft tissue and yields a soft, white substance comprising predominantly fatty acids (Fiedler and Graw, 2003). Although the chemical process is not completely understood, it is generally agreed that hydrolysis and hydrogenation of the neutral fats occurs to form adipocere. It will initially form in the regions of the body containing higher fat content, but may also be observed in leaner tissue due to the translocation of liquefied fats. Many studies suggest that adipocere formation requires bacterial action, excess moisture, and an anaerobic environment (Yan et al., 2001). While these factors will enhance the formation of adipocere, recent studies have shown that there is sufficient moisture in soft tissue for adipocere to form even in a dry environment (Forbes et al., 2005a). Furthermore, adipocere formation has been reported to form as a result of both biotic and abiotic processes (Makristathis et al., 2002). As with all decomposition and preservation processes, its formation can be accelerated or retarded by the circumstances and environment surrounding the body.

Adipocere formation has been reported in various environments, including soil burials (Forbes et al., 2002, 2003; Vane and Trick, 2005; Fiedler et al., 2004), water bodies (Kahana et al., 1999; Mellen et al., 1993), and glaciers (Makristathis et al., 2002; Mayer et al., 1997; Bereuter et al., 1997). Its formation acts to preserve the underlying tissue or organ and, in some cases, may even preserve recognizable facial characteristics of the body (Fielder et al., 2003). Once formed, adipocere is a relatively stable product; however, decomposition can eventually occur. The process of adipocere decomposition has not yet been determined, and it is unclear which environmental factors will cause it to disintegrate. Studies suggest that aerobic conditions and the presence of gram positive bacteria may aid the decomposition of adipocere (Pfeiffer et al., 1998).

Rate of Decomposition: Surface versus Burial

Decomposition rates on the surface and in burials are extremely diverse due to the different conditions present in both types of environment. When measured under the same ecological conditions, the rate of decomposition of buried bodies is generally much slower than that of a body exposed to the air (Rodriguez, 1997). The two major factors that contribute to accelerated decomposition on the surface are insect and carnivore activity. Insect activity accounts for the majority of soft tissue degradation due to the secretion of enzymes and bacteria by fly larvae, which facilitates consumption of the liquefied soft tissue. Carnivore activity can also be extensive and will result in the modification and consumption of soft tissue and bone, as well as scattering of the remains (Haglund, 1997).

Numerous other factors will affect the rate of surface decomposition, and some of these factors can be linked to the activity associated with decomposing remains. Ambient temperature is by far the most important ecological factor to affect decomposition rates. Warm environments are more conducive to the rapid degradation of soft tissue because they allow for the survival and growth of the bacteria associated with decomposition (Polson et al., 1985; Micozzi, 1997). Additionally, fly activity on a cadaver is more prominent in warmer temperatures because fly eggs and larvae are unable to survive at temperatures much below freezing (0°C) (Mann et al., 1990). Similarly, a lack of humidity or moisture is detrimental to both decomposition bacteria and fly larvae (Bass, 1997). Both require moisture to survive, and as a result, arid environments are more commonly associated with mummification, rather than decomposition.

Characteristics of the body at the time of death will also influence the rate of decomposition. Trauma or wounds on a body will accelerate decomposition or result in localized skeletonization due to enhanced maggot activity. Flies are quickly attracted to the site because it represents an ideal location for laying eggs due to the exposed moisture (Anderson and Cervenka, 2002). Clothing present on the body may alter the rate of decomposition by providing protection from sunlight for fly larvae (Mann et al., 1990), or by absorbing moisture, which may result in differential decomposition patterns. Other coverings, such as carpet and plastic, may accelerate decomposition if the remains are still accessible by insects (Morton and Lord, 2002), or may completely preserve the body by restricting insect access (Rodriguez, 1997).

The slowed rate of decomposition in a burial environment can also be attributed to a lack of the two major contributors: fly and carnivore activity (Rodriguez, 1997). The burial of a body will significantly limit or completely prohibit access to the body by insects and carnivores. The degree to which this access is limited will generally correlate with the burial depth. In deeper burials (more than a foot), the odors associated with decomposition are not detectable aboveground, and decomposition will proceed without attracting insects and other animals. In shallow burials (less than a foot), the odors are capable of penetrating the soil, thus attracting insects and carnivores (Rodriguez and Bass, 1985). Insects gain access to the body by burrowing through cracks in the grave surface, which result from soil compaction. Carnivores will instead expose the remains in order to feed on the soft tissue or skeletal material. Exposure of the remains by carnivore activity will subsequently provide additional access for insects and other arthropods (Turner and Wiltshire, 1999). For these reasons, decomposition in shallow burials is often faster than in deep burials, but slower than on the surface. The other major factor that accounts for the reduced rate of decomposition in a burial site is the soil environment.

Soil Environment

Soil as the Matrix

In burial environments, soil acts as the matrix and represents a complex assemblage of minerals, organic matter, and various microscopic and macroscopic organisms (Dent et al., 2004). The soil system is composed of three phases: solid, liquid, and gas. The solid phase is a mixture of inorganic and organic material that makes up the skeletal framework of soils. In mineral soils the inorganic fraction is present in significant quantities, while the organic fraction is found in substantially smaller amounts (Tan, 1994). The inorganic soil fraction can be distinguished into three major soil fractions: sand, silt, and clay, according to size. Sand grains are irregular in size and shape and are not sticky or plastic when wet. Their presence in soil produces a loose and friable condition, allowing rapid water and air movement in all directions. Silt particles are intermediate in size and possess characteristics between those of sand and clay. Silt may exhibit some plasticity due to clay film coatings on the silt particles, promoting an absorptive capacity for water and cations. Clay is the smallest particle in soil and has colloidal properties. It is chemically the most active inorganic constituent in soils, as it typically carries an excess negative charge. The presence of clay in a soil increases the water-holding and cation exchange capacities, and promotes a sticky plasticity when wet (Grim, 1968).

Soil Factors Affecting Decomposition

The soil type will have a significant influence on the process of decomposition based on the moisture it retains and the air movement it permits. A burial in a sand type soil is unlikely to retain significant amounts of water and, when coupled with extreme temperatures such as desert environments, will most likely lead to mummification of the soft tissue. However, a sandy burial containing sufficient moisture, such as a beach environment, will allow decomposition to proceed and may also encourage partial adipocere formation (Forbes et al., 2005a). A clay type soil environment is also conducive to adipocere formation due to the excess moisture retained by the clay particles. Bodies recovered from these types of soil often show extensive adipocere formation of almost the entire body (Rodriguez, 1997).

Regardless of soil type, the amount of oxygen present in a burial site is limited, and its use in the decomposition process depends on gas diffusion in the attendant soil (Dent et al., 2004). Initially, the oxygen available in a burial will be in excess due to the aeration of the soil through the processes of digging and backfilling into the grave. However, soil compaction as a result of backfilling, rainfall, and subsequent water percolation (Hunter and Martin, 1996) will reduce the amount of oxygen available for the chemical processes of decomposition. Calculations suggest that as little as 150 to 200 g of oxygen gas are available in a deep burial, and even less in a shallow burial, thus demonstrating that the decomposition processes in a burial environment are predominantly anaerobic (Dent et al., 2004). An anaerobic environment will affect decomposition by restricting the types of bacteria and fungi that can survive in the gravesite.

In addition to restricting insect and animal activity, the burial depth can also affect decomposition through variations in temperature (Weitzel, 2005). In a shallow burial, the body will undergo temperature fluctuations similar to those experienced aboveground, and decomposition will be accelerated when compared to a burial at lower depths. In addition to soil temperature fluctuations, the body temperature will also fluctuate and correlate directly with the burial depth (Rodriguez and Bass, 1985). Furthermore, the remains will be subjected to the activity of soil organisms and plant growth due to the nutrient-rich upper layer of soil (Rodriguez, 1997).

In the short term, soil pH is not a particularly important factor governing decomposition; however, over time it can have an influence on the extent of preservation of soft tissue and associated materials (Janaway, 1996). Soil pH values normally range from 3 to 9, although occasionally values outside this range are found. Generally, neutral or alkaline soils will demonstrate better soft tissue and bone preservation. Highly acidic sandy and gravelly soils have the ability to reduce a body to a dark stain (Bethell and Carver, 1987). It has been shown that even the skeletal material may be partially or completely dissolved under highly acidic conditions (Merbs, 1997). Conversely, in highly alkaline environments such as those containing calcium oxide, preservation of soft tissue may be extensive due to the inhibitory effect it has on bacterial survival (Forbes et al., 2005b). Unfortunately, skeletal remains do not fare as well in highly alkaline soils (Rodriguez, 1997).

Soil pH can also vary as a result of the decomposition process. During the early stages of decomposition, an acidic environment is created around the body due to the fermentation of anaerobic bacteria in the soil and the large bowel (Gill-King, 1997). A lowered soil pH will enhance the growth of ubiquitous fungi and increase plant activity in the vicinity, both of which may act upon the remains and alter the process of decomposition. As decomposition progresses, the soil pH may increase in alkalinity by a value between 0.5 and 2.1 as a result of proteolysis (Rodriguez and Bass, 1985). Small pH changes are unlikely to affect decomposition; however, large increases in alkalinity may affect the proliferation of certain bacteria species.

Soil Microorganisms Affecting Decomposition

As demonstrated previously, soil factors present in a burial environment can alter the rate of decomposition due to their effect on soil microorganisms. Soil microorganisms as a group include bacteria, fungi, algae and protozoa (Janaway, 1996). However, those organisms that are autotrophic and do not require an organic substrate to survive (e.g., algae and protozoa) do not have a significant bearing on the decomposition process. For this reason, only bacteria and fungi will be considered here.

Soil bacteria are microscopically small and can be found in abundance in food-rich soils. Under favorable conditions they can multiply rapidly, producing several billion in 1 g of soil. Their distribution is not uniform and is determined by the amount of organic matter in the soil, which can originate from dead plant material, fecal matter, or a decomposing body or other animal. As a result, bacterial distribution can be correlated to burial depth in an undisturbed soil (Janaway, 1996). Bacterial growth is dependent on the availability of a suitable food source and can be affected by soil conditions, including temperature, moisture, oxygen content, and pH.

Temperature will directly affect bacterial growth, especially in extremes of heat and cold. At temperatures below 4°C, bacterial growth is inhibited, which subsequently leads to preservation of the soft tissue for a period of time (Micozzi, 1997). At temperatures above 4°C but below 10°C, bacterial reproduction is severely slowed but does not halt altogether; hence, decomposition can still occur. Bacterial proliferation can occur between 10 and 40°C but is greatest between 15 and 37°C, as this temperature range allows for the highest rate of bacterial cell division. Decomposition is therefore rapid in an environment that falls within this temperature range. At temperatures significantly above this, bacterial growth will once again be inhibited, and the soft tissue will be preserved through the process of mummification (Galloway, 1997).

As has been previously noted, bacteria require moisture to survive and reproduce. As soil becomes dry, the microbial population will decrease considerably, which will ultimately affect their ability to decompose soft tissue (Janaway, 1996). Similarly, an environment with excess moisture, such as a waterlogged burial, will also inhibit bacterial growth through the reduction of available oxygen. The optimum pH value for bacteria proliferation is greater than 7, although soil bacteria can tolerate values between 3 and 10. Bacterial presence is often limited by both low pH and low oxygen content. In an acidic soil with limited aeration, anaerobic bacteria will be limited by pH, while aerobic bacteria will be limited by the low oxygen content.

Fungi are heterotrophic organisms that require an organic substrate for growth and development. They can obtain their food from nonliving organic matter, such as dead plants and animals, or from living organic matter by acting as parasites. Microfungi are the most active fungi in the decomposition process due to their ubiquitous nature. Their ability to produce a broad range of degradative enzymes makes them powerful decomposers of all types of organic matter. Microfungi use successional behavior to break down complex organic substrates. Primary colonizers will grow and sporulate rapidly but are only capable of degrading relatively simple carbohydrates. Secondary colonizers will subsequently take over and are capable of degrading complex carbohydrates and other compounds. In addition to these colonizers, other specific microfungi will act to degrade difficult organic substrates, such as hair and nails. Similar to bacteria, the microfungi succession associated with a cadaver will be determined by the surrounding environmental conditions, including temperature, moisture, and oxygen content (Janaway, 1996).

Microfungi that feed on decaying remains require a relatively high moisture content in both their food source and the surrounding atmosphere. It has been reported that the optimal moisture content should be greater than 20% in the food substrate, and the humidity of the atmosphere should be greater than 65% to minimize desiccation of microfungi (Cronyn, 1990). Many microfungi also require oxygen to function and, as a result, cannot survive in anaerobic environments. In a burial environment, the soil environment will initially be dominated by bacteria during the early stages of decomposition because microfungi are unable to compete with the rapidly colonizing bacteria (Carter et al., 2006b). However, in the later stages of decomposition, microfungi will appear and start to dominate the soil environment.

Macrofungi have also been identified as having a potential role in cadaver decomposition. It has been suggested that the advanced stage of decomposition may be associated with postputrefaction fungi (Sagara, 1981). It is thought that postputrefaction fungi will fruit in response to the large influx of ammonia that a decomposing cadaver adds to the soil (Carter and Tibbett, 2003). During the skeltonization stage, postputrefaction fungi may also fruit in response to the influx of organic nitrogen, ammonium, and nitrate. Research in this area suggests that macrofungi associated with cadaver decomposition may be useful for locating clandestine gravesites, or with further research, may demonstrate a predictable fruiting sequence that can be used to estimate postburial interval (PBI) (Carter et al., 2006a).

Role of Soil Microbial Community in Decomposition

When a cadaver is placed in the ground, soil microbes will respond to it rapidly, and microbial activity has been recorded as soon as 24 hours after deposition (Tibbett et al., 2004). During the early and intermediate stages of decomposition, including autolysis and putrefaction, purging of fluids into the surrounding soil will result in an increase in soil microbial biomass, accompanied by changes to the soil faunal community and an increase in the soil nutrient level (Carter et al., 2006b). At the same time, maggot activity will increase the amount of energy and nutrients that enter the soil, and peak levels of microbial activity will be observed during this phase. Through the degradation of soft tissue and rupturing of the skin, soil microbial propagules can enter the cadaver and may contribute to the decomposition process. As nutrients and moisture are depleted and advanced decay predominates, the activity of decomposer microorganisms will also start to decline. Some microfungi that are capable of degrading keratinous material, such as hair and nails, will remain. Changes in soil nematode community may also be observed during this stage (Carter et al., 2006a).

This later stage of decomposition has been associated with an increase in bacteria-feeding nematodes, followed by fungal-feeding nematodes. It is also predicted that a concomitant increase in nematophagous fungi will result from the bacterial-fungal succession (Jaffee et al., 1992). Research in this area has the potential to provide additional insight into the soil microbial activity that results from cadaver decomposition. Further research in the area of soil microbial community analysis may also be useful in identifying decomposition bio-markers that are indicative of burial environments.

Soil Fauna and Decomposition

Role of Invertebrates in Decomposition

The invertebrate assemblage associated with carrion is composed mainly of insects (Smith, 1986). The role of insects in the decomposition process has become a well-recognized area of forensic investigation, with forensic entomology commonly employed in homicide investigations to estimate postmortem interval (Morris and Dadour, 2005). The estimation of postmortem interval (PMI) is based upon knowledge of the locality-specific succession of insects occurring on a corpse following death. This predictable succession may be used, in conjunction with temperature-dependent developmental data for carrion-frequenting insects, to estimate PMI based on the minimum amount of time required for insect development. The use of successional data in estimation of PMI assumes that following death, an orderly and predictable succession of insect species occurs on a corpse. The insects may be classified as adventive or incidental species, simply visiting by chance and having little forensic relevance (Goff and Catts, 1990); predators that may feed on other species that have already colonized the body; or necrophages that feed from the body itself and are most useful in estimation of PMI.

If decomposition occurs in an insect-accessible area, the process of autolysis will provide a suitable substrate for fly colonization (Carter et al., 2006b). Insects are attracted to a body within minutes after death, and blow flies (Diptera: Calliphoridae) are often observed as the first colonizers on a cadaver (Anderson, 2001; Haskell et al., 1997). During the early stages of decomposition, oviposition will predominate and minimal feeding will occur, as the tissue has not yet begun to disintegrate. Egg laying will occur in the facial orifices, including nasal openings, ears, mouth, and eyes (Galloway, 1997), the genital region (Rodriguez and Bass, 1983), or near a wound site. The site of oviposition is important to ensure that the hatched first instar has access to a suitable liquid protein meal (Anderson and Cervenka, 2002). Oviposition by one female attracts many more females to lay eggs on the cadaver, thus producing large numbers of eggs in a concentrated region (Anderson, 2001). Over time, these eggs will yield large maggot colonies, also referred to as maggot masses, which will ultimately assist in the decomposition of the body.

In addition to blow flies, flesh flies (Diptera: Sarcophagidae) will be seen during the early stages of decomposition. The flesh flies have an advantage over blow flies because the female will larviposit directly onto the cadaver, thus increasing its chances of acquiring a food source before other competitors (Haskell et al., 1997). Beetle species, including rove beetles (Coleoptera: Staphylinidae), carrion beetles (Silphidae), hister beetles (Histeridae), and hydrophilid beetles (Hydrophilidae), may also be observed feeding on the fly eggs and

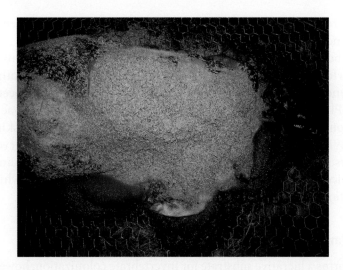

Figure 11.3 Pig (*Sus domesticus*) carcass showing frothing of the decomposition fluids as a result of maggot mass movement and digestion following placement on the soil surface in a southern Ontario environment during summer 2007. (Photo courtesy of Dr. Shari Forbes.)

larvae. In the presence of high beetle populations, the fly population on the cadaver may be reduced, thus affecting the decomposition process.

Following oviposition during the autolysis stage, the eggs will hatch into first instar (or may already be first instar due to larviposition) and will subsequently mature through the second and third stages of larval development as putrefaction progresses (Anderson and Cervenka, 2002). The gases emanating from the cadaver will attract further colonizers, which will subsequently oviposit or larviposit the cadaver. The odors vary in attractiveness to different insects, and as the body decomposes and various resources are depleted, new insect types will colonize, being more suited to the current decompositional stage. These insect taxa reflect the physical changes in the body and are therefore predictable and useful in estimation of PMI. Fly larvae will feed extensively on the liquefying soft tissue, and peak maggot activity will be observed during the putrefaction stage (Carter et al., 2006b). The large maggot masses that form during this stage may greatly increase the temperature of the cadaver, thus accelerating the decomposition process. The movement and digestion of maggot masses can produce frothing of the decomposition fluids (Figure 11.3). The size of the maggot masses present on the cadaver will contribute to the lateral extent of the cadaver decomposition island that forms. During the putrefaction stage, carrion and hide beetles will be observed feeding on the fly larvae, while checkered beetles (Coleoptera: Cleridae) will feed directly on the decomposing tissue (Wolff et al., 2001; Rodriguez and Bass, 1983; Haskell et al., 1997). Putrefaction will continue until the fly larvae have migrated from the cadaver, at which point they enter their pupal stage.

Following putrefaction, the remains start to desiccate as the body enters the advanced decay stage. Carrion flies often observed during the advanced stages of decomposition include black soldier flies (Diptera: Stratiomyidae), house flies (Diptera: Muscidae), and skipper flies (Diptera: Piophilidae). As the tissue desiccates, the remains become less attractive to fly colonizers, and significant numbers of beetles and beetle larvae may be observed feeding on the dry remains (Anderson and VanLaerhoven, 1996). Predaceous beetle species include the rove beetles, carrion beetles, hister beetles, and checkered beetles. Beetle activity on dry remains can produce a pitting of the skin, which is often confused with perimortem or postmortem trauma (Haskell et al., 1997).

Dermestid beetles (Coleoptera: Dermestidae) are common during this stage and have even been reported on skeletal remains (Wolff et al., 2001). A variety of other beetle families may be found in association with the skeleton, including carcass beetles (Trogidae), spider beetles (Ptinidae), the red-legged ham beetles (Cleridae), and larvae of some of the above beetle groups, which may be observed feeding on remains of hair, skin, and clothing (Smith, 1986). With the exception of beetles, very few insects will be observed on a cadaver once it becomes skeletal (Rodriguez and Bass, 1983). However, close examination of the remaining crevices may yield the remains of early successional insects, which can provide useful information to the forensic entomologist.

Role of Soil Fauna in Decomposition

Carrion itself forms a lucrative substrate for invertebrate colonization; however, in outdoor situations, the interface between this substrate and the underlying soil also forms attractive environments for certain organisms. The seepage of nutrient-rich fluids into the soil beneath the cadaver significantly alters the microenvironment, affecting the inhabitant fauna. The arthropod assemblage may thus be considered to be affected by and reflective of the decomposition of the cadaver, and may therefore have some potential in contributing to forensic investigations (Bornemissza, 1957).

While the body itself forms the primary decompositional site, the soil beneath it may be equally important. Cadavers located in outdoor environments on a terrestrial surface create an interface within which soil fauna and carrion-dwelling organisms interact. The interactions in this zone are affected by soil type, vegetation, decomposition of the body, and a variety of environmental factors. Apart from the work by Bornemissza (1957) and Lundt (1964), the succession of insects in this interface, and within the soil itself, has been largely overlooked in the literature, and the forensic implications have yet to be considered.

A decomposing body in a terrestrial environment will considerably alter the substrate beneath it. This initiates a series of changes in vegetation and fauna, beginning a succession of arthropods affected by the decomposing carrion above. Bornemissza (1957) observed the greatest effect of the decomposing cadaver on the soil beneath to occur during the autolytic and putrefactive stages. Fluid seepage contributes to development of a crust of hair, plant matter, and the upper-most soil layer beneath the body. During putrefaction, the decomposition fluids released from the body, along with the waste products excreted by the insects feeding on the body, combine to kill the plants beneath the body (Figure 11.4) and the soil fauna, altering the microenvironment. Anderson and VanLaerhoven (1996) found that vegetation under and around the body for 20 to 30 cm was killed by fluids released during active and advanced decay stages, and the number of arthropod species in the soil was reduced from thirty species before carrion placement to two species by 15 days following placement in British Columbia.

During the active decay stage the soil beneath the carrion may become disturbed to a depth of approximately 1 inch by the action of arthropods, particularly dipteran larvae, burrowing (Reed, 1958). Decomposition fluids and associated arthropods are reported to affect the soil to a depth of 14 cm, with the most effect in the upper soil layers. The area directly beneath the body, the "carrion zone," serves as a decompositional zone occupied by carrion dwellers, distinct from a surrounding area of approximately 10 cm, which

Figure 11.4 Alteration of the microenvironment surrounding a decomposing pig (*Sus domesticus*) carcass following placement on the soil surface near Mead, Nebraska, during summer 2005. (Photo courtesy of Dr. David Carter.)

provides an "intermediate zone" of both carrion and regular soil-dwelling invertebrates (Bornemissza, 1957). At a distance of 10 to 20 cm away from the body, the soil fauna is typical of general litter-dwelling fauna, but perhaps the size of these zones may be dependent on the size of the carrion, as the work of Bornemissza (1957) was based on guinea pigs, and human decomposition may produce greater amounts of fluid.

Payne (1965) indicated that in South Carolina, once carrion reached the dry stage of decomposition, there began an overlap of carrion and soil-dwelling insects on and around the corpse. Bornemissza (1957) recorded that one year following the placement of carrion, the soil arthropod assemblage had not yet returned to its predecompositional state. The soil-surface and litter-dwelling arthropods took longer to recolonize than subterranean taxa. The crust formed by hair, vegetation, and fluids persisted long after placement, and was most pronounced beneath oral and anal regions, where seepage was greatest. Restoration of the soil community occurred only following heavy rains. Anderson and VanLaerhoven (1996) similarly observed that 271 days following death, the vegetation and soil fauna had not returned to normal. This alteration to the soil may perhaps be informative in PMI estimation for an extended period of time following death.

Invertebrates in Gravesites

Buried remains, while excluding most of the insect fauna generally seen on surface remains, may still contain their own unique set of insect fauna within the burial environment. The most important effect of burial on decomposition is the increase in time required for biomass reduction, relative to exposed carrion. Although a burial environment will retard the rate of decomposition by limiting insect activity, many invertebrates still have the ability to access buried remains. In a burial situation, certain insects, such as the flies *Muscina* spp. (family Muscidae) and *Morpholeria kerteszi* (family Heliomyzidae), will lay eggs on the soil surface, and once hatched, larvae will burrow through the soil to the carrion (Smith, 1986). Other adult insects, including the Staphylinidae (Coleoptera) and Phoridae (Diptera), will burrow down through the soil to oviposit directly onto carrion. Cheese skipper larvae (Piophilidae) have been recorded on buried carrion in rural areas of Nova Scotia (Simpson and Strongman, 2002), and Sphaeroceridae (Diptera) may also be observed in burial environments (Bourel et al., 2004). Although buried remains can still be colonized by insects, the sequence of colonization, species involved, and rate of decomposition may be altered due to influences from the burial environment (Anderson, 2001).

The burial depth plays a distinct role in determining the type and number of arthropods observed in a gravesite. One of the earliest studies of arthropod succession on buried pigs identified both fly (Sphaerocerids, Leptocera, Phorids, and Psychodids) and beetle (Staphylinids) species at depths of 50 to 100 cm. Additional arthropods observed feeding on the carrion included ants, mites, collembola, cryptophagids, and millipedes (Payne and King, 1968). Of the forty-eight arthropod species identified in the study, twenty-six species were associated only with the buried pigs and were not identified in succession studies on the surface (Payne, 1965). However, the pig carcasses were placed in mock coffins before being buried in the shallow graves, and the study is therefore not representative of most clandestine burials encountered in forensic investigations.

Although the arthropod species observed in a grave may be similar, the burial conditions reported can vary considerably. Smith (1986) suggests that blow flies, responsible for the majority of biomass reduction on carrion, are excluded from the corpse at a depth of just 2.5 cm. Lundt (1964) also observed the exclusion of *Calliphora* and *Lucilia* sp. at a depth of 1 to 2 cm. However, Simpson and Strongman (2002) report the occurrence of the blow fly *Cynomyopsis cadaverina* on carrion buried at a depth of 30 cm. Lundt (1964) observed *Muscina pabulorum* and *Ophyra leucostoma* at 2.5 to 10 cm depth, and Phoridae and Staphylinidae at 25 to 50 cm below ground. Rodriguez and Bass (1985) observed Sarcophagidae (flesh fly) and Calliphoridae (blow fly) larvae on burials of human cadavers at a depth of 1 foot (~30 cm). They also reported temperature increases in the cadaver when compared with the surrounding soil temperature, presumably as a result of insect and bacterial activity.

Studies by VanLaerhoven (1997) and VanLaerhoven and Anderson (1999) also observed the presence of flesh flies and blow flies in pig burials at a depth of approximately 1 foot; however, maggot masses did not form, and carcass temperatures were found to be comparable to soil temperatures. Insect colonization of buried remains was distinctly different from exposed remains, which were used as a control group in the studies. In general, muscid flies were more common on buried remains than on surface carcasses. Additionally, species that colonized both buried and exposed remains did so at different elapsed times since death (VanLaerhoven, 1997; VanLaerhoven and Anderson, 1999). In the latter study, adult flies were observed moving through cracks in the soil surface in an attempt to reach the body, particularly following heavy rain. Eggs were also laid on the surface following rain, and on hatching, larvae moved down through the soil cracks.

Invertebrates have also been observed in much deeper burials, usually following exhumation of cemetery burials. In 1898, Motter reported on the various insect fauna associated with human remains following 150 disinterments (Motter, 1898). The majority of species identified were from the *Diptera* and *Coleoptera* orders; however, a comprehensive list of other identified species was also reported for each disinterment. Colyer has also reported on fly species associated with human remains buried in coffins (Colyer, 1954). *Conicera tibialis* Schmitz (Diptera: Phoridae) was found to commonly inhabit coffins and is believed to burrow to the cadaver in order to lay eggs closer to the body (Anderson, 2001).

A recent exhumation of twenty-two cadavers in the Lille area of northern France also identified *Conicera tibialis* along with other fly species, including *Leptocera caenosa*, *Ophyra capensis*, and *Triphleba hyalinata* (Bourel et al., 2004). It is believed that these species were found in the burials because of their preference for underground or closed environments. *C. tibialis* is often referred to as the coffin fly because of its ability to burrow into the soil to a depth of 2 m and oviposit directly on a cadaver enclosed in a coffin. All stages of development for this species were observed in this study, indicating that several generations had occurred within the confined spaces. *L. caenosa* is often associated with organic materials found in underground environments, such as sewers and cracked soil pipes. *Leptocera* sp. is known to occur on buried pig carcasses (Payne, 1965; VanLaerhoven and Anderson, 1999) and was regularly identified inside the coffins of the exhumed remains. *O. capensis* can colonize confined bodies by laying its eggs on the surface of coffins. It is usually associated with feces and carrion, and has been found on human cadavers located indoors for several years. It was the most abundant fly found on the exhumed cadavers in the study. In total, eight *Diptera* and two *Coleoptera* species were identified as being associated with the coffined cadavers (Bourel et al., 2004).

Beetles may also be found on buried carrion. Payne and King (1969) discovered a carabid beetle species *Anillinus fortis* Horn, associated with buried carrion. Simpson and Strongman (2002) also reported the sylphid *Necrophila americana*, and some histerid, carabid, and staphylinid species on buried carrion in rural areas of Nova Scotia, but not in urban areas. Bourel et al. (2004) reported the occurrence of *Omalium rivulare* and *Philonthus* sp. in the coffins of exhumed remains. These species are predaceous beetles usually found on the surface of the soil, but they have been found feeding on Diptera larvae associated with buried carrion (Bourel et al., 2004). VanLaerhoven and Anderson (1999) report the occurrence of many beetle species, including silphids, on buried carrion, but suggest these beetles to be unreliable in a successional sense, as they are generalized predators and their arrival is not predictive.

Determination of Postburial Interval Using Insect Succession

The most recognized application of forensic entomology involves the estimation of postmortem interval and circumstances surrounding a death. Since this area has been covered extensively in previous chapters, it will not be considered here. Instead, the potential application of forensic entomology to the estimation of postburial interval will be discussed. A thorough knowledge of the succession occurring on buried carrion is required if it is to be employed in postburial interval (PBI) estimation. As with surface carrion, a predictable succession may be used to estimate PBI, and in cases where dipteran larvae are still present, the use of temperature-dependent developmental data may be applied. The use of such data obviously requires consideration of season and temperature.

VanLaerhoven and Anderson (1999) recorded the occurrence of some species on buried carrion to be predictable, and suggested that burial successions can be used in estimation of PBI. Dipteran species, with the exception of the family Calliphoridae, were identified as the most useful indicator species, allowing for an estimation of the minimum PBI in a shallow burial environment. This assumption is based on the hypotheses that once insects are able to locate buried remains, they will colonize, feed, and develop in a normal, predictable sequence. The study also determined soil temperature to be a better predictor of internal temperature of buried carrion than ambient temperature, and suggested that soil temperature should be used for estimation of insect development (VanLaerhoven and Anderson, 1999). Although useful in British Columbia, this research requires further studies in each specific locality where such data are to be employed. The variability in species present and differences in behavior and developmental rates make the development of a generalized successional database impractical, as with the current successional data for aboveground carrion. However, further research in this area has the potential to identify a new method for estimating PBI.

Successional behavior can also be used to estimate the season in which a body was buried. One of the first applications of burial entomofauna occurred in the field of archaeology and was used to identify Calliphoridae (blow flies) and Sarcophagidae (flesh flies) associated with 130- to 160-year-old burials (Gilbert and Bass, 1967). Regardless of the extended timeframe between burial and exhumation of the remains, pupal cases had survived in several of the burials due to their tough exoskeletal coating. The presence of pupae in the burial environments seasonally dated the burials from late March to mid-October based on the arrival and disappearance of these flies in South Dakota.

However, the effect of soil characteristics on insect succession in a burial environment can prove problematic to a forensic investigation, as demonstrated by a case study in the United Kingdom (Turner and Wiltshire, 1999). The case involved the discovery of a male corpse in a shallow grave that had been partly exposed by large scavengers. Entomological evidence collected from the cadaver provided a PMI estimate of 6 to 7 weeks; however, this estimate was in contradiction to other evidence in possession of the police, which suggested a PMI of at least 17 weeks. The investigation provided an opportunity to conduct an experimental validation on the accuracy of estimating PMI of buried remains using entomological evidence. Replication of the homicide and decomposition process was produced through the use of pig carcasses buried in close proximity to the original shallow gravesite, and regular observations were made until the corpses were exhumed. The study showed that insects played no role in the decomposition process until the remains were exposed by scavengers, at which point blow flies oviposited on the exposed tissues. As a result, the estimation of PMI based on entomological evidence was only reflective of the time the cadaver was exposed and was not able to provide information specific to the time of death (Turner and Wiltshire, 1999).

Although the field of forensic entomology is regularly employed to estimate time since death of exposed remains in forensic investigations, the few studies that have investigated arthropod succession on buried corpses suggest that it cannot be used as accurately for estimating PBI or PMI of buried remains. Several insect species have been identified as possible indicators of time since death, and entomological evidence has been successfully used in the seasonal dating of buried remains. Although the research conducted thus far is only preliminary in nature, the use of entomological evidence in estimating PBI holds great potential.

Conclusion

Decomposition is an inherently difficult process to predict because of the many environmental factors that can affect its rate and progression. In an outdoor environment, insect activity will contribute to the majority of soft tissue decomposition. However, the soil environment will have a major influence on the process as well. Hence, it is important to understand the effect of the soil environment on the insect fauna associated with decomposition. This chapter has outlined the potential influences the soil environment may have on insect succession. It has also discussed the potential of forensic entomology to assist in estimating postburial interval (PBI). The use of soil succession in PBI estimation will require development of successional databases for locations where it is to be applied. Invertebrates play an important role in decomposition in a soil environment, and further study in this area is clearly warranted.

References

Anderson, G. S. 2001. Insect succession on carrion and its relationship to determining time of death. In *Forensic entomology: The utility of arthropods in legal investigations*, Byrd, J. H., and Castner, J. L., Eds. Boca Raton, FL: CRC Press, 143.

Anderson, G. S., and Cervenka, V. J. 2002. Insects associated with the body: Their use and analyses. In *Advances in forensic taphonomy: Method, theory, and archaeological perspectives*, Haglund, W. D., and Sorg, M. H., Eds. Boca Raton, FL: CRC Press, 173.

Anderson, G. S., and VanLaerhoven, S. L. 1996. Initial studies on insect succession on carrion in south western British Columbia. *J. Forensic Sci.* 41:617.

Bass, W. M. 1997. Outdoor decomposition rates in Tennessee. In *Forensic taphonomy: The postmortem fate of human remains*, Haglund, W. D., and Sorg, M. H., Eds. Boca Raton, FL: CRC Press, 181.

Bereuter, T. L., Mikenda, W., and Reiter, C. 1997. Iceman's mummification—Implications from infrared spectroscopical and histological studies. *Chem. Eur. J.* 3:1032.

Bethell, P. H., and Carver, M. O. H. 1987. Detection and enhancement of decayed inhumations at Sutton Hoo. In *Death, decay and reconstruction: Approaches to archaeology and forensic science*, Boddington, A., Garland, A. N., and Janaway, R. C., Eds. Manchester, UK: University Press, 10.

Bornemissza, G. F. 1957. An analysis of arthropod succession in carrion and the effect of its decomposition on the soil fauna. *Aust. J. Zool.* 5:1.

Bourel, B., et al. 2004. Entomofauna of buried bodies in northern France. *Int. J. Legal Med.* 118:215.

Carter, D. O., and Tibbett, M. 2003. Taphonomic mycota: Fungi with forensic potential. *J. Forensic Sci.* 48:168.

Carter, D. O., et al. 2006a. Nematode community dynamics associated with cadaver (*Sus scrofa* L.) decomposition and insect activity on the soil surface. Paper presented at Proceedings of the Annual Meeting of the American Academy of Forensic Sciences, Seattle.

Carter, D. O., Yellowlees, D., and Tibbett, M. 2006b. Cadaver decomposition in terrestrial ecosystems. *Naturwissenschaften*, epub.

Clark, M. A., Worrell, M. B., and Pless, J. E. 1997. Postmortem changes in soft tissue. In *Forensic taphonomy: The postmortem fate of human remains*, Haglund, W. D., and Sorg, M. H., Eds. Boca Raton, FL: CRC Press, 151.

Colyer, C. N. 1954. The "coffin fly" *Conicera tibialis* Schmitz (Diptera: Phoridae). *J. Br. Entomol.* 4:203.

Cronyn, J. M. 1990. *The elements of archaeological conservation.* London: Routledge.

Dent, B. B., Forbes, S. L., and Stuart, B. H. 2004. Review of human decomposition processes in soil. *Env. Geol.* 45:576.

Dix, J., and Graham, M. 2000. *Time of death, decomposition, and identification: An atlas.* Boca Raton, FL: CRC Press, 8.

Fiedler, S., and Graw, M. 2003. Decomposition of buried corpses, with special reference to the formation of adipocere. *Naturwissenschaften* 90:291.

Fiedler, S., Schneckenberger, K., and Graw, M. 2004. Characterization of soils containing adipocere. *Arch. Environ. Contam. Toxicol.* 47:561.

Forbes, S. L., et al. 2003. Development of a GCMS method for the detection of adipocere in grave soils. *Eur. J. Lipid Sci. Tech.* 105:761.

Forbes, S. L., Stuart, B. H., and Dent, B. B. 2002. The identification of adipocere in grave soils. *Forensic Sci. Int.* 127:225.

Forbes, S. L., Stuart, B. H., and Dent, B. B. 2005a. The effect of soil type on adipocere formation. *Forensic Sci. Int.* 154:35.

Forbes, S. L., Stuart, B. H., and Dent, B. B. 2005b. The effect of the burial environment on adipocere formation. *Forensic Sci. Int.* 154:24.

Galloway, A. 1997. The process of decomposition: A model from the Arizona-Sonoran desert. In *Forensic taphonomy: The postmortem fate of human remains*, Haglund, W. D., and Sorg, M. H., Eds. Boca Raton, FL: CRC Press, 139.

Gilbert, B. M., and Bass, W. M. 1967. Seasonal dating of burials from the presence of fly pupae. *Am. Antiquity* 32:534.

Gill-King, H. 1997. Chemical and ultrastructural aspects of decomposition. In *Forensic taphonomy: The postmortem fate of human remains*, Haglund, W. D., and Sorg, M. H., Eds. Boca Raton, FL: CRC Press, 93.

Gniadecka, M., et al. 1999. Near-infrared Fourier transform Raman spectroscopy of the mummified skin of the alpine Iceman, Qilakitsoq Greenland mummies and Chiribaya mummies from Peru. *J. Raman Spectrosc.* 30:147.

Goff, M. L., and Catts, E. P. 1990. Arthropod basics—Structure and biology. In *Entomology and death: A procedural guide*, Catts, E. P., and Haskell, N. H., Eds. Clemson, SC: Joyce's Print Shop, 41–71.

Grim, R. E. 1968. *Clay mineralogy*. New York: McGraw-Hill.

Haglund, W. D. 1997. Dogs and coyotes: Postmortem involvement with human remains. In *Forensic taphonomy: The postmortem fate of human remains*, Haglund, W. D., and Sorg, M. H., Eds. Boca Raton, FL: CRC Press, 367.

Haskell, N. H., et al. 1997. On the body: Insects' life stage presence and their postmortem artifacts. In *Forensic taphonomy: The postmortem fate of human remains*, Haglund, W. D., and Sorg, M. H., Eds. Boca Raton, FL: CRC Press, 415.

Hunter, J. R., and Martin, A. L. 1996. Locating buried remains. In *Studies in crime: An introduction to forensic archaeology*, Hunter, J., Roberts, C., and Martin, A., Eds. London, UK: B.T. Batsford, 86.

Jaffee, B., et al. 1992. Density-dependent host-pathogen dynamics in soil microcosms. *Ecology* 73:495.

Janaway, R. C. 1996. The decay of buried remains and their associated materials. In *Studies in crime: An introduction to forensic archaeology*, Hunter, J., Roberts, C., and Martin, A., Eds. London, UK: B.T. Batsford, 58.

Lundt, H. 1964. Ecological observations about the invasion of insects into carcasses buried in soil. *Pedobiologia* 4:158.

Makristathis, A., et al. 2002. Fatty acid composition and preservation of the Tyrolean Iceman and other mummies. *J. Lipid Res.* 43:2056.

Mann, R. W., Bass, W. M., and Meadows, L. 1990. Time since death and decomposition of the human body: Variables and observations in case and experimental field studies. *J. Forensic Sci.* 35:103.

Mayer, B. X., Reiter, C., and Bereuter, T. L. 1997. Investigation of the triacylglycerols composition of iceman's mummified tissue by high-temperature gas chromatography. *J. Chromatgr. B* 692:1.

Mellen, P. F. M., Lowry, M. A., and Micozzi, M. S. 1993. Experimental observations on adipocere formation. *J. Forensic Sci.* 38:91.

Merbs, C. F. 1997. Eskimo skeleton taphonomy with identification of possible polar bear victims. In *Forensic taphonomy: The postmortem fate of human remains*, Haglund, W. D., and Sorg, M. H., Eds. Boca Raton, FL: CRC Press, 249.

Micozzi, M. S. 1997. Frozen environments and soft tissue preservation. In *Forensic taphonomy: The postmortem fate of human remains*, Haglund, W. D., and Sorg, M. H., Eds. Boca Raton, FL: CRC Press, 171.

Morris, B., and Dadour, I. R. 2005. Forensic entomology: The use of insects in legal cases. In *Expert evidence*, Freckelton, I., and Selby, H., Eds. Hampshire, UK: Thomson Publishing Services.

Morton, R. J., and Lord, W. D. 2002. Detection and recovery of abducted and murdered children: Behavioral and taphonomic influences. In *Advances in forensic taphonomy: Method, theory, and archaeological perspectives*, Haglund, W. D., and Sorg, M. H., Eds. Boca Raton, FL: CRC Press, 151.

Motter, M. G. 1898. A contribution to the study of the fauna of the grave. A study of one hundred and fifty disinterments, with some additional experimental observations. *J. New York Entomol. Soc.* 6:201.

Payne, J. A. 1965. A summer carrion study of the baby pig *Sus scrofa* Linnaeus. *Ecology* 46:592.

Payne, J. A., and King, E. W. 1968. Arthropod succession and decomposition of buried pigs. *Nature* 219:1180.

Payne, J. A., and King, E. W. 1969. Coleoptera associated with pig carrion. *Entomol. Mon. Mag.* 105:224.

Pfeiffer, S., Milne, S., and Stevenson, R. M. 1998. The natural decomposition of adipocere. *J. Forensic Sci.* 43:368.

Polson, C. J., Gee, D. J., and Knight, B. 1985. *The essentials of forensic medicine.* 4th ed. Oxford: Pergamon Press.

Reed, H. B. 1958. A study of dog carcass communities in Tennessee, with special reference to the insects. *Am. Mid. Nat.* 59:213.

Rodriguez, W. C. 1997. Decomposition of buried and submerged bodies. In *Forensic taphonomy: The postmortem fate of human remains*, Haglund, W. D., and Sorg, M. H., Eds. Boca Raton, FL: CRC Press, 459.

Rodriguez, W. C., and Bass, W. M. 1983. Insect acitivity and its relationship to decay rates of human cadavers in East Tennessee. *J. Forensic Sci.* 28:423.

Rodriguez, W. C., and Bass, W. M. 1985. Decomposition of buried bodies and methods that may aid their location. *J. Forensic Sci.* 30:836.

Roksandic, M. 2002. Position of skeletal remains as a key to understanding mortuary behaviour. In *Advances in forensic taphonomy: Method, theory, and archaeological perspectives*, Haglund, W. D., and Sorg, M. H., Eds. Boca Raton, FL: CRC Press, 99.

Sagara, N. 1981. Occurrence of *Laccaria proxima* in the grave site of a cat. *Trans. Mycol. Soc. Jpn.* 22:271.

Simpson, G., and Strongman, D. B. 2002. Carrion insects on pig carcasses at a rural and an urban site in Nova Scotia. *Can. Soc. Forensic Sci. J.* 35:123.

Sledzik, P. S., and Micozzi, M. S. 1997. Autopsied, embalmed, and preserved human remains: Distinguishing features in forensic and historic contexts. In *Forensic taphonomy: The postmortem fate of human remains*, Haglund, W. D., and Sorg, M. H., Eds. Boca Raton, FL: CRC Press, 483.

Smith, K. G. V. 1986. *A manual of forensic entomology.* Oxford: British Museum (Natural History) and Cornell University Press.

Tan, K. H. 1994. *Environmental soil science.* 1st ed. New York: Dekker.

Tibbett, M., et al. 2004. A laboratory incubation method for determining the rate of microbiological degradation of skeletal muscle tissue in soil. *J. Forensic Sci.* 49:560.

Turner, B., and Wiltshire, P. 1999. Experimental validation of forensic evidence: A study of the decomposition of buried pigs in a heavy clay soil. *Forensic Sci. Int.* 101:113.

Vane, C. H., and Trick, J. K. 2005. Evidence of adipocere in a burial pit from the foot and mouth epidemic of 1967 using gas chromatography–mass spectrometry. *Forensic Sci. Int.* 154:19.

VanLaerhoven, S. L. 1997. Successional biodiversity in insect species on buried carrion in the Vancouver and Caribou regions of British Columbia, MPM thesis, Department of Biological Sciences, Simon Fraser University, Burnaby, British Columbia.

VanLaerhoven, S. L., and Anderson, G. S. 1999. Insect succession on buried carrion in two biogeoclimatic zones of British Columbia. *J. Forensic Sci.* 44:32.

Vass, A. A., et al. 2002. Decomposition chemistry of human remains: A new methodology for determining postmortem interval. *J. Forensic Sci.* 47:542.

Weitzel, M. A. 2005. A report of decomposition rates of a special burial type in Edmonton, Alberta from an experimental field study. *J. Forensic Sci.* 50:641.

Wolff, M., et al. 2001. A preliminary study of forensic entomology in Medellin, Colmbia. *Forensic Sci. Int.* 120:53.

Yan, F., et al. 2001. Preliminary quantitative investigation of postmortem adipocere formation. *J. Forensic Sci.* 46:609.

Entomotoxicology
Insects as Toxicological Indicators and the Impact of Drugs and Toxins on Insect Development

12

M. L. GOFF
WAYNE D. LORD

Contents

Introduction

Careful analyses of the community of insects encountered on a decomposing body, combined with knowledge of insect biology, ecology, and local environmental conditions, can often provide valuable forensic insights. These can include the estimation of time since death, movement of the remains after death, indication of antemortem injuries, and the presence of drugs or toxins.

Over the past two decades, there has been an apparent increase in the incidence of drug-related deaths reported within the United States and other countries. Decedents in such cases are, in many instances, not discovered for a substantial period of time (days or weeks). The resulting state of advanced decomposition and environmental recycling typically encountered in these situations often dictates the employment of various entomological methodologies. The entomological techniques most frequently utilized are based on comprehensive analyses of the insects and other arthropods associated with the remains, their development, and patterns of succession (Goff and Flynn 1991, Goff and Odom 1987, Lord et al. 1986).

The accuracy of entomological estimates in deaths involving narcotic intoxication has been subject to debate in recent years, as few available studies have explored the effects of drugs contained in decomposing tissues on fly colonization and ovipositional behavior, or on the rates of development of carrion-frequenting insects feeding on such food sources (Goff 1993). Additionally, relatively few studies have examined the effects of other tissue contaminants, such as toxins or environmental pollutants on these behaviors or the developmental patterns of the insects colonizing such tissues.

In recent years, interest has also focused on the potential use of carrion-frequenting insects as alternative toxicological specimens in situations where traditional toxicological sources, such as blood, urine, or solid tissues, are unavailable or not suitable for analysis. The use of anthropophagic fly larvae (maggots) as alternate toxicological specimens is well documented in the entomological and forensic science literature (Miller et al. 1994). Detection of various toxins and controlled substances in insects found on decomposing human remains has contributed to the assessment of both cause and manner of death (Lord 1990, Goff and Lord 1994, Nolte et al. 1992). With the development of hair extraction technologies, attention has recently focused on the analysis of chitinized insect remnants that are frequently encountered with mummified and skeletonized remains (Miller et al. 1994). In such cases, the standard toxicological specimens are often absent.

Studies of the use of carrion-feeding arthropods as alternative toxicological specimens, and of the impact that tissue toxins and contaminants have on the development of immature insects feeding on these substances, currently comprise the major avenues of exploration in the emerging field of entomotoxicology.

Detection of Drugs and Toxins in Carrion-Feeding Insects

As previously mentioned, it is not unusual for human remains to be discovered in a highly decomposed or skeletonized state. Historically, it has been difficult to obtain toxicological information in cases of such advanced decay due largely to a lack of sufficient, analyzable tissue. A variety of arthropods and their cast larval and puparial skins, however, are commonly encountered on putrefied, mummified, and skeletonized remains.

Several recent studies have detailed the detection of toxins and controlled substances in both the insects and chitinized remnants recovered from badly decomposed victims. In these reports, the recovered arthropods have generally been homogenized and subsequently processed in a manner similar to that for other, more traditional tissues and fluids, or subjected to extraction techniques developed for the analysis of rigorous tissues, such as hair and nails. Analytic procedures have included radioimmunoassay (RAI), gas chromatography (GC), gas chromatography–mass spectrometry (GC/MS), thin-layer chromatography (TLC), or high-performance liquid chromatography–mass spectrometry (HPLC/MS).

Nuorteva and Nuorteva (1982) described the successful recovery of mercury from various species of blow fly larvae (Calliphoridae) reared on fish tissues containing known concentrations of the heavy metal. Mercury was observed to bioaccumulate in the developing larvae as they fed on the contaminated tissues, and the recovered concentrations within the larvae increased with the duration of the feeding period. The accumulation of mercury in the larvae was also observed to be directly related to the presence of mercury in the methylated form. In those larvae reared on tissues in which 94% of the mercury was methylated, a 4.3 times greater concentration was found than in the tissues upon which they had fed. In tissues where a lesser percentage of methylated mercury was present, recovered larval concentrations were only 1.5 times greater.

The mercury ingested by the developing calliphorid larvae was retained through the puparial stage and detectable in the emerging adult flies. Upon reaching adulthood, however, the flies rapidly eliminated the mercury. Two days following emergence, adult flies contained only 50% of the mercuric concentrations detected in the developing larvae. More detailed observations revealed that mercury was excreted into the meconium of the hindgut during the process of pupariation. No adverse effects from the bioaccumulated mercury were detected in either the adult flies or the developing larvae. In some instances, however, difficulties were observed in pupariation.

The pioneering work of Nuorteva and Nuorteva (1982) clearly demonstrated that substances contained in food sources exploited by carrion-feeding insects could be detected in both immatures and adults, and that their levels could be quantified by toxicological means. As an extension of this study, larvae that had been reared on mercury-laden tissues were fed to adult staphylinid beetles, *Creophilus maxillosus* (L.). Secondary bioaccumulation of mercury was also seen in these predaceous beetles. However, no adverse effects were detected in the adult staphylinids studied. But Nuorteva and Nuorteva (1982) observed "minimata-like symptoms," consisting primarily of irregularities in motor control in adult tenebrionid beetles, when *Tenebrio molitor* L. fed on fly larvae containing mercury. Subsequently, Schott and Nuorteva (1983) demonstrated decreased levels of activity in the same species of tenebrionid beetles when fed on a diet consisting of dried fly larvae contaminated with a high mercury content. These studies further demonstrated the ability to detect and quantify levels of food-borne contaminants in both adult and immature carrion-feeding insects, and the bioaccumulation of these substances within the insects. Additionally, the detection of such substances in predacious beetles feeding on contaminated fly larvae and the resulting development of potentially negative side effects were illustrated.

In a similar manner, Sohal and Lamb (1977, 1979) demonstrated the accumulation of various metals (including copper, iron, and zinc) and calcium in the tissues of adult house flies, *Musca domestica* L. No detrimental effects to the adult flies were associated with the bioaccumulation of these metals. Utsumi (1958) observed that rat carcasses varied in their attractiveness to adult flies, depending on the poison causing death. This research, however, did not include any attempt to detect the toxins in the maggots subsequently developing on the rat tissues.

Goff et al. (1997) reported a concentration of 3,4-methylenedioxymethamphetamine in puparial casings of the sarcophagid fly, *Parasarcophaga ruficornis* (Fabricius), which was higher than that detected in the tissues on which the developing larvae of this fly had been feeding. While this observation can be interpreted as yet another example of the phenomenon of dipteran bioaccumulation, the reported deposition of toxins in the cuticle of insects as a method of excretion suggests other potential mechanisms for the accumulation of ingested materials in nonliving insect tissues.

Sadler et al. (1997) failed to observe bioaccumulation of barbiturates in blow fly larvae, *Calliphora vicina* (Rodineau-Desvoidy), reared on an artificial food medium. Additionally, they observed that even closely related chemical compounds appeared to be processed differently by similar cohorts of developing *C. vicina* larvae. They cautioned against the quantitative interpretation of entomotoxicological findings given the current limited knowledge of the ways that drugs and toxins are handled by immature insects. Clearly, much more research into the mechanisms and processes by which carrion-feeding insects incorporate ingested drugs and toxins into their living and nonliving tissues is needed before the full forensic potential of entomotoxicology is realized.

Researchers have employed various methods of drug administration in their attempts to accurately duplicate tissue concentrations of drugs and toxins seen in human overdose/poisoning deaths. Sadler et al. (1997), for example, fed developing fly larvae an artificial food medium spiked with a known concentration of the drug being tested. Other researchers have employed alternative methodologies wherein known quantities of a drug or toxin are administered orally and by infusion to a live animal model, and the resulting vertebrate tissues are then utilized as the insect food source. The latter method allows for the drugs and toxins to be metabolized by the vertebrate host prior to insect ingestion. Goff et al. (1997) have suggested that many drugs, such as cocaine, exert their effects as the mammalian metabolite rather than the parent compound, and that live animal models present a scenario more closely related to that seen in actual drug-related deaths. However, further research is needed before a clear understanding of the mechanisms underlying these processes and the optimum research model is elucidated.

An example of the potential forensic application of these types of observations, studies, and data was detailed by Nuorteva (1977). In this case, fly larvae were collected and reared from the decomposing body of an unidentified woman discovered in a rural area of Inkoo, Finland. An analysis for mercury was performed on the emerging adults in an effort to determine the geographic origin of the unidentified victim. The low mercury content of the emerging adult flies indicated that the victim came from an area relatively free of mercury pollution. When the victim's identity was eventually determined, she proved to be a student from the city of Turku, an area relatively free of mercury pollution. In this case, the entomotoxicological analysis allowed police to focus their investigative efforts in a more limited geographic area, thereby enhancing the chances of successful victim identification and case resolution.

Leclercq and Brahy (1985) successfully detected the presence of arsenic through the toxicological analysis of species in the families Piophilidae, Psychodidae, and Muscidae in a case of untimely death in France. Detection of the organophosphate insecticide Malathion in fly larvae was reported by Gunatilake and Goff (1989). In this case, a 58-year-old male with a previous history of suicide attempts was discovered in the crawl space under his mother's home in Honolulu. Adjacent to the victim's body was a bottle of Malathion with approximately 117 ml missing. Toxicological tests (GC) for Malathion were conducted on a variety of the victim's tissues and on two species of fly larvae (Calliphoridae) collected from the remains. Fat tissues from the victim revealed Malathion in a concentration of 17 mg/kg. A combined sample of the two species of fly larvae, *Chrysomya megacephala* (F.) and *Chrysomya rufifacies* (Macquart), revealed Malathion in a concentration of 2,050 µg/g. It is significant to note that the ages of both species of *Chrysomya* larvae were indicative of a postmortem interval of approximately 5 days. The victim was last reliably seen alive 8 days prior to the discovery of his body. The Malathion in the victim's tissues may have served, in this case, to delay oviposition by adult female *Chrysomya* for a period of several days. Additionally, a far greater diversity of both carrion fly and predatory beetle species would have been expected to have been encountered on remains present for 5 to 8 days in an outdoor Hawaiian habitat. The absence of a well-developed insect community on the remains of the decedent lends support to the notion that the ingested Malathion had a negative influence on carrion insect colonization.

Beyer et al. (1980) detailed the case of a 22-year-old female whose decomposed remains were discovered 14 days following her disappearance. The young woman had a lengthy history of mental illness, including multiple attempts at suicide. An empty bottle from a prescription for phenobarbital tablets filled 2 days prior to her disappearance was found in a purse adjacent to her body. As there were no soft tissues suitable for classic toxicological testing on her almost skeletonized remains, larvae of the calliphorid fly *Cochliomyia macellaria* (F.), found feeding on the remains, were collected and analyzed for drug content. Phenobarbital was subsequently detected by gas chromatography and confirmed by thin-layer chromatography at a concentration of 100 µg/g.

Kintz et al. (1990a, 1990b) have demonstrated further detection of prescription drugs through the analyses of fly larvae feeding upon human remains. In one case, toxicological tests were performed on the remains of a male decedent having a known postmortem interval of 67 days. Liquid chromatography was employed in the analysis of heart, liver, lung, spleen, and kidney tissues as well as calliphorid fly larvae collected from the victim. Results of these analyses revealed the presence of five drugs (triazolam, oxazepam, phenobarbital, alimemazine, and clomipramine) in both the tissues and fly larvae examined. Triazolam was not detected in either the spleen or kidney samples, although the other drugs were present. All five drugs were, however, isolated from the developing fly larvae. In this case, it was not possible to establish any quantitative correlations between the concentrations of the drugs detected in the fly larvae and the human tissues. It is of interest to note, however, that Kintz et al. (1990b) observed fewer endogenous peaks in the chromatograms obtained from the maggot extractions than in those from the human tissues. Further evidence of the toxicological potential of insect specimens is provided by Kintz et al. (1990b) in the subsequent successful recovery of bromazepam and levomepromzind from fly larvae obtained from decomposed human remains.

Introna et al. (1990) present the results of toxicological analyses conducted on fly larvae reared on human liver tissues collected from 40 cases in which opiates were detected during routine postmortem examinations. In this study, opiates were effectively identified in the fly larvae through the use of radioimmunoassay (RIA) techniques. A significant correlation was reported between the concentrations of opiates observed in the liver tissues and the concentrations detected within the fly larvae tested. While the qualitative drug findings were quite clear, the quantitative relationships between human host and insect consumer concentrations were less dramatic.

Goff et al. (1989, 1991) detailed similar qualitative results in studies of known dosages of cocaine and heroin administered to laboratory rabbits. Fly larvae, subsequently fed on the tissues of these animals and analyzed for drug concentrations, demonstrated clear evidence of both drug presence and bioaccumulation. Quantitative relationships between the tested rabbit tissues and the developing fly larvae were suggestive but not definitive. Introna et al. (2001) and Bourel at al. (2001) noted similar relationships between opiate concentrations in substrate and the developing larvae, but noted a lack of such correlations in other studies. In these studies, the larvae were fed on a substrate with a known, somewhat uniform concentration of the drug. This is obviously not the case in life, and there is considerable variation between different organs in the body in drug and metabolite concentration. This is further complicated by postmortem redistribution of substances and the various processes involved in decomposition. As noted by Tracqui et al. (2004), at the present time, it does not appear that reliable correlations can be drawn between concentrations of drugs and metabolites in a decomposing body and the dosage administered.

Nolte et al. (1992) further illustrate the forensic applicability of toxicological information obtained through the analysis of insects collected from badly decomposed human remains. In this instance, the nearly skeletonized body of a 29-year-old intravenous drug user was discovered in a wooded area 5 months after his disappearance. Friends of the victim reported that he had used a substantial quantity of intravenous cocaine immediately prior to wandering away from a rural residence. Associated with his remains were numerous fly larvae and puparia. Skeletal muscle was submitted for toxicological analysis along with samples of the collected insects. Both the victim's tissues and the larval insects tested positive for cocaine and its major mammalian metabolite, benzoylecognine, using gas chromatography. The empty puparial cases were also subjected to analysis and found to be weakly positive for both the parent drug and the metabolite. The prospect of effectively extracting drugs or toxins from the chitin matrix of insect puparia opens new avenues of entomotoxicology, as these and other chitinized insect remnants often remain unaltered in the environment for extremely long periods of time.

With the advent of hair extraction techniques (Baumgartner et al. 1989), further interest in the use of chitinized insect remnants as potential sources of toxicological information emerged. Manhoff et al. (1988) successfully extracted cocaine from fly larvae and beetle fecal material collected from human remains using gas chromatography and mass spectrometry. Miller et al. (1994) isolated both amitriptyline and nortriptyline from empty fly puparia, dermestid beetle exuviae and frass, and mummified human tissues associated with the body of a middle-aged female discovered in her residence more than 2 years following her death. Strong acid and base extraction procedures, originally developed for human hair analyses, were employed to successfully release the drugs from the chitin/protein matrix of the puparia and beetle remnants. Similarly, Goff et al. (1997) were able to demonstrate 3,4,-methylenedioxymethamphetamine in both fly larvae and spent puparia reared on infused rabbit tissues.

The potential value of larval and adult carrion-feeding insects, and their chitinous remnants, as alternative sources of toxicological information has been clearly demonstrated. As with other emerging technologies, however, great care must be taken in the interpretation and use of such data, particularly within the forensic arena. Given recent advances in analytical procedures, it has become more practical to use even decomposed tissues for analysis (Tracqui et al. 2004). The situation may still be encountered where for various reasons there are no tissues remaining and the arthropods remain the only available material for analyses. In these instances, a qualitative analysis will be of value, but any attempt at quantitation must be viewed with skepticism. Much more research is required before the full potential of this discipline can be recognized.

Impact of Drugs and Toxins on Insect Development

While many of the studies mentioned earlier documented the potential for use of maggots and puparia as alternate specimens for toxicological analyses, few were concerned with the potential effects of these drugs on the development of the insects ingesting them. In providing an estimate of the postmortem interval, particularly within the first 2 to 4 weeks of decomposition, it is assumed that the insects will develop at predictable rates for given environmental conditions. That this might not always be the case was first demonstrated by Goff et al. (1989) in their studies on the effects of cocaine on development of the sarcophagid fly *Boettcherisca peregrina* (Rodineau-Desvoidy). In this study, maggots were reared on tissues from rabbits that had received known dosages of cocaine, corresponding to 0.5, 1.0, and 2.0 times median lethal dosage by weight. Two patterns of development were noted. Control and sublethal dosage colonies developed at approximately the same rate, as indicated by total body length. By contrast, the colonies fed on tissues from the lethal and twice lethal dosages developed more rapidly. This difference continued until maximum size was attained and the postfeeding portion of the third instar was reached. Due to the increased rate of development during the feeding stages, pupariation occurred first in the lethal and twice lethal colonies, but the actual duration of the puparial period was the same for all colonies, and there were no detectable differences in puparial mortality.

Similar studies were conducted by Goff et al. (1991) for heroin using *B. peregrina* maggots. In these studies, heroin, in tissues as morphine, resulted in more rapid development of maggots and production of larger maggots in all the treated colonies, until maximum size was attained. The puparial period was longer in duration for all colonies fed on tissues from the test animals and appeared proportional to the amount of the drug administered. This study demonstrated that an error of up to 29 hours could occur if the estimated post-mortem interval was based on larval development, and 18 to 38 hours if based on the duration of the puparial stage.

In studies on methamphetamine, the situation became more complicated. There were observed increases in rates of development for the sarcophagid *P. ruficornis* for colonies fed on tissues containing lethal and twice median lethal dosages of methamphetamine, while colonies from the control and half median lethal dosage animals developed at approximately the same rate. There was increased puparial mortality in the half and median lethal dosage colonies, and both median lethal and twice median lethal dosage colonies failed to produce viable offspring by the second generation. By contrast, for 3,4-methyl-enedioxymethamphetamine (MDMA), larval and puparial mortality was highest for the control and half median lethal dosage colonies and lowest for the twice median lethal dosage colony. Additionally, emerging adults produced viable larvae.

The tricyclic antidepressant amitriptyline was tested in a similar manner on *P. ruficornis* (Goff et al. 1993). Here there were no significant differences in rate of development related to the concentrations of the drug administered until the postfeeding third instar. The duration of the postfeeding stage was longer for all colonies fed on the drug, and larval mortality was greater in the treated colonies. No significant differences were observed in puparial mortality among the treatments; however, the puparial stage was longer for colonies fed on tissues from the rabbits receiving the median and twice median lethal dosages.

Recently, Hedouin et al. (1999) have demonstrated the potential underestimation of postmortem interval based on the developmental analysis of necrophagous fly larvae (*Lucilia sericata* Meigen) fed on the tissues of deceased rabbits previously perfused with various concentrations of morphine. In this study, a possible error in the estimation of time since death of up to 24 hours was noted if the presence of morphine and its resultant effects on fly development were not considered. During this study, a new experimental model was used to obtain concentrations of drugs in rabbit tissues, which were similar to those encountered in humans who had expired as the result of a drug overdose. This model employed the administration of morphine hydrochloride via ear artery perfusion through a plastic catheter placed into the main ear artery of experimental rabbits. This experimental methodology allowed for a more precise control of both blood and tissue levels of the drug, and facilitated more accurate duplication of visceral concentrations similar to those encountered in human cases.

While there have been relatively few applications of these data to cases, Lord (1990) details a case that serves to illustrate the potential significance of these alterations to larval and puparial development. The body of a Caucasian woman, approximately 20 years old, was discovered in a pine woods area northeast of Spokane, Washington. The body was physically in the early bloated stage of decomposition and had extensive populations of maggots on the face and upper torso. Maggots were submitted to the entomologist after being refrigerated for 5 days, and reared to the adult stage. Two species were identified from the adults: *Cynomyopsis cadaverina* (Rodineau-Desvoidy) and *Phaenicia sericata* (Meigen).

Typically, *P. sericata* oviposits within 24 hours following death, while *C. cadaverina* oviposits 1½ days following death. Three classes of maggots were present in the corpse, based on size. The first consisted of maggots measuring 6 to 9 mm in length that were consistent with a period of development of approximately 7 days. The second consisted of smaller maggots, consistent with continued oviposition by adult flies. The third consisted of a single maggot measuring 17.7 mm in length and indicative of a developmental period, under prevailing conditions at the scene, of approximately 3 weeks.

Given the other data associated with the case, this period did not seem possible. The possibility that this maggot had migrated from another nearby source was eliminated, as no carrion could be located nearby, and the probability of only a single maggot migrating was low. The alternate explanation was that the maggot's growth rate had been accelerated in some manner. It was learned that the victim had a history of cocaine abuse, and that she had snorted cocaine shortly before her death. This maggot had most probably developed in a particular pocket in the nasal region containing a significant amount of cocaine.

Conclusions

Insects and other arthropods can prove to be valuable tools in the investigation of homicides, suicides, and other unattended human deaths. In addition to the recognized applications to the estimation of postmortem interval, remains relocation, and assessment of antemortem injury, insects may also serve as reliable alternative specimens for qualitative toxicological analyses in the absence of tissues and body fluids normally sampled for such purposes. In cases of badly decomposed or skeletonized remains, analyses of collected carrion-feeding insects may provide the most accurate qualitative sources of toxicological information; however, attempts to determine dosage from analyses of arthropods must be viewed with skepticism.

While the data reviewed above concerning the potential effects of drugs and toxins on rates of insect development are limited in scope and no adverse impacts on case analyses have been reported to date, it is not unreasonable to assume that such substances, contained in tissues fed upon by carrion insects, have the potential for altering developmental patterns. Any factors mitigating insect development have the potential of affecting subsequent insect-based estimates of postmortem interval. Until more comprehensive studies and appropriate baseline data are available, care must be taken in interpretations of arthropod developmental patterns in cases where drugs or toxins may be a factor. In addition, it becomes essential that forensic entomologists be made aware of any information concerning the presence of these substances in remains.

Entomotoxicology may prove to be another valuable tool in the forensic science arsenal. More detailed and comprehensive research is required, however, before the full potential of this emerging discipline can be recognized.

References

Baumgartner, W. A., V. A. Hill, and W. H. Blahd. 1989. Hair analysis for drugs of abuse. *Journal of Forensic Sciences* 34:1433–53.

Beyer, J. C., W. F. Enos, and M. Stajic. 1980. Drug identification through analysis of maggots. *Journal of Forensic Sciences* 25:411–12.

Bourel, B., G. Tournel, V. Hedouin, M. Deveaux, M. L. Goff, and D. Gosset. 2001. Morphine extraction in necrophagous insects remains for determining ante-mortem opiate intoxication. *Forensic Science International* 120:127–31.

Goff, M. L. 1993. Estimation of postmortem interval using arthropod development and successional patterns. *Forensic Science Review* 5:81–94.

Goff, M. L., W. A. Brown, K. A. Hewadikaram, and A. I. Omori. 1991. Effects of heroin in decomposing tissues on the development rate of *Boettcherisca peregrina* (Diptera: Sarcophagidae) and implications of this effect on estimations of postmortem intervals using arthropod development patterns. *Journal of Forensic Sciences* 36:537–42.

Goff, M. L., W. A. Brown, A. I. Omori, and D. A. LaPointe. 1993. Preliminary observations of the effect of amitriptyline in decomposing tissues on the development of *Parasarcophaga ruficornis* (Diptera: Saracophagidae) and implications of this effect on the estimations of postmortem intervals. *Journal of Forensic Sciences* 38:316–22.

Goff, M. L., and M. M. Flynn. 1991. Determination of postmortem interval by arthropod succession: A case study from the Hawaiian Islands. *Journal of Forensic Sciences* 36:607–14.

Goff, M. L., M. L. Miller, J. D. Paulson, W. D. Lord, E. Richards, and A. I. Omori. 1997. Effects of 3,4-methylenedioxymethamphetamine in decomposing tissues on the development of *Parasarcophage ruficornis* (Diptera: Sarcophagidae) and detection of the drug in postmortem blood, liver tissue, larvae and puparia. *Journal of Forensic Sciences* 42:276–80.

Goff, M. L., and W. D. Lord. 1994. Entomotoxicology: A new area for forensic investigation. *American Journal of Forensic Medicine and Pathology* 15:51–57.

Goff, M. L., and C. B. Odom. 1987. Forensic entomology in the Hawaiian Islands: Three case studies. *American Journal of Forensic Medicine and Pathology* 8:45–50.

Goff, M. L., A. I. Omori, and J. R. Goodbrod. 1989. Effect of cocaine in tissues on the rate of development of *Boettcherisca peregrina* (Diptera: Sarcophagidae). *Journal of Medical Entomology* 26:91–93.

Gunatilake, K., and M. L. Goff. 1989. Detection of organophosphate poisoning in a putrefying body by analyzing arthropod larvae. *Journal of Forensic Sciences* 34:714–16.

Hedouin, V., B. Bourel, L. Martin-Bouyer, A. Becart, G. Tournel, M. Deveaux, and D. Gossett. 1999. Morphine perfused rabbits: A tool for experiments in forensic entomology. *Journal of Forensic Sciences* 44:347–50.

Introna, F., Jr., C. P. Campobasso, and M. L. Goff. 2001. Entomotoxicology. *Forensic Science International* 120:42–47.

Introna, F., Jr., C. LoDico, Y. H. Caplan, and J. E. Samlek. 1990. Opiate analysis of cadaveric blow fly larvae as an indicator of narcotic intoxication. *Journal of Forensic Sciences* 35:118–22.

Kintz, P., A. Godelar, A. Tracqui, P. Mangin, A. A. Lugnier, and A. J. Chaumont. 1990a. Fly larvae: A new toxicological method of investigation in forensic medicine. *Journal of Forensic Sciences* 35:204–7.

Kintz, P., A. Tracqui, B. Ludes, et al. 1990b. Fly larvae and their relevance to forensic toxicology. *American Journal of Forensic Medicine and Pathology* 11:63.

Leclercq, M., and G. Brahy. 1985. Entomologie et medecine legale: datation de la mort. *Journal de Medecine Legal* 28:271–78.

Lord, W. D. 1990. Case studies in the use of insects in investigations. In *Entomology and death: A procedural guide*, Catts, E. P., and N. H. Haskell, Eds. Clemson, SC: Joyce's Print Shop, 9–37.

Lord, W. D., E. P. Catts, D. A. Scarboro, and D. B. Hadfiled. 1986. The green blow fly, *Lucilia illustris* (Meigen), as an indicator of human post mortem interval: A case of homicide from Fort Lewis, Washington. *Bulletin of the Society for Vector Ecology* 11:271–75.

Manhoff, D. T., I. Hood, F. Caputo, J. Perry, S. Rosen, and H. G. Mirchandani. 1988. Cocaine in decomposed human remains. *Journal of Forensic Sciences* 36:1732–35.

Miller, M. L., W. D. Lord, M. L. Goff, D. Donnelly, E. T. McDonough, and J. C. Alexis. 1994. Isolation of amitriptyline and nortriptyline from fly puparia (Phoridae) and beetle exuviae (Dermestidae) associated with mummified human remains. *Journal of Forensic Sciences* 39:1305–13.

Nolte, K. B., R. D. Pinder, and W. D. Lord. 1992. Insect larvae used to detect cocaine poisoning in a decomposed body. *Journal of Forensic Sciences* 37:1179–85.

Nuorteva, P. 1977. Saracosaprophagous insects as forensic indicators. In *Forensic medicine: A study in trauma and environmental hazards*, Tedeshi, G. C., W. G. Eckert, and L. G. Tedeshi, Eds. Vol. 2. Philadelphia: W.B. Saunders, 1072–95.

Nuorteva, P., and S. L. Nuorteva. 1982. The fate of mercury in sarcosaprophagous flies and in insects eating them. *Ambio* 11:34–37.

Sadler, D. W., L. Roberson, G. Brown, C. Fuke, and D. J. Pounder. 1997. Barbituates and analgesics in *Calliphora vicina* larvae. *Journal of Forensic Sciences* 42:481–85.

Schott, S., and P. Nuorteva. 1983. Metylkvicksilvrets inverkan pa aktiviet hos *Tenebrio molitor* (L.) (Col. Tenebrionidae). *Acta Entomologica Fennica* 42:78–81.

Sohal, R. S., and R. E. Lamb. 1977. Intracellular deposition of metals in the midgut of the adult housefly, *Musca domestica. Journal of Insect Physiology* 23:1349–54.

Sohal, R. S., and R. E. Lamb. 1979. Storage excretion of metallic cations in the adult housefly, *Musca domestica. Journal of Insect Physiology* 25:119–24.

Tracqui, A., C. Keyser-Tracqui, P. Kintz, and B. Ludes. 2004. Entomotoxicology for the forensic toxicologist: Much ado about nothing? *International Journal of Legal Medicine* 118:194–96.

Utsumi, K. 1958. Studies on arthropods congregating to animal carcasses with regard to the estimation of postmortem interval. *Ochanomizu Journal of Medicine* 7:202–23.

Molecular Methods for Forensic Entomology

13

JEFFREY D. WELLS
JAMIE R. STEVENS

Contents

Practical Advice for a Death Investigator

Molecular biology is a very technical field. Although we have tried to explain some of the basic concepts and avoid the worst jargon, we suspect that a reader will need a fair amount of biology or chemistry training in order to follow all of the details included within this chapter. It will be a lost opportunity, though, if we fail to reach the crime scene investigator who lacks a bench science background. To state the obvious, if the evidence is not collected in the first place, and then properly handled, the analyses described in this chapter will not be possible.

In our experience, most investigators now know that a forensic entomologist can help pinpoint the time of death. Please keep in mind that deoxyribonucleic acid (DNA)-based identification of the insect specimens may be needed to produce the most useful postmortem interval estimate that can be made. Also, insects associated with a corpse may, in rare instances, indicate that a body was moved after death, and an insect found at the scene after a corpse was moved may help identify the victim.

Perhaps the most important step for a death investigator is to kill and preserve a representative sample of insects associated with the corpse as soon as possible. There is some flexibility concerning preservation methods, but good DNA can usually be recovered from an insect preserved using 95% ethanol, or by freezing, or both. A preservative solution containing formaldehyde (formalin) should not be used if it can be avoided, because it makes it difficult to do DNA analysis. If ethanol or a freezer is not immediately available, the specimens can be chilled, e.g., with ice from the nearest food store or soft drink cans from a vending machine, and kept cold for transport to the lab. Some forensic entomologists recommend killing maggots by blanching in hot water. This technique will not hinder any subsequent DNA analysis.

Maggots found in the absence of a corpse may still have the victim's tissue in their gut. Such specimens must be killed and preserved (or put on ice) right away. Otherwise, the evidence will be digested and lost.

Background

Pre–polymerase chain reaction (PCR) genotyping technology generated mitochondrial DNA (mtDNA) sequence data for higher flies, such as *Drosophila yakuba* (Clary and Wohlstenholme, 1985), *Cochliomyia hominivorax* (Roehrdanz and Johnson, 1988), and *Phormia regina* (Goldenthal et al., 1991). However, the advent of PCR (Sakai et al., 1988; Figure 13.1), direct sequencing of PCR product (Hillis et al., 1996), and universal PCR primers, those that work for a range of taxa (e.g., Kocher et al., 1989), made it relatively easy to produce molecular genetic data for many species that had never before been examined.

Sperling et al. (1994) were the first to propose a DNA-based species diagnostic test for forensically important insects. They directly sequenced the mtDNA genes for cytochrome c oxidase subunits one and two (COI+II). They also searched their sequence data for differences between species in the location of restriction enzyme sites, and used this information to design a fast and inexpensive identification procedure based on digested PCR product restriction fragment lengths (PCR-RFLPs; Figures 13.2 and 13.3). Because they characterized only three blow fly species, *Protophormia terraenovae*, *Lucilia* (*Phaenicia*) *sericata*, and *Lucilia illustris*, theirs was more of a proof of concept than a practical test. However, the methods they developed or demonstrated are remarkably robust, and are the basis of most of the papers that have since been published on this topic.

When the first edition of *Forensic Entomology* (Byrd and Castner, 2001) was published, related molecular biology research and practice was overwhelmingly directed at calliphorid (blow fly) specimen identification. At the time of this writing that is still the case. However, since 2001 there has been a huge increase in published genotype data, allowing for a reappraisal of DNA-based species determination for forensically important insects. There have also been several exciting technological developments that are either beginning to affect casework or seem likely to in the near future.

Specimen Identification

There is little doubt about the need for correct specimen identification in forensic entomology. Under many circumstances this can be difficult to do for immature stages, damaged specimens, or even complete adults of some taxa, such as the Sarcophagidae (Smith, 1986). A molecular genotype offers an obvious alternative, or complement, to classical methods (Wells and Stevens, 2008). A genotype obtained from an unknown evidence specimen is matched (see below) to a genotype from an identified reference specimen. Similar analyses occur in other areas of forensic science, in which identifying a tissue sample aids in the enforcement of hunting and conservation laws (Cronin et al., 1991; Dizon et al.,

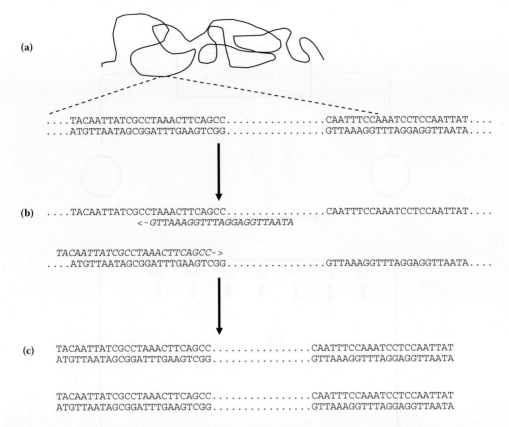

Figure 13.1 Schematic diagram of the polymerase chain reaction process. (a) Long molecules of DNA, extracted from the experimental sample, are mixed in solution with the chemical ingredients for DNA replication. The typical target is a small section of the DNA, indicated by the dashed lines. (b) The double-stranded sample DNA is split into single strands by heat, allowing short strands of DNA, called primers, to anneal to the longer strands. The only DNA sequence shown is the primers and primer annealing sites. Usually the two primer sites are a few hundred bases apart. A DNA-synthesizing enzyme, indicated by an arrow, will build a new DNA molecule from the primer, using the sample strand as a template. (c) At the end of the previous step, two copies have been made of the DNA region defined by the primers. The process is repeated perhaps twenty-five to thirty-five times, with the amount of target DNA approximately doubled each time.

2000), fraudulent labeling of high-value foods (Bottero et al., 2002; Marko et al., 2004), and identification of corpse stomach contents (Zehner et al., 1998).

Current Methods

Most molecular identification tests proposed for forensic entomology use some portion of COI+II (e.g., Malgorn and Coquoz, 1999; Sperling et al., 1994; Vincent et al., 2000; Wallman and Donnellan, 2001; Wells et al., 2001; Wells and Sperling, 2001; Harvey et al., 2003, Chen et al., 2004; Ames et al., 2006). MtDNA is haploid, occurs in a much greater number of copies per cell compared to nuclear DNA (nuDNA), and animal mitochondrial DNA has a relatively stable arrangement of polymorphic protein-coding genes flanked by highly conserved transfer RNA genes (Avise, 1994). Thus, mtDNA quickly became easy to genotype because it can be recovered from even degraded tissue (there is no need to separate different alleles) and because of the early availability of universal PCR primers based on tRNA annealing

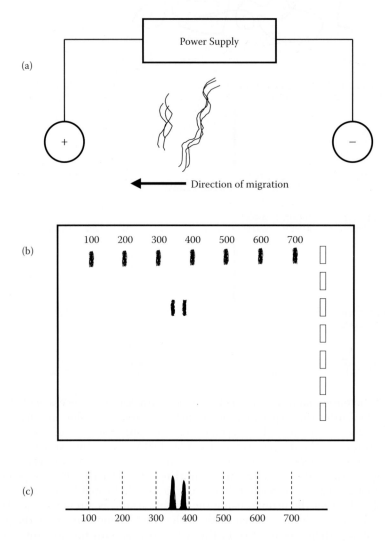

Figure 13.2 Diagrams of electrophoresis, in which a mix of DNA of two different lengths is separated so that those lengths can be estimated. (a) DNA is negatively charged and will move toward the positive electrode in solution. If this occurs within a medium such as gel-atin-like agarose, small DNA molecules will migrate more quickly than large molecules. (b) Arrangement of an electrophoresis gel image. DNA molecules are stained to form bands on the gel. The small rectangles on the right indicate wells in the gel into which DNA was loaded. The voltage was turned off after DNA fragments were separated during migration from right to left. The length, in bases, of the two DNA fragments loaded in the center is estimated by compari-son to the 100- to 700-base set of DNA size standards in the top lane. (c) DNA electrophoresis is also performed within tiny capillary tubes. Rather than a gel image, DNA is recorded as it passes a detector at the end of the capillary. This signal is converted to a graphics file with DNA peaks rather than bands. In this case, the size standard is injected with the sample DNA. Although the size standard is shown here as dashed lines, it would actually be a series of peaks as well, but a different color than the sample peaks.

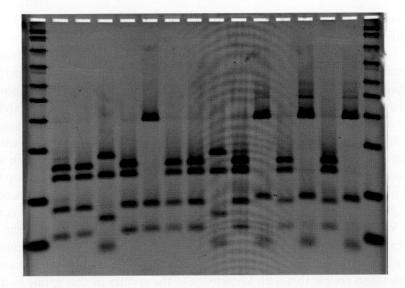

Figure 13.3 PCR-RFLP profiles of Sarcophagidae produced by *Alu*I digestion of a portion of COI. The ethidium bromide-stained agarose gel is shown as a negative image. From left to right, lanes 1 and 16 = size standard; lanes 2, 3, 5, 7, 8, 10, and 12 = *Sarcophaga africa*; lanes 4 and 9 = *Ravinia sueta*; lane 6 = *S. mimoris*; lanes 11, 13, and 15 = *Blaexosipha plinthopyga*; and lane 14 = *S. crassipalpis*. (Courtesy of J. Linville, University of Alabama at Birmingham.)

sites (Kocher et al., 1989; Simon et al., 1994). The 5' end of COI is also the proposed universal "DNA barcode" for animals (Herbert et al., 2003; Rubinoff et al., 2006).

However, various authors have suggested the use of loci other than COI+II for identifying carrion insects. These included nuDNA randomly amplified polymorphic DNA (RAPD; Benecke, 1998; Skoda et al., 2002; Stevens and Wall, 1995, 1996), the gene for 28S ribosomal RNA (Stevens and Wall, 2001), the ribosomal internal transcribed spacer regions (Ratcliffe et al., 2003; Nelson et al., 2007), or alternative mtDNA protein coding genes (Wallman et al., 2005; Zehner et al., 2004). A few studies have included both mtDNA and nuDNA loci in a single analysis (Stevens, 2003; Stevens et al., 2002; Nelson et al., 2007).

Data interpretation varies somewhat according to the method. PCR-RFLP produces a particular set of DNA fragment sizes that can be measured on an agarose or polyacrylamide gel (Dowling et al., 1996; Figure 13.3). This method requires an exact match between the fragment profile of an unknown specimen and the profile of a known reference specimen.

A DNA sequence is usually inferred from a complex computer-generated file called an electropherogram or chromatogram (Figure 13.4), representing fluorescently labeled DNA

TA TA G TA G A A A A CG G A G C T G G A ACA G G A T G A ACA G T T TA

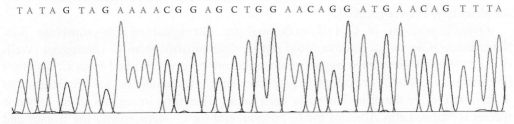

Figure 13.4 A DNA sequence data file, showing a portion of the COI gene in *Lucilia coeruleiviridis*.

molecules separated by electrophoresis. See Hillis et al. (1996) for further details about this process. This cycle sequencing technology has been shown by vast experience to work well for a wide variety of sample types, even very degraded tissue, and it is available at most research universities, many crime labs, and as a commercial service from private companies. Although sending samples to a commercial service may present chain of custody problems for a criminal investigation, it can be an economical option for research.

Intraspecific variation in DNA sequence is commonly observed, so an unknown specimen will often not exactly match the genotype of a reference specimen. Some authors who used sequence data for this purpose were a bit vague about the logic of making identifications in this manner. However, many authors use phylogenetic analysis to group the genotype of an unknown specimen with its closest relative in the reference database (Stevens and Wall, 2001; Wallman and Donnellan, 2001; Harvey et al., 2003; Chen et al., 2004; Wells and Williams, 2007; Nelson et al., 2007; Wells et al., 2007). Even if the evidence specimen cannot be precisely identified by conventional means, a forensic entomologist will usually be able to narrow the taxonomic possibilities, i.e., identify the specimen to the level of genus, and couple this with knowledge of the species in that genus found at that site. If all candidate species are represented in the reference database, then the reference genotype found to be the closest relative of the evidence specimen genotype will be the same species.

How Well Do DNA Sequence Data Work for Distinguishing Closely Related Carrion Insects?

On average, it is closely rather than distantly related species that most resemble each other in appearance. Therefore, if the performance of a DNA-based identification test is to be judged, we should consider how well it distinguishes the most closely related species of interest. Using sequence data for species determination will yield unambiguous results only if the locus shows reciprocal monophyly (Moritz, 1996; see Wells and Williams, 2007). It is thought that over evolutionary time following the split of one species into two, reciprocal monophyly is likely to develop for every locus (Avise, 2000), but recently diverged species may still share alleles (Funk and Omland, 2003).

The group of carrion-feeding insects for which we have the most genotype data is the blow fly subfamily Chrysomyinae, which includes the familiar and forensically useful genera *Chrysomya*, *Phormia*, *Protophormia*, and *Cochliomyia*. There are now published sequence data for at least twenty of the carrion-feeding species (Wells and Sperling, 2001; Harvey et al, 2003; Chen et al., 2004; Wallman et al., 2005; Wells and Williams, 2007), plus some closely related vertebrate parasites (Wells and Sperling, 2001; Lessinger et al., 2000, Whitworth et al., 2007). The specimens represent all of the continents except Antarctica. For the more common species, those most used in casework, tens of specimens from many countries have been sequenced.

Overall, portions of COI+II work well for distinguishing Chrysomyinae flies. Nevertheless, *Chrysomya putoria* could not be distinguished from *C. chloropyga* (Wells et al., 2004), and *Chrysomya megacephala* could not be distinguished from *C. saffranea* (Wallman et al., 2005). For both of these species pairs, one member occurs over large areas of the earth where the other has never been found. Therefore, in regions where only one species is found, Latin America for *C. putoria*, and for *C. megacephala* the tropics and warm regions around the world, outside of Australia and nearby islands, it can be identified with DNA.

The next best-studied group is the greenbottle flies *Lucilia* (including *Phaenicia*) and some closely related genera. Portions of the COI+II sequence are available for fourteen *Lucilia* spp. (Wells et al., 2002; Chen et al., 2004; Wallman et al., 2005; Wells and Stevens, 2008), two *Hemipyrellia* spp. (Chen et al., 2004; Wallman et al., 2005), all carrion flies, and a genus, *Dyscritomyia*, the biology of which is not well understood (Wells et al., 2002). The number of individuals examined per species is not yet as high as for the Chrysomyinae, but preliminary analysis suggests that COI works less well for distinguishing greenbottle species (Wells et al., 2007). In fact, mtDNA reciprocal monophyly has yet to be found for any greenbottle flies. However, as described above, some pairs of species that cannot be distinguished by DNA (e.g., *Lucilia illustris* and *L. caesar*) have unequal geographical distributions. Therefore, COI should still be useful in regions where only one of the species occurs (e.g., *L. illustris* in North America).

Ribosomal RNA (rRNA) genes, including the flanking internal transcribed spacer (ITS) regions, are a common tool for molecular systematics and species identification, and a few studies have looked at carrion flies. Ribosomal genes occur in both nuDNA and mtDNA. In contrast to many other nuDNA genes, nuclear rRNA genes exist in multiple copies, making them abundant and relatively easy to amplify and analyze. The ITS regions between the nuclear rRNA genes are noncoding. Such loci tend to be hypervariable, and perhaps useful for species diagnosis. They have been extensively used for this purpose with mosquitoes (e.g., Beebe et al., 2007). A large number of calliphorid ITS sequences have been deposited in public sequence databases, such as GenBank (http://www.ncbi.nlm.nih.gov/Genbank/index.html), but so far, few of these have been included in a scientific publication. A comparison of ITS and COI sequences was used to infer that some blow fly specimens were a hybrid of two species (Nelson et al., 2007), a condition that could throw off a diagnostic test if unrecognized.

Regions coding for ribosomal subunits have evolved slowly, and in general they appear more useful for resolving deeper evolutionary patterns than for separating close relatives (Stevens and Wall, 1997, 2001). However, as with the ITS example above, they can be a vital independent source of data for interpreting the results of an mtDNA analysis (Stevens et al., 2002).

Future Technology

Technological progress within molecular biology is extremely rapid, and it is difficult to foresee the methods that will be common practice only a few years hence. However, two new genotyping techniques, already routinely used in other fields such as medical genetics, seem particularly suitable for forensic insect identification. We will therefore briefly describe them here.

Pyrosequencing is a method for determining the sequence of a short (<100 bases) segment of DNA. The specimen DNA is copied, and the identity of each nucleotide (i.e., the sequence) incorporated into the synthesized strand is recorded in real time (Ahmadian et al., 2000; Alderborn et al., 2000). No PCR or electrophoresis step is required, and once the specimen DNA has been extracted, a large number of pyrosequencing analyses can be performed within a few minutes for a relatively low cost. For this to become a practical method for forensic entomology, it will probably be necessary to find a small number of short diagnostic DNA regions, ideally a single region that could distinguish every species of interest, flanked by a universal primer annealing site. This would make a relatively simple and universal protocol possible.

A DNA microarray is a glass slide or silicon chip to which one or more types of oligo-nucleotide probe have been bound (Chee et al., 1996). An oligonucleotide probe is usually single-stranded DNA, and it is designed to bind a specific target sample sequence (= allele) according to the rules of DNA bases pairing (e.g., an A in the probe molecule binds to T in the target molecule). The sample DNA is amplified by PCR, the PCR product is denatured (split into single strands), and flooded onto the microarray. Because the probes for a particular target sequence are attached to only one spot on the array, the binding of sample DNA at a given location reveals its sequence. See Kwok (2001) for an explanation of ways in which a microarray is "read" by the analyst.

Although there are other reliable probe-based genotyping methods, such as real-time PCR (Livak, 1999; Giesendorf et al., 1998), they are limited in the number of different genotypes that can be simultaneously screened. In contrast, a microarray can be equipped with thousands of different probes (Mir, 2000), and because a PCR can be multiplexed to amplify many different genes at once (Henegariu et al., 1997), it might be possible to develop a single PCR/DNA chip combination that could identify every common species of carrion insect. An obstacle to this goal is that a great deal of research and development effort would be required to make it work.

Finally, one very active area of research will almost certainly have a major impact on all areas of DNA diagnostics in the near future. Molecular biology is currently undergoing a technological revolution resulting from newly available ultra-high-throughput DNA sequencing methods (Margulies et al., 2005; Bentley, 2006). As a result, the price of a large-scale sequencing project has dropped precipitously, and this trend will continue, perhaps to the point where the average researcher can afford to order complete copies of the genomes of his or her study organisms. Should that happen, it would be possible to search those genomic data with a computer to design a new DNA test that is just right for the task at hand.

Is a Particular Method Valid?

Validation is the applied science term for the determination that an analytical procedure is accurate and reliable (Technical Working Group on DNA Analysis Methods, 1995). In contrast to forensic DNA human identity testing (Butler, 2005), there are no official validation standards for DNA-based species determination. Although we are not ready to propose such standards, we have considered two aspects of validation study design that we think are crucial, and that are not often addressed (Wells and Stevens, 2008). The first is replication. How many times must the proposed method be shown to produce the correct answer before it can be trusted? This is a new area of statistical theory, and the answer is not yet clear. However, simple but conservative calculations indicate that a validation study for a DNA-based diagnostic test should include hundreds of specimens (Wells and Stevens, 2008).

The second issue is whether or not the reference database includes a useful selection of species. In other words, if genotypes for all likely candidate species are not available for comparison, a proposed test could produce a misleading match with an incorrect species, leading to misidentification of the evidence specimen. Determining that the reference database is adequate for a given investigation requires expert knowledge of the species involved. For example, an experienced forensic entomologist might be unable to identify a sarcophagid larva from a corpse using anatomical characters, but he or she could be sure that it was Sarcophagidae. Furthermore, if the entomologist knows the local carrion fauna well, it might be possible to narrow the choice to only a few sarcophagid species found in that location at that time of year. If unique, trusted, reference genotype data were available for each of those candidate species, then a match, however defined (see above), between evidence and reference specimen would provide an unambiguous identification.

Population Genetics

Very little is known about the extent to which forensically important insects show geographic genetic variation. There are at least two practical reasons that a death investigator might care about this subject. First, if geographic variation does exist, it might be useful because it would be possible to reconstruct the postmortem movement of a corpse. The hypothetical scenario is that a maggot from the victim has a genotype uncharacteristic of the location where the body is discovered. Using a genotype to infer the geographic source of an individual is called assignment, and it is now routine in scientific fields such as fisheries management or conservation biology (Manel et al., 2005). Second, if regional genetic variation were found to exist, investigators would want to know if these populations differ in any forensically important aspect of their biology, such as development rate.

These issues have barely been investigated. Development studies on the same species performed independently at separate locations sometimes suggest variation in growth rate, but this interpretation is confounded by differences in experimental methods (Wells and Kurahashi, 1994; Tarone and Foran, 2006). Single experiments in which one rearing protocol was used for the same species from different locations failed to find a geographic effect across the United States on *Phormia regina* development (Cyr, 1993), but another did find that *Lucilia sericata* originating from California fed longer, and therefore reached a larger size, than strains from Michigan or West Virginia (Tarone, 2007).

Molecular studies have given us a mixed picture. Byrne et al. (1995) reported population differentiation on a very fine geographic scale for the widespread calliphorid *Phormia regina*. The authors measured cuticular hydrocarbon profiles, a phenotype that is almost entirely determined by genotype. Regional surveys using mtDNA found geographic variation in Brazilian *Cochliomyia macellaria* (Valle and Azeredo-Espin, 1995), but no regional variation across the United States for several species of Chrysomyinae blow flies (Böhme, 2006; Wells and Williams, 2007). The association of distinct mtDNA lineages with geographic and phenotypic variants of *Chrysomya rufifacies* and *Lucilia cuprina* may be evidence for cryptic species (Wallman et al., 2005; Wells et al., 2007).

However, in most of the work just listed, the authors were analyzing data more appropriate for systematics and species discrimination than for detecting population variation. For animal taxa, nuDNA loci rather than mtDNA are most useful for population genetic studies, and there is a great need to develop such methods for carrion insects. Judging from what has worked well for other taxonomic groups, what we would most like to have would be a genotyping protocol for the most common carrion-feeding calliphorids based on several microsatellite loci. Microsatellites, also called short tandem repeats (STRs), are widely used for measuring such things as gene flow (Dowling et al., 1996), as well as for the standard forensic genetic questions of individual identity and paternity (Butler, 2005). A microsatellite locus can possess many different alleles (i.e., genetic forms of slightly different but distinctive size), and measuring an individual's genotype is a relatively simple matter because the alleles differ by length. If the locus can

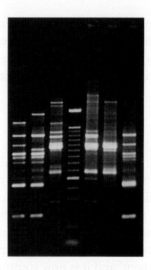

Figure 13.5 RAPD profiles from an agarose gel stained with ethidium bromide. From left to right, lanes 1, 2, and 7 are from the DNA of *Lucilia sericata* specimens. Lanes 3, 5, and 6 are *L. cuprina*. Lane 4 is the size standard (Stevens and Wall, 1996).

be amplified by PCR, then the length of the alleles, and therefore the genotype, can be determined by electrophoresis.

Development of a new microsatellite test is currently very labor-intensive. Unfortunately, the genomes of arthropods appear to contain far fewer microsatellites than those of vertebrates (Hammond et al., 1998; Ji et el., 2003; Navajas et al., 1998). Nevertheless, microsatellite protocols were developed for European *Lucilia* spp. (Florin and Gyllenstrand, 2002), and Brazilian *Cochliomyia* and *Chrysomya* spp. (Torres and Azeredo-Espin, 2005; Torres et al., 2004). Unfortunately, one of us (JDW) found the application of these methods to U.S. blow flies to be difficult to perform, or the resulting genotypes did not show variation between individuals. Future research may show that these methods can be adapted to the North American fauna.

An alternate approach to producing highly discriminating genotypes involves procedures that can be applied with little prior development to any species. RAPD analysis (see above) involves PCR primers so short that they bind to the specimen DNA in many places just because of a random match between primer and genomic DNA (Williams et al., 1990). The result is a multilocus profile that can be used for many kinds of population genetic research (Figure 13.5). Stevens and Wall (1995) were able to distinguish some UK populations of *Lucilia sericata* using RAPD profiles. The problem with RAPD is that the profile is extremely sensitive to experimental conditions; that is, the results are difficult to independently reproduce (MacPherson et al., 1993; Pérez, 1998). A newer, more reproducible method is amplified fragment length polymorphism (AFLP; Vos et al., 1995). Briefly, an AFLP profile is generated by cutting the sample DNA into many pieces, and then attaching PCR primer annealing sites to the cut ends. These fragments are amplified to produce a complex genetic profile (Figure 13.6). Again, the advantages of AFLP are that it can be reliably used on a new species with very little time spent developing the methods, and the resulting genotype can work very well for detecting population differentiation (Campbell et al., 2003). A disadvantage, compared to microsatellites, is that we do not actually know the genetic loci and the alleles that make up the profile, and this reduces the number of statistical analyses that can be performed (Campbell et al., 2003).

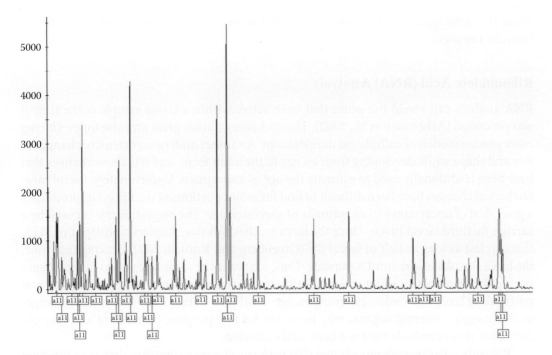

Figure 13.6 AFLP profile for a specimen of *Phormia regina*. A label below a peak indicates that it has been recorded as an allele according to a height threshold set by the operator. (Courtesy of C. Picard, West Virginia University.)

AFLP has been performed on blow flies that parasitize birds (Baudry et al., 2003; Whitworth et al., 2007), and we have successfully applied it to several forensic calliphorid species (unpublished; Figure 13.6).

Gut Content Analysis

It is possible to identify the food in an insect's gut (see review by Campobasso et al., 2005). Forensic reasons for doing so include identifying a missing body from a maggot that dropped off of it, or identifying a sexual assailant from a pubic louse transferred to the victim with the assailant's blood still inside. The laboratory protocol for such a case would be one of the vertebrate genotyping methods widely used in forensic science. The species of the tissue could be determined by sequencing the cytochrome b gene of mtDNA (Bartlett and Davidson, 1992; Zehner et al., 1998). If the source were human, this could be shown by one of several specialized tests, and the individual on whom the insect fed could be identified using STRs or the D-loop region of mtDNA (Schiro, 2001; Wells et al., 2001).

In the case of maggot gut content analysis, the most common barrier to success is the fact that the insect's gut empties quickly once the larva is off of the food (Linville, 2003). Therefore, it is essential that live larvae that might be analyzed in this way, e.g., larvae found in the absence of a corpse, be killed and preserved immediately. Another way to approach this problem may emerge from population genetic research (see above). The tools for comparing geographic populations can also be used for kinship analysis, e.g., to demonstrate that two larvae are full siblings. A blow fly female typically deposits all of the eggs she has within a few minutes, and then at least several days are required for the next batch of eggs to be produced. Because of this, the discovery of larvae that are full siblings, and about the same age, at separate locations would strongly suggest that a corpse was moved between the sites.

Ribonucleic Acid (RNA) Analysis

RNA analysis can reveal the genes that were active within a tissue sample at the time it was processed (Arbeitman et al., 2002). This technology holds great promise for producing more precise models of calliphorid development. An insect undergoes extensive changes in size and shape while developing from an egg to the adult form, and it is those changes that have been traditionally used to estimate the age of a specimen. Unfortunately, useful morphological changes have been difficult to find for major portions of the life cycle, producing a great deal of uncertainty in an estimate of specimen age. The biggest puzzle is posed by a carrion fly third larval instar. Once the larva reaches full size, it enters a postfeeding stage that can last as long as half of larval life (Greenberg and Kunich, 2002). During that time, the larva shrinks in size until it pupates. Thus, a larva is the same size at two different ages during the third instar, and it is our inability to clearly distinguish between a feeding and postfeeding larva that makes assigning an age so difficult. Other evidence, such as changes in the maggot's internal organs, may be useful for this purpose (Greenberg and Kunich, 2002), but these methods have not been well evaluated.

Recently it has been demonstrated that patterns of gene expression that are a function of age can be tracked in *Lucilia sericata* by measuring messenger RNA (mRNA) levels (Tarone, 2007; Tarone et al., in press). For example, the authors found clear differences between feeding and postfeeding larvae in the levels of mRNA produced by two genes. This advance has the potential for greatly increasing the time period in which maggot evidence can yield a precise minimum PMI estimate. RNA is chemically unstable compared to DNA, so a critical area of future research will be the evaluation of field protocols for preserving specimens suitable for RNA analysis.

Acknowledgments

Our research programs on this topic were made possible by the support of the U.S. National Institute of Justice (JDW) and the UK Wellcome Trust and Natural Environmental Research Council (JRS). The views we have expressed are not necessarily those of the granting agencies.

References

Ahmadian, A., B. Gharizadeh, A. C. Gustafsson, F. Sterky, P. Nyren, M. Uhlen, and J. Lundeberg. 2000. Single-nucleotide polymorphism analysis by pyrosequencing. *Anal. Biochem.* 280:103–10.

Alderborn, A., A. Kristofferson, and U. Hammerling. 2000. Determination of single nucleotide polymorphisms by real-time pyrophosphate DNA sequencing. *Genome Res.* 10:1249–58.

Ames, C., B. Turner, and B. Daniel. 2006. The use of mitochondrial cytochrome oxidase I gene (COI) to differentiate two UK blowfly species—*Calliphora vicina* and *Calliphora vomitoria*. *Forensic Sci. Int.* 164:179–82.

Arbeitman, M. N., E. E. M. Furlong, F. Imam, E. Johnson, B. H. Null, B.S. Baker, M. A. Krasnow, M. P. Scott, R. W. Davis, and K. P. White. 2002. Gene expression during the life cycle of *Drosophila melanogaster*. *Science* 297:2270–75.

Avise, J. 2000. *Phylogeography. The history and formation of species.* Cambridge, MA: Harvard University Press.

Bartlett, S. E., and W. S. Davidson. 1992. FINS (forensically informative nucleotide sequencing): A procedure for identifying the animal origin of biological specimens. *Biotechniques* 12:408–11.

Baudry, E., J. Bartos, K. Emerson, T. Whitworth, and J. H. Werren. 2003. *Wolbachia* and genetic variability in the birdnest blowfly *Protocalliphora sialia*. *Mol. Ecol.* 12:1843–54.

Beebe, N. W., P. I. Whelan, A. F. Van den Hurk, S. A. Ritchie, S. Corcoran, and R. D. Cooper. 2007. A polymerase chain reaction-based diagnostic to identify larvae and eggs of container mosquito species from the Australian region. *J Med. Entomol.* 44:376–80.

Benecke, M. 1998. Random amplified polymorphic DNA (RAPD) typing of necrophageous insects (Diptera, Coleoptera) in criminal forensic studies: Validation and use in praxi. *Forensic Sci. Int.* 98:157–68.

Bentley, D. R. 2006. Whole-genome re-sequencing. *Curr. Opin. Genet. Dev.* 16:545–52.

Böhme, P. 2006. Population genetics of forensically important North American blow flies (Diptera: Calliphoridae) using the A+T-rich region of mitochondrial DNA. Diploma thesis, Universität Bonn.

Bottero, M. T., T. Civera, A. Anastasio, R. M. Turi, and S. Rosati. 2002. Identification of cow's milk in "buffalo" cheese by duplex polymerase chain reaction. *J. Food Prot.* 65:362–66.

Butler, J. M. 2005. *Forensic DNA typing.* 2nd ed. Amsterdam: Elsevier.

Byrd, J. H., and J. L. Castner, Eds. 2001. Forensic entomology. In *The utility of arthropods in legal investigations.* 1st ed. Boca Raton, FL: CRC Press.

Byrne, A. L., M. A. Camann, T. L. Cyr, E. P. Catts, and K. E. Espelie. 1995. Forensic implications of biochemical differences among geographic populations of the black blow fly, *Phormia regina*. *J. Forensic Sci.* 40:372–77.

Campbell, D., P. Duchesne, and L. Bernatchez. 2003. AFLP utility for population assignment studies: Analytical investigation and empirical comparison with microsatellites. *Mol. Ecol.* 12:1979–91.

Campobasso, C. P., J. G. Linville, J. D. Wells, and F. Introna. 2005. Forensic genetic analysis of insect gut contents. *Am. J. Forensic Med. Pathol.* 26:161–65.

Chee, M., R. Yang, E. Hubbell, A. Berno, X. C. Huang, D. Stern, J. Winkler, D. J. Lockhart, M. S. Morris, and S. P. A. Fodor. 1996. Accessing genetic information with high-density DNA arrays. *Science* 274:610–14.

Chen, W. Y., T. H. Hung, and S. F. Shiao. 2004. Molecular identification of forensically important blow fly species (Diptera: Calliphoridae) in Taiwan. *J. Med. Entomol.* 41:47–57.

Clary, D. O., and D. R. Wohlstenholme. 1985. The mitochondrial DNA molecule of *Drosophila yakuba*: Nucleotide sequence, gene organization, and the genetic code. *J. Mol. Evol.* 22:252–71.

Cronin, M. E., D. A. Palmisciano, E. R. Vyse, and D. G. Cameron. 1991. Mitochondrial DNA in wildlife forensic science: Species identification of tissues. *Wildl. Soc. Bull.* 19:94–105.

Dizon, A., C. S. Baker, F. Cipriano, G. Lento, P. Palsboll, and R. Reeves, Eds. 2000. Molecular genetic identification of whales, dolphins, and porpoises. In *Proceedings of Workshop on the Forensic Use of Molecular Techniques to Identify Wildlife Products in the Marketplace,* La Jolla, CA, June 14–16, 1999. NOAA Technical Memorandum NOAA-TM-NMFS-SWFSC-286.

Dowling, T. E., C. Moritz, J. D. Palmer, and L. H. Rieseberg. 1996. Nucleic acids III: Analysis of fragments and restriction sites. In *Molecular systematics,* D. M. Hillis, C. Moritz, and B. K. Mable, Eds. Sunderland, UK: Sinauer, 249–320.

Florin, A. B., and N. Gyllenstrand. 2002. Isolation and characterization of polymorphic microsatellite markers in the blowflies *Lucilia illustris* and *Lucilia sericata*. *Mol. Ecol. Notes* 2:113–16.

Funk, D. J., and K. E. Omland. 2003. Species-level paraphyly and polyphyly: Frequency, causes, and consequences, with insights from animal mitochondrial DNA. *Annu. Rev. Ecol. Syst.* 34:397–423.

Giesendorf, B. A., J. A. Vet, S. Tyagi, E. J. Mensink, F. J. Trijbels, and H. J. Blom. 1998. Molecular beacons: A new approach for semiautomated mutation analysis. *Clin. Chem.* 44:482–86.

Goldenthal, M. J., K. A. McKenna, and D. J. Joslyn. 1991. Mitochondrial DNA of the blowfly *Phormia regina*: Restriction analysis and gene localization. *Biochem. Genet.* 29:1–11.

Greenberg, B., and J. C. Kunich. 2002. *Entomology and the law*. Cambridge, UK: Cambridge University Press.

Hammond, R. L., I. J. Saccheri, C. Ciofi, T. Coote, S. M. Funk, W. O. McMillan, M. K. Bayes, E. Taylor, and M. W. Bruford. 1998. Isolation of microsatellite markers in animals. In *Molecular tools for screening biodiversity*, A. Karp, P. G. Isaac, and D. S. Ingram, Eds. London: Chapman & Hall, 279–85.

Harvey, M. L., M. W. Mansell, M. H. Villet, and I. R. Dadour. 2003. Molecular identification of some forensically important blowflies of southern Africa and Australia. *Med. Vet. Entomol.* 17:363–69.

Henegariu, O., N. A. Heerema, S. R. Dlouhy, G. H. Vance, and P. H. Vogt. 1997. Multiplex PCR: Critical parameters and step-by-step protocol. *Biotechniques* 23:504–11.

Herbert, P. D., A. Cywinska, S. L. Ball, and J. R. de Waard. 2003. Biological identifications through DNA barcodes. *Proc. Roy. Soc. London B* 270:313–21.

Hillis, D. M., B. K. Mable, A. Larson, S. K. Davis, and E. A. Zimmer. 1996. Nucleic acids. IV. Sequencing and cloning. In *Molecular systematics*, D. M. Hillis, C. Moritz, and B. K. Mable, Eds. Sunderland, UK: Sinauer, 321–41.

Ji, Y., J. D. X. Zhang, G. M. Hewitt, L. Kang, and D. M. Li. 2003. Polymorphic microsatellite loci for the cotton bollworm *Helicoverpa armigera* (Lepidoptera: Noctuidae) and some remarks on their isolation. *Mol. Ecol. Notes* 3:102–4.

Kahana, T., J. Almog, J. Levy, E. Shmeltzer, Y. Spier, and J. Hiss. 1999. Marine taphony: Adipocere formation in a series of bodies recovered from a single shipwreck. *J. Forensic Sci.* 44:897.

Kocher, T. D., W. K. Thomas, A. Meyer, S. V. Edwards, S. Pääbo, F. X. Villablanca, and A. C. Wilson. 1989. Dynamics of mitochondrial DNA evolution in animals: Amplification and sequencing with conserved primers. *Proc Natl. Acad. Sci. U.S.A.* 86:6196–200.

Kwok, P.-Y. 2001. Methods for genotyping single nucleotide polymorphisms. *Annu. Rev. Genom. Hum. Genet.* 2:235–58.

Lessinger, A. C., A. C. Martins Junqueira, T. A. Lemos, E. L. Kemper, F. R. da Silva, A. L. Vettore, P. Arruda, and A. M. Azeredo-Espin. 2000. The mitochondrial genome of the primary screw-worm fly *Cochliomyia hominivorax* (Diptera: Calliphoridae). *Insect Mol. Biol.* 9:521–29.

Linville, J. G. 2003. The recovery and characterization of vertebrate DNA from forensically important fly larvae: An optimization study. PhD thesis, University of Alabama, Birmingham.

Livak, K. J. 1999. Allelic discrimination using fluorogenic probes and the 5' nuclease assay. *Genet. Anal.* 14:143–49.

MacPherson, J. M., P. E. Eckstein, G. J. Scoles, and A. A. Gajadhar. 1993. Variability of the random amplified polymorphic DNA assay among thermal cyclers, and effects of primer and DNA concentration. *Mol. Cell Probes* 7:293–99.

Malgorn, Y., and R. Coquoz. 1999. DNA typing for identification of some species of Calliphoridae. An interest in forensic entomology. *Forensic Sci. Int.* 102:111–19.

Manel, S., E. O. Gaggiotti, and R. S. Waples. 2005. Assignment methods: Matching biological questions with appropriate techniques. *Trends Ecol. Evol.* 20:136–42.

Margulies, M., M. Egholm, W. E. Altman, S. Attiya, J. S. Bader, L. A. Bemben, J. Berka, M. S. Braverman, Y. J Chen, Z. Chen, S. B. Dewell, L. Du, J. M. Fierro, X. V. Gomes, B. C. Godwin, W. He, S. Helgesen, C. H. Ho, G. P. Irzyk, S. C. Jando, M. L. Alenquer, T. P. Jarvie, K. B. Jirage, J. B. Kim, J. R. Knight, J. R. Lanza, J. H. Leamon, S. M. Lefkowitz, M. Lei, J. Li, K. L. Lohman, H. Lu, V. B. Makhijani, K. E. McDade, M. P. McKenna, E. W. Myers, E. Nickerson, J. R. Nobile, R. Plant, B. P. Puc, M. T. Ronan, G. T. Roth, G. J. Sarkis, J. F. Simons, J. W. Simpson, M. Srinivasan, K. R. Tartaro, A. Tomasz, K. A. Vogt, G. A. Volkmer, S. H. Wang, Y. Wang, M. P. Weiner, P. Yu, R. F. Begley, and J. M. Rothberg. 2005. Genome sequencing in microfabricated high-density picolitre reactors. *Nature* 437:376–80.

Marko, P. B., S. C. Lee, A. M. Rice, J. M. Gramling, T. M. Fitxhenry, J. S. McAllister, G. R. Harper, and A. Moran. 2004. Mislabelling of a depleted reef fish. *Nature* 430:309–10.

Mir, K. U. 2000. The microarray meeting, Scottsdale, Arizona, USA, 22–25 September 1999. *Trends Genet.* 16:63–64.

Moritz, C. 1996. The uses of molecular phylogenies for conservation. In *New uses for new phylogenies*, P. H. Harvey, A. J. Leigh Brown, J. Maynard Smith, and S. Nee, Eds. Oxford: Oxford University Press, 203–16.

Navajas, M. J., H. M. A. Thistlewood, J. Lagnel, and C. Hughes. 1998. Microsatellite sequences are under-represented in two mite genomes. *Insect Mol. Biol.* 7:249–56.

Nelson, L. A., J. F. Wallman, and M. Dowton. 2007. Using COI barcodes to identify forensically and medically important blowflies. *Med. Vet. Entomol.* 21:44–52.

Pérez, T., J. Albornoz, and A. Domínguez. 1998. An evaluation of RAPD fragment reproducibility and nature. *Mol. Ecol.* 7:1347–57.

Ratcliffe, S. T., D. W. Webb, R. A. Weinzievr, and H. M. Robertson. 2003. PCR-RFLP identification of Diptera (Calliphoridae, Muscidae and Sarcophagidae)—A generally applicable method. *J. Forensic Sci.* 48:783–85.

Roehrdanz, R. L., and D. A. Johnson. 1988. Mitochondrial DNA variation among geographical populations of the screwworm fly *Cochliomyia hominivorax*. *J. Med. Entomol.* 25:136–41.

Rubinoff, D., S. Cameron, and K. Will. 2006. A genomic perspective on the shortcomings of mitochondrial DNA for "barcoding" identification. *J. Hered.* 97:581–94.

Sakai, R. K., D. H. Gelfand, S. Stoffel, S. J. Scharf, R. Higuchi, G. T. Horn, K. B. Mullis, and H. A. Erlich. 1988. Primer-directed enzymatic amplification of DNA with a thermostable DNA polymerase. *Science* 239:487–91.

Schiro, G. J. 2001. Extraction and quantification of human deoxyribonucleic acid, and the amplification of human short tandem repeats and a sex identification marker from fly larvae found on decomposing tissue. MS thesis, University of Central Florida, Orlando, FL.

Simon, C., F. Frati, A. Beckenbach, B. Crespi, H. Liu, and P. Flook. 1994. Evolution, weighting and phylogenetic utility of mitochondrial gene sequences and a compilation of conserved polymerase chain reaction primers. *Ann. Entomol. Soc. Am.* 87:651–70.

Skoda, S. R., S. Pornkulwat, and J. E. Foster. 2002. Random amplified polymorphic DNA markers for discriminating *Cochliomyia hominivorax* from *C. macellaria* (Diptera: Calliphoridae). *Bull. Entomol. Res.* 92:89–96.

Smith, K. G. V. 1986. *A manual of forensic entomology*. London: British Museum (Natural History).

Sperling, F. A. H., G. S. Anderson, and D. A. Hickey. 1994. A DNA-based approach to the identification of insect species used for postmortem interval estimation. *J. Forensic Sci.* 39:418–27. [Published erratum appears in Wells and Sperling (2000).]

Stevens, J. R. 2003. The evolution of myiasis in blowflies (Calliphoridae). *Int. J. Parasitol.* 33:1105–13.

Stevens, J., and R. Wall. 1995. The use of random amplified polymorphic DNA (RAPD) analysis for studies of genetic variation in populations of the blowfly *Lucilia sericata* (Diptera: Calliphoridae) in southern England. *Bull. Entomol. Res.* 85:549–55.

Stevens, J., and R. Wall. 1996. Species, sub-species and hybrid populations of the blowflies *Lucilia cuprina* and *Lucilia sericata* (Diptera: Calliphoridae). *Proc. Biol. Sci.* 263:1335–41.

Stevens, J., and R. Wall. 2001. Genetic relationships between blowflies (Calliphoridae) of forensic importance. *Forensic Sci. Int.* 120:116–23.

Stevens, J. R., R. Wall, and J. D. Wells. 2002. Paraphyly in Hawaiian hybrid blowfly populations and the evolutionary history of anthropophilic species. *Insect Mol. Biol.* 11:141–48.

Tarone A. 2007. *Lucilia sericata* development: Plasticity, population differences, and gene expression. PhD thesis, Michigan State University, East Lansing, MI.

Tarone, A., and D. R. Foran. 2006. Components of developmental plasticity in a Michigan population of *Lucilia sericata* (Diptera: Calliphoridae). *J. Med. Entomol.* 43:1023–33.

Tarone, A., K. C. Jennings, and D. R. Foran. 2007. Aging blow fly eggs using gene expression: A feasibility study. *J. Forensic Sci.* 52(6):1350–54.

Technical Working Group on DNA Analysis Methods. 1995. Guidelines for a quality assurance program for DNA analysis. *Crime Lab Digest.* 22:21–43.

Torres, T. T., and A. M. L. Azeredo-Espin. 2005. Development of new polymorphic microsatellite markers for the New World screw-worm *Cochliomyia hominivorax* (Diptera: Calliphoridae). *Mol. Ecol. Notes* 5:815–17.

Torres, T. T., R. P. V. Brondani, J. E. Garcia, and A. M. L. Azeredo-Espin. 2004. Isolation and characterization of microsatellite markers in the new world screw-worm *Cochliomyia hominivorax* (Diptera: Calliphoridae). *Mol. Ecol. Notes* 4:182–84.

Valle, J. S., and A. M. L. Azeredo-Espin. 1995. Mitochondrial DNA variation in two Brazilian populations of *Cochliomyia macellaria* (Diptera: Calliphoridae). *Braz. J. Genet.* 18:521–26.

Vincent, S., J. M. Vian, and M. P. Carlotti. 2000. Partial sequencing of the cytochrome oxydase b subunit gene I: A tool for the identification of the European species of blow flies for postmortem interval estimation. *J. Forensic Sci.* 4:820–23. [Published erratum appears in Wells and Sperling (2000).]

Vos, P., R. Hogers, M. Bleeker, M. Reijans, T. van de Lee, M. Hornes, A. Frijters, J. Pot, J. Peleman, and M. Kuiper. 1995. AFLP: A new technique for DNA fingerprinting. *Nucleic Acids Res.* 23:4407–14.

Wallman, J. F., and S. C. Donnellan. 2001. The utility of mitochondrial DNA sequences for the identification of forensically important blowflies (Diptera: Calliphoridae) in southeastern Australia. *Forensic Sci. Int.* 120:60–67.

Wallman, J. F., R. Leys, and K. Hogendoom. 2005. Molecular systematics of Australian carrion-breeding blowflies (Diptera: Calliphoridae) based on mitochondrial DNA. *Invert. Syst.* 19:1–15.

Wells, J. D., M. L. Goff, J. K. Tomberlin, and H. Kurahashi. 2002. Molecular systematics of the endemic Hawaiian blowfly genus *Dyscritomyia* Grimshaw (Diptera: Calliphoridae). *Med. Entomol. Zool.* 53(Suppl. 2):231–38.

Wells, J. D., F. G. Introna, G. Di Vella, C. P. Campobasso, J. Hayes, and F. A. H. Sperling. 2001. Human and insect mitochondrial DNA analysis from maggots. *J. Forensic Sci.* 46:685–87.

Wells, J. D., and H. Kurahashi. 1994. *Chrysomya megacephala* (F.) development: Rate, variation and the implications for forensic entomology. *Jpn. J. Sanit. Zool.* 45:303–9.

Wells, J. D., N. Lunt, and M. H. Villet. 2004. Recent African derivation of *Chrysomya putoria* from *C. chloropyga* and mitochondrial DNA paraphyly of cytochrome oxidase subunit one in blowflies of forensic importance. *Med. Vet. Entomol.* 18:445–48.

Wells, J. D., and F. A. H. Sperling. 2000. Commentary: A DNA-based approach to the identification of insect species used for postmortem interval estimation and partial sequencing of the cytochrome oxydase b subunit gene I: A tool for the identification of European species of blow flies for postmortem interval estimation. *J. Forensic Sci.* 45:1358–59.

Wells, J. D., and F. A. H. Sperling. 2001. DNA-based identification of forensically important Chrysomyinae (Diptera: Calliphoridae). *Forensic Sci. Int.* 120:110–15.

Wells, J. D., and J. R. Stevens. 2008. The application of DNA-based methods in forensic entomology. *Annu. Rev. Entomol.* 53:103–20.

Wells, J. D., R. Wall, and J. R. Stevens. 2007. Phylogenetic analysis of forensically important *Lucilia* flies based on cytochrome oxidase 1: A cautionary tale for forensic species determination. *Int. J. Leg. Med.* 121:1–8.

Wells, J. D., and D. W. Williams. 2007. Validation of a DNA-based method for identifying Chrysomyinae (Diptera: Calliphoridae) used in a death investigation. *Int. J. Leg. Med.* 121:1–8.

Whitworth, T. L., R. D. Dawson, H. Magalon, and E. Baudry. 2007. DNA barcoding cannot reliably identify species of the blowfly genus *Protocalliphora* (Diptera: Calliphoridae). *Proc. Biol. Sci.* 274:1731–39.

Williams, J. G., A. R. Kubelik, K. J. Livak, J. A. Rafalski, and S. V. Tingey. 1990. DNA polymorphisms amplified by arbitrary primers are useful as genetic markers. *Nucleic Acids Res.* 18:6531–35.

Zehner, R., J. Amendt, S. Schutt, J. Sauer, R. Krettek, and D. Povolny. 2004. Genetic identification of forensically important flesh flies (Diptera: Sarcophagidae). *Int. J. Legal Med.* 118:245–47.

Zehner, R., S. Zimmerman, and D. Mebs. 1998. RFLP and sequence analysis of the cytochrome b gene of selected animals and man: Methodology and forensic application. *Int. J. Legal Med.* 111:323–27.

The Forensic Entomologist as Expert Witness

14

ROBERT D. HALL

Contents

Introduction

Evidence from medicocriminal entomology can affect investigative or legal proceedings in various ways. Oral and written anecdotes pertaining to insects may be useful as investigators piece together a present or retrospective look at pertinent circumstances. Occasionally, insect evidence may lead to other lines of investigation, which in turn may reveal the truth. Insect-derived data may simply corroborate other real or testimonial evidence. In fact, litigation seldom turns solely on insect evidence. Sometimes it does.

As the medicocriminal entomologist becomes increasingly involved with the judicial system, the procedures governing such involvement become progressively more stringent. For example, the aforementioned anecdotes, verbal comments, or suggestions during the processing of a crime scene may represent informal involvement. However, entomological opinions by written report (usually discoverable by the opposition), affidavit, deposition, or in-court testimony represent formal procedures ultimately characterized by testimony under oath, where prescribed penalties for perjury attach.

If entomological evidence is to have any impact on the outcome of litigation, it somehow must find its way into the proceedings. In virtually all instances, insect evidence is evaluated by state or federal courts under the general rubric of scientific "expert testimony." In brief, this means that the entomological data must be analyzed by someone qualified to render an opinion regarding how such evidence fits the facts of the case at hand. Such expert testimony is governed by several Federal Rules of Evidence, or the codification thereof that exist in state jurisdictions (see Giannelli and Imwinkelried 1993).

Theories of Admissibility

The guidelines governing whether or not evidence will be admitted—that is, evaluated by the trier of fact (a jury in criminal proceedings)—depend upon several fundamental tests that appear deceptively simple (see Imwinkelried 1992). First, the proffered evidence must be *relevant*—both logically and legally. Irrelevant evidence is not admissible, because it would not serve to make the trier of fact believe that any certain set of circumstances was more or less likely. Therefore, evidence about the presence of a certain species of butterfly at a crime scene would be considered logically irrelevant if such information had no scientific or other bearing on the analysis of the case. The issue of legal relevance is treated under Federal Rule of Evidence 403, where evidence may be excluded by the trial judge if "its probative value is substantially outweighed by the danger of unfair prejudice, confusion of the issues, or misleading the jury, or by considerations of undue delay, waste of time, or needless presentation of cumulative evidence." This powerful rule obviously gives the judge wide latitude.

In addition to being relevant, proffered evidence and witnesses must also be *competent*. Witnesses may be declared incompetent for violation of various procedural rules. As an example, if the court has adopted an exclusionary rule, witnesses must remain outside the courtroom until called. If an expert violates this procedure, he or she may be barred from testifying in that proceeding.

To qualify as competent, witnesses and evidence must not be protected by any of the various common law, statutory, or constitutional *privileges* (such as spousal privilege, attorney-client privilege, privilege against self-incrimination, and so forth), and evidence must be *trustworthy*. It is argument over trustworthiness that absorbs the most time when admissibility, or weight of insect evidence, is contested. If it can be shown that evidence is untrustworthy, it follows as a matter of law that it is incompetent, and if incompetent, it should not factor into the trier of fact's decision-making process. Such a straightforward analysis, however, can lead to emotionally charged argument because the legal terms of art employed can be interpreted as hurtful when perceived as an *ad hominem* (that is, against a person rather than against the evidence) attack. In fact, most affidavits, pretrial motions to exclude, and oral argument directed toward exclusion of evidence affirmatively use such terminology as the legally proven way to achieve the most direct defense against entomological, and indeed all, expert testimony: by simply keeping it out of the courtroom altogether.

Federal Rule of Evidence 702, followed in federal courts and virtually all states, is entitled "Testimony by Experts." It is remarkably broad, stating that "if scientific, technical, or other specialized knowledge will assist the trier of fact to understand the evidence or to determine a fact in issue, a witness qualified as an expert by knowledge, skill, experience, training, or education, may testify thereto in the form of an opinion or otherwise." The question, then, becomes "Who is an expert?" Whereas educational credentials may not be important in qualifying a witness whose "expertise" was gained through experience—such as carpentry or driving a truck—they are critical in the qualifications of an expert in science. An expert lacking such credentials may be barred from testifying.

Legal Tests of Admissibility

Whether the analysis is under a federal *Daubert** standard or any of the various state standards typified by *Frye*,† the same four broad elements seem to shape decisions regarding whether or not scientific, including entomological, evidence will be admitted (Cantor 1994). First, is the "expert" qualified? Second, is the opinion supported by scientific principles? Third, is the opinion based on reliable data? Fourth, is the opinion so confusing or prejudicial that it should be excluded? The first three questions may be restated in another way, which can be regarded as a three-pronged test for admissibility. Is it (first) good science, incorporating (second) reliable data, as applied to the facts of the case at hand, by (third) a qualified expert? If the answer to any of these is no, a good legal argument can be made to keep the jury from hearing the evidence.

Good Science

Although the federal and state standards are at variance with each other on minor points, the issue of whether a field, technique, or procedure constitutes "good science" is fairly straightforward. Under the *Frye* general acceptance principle, in order to qualify as good science, the relevant scientific community must have generally accepted a theory. Publication, peer review, and reliance all are elements to be examined and weighed when making a decision about whether something constitutes such good science. Additional elements under *Daubert* include evaluation of known error rates, and any standards or indices of reliability. The latter decision makes the trial judge the "gatekeeper," deciding what testimony enters the courtroom. While a judge has arguably more discretion under the latter standard—and many have used it since *Daubert* was announced—most scientists, attorneys, and the judiciary seem to understand obvious differences between documented and accepted science, scientific uncertainty, and pure speculation.

From a practical standpoint, medicocriminal entomology has enjoyed more than a century of judicial acceptance in regard to its fundamental scientific principles, including temperature-dependent insect development, necrophilous ecological behavior patterns, and the generally predictable succession of decomposer fauna. These have also been well documented in the scientific literature. It is therefore highly unlikely an assertion that medicocriminal entomology does not constitute good science per se and thus should be excluded would prevail in any U.S. jurisdiction today. When entomological evidence is

* *Daubert v. Merrell Dow Pharmaceuticals*, 509 U.S. 579 (1993).
† *Frye v. United States*, 293 F. 1013, D.C. Cir. (1923).

important to a case, then the first prong of the three-part admissibility test is essentially unassailable. The same cannot be said for the remaining two.

Qualifications of the Expert

Whether or not an individual is qualified to render an opinion in a case involving medico-criminal entomology depends principally on his or her educational and experiential background. As described in the introduction of this book, the modern trend in U.S. courts is to view postgraduate education in entomology, such as the master of science (MS) or master of arts (MA), or doctoral degrees such as the doctor of philosophy (PhD) or doctor of science (ScD), as the educational minima to qualify an expert. Organizations such as the American Board of Forensic Entomology require that applicants possess an earned (in contrast to honorary) doctoral degree with a major in medical entomology, and this seems to be recognized as reasonable by the judiciary. However, testimony by masters-prepared "experts" is still frequently admitted in many state courts. The progression, however, is inexorable: the expectation that the best-credentialed expert will be best received by the jury militates toward the highest academic degrees. There is, of course, ample precedent for this pattern in expert medical testimony.

In addition to demonstrating educational qualifications, the putative expert, to be considered fully qualified, must also show consistent involvement with research or teaching of forensic entomology. This can be done by exhibiting publication lists of the expert's scientific papers germane to the subject, and documenting involvement with classroom teaching or workshops. The latter sort of hands-on association is important to show that the expert is involved day to day with the so-called "wet work" in a manner analogous to medical experts in surgery and related clinical areas. The goal here is to identify and eliminate so-called experts who are "book smart" but without actual experience.

Another important aspect of expert qualification is absence of genuine or perceived bias. That is, the jury may perceive, rightly or wrongly, an expert whose record shows testimony only for the defense in criminal cases as biased toward the defense in all cases. Such a pattern is certain to be brought up during cross-examination of qualifications. On the other hand, a record of cases where the testimonial history shows involvement with either prosecution or defense, without regard for the type of entomological evidence, tends to demonstrate a lack of bias.

In practicality today, unless an individual offered as an expert witness in medicocriminal entomology can be shown to be patently unqualified on educational grounds (such as no college degree altogether), or to be unconscionably biased, it is likely he or she will be allowed to testify. This is only one reason why expert testimony on one side must usually be countered by expert testimony on the other, resulting in a predictable "battle of the experts."

Application of Reliable Data to the Facts

It is therefore the third and last prong of the three-part test, that is, the application of entomological science to the facts of the case, where most arguable points arise and most controversy centers. In addition to presenting as tight and scientifically valid an entomological analysis as possible, another function of the expert witness is to identify weak points in the opponent's analysis and educate counsel so that they may be thoroughly attacked. As will be explained later, this involvement does not constitute advocacy. Although the science of medicocriminal entomology per se will be considered "good science" by virtually all courts, application of scientific principles to individual cases can readily be seen as fraught with opportunity for error. Absent measurements or evidence of repeatable and documented phenomena, the expert witness quickly moves from science to guesswork. While a prerogative of the expert is to render an opinion, such an opinion must be supported by good science, and guesswork is speculation—not science. One of the tasks of our adversarial legal system is to identify and attack such speculation, thus exposing shortcomings and showing the trier of fact why it is unreliable and incompetent. Commonly argued weak points in the application of reliable entomological data to the facts of a case include, but are by no means limited to, the following:

Were the insects or other arthropods identified correctly? Accurate taxonomic identification is the foundation upon which the remainder of the entomological analysis rests, yet it is seldom challenged. It is obvious that errors here can have a fundamental impact on the validity of any estimates made, because different species of flies, for example, may have different developmental times and different proclivities. If a postmortem interval is calculated on the mistaken assumption that it was species X developing, but it can be shown that it was in fact species Y, it is easy for the jury to conclude that the entomological analysis is flawed. Further, demonstration of such misidentification can be damaging to the credibility of the expert who made the mistake. "Why, you couldn't identify species Y correctly when it was right in front of you, could you, Doctor?" can be fatal to an entomological analysis on re-cross-examination, in addition to being personally humiliating. To reduce the chances of this sort of attack being successful, prudent medicocriminal entomologists "backstop" identifications by getting second opinions from other specialists before completing analysis of the case. In the future, it is likely that molecular techniques such as polymerase chain reaction or DNA sequencing may be useful as taxonomic "litmus tests" in species identification of forensically important insects, especially hard-to-identify eggs, and larvae. At present, attempts to rear a subsample of specimens to the adult stage, which is most reliably identified, represent a prudent course of action.

Temperatures from remote sites do not accurately reflect temperatures at the scene. When temperature-dependent insect development is used to gauge postmortem interval, so-called retrospective weather records usually have to be employed. What this means is that the entomologist or investigator will get weather records from the nearest, or several nearest, weather recording stations for the timeframe in question. Weather conditions are recorded routinely at many sites, but most often at airports, agricultural experiment stations, water treatment facilities, and other municipal facilities. Such data may range from complete hourly temperatures, precipitation, relative humidity, wind speed, and cloud cover, to the starker record of daily maximum and minimum temperatures. The entomologist uses these temperatures, known to have prevailed during the time when the case at hand was transpiring, to estimate the temperatures under which the insect evidence developed. It is immediately apparent to any lay observer that the basic assumption—that the remote, retrospective temperatures in fact represent what actually prevailed at the crime scene—is tenuous at best.

If the weather recording station was generally proximate—say, within 20 miles or so—to the crime scene, it is reasonable to assume that prevailing conditions were at least similar at the two locations. As one judge remarked, "When it's hot in Oklahoma City, it's hot in Stroud [a small town about 60 miles away], and when it's cold in Oklahoma City, it's cold in Stroud." This proposition is generally true when two locations are fairly close together, as long as there is no major difference in elevation, no big geographical feature, such as a mountain range or body of water, separating the two, and the situation under evaluation is out-of-doors. Trivially, everyone has witnessed rain on one side of a street while the other side was sunny, or rapid changes in temperature as a cold front moved through.

The weakness in using remote temperature data is when microclimatic factors introduce unknown but consistent variables. Weather stations are usually designed to measure air temperatures in unshaded sites, with the thermal equipment protected from direct solar radiation. Suppose a body was found hidden in a deep, wooded ravine where midsummer overstory created intense shade? The combination of shade with cool air currents, especially in the evening, might produce a microclimate measurably cooler than the temperatures measured at an airport several miles away. On the other hand, a body found inside a closed car trunk, when the car was parked in direct summer sunlight, might have been exposed to temperatures much higher than those recorded at a remote weather station.

Because temperatures are so important to many entomological analyses, and because in most cases there is no way to escape using retrospective, remote temperature data, it is important to measure concordance, or lack thereof, between the remote site and the crime scene. This can simply involve taking temperatures, hourly or daily maximum-minimums, for about a week at the crime scene, at the same time of year and under conditions identical to those prevailing when the crime occurred, and subsequently co-locating the thermometer with the site producing the retrospective weather data. This technique will determine if the two instruments measure identically, and allow correction if not. Furthermore, it will allow correction of the appropriate thermal units for any consistent microclimatic differences noted between the crime scene and the remote temperature site. Absent such measurements, questions on cross-examination can draw attention to flaws.

Q. *Doctor, you used the daily maximum and minimum temperatures from City Airport when you calculated the fly development in this case, is that right?*
A. *Yes.*
Q. *How far is City Airport from where the body was found?*
A. *Twenty miles, more or less.*
Q. *Well, Doctor, we've all experienced situations where the weather was different over a 20-mile distance, right? How do you know they were the same?*
A. *You're right about the weather changes, but in this case I'm quite sure that the temperatures were similar.*
Q. *Did you make any measurements?*
A. *No.*

Q. *Did you have any temperature data available to you other than those derived from City Airport, 20 miles from the crime scene?*

A. *No.*

Q. *Then you cannot say for sure that the temperatures were identical, can you?*

A. *No, but...*

Q. *In fact, you had to guess they were the same, isn't that correct?*

A. *Yes.*

Maggot Mass May Have Affected the Temperature

A fascinating aspect of studying the thermal development of maggots is that under certain circumstances they can generate heat. Although insects are cold-blooded and generally must develop as a function of ambient temperature, the teeming, writhing mass of maggots that sometimes occurs during decomposition may produce significantly elevated temperatures (Greenberg 1991). These higher temperatures have the effect of shortening the developmental time of the insects themselves in relation to what analysis of ambient temperatures would suggest. Obviously, the occurrence of any maggot mass is important and should be taken into account. Problems arise, however, when there is no evidence that a maggot mass occurred. Absent eye-witness testimony of the existence of fly pupal cases in large numbers (resulting from the maggots in the mass), or in some cases, the disarticulation of skeletal remains in the absence of scavenging animals, conjecture about the effect of maggot mass temperatures may represent speculation. Conversely, failure to account for maggot mass temperatures when one obviously occurred (documented in scene photographs, for instance) also represents error. When actual measurements of maggot mass temperature are absent, such "refinement" of ambient thermal data generally involves some amount of guesswork. The proper way to reflect this uncertainty is to present conclusions describing a range, rather than a discrete number.

Accumulated Degree-Hours May Give the Impression of Precision without Substance

The thermal units necessary for well-known necrophilous flies to progress through their various life stages have been recorded in various data sets. The theory of application is simple: if one knows how many so-called thermal units were available at a crime scene, and also knows the thermal units necessary for a given species of fly to develop from the stage deposited by the female (generally egg or first-instar larva) to the stage collected, by putting the two together, one can infer that the decedent must have been dead *for at least that period of time.* This interval is termed the minimum postmortem interval, or minimum PMI.

The climatological theory of accumulating degree-hours or degree-days is widely accepted in both agricultural science and heating and cooling engineering, and is readily applied to insect development. In brief, it involves cumulative additions of temperatures (as either degrees Celsius or Fahrenheit) on an hourly or daily basis (see Chapter 9). The attractiveness of such an approach is that it can be quite precise if applied correctly. From the standpoint of litigation, it typically *appears* to be quite precise when presented to a jury. Some entomologists count only those thermal units over a base of, for example, 6 or 10°C, depending on the fly species involved. Accumulated degree-hours (ADH) are calculated appropriately only from hourly temperature data. Daily maximum and minimum temperatures can be used to calculate a daily average and accumulated degree-days (ADD), but no more. Some analysts have attempted to calculate ADH by multiplying the daily average (calculated from a maximum and minimum temperature) by 24. This calculation is valid only if the daily temperature change from warm in the afternoon to cool in the early morning fluctuates as a sine curve about that average. While this presumption may have some validity when considered over a lengthy time span, it is obviously problematic over shorter times, such as the few days generally considered in most crime analyses. It is, of course, possible for temperature to "dwell" more on the hot or cool side of a daily average, and this cannot be calculated when only maximum and minimum temperatures are analyzed. Misuse of the ADH developmental models thus puts the entomologist at risk of attack.

Did Other Temperatures Affect Insect Development?

Bodies exposed outdoors soon become an insect habitat thermally affected by ambient temperature and perhaps by maggot mass. However, complications can arise when bodies, along with the maggots on them, are refrigerated. As temperature falls, maggot development slows. When the temperature is low enough, for example, under 6 to 10°C for many common blow flies, development—for all practical purposes—ceases. Therefore, prolonged periods in air-conditioned ambulances, morgues, and so forth can have major effects on thermal calculations. Whereas body temperature drops in fairly predictable fashion shortly following death, when actual temperature measurements (which constitute recommended procedure) are absent, there is no way to know for sure how rapidly a corpse's temperature dropped. In some cases, engineers have calculated cooling curves applied to this question on an a posteriori basis, but they remain vulnerable to criticism for possible inaccuracy.

When Did the Insects Arrive at the Decedent? When Did They Lay Their Eggs or Larvae?

The most often asked question in relation to medicocriminal entomology is: "How long does it take for flies to arrive at a corpse?" The typical answer is: "Within minutes." This seemingly incredible ability of necrophilous flies has been documented repeatedly over many years and constitutes a repeatable phenomenon widely accepted by forensic entomologists.

The problem is that when this observation is applied to the facts of a particular case, the assumptions necessary may not be met. Three assumptions are required to support the assertion that flies will arrive "within minutes." First, it must be the season of the year when flies are active. To assume that flies will arrive within minutes at a decedent during wintertime depends upon the climatic conditions prevailing at the moment. Cold weather hinders and finally stops fly activity. Second, it must be during daylight hours. Necrophilous flies are generally inactive nocturnally, and so the arrival "within minutes" must usually occur during daylight. Third, the flies must have ready access to the corpse.

The repeated observations regarding prompt arrival of necrophilous flies have inevitably been made when the corpse or surrogate was exposed in open air to fly activity. Thus, a decedent lying on the ground surface, such as on the side of a rural road or in a city park, would meet this criterion. The basic assumption regarding the out-of-doors decedent is that there are "no barriers" to fly access and "no barriers" to dispersal of the decomposition odors recognized as an attractant by the flies. At the other extreme, it is easy to conceive of a decedent, otherwise highly attractive to necrophilous flies, that would never exhibit any fly activity at all. A body sealed within a closed casket, zipped inside a tight body bag, or stuffed inside a tightly sealed automobile trunk would qualify. This is one of the reasons why dismembered bodies wrapped in plastic garbage bags often prove refractory to successful entomological analysis. Frozen bodies constitute a similar problem.

Between the two extremes, one of complete exposure and the other of complete protection, lies an infinite number of gradations in accessibility of flies to corpses. What about corpses inside houses with doors and windows closed? With a window open? What about corpses wrapped in various numbers of blankets? Waffle-weave blankets versus thick wool blankets? It will quickly be seen that evidence of any major "barrier" to fly access is liable to introduce insurmountable uncertainty to estimates made therefrom. In many such cases, estimates about when flies actually accessed decedents represent nothing more than guesswork, although some experiments have been conducted (Goff 1992).

Further, access to corpses by flies does not necessarily mean that oviposition or larviposition occurred at that time. Given the major assumptions enumerated earlier—season, time of day, and access—it is generally accepted that necrophilous flies make access to and utilization of the corpse for their offspring reasonably contemporaneous. This has not been demonstrated equally convincingly when major barriers come into play. Thus, estimates derived under these circumstances often may be challenged as speculation, as in the following illustration of cross-examination:

Q. *Now, the decedents in this case were found inside a closed building, correct?*
A. *Yes.*
Q. *The screens were shut, the blinds were down, and the doors were all shut and locked, is that right?*
A. *Yes.*
Q. *And your assumption, if I understood you correctly, is that the flies—the ones that laid the eggs producing the maggots you used to estimate the PMI in this case—arrived at the decedents' within 1 hour after the decedents had died, correct?*
A. *Yes, that was my conclusion.*
Q. *Doctor, do you think the decedents in this case had any maggots on them before they were dead?*
A. *No, that's very unlikely.*
Q. *And the crime scene report reflects that the decedents were covered with maggots and flies when they were found, do you remember that?*
A. *Yes.*
Q. *So what we know for sure is that the decedents didn't have any maggots when they died, and they had plenty of maggots when they were found, right?*
A. *Yes.*
Q. *Now, Doctor, were you in the house when the flies arrived on the decedents?*
A. *Of course not.*
Q. *So you don't know exactly when the flies actually got there?*
A. *No.*
Q. *So your statement that the flies arrived within an hour is really a guess, then, isn't it?*
A. *Yes.*

When Were the Insect Specimens Actually Preserved?

The theory behind using temperature-dependent insect developmental times to calculate an entomological-based PMI is to "work back" from a known point in the insect's life history. That is, if the insects collected were mid-stage third instars, as evidenced by gross length, length-to-crop ratio, or other factors, it is possible, by knowing prevailing temperatures, to estimate how long it would have taken the species in question to grow from the stage (egg or larva) deposited to the stage collected. This minimum PMI, as mentioned previously, is the length of time that the decedent in question *had to be dead*. Infestation of living humans with necrophilous fly maggots (myiasis) is comparatively rare today. Such PMI estimates are routinely analyzed by the entomologist in accordance with time of day, because nocturnal oviposition by necrophilous flies seldom occurs.

The validity of these "had to be dead at least so long" estimates depends upon knowing, as medicocriminal entomologists are fond of saying, when the "clock was stopped" on the insects. That is, when were they actually killed and preserved so that their development terminated? Written records by crime scene investigators or medical examiners to the effect "maggot samples obtained and preserved in 80% alcohol at 0900 hours this date" leave little room for argument. Fluids as exotic as special entomological fixatives or as simple as embalming preservative have a similar effect. Perhaps the best maggot specimens are obtained by dropping them alive into boiling water, which causes them to extend full length and kills them, after which they are moved to preservative. Freezing the samples, although not recommended, will also serve to kill them quickly and preserve them. The common denominator documenting all these techniques is the written record.

Absent specimens killed and preserved at known times, the resulting entomological analysis can be fatally confused. In some cases, investigators retain insect evidence with no attempt to kill or preserve it. As an example, maggots collected into empty plastic 35 mm film containers can continue to develop for variable times afterward, even molting to the next stage. With this sort of evidence, the entomologist can only guess at the time that the insects developed to the stage identified, and this sort of speculation is vulnerable to attack. Similarly, insect evidence retained in paper bags with no preservative may be examined years later. Attempts at speculation that specimens were "crushed" and thus "preserved" at known times, such as when a decedent was moved from the crime scene to autopsy, represent no verifiable phenomena and are rightly attacked in court.

In addition, mistakes are frequently made by misinterpreting the time when the specimens were actually preserved. Bodies may be found but not autopsied until a day or so later. It is an obvious error to calculate the portion of the postmortem interval based on insect activity by using an insect specimen collected and preserved at autopsy, but employing the date the body was found as the starting point for analysis.

How Were the Collections Made? Were These the Oldest Specimens?

From the foregoing, it should be clear that reliable estimation of the minimum PMI from insect evidence is contingent upon several assumptions. First, it must be known when the specimens were preserved, so that one knows when to start to "work backwards" in time. Second, there must be concordance between the temperatures used and those prevailing at the scene. Finally, the "minimum" PMI represents the time it would have taken the *oldest specimens on the body* to get to the stage identified. Of course, if a body has 4-day-old insects on it, it is possible for that same body to have younger, say 3-day-old, insects on it also. Because necrophilous flies ovi- or larviposit over a period of time, an accurate estimation of minimum PMI rests upon analysis of the oldest insects associated with the corpse. All else being equal, "oldest" in the context of necrophilous flies generally means "largest." Thus, the procedure at crime scenes is to collect the largest specimens available. Because this typically is done by personnel other than the medicocriminal entomologist, the latter must usually depend on the skill of the collector as one fundamental assumption in the analysis. Shortcomings may come out during cross-examination.

Q. *So, Doctor, you identified the insects in this case as third-instar* Phormia regina, *is that correct?*
A. *Yes.*
Q. *And you used the temperatures available to you to calculate that these maggots were 4 days old, is that right?*
A. *Yes, 4 days.*
Q. *And did you base your estimate of how long the decedent had been dead on this 4-day interval?*
A. *Yes.*
Q. *Now, Doctor, if the maggots you examined were not the oldest available on the body, then your estimate would be incorrect, would it not?*
A. *Yes, it would tend to underestimate the PMI.*
Q. *And you don't know for sure that the specimens you examined were in fact the oldest available.*
A. *They were the oldest of the ones I looked at.*
Q. *My point is, there might have been older ones on the body that were not available to you, isn't that possible?*
A. *Well, I'm sure that the investigator who collected them followed proper procedure.*
Q. *But you didn't collect the specimens yourself did you, Doctor?*
A. *No.*
Q. *In fact, you never actually examined the decedent in this case at all, did you?*
A. *No.*

Another shortcoming in basing analyses on evidence collected by someone else is the appearance of unfamiliarity with the crime scene in question. If possible, it is always best to visit the scene during the acquisition of evidence, and if this is impossible, to view the scene personally before rendering a final opinion. As a last resort, photographs of the scene may be examined, and this has become more convenient with the advent of CD-ROM computer discs containing many image files. A forced admission of having not visited the crime scene can be especially damaging if the opposing expert has done so.

Q. *Now, Doctor, have you had a chance to examine the basement where the decedent was found and where Dr. X made the extensive insect collections?*
A. *I've seen some photographs.*
Q. *Have you been to the house where all this took place?*
A. *Not personally, no.*

Q. *So you haven't seen with your own eyes any of the things we're talking about here today, have you?*

A. *No.*

Serving as Expert Witness

The medicocriminal entomologist must be prepared to function in the adversarial legal system if his or her analyses are to affect the outcome of litigation. The initial contact may stem from being part of a crime scene investigation team, where the entomologist physically collects insect evidence, documents and analyzes it, or delivers it to someone else for analysis. In other cases, the entomologist may be called to the necropsy and be involved with collection and preservation of specimens. Most frequently, however, the entomologist is contacted at some point during the investigation when it has become apparent that insect evidence is important. In times past, and fortunately becoming less frequent today, the entomologist might not be contacted until insect evidence surfaced during a new trial, often years later.

The rubric of "expert witness" attaches when the attorneys handling either side of a case make an oral or written formal agreement to retain an individual in that capacity. A lawyer would say that the legal theory here "sounds in contract law" because of the actual or implied contractual relationship. This may be either as a testifying or nontestifying expert, the distinction being whether the identity of the expert need be disclosed to the opposing side. Whereas experts anticipated to testify in court must be disclosed, occasionally an expert will be retained in a nontestifying status as a second-opinion backup or to deny his or her services to the opposing side. In any event, the attorneys involved should make the status clear. During initial dialog, the actual work to be performed should be discussed and the rate of compensation, payment schedules, and billing particulars set (Cantor 1997). Expert witnesses can expect reasonable compensation for their services, but exorbitant fees affect credibility. The amount of fees, expenses, and other compensation is fair game during cross-examination, and it is important for the expert to keep in mind that such fees represent compensation for his or her expertise and expenses—not for the testimony itself. It is good practice for the expert to keep stringent records of time spent on a case, usually to a fraction of an hour. "Book" billings should always be avoided, such as "I always charge $1,000 for an opinion in a murder case," or "I always bill for 3 hours for an insect identification, no matter how long it takes me." Unless the retaining side represents a governmental or other entity likely good for any amount due, it is prudent for an expert to ask for advance payment sufficient to cover initial expenses. Further, if travel is anticipated, agreement about reimbursable expenses should be reached early in the relationship. As trial nears, the time value of money often makes it expedient to request, for example, that airline tickets be forwarded directly to the expert, or that hotel bookings be billed directly to the retaining party, rather than to depend on reimbursement at some time in the future.

It is unethical for expert witnesses to enter into fee agreements contingent upon the outcome of the case or for a percentage of any settlement. Therefore, an arrangement where "we'll pay your fee, but only if we win" is forbidden as unconscionably biased, as is an arrangement where the retaining side says, "We'll pay you 20% of whatever damages we recover." Such arrangements put lawyers in jeopardy of disciplinary action. No ethical attorney will suggest such a fee arrangement, and no expert witness should accept one.

From a purely practical standpoint, such agreements are considered void as against public policy; thus, they are unenforceable and an aggrieved expert will have no legal remedy.

Experts Are Not Advocates

Perhaps no other line becomes blurred so easily as that between detached, scientific expert and biased advocate for one side or the other. The function of the expert is to assist the trier of fact in understanding the circumstances of the case, and this cannot be done except in a neutral and detached capacity. This is one rationale for the court-appointed expert, but in U.S. courts litigants have the right to retain experts of their own choosing. The public perception of "hired gun" expert witnesses who will say anything if the price is right has some basis in reality, although by far the majority of scientific experts do their best to present unbiased reports and testimony.

When contacted initially by law enforcement authorities, the prosecuting attorney, or defense counsel, the responsibility of the medicocriminal entomology expert is to gather available information, perform an entomological analysis, and apply the analysis to the facts of the case as they are known—"take it or leave it." The same can be said for civil matters. Whether or not the expert's analysis supports the plaintiff's or defense's position should be immaterial in the analysis itself. Whether it is scientifically valid is everything. Most cases are settled before trial, and the support of experts, or the lack of such support, is often critical in decisions to settle. When performing this initial analysis, it is probably best for the expert to know only the minimum necessary about unrelated facts of the case, so that potential sources of bias can be avoided. At this point, if the entomological analysis runs counter to the case being made by counsel, the expert can be compensated for time spent with no further involvement anticipated. Alternatively, the expert may be put "on hold" as a nontestifying witness. In either event, the information generated by the expert to that point might be protected under various confidentiality doctrines. Therefore, the expert should not discuss the case with outside parties until it has been resolved. In particular, experts should not discuss a case in the presence of an opposing expert except during formal proceedings with counsel present.

The danger lurking in the background here is twofold: one is that the expert witness will tend to identify excessively with the particulars of the case as they favor his side, and the other is that prospective remuneration will be enough to sway objectivity. Probably analogous to the psychological identification of prisoners with their captors, it is not unusual to see scientific experts adopt biases as they increasingly identify with "their side" in a case. Highly insidious, such bias is unethical and prejudicial to fair resolution of controversies. As an example, an entomologist routinely contacted by law enforcement agencies may begin to adopt the "us or them" attitude frequently a product of street survival. "Help us put this dirtbag behind bars, where we know he belongs" may be enough to induce an impressionable expert to stretch his analysis to fit a theory that will do just that. Similarly, it is a rare defense team unable to make the compassionate statement, "There's just no way that this guy could have done that crime. We're sure of that—all we have to do is prove it." Under this type of pressure, skewed analyses may result and thus negate the theory behind admissibility of entomological, and indeed all expert, testimony: that it will lead to discovery of the truth.

Under the ethical codes of virtually every scientific society, including the American Academy of Forensic Sciences and the Board Certified Entomologist category of the

Entomological Society of America, slanting or skewing analyses to favor one side or the other, without a valid scientific reason, is actionable. On the other hand, the expert will be expected to educate those responsible for retaining him or her in regard to weak points in the opposition's case. As will be amplified later, this requires walking a fine line between advocacy and education. If the expert remembers that his or her job is to perform an unbiased analysis and apply it to the facts of the case, and that it is the job of counsel to present that application in the light most favorable to one side or the other, things generally will go smoothly.

Occasionally, an expert witness may be retained in the capacity of *consultant*. In this instance, there should be no expectation of actual testimony. The job of the consultant is to assist counsel in putting together the best possible case. An expert functioning as a consultant may properly be regarded as an advocate, and bias in such cases is not problematical because the expert will not testify or give sworn statements in the proceedings.

The Entomology Expert and Formal Legal Proceedings

As the entomologist deploys his scientific expertise toward the resolution of a litigated controversy, there are three principal ways in which such expert opinion can be documented. These include the filing of an *affidavit*, the giving of a *deposition*, and *courtroom testimony*. The common denominator is that all constitute sworn statements. In addition to the expectation of intellectual honesty (which of course pervades day-to-day activities in science and scholarship), penalties for perjury attach.

Affidavit

An affidavit is a written, voluntary declaration of fact or opinion made before one authorized to administer an oath. Most entomology experts become associated with affidavits as they are used in support of pretrial motions. Typically, discussion between the expert and attorneys results in some consensus regarding the "fit" between the expert's opinion and the fact pattern of the case. When the expert's opinion can be used to support any of the many pretrial motions possible in legal proceedings, he or she may be asked to sign a notarized affidavit, which then accompanies the motion. In essence, the expert's opinion is used as testimony in support of the motion. Perhaps the example most common in medicocriminal entomology is the pretrial motion to exclude expert testimony. The sequence of events usually takes the following pattern. If one side or the other proposes to use entomology evidence in support of their case, that intent plus evidence accumulated and analyzed to that point must be disclosed. This disclosure may involve the name of the entomologist, any reports he or she has filed, and an accounting of the entomological evidence, such as specimens and weather data.

Upon such disclosure, the opposing side has the opportunity to respond to such evidence, and often does so by retaining a separate expert to review it. If the latter expert's analysis points out flaws or deficiencies in the entomological analysis, then a motion to exclude can be made under the legal standard appropriate in that jurisdiction. The attorneys making the motion to exclude will draft it. What often comes as a surprise to the entomologist is that they will also draft an affidavit (a voluntary, written statement of facts made under oath) for his or her signature. This procedure is efficient in several ways, because the affidavit will reflect the entomological results and conclusions as they have been discussed between the expert and the attorneys and will be in the proper format. Most entomologists have no background in preparation of affidavits.

However, entomologists should be keenly aware that the affidavit they are asked to sign becomes their own statement. It will be drafted by the attorneys to give the best support to the motion, and will typically contain legal terms of art. As mentioned previously, such terms are included solely to achieve a desired legal result. The entomologist should read the affidavit carefully before signing it, to ensure that it presents a scientific analysis he or she can support in good conscience. If there are misstatements or other problems, the affidavit should be edited and revised so that the entomologist—the one "making" the statement—is completely in agreement with it. Once notarized and filed, affidavits become part of the official record and may be used to impeach the affiant in the present or future proceedings.

Q. *Doctor, you have testified just now on direct examination that in your opinion it is impossible to perform an entomological analysis solely from photographic evidence, is that correct?*

A. *Yes.*

Q. *And that is your expert opinion in this case?*

A. *Yes, it is.*

Q. *I have here your affidavit filed in the case of State v. Smith. Do you remember that case, Doctor?*

A. *Well, I think that's been several years ago.*

Q. *Paragraph 4 of this affidavit, which you signed under oath, Doctor, reads as follows, "My conclusion from examination of the crime scene photographs, which were all the entomology evidence available, is that the insects present were probably migrating third-instar black blow flies." Do you remember that assertion?*

A. *Yes.*

Q. *So your conclusion in State v. Smith was in fact based solely on photographic evidence, wasn't it?*

A. *Yes.*

Q. *And that's directly counter to what you've just said in this courtroom, isn't it?*

A. *Yes.*

Depositions

There is considerable misunderstanding surrounding depositions and the entire associated process by those outside the legal system. Current federal rules actually limit the number of depositions in civil cases. A deposition is testimony—out of court, to be sure, but under oath—in response to questions posed, most often, by attorneys for the opposing side. It is part of the carefully regulated exchange of information called discovery. A deposition may be taken in the attorney's office, but more often is conducted in the expert's office or conference room. An authorized court reporter will administer the oath and will record the deposition word for word for transcription. The expert, as deponent, has the right to read and correct the transcription before affixing his notarized signature. Occasionally, a deposition is so straightforward that review is waived. Videotaping of depositions is becoming increasingly popular and places an additional burden on the deponent—that of a visual, in addition to a written, record.

A deposition by a medicocriminal entomologist typically focuses on the entomology report and conclusions. At the beginning, prefatory remarks usually include a statement and spelling of the expert's name, introduction of the attorney who will be asking the questions, and a reminder that clarification of questions may be sought and that responses must be verbal (court reporters cannot record grunts, nods, or shakes of the head).

Because the questioning attorney will generally have studied the entomology report and the expert's curriculum vitae, the trend of medicocriminal entomology depositions is fairly predictable. They usually begin with background questions in regard to the expert's academic qualifications and performance history. This may be glossed over, or may become the subject of in-depth questioning. In some cases, questions may be asked about the particulars of every scientific article the expert has published. It is thus important to be well prepared.

Sooner or later, though, questioning will turn to the entomology report and the application of entomological science to the facts of the case. Because responses on deposition can be used to impeach subsequent in-court testimony, the deponent must take care to provide accurate and supportable responses to questions asked. Although the expert will often be allowed to provide a discourse (narrative) on a particular topic, it is good advice to limit one's responses to the best and most straightforward answer to the question asked. The more one rambles on, the greater chance that such information will be inconsistent with later statements. Although minor inconsistencies and fine points may be clearly understood as such by fellow entomologists, they may be perceived by the jury as undermining the expert's credibility.

Most depositions are straightforward, although they may be lengthy. If a break is needed, simply make a request. One major difference between depositions and courtroom testimony is that no judge is present. Therefore, objections cannot be ruled upon at the moment. Still, if the questioning attorney asks an objectionable question, the attorney representing the side retaining the expert (a critical point is that he or she does not represent the expert) may register an objection, saying "subject to that, you can answer." The effect of this give-and-take is that much information usually comes out during a deposition, and each bit can be fairly brought up during future testimony, where objections may have to be argued, or where it may be inconsistent with responses offered at that time.

Depositions may or may not be done under *subpoena*. A subpoena, especially a *subpoena duces tecum*, commands a party to appear in court or for deposition at a certain date and time, and to bring relevant documents. This court order must be obeyed, or contempt sanctions may result. Most expert witnesses are willing to have depositions scheduled without compulsion, and the subpoena is often dispensed with. If this is the case, it is wise to ensure that payment for time and expenses will be forthcoming despite the lack of a subpoena. In some cases, an expert may be compelled to testify under subpoena with only the statutory compensation for travel expenses and appearance in court. This situation is far from ideal and is comparatively rare, but it underscores the wisdom of arriving at written expert fee and compensation agreements early in the relationship. The party requesting the deposition, either the prosecution or defense in criminal matters, is responsible for paying for the deponent's time and travel expenses, if any. Needless to say, they are also responsible for the transcription fee, which is often greater than the expert's charges.

In addition, the attorney with whom the expert is associated should examine all documents the expert expects to bring to the deposition. Often, this is not done and can constitute a major mistake. Such documents are discoverable, and stray notes, memoranda, and correspondence may contain statements or information adverse to one's party. Inadvertently bringing harmful documents may constitute negligence. The best procedure is to avoid creating these in the first place; thus, the expert should give much thought before making any notes, memoranda, letters, reports, or other written materials.

Courtroom Testimony

When entomological evidence is going to be argued in court, whether at trial or in relation to pretrial motions, the expert witness will generally be needed on the stand. As with affidavits and depositions, courtroom testimony is under oath. Also similar to other statements under oath, trial transcripts are historical documents available to future litigants. What this means in practice is that testimony from one trial can be used to impeach a witness in a second proceeding. Therefore, the expert should be aware of and avoid inconsistent statements, and if these become necessary, for instance, because of new scientific knowledge, the expert should be prepared to explain inconsistencies.

Good preparation is fundamental to success in litigation, and expert testimony is no exception. Prior to trial, the expert should review all reports, notes, and associated documents relevant to analysis of the case. Further, there should be a pretestimony meeting between the expert and the attorney who will be representing the side for which the expert is testifying. A trial lawyer's maxim is "never ask a question when you don't know what the answer will be." This applies to expert testimony and represents good preparation and rehearsal. It does not constitute advocacy for the expert to review the salient scientific aspects of his or her analysis and alert the attorney as to which points need to be made, and to go through a question-and-answer session to ensure that both are in accord. Similar to depositions, all documents that the expert expects to bring into court should be screened beforehand. Like the situation with depositions, a subpoena may or may not be issued by the court; if one is issued, absent an agreement to the contrary with appropriate counsel, the expert is entitled only to the statutory compensation for his or her appearance.

On the day of courtroom testimony, the expert should arrive with sufficient lead time for a final session with counsel, if required. Sometimes, an expert can request to be put "on call" so that it is not necessary to wait at the courthouse. If an exclusionary rule is in effect, it will be necessary to wait outside the courtroom (in a hallway, or in a witness room) until called. This time can be used for final review of relevant documents. If the exclusionary rule has been waived, the expert may and should listen to the testimony of opposing experts. If in doubt, consult with the appropriate attorney.

While an expert witness need not adopt the formal attire of trial lawyers, he or she should present a professional appearance. This means, at a minimum, coat and tie for men and its equivalent for women. Remember that expert testimony is valueless unless it affects the decision-making process of the trier of fact. The goal is for the jury to believe the scientific opinion of the expert. Therefore, inappropriate attire or mannerisms can have an adverse effect by causing the jury to reject the expert and his or her theories. In extreme circumstances, attorneys may request a recess and have a clerk or paralegal purchase a change of clothing for an expert—to be paid for from the fee owed.

While on the witness stand, the expert should strive to maintain composure even under pressure. Anger, argumentative responses, and annoying mannerisms are usually counterproductive. The best results are obtained by assuming a relaxed but alert attitude and making eye contact with the questioning attorney, judge, or jury as appropriate. It is especially important that the expert not "look to" the attorney with whom he or she has been working for support when difficult questions are asked on cross-examination.

The initial questioning of the expert in court is called *direct examination* and is done by the attorney representing the party calling the expert. This critical period is relatively friendly, because it consists of the attorney and expert who recently rehearsed precisely for this occasion. The initial portion of the question-and-answer period will be devoted to the expert's qualifications, in order to convince the trial judge to admit the testimony under the appropriate rules of evidence pertaining to such expert witnesses. In the case of medicocriminal entomologists, the focus will be on academic preparation and degrees, academic appointments and other professional positions, and research contributions, including papers published, students advised, and grants awarded. It is important to document qualifications in the field of forensic entomology—the fact that one is an expert in another entomological field, control of insect pests on corn, for example, is irrelevant when seeking qualification as a medicocriminal entomologist. A medicocriminal entomologist with experience testifying as an expert witness can effectively point out his or her academic and professional background in narrative form. If one is not so experienced, it is best to allow the attorney to take the lead by asking pertinent questions. In either event, it is important to ensure that the judge and jury understand the full impact of the expert's credentials. Whereas misrepresentation of one's credentials, by affirmative misstatement or by omission, is an ethical violation and not tolerated, this is no time to be modest about one's honestly earned background. The issue here is believability, and the expert the jury considers best qualified is often the one believed. Thus, one's professional title should be stated and all academic degrees, along with the institution where each was earned. After that, a coherent presentation covering professional stature, number and type of publications, major grants or endowments, membership in professional and scientific societies, and significant honors should be provided by narrative or questioning. In particular, the manner in which the expert's background makes him or her uniquely qualified to enlighten the jury should be emphasized. Frequently, opposing counsel will attempt the old ploy of stipulating to the expert's credentials. The purpose of this tactic is to cut short the litany so the jury will not hear it. Inevitably, the attorney seeking to qualify the expert will request permission to proceed—to preserve the matter on the record—and this is typically granted.

After the expert has been qualified by the court, the next function of the direct examination is to present the expert's theory of the case to the jury. This will usually start with the expert's written report, which has earlier been disclosed. As in the qualification phase, the expert may proceed by testifying in narrative form, or in direct response to questions from counsel. Typically, direct examination consists of responses elicited by nonleading direct questioning, and therefore, in some instances, opposing counsel may object to narrative testimony. If this objection is sustained, it constitutes a major reason for adequate preparation between the expert and associated counsel. The goal is for the expert to be able to teach the jury and instruct them why his or her theory of the case is correct and should be believed. Two points become important here. One is that the expert may refresh his or her memory by referring to a wide range of materials. Because accuracy is critical to expert testimony, it is not a sign of weakness to ask permission to refer to notes, documents, or other written sources to ensure that testimony is factually correct. The second point is that responses during direct examination impact the scope of cross-examination, in that the latter phase is a derivative of those questions answered during direct examination.

As the expert testifies in narrative form or in response to questions, it is important to include the jury in the discussion by eye contact. It is not necessary to look at the jury to the exclusion of everyone else in the courtroom—indeed, this would appear awkward—but a relaxed demeanor in which the expert looks at counsel when questions are asked and at the jury as they are answered is often effective. Within the strictures of good science and ethical limits, the expert should appear positive and forthright, and able to explain the biological variability limiting the precision of his or her answer.

Another often fatal error of scientific experts is to infuse their responses with excessive technical jargon. It is a mistake to assume that reliance on mystical-sounding terminology will be taken by the jury to represent education or wisdom. In fact, often the opposite occurs. If an expert confuses responses with arcane jargon, thinking that it sounds technical, and that the jury will believe that someone who knows so many technical terms must also know the correct analysis of the case, a fundamental mistake has been committed. While the use of some jargon or technical language may be unavoidable, it is best to couch answers in terminology that anyone can understand. Remember that the job of the expert is to educate the trier of fact, and it is impossible to provide doctoral-level education in an hour or even many hours on the stand. Usually, exactly the opposite results: the jury becomes bored with the testimony and simply ignores it. Therefore, the key is to use clear language while avoiding "talking down" to the jury. This sort of presentation involves craftsmanship and can be learned with practice.

Q. *Now, Doctor, can you tell us what your entomological analysis of this case was?*
A. *Yes, the climatological data were applied retrospectively to the thermal developmental profile for putatively late but premigratory third-instar Phormia regina collected in this instance by the medical examiner at necropsy. At least 3,472 accumulated degreehours are required for this species to enter the prepupa; thus, I calculated that.*

Whereas this response would be intelligible to another medicocriminal entomologist, it is one only an entomology graduate student could love. Unless much time is taken to define each term and make it understandable, the jury will fail to learn much, if anything, from it. Without talking down to the jury, a better response might be as follows:

A. *The insects tell us a good bit about how long Mr. Smith had been dead. I identified the flies found on his body and their stage of development. I also checked the temperatures for that time from Central City Airport and performed a short test that showed it was valid to use them. Knowing that flies grow up at different rates, depending on how warm it is, the insect evidence here tells me that Mr. Smith had to have been dead for at least 4 days when he was found.*

If use of scientific terminology is unavoidable, such as when discussing the various species of flies, it may be useful to prepare a list of arcane terms to hand to the court reporter before testifying. This will at least ensure that the terms are spelled correctly in the trial transcript. Although the fine points of the analysis, such as latinized names of species, use of thermal data like ADH, and so forth, will surely be argued on cross-examination, the effect of simple initial responses is for the entomologist to transmit their result to the jury in a fashion they will understand and remember.

It has been well documented that retention and learning improve with visual, in addition to auditory, input. Therefore, many experts enjoy success with well-designed courtroom presentations, which fall under the category of *demonstrative*, rather than testimonial, evidence. These may be as simple as chalkboards or flip charts, as straightforward as slide presentations from a projector or television, or as sophisticated as preserved or living exhibits, computer imagery, or videotapes. Although possibly smacking a bit of theater, there is no question that thoughtful visual aids are very effective in getting the expert's point across. Often, they can be left in place after the expert testifies, and thus serve to remind and reeducate the jury as the proceedings continue. Because these sorts of aids or exhibits must be disclosed before they will be permitted in the court, be sure to advise counsel of what you intend to present. Do not wait until trial, or opposing counsel will likely object to the surprise and will probably be successful in keeping such demonstrative evidence out of court. This can be especially damaging if the expert is building a critical presentation around the visuals.

At the conclusion of the direct examination, opposing counsel is given the opportunity to *cross-examine* the expert witness. A principal difference between the form of questions on direct versus cross-examination is that the latter may be *leading*; that is, the question itself may suggest the answer. The expert must be especially alert during cross-examination, because an experienced opposing attorney will have identified all possible inconsistencies arising as a result of the direct examination, affidavits, depositions, or former testimony. These inconsistencies may then be pointed out to the jury and serve to impeach (to make less credible) the expert's opinion. Experts should pay particular attention to the following points because cross-examination is such a critical phase of testimony.

Listen carefully to the question asked. If the question is not clear, ask for clarification. It is good practice to develop a habit of pausing deliberately before responding to any question on cross-examination, in order to give counsel on your side time to recognize and register an objection. Generally, if the question is unclear, poorly phrased, argumentative, or exhibits similar defects, counsel will register a timely objection. If the question is not legally objectionable but is unclear scientifically, it is appropriate to request restatement or clarification.

Q. *So, as I understand it, it is your opinion that the insect larvae in this case support a PMI estimate of four days, is that right?*
A. *I'm sorry, there were three insect species involved: two species on one decedent and one species on the other. I am not sure which species you mean.*
Q. *I'm talking about the species that was found on Mr. Jones. I think that decedent had only one kind of fly maggot on him.*

On cross-examination, an experienced expert will respond as truthfully and briefly as possible to the questions asked. While it may be tempting to offer additional explanation, it is best to refrain from doing so. The opportunity may seem particularly tempting when a well-phrased question has apparently exposed some weakness in the expert's response. The best way to handle this is to permit the attorney representing the side retaining the expert to *rehabilitate* the expert on *re-direct examination*. Of course, a re-direct may be followed by *re-cross-examination*, but the number of iterations of these decreases rapidly as the scope of possible questions narrows.

It is imperative that the expert retains his or her composure during cross-examination. This may be difficult to do when the questions posed constitute a direct attack on the expert's credentials, scope of practice, and professional competence. Answers given in anger are often regretted. Resist the temptation to "match wits" with the attorney asking the questions, and remember that the courtroom is his or her professional habitat. Making a witness angry is only one of many strategies employed by trial lawyers. An experienced attorney will never try to match an expert one on one when arguing the fine points of entomological science. Similarly, an expert who tries to "outsmart" the attorney in the courtroom is generally doomed to failure.

Q. *Doctor, the data you used to analyze this case were generated at the "Body Farm" in Tennessee, weren't they?*
A. *I see what you're getting at. There's no proof that data from the Body Farm reflect faunal enrichment or are otherwise unreliable.*
Q. *Doctor, did I ask you about faunal enrichment?*
A. *No.*
Q. *Did I ask whether or not such data were unreliable?*
A. *No.*
Q. *Well, would you please simply respond to the questions that I ask?*

In this manner, the attorney has made it clear to the jury who is in charge during the cross-examination. The expert in this example has not educated the jury; worse, his or her esteem has been lessened because the jury will perceive that he or she "lost" this confrontation.

Another way in which expert witnesses get into difficulty is by venturing outside their discipline. What this means is that experts must be highly cognizant of the boundaries demarcating their scientific specialty. The medicocriminal entomologist is an arthropod expert qualified to render an opinion about the identity of insects and related species, their biology, including reproductive behavior, successional occurrence in relation to geography, season and time of day, rate of development, and so forth. The entomologist may be extremely familiar with closely allied fields, such as forensic pathology, but must be sensitive to questions that call for an opinion outside the area of qualification.

Q. *Doctor, may I refer you to the photograph marked "State's Exhibit 43," which you have previously testified depicts third-instar larvae of the black blow fly?*
A. *Yes, I have that photograph.*
Q. *If you will examine the decedent's forehead in the photo. Can you see the forehead, Doctor?*
A. *Yes, I can see it.*
Q. *What appear to me to be fly maggots are depicted crawling around a hole in the forehead, are they not?*
A. *I see them around a hole, yes.*
Q. *Doctor, does that hole appear to have been made by a 9 mm bullet?*
A. *I don't know. I'm not an expert in regard to bullet holes.*

Venturing outside one's area of expertise can be tempting, especially during heady moments on the witness stand. Be assured that experienced trial attorneys will know this and may lead into it simply to "dull" the expert's luster and erode his or her credibility with the jury.

An important role of the expert witness, as emphasized previously, is to educate the attorneys involved so that they can elicit the truth during the direct and cross-examinations for which they are responsible. This role can become especially interesting if the exclusionary rule is waived. Then, the opposing experts are present in the courtroom during testimony and are expected to provide expert insight into the responses provided. In some cases, the expert will sit at the appropriate counsel's table and take notes as his or her counterexpert undergoes direct or cross-examination. While this appears close to advocacy (it certainly has all the visual trappings of it), it is important to remember that the expert's role continues to be one of education. That is, the expert is used to alert the appropriate counsel to inaccurate statements of fact, misrepresentations, subtly artful responses, and so forth. This allows clarification on re-direct or re-cross examinations. As might be expected, the plainly adversarial nature of such participation can elicit hard feelings and misunderstandings between experts. If this happens, it is a good idea to deal with it immediately so that interpersonal bad feelings do not become a major issue. In fields such as medicocriminal entomology, where there are relatively few experts, long-term associations with colleagues work best on a positive, rather than negative, note.

Malpractice

An area seldom considered by expert witnesses is malpractice liability. Whereas the attorneys involved in a case are invariably well insured, most experts are not. While it is true that certain immunities attach to testimony under oath, such as immunity from slander and similar charges, many vulnerable areas remain. The theory most commonly applied to malpractice of expert witnesses is common law *negligence*, which is a tort. In order to establish a negligence cause of action, the plaintiff must prove the existence of a duty, breach of such duty, cause-in-fact and proximate cause, and damages. Often, these elements are not difficult to establish in malpractice cases.

The agreement in contract between the expert and the side employing him or her establishes duties owed, and the expert's appearance in court is evidence of awareness. The issue of causation is also straightforward: whether or not the expert's actions caused the loss of a criminal case or civil lawsuit, for instance, and whether or not it was reasonably foreseeable that such a result would occur. The issue of damages is often easily determined, in that criminal penalties or civil monetary awards are clear. Most argument centers on whether or not the expert in fact breached his or her duty to the side retaining him.

Common mistakes by expert witnesses that can be considered breach of duty and thus incur malpractice liability include factual misstatements or errors, breaches of confidentiality, and inadvertent disclosure of documents. The expert should thus take care to ensure that all testimony is factually accurate. The truth is a powerful defense against malpractice. Further, the expert should not discuss an ongoing case outside the courtroom with anyone other than counsel for the side retaining him. Narrow exceptions exist for personnel coming under the umbrella of the confidentiality doctrine, such as technicians and other employees of the expert. Expectations of confidentiality extend to these personnel, and any breaches may be actionable against them or against the expert himself or herself under the theory of *respondeat superior*. As stated earlier, negligent disclosure of damaging documents may be considered malpractice. In addition to following good practice, the prudent medicocriminal entomologist will carry adequate malpractice insurance and may seek legal advice regarding methods to limit potential exposure by skillful use of certain business organizations.

References

Cantor, B. J. 1994. *Reference manual on scientific evidence*. Federal Judicial Center. New York: Matthew Bender.

Cantor, B. J. 1997. The role of the expert witness in a court trial. Civil evidence, photo. Seminars, Belmont, MA.

Giannelli, P. C., and E. J. Imwinkelried. 1993. *Scientific evidence*. 2nd ed. Vols. 1 and 2. Charlottesville, VA: Michie Co.

Goff, M. L. 1992. Problems in estimation of postmortem interval resulting from wrapping of the corpse: A case study from Hawaii. *Journal of Agricultural Entomology* 9:237–43.

Greenberg, B. 1991. Flies as forensic indicators. *Journal of Medical Entomology* 28:565–77.

Imwinkelried, E. J. 1992. *The methods of attacking scientific evidence*. Charlottesville, VA: Michie Co.

Livestock Entomology

15

JEFFERY K. TOMBERLIN
JUSTIN TALLEY

Contents

Introduction

Many arthropods in livestock can negatively impact both the people and property they encounter. Arthropod encounters with people can range from nonlethal bites and stings to life-threatening events of severe reactions. When these encounters occur with personal property, interactions can include anything from infestation to complete destruction.

Livestock and poultry are essential parts of our society. These animals provide both consumable food items and nonconsumable commodities. Most livestock and poultry are produced in confined animal facilities (CAFOs). These systems have large numbers of animals restricted to one location and are efficient at producing the raw products necessary to meet the demands of society. This concentration of animals results in the production of large quantities of wastes in these facilities. Regulations, such as monitoring the exportation of wastes from CAFOs by the Environmental Protection Agency (EPA), are in place to protect environmental integrity and prevent potential pollution issues from occurring.

477

Fifty-six percent of the world's population and 80% of the resident populations in Germany, France, and the United States are expected to be located in urban environments by 2010 (Rust 1999). Because of urbanization and globalization, many arthropod pests have been exported outside of their native range and occur in areas without natural measures to keep their populations under control. Such situations allow for populations of these pests to grow unchecked. Furthermore, contact between urban and agricultural regions has resulted in the increased movement of arthropods from one environment to the other.

Urban encroachment into areas occupied by CAFOs in some instances has resulted in civil litigation against these operations. These lawsuits are commonly due to complaints about dust, odors, and arthropods in some cases. There are instances where these lawsuits are warranted; however, some of these cases are due to individuals relocating from urban areas into an agricultural setting, and consequently having more contact with "nature." Because of the increase in the number of litigation issues against CAFOs, states have adopted statues known as right-to-farm laws (Horne 2000). These vary from state to state but generally focus on protecting the agricultural practice.

Impacts of Arthropods on People

Many arthropods are naturally attracted to homes and businesses, while others are purely incidental. Regardless of how an arthropod invades a human habitation, an individual's tolerance level typically varies depending on past experiences. In most cases, any arthropod in a home or business is viewed negatively. The types of arthropods that are often encountered range from easily noticed species, such as mosquitoes, flies, and moths, to cryptic species, such as bed bugs and mites (Figure 15.1). Additionally, many interactions between arthropods and people vary from simple annoyance to pain via a bite or sting. The question regarding the origin of these arthropods often arises when an encounter occurs. In most cases, these arthropods are commonly associated with people, while in other cases they originate from resources surrounding an urban area. For example, aquatic habitats,

Figure 15.1 Biting arthropods such as the cryptic bed bug *Cimex lectularius*, once thought to be a pest of the past, are making a dramatic resurgence in the United States. Undoubtedly, the future of urban entomology will include cases of bed bug infestations in homes and the hospitality industry. (Photo courtesy of Bart Drees, Texas Agrilife Extension, Texas A&M University.)

such as streams, ponds, and lakes, support populations of certain mosquito species and other biting arthropods not normally encountered.

Invasion of human habitations can also include arthropods that colonize structures. Termites, beetles, and many other arthropods arrive, colonize, and feed on wood structures. Potter (2004) concluded that no other structural pest has received more attention than termites. This attention is primarily due to the level of damage that can occur due to their presence as well as the lack of satisfaction with current control measures (Potter 2004). Other insects, such as beetles, will attack various wood products, including living, dead, and manufactured goods (Gulmahamad 2004).

Although most food naturally contains insect parts, many food items, such as dried goods, will be colonized after processing. Arthropods often shed cuticles and scales during their life, and some of these items will find their way into food. In other instances, these arthropods will pick up various microscopic organisms on their tarsi and legs that can be transferred from one environment to another; however, the question that typically arises is: Was the item colonized prior to or after being purchased? Such contaminations are often benign; however, cases do occur where contamination levels result in allergic reactions and possibly the transmission of pathogens. Gorham (1975) provides an excellent review of arthropods and food contamination.

Individuals claiming to have encountered an arthropod in their home are not uncommon; however, cases do occur where individuals suspect an encounter when one did not occur. Delusory parasitosis, or Ekborn's syndrome (Altschuler et al. 2004), is a psychiatric condition where individuals are convinced without any supporting empirical evidence that they are infested with arthropods or parasites (Poorbaugh 1993). Individuals with delusory parasitosis can experience itching, redness, crawling sensation, and other skin-related conditions (Altschuler et al. 2004). Additionally, physical evidence recovered from target sites includes white flaky materials, fibers, and colored fragmented materials (Altschuler et al. 2004). These individuals will often consult a medical doctor or specialist, such as a dermatologist. These consultations often yield no evidence of arthropod colonization of the patient; thus, the afflicted person contacts an entomologist. Although most cases for the entomologist result in findings similar to those recorded by the physician, all cases should be taken seriously and professionally. Individuals suspecting infestations of their body will often resort to home treatment with over-the-counter products or even more extreme treatments, such as pesticides and other compounds that can be harmful to the individual (Altschuler et al. 2004).

Allergic reactions to insects and arthropods are common. Most reactions are limited to a mild irritation, redness, or swelling, but some can be more severe. Respiratory reactions due to the inhalation and contact with airborne arthropod materials such as feces can occur. Dust mites (Pyroglyphidae: *Dermatophagoides* sp.) (Shin et al. 2005) and house flies, *Musca domestica* L. (Diptera: Muscidae) (Focke et al. 2003), are two examples of arthropods that have been linked to asthma and other respiratory illnesses (Figure 15.2). Inhalation of materials related to many other arthropods, including roaches (Blattodea), caterpillars (Lepidoptera), and fleas (Siphonaptera), can also result in asthmatic reaction (Heyworth 1999) (Figure 15.3).

Other reactions to arthropods are due to direct contact and envenomation. Presently, it is estimated that 5% of the U.S. population is allergic to hymenopteran stings (Klotz et al. 2005). Fifty percent of some populations living in areas with fire ants have been stung, and 17% are sensitized to their venom (Klotz et al. 2005). Certain species of blister beetles (Coleoptera:

Figure 15.2 Worldwide in distribution, the house fly has an extensive association with human activity and their dwellings. It is a notorious pest that has proved difficult to control despite extensive efforts directed toward its eradication. (Photo courtesy of Bradley Mullens, University of California at Riverside.)

Figure 15.3 Roaches are a familiar urban pest species to most individuals. In addition to being a revolting urban pest, exposure to roach populations may trigger respiratory distress, and in many cases of medicolegal entomology, they feed on the epidermis of living and deceased humans, producing scavenging artifacts that may mimic perimortem trauma. (Photo courtesy of Bart Drees, Texas Agrilife Extension, Texas A&M University.)

Meloidae) are capable of releasing compounds that cause a chemical burn. Accordingly, contact with caterpillars that have urticating hairs can also result in a skin reaction.

Mullen and Durden (2002) indicate people and animals can react in one of two ways to contact with envenomation or inhalation of arthropod artifacts. Individuals can either become desensitized to arthropod components or develop an allergic reaction to them. They also outline five steps that will occur when an individual is repeatedly exposed to a potential allergen: (1) no skin reaction but hypersensitivity occurs over time, (2) a delayed hypersensitivity occurs, (3) immediate sensitivity reaction and then a hypersensitivity reaction, (4) immediate reaction, and (5) no reaction due to desensitization. Individuals appearing to have a severe reaction to contact with an arthropod should be taken to a physician for immediate treatment.

Current Status of Livestock Production in the United States

Livestock production is an essential part of our society. Many products outside of food items are derived from these animals. As mentioned previously, most livestock and poultry are produced in CAFOs. Although these operations result in large numbers of animals restricted to one location, they are efficient at producing the raw products necessary to meet the demands of society. On the other hand, large quantities of wastes are also produced in these operations. Regulations, such as monitoring the exportation of wastes from CAFOs, by the EPA, are in place to protect environmental integrity and prevent potential pollution issues from occurring. The following sections review the current production of primary livestock and poultry production industries in the United States.

Dairy

The dairy industry has increased milk production by 15% in the past 10 years, with a 40% decrease in the number of dairy operations (NASS 2006a). This trend indicates that production is becoming more efficient as well as concentrated into fewer operations with greater numbers of animals. Another fact that supports this trend of higher efficiency is that milk production per cow has increased 19% in the past 10 years (NASS 2006a). The dairy industry alone increased the total value of production from $24 billion in 2001 to $26 billion in 2005, representing an 8% increase in cash receipts (NASS 2006a). The top five dairy-producing states in 2005 were (1) California, (2) Wisconsin, (3) New York, (4) Pennsylvania, and (5) Idaho, combining to produce approximately 93 billion pounds of dairy products (NASS 2006a). Every top dairy-producing state supports a significant urban population that continues to expand into historically agricultural land. In fact, California is ranked number 1 out of all states in estimated population, with a 6.7% increase in population from 2000 to 2005 (Population Division, United States Census Bureau 2005).

The dairy industry is experiencing a major restructuring period with more dairy managers incorporating modern technologies to improve their efficiency and environmental stewardship. Food safety is a major concern, and ongoing efforts to implement best management practices that reduce milk contamination are a primary focus for the dairy industry. Within the realm of these best management practices are techniques that reduce insect pest populations without contaminating dairy products. The dairy industry is striving to implement pest management plans, and the awareness of an integrated approach to pest management is a priority since waste management and pest control go hand in hand.

Beef

The beef industry has experienced a steady increase in total inventory during the last 3 years, with a 1% increase of all cattle and calves from 2005 to 2006, totaling 105.7 million head (NASS 2006a). The dynamics of beef CAFOs can vary significantly as far as size and total beef animals present. A good rule of thumb is that a beef CAFO has a 2.5 turnover rate of its capacity in animals on an annual basis. For example, if a beef CAFO has a carrying capacity of 15,000 head, then the total amount of beef animals being housed at this facility on an annual basis would be about 37,500 head. These numbers are estimates, and the actual number of head vary based on market trends and adverse environmental factors, such as drought. The top five beef-producing states in 2006 were (1) Texas, (2) Kansas, (3) Nebraska, (4) California, and (5) Oklahoma, and they produced 38.3 million head of cattle (NASS 2006a). While Texas is vast in land area size (695,673 km^2) it also supports the nation's second largest human population, with a 9.6% increase in population from 2000 to 2005 (Population Division, United States Census Bureau 2005).

Urban encroachment into areas solely occupied by beef CAFOs has brought on legal litigation against the CAFO mainly from dust or odors, but flies also have been implicated as a nuisance. The increase in the number of litigation issues against CAFOs has led to the adoption of state statutes known as right-to-farm laws (Horne 2000). These vary from state to state but generally focus on protecting the agricultural interest. Another issue is the classification of CAFOs by either the standard size component or the manner in which waste is disposed. For instance, the EPA defines a CAFO as any facility with more than 1,000 animal units (AU) (1 AU is equivalent to one beef cow), or more than 300 AU per facility when pollutants are discharged into navigable waters through a man-made device, and if pollutants are discharged into waters that are in direct contact with the CAFO under the Clean Water Act (USDA-EPA 1999). While the first aspect applies to the larger commercial type operation that can implement environmental controls without decreasing profits, the second deals with smaller operations, such as preconditioning facilities. The issue that arises in the second category is that it requires smaller facilities to submit comprehensive nutrient management plans (CNMPs), which can be costly and labor-intensive. A CNMP is best described by the U.S. Department of Agriculture and U.S. Environmental Protection Agency *Unified National Strategy for Animal Feeding Operations* in the following statement:

> CNMPs should address feed management, manure handling and storage, land application of manure, land management, record keeping, and other utilization options. While nutrients are often the major pollutants of concern, the plan should address risks from other pollutants, such as pathogens, to minimize water quality and public health impacts from animal feeding operations. (USDA-EPA 1999)

While the creation and implementation of CNMPs can be costly and vast in scope, one potential benefit for producers is that it provides a framework for record keeping of all pest management decisions for use in possible future litigation.

The beef CAFO industry as a whole is undergoing a major effort in implementing and adopting quality assurance programs that focus on both animal husbandry and environmental stewardship. The National Cattleman's Beef Association (NCBA) provides guidelines for quality assurance that focus on keeping records on pesticide use (NCBA 2001). These and other organizations on the state and national level are focusing on providing a safe product that focuses on environmental stewardship.

Swine

The total number of hogs and pigs in the United States experienced steady to slight increases from 2001 to 2005, with an approximate increase of 2.7%, with 61.3 million head in 2005

(NASS 2006a). Cash receipts from the swine industry totaled $15 billion in 200,4 which represents a 21.3% increase in cash receipts from 2000 (NASS 2006a). The top swine-producing states in 2006 were (1) Iowa, (2) North Carolina, (3) Minnesota, (4) Illinois, and (5) Indiana. These states produced 40.2 million head of swine, which represents 52.8% of the total swine in the United States (NASS 2006a). These states also support a significant human population representing 12% of the U.S. population in 2005, with an average increase in population of 3.9% from 2000 to 2005 (Population Division, United States Census Bureau 2005).

The swine industry has undergone vast changes over the past decade, with more animals being produced in fewer facilities. Swine production has become vertically integrated, with corporations controlling all production aspects from farrowing to marketing. This trend of higher commercialization has led many states to redefine their right-to-farm laws to either include these types of operations as agriculture entities or define them separately. In the past decade, confined swine facilities have experienced a significant amount of litigation, with odor being a main problem. For example, North Carolina was one of the first states to enact a right-to-farm law in 1979, but in a court's interpretation in a case titled *Mayes v. Tabor*, the court ruled in favor of a campground that labeled a hog farm a nuisance due to the odors coming from the operation (Horne 1999). While most litigation in the swine industry is focused on odor, the issue of what is considered a nuisance can vary, and in most cases is left up to the courts to interpret.

Poultry

The poultry industry is vast in magnitude. It requires extensive housing for the birds and their associated wastes. Modern poultry production systems are large operations with high densities of animals. These facilities often are financed and managed by large companies (Axtell 1999). These large companies are known as integrators that utilize a contracting system with individual producers. Facilities are usually located in close proximity to an integrator's feed mill because they provide the feed for the growers. The integrator usually owns the feed mills, hatcheries, processing plants, and transportation (Axtell 1999). The adoption of the fully integrated production system has allowed for the production of a large number of birds for meat or eggs on a small amount of land (Axtell 1999). Poultry production can be grouped into five main production systems: broiler breeders, turkey breeders, growout (broilers or turkeys), caged layers, and pullets (Axtell 1999). For a complete overview of poultry production systems, consult North and Bell (1990).

Poultry production can be separated into either broiler production (chickens produced for meat consumption) or layer production (chickens utilized for the production of eggs). The broiler industry in the United States produced 8,870,350,000 head of chickens for meat consumption in 2005, which represents a 1.5% increase from the previous year (NASS 2006b). The top five broiler-producing states in 2005 were all located in the southeastern United States, with rankings as follows: (1) Georgia, (2) Arkansas, (3) Alabama, (4) Mississippi, and (5) North Carolina (NASS 2006b). These states alone accounted for 71% of the total value of production in the United States in 2005 (NASS 2006b). The layer industry in the United States produced 89,960,000,000 eggs in 2005, which is slightly higher than production in 2004 (NASS 2006b). The top five layer-producing states in the United States in 2005 were (1) Iowa, (2) Ohio, (3) Pennsylvania, (4) Indiana, and (5) California, accounting for 43% of all eggs produced in the United States (NASS 2006b). The states representing the greatest amount of production also support a significant human population that represented approximately 32% of the U.S. population in 2005 (Population Division, United States Census Bureau 2005).

The turkey industry is similar in structure to the broiler industry, with the majority of production being concentrated into large facilities that occupy a small area of land. Production in the United States totaled approximately 256 million birds in 2005, which represents a 2.7% decrease from 2004 in the total number of turkeys (NASS 2006b). Although there was a decrease in turkey production, the value of production increased from 2004 to 2005 by approximately 6%, with the total value of production at $3,232,576,000 (NASS 2006b). The top turkey-producing states in 2006 were (1) Minnesota, (2) North Carolina, (3) Arkansas, (4) Virginia, and (5) Missouri, with production from these states accounting for approximately 70% of all turkeys being produced in the United States (NASS 2006b).

Sheep and Goats

The sheep industry has been experiencing a downtrend in the number of animals for the past 15 years, but seems to have stabilized recently (NASS 2006a). Total number of sheep increased slightly from 2005 to 2006, with all sheep categories totaling 6,230,000 (NASS 2006a). The top sheep-producing states in 2005 were (1) Texas, (2) California, (3) Wyoming, (4) Colorado, and (5) South Dakota, and they represent 48% of all sheep production in the United States (NASS 2006a). While most sheep production is on range land, there are a few operations that confine sheep. The sheep industry's decline over the past decade can be attributed to several reasons: a shift from wool-based products to synthetic, losses to predators, inability to compete with imports, and consumer's preference for other meats rather than lamb. Recent stabilization could be attributed to a smaller niche market, where low-income families can produce and raise sheep effectively without having to worry about high-input costs.

The goat industry has prospered in recent years, with increased interest in meat production. Once an industry heavily focused on angora mohair production, it has now shifted to focus on meat production. One reason for such a shift is that the mohair incentive program was phased out in 1996, and the introduction of the Boer goat from South Africa has improved performance through genetics. The Boer goat has brought higher prices to the goat industry and has increased embryo transfers and artificial insemination practices to improve breeding lines. The total number of goats in the United States was 2,826,000 in 2006, representing a 4% increase from 2005 (NASS 2006a). In fact, Texas accounts for 42% of the goat inventory in the United States, totaling 1,200,000 (NASS 2006a).

Muscoid Flies

The dipteran family Muscidae contains 700 different species in forty-six genera in North America, with only a few of the genera considered pests. Flies in the Muscidae family are diverse, ranging from blood-feeding facultative parasites to fly species that are a nuisance to the general public (Figure 15.2). Muscoid flies are commonly referred to as filth flies due to the type of substrates in which they develop.

Figure 15.4 The pupal casings (puparia) of most flies can be used for species identification. They are durable, and can be found months and sometimes years after the adult has emerged. This life stage can also be utilized as an alternative specimen for postmortem toxicology in medicolegal cases. (Photo courtesy of Dr. Jason H. Byrd, University of Florida.)

Muscoid flies have a holometabolous life cycle consisting of the egg, larva, pupa, and adult stages. Eggs are creamy white, elongate in shape, range in length from 0.8 to 2.0 mm, and can be laid singly or in aggregated clumps. Muscoid fly larvae are more commonly known as maggots. They have three instars and develop in moist decomposing organic matter. Pupae occur in cases known as puparia (singular: puparium) that form during a process called pupariation (Frankel and Bhaskaran 1973) (Figure 15.4). Puparia can vary in size and color and retain the form of the posterior spiracles developed in the larval stages, which can be used for identification of specific species. Adult muscoid flies can range in size from 4 to 12 mm in length, and all have wings that are longer than their abdomen. Mouthparts of nonbiting and biting muscoid flies vary (Elzinga and Broce 1986). The nonbiting flies have enlarged labellae for sponge feeding (Elzinga and Broce 1986). In contrast, the biting flies have smaller labellar lobes, but with highly sclerotized and enlarged prestomal teeth for piercing skin (Elzinga and Broce 1986).

In the field of veterinary medicine, important muscoid species are mostly oviparous, which means that the females deposit fertilized eggs into or onto the ovipositional substrates. Larvae then hatch and begin feeding in the location in which they were placed as eggs. Once the larvae have completed most of the third-instar stage, they will empty their alimentary tract and enter a dispersal phase and search for adequate sites to pupate. Developmental times vary between species, but generally time of development from egg to adult ranges from 1 to 6 weeks during optimal conditions. Optimal development occurs between 27 and 34°C.

Upon completion of the larval and pupal stages, most muscoid flies contain little or no nutrients when they emerge from the puparium. Therefore, it is imperative that newly emerged flies locate water and carbohydrates in order to survive. The feeding behavior for a majority of nonbiting flies can be described as a process of salivating, scrubbing, and sucking (Moon 2002). Nonbiting flies obtain most carbohydrates off host sites such as plant nectar or aphid secretions. Biting fly-feeding behavior can be described as the process of excoriating the skin surface, thus forming an opening that fills with blood (hematoma) and sucking the blood up with a long tube-like proboscis. Females of all muscoid flies require a protein source to produce each egg clutch.

Mating behavior in muscoid flies varies, but the majority of males perch on resting surfaces and dart out at females flying in the area. Most of the veterinarily important muscoid flies are active during the daytime, with differing time intervals of peak activity. Daily activities can include flying, mating, host location, feeding, and ovipositing (Moon 2002). Excluding the horn fly, *Haematobia irritans irritans* (L.), most biting flies only spend a short time on their host, with the remainder of their time being spent digesting their meal in the surrounding environment. Flight ranges of muscoid flies vary between individual species. Some species can travel for more than 25 km from known breeding habitats. Seasonal patterns of muscoid flies will vary depending on geographical location, species, and years. Most muscoid flies are multivoltine, meaning they can develop through at least two or more generations per breeding season (Moon 2002). Muscoid flies have different overwintering mechanisms, with some species adapting behaviorally and others going into a diapause state.

House Fly (*Musca domestica* L.)

The house fly has long been associated with humans, and despite this extensive association, it still remains one of the most difficult pests to control. It is a nonbiting fly that is cosmopolitan in distribution. Adults are 6 to 9 mm in length, with four longitudinal black stripes on their thorax. A female can deposit several hundred eggs in her lifetime. Development from egg to adult can occur in 15 days under optimal conditions. Immature house flies can be found in a variety of decaying organic substrates located in dairy, swine, or poultry barns, garbage cans, and animal kennels. As far as public health, the major issue that house flies cause is annoyance; however, recent research has implicated house flies as a potential disseminating agent of *E. coli* O157:H7 from cattle farms (Alam and Zurek 2004). This research, along with others, is continually providing evidence that house flies serve as routes for the spread of major enteric pathogens.

Little House Fly (*Fannia canicularis* [L.])

The little house fly is a nonbiting fly that is primarily a pest in or around poultry houses. Adults swarm at head or shoulder height of an average person, and therefore can be a considerable nuisance. Adults are 5 to 8 mm long with a dark thorax and an abdomen that has yellow markings. The little house fly when at rest will hold its wings over its abdomen, which creates a narrow V shape to the wing outline. Larvae develop in poultry manure and are brown and somewhat flattened with fleshy spines. Larvae can tolerate various moisture ranges, which make them particularly difficult to control. But the most effective control strategy is sanitation of breeding grounds. The development time from egg to adult is longer than the house fly at all temperature ranges.

False Stable Fly (*Muscina stabulans* [Fallen])

The false stable fly is a nonbiting and stout-bodied fly measuring 8 to 12 mm in length. It has thoracic markings similar to that of the stable fly, with three longitudinal black stripes; however, the tip of the scutellum is red to orange in color. Newly hatched larvae are white and 1.5 mm in length. Resulting larvae gradually change from a gray to cream color and measure 12 to 18 mm long when mature. This fly species has been implicated in causing urinary tract myiasis in humans (Moon 2002), mainly due to its feeding and defecation patterns on fresh

Figure 15.5 The stable fly (*Stomoxys calcitrans*) is not considered a primary urban pest by many since it primarily feeds on livestock. However, it will feed on most any warm-blooded animal, and humans often experience their annoyance and painful bite when attempting to enjoy outdoor activities. Many people are familiar with this species as the "dog fly," and have encountered it during a summer visit to a beach. (Photo courtesy of Bradley Mullens, University of California at Riverside).

fruit. This can pose a particular problem when abundant breeding habitats, such as those in animal production systems, are located near fresh outdoor produce markets.

Stable Fly (*Stomoxys calcitrans* [L.])

Stable flies have been associated with CAFOs for decades (see Figure 15.5). It has been referred to by many common names, such as the beach fly, lawnmower fly, dog fly, and biting house fly. It is a biting fly that is cosmopolitan in distribution. Adults are 5 to 7 mm in length, with a protruding proboscis that resembles a bayonet sticking directly forward from the head. Females require several bloodmeals to produce fertile eggs, and females can deposit several hundred eggs in their lifetime (Morrison et al. 1982). Generation time can occur in as few as 20 days under optimal conditions, with temperatures between 28 and 30°C (Lysyk 1998). Most immature development occurs in substrates that contain manure that is mixed with organic matter (spilled feed or hay). Adults of both sexes primarily blood-feed on beef or dairy cattle, and other warm-blooded animals secondarily. The main public health issue is the annoyance they cause when biting humans, especially in recreation areas such as beaches. Stable flies have been implicated in transmitting retroviruses to horses and cattle, but in nature their role as vectors has been negligible (Moon 2002).

Horn Fly (*Haematobia irritans irritans* [L.])

The horn fly is an obligate blood-sucking parasite of cattle. Adults spend most of their time on the animal, with females leaving only to deposit eggs in fresh manure. They can be found mainly on the withers, back, and sides of beef animals, but can move to the underside of the animal during high temperatures. This fly can feed up to thirty-eight times per day, thus causing continued annoyance to cattle (Harris et al. 1974). Horn flies can vary in activity by different locations, with northern states experiencing significant numbers from 3 to 4 months in the summer, whereas in the southern states activity can be throughout the entire year. Adults are 3 to 5 mm in length and have a piercing type proboscis for sucking blood. Immature development occurs solely in dung pats and is optimal at 25 to 27°C.

Beetles

Beetles (Coleoptera) encountered in urban structures are often identified by artifacts, such as exit holes, frass (feces), and gallery-feeding sites (Gulmahamad 2004). Additionally, in cases involving infestations by these beetles, adults and immatures are often not available for identification. The beetles discussed below colonize CAFOs, and when facilities are cleaned, associated beetles will often migrate into urban areas.

Lesser Mealworm (*Alphitobius diaperinus* [Panzer])

The lesser mealworm or darkling beetle, *Alphitobius diaperinus* (Panzer), is considered the most damaging pest in broiler operations. The lesser mealworm life cycle includes eggs, larvae (six to nine stages), pupae, and adults, all of which are found in litter or manure at poultry houses (Figure 15.6a and b). A complete life cycle from egg to adult usually requires about 5 weeks, depending on the temperature (Rueda and Axtell 1996). Adults can live for long periods of time, usually living at least a year (Axtell 1999). Lesser mealworm beetles are cosmopolitan in distribution and can be associated with stored product production. Adults are broadly oval, moderately convex, 5 to 7 mm in length, and usually shiny black in appearance. Eggs are deposited in cracks or crevices in the poultry house. Larvae exhibit segmented bodies that taper posteriorly, three pairs of legs, and range from 7 to 11 mm in length. Pupae are approximately 6 to 8 mm in length, creamy white to tan colored, with legs tucked alongside the body.

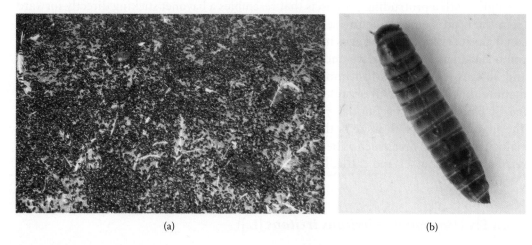

(a)   (b)

Figure 15.6 (a) The lesser mealworm (*Alphitobius diaperinus*) is a damaging pest commonly found in litter or manure in poultry operations. The longevity of the adult, often a year or more, makes this species a particularly difficult pest to control. (b) The larvae are a common pest of stored products. (Photo courtesy of Aubree Roche, University of Georgia, Athens.)

Control efforts are usually focused on spraying the litter with insecticides between flocks (Weaver 1996).

Larder Beetles (*Dermestes* sp.)

The two dermestid beetles (covered in other chapters) most commonly associated with the litter beetle complex are the hide beetle, *Dermestes maculates* (De Geer), and the larder beetle, *Dermestes lardarius* (L.). These beetles have been reviewed in previous chapters in regards to their colonization of human remains. Both species are less abundant than the darkling beetle but still are considered serious pests (Axtell 1999). The biology of these beetles is very similar to that of the darkling beetle. Adults of hide beetles are generally larger than the darkling beetle, and the larvae are covered with long brown hairs (setae). The larvae cause the most damage when they feed on insulation or even wood in poultry houses. The adult larder beetle can be distinguished by a dense covering of yellow hairs on the basal half of the elytra. This pest has historically been associated with homes or urban settings and causes damage similar to that of the hide beetle in poultry houses.

Red and Black Imported Fire Ants

A tremendous amount of literature is available on red and black imported fire ants, *Solenopsis* sp. Red and black imported fire ants were introduced into the United States through Mobile, Alabama, between 1933 and 1945 (Drees 2006). Since their introduction, they have spread throughout the southern United States, including southern California. Red imported fire ants are highly aggressive and will severely impact local fauna by outcompeting them for resources through predatory actions. Fire ants will nest close to or in electrical devices, which could result in short circuitry. Identification of red and black imported fire ants is difficult without a hand lens or dissecting scope; however, their mounds and aggression toward others are key biological characteristics. Fire ants are notorious for invading homes and other structures. They will also feed on decomposing human remains or those unable to care for themselves.

Conclusion

Due to the economic and health-related impact of arthropods associated with livestock and poultry, livestock entomology will continue to represent a major section in forensic entomology. As urbanization continues to grow and spread into traditionally agricultural settings, it is expected that the role of the livestock entomologist in the courtroom will also increase. However, it is hoped that through educating the urban and agricultural populous on these arthropods and their associated roles in the environment, the growing pains of our society can be eased.

Acknowledgments

The authors thank Gary Mullen, Department of Entomology and Plant Pathology, Auburn University; Michelle Sanford, Department of Entomology, Texas A&M University; Patricia

Zingoli, Department of Entomology, Soils, and Plant Sciences, Clemson University; and Dr. Roger Gold, Department of Entomology, Texas A&M University, for comments on earlier drafts of this chapter.

References

Alam, M. J., and L. Zurek. 2004. Association of *Escherichia coli* O157:H7 with houseflies on a cattle farm. *Appl. Environ. Microbiol.* 70:7578–80.

Altschuler, D. Z., M. Crutcher, N. Dulceanu, B. A. Cervantes, C. Terinte, and L. N. Sorkin. 2004. Collembola (Sprintails) (Arthropoda: Hexapoda: Entognatha) found in scrapings from individuals diagnosed with delusory parasitosis. *J. New York Entomol. Soc.* 112:87–95.

Axtell, R. C. 1999. Poultry integrated pest management: Status and future. *Integrated Pest Management Rev.* 4:53–73.

Drees, B. 2006. Managing red imported fire ants in urban area. B-6043, 8-06, Texas Cooperative Extension, Texas A&M University.

Elzinga, R. J., and A. B. Broce. 1986. Labellar modifications of Muscomorpha flies (Diptera). *Ann. Entomol. Soc. Am.* 79:150–209.

Focke, M., W. Hemmer, S. Wöhrl, M. Götz, R. Jarisch, and H. Kofler. 2003. Specific sensitization to the common housefly (*Musca domestica*) not related to insect panallergy. *Allergy* 58:448–51.

Frankel, G., and G. Bhaskaran. 1973. Pupariation and pupation in cyclorrhaphous flies (Diptera): Terminology and interpretation. *Ann. Entomol. Soc. Am.* 66:418–22.

Gorham, J. R. 1975. Filth in foods: Implications for health. *J. Milk Food Technol.* 38:409–18.

Gulmahamad, H. 2004. Wood-boring beetles. In *Handbook of pest control*, Mallis, A., Ed. 9th ed. Richfield, OH: GIE Media, 381–430.

Harris, R. L., J. A. Miller, and E. D. Frazar. 1974. Horn flies and stable flies: Feeding activity. *Ann. Entomol. Soc. Am.* 67:891–94.

Heyworth, M. F. 1999. Importance of insects in asthma. *J. Med. Entomol.* 36:131–32.

Horne, J. E. 2000. *Rural communities and CAFOs: New ideas for resolving conflict.* Kerr Center for Sustainable Agriculture, Poteau, OK. http://www.kerrcenter.com/publications/CAFO.pdf.

Klotz, J. H., H. C. Field, S. A. Klotz, and J. L. Pinnas. 2005. Ants and public health. *Pest Control Tech.* 34:40–60.

Lysyk, T. J. 1998. Relationships between temperature and life-history parameters of *Stomoxys calcitrans* (Diptera: Muscidae). *J. Med. Entomol.* 35:107–19.

Moon, R. D. 2002. Muscid flies (Muscidae). In *Medical and veterinary entomology*, Mullen, G. R., and L. A. Durden, Eds. New York: Academic Press, 279–301.

Morrison, P. E., K. Venkatesh, and B. Thompson. 1982. The role of male accessory gland substance on female reproduction with some observations of spermatogenesis in the stable fly. *J. Insect Physiol.* 28:607–14.

Mullen, G. R., and L. A. Durden. 2002. Introduction. In *Medical and veterinary entomology*, Mullen, G. R., and L. A. Durden, Eds. New York: Academic Press, 1–13.

National Agricultural Statistics Service (NASS). 2006a. Statistical highlights of U.S. agriculture: 2005 and 2006. USDA Statistical Bulletin. http://www.nass.usda.gov/Publications/Statistical_Highlights/index.asp.

National Agricultural Statistics Service (NASS). 2006b. Poultry production and value: 2005 summary. http://usda.mannlib.cornell.edu/usda/current/PoulProdVa/PoulProdVa-05-18-2006_revision.pdf.

National Cattlemen's Beef Association (NCBA). 2001. Beef quality assurance national guidelines. http://www.beefusa.org/uDocs/NCBA_QA_Guidelines_August_2001_color.doc.

North, M. O., and D. D. Bell. 1990. *Commercial chicken production manual.* 4th ed. New York: Chapman & Hall.

Poorbaugh, J. H. 1993. Cryptic arthropod infestations: Separating fact from fiction. *Bull. Soc. Vector Ecol.* 18:3–5.

Population Division, U.S. Census Bureau. 2005. Table 2: Cumulative estimates of population change for the United States and states, and for Puerto Rico and state rankings: April 1, 2000 to July 1, 2005. NST-EST2005-02. Available from http://www.census.gov/popest/states/tables/NST-EST2005-02.xls.

Potter, M. F. 2004. Termites. In *Handbook of pest control*, A. Mallis, Ed. Richfield, OH: GIE Media, 217–361.

Rueda, L. M., and R. C. Axtell. 1996. Temperature-dependent development and survival of the lesser mealworm, *Alphitobius diaperinus*. *Med. Vet. Entomol.* 10:80–86.

Rust, M. K. 1999. Urban entomology: Past and present. Paper presented at Proceeding of the 3rd International Conference on Urban Pests. Prague, Czech Republic.

Shin, J. W., J. H. Sue, T. W. Song, K. W. Kim, E. S. Kim, M. H. Sohn, and K. E. Kim. 2005. Atopy and house dust mite sensitization as risk factors for asthma in children. *Yonsei Med. J.* 46:629–34.

U.S. Department of Agriculture and U.S. Environmental Protection Agency (USDA-EPA). 1999. *Unified national strategy for animal feeding operations.*

Weaver, J. E. 1996. The lesser mealworm, *Alphitobius diaperinus*: Field trials for control in a broiler house with insect growth regulators and pyrethroids. *J. Agric. Entomol.* 13:93–97.

Ecological Theory and Its Application in Forensic Entomology

16

SHERAH L. VANLAERHOVEN

Contents

Introduction

In forensic entomology, we use two types of insect evidence to estimate the time of insect colonization, or the minimum postmortem interval of dead bodies. One method utilizes the developmental rate of the first generation of flies to colonize the remains to estimate postmortem interval. The other method utilizes the sequence of colonization and extinction of populations of different insect species feeding and coexisting on a body as it decomposes from fresh to skeletal remains. This latter method uses the presence and absence of particular insect species, as well as the evidence that remains from particular insect species that were previously present, like the numbers on a clock marking the passage of time. The current paradigm in forensic entomology is that the successional pattern of insect species is predictable and depends on the specific biogeoclimatic region and habitat that the remains are in, as well as the specific circumstances (i.e., hanging, clothed, scavenged, etc.) of the remains (Anderson and VanLaerhoven, 1996; Catts and Goff, 1992; Grassberger and Frank, 2004; Tabor et al., 2004; Tenorio et al., 2003; Watson and Carlton, 2003). Thus, the majority of research that has described insect succession in decomposing remains has focused on testing different biogeoclimatic regions (Early and Goff, 1986; VanLaerhoven and Anderson, 1999), different habitats (Joy et al., 2006; Shean et al., 1993), or the varied condition of the remains, such as carcass size (Hewadikaram and Goff, 1991; Kuusela and Hanski, 1982), carcass type (Watson and Carlton, 2005), presence of clothing or wrappings (Goff, 1992; Komar and Beattie, 1998), and other variables. Although these studies have demonstrated that patterns of insect assembly exist under these conditions, and that patterns change with different conditions, as stated by Keddy and Weiher (2004), asking if patterns exist in nature is akin to asking if bears shit in the woods. Few of these studies provide insight into the actual mechanisms that shape assembly of the insect community on remains over time and space. Without an understanding of the mechanisms that influence community assembly, and therefore modify community assembly, accurate and precise prediction of insect assembly on carrion is tenuous and should encompass wide margins of error to account for potential variability in assembly.

The field of ecology has been asking the question "What drives particular species to colonize, coexist, or go extinct in particular patterns?" for over 100 years. The process of community assembly often exhibits particular patterns; however, these patterns may not be clearly related to particular mechanisms. Some patterns arise from multiple mechanisms, and the important mechanisms may be difficult to identify by observational experiments. In some cases, what initially appears to be an important community pattern eventually is proved be indistinguishable from a random pattern. Despite years of research into this question, how communities assemble and change remains one of the most crucial questions in ecology. This chapter will discuss the development of ecological theory, from the pioneering studies in plant and animal ecology, to the modern theoretical framework that describes our current understanding of how communities are formed and evolve over space and time. Although it is not possible to fully explore all of the tenets of community assembly, as there are numerous books and papers on the subject, the goal of this chapter will be to introduce the ecological framework and relate it to current knowledge and assumptions in forensic entomology. This should allow the reader to pursue further investigation into relevant concepts. The idea will be explored that although community assembly of insects on decomposing remains exhibits patterns, these patterns may only be predictable to a certain degree. The degree of variability within community assembly in a given biogeoclimatic zone, habitat, and circumstances must be considered when estimates of postmortem interval are made. Suggestions for future research directions will be made that would provide forensic entomologists with the data needed to define how much variation exists in community assembly, and how much accuracy can be assigned to estimates of postmortem interval based on community assembly.

Development of Ecological Theory of Community Assembly

Historical Views on Community Assembly

In the early part of the twentieth century, there were two views as to the structure of communities. The Clementsian deterministic view stated that the community is a superorganism (Clements, 1916). The species within the community share a common evolutionary history. Therefore, if a community is an organism, then its populations of species and individuals are the tissues and cells. In this view, community development, or assembly, is controlled by climate and is defined by the dominant species. The species composition of a particular community is predictable. Clements defined "pioneer species" as those that are the first to arrive, and they modify the environment to make it suitable for subsequent colonizing species. In contrast, the Gleasonian individualistic view stated that the community is a relationship of coexisting species due to the result of similar tolerances, resource requirements, and chance (Gleason, 1917). Associations of species are less predictable than stated in the Clementsian view and community boundaries are less defined, with species distributed along a gradient such as temperature or particular nutrients. A particular community may be defined by a characteristic species composition, but not necessarily the dominant species. Gleason's view was the first of the niche-based community assembly theories.

The first to use the term *niche* was Grinnell (1917). He defined it to be all the sites where a species could live as determined by habitat conditions. Elton (1927) defined niche as the function a species performs in its community, such as the role a blow fly performs by eating the remains of deer left by wolves. Miller (1967) stated that the Grinnellian niche is a species' address, while Elton's niche is its profession. Hutchinson (1957) proposed that a niche is the totality of sets of conditions that are compatible with a species' persistence and success, and that this forms an n-dimensional hypervolume bounded by these limits. An example of an n-dimensional hypervolume using three dimensions is the space encompassed by the minimum and maximum limits of temperature and humidity, and the minimum resources required by a particular insect species to survive and reproduce over time. This n-dimensional hypervolume is the fundamental niche of a species, while the realized niche is all the conditions in which a species succeeds in the presence of competitors. MacMahon and coauthors (1981) redefined fundamental niche as the maturationally restricted range of states that an organism could endure, while the realized niche is the state of an organism at any instant during its life. Grace and Wetzel (1981) demonstrated the difference between fundamental and realized niches on the basis of competition. Therefore, the definition of niche has changed from focusing on habitat, to food, to n-dimensional, and from the species level to the individual organism level. The debate over the definition of niche continues, as James and coauthors (1984) recommend the Grinnellian approach over the Hutchinsonian approach.

Odum (1969) expanded on niche-based community assembly theory by examining r- and K-selected species, energetics, community structure, and nutrient cycling. Pioneer species are r-selected, as characterized by rapid development, poor competitors, high rate of reproduction, early, single reproduction, small body size, and a short life span. Later successional species, or those species arriving later in community assembly, are K-selected, which are characterized by slow development, good competitors, delayed, repeated reproduction, large body size, and a longer life span. Odum stated that net primary productivity, or conversion of sunlight to energy and growth, is greater in the early successional stages and low in climax stages, but that biomass of the community is greater in the climax stage. Food chains are linear in early successional stages, but weblike at the climax, with more species interactions. Species diversity, niche specialization, and spatial diversity are greater at the climax stage than during successional stages. He stated that nutrient cycles are closed, with slow exchange between organisms and environment, and a greater role of detritivores in nutrient cycling during the climax, compared to earlier successional stages.

Connell and Slatyer (1977) further expanded the concept of niche-based community assembly over time, into three models. While the first model is Clements' (1916) concept of changes in community assembly over time (= succession), renamed the facilitation model, the other two models, tolerance and inhibition, are new concepts. The tolerance and inhibition models differ from the facilitation model by assuming that any arriving species may be able to colonize, instead of only pioneer or early successional species colonizing immediately after the disturbance. This illustrates that the main difference between the models is the mechanisms by which new species appear in the assembly sequence. While early successional species modify the environment to make it more suitable for subsequent species in the facilitation model, in the tolerance model, early successional species have no impact on later successional species. Instead, the later successional species are those that may have arrived either at the beginning or later, but grew slowly compared to early successional species. As community assembly proceeds, tolerance to environmental factors such as shade, moisture, nutrients, or grazing limits the growth of species until only the most tolerant survive, resulting in the climax stage. In the inhibition model, the earliest colonists inhibit the invasion of subsequent colonists or suppress the growth of other colonists by securing the space or resources. Only when an early colonist dies does this give another of the same species or a later successional species a chance to grow in its place. Because late successional species tend to have longer life spans than early successional species, they will likely replace the early successional species at a higher rate than the replacement of late successional species. This gradually increases the relative abundance of the late successional species. In effect, the succession is from short-lived to long-lived species. One final difference between the three models is the cause of death of the early colonists. In the facilitation and tolerance models, early colonists are killed in competition with the later species, as they are deprived of nutrients, light, or other resources. However, in the inhibition model, the early colonists are killed by local disturbances such as herbivores, parasites, or pathogens, or by physical extremes.

Diamond (1975) suggested that certain species are found only in communities with certain properties or with certain values of species richness and coined the term *assembly rules*. He described incidence functions to describe the probability of a particular species occurring in a particular community based on properties of that community. Some species, called high-S species, occur only in communities with high species richness. High-S species require more specialized properties found in communities that support a variety of other species. Other species, called tramp species, occur in a broad range of communities, including those of low species richness. Tramps species are good colonizers with generalized requirements that allow them to persist in relatively simple communities. Diamond's incidence functions are not mechanistic, in that they do not say why certain species appear more or less often than others in communities that contain a particular number of species. If species do not interact, such that they colonize independently, then incidence functions for each species could be used to predict the probability that two species will coexist. The idea the species colonize independently, with no interaction between

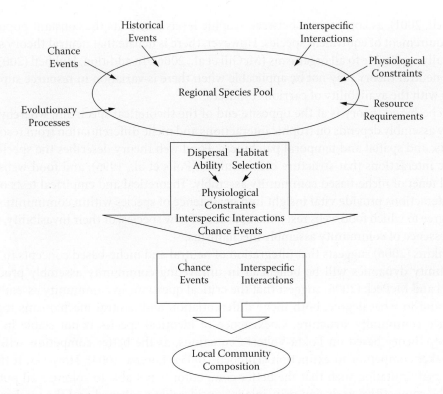

Figure 16.1 Hierarchical processes modifying community assembly. (Adapted from Morin 1999.)

species, leads to the current dichotomy between neutral theory and niche-based theory of community assembly.

Current Views on Community Assembly

Ecologists recognize that many factors affect the species composition of a given community, with no single factor providing a complete explanation for observed patterns (Schoener, 1986). Community assembly is a consequence of a hierarchy of interacting processes (Figure 16.1) (Morin, 1999). The composition of the regional species pool depends on evolutionary processes such as speciation, gene flow, and mutation; physiological constraints in terms of tolerances of particular species to the climatic conditions of the region; and historical events such as climatic shifts, natural disasters, or chance introductions of exotic species. Dispersal and habitat selection filter species from the regional species pool to identify those species available to colonize a given community (Keddy, 1992). Thus, communities are nonrandom subsets of the regional species pool. It is at this stage that there are two distinct branches of current theory on community assembly. Neutral theory (Hubbell, 2001) states that community assembly is a random process, whereby the local community is randomly composed of species from the regional species pool. Neutral theory assumes that within the regional species pool, equivalent species share similar characteristics, making them equally likely to colonize the community. Thus, fluctuations in species composition during assembly are due to dispersal limitations, not differences in competitive ability. Neutral theory tends to be applied on single functional groups or trophic levels

(Hubbell, 2001), as comparison between trophic levels rarely fits the constant population size requirement of equivalent species. However, there is debate that neutral theory should be equally applicable to all organisms (McGill et al., 2006). In addition, Hubbell (2001) suggested neutral theory may not be applicable when there is variation in resource supply, as occurs with the availability of carrion resources.

Niche-based theory is at the opposite end of the theoretical spectrum, whereby community assembly depends on trophic interactions and niche differentiation from resources, habitats, and spatial and temporal patchiness. Food web theory describes the species and trophic interactions that structure communities (Polis et al., 1996), and food webs are a central tenet of niche-based community assembly. Theoretical and empirical tests of food web interactions provide vital insight into persistence of species within communities, and the degree to which food webs resist the addition of new species, or their invasibility, which is the essence of community assembly mechanisms.

Jenkins (2006) suggests that integration of neutral and niche-based concepts in metacommunity dynamics will be beneficial in unraveling community assembly processes. Leibold and McPeek (2006) suggest that the critical questions in community assembly are when, and to what degree, both niche differentiation and neutral mechanisms together influence community structure. Coexistence of identical species is not stable in niche assembly theory based on Lokta-Volterra equations, as the better competitor will drive the weaker competitor to extinction (Hubbell, 2005a; Loreau, 2004). However, if there is a dispersal limitation such that the better competitor is not able to colonize all potential sites, the competitive exclusion principle is considerably weakened and the weaker competitor can persist in these sites (Hubbell, 2005b). Dispersal limitation can be defined as the failure of a species to disperse individuals to all sites suitable for their colonization, survival, and growth, and recruitment limitation is the failure to recruit immatures, and ultimately reproductive adults, in all such sites. Under dispersal limitation, many sites are won by default by an inferior competitor because the best competitor for the site did not disperse or recruit to the site (Hubbell, 2006), such as in the case of fugitive species. In the absence of a competition-colonization trade-off, large numbers of species could coexist if they were sufficiently dispersal and recruitment limited (Hurtt and Pacala, 1995). This is not to say that the presence of equivalent species can only be a result of neutral mechanisms. Niche-based mechanisms can result in ecological equivalence of species (Chave, 2004). This occurs because species that have spatial or temporal niche partitioning become equivalent in their competitive abilities at some spatial or temporal scales.

Mechanisms That Define the Regional Species Pool

Importance of History

History shapes every community to some extent, either by influencing which species can colonize developing communities or by setting the sequence of species arrival as species accumulate and interact. Models of competitive interactions (Holt et al., 1994) predict that small historical differences, such as differences in the initial abundances of two competing species, can produce very different communities. In Lotka-Volterra models of interspecific competition, communities will reach the same final equilibrium composition, regardless of historical differences in the initial species abundances, when conditions for a stable two-species equilibria occur. In contrast, when different parameters create an unstable equilibrium, historical effects become very important because initial differences in the species abundances can determine which species will competitively exclude the other. It is unclear to what extent natural communities are mostly shaped by history-free processes, compared with those communities that are strongly influenced by historical processes (Morin, 1999).

Timescale is very important when considering community assembly. Differences in just a few days in the arrival times of mycophagous *Drosophila* change the outcome of competitive interactions in decaying mushrooms (Shorrocks and Bingley, 1994). Temporal abundances change arrival patterns, and changes due to climate change and natural disturbance can be especially important if earlier species either inhibit or facilitate the species that follow. Given that for any particular species, its timing of first arrival or maximum activity in communities often differs, changes in the temporal abundance of species create situations in which the outcomes of interspecific interactions may depend on temporal patterns of species (Morin, 1999). For example, with two species arriving at different times, there are three possible temporal interactions. The earlier species may facilitate, inhibit, or have no effect on the later-arriving species (Connell and Slatyer, 1977). When earlier-arriving species influence later-arriving species, these are known as priority effects.

Chance Events

Chance events can influence community assembly at multiple scales. Whether or not a species becomes a member of a regional pool may depend on chance introductions to a region. Stochastic events can limit when a species is available by preventing dispersal or reducing the number of individuals available through mortality (i.e., flood), but unless these events result in extinction from the regional species pool, their effect on availability of particular species is transient and difficult to predict.

Temporal Availability

Species may only be available to colonize a new resource and become a member of a new community part of the year. There are various mechanisms that explain a particular species' seasonal availability, and in order to use temporal availability as a predictor of the probability of a species becoming a member of a community, the mechanisms involved must be determined for that particular species.

Physiological Constraints

The species may only be able to tolerate the climate for a particular portion of the year; thus, physiological constraints may evolve into traits such as hibernation, diapause, or migration to allow the species to avoid the unfavorable conditions. Thus, research designed to test the physiological constraints of a species, such as developmental rate, reproduction, dispersal, and activity patterns, would determine what temperature, diel cycle, and humidity requirements limit the availability of particular species. Temperature, daylength, or the interaction between the two may induce diapause, which is a mechanism that may make particular species of blow flies unavailable to colonize a resource at different times of the year, as has been demonstrated for *Protophormia terraenovae* Robineau-Desvoidy (Diptera: Calliphoridae) (Numata and Shiga, 1995) and *Lucilia sericata* (Meigen) (Tachibana and Numata, 2004). By measuring what specific environmental conditions induce and break diapause for a particular carrion species, one can calculate the likelihood that a species may be available to colonize a carrion resource.

On a much shorter timescale, one factor that likely limits nocturnal oviposition by blow flies is their diel activity pattern. By specifically testing the hypothesis of whether light level (as well as how much light), time of day, or an interaction of both is the mechanism that determines the activity pattern and oviposition behavior of particular blow fly species, one would be able to quantify the probability that any given species is available to colonize a body at night. Of course, this assumes that the probability that the species is available in the region is already known. Another potential mechanism limiting nocturnal oviposition is the reproductive state of the female, as eggload may make females more or less willing to oviposit, or to respond to attractive odors. Although several studies have placed baits in nocturnal and nocturnal with artificial light scenarios to attempt to determine the likelihood of nocturnal oviposition (Baldridge et al., 2006; Greenberg, 1990; Singh and Bharti, 2001; Tessmer et al., 1995), none have yet conducted laboratory tests with variation of light level (equivalent of full sun, down to full dark) and time of day in a full factorial experiment. By testing willingness to fly, walk, and oviposit when fully gravid on attractive media under these controlled laboratory conditions, one would be able to assign a probability of oviposition occurring, assuming that the species is present.

Adaptive Responses to Interactions with Other Species

Temporal availability may be a result of adaptive responses to interactions with other species, in which case, these responses will depend on the presence or absence of other species. If these responses are consistently demonstrated through experimentation, then these responses can be built into a predictive model to allow calculation of the probability that particular species will be available for membership in a community.

Mechanisms to Reduce Competition

Adaptive responses to interactions with other species may include mechanisms to reduce competition. Gause's competitive exclusion principle (1934) states that it is impossible for two or more species to coexist indefinitely in the same region if they occupy the same niche and resources are limited. However, Lotka-Volterra models of competition demonstrate that if intraspecific competition is greater than interspecific competitions, both species can coexist (Lotka, 1925; Volterra, 1926). Schoener (1974) described various ways that species can separate their environment to reduce niche overlap, including diurnal variation and seasonal variation. Both of these are examples of temporal resource partitioning, in which ecologically similar species coexist by using the same limited resource at different times. This assumes that competition among species that are active at different times is less intense than if the same set of species all attempted to use the resource at the same time. This scenario is most plausible for situations where the resource rapidly recovers from utilization or is constantly renewed, as is the case in the carrion system over a larger spatial scale. Otherwise, the first species to exploit the resource effectively depletes the resource availability for later species. This occurs at the local scale on carrion, but across the landscape, new carrion resources become available. In a carrion resource system, blow flies are a model organism of resource exploitation, with adults as the dispersal stage, and larvae comprise the nondispersing, carrion-feeding stage. Mechanisms of coexistence among blow flies have not yet been fully explained, although several researchers have explored the issue. Some have suggested that niche overlap of different fly species is minimized, due to temporal resource partitioning over season (Denno and Cothran, 1975; Hanski and Kuusela, 1980; Johnson, 1975). Mechanisms such as optimal temperatures for development and adult preferences may interact with interspecific competition (Denno and Cothran, 1976; Hanski and Kuusela, 1977; Kneidel, 1984a; Wells and Greenberg, 1994a) to promote temporal resource partitioning over season.

Tracking Resource Abundance

Tracking resources, which vary in abundance over time, is another potential adaptive response to interactions with other species that results in temporal species availability. One of the ways that Schoener (1974) described that species can separate their environment to reduce niche overlap is variation in resources. VanValen (1965) examined niche breadth and determined that within-species phenotypic variation is greater in a community of generalist consumers than in a community of specialist consumers, but between-species phenotype variation is greater in a community of specialists. This adjustment of niche breadth between specialists and generalists is another proposed mechanism that allows species to coexist. Many decomposers feed on a variety of resources; thus, experimentally measuring the diet breadth would allow one to assign a probability that a particular species would consume a particular resource within a particular community based on the presence of other resources within the area during that timeframe. For example, the presence of dung may increase the likelihood that both dung and carrion-feeding insects would colonize carrion resources. The absence of a required resource for a particular species would also indicate the likelihood of the presence of that species within the community. This is somewhat complicated by the fact that many species that are reported to feed on one type of resource may have a larger diet breadth. Skin is typically consumed by dermestid beetles (Coleoptera: Dermestidae), and one might assume that these beetles would not arrive until the skin had reached an appropriate state of dryness. However, adult *Dermestes maculatus* (DeGeer) prefer moist muscle tissue and ligamentous remains to skin, occasionally act as predators on blow fly larvae, and also consume dead insects (Braack, 1987). Thus, the arrival time of *D. maculatus* is not necessarily tied to the condition of the carrion resource.

Predator Avoidance

Predator avoidance is another potential adaptive response to interactions with other species that results in temporal species availability. Predators and other natural enemies can vary in abundance over time, and in turn affect the temporal abundance of their prey. Prey may attempt to avoid seasonally intense predation by evolving temporal and behavioral changes in their availability within a community. Community composition is likely to depend, in a complex interaction, on how long predators and prey interact, as predators may enhance the survival of some species by reducing the survival of others. Differences in the timing of arrival of the interacting species may determine whether those species interact as competitors or as predators and prey. *Chrysomya rufifacies* (Macquart) (Diptera: Calliphoridae) is an invasive blow fly species that feeds on carrion. However, unlike native species that strictly feed on carrion (sarcosaprophytic feeding strategy), this invasive blow fly engages in omnivory in its later larval stages, feeding on both the carrion resource and conspecifics and larvae of other blow fly species (Baumgartner, 1993; Wells and Greenberg,

1992a, 1992b), resulting in reduced blow fly species richness (Rosati and VanLaerhoven, unpublished data). One could hypothesize that if *C. rufifacies* arrives and colonizes carrion earlier than other sarcosaprophytic flies, the older instars of *C. rufifacies* would be able to use the younger instars of other species as another food source in addition to the carrion. However, if *C. rufifacies* arrives later than other sarcosaprophytic flies, then it must compete with larger instars of other species for carrion resources until its larvae have developed to a sufficient size to be able to prey on other species. By this point, the oldest larvae from other species may have already migrated from the carrion resource to pupate. Early migration of other species of blow flies has been observed in the presence of *C. rufifacies*, suggesting a predator avoidance mechanism by native blow fly species (Rosati and VanLaerhoven, 2007; Tillyard and Seddon, 1933; Watson and Carlton, 2005). It is possible that *C. rufifacies* larvae that prey on other species as well as the carrion resource may have a higher fitness due to some benefits derived from diet mixing, as has been observed in generalists (Bernays et al., 1994), compared with *C. rufifacies* that develop without other species to prey upon. This may be a mechanism that drives arrival patterns of *C. rufifacies* on carrion.

Priority Effects

Some species already present in a community may facilitate the establishment of a new arrival; thus, positive interactions can result in adaptive responses that are the mechanism for temporal patterns of species abundance. As an example, the facilitating species may change the habitat in ways that promote the successful colonization by other species, or the facilitating species may be a resource that is used by later colonists. Although the mechanism is obvious, the presence of prey facilitates the arrival of predators. This temporal pattern is one, although perhaps trival, example of assembly rules.

Priority effects occur when a species that is already present in a community either inhibits or facilitates other species that subsequently arrive in the community. Priority effects do not need to be tightly linked to temporal abundance differences among species to affect community structure. In situations where resources for community formation become available at unpredictable times and resources are patchy, priority effects can have strong impacts on community structure. The lottery model (Chesson and Warner, 1981) states that when two species are competing, environmental fluctuation prevents extinction as first one, then the other, is favored by the environmental fluctuation. In these models, species that happened to have more offspring available when a resource opens up have an increased probability of utilizing that resource. To the extent that resource availability is unpredictable, winning a resource is like winning a lottery. *Drosophila phalerate* and *Drosophila subobscura* reproduce in decaying mushrooms (Shorrocks and Bingley, 1994). When one species arrives a few days after the first, the later-arriving species has a lower survival rate, smaller adult size, longer development time, and lower fitness. When both species arrive together, *D. phalerate* outcompetes *D. subobscura*. *D. subobscura* manages to persist as a fugitive species under conditions in which it discovers and exploits new mushrooms before they are found by *D. phalerate*.

Priority effects have been demonstrated for insects on carrion. The competitive advantage in arriving first has been recognized several times (Hanski, 1987; Hanski and Kuusela, 1977; Kneidel, 1983). Although facilitation of subsequent species has been suggested (Schoenly and Reid, 1987) for carrion species, it has never been experimentally demonstrated except in the most basic sense (i.e., prey must be present before predators or parasitoids can assemble in the community). In order to demonstrate that facilitation is a mechanism by which some species assemble in the carrion community, one must show that a species either cannot colonize without the presence of the facilitating species, or has a lower survivorship or fitness without the presence of the facilitating species. Although it has been assumed in forensic entomology literature that consumption of the carrion resource by blow flies facilitates the colonization of dermestid beetles (Schoenly and Reid, 1987), this is contrary to the observation that dermestid beetles such as *D. maculatus* prefer moist muscle tissue to dried skin (Braack, 1987). Upon further examination, it may prove to be true that blow flies competitively exclude *D. maculatus* until later in community assembly, or that *D. maculatus* is a poor disperser and as such is unable to reach the carrion resource until later in community assembly. In contrast, there is preliminary evidence that *Phormia regina* (Meigen) (Diptera: Calliphoridae) is facilitated by the presence of other blow fly species. In the absence of other blow fly species, *P. regina* is able to oviposit on a fresh carrion resource immediately, but has a higher offspring mortality rate, earlier prepupal migration, and smaller adult size than when it is allowed to oviposit concurrently or after *L. sericata* (Rosati and VanLaerhoven, unpublished data). As smaller adults have a reduced fitness (MacKerras, 1933; Prinkkila and Hanski, 1995), this likely explains the observation that *P. regina* may sometimes oviposit after other blow fly species have colonized the carrion resource (Anderson and VanLaerhoven, 1996; Denno and Cothran, 1976; Goddard and Lago, 1985; Hall and Doisy, 1993; Shean et al., 1993; Watson and Carlton, 2003).

Filtering Mechanisms between the Regional Species Pool and the Local Community

Spatial Scale and Dispersal

The process of community assembly depends on the population sizes of species in the regional species pool, their ability to disperse into the community, the area of the community, and its distance from the species pool. Thus, one cannot consider community assembly without considering the importance of spatial scale. Heterogeneity in spatial distribution of species can influence the composition and dynamics of communities. Carrion resources represent a fragmented landscape of habitat islands for decomposers. This spatial complexity influences the structure of insect communities on carrion. Spatial dynamics that influence saprophytic insects, such as size of resource and distance between resource islands, can indirectly imply spatial effects for species at higher trophic levels based on an island biogeographic theory for food chains developed by Holt (1993).

Species are seldom spatially distributed in homogeneous or random patterns. Clumping, or spatial heterogeneity in abundance can influence persistence of interspecific interactions and resulting community patterns in patchy or subdivided habitats. Partial heterogeneity in the influx of individuals into local communities can set the stage for increases or decreases in density-dependent interactions. The relative importance of density-dependent interactions at a particular site may therefore be determined by processes at other locations that influence the supply of colonizing organisms that reach the site. Spatial heterogeneity in web structure may exist in the absence of environmental heterogeneity, due to interactions resulting from chance colonization events. Carrion resources exhibit both spatial and temporal variability and are an unpredictable resource for the species that colonize it. Spatial variation can occur at very different scales, and spatial scale is important when considering species interactions. Community assembly on carrion exhibits variation in the presence of particular species. Although many species of decomposers seem to be ubiquitous across large spatial scales, even common individual species may not be present in all resource patches within a habitat type (Beaver, 1977; Hwang and Turner, 2005; Kneidel, 1983; Kuusela and Hanski, 1982). Beaver (1977) found considerable differences in species composition between individual snails; most species occurred in less than one-third of the snails, and no species occurred in all of them. In contrast, there was relative stability at the regional level, with the total numbers of species and individuals being similar in each collection. According to the fugitive principle (Beaver, 1977), species that are poor competitors survive within the regional species pool because of superior dispersal ability. This allows them to find and colonize resource patches that are missed by the better competitors. Once they colonize the patch, they inhibit the colonization of other competitors through a priority effect (Hanski and Kuusela, 1977). Depending on the species involved, this could lead to alternative community assembly compared with the assembly that occurs when the better competitors colonize the resource first.

Temporal and spatial variation in species abundance can increase or decrease the impact of species interactions on community composition. A tendency for individuals of a species to aggregate within spatially isolated habitats can promote coexistence of competitors within a mosaic of habitat patches (Atkinson and Shorrocks, 1981; Ives and May, 1985). These models predict that two or more competing species can coexist if they have independently aggregated distributions across resource patches. The role of resource patchiness on the coexistence of two species of carrion-breeding flies was investigated by Kneidel (1985), who demonstrated that increased resource patchiness reduced interspecific competition. Using multiple carrion-breeding flies, Hanski (1987) suggested that both resource patchiness and seasonal changes in population sizes are important in carrion fly community dynamics.

Distribution of organisms among subdivided habitats can stabilize interactions that are unstable in undivided habitats (Caswell, 1978; Holyoak and Lawler, 1996a, 1996b; Huffaker, 1958). Persistence of some predator–prey interactions depends on a complex spatial framework of patchy habitats that create a shifting patchwork of temporary prey refugia. Without these refugia, the interaction is unstable and the predator drives the prey population to extinction. It is not clear whether natural patchy systems exhibit the types of behavior predicted by the models of Huffaker (1958), Caswell (1978), or Holyoak and Lawler (1996a, 1996b); however, the patchy resource distribution of carrion would provide the means to test this hypothesis.

At larger scales, spatial variation in recruitment influences the intensity of postrecruitment density-dependent interactions. Size and isolation of patchy resource islands influence the number of species that coexist. Departure of predators from sites with few prey mean low prey recruitment can create density-dependent refugia from predators, since predators tend to overlook such sites while foraging elsewhere. Density of interacting organisms is set by processes that are extrinsic to the site where the interactions ultimately take place. Thus, such communities must be studied at spatial or temporal scales that are sufficient to reveal variation in recruitment to fully understand why phenomena like keystone predation or priority effects vary in importance among locations. Although alpha diversity (Whittaker, 1975), or the number of species within a community, is often measured in experiments describing community structure, it is beta diversity, or the variation in species composition among comparable habitats, that provides some insight into the potential variation in community assembly for a given region, habitat, and circumstance. Beta diversity may depend on when sites or resources first become available for colonization through mechanisms of temporal species availability, the order in which colonists arrive and successfully establish due to priority effects and dispersal, and chance events that impact the availability and success of potential colonists. Within the forensic entomology literature, numerous studies have measured alpha diversity and compared it between habitat types and between biogeoclimatic regions (for example, Early and Goff, 1986; VanLaerhoven and Anderson, 1999). However, insect decomposers such as blow flies can have a large dispersal range. Marked *Chrysomya* sp. have been recaptured in baited traps up to 64 km from their release point (Braack, 1981). A single biogeoclimatic zone, as defined by the climatic conditions, generalized soil type, and dominant vegetation, often spans hundreds to thousands of hectares, depending on the topography. The vast majority of published insect community assembly studies on carrion for a particular biogeoclimatic region have been conducted within relatively short spatial scales (<1 km) (for example, Anderson and VanLaerhoven, 1996; Easton and Smith, 1970; Gruner et al., 2007; Joy et al., 2006; Martinez et al., 2007; Payne, 1965; Reed, 1958; Rodriguez and Bass, 1983; Tabor et al., 2005), which is well within the dispersal range of a single local population of many of the insect decomposers. As such, they have failed to measure beta diversity, and so do not represent the potential range of community dynamics present within a single biogeoclimatic zone or climatic region. Instead, studies need to be conducted where the unit for replication is based on a spatial scale larger than that of the dispersal range of a single population of the decomposer species with the greatest dispersal ability for the community of interest.

Habitat Selection

Habitat selection can influence patterns of community assembly. Habitat selection provides one possible explanation for the conspicuous absence of highly mobile, readily dispersing species from an apparently suitable community. However, chance events can produce similar results and must be ruled out as an alternative hypothesis before assuming habitat selection is at work. Habitat selection can function as a filter between a developing community and the regional species pool by sorting among species that can actively avoid, or choose to colonize, a particular place.

Habitat selection may be a mechanism for some species to avoid physiological stress, by choosing a habitat that is more optimal in terms of life history traits such as development, growth, and reproduction. Some species selectively use habitats in ways that minimize strong negative interactions such as predation and competition, or maximize strong positive interactions such as prey availability or the availability of other resources. Schoener (1974) described several ways that species can separate their environment to reduce niche overlap, including macrohabitat variation and microhabitat variation. There is often a trade-off between availability of resources and risk of predation. Habitat selection models predict that this trade-off depends on the benefits of foraging in a particular place, discounted by the risk of mortality in that location.

Evidence for habitat selection can be found in natural history observations that associate the presence of particular species with biotic or abiotic features of a habitat. Correlative studies that link the abundance of species to particular habitat features can be highly qualitative, often using multivariate statistics of community patterns. These studies assume that any associations between animals and particular habitats are a consequence of habitat selection, since mobile species can move freely among habitats. However, these studies do not provide insight into causal mechanisms of habitat selection. Assumptions have been made that certain species on carrion are found preferentially in urban habitats, that other species are preferentially found in shaded habitats, and that still other species are preferentially found in open sunny habitats. *Lucilia ceasar* is said to prefer shaded habitats (Cragg and Ramage, 1945; Hanski, 1976; Lane, 1975; MacLeod and Donnelly, 1957; Nuorteva, 1963), but it has also been said to prefer intermediate habitats (MacLeod and Donnelly, 1957). *Calliphora vicina* is said to prefer shaded habitats (Hanski, 1976; Hanski and Kuusela, 1980; Nuorteva, 1963), but others have stated that it prefers sunny habitats (MacLeod and Donnelly, 1957) or both types of habitats (Lane, 1975). These assumptions are based on natural history observations and correlative studies, but could be due to various mechanisms, such as chance events, physiological constraints, dispersal, or interactions between species. If, for example, the mechanism that drives habitat selection is the optimal temperature for development, then habitat selection by a particular species would vary between seasons, latitudes, and biogeoclimatic regions. For example, in the fall, *Lucilia cluvia* and *Cochliomyia macellaria* colonized pig carrion in both sun and shade, whereas in the winter, they only colonized pig carrion in the shade in Argentina (Centeno et al., 2002). It may also be the case that flaws in experimental design, such as trap placement, sampling size, or sampling effort, result in invalid assumptions, and that some species do not exhibit preferences. This becomes extremely perilous when these assumptions are used to state that a body has been moved after death from one location to another. Instead, a more direct approach is required to draw a conclusion that a particular species preferentially selects a particular habitat type. This approach involves experimental manipulation of factors thought to influence habitat choice and subsequent observation of whether particular species respond to habitat alteration.

Mechanisms That Define the Local Community

Food Web Interactions

Food web patterns may be explained by the dynamics of community assembly (Diamond, 1975). Food webs are often depicted as static representations of communities (Polis et al., 1996); however, the ephemeral nature of carrion resources should result in dynamic trophic relationships. Temporal fluctuations in food web complexity have been observed during succession of aquatic communities in artificial tree holes (Pimm and Kitching, 1987) and occur during community assembly on carrion (Braack, 1987). Species colonization is a qualitative process, determined by a species' presence or absence within a community. Once present, its local dynamics are quantitatively influenced (Polis et al., 1996) through interactions with resources, or "bottom-up regulation" (Hairston et al., 1960); interactions within the same trophic level, such as competition; and interactions with higher trophic levels, or "top-down regulation" by natural enemies (Hairston et al., 1960). The degree to which a particular species is impacted by interactions within the food web will depend on the trophic position of the species, how many possible interactions are present (i.e., connectance) with other species, and the strength of each interaction. Trophic linkages provide the potential for species interactions resulting in exclusion of particular species, such as through competition (Polis et al., 1996) or intraguild predation (Polis and Holt, 1992). Together with chance events in colonization, this may result in different pathways for community assembly, leading to noninvasible, alternative community status. In addition, decomposers avoid concurrent predator-prey coevolutionary arms races with carrion, as dead animals do not develop defenses against decomposers. Instead, natural selection may act through competition between decomposers. The carrion resource is always recycled, and no one species or taxon completely dominates its use (DeVault et al., 2003).

Top-Down Regulation: Predation, Parasitism, and Omnivory

Simple food webs have been constructed for insect communities on carrion (Braack, 1987), but the degree of interaction of many species in these communities has yet to be determined. As an example, the older larvae of some blow fly and house fly species are known to be both predatory and saprophytic (Goodbrod and Goff, 1990). The degree of omnivory exhibited by these larvae, and under what conditions they chose to prey on other members of the community compared with feeding on the carrion resources, is unknown. Omnivory can be defined as feeding at more than one trophic level (Pimm and Lawton, 1978), including intraguild predation, which can have various effects on food web dynamics (Polis et al., 1989). The unifying benefit of this feeding strategy is that diet breadth allows for a degree of plasticity when feeding, allowing the omnivore to change its diet when its preferred food is reduced to low levels. This strategy should allow persistence of omnivore populations when specialist feeders may be excluded (Coll and Guershon, 2002; Polis and Holt, 1992). Unlike native North American species, which strictly feed on carrion, the invasive blow fly species *C. rufifacies* engages in omnivory in its later larval stages, feeding on both the carrion resource and conspecifics and larvae of other blow fly species (Baumgartner, 1993; Wells and Greenberg, 1992a, 1992b). Research has demonstrated that the presence of *C. rufifacies* results in reduced diversity of species on carrion resources and reduction in the relative abundance of other blow fly species (Baumgartner, 1993; Wells and Greenberg, 1992a, 1992b) as well as other omnivores and predators such as rove beetles and carrion beetles (Watson and Carlton, 2005). In a comparison of pig carrion where *C. rufifacies* was either present or absent, the number of blow fly species (i.e., species richness) was reduced when *C. rufifacies* was present (Rosati and VanLaerhoven, unpublished data). Other possible trait-mediated effects have been observed with the presence of *C. rufifacies*, as other blow flies may exhibit changes in larval feeding and larval migration behavior (Watson and Carlton, 2005), which may make them more susceptible to predators or parasitoids or reduce their fitness.

Other predators can have important interactions with the carrion food web, such as fire ants, which are capable of completely removing the saprophytic insects on carrion (Wells and Greenberg, 1994b). As these examples have demonstrated, top-down regulation can change the dynamics of community assembly through various mechanisms. It is vital to fully explore these mechanisms so that once the likelihood of the presence of upper trophic guild members within a local community can be predicted, the impact of these members on assembly can be taken into account.

Within Trophic Level Regulation: Competition and Other Interactions

One way to define competition is a mutually negative interaction between two or more organisms within the same guild or trophic level. Intraspecific competition occurs between individuals of the same species, whereas interspecific competition occurs between two or more different species. These negative interactions usually are manifested by reduced abundance, decreased fitness, or some component of fitness, such as body size, growth rate, fecundity, or survivorship. It is assumed that decreased fitness eventually translates to reduced abundance of that species. There are various mechanisms proposed for competitive interactions. Interference competition is a direct interaction, such as territorial interactions or chemical interference. Contest competition occurs when resource utilization is uneven between species such that some species obtain all the resources they require while others do not. Scramble competition occurs when resource utilization is even between species, but there are not enough resources for all. In carrion systems, all of these mechanisms are present. Blow flies interact with bacteria by secreting antibiotics (Pavillard and Wright, 1957), which inhibits bacteria from fully utilizing the carrion resource in an example of interference competition. Within the guild of carrion-feeding insects, scramble competition is the mechanism that likely best describes the overall interaction, as carrion resources are finite and do not support successful development of all the immatures that attempt to utilize it (Denno and Cothran, 1976). However, within any particular pair of carrion-feeding species, contest competition is the mechanism that likely describes the interaction, as each species differs in its ability to utilize the resource. Between flesh flies and blow flies, the first flesh flies to develop obtain the carrion resources they require to reach adult, whereas resource use by blow flies usually prevents full development of late deposited flesh fly larvae (Denno and Cothran, 1976).

Although not yet studied in carrion systems, other types of interactions are also likely of importance in determining the local community assembly. Mutualism is an interaction where both species benefit. Commensalism is an interaction where one species benefits, but there is a neutral impact on the other species. Amensalism is an interaction where one species is negatively impacted, with a neutral impact on the other species. As discussed previously, the facilitation of *P. regina* by another blow fly species may be an example of commensalism, as *P. regina* appears to have a higher fitness when it colonizes at the same time or after another blow fly species, and *L. sericata* does not seem to be impacted by the colonization of *P. regina* (Rosati and VanLaerhoven, unpublished data). It has been assumed that *P. regina* colonizes after other blow fly species based on its delayed colonization in some succession studies (Anderson and VanLaerhoven, 1996; Hall and Doisy, 1993; Shean et al., 1993; Watson and Carlton, 2003). However, it is clearly able to colonize within the first 24 h, as has been demonstrated in both laboratory and field experiments (Rosati and VanLaerhoven, unpublished data).

Bottom-Up Regulation: Resource Type, Quality, and Size

Resources influence the nature and intensity of interactions of species within communities, altering the structure of those communities (Kagata and Ohgushi, 2006). These resource-mediated interguild interactions are fundamental processes in community ecology. The strength and outcome of top-down effects of natural enemies on lower-level consumers can be moderated by intraguild interactions between natural enemies (Schmitz and Suttle, 2001). For omnivores, basal resources are more than just a substrate on which to search for prey or intraguild prey; the basal resource is another food source, thus adding an additional layer of complexity to the interaction.

Using a theoretical model of a plant system, Gillespie and Roitberg (2006) demonstrated that resource nutrient quality can affect intraguild predation by mediating prey densities and omnivore growth rates, and shifting omnivore feeding rates, thereby changing encounters with, and attacks on, intraguild prey. As resource nutrient quality increases, prey and omnivores should have increased fitness returns from resource feeding. Recruitment of pure predators (intraguild prey) should increase with resource quality due to effects of resource nutrition on prey quality and quantity (Gillespie and Roitberg, 2006). In plants, nitrogen fertilization is positively correlated with recruitment and numerical increases in herbivores (Moon and Stiling, 2002). Parasitism of aphids increased on N-fertilized plants (Moon and Stiling, 2002). The presence of more nutritious prey might divert foraging effort away from potential intraguild prey; however, the interaction would be complicated by apparent competition between the omnivore and the intraguild prey, making the resulting dynamics uncertain (Gillespie and Roitberg, 2006). Recruitment of omnivores should increase with increasing resource quality because of the inherent value of the resource as a food source, but they may or may not be impacted by the indirect effects of resource quality on prey quality (Gillespie and Roitberg, 2006). If the nutritional needs can increasingly be met by resource feeding, then increased resource quality should reduce the per capita intraguild predation by the omnivore. Although not studied for the upper trophic levels on carrion, it is likely that carrion type, size, and quality influence interactions and assembly of these species.

As discussed previously, in the presence of patchy, unpredictable resources, generalists are expected to persist longer and have a more stable population structure than specialists. In contrast, when resources vary in quality, specialization allows a species to improve its efficiency in utilizing a resource beyond the performance of generalists (Kuusela and Hanski, 1982). Carrion resources vary in their attractiveness and nutrient value to those species that feed directly on the resource (Rosati and VanLaerhoven, 2007). Cholesterol is an essential growth factor in the blow fly *L. sericata*, and ergosterol included in yeast and sitosterol included in wheat germ also promote larval development, but are less effective than cholesterol (Hobson, 1935). No differences were observed in mortality during the larval and pupal stages, and in pupal weight between *L. sericata* larvae reared on a diet of yeast, whole milk, and wheat germ, and those reared on beef liver, although duration of the larval stage on the diet was longer than that on beef liver (Tachibana and Numata, 2001). Although other impacts of resource quality are also likely, this demonstrates that resource quality can impact developmental rates of carrion-feeding insects. Although this has not been studied, this has the potential to change interactions between species and community assembly.

It has been commonly stated in the literature that flies prefer different types (Hennig, 1950; Nuorteva, 1959; Schoof and Savage, 1955; Savage and Schoof, 1955; Uecker and Keilbach, 1966; Williams, 1954) and sizes (Davies, 1990; Denno and Cothran, 1975; Nuorteva, 1959; Palmer, 1980) of carrion. *Phaenicia caeruleiviridis* (Diptera: Calliphoridae) was bred from mammal, snake, toad, and frog carrion, but was absent in slug, other snail, arthropod, and salamander carrion (Kneidel, 1984b). It has been stated that *Calliphora vomitoria* (L.) (Diptera: Calliphoridae) prefers large carrion (Nuorteva, 1959), whereas *Lucilia richardsi* Collin (Diptera: Calliphoridae) and flesh flies (Diptera: Sarcophagidae) prefer small carrion (Denno and Cothran, 1975; Nuorteva, 1959). The majority of these studies conducted trapping experiments or set out carrion of different types and sizes. These types of studies confuse the variability associated with community assembly between patches with preference for particular carrion types or sizes. A study by Kuusela and Hanski (1982) noted that considerable variation in species composition between individual carcasses exists, and variation between different types or sizes was not greater than variation between identical pieces of carrion. In addition, the association between particular species and carrion type or size has been observed to vary between weeks and season for individual fly species (Cragg, 1955; Watson and Carlton, 2005; Williams, 1954). It has been demonstrated that habitat or temporal availability of species is more important than carrion type in determining species composition (Cornaby 1974; Suenaga, 1959). More experimentation under carefully controlled conditions is required to determine if particular carrion-feeding insects exhibit preferences for particular carrion types or sizes, or if these observations are the result of variability in community assembly.

Ullyett (1950) states that specific preferences for carrion type would reduce interspecific competition overall, and intraspecific competition would be more severe. It has been suggested that flesh flies prefer smaller carrion to avoid niche overlap with blow flies (Denno and Cothran, 1975, 1976; Hanski and Kuusela, 1980). Some species may have evolved to become better in utilizing smaller pieces of carrion by becoming better dispersers, and by laying only a small number of offspring at one time (Kuusela and Hanski, 1982). Other species might develop toward laying all the eggs from one batch on a single piece of carrion, which should be large enough to provide sufficient food for at least one developing batch of offspring. Differences in the egg-laying behavior of carrion flies do occur (Beaver, 1972; MacKerras, 1933), leading to differences in spatial variance (Hanski and Kuusela, 1980), but it is not clear whether these differences are due to adaptations to the size of carrion (Hanski and Kuusela, 1980; Ullyett, 1950). Thus, the mechanisms driving association of particular species with particular types or sizes of carrion need to be determined in order to quantify the likelihood that particular species will colonize particular carrion.

Research into the actual mechanisms of interactions between species during community assembly is vital to ensure that assumptions made in forensic entomology regarding community assembly are valid, not simply the result of misinterpretation of life history observations, small sample sizes, small spatial scales, or other experimental design flaws. Once we fully understand the mechanisms, we can then predict under what circumstances particular patterns are likely to occur, and quantify the likelihood of occurrence.

Future Directions in Forensic Entomology Research

Carrion systems are ideal models for testing community assembly due to the rapid community turnover (days to weeks) in comparison with more traditional model systems. Carrion systems have the added advantage of manipulation (i.e., size, habitat placement, and distance between carrion resources) and replication (available in large numbers of consistent size). Community members can be added or excluded readily by manipulating access to the resource, making carrion systems invaluable for community assembly studies. Carrion is ideal for testing food web dynamics, as well as nonequilibrium versus equilibrium dynamics as carrion systems exhibit elements of both. Thus, research in forensic entomology and carrion systems has the potential to make important contributions to ecology and advance ecological theory, as has been demonstrated previously (for example, Denno and Cothran, 1975; Hanski and Kuusela, 1977; Kneidel, 1983, 1984a).

Framing forensic entomology research on community assembly within the context of ecological theory is vital to advance our understanding of how insects assemble on carrion, as well as why and when assembly varies. Future research needs to address the mechanisms of community assembly in a hierarchical manner (Figure 16.1):

1. In order to assign a probability that a species is available (i.e., within the regional species pool) to colonize a carrion resource, research must test mechanisms that limit:
   a. A species' temporal availability
   b. Physiological constraints
   c. Resource requirements, diet breadth, and foraging preferences
   d. Interactions with other species (including competition, predation, and priority effects)
   e. Chance events through measuring variability in replication
2. Once the likelihood that a species is available to become a member of a local carrion community can be predicted, the next step is to determine the probability that a species can get to a local carrion community. Research must test mechanisms that limit:
   a. A species' dispersal and habitat selection
   b. Physiological constraints
   c. Interactions with other species
   d. Chance events
3. After the probability of a species arriving at a local carrion community can be predicted, the final step is to determine whether the species will be able to successfully become a member of the local carrion community. Research must test mechanisms that limit:
   a. A species' ability to colonize a resource
   b. Interactions with higher trophic levels (predator, parasitoids, and omnivores)
   c. Interactions within the same trophic level (competition, mutualism)
   d. Interactions with lower trophic levels (carrion resource, prey)
   e. Chance events

It is apparent that some mechanisms influence community assembly at several scales. Chance events influence who is available to become a member of a particular community, and who manages to stay within that community. The end result is that some species make it, while others do not. Extinction, colonization, and migration of species impart changes in community structure over time.

The focus in forensic entomology has predominantly been to document patterns (for example, Anderson and VanLaerhoven, 1996; Catts and Goff, 1992; Grassberger and Frank, 2004; Payne, 1965; Tabor et al., 2004; Tenorio et al., 2003; Watson and Carlton, 2003). Simple descriptive food webs have been constructed (Braack, 1987). Niche partitioning over habitat (Early and Goff, 1986; Hanski, 1976; MacLeod and Donnelly, 1957), season (Denno and Cothran, 1975; Hanski and Kuusela, 1980; Johnson, 1975), arrival time on carrion (Denno and Cothran, 1975; Schoenly and Reid, 1987), and resource type or size (Kneidel, 1984b; Schoenly and Reid, 1983) have been suggested as mechanisms for coexistence among blow flies; however, much more research is required. For the most part, mechanisms shaping community assembly processes have not been identified or their relative importance quantified. The degree of interaction of many species within carrion communities has yet to be determined. Current ecological data for insect communities on carrion are not sufficient to permit explanation of underlying community processes (Schoenly and Reid, 1987). If forensic entomologists are to make accurate and precise predictions of insect assembly on carrion, it is vital that research into the mechanisms of community assembly be conducted so that estimates of postmortem interval are based on sound ecological theory, thereby providing the means to calculate the variability associated with these estimates.

Acknowledgments

I thank my former and current graduate students, J. Bennett, A. Brommit, C. Fitzpatrick, N. Lamont, H. Murillo, J. Rosati-Deslippe, and L. Sohail, and undergraduate thesis students, S. Edwards, C. Hughes, M. Kereliuk, A. Laschuk, I. Svilans, and K. Yap, for their work on and relating to forensic science, community assembly, and trophic interactions in multiple systems. My research program would not be possible without the assistance from my undergraduate research assistants, L. Alexander, S. Almosawi, O. Billy, E. Chew, J. Davies, J. Ferenz, K. Figgens, J. Fournier, S. Gignac, D. Granados, A. Hasselhurst, L. Holmes, J. Hu, M. Kilani, A. Koss, A. Longmore, J. Manchee, M. Marshall, M. Martin, K. McGuffin, A. Michaud, S. Najafi, T. Nguyen, M. Novis, F. Omar, S. Parlee, N. Suriyakumar, A. Trifonov, L. Whelton, J. Willert, C. Willock, and H. Veres. I thank D. R. Gillespie for allowing me to use him as a sounding board for my ideas. I thank the Canadian Police Research Centre, Canadian Foundation for Innovation, Ontario Innovation Trust, Ford Motor Co., HJ Heinz Co. Canada Ltd., Molds Are Us, Romanos Specialty Italian Deli, Bart DiGiovanni Construction, Essex Regional Conservation Authority, Windsor Airport, Ojibway Nature Centre, A. Rosati, D. Molnar, D. Cotter, J. D'Alosio, H. Fackrell, M. Pollard, D. Rizzo, T. Rizzo, R. White, Ontario Police College, Office of the Chief Coroner for Ontario, Ontario Provincial Police, B. Yamashita and the Royal Canadian Mounted Police, Toronto Police Service, York Regional Police, Windsor Police Service, G. Olson and the Office of the Fire Marshal for Ontario, and the University of Windsor for supporting forensic entomology and forensic entomology research in Canada.

References

Anderson, G. S., and S. L. VanLaerhoven. 1996. Initial studies on insect succession on carrion in southwestern British Columbia. *J. Forensic Sci.* 41:617–25.

Atkinson, W. D., and B. Shorrocks. 1981. Competition on a divided and ephemeral resource: A simulation model. *J. Anim. Ecol.* 50:461–71.

Baldridge, R. S., S. G. Wallace, and R. Kirkpatrick. 2006. Investigation of nocturnal oviposition by necrophilous flies in central Texas. *J. Forensic Sci.* 51:125–26.

Baumgartner, D. L. 1993. Review of *Chrysomya rufifacies* (Diptera: Calliphoridae). *J. Med. Entomol.* 30:338–52.

Beaver, R. A. 1972. Ecological studies on Diptera breeding in dead snails. I. Biology of the species found in *Cepaea nemoralis* (L.). *Entomology* 105:41–52.

Beaver, R. A. 1977. Non-equilibrium "island" communities: Diptera breeding in dead snails. *J. Anim. Ecol.* 46:783–98.

Bernays, E. A., K. L. Brigh, N. Gonzalez, and J. Angel. 1994. Dietary mixing in a generalist herbivore: Tests of two hypotheses. *Ecology* 75:1997–2006.

Braack, L. E. O. 1981. Visitation patterns of principal species of the insect-complex at carcasses in the Kruger National Park. *Koedoe* 24:33–49.

Braack, L. E. O. 1987. Community dynamics of carrion-attendant arthropods in tropical African woodland. *Oecologia* 72:402–9.

Caswell, H. 1978. Predator mediated coexistence: A nonequilibrium model. *Am. Nat.* 112:127–54.

Catts, E. P., and M. L. Goff. 1992. Forensic entomology in criminal investigations. *Ann. Rev. Entomol.* 37:253–72.

Centeno, N., M. Maldonado, and A. Oliva. 2002. Seasonal patterns of arthropods occurring on sheltered and unsheltered pig carcasses in Buenos Aires Province (Argentina). *Forensic Sci. Int.* 126:63–70.

Chave, J. 2004. Neutral theory and community ecology. *Ecol. Lett.* 7:241–253.

Chesson, P. L., and R. R. Warner. 1981. Environmental variability promotes coexistence in lottery competitive systems. *Am. Nat.* 117:923–43.

Clements, F. E. 1916. *Plant succession.* Carnegie Institute of Washington Publication 242.

Coll, M., and M. Guershon. 2002. Omnivory in terrestrial arthropods: Mixing plant and prey diets. *Ann. Rev. Entomol.* 47:267–97.

Connell, J. H., and R. O. Slatyer. 1977. Mechanisms of succession in natural communities and their role in community stability and organization. *Am. Nat.* 111:1119–44.

Cornaby, B. W. 1974. Carrion reduction by animals in contrasting tropical habitats. *Biotropica* 6:51–63.

Cragg, J. B. 1955. The natural history of sheep blow flies in Britain. *Ann. Appl. Biol.* 42:197–207.

Cragg, J. B., and G. R. Ramage. 1945. Chemotropic studies on the blowflies *Lucilia sericata* (Mg.) and *Lucilia caesar* (L.). *J. Parasitol.* 36:168–75.

Davies, L. 1990. Species composition and larval habitats of blowfly (Calliphoridae) populations in upland areas in England and Wales. *Med. Vet. Entomol.* 4:61–68.

Denno, R. F., and W. R. Cothran. 1975. Niche relationships of a guild of necrophagous flies. *Ann. Entomol. Soc. Am.* 68:741–54.

Denno, R. F., and W. R. Cothran. 1976. Competitive interactions and ecological strategies of sarcophagid and calliphorid flies inhabiting rabbit carrion. *Ann. Entomol. Soc. Am.* 69:109–13.

DeVault, T. L., O. E. Rhodes, and J. A. Shivik. 2003. Scavenging by vertebrates: Behavioral, ecological, and evolutionary perspectives on an important energy transfer pathway in terrestrial ecosystems. *Oikos* 102:225–34.

Diamond, J. M. 1975. Assembly of species communities. In *Ecology and evolution of communities*, ed. Cody, M. L., and J. M. Diamond. Cambridge, MA: Harvard University Press, 342–44.

Early, M., and M. L. Goff. 1986. Arthropod succession patterns in exposed carrion on the Island of Oahu, Hawaii Island, USA. *J. Med. Entomol.* 23:520–31.

Easton, A. M., and K. G. V. Smith. 1970. The entomology of the cadaver. *Med. Sci. Law* 10:208–15.

Elton, C. 1927. *Animal ecology.* London: Sidgwick and Jackson.

Gause, G. F. 1934. *The struggle for existence.* Baltimore: Williams and Wilkins.

Gillespie, D. R., and B. D. Roitberg. 2006. Inter-guild influences on intra-guild predation in plant-feeding omnivores. In *Trophic and guild interactions in biological control*, ed. Brodeur, J., and G. Boivin. Dordrecht, The Netherlands: Springer, 71–100.

Gleason, H. A. 1917. The structure and development of the plant association. *Bull. Torrey Bot. Club* 44:463–81.

Goddard, J., and P. K. Lago. 1985. Notes on blowfly (Diptera: Calliphoridae) succession on carrion in Northern Mississippi. *J. Entomol. Sci.* 20:312–17.

Goff, M. L. 1992. Problems in estimation of postmortem interval resulting from wrapping of the corpse: A case study from Hawaii. *J. Agric. Entomol.* 9:237–43.

Goodbrod, J. R., and M. L. Goff. 1990. Effects of larval population density on rates of development and interactions between two species of *Chrysoma* (Diptera: Calliphoridae) in laboratory culture. *J. Med. Entomol.* 27:338–43.

Grace, J. B., and R. G. Wetzel. 1981. Habitat partitioning and competitive displacement in cattails (*Typha*): Experimental field studies of the intensity of competition. *Am. Nat.* 118:463–74.

Grassberger, M., and C. Frank. 2004. Initial study of arthropod succession on pig carrion in a central European urban habitat. *J. Med. Entomol.* 41:511–23.

Greenberg, B. 1990. Nocturnal oviposition behaviour of blow flies (Diptera: Calliphoridae). *J. Med. Entomol.* 27:807–10.

Grinnell, J. 1917. Field tests of theories concerning distributional control. *Am. Nat.* 51:115–28.

Gruner, S. V., D. H. Slone, and J. L. Capinera. 2007. Forensically important Calliphoridae (Diptera) associated with pig carrion in rural north-central Florida. *J. Med. Entomol.* 44:509–15.

Hairston, N. G., F. E. Smith, and L. B. Slobodkin. 1960. Community structure, population control, and competition. *Am. Nat.* 94:421–25.

Hall, R. D., and K. E. Doisy. 1993. Length of time after death: Effect on attraction and oviposition or larviposition of midsummer blow flies (Diptera: Calliphoridae) and flesh flies (Diptera: Sarcophagidae) of medicolegal importance in Missouri. *Ann. Entomol. Soc. Am.* 86:589–93.

Hanski, I. 1976. Breeding experiments with carrion flies (Diptera) in natural conditions. *Ann. Entomol. Fenn.* 42:113–21.

Hanski, I. 1987. Carrion fly community dynamics: Patchiness, seasonality and coexistence. *Ecol. Entomol.* 12:257–66.

Hanski, I., and S. Kuusela. 1977. An experiment on competition and diversity in the carrion fly community. *Ann. Entomol. Fenn.* 43:108–15.

Hanski, I., and S. Kuusela. 1980. The structure of carrion fly communities: Differences in breeding seasons. *Ann. Zool. Fenn.* 17:185–90.

Hennig, W. 1950. Entomologishe Beobachtungen an kleinen Wirbeltierleichen. *Z. Hyg. Zool.* 38:33–88.

Hewadikaram, K. A., and M. L. Goff. 1991. Effect of carcass size on rate of decomposition and arthropod succession patterns. *Am. J. Forensic Med. Pathol.* 12:235–240.

Hobson, R. P. 1935. On a fat-soluble growth factor required by blow-fly larvae. II. Identity of the growth factor with cholesterol. *Biochem. J.* 29:2023–26.

Holt, R. D. 1993. Ecology at the mesoscale: The influence of regional processes on local communities. In *Community diversity in ecological communities, historical and biogeographical perspectives*, ed. Ricklefs, R. E., and D. Schluter. Chicago: University of Chicago Press, 77–88.

Holt, R. D., J. Grover, and D. Tilman. 1994. Simple rules for interspecific dominance in systems with exploitative and apparent competition. *Am. Nat.* 144:741–71.

Holyoak, M., and S. P. Lawler. 1996a. Persistence of an extinction-prone predator-prey interaction through metapopulation dynamics. *Ecology* 77:1867–79.

Holyoak, M., and S. P. Lawler. 1996b. The role of dispersal in predator-prey metapopulation dynamics. *J. Anim. Ecol.* 65:640–52.

Hubbell, S. P. 2001. *The unified neutral theory of biodiversity and biogeography.* Princeton, NJ: Princeton University Press.

Hubbell, S. P. 2005a. Neutral theory in community ecology and the hypothesis of functional equivalence. *Functional Ecol.* 19:166–72.

Hubbell, S. P. 2005b. The neutral theory of biodiversity and biogeography and Stephen Jay Gould. *Paleobiology* 31:122–32.

Hubbell, S. P. 2006. Neutral theory and the evolution of ecological equivalence. *Ecology* 87:1387–98.

Huffaker, C. B. 1958. Experimental studies on predation: Dispersion factors and predator-prey oscillations. *Hilgardia* 27:343–83.

Hurtt, G. C., and S. W. Pacala. 1995. The consequences of recruitment limitation: Reconciling chance, history and competitive differences between plants. *J. Theor. Biol.* 176:1–12.

Hutchinson, G. E. 1957. Concluding remarks. *Cold Spring Harbor Symp. Quantitative Biol.* 22:415–27.

Hwang, C., and B. D. Turner. 2005. Spatial and temporal variability of necrophagous Diptera from urban to rural areas. *Med. Vet. Entomol.* 19:379–91.

Ives, A. R., and R. M. May. 1985. Competition within and between species in a patchy environment: Relations between microscopic and macroscopic models. *J. Theor. Biol.* 115:65–92.

James, F. C., R. F. Johnston, N. O. Wamer, G. J. Niemi, and W. J. Boecklen. 1984. The Grinnellian niche of the wood thrush. *Am. Nat.* 124:17–30.

Jenkins, D. G. 2006. In search of quorum effects in metacommunity structure: Species co-occurrence analyses. *Ecology* 87:1523–31.

Johnson, M. D. 1975. Seasonal and microseral variations in the insect populations on carrion. *Am. Midland Nat.* 93:79–90.

Joy, J. E., N. L. Liette, and H. L. Harrah. 2006. Carrion fly (Diptera: Calliphoridae) larval colonization of sunlight and shaded pig carcasses in West Virginia, USA. *Forensic Sci. Int.* 164:183–92.

Kagata, H., and T. Ohgushi. 2006. Bottom-up trophic cascades and material transfer in terrestrial food webs. *Ecol. Res.* 21:26–34.

Keddy, P. 1992. Assembly and response rules: Two goals for predictive community ecology. *J. Veg. Sci.* 3:157–64.

Keddy, P., and E. Weiher. 2004. Introduction: The scope and goals of research on assembly rules. In *Ecological assembly rules: Perspectives, advances, retreats*, ed. Keddy, P., and E. Weiher. Cambridge, UK: Cambridge University Press, 1–22.

Kneidel, K. A. 1983. Fugitive species and priority during colonization in carrion-breeding Diptera communities. *Ecol. Entomol.* 8:163–69.

Kneidel, K. A. 1984a. Competition and disturbance in communities of carrion-breeding Diptera. *J. Anim. Ecol.* 53:849–65.

Kneidel, K. A. 1984b. The influence of carcass taxon and size on species composition of carrion-breeding Diptera. *Am. Midland Nat.* 111:57–63.

Kneidel, K. A. 1985. Patchiness, aggregation, and the coexistence of competitors for ephemeral resources. *Ecol. Entomol.* 10:441–48.

Komar, D., and O. Beattie. 1998. Postmortem insect activity may mimic perimortem sexual assault clothing patterns. *J. Forensic Sci.* 43:792–96.

Kuusela, S., and I. Hanski. 1982. The structure of carrion fly communities: The size and type of carrion. *Holarctic Ecol.* 5:337–48.

Lane, R. P. 1975. An investigation into blowfly (Diptera: Calliphoridae) succession on corpses. *J. Nat. Hist.* 9:581–98.

Leibold, M. A., and M. A. McPeek. 2006. Coexistence of the niche and neutral perspectives in community ecology. *Ecology* 87:1399–410.

Loreau, M. 2004. Does functional redundancy exist? *Oikos* 104:606–11.

Lotka, A. J. 1925. *Elements of physical biology*. Baltimore: Williams and Wilkins.

MacKerras, M. J. 1933. Observation on the life histories, nutritional requirements and fecundity of blowflies. *Bull. Entomol. Res.* 24:353–61.

MacLeod, J., and J. Donnelly. 1957. Some ecological relationships of natural populations of calliphorine blowflies. *J. Anim. Ecol.* 26:135–70.

MacMahon, A., D. J. Schimpf, D. C. Anderson, K. G. Smith, and R. L. Bayn, Jr. 1981. An organism-centered approach to some community and ecosystem concepts. *J. Theor. Biol.* 88:287–307.

Martinez, E., P. Duque, and M. Wolff. 2007. Succession pattern of carrion-feeding insects in Paramo, Columbia. *Forensic Sci. Int.* 166:182–89.

McGill, B. J., B. A. Maurer, and M. D. Weiser. 2006. Empirical evaluation of neutral theory. *Ecology* 87:1411–23.

Miller, R. S. 1967. Pattern and process in competition. *Adv. Ecol. Res.* 4:1–74.

Moon, D. C., and P. Stiling. 2002. The effects of salinity and nutrients on a tritrophic salt-marsh system. *Ecology* 83:2465–76.

Morin, P. J. 1999. *Community ecology*. Malden, MA: Blackwell Publishing.

Numata, H., and S. Shiga. 1995. Induction of adult diapause by photo-period and temperature in *Protophormia terraenovae* (Diptera: Calliphoridae) in Japan. *Environ. Entomol.* 24:1633–39.

Nuorteva, P. 1959. Studies on the significance of flies in the transmission of poliomyelitis. III. The composition of the blow fly fauna and the activity of the flies in relation to the weather during the epidemic season of poliomyelitis in South Finland. *Ann. Entomol. Fenn.* 25:135–36.

Nuorteva, P. 1963. The flying activity of blow flies (Dipt., Calliphoridae) in Finland. *Ann. Entomol. Fenn.* 29:1–49.

Odum, E. P. 1969. The strategy of ecosystem development. *Science* 164:262–70.

Palmer, D. H. 1980. Partitioning of the carrion resource by sympatric Calliphoridae (Diptera) near Melbourne. Ph.D. thesis, La Trobe University, Melbourne, Australia.

Pavillard, E. R., and E. A. Wright. 1957. An antibiotic from maggots. *Nature* 180:916–17.

Payne, J. A. 1965. A summer carrion study of the baby pig *Sus scrofa* Linnaeus. *Ecology* 46:592–602.

Pimm, S. L., and J. H. Lawton. 1978. On feeding on more than one trophic level. *Nature* 275:542–44.

Pimm, S. L., and R. L. Kitching. 1987. The determinants of food chain lengths. *Oikos* 50:302–7.

Polis, G. A., and R. D. Holt. 1992. Intraguild predation: The dynamics of complex trophic interactions. *TREE* 7:151–54.

Polis, G. A., R. D. Holt, B. A. Menge, and K. O. Winemiller. 1996. Time, space, and life history: Influences on food webs. In *Food webs: Contemporary perspectives*, ed. Polis, G. A., and K. O. Winemiller. London: Chapman and Hall, 435–60.

Polis, G. A., C. A. Myers, and R. D. Holt. 1989. The ecology and evolution of intraguild predation. *Ann. Rev. Ecol. Syst.* 20:297–330.

Prinkkila, M. L., and I. Hanski. 1995. Complex competitive interactions in 4 species of *Lucilia* blow flies. *Ecol. Entomol.* 20:261–72.

Reed, H. B., Jr. 1958. A study of dog carcass communities in Tennessee, with special reference to the insects. *Am. Midl. Nat.* 59:213–45.

Rodriguez, W. C., and W. M. Bass. 1983. Insect activity and its relationship to decay rates of human cadavers in east Tennessee. *J. Forensic Sci.* 28:423–32.

Rosati, J. Y. and S. L. VanLaerhoven. 2006. Seasonal blow fly species diversity and potential effects of an invasive species on diversity and food web dynamics in carrion systems. Canadian Society for Ecology and Evolution, Montreal, Canada. (unpublished data)

Rosati, J., and S. L. VanLaerhoven. 2007. New record of *Chrysomya rufifacies* (Macquart) in Canada: Current distribution and the possibility of future range expansion. *Can. Entomol.* 139:670–77.

Savage, E. P., and H. F. Schoof. 1955. The species composition of fly populations at several types of problem sites in urban areas. *Ann. Entomol. Soc. Am.* 48:251–57.

Schmitz, O. J., and K. B. Suttle. 2001. Effects of top predator species on direct and indirect interactions in a food web. *Ecology* 82:2071–81.

Schoener, T. W. 1974. Resource partitioning in ecological communities. *Science* 185:27–39.

Schoener, T. W. 1986. Overview: Kinds of ecological communities—Ecology becomes pluralistic. In *Community ecology*, ed. Diamond, J., and T. J. Case. New York: Harper and Row, 467–79.

Schoenly, K., and W. Reid. 1983. Community structure of carrion arthropods in the Chihuahuan Desert. *J. Arid Environ.* 6:253–63.

Schoenly, K., and W. Reid. 1987. Dynamics of heterotrophic succession in carrion arthropod assemblages: Discrete seres or a continuum of change? *Oecologia* 73:192–202.

Schoof, H. F., and E. P. Savage. 1955. Comparative studies of urban fly populations in Arizona, Kansas, Michigan, New York and West Virginia. *Ann. Entomol. Soc. Am.* 48:1–12.

Shean, B. S., L. Messinger, and M. Papworth. 1993. Observations of differential decomposition on sun exposed vs. shaded pig carrion in coastal Washington State. *J. Forensic Sci.* 38:938–49.

Shorrocks, B., and M. Bingley. 1994. Priority effects and species coexistence: Experiments with fungal-breeding *Drosophila*. *J. Anim. Ecol.* 63:799–806.

Singh, D., and M. Bharti. 2001. Further observations on the nocturnal oviposition behaviour of blow flies (Diptera: Calliphoridae). *Forensic Sci. Int.* 120:124–26.

Suenaga, O. 1959. Ecological studies of flies. V. On the amount of flies breeding out from several kinds of small dead animals. *Endem. Dis. Bull. Nagasaki Univ.* 1:407–23.

Tabor, K. L., C. C. Brewster, and R. D. Fell. 2004. Analysis of the successional patterns of insects on carrion in southwest Virginia. *J. Med. Entomol.* 41:785–95.

Tabor, K. L., R. D. Fell, and C. C. Brewster. 2005. Insect fauna visiting carrion in Southwest Virginia. *Forensic Sci. Int.* 150:73–80.

Tachibana, S., and H. Numata. 2001. An artificial diet for blow fly larvae, *Lucilia sericata* (Meigen) (Diptera: Calliphoridae). *Appl. Entomol. Zool.* 36:521–23.

Tachibana, S., and H. Numata. 2004. Effects of temperature and photoperiod on the termination of larval diapause in Lucilia sericata (Diptera: Calliphoridae). *Zool. Sci.* 21:197–202.

Tenorio, F. M., J. K. Olson, and C. J. Coates. 2003. Decomposition studies, with a catalog and description of forensically important blow flies (Diptera: Calliphoridae) in central Texas. *Southwest. Entomol.* 28:37–45.

Tessmer, J. W., C. L. Meek, and V. W. Wright. 1995. Circadian patterns of oviposition by necrophilous flies (Diptera: Calliphoridae). *Southwestern Entomol.* 24:439–45.

Tillyard, R. J., and H. R. Seddon. 1933. *The sheep blowfly problem in Australia.* Report 1, Australian Council of Scientific and Industrial Research Pamphlet 37. New South Wales Department of Agriculture Science Bulletin 40. Melbourne.

Uecker, C. von, and R. Keilbach. 1966. Die Ernahrungsweise synantroper Fliegen in ihrem natiirlichen Milieu. *Angew. Parasitol.* 7:259–70.

Ullyett, G. C. 1950. Competition for food and allied phenomena in sheep-blowfly populations. *Philos. Trans. R. Soc. London B* 234:77–174.

VanLaerhoven, S. L., and G. S. Anderson. 1999. Insect succession on buried carrion in two biogeoclimatic zones of British Columbia. *J. Forensic Sci.* 44:31–41.

VanValen, L. 1965. Morphological variation and width of ecological niche. *Am. Nat.* 99:377–90.

Volterra, V. 1926. Variations and fluctuations in the numbers of individuals in animal species living together. In *Animal ecology*, ed. Chapman, R. N. New York: McGraw-Hill, 409–48. Reprinted in 1931.

Watson, E. J., and C. E. Carlton. 2003. Spring succession of necrophilous insects on wildlife carcasses in Louisiana. *J. Med. Entomol.* 4:338–47.

Watson, E. J., and C. E. Carlton. 2005. Insect succession and decomposition of wildlife carcasses during fall and winter in Louisiana. *J. Med. Entomol.* 42:193–203.

Wells, J. D., and B. Greenberg. 1992a. Laboratory interaction between introduced *Chrysomya rufifacies* and native *Cochliomyia macellaria* (Diptera: Calliphoridae). *Environ. Entomol.* 21:640–45.

Wells, J. D., and B. Greenberg. 1992b. Rates of predation by *Chrysomya rufifacies* (Macquart) on *Cochliomyia macellaria* (Fabr) (Diptera: Calliphoridae) in the laboratory—Effect of predator and prey development. *Pan Pacific Entomol.* 68:12–14.

Wells, J. D., and B. Greenberg. 1994a. Resource use by an introduced and native carrion flies. *Oecologia* 99:181–87.

Wells, J. D., and B. Greenberg. 1994b. Effect of the red imported fire ant (Hymenoptera: Formicidae) and carcass type on the daily occurrence of postfeeding carrion-fly larvae (Diptera: Calliphoridae, Sarcophagidae). *J. Med. Entomol.* 31:171–74.

Whittaker, R. H. 1975. *Communities and ecosystems.* 2nd ed. New York: MacMillan.

Williams, R. W. 1954. A study of the filth flies in New York City. *J. Econ. Entomol.* 47:556–63.

Forensic Meteorology
The Application of Weather and Climate

17

JOHN R. SCALA
JOHN R. WALLACE

Contents

Introduction

The science of forensics draws upon a body of knowledge gained from observation, experiment, and experience, and applies that understanding to questions of interest within the legal system. Forensic meteorology utilizes the climate record as well as local meteorological observations in the context of atmospheric science to reconstruct key weather events for use in criminal or civil litigation. Weather reconstruction has proven to be an invaluable tool in today's litigious world, to the extent that many legal debates cannot be adjudicated without knowledge of the local meteorology. These investigations include, but are not limited to, lightning strikes, obscurations to visibility, sun or illumination issues, crop and perishable product loss, flooding, property loss, motor vehicle accidents, transportation interruptions, construction delays, questionable death, and personal injury. Weather and climate may behave in a stealthy manner, providing little clue as to its importance to the forensic scientist. The influence of the atmosphere may also be swift and uncompromising in its veracity. Regardless, the forensic scientist and his or her legal complement must be cognizant of the limits as well as the potential sources of error associated with meteorological data.

On occasion, the domain of the forensic meteorologist may be expanded beyond what is clearly a meteorological or even climatological context to address an astronomical state. For example, a celestial reconstruction might include the presence or absence of twilight or the phase of the moon. A famous application of a celestial observation in a court of law occurred in 1858 in Beardstown, Illinois. The prepresidential Abraham Lincoln defended

William "Duff" Armstrong, the son of a longtime friend who, along with another assail-ant, was accused of murder (Walsh 2000). The alleged attack followed a nighttime brawl. A witness claimed to have seen Armstrong and another man launch the attack beneath a full moon, the source of the illumination utilized for the identification. Lincoln obtained an 1857 copy of *Jayne's Almanac* to show the three-quarter moon was 3 minutes short of setting at the time of the attack (Potter 2004). The alleged amount of moonlight was impossible given its position just above the horizon, and Armstrong was acquitted on May 7, 1858.

Personal injuries, vehicle accidents, even homicides often require an assessment of the natural as well as the artificial lighting present at the time of an incident. The forensic meteorologist may be required to reconstruct the moon phase, its azimuth and elevation, or the amount of direct and indirect sunlight to assess questions of illumination, par-ticularly during the early morning and early evening. The U.S. Naval Observatory defines twilight as the period preceding sunrise and following sunset when indirect lighting of the Earth's surface is provided by the atmosphere through the process of reflection. The amount of natural light during twilight is determined by the state of the atmosphere gen-erally and local weather conditions in particular. Civil twilight, defined by the position of the center of the sun relative to the horizon, is recognized as the limit of natural illumi-nation. Surface-based objects may be identified without the use of artificial lighting, and in the presence of good weather conditions during civil twilight. The Commonwealth of Pennsylvania uses civil twilight to define the legal commencement of hunting in the pre-dawn hours, and cessation in the evening in an effort to increase safety and situational awareness. It is the assessment of the local weather conditions at the time of a questionable death, like a hunting accident or a possible suicide, that drives the working relationship between the forensic meteorologist and the forensic entomologist. Litigation has been won or lost on the basis of the accuracy of available weather data and its appropriate application to cases such as these.

Perhaps, the common thread linking the forensic meteorologist to most, if not all, legal investigations is the measurement or reconstruction of a single meteorological vari-able: temperature. The simultaneous existence of water in all three phases on the Earth's surface at commonly observed temperatures and pressures often accounts for the complex weather reconstructions required in a case. Similarly, human-perceived discomfort as well as postmortem decay is critically dependent upon the combination of sensible tempera-ture, wind, and humidity. For example, a high-profile homicide investigation questioned the legitimacy of the defendant's claim that he was lounging at poolside after midnight in shorts and a t-shirt at the time of his wife's death. A forensic meteorologist working for the prosecution determined that the early morning conditions during this period of alleged relaxation included a sensible temperature in the lower 50s, conditions hardly conducive for the admitted warm weather attire of shorts and a shirt.

Quality-controlled observations are rarely, if ever, available at a crime or accident scene. Only in cases of extreme serendipity can a forensic meteorologist utilize *in situ* data without accounting for the impact of local climatic influences, elevation differences, or exposure. Even in the best cases, where the instrumentation is sited in close proximity to the scene of interest, poor calibration, site contamination, or even instrument failure may render the existing data useless. The catastrophic losses experienced by residents of the Louisiana and Mississippi Gulf Coast in August 2005 from Hurricane Katrina triggered tens of thousands of legal confrontations between homeowners and insurance carriers. Settlements were delayed due to the laborious process of reconstructing the critical sequence of events leading to individual losses and comparing anecdotal information with archived weather observations. Loss of electrical power and weather instrument failure several hours prior to the observed catastrophic property loss were common along the immediate Gulf Coast. Consequently, distinguishing between damage associated with storm surge flooding and damage from wind-driven rain became a complicated, if not impossible, task in most of these cases. Thus, it is the responsibility of the forensic meteorologist to: (1) recognize the limitations of the temperature record, (2) identify periods of questionable sampling, and (3) convey the uncertainty implicit in the data to all parties. Irrevocable damage could be done to a forensic entomological study by poorly understood and applied temperature data.

Temperature observations are impacted not only by instrument quality, precision, and design, but also by the local environmental influences exerted by wind, precipitation, exposure, insolation, surface characteristics, vegetative cover, and proximity to water. An accurate reconstruction of daily and more often hourly temperature data surrounding the event of interest characterizes the working relationship between a forensic meteorologist and a forensic entomologist (Figure 17.1). The latter seeks to estimate the minimum *post-mortem interval* (PMI), defined as the time between insect colonization on a body and the time of corpse discovery (Tomberlin et al. 2006). The minimum PMI is based on the empirical relationship between insect growth rates and ambient temperatures (Tomberlin et al. 2006; Hart and Whitaker 2006). The former relies on available climate records and crime scene temperature data (if available) while taking into account the known influences cited earlier to reconstruct the local temperature history. The strength of the resulting collaboration rests equally upon the true measure of the ambient temperature record as well as the experiential knowledge of forensically important blow fly species where some degree of uncertainty exists (Davies 2006). The current literature suggests that additional developmental growth data are required, particularly at the low-temperature thresholds in conjunction with accumulated degree-day (ADD) assumptions, to ensure accuracy of a minimum PMI estimate (Nabity et al. 2006).

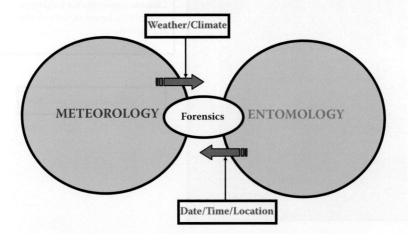

Figure 17.1 Flow diagram illustrating how the field of forensics lies at the heart of the transfer of information between a meteorologist and an entomologist. The interface is defined by the type of data required by the forensic entomologist and the ability of the forensic meteorologist to apply that which is known about the local climate to the case at hand.

This chapter seeks to impress upon the reader the importance of ensuring that the temperature data utilized by the forensic entomologist is as accurate as possible and representative of the crime scene. The material will be developed by first exploring the controls on temperature. The content will evolve into a discussion of field measurements and instrumentation, the importance of understanding the impact of microclimates, and pertinent legal issues. The chapter concludes with the presentation of case histories to illustrate the pertinence of the developed material.

Fundamentals of Meteorology: Controls on Temperature

The difference between weather and climate is an important distinction that warrants special attention. Weather defines the dynamic state of the atmosphere at a specific place, and at a specific time. For example, the statement "A devastating tornado struck Moore, Oklahoma, at 6:15 p.m. on May 3, 1999" describes a weather event, albeit a tragic one. The state of the atmosphere at a given instant in time is assessed through a suite of routine observations. The most common of these are air temperature, station pressure, wind, moisture content, cloud cover, and precipitation. Climate is the long-term variability inherent in the weather, or alternatively stated, it is the average of extremes (Figure 17.2). The historical archive allows a meteorologist to compare a single observation or period of record with statistical averages. For example, the long-term seasonal snowfall average in the vicinity of Lancaster, Pennsylvania, is 26.5 inches. A comparison of December, January, and February

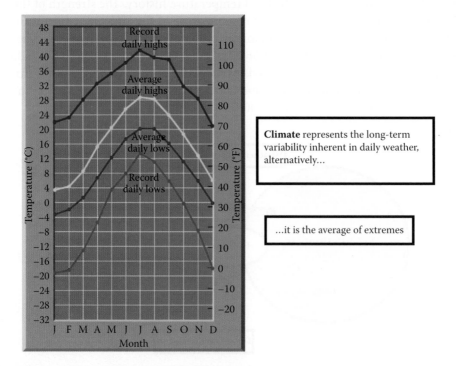

Figure 17.2 The local climatology provides a measure of the variability inherent in the long-term temperature record. Average daily high and low temperatures lie within the historical range of observed maximum and minimum values, in this case for New York City. (Adapted from Lutgens and Tarbuck 2004. With permission of Pearson Education.)

snowfall for any winter with this seasonal average can provide a measure of the seasonal snowfall anomaly in central Lancaster County.

Weather can influence the nature of a crime scene and impact the ability of the forensic expert to reconstruct the critical events leading up to the event of interest. From an entomological perspective, the rate of decay, the recovery of evidence, and the accurate application of insect indicator species to determine the minimum PMI are selectively determined by the exposure of the corpse to the local weather elements. The size and developmental stage of forensically important species of blow flies are well known to be a function of time and temperature (Slone and Gruner 2007; Donovan et al. 2006), and are utilized by forensic entomologists in this regard. The physical controls on temperature will ultimately determine how well those indicator species can be utilized to derive an estimate of the accumulated degree-hours (ADH) and a measure of the minimum PMI.

A suitable starting point for a discussion of the physical influences on temperature is the recognition that the heating of the Earth's surface is spatially nonuniform. This fact is well known to anyone who has traveled beyond the boundaries of his or her local community. Maximum and minimum temperatures, and the resultant diurnal temperature ranges, are a direct consequence of the amount of incoming solar radiation, or insolation, received at the Earth's surface. Insolation varies according to the time of day, latitude, and season in a general sense, but is also influenced locally by exposure (direct or indirect) to that radiation. Since the atmosphere is warmed from below, any reduction in the maximum insolation by local influences like topography, vegetation, and surface character will lower the amount absorbed at the Earth's surface and subsequently radiated to the overlying atmosphere. A balance must exist between the Earth's surface and the atmosphere at any latitude; otherwise, energy gained or lost by the Earth-atmosphere system would result in a runaway temperature such that the equator would continually warm and the poles cool. This equilibrium state permits only small changes in the Earth's average temperature from one year to the next.

The Earth follows an elliptical orbit about the sun, leading to its closest distance (called opposition) on or about January 3, and its farthest distance about July 4. If this distance were the sole control on temperature, then the coldest time of the year would occur in early July, not in January. The apparent paradox is resolved by recognizing that the Earth revolves on an axis tilted 23 1/2° from the perpendicular relative to its orbital path around the sun. Consequently, the northern hemisphere is tilted away from the direct rays of the sun in December and toward them in June (Figure 17.3). The amount of daylight in the northern hemisphere increases between the winter solstice (December 21 or 22) and the summer solstice (June 21 or 22), then decreases for the next 6 months as the Earth's orbit progresses to its winter position.

The sun follows a much lower trajectory across the sky in winter than in summer, extending nighttime to 24 hours north of the Arctic Circle, while residents of the southern hemisphere south of 66 1/2° experience the exact opposite: 24 hours of daylight! The low sun angle in winter not only reduces the intensity of the solar radiation, but also forces that radiation to pass through a greater depth of the atmosphere, where the processes of reflection, refraction, and absorption spread the available sunlight over a greater surface area (Figure 17.4).

Latitude is not the only control on temperature. Local seasonal variations develop as a consequence of geography, elevation, and exposure to solar radiation. North-facing slopes receive less direct sunlight through the course of a year than the south side of a hill

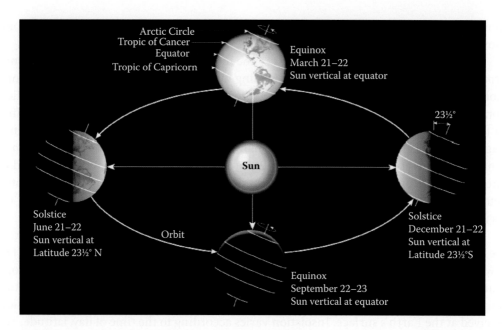

Figure 17.3 The Earth revolves around the sun on an axis tilted 23.5 degrees from the perpendicular relative to the plane of the elliptic exposing the Northern hemisphere to the most intense sunlight at the summer solstice. (Adapted from Lutgens and Tarbuck 2004. With permission of Pearson Education.)

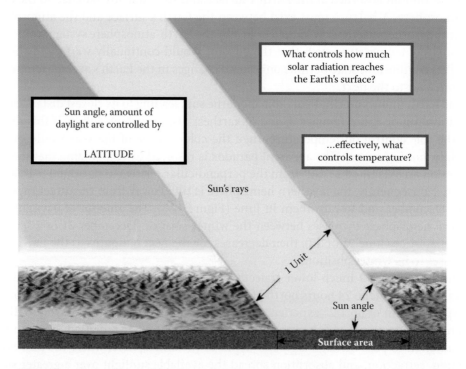

Figure 17.4 The effect of latitude is to spread the incoming solar radiation over a greater surface area as one moves north and south of the sun's position at solar noon. The latitudinal influence is analogous to the lengthening of shadows as the day progresses. Longer shadows indicate passage of the sun's radiant energy through a greater depth of the atmosphere, reducing its intensity. (Adapted from Lutgens and Tarbuck 2004. With permission of Pearson Education.)

or mountain. Snow cover remains longer on a north-facing slope, where cooler temperatures prevail and the soil tends to contain more moisture. Similarly, a south-facing side is typically drier and may be characterized by lower-density vegetation cover due to warmer temperatures and higher evaporation rates. It is reasonable to expect this combination of exposure, temperature, and moisture would increase the decay rates on the south-facing slope and reduce the rate of decay on the north-facing side.

The daily cycle of warming and cooling at the Earth's surface is controlled by a variety of inputs, in addition to the most important one, the amount of incoming solar radiation. Surface measurements of insolation reveal a peak around solar noon in the absence of clouds. Yet, the warmest time of the day is not at noon, but typically 3 to 5 hours later (Figure 17.5). This apparent paradox is easily explained by the fact that the atmosphere continues to be heated even though the intensity of the incoming radiation is on the wane. Since the atmosphere is heated from below, the near-surface layer of air will continue to be warmed as long as the incoming radiation (insolation) exceeds that which is lost by the Earth's surface (terrestrial radiation). The warmest temperatures are usually observed directly on the ground, and then decrease rapidly above it. For this reason, the World Meteorological Organization mandates that surface temperature observations should be made at a standardized height of 1.2–2.0 m (3.9–6.5 ft) to ensure a representative sampling of the unobstructed atmosphere in the absence of near-surface influences. It is important to note that in the presence of a temperature inversion, an elevated warm layer (i.e., the local temperature increases with height in the lowest levels of the atmosphere, rather than decreases) will often trap cooler air at the surface. The resultant weather is often fog, which lowers the local visibility and leaves the near-surface layer cool and moist. Thus, it behooves

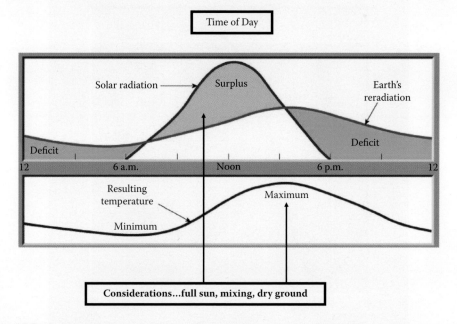

Figure 17.5 Peak insolation occurs at solar noon; however, the atmosphere continues to warm well into the afternoon, in the absence of clouds and other environmental influences, until the amount of incoming solar radiation no longer exceeds that which is lost by the Earth's surface. A deficit in the local energy budget occurs at this time; consequently, the atmosphere begins to cool until shortly after sunrise of the following day, when the process is repeated. (Adapted from Lutgens and Tarbuck 2004. With permission of Pearson Education.)

crime scene investigators and forensic entomologists to catalog all available weather information that may impact the local temperature data used to determine a minimum PMI.

Peak diurnal temperature ranges occur in the presence of clear skies, low humidity, and dry ground. These conditions are typical in the desert southwest of the United States, where the difference between the daily minimum and maximum temperatures can exceed 22°C (40°F) routinely in the summer months, in the absence of the monsoonal rains of July and August. The dew point depression, the difference between the ambient temperature and the temperature at which condensation would occur through cooling alone, is an excellent measure of humidity. In fact, dew point depressions in the southwestern United States may approach 35°C (63°F) during early summer. The coexistence of exceptionally dry air and high temperatures in desert environments (e.g., Las Vegas) may lead to an accelerated rate of body mummification. Consequently, some forensically important insects (e.g., blow flies) may be absent from mummified remains due to the limited opportunity for colonization.

The addition of moisture to the atmospheric column in the form of water vapor (i.e., the gaseous phase of liquid water) will tend to reduce daytime temperatures and increase overnight temperatures due to its selective absorption of radiant energy. Some degree of cloud cover, haze, and fog is usually present in most locations and acts to reduce the amount of insolation and the observed diurnal temperature range (Figure 17.6). Consequently, the warmest time of day may occur closer to solar noon or even in the morning, rather than later in the afternoon, as is typically observed.

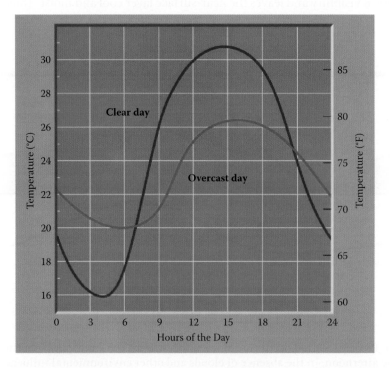

Figure 17.6 A comparison of the diurnal temperature distribution for Peoria, Illinois, representative of clear and cloudy days in July. Note the impact of clouds is to not only reduce the temperature range, but also slow the rate at which warming and cooling is observed. (Adapted from Lutgens and Tarbuck 2004. With permission of Pearson Education.)

Experience indicates that temperature is affected by more than simply latitude and cloud cover. San Francisco, California (SFO), and Wichita, Kansas (ICT), are at approximately the same latitude and exhibit similar average annual temperatures: 14.1°C (57.3°F) and 13.4°C (56.2°F), respectively. However, extraordinarily different near-surface conditions prevail along the Pacific Coast versus the Central Plains, and not surprisingly, these are revealed in the climate record. You may recall that at the beginning of this section, climate was defined as the average of extremes. The record low and high temperatures at SFO range from –3°C (27°F) to 39°C (103°F), as compared to the much broader –30°C (–22°F) to 46°C (114°F) at ICT. A partial explanation for this remarkable contrast is the difference in elevation (3 m compared to 407 m above mean sea level for SFO and ICT, respectively). However, the disparity more accurately reflects the control that water, cloud cover, geography, and surface characteristics (e.g., soil type, land use, vegetation) exert on the ambient temperature.

The relevant concept is the variation in the specific heat capacity present in air, water, soil, trees, and other land surfaces. The degree to which a substance reflects or absorbs solar radiation, and to what depth that absorption penetrates, dictates how rapidly its surface warms or cools. Consider the surface temperature of asphalt compared to fresh snow, or an open field compared to a forest floor residing beneath a dense vegetative canopy. The opaque asphalt absorbs most of the incoming radiation while reflecting very little. A fresh snow cover absorbs little of the sunlight and reflects a much greater percentage of the incoming radiation. Similarly, an open field (particularly one that is sloped) will respond dramatically to the drying effects of sun and wind, or the moistening influence of rain, while a shaded location will accumulate organic matter, retain surface moisture, and experience a reduced diurnal range in temperature.

The ratio of reflected to incoming solar radiation is termed the albedo. The albedo of snow cover, sandy soil, and a field of grain is much greater than that of a parking lot, plowed field, or rooftop. Consequently, air in direct contact with a dark substance will warm faster and to a greater degree than air overlying a white surface. For this reason, thermometers are housed within a white covered shelter away from direct sunlight and surfaces that may introduce biases in the temperature data.

The presence of water in the form of small droplets, the prime constituent of clouds, exerts a strong forcing on the environment by contributing to the heating and cooling of the Earth–atmosphere system. This forcing is accomplished primarily through the reflection of incoming solar radiation and absorption of outgoing terrestrial radiation (Figure 17.7). Nighttime temperatures tend to be warmer in the presence of clouds as a consequence of the absorption of terrestrial radiation and its emission back toward the Earth's surface. Daytime temperatures are generally lower in the presence of clouds due to the reduction in insolation.

Local geography contributes greatly to the observed diurnal temperature range by limiting the degree of heating and cooling. Coastal locations, for example, are strongly influenced by wind and the presence of water. These two factors act to moderate the severity of the climate by reducing maximum and raising the minimum surface temperatures. Locations downwind (leeward) of mountains are typically warmer and drier than their upstream (windward) counterparts due to the lack of clouds and the presence of subsiding (sinking) air.

Additional controls on temperature are linked to the land use and land cover characteristics surrounding a local observation site (Gallo et al. 1996). Clearly, the observed

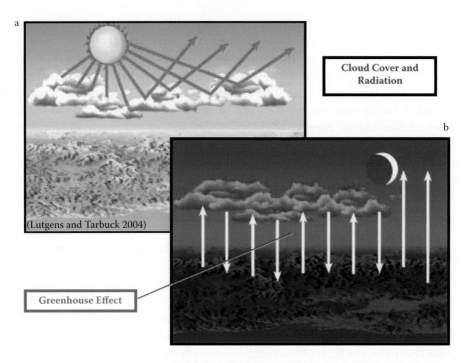

Figure 17.7 (a) Clouds absorb and reflect about 20% of the incoming solar radiation, with reflectance dominating. Consequently, daily maximum temperatures are reduced in the presence of cloud cover. (b) Alternatively, a nocturnal cloud cover is a relatively effective absorber of outgoing terrestrial radiation. The subsequent emission of infrared radiation back to the Earth raises overnight temperatures in the absence of additional influences. (Adapted from Lutgens and Tarbuck 2004. With permission of Pearson Education.)

climate is different in an urbanized area dominated by asphalt and concrete, compared to farmland or forest. The emphasis is on the dramatic contrast that exists between a natural environment, where soil type, vegetation, and water are the primary controls on temperature, and a population center dominated by roads and buildings, which generate and retain heat more effectively than the natural landscape, and often enhance street-level wind circulations. This effect, known as the urban heat island, leads to warmer overnight temperatures, less diurnal temperature range, and a greater potential for clouds and precipitation downwind of the urban center. The abundance of impermeable surfaces supports rapid runoff and greater evaporation within a zone of elevated temperatures. The cooling effect of trees is lost within a city, while the release of aerosols adds to the absorption of radiation. Forensic entomologists using temperature data obtained from urban death scenes to determine accumulated degree-days (or degree-hours) should consider these microclimatic influences, specifically maximum and minimum temperatures, when adjusting between quality-controlled data sites and an urban death scene environment.

As noted previously, land surface characteristics modulate the ratio of reflected to incoming solar radiation contributing directly to the degree of surface heating or cooling. The roughness of the land's surface, as pointed out previously, will also influence temperature through the magnitude and direction of low-level circulations. A forensic entomologist in consultation with a forensic meteorologist must remain cognizant of these local controls on temperature when evaluating exposure of a decaying corpse.

Field Measurement Techniques and Sampling Fundamentals

National politics and international relations occupied much of the nation's first secretary of state in 1790, yet Thomas Jefferson pursued his interest in meteorology with the same vigor he accorded his political appointments. Jefferson and his close friend and fellow Virginian James Madison recorded temperature measurements from 1777 through 1816 from their respective plantations, located 21 miles apart. In 1803, both men turned their backs on the Royal Meteorological Society's recommendation of an unheated room for measuring temperature and began to report their readings from an outdoor thermometer (Solomon et al. 2007). This new location revealed the profound influence that environmental exposure can exert on temperature. Jefferson also realized that the local elevation as well as the shelter design that housed the instrument influenced the measurement. The challenge these two men faced more than 200 years ago remains one of great importance today for forensic entomologists: how to maximize the accuracy of the temperature measurement with a minimum of environmental influence.

A crime scene investigation often requires temperature sampling at several positions above the surface in addition to ground and subsurface values. These measurements may also include the temperatures of the deceased, the underlying soil (at various depths), within streams and lakes, or along a shoreline. The accuracy of the ambient measurement rests to some degree on the choice of instrument, a selection based on several considerations, including its performance, precision, design, durability, purchase cost, and maintainance cost. The forensic entomologist is encouraged to consult DeFelice (1998) for a comprehensive treatment of meteorological instrumentation and measurement, and the 2006 publication by the World Meteorological Organization (WMO) for further details on instrumentation and the measurement of air temperature.

Erroneous temperature data resulting from inaccurate sampling or inappropriate application of non–*in situ* measurements are recognized as the single greatest source of error in the calculation of accumulated degree-days in death scene investigations (Nabity et al. 2006; Higley et al. 1986). Therefore, the measurement of air temperature should reflect just that: the ambient atmosphere, not the temperature of the device housing the instrument or the hand holding it. Care must be exercised to avoid direct contact with the thermometer while permitting the instrument to achieve thermal equilibrium with the environment, rendering the temperature of the thermometer equal to the air it is sampling.

As we noted earlier in this chapter, exposure (in general terms, the local microclimate) will exert a strong influence on temperature. The observations (either *in situ* or obtained from a nearby observing site) should be accompanied by a description of the local environment and ambient conditions present at the time of sampling (clear, overcast, windy, raining, etc.). The height and distribution of foliage, surface character, soil composition, compass heading, orientation relative to the prevailing wind, topography and elevation, proximity to water, and the presence of buildings can affect the local temperature, essentially biasing the crime scene measurement while complicating comparisons to nearby observations. Crime scenes are rarely collocated with quality-controlled instrumentation or in sites devoid of local singularities. Standardizing the elevation above ground level (~1.5 m), at which the air is sampled, minimizes the impact of the vertical temperature gradient while achieving the representative results required for comparisons with different locations and at different times. Of course, additional data are often obtained at nonstandard levels to achieve a characterization of the local environment.

Routine field measurements of air temperature are typically obtained with a mechanical (mercury- or alcohol-filled thermometer) or electronic (thermocouple, resistance element, or thermistor) device. Mechanical thermometers tend to be stable and capable of excellent measurement accuracy. Electronic thermometers utilize variations in an electric signal through the generation of an electric current or application of an external signal to derive temperature. The selection of an electric thermometer should consider the expected temperature range that will be sampled, the required sensitivity, and the anticipated accuracy.

The forensic entomologist should be aware of the common errors associated with ambient air temperature measurements. The true air temperature can only be obtained when the instrument is in equilibrium with the atmospheric environment that it is sampling. The sensor must be shielded from direct radiation from the sun as well as its surroundings. Care must also be taken to minimize exposure to wind and direct contact with a surface of contrasting temperature. A thermometer should be calibrated against a laboratory standard before taking it into the field, with the derived corrections applied accordingly and the inherent uncertainty (e.g., 0.2°C) noted. The required response time of the measurement will determine instrument selection. Slow response times (on the order of minutes) are typically well suited for *in situ*, stationary measurements.

Most investigations will necessitate acquiring nearby observations from a network of Automated Surface Observing System (ASOS) instruments maintained by the National Weather Service (NWS). Each ASOS is designed to measure the typical atmospheric variables: temperature, pressure, dew point, wind speed, wind direction, visibility, cloud height, and precipitation. The ASOS instrument will also report sky condition, obscurations to visibility, pressure tendency, peak winds, and other important attributes of the weather. An ASOS will not report short- and long-wave radiation or surface moisture content, nor is it designed to see the horizon since measurements are taken directly overhead. This information is quality controlled and archived at the National Climatic Data Center (NCDC) for future use. The forensic entomologist will rely solely on this climatic record in the absence of data obtained from the scene to reconstruct the weather. Local data, where available, may be obtained from the Office of the State Climatologist. Either repository can designate the retrieved data as official, thereby increasing its acceptance in a courtroom environment.

Microclimates

A forensic entomologist may rely on a data logger placed at a crime scene to complement meteorological observations obtained from the nearest NWS reporting station. The objective is to reconstruct an ambient temperature record for use in determining the minimum PMI. A regression analysis is often employed to develop a statistically robust relationship between the two data sets and enhance the credibility of the minimum PMI estimate. That relationship gains additional credence when sources of instrument error are identified and eliminated. For example, nonclimatic biases may be introduced into the data record as a consequence of changes in observation time, location, instrumentation, or poor instrument siting (Karl et al. 1986; Gallo et al. 1996; Peterson 2006). The analysis of any temperature data must consider the accuracy of the observations as well as sources of potential error. A recent study by Sublette (2006) identified the relocation of an airport-based automated surface observing system, and not a local climatic trend, as being responsible for a decrease in the observed mean low temperature.

The U.S. Historical Climate Network (USHCN) incorporates automated NWS stations and cooperative observing sites. The associated instrumentation must adhere to internationally accepted standards prepared by the WMO regarding siting and exposure. These standards must also be maintained throughout the record of observation to ensure the data are representative of the local surroundings. Site exposure violations may misrepresent long-term climatic trends by introducing temperature biases into regional averages. These biases are often associated with changes in land use, demographics, vegetation cover, or simply inadequate temperature sensor ventilation (Davey and Pielke 2005).

Temperature sensor exposure violations can result from the siting of an instrument in close proximity to a building or ventilation system, over a patchy land surface, within a narrow river valley, or positioned at heights other than the WMO accepted standard of about ~1.5 m above ground level. Erroneously cold or warm temperatures introduced as a consequence of sampling errors could result in an inaccurate ADD and an artificially lengthened or shortened estimate of the minimum PMI. Spurious trends associated with poorly sited stations may be adjusted such that the resulting time series could compare favorably with a properly sited station (Peterson 2006).

NWS ASOS observations are expected to be quality controlled. However, an assurance of accuracy does not equate to a license for applying archived climate data to a crime scene without first determining if differences exist with respect to elevation and exposure. Calibrated and quality-controlled ASOS instrumentation can be influenced by local land use, geography, and proximity to water, as illustrated in Figure 17.8. Abrupt

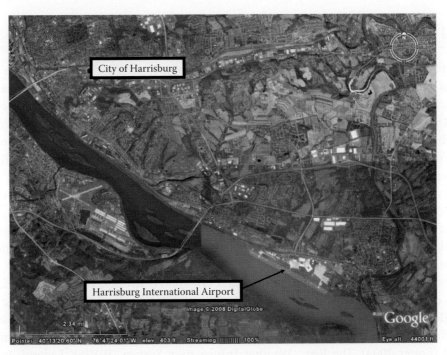

Figure 17.8 Aerial view of Harrisburg International Airport (MDT), the Susquehanna River, the city of Harrisburg, Pennsylvanina, and surrounding areas. The NWS ASOS instruments are located on the grounds of MDT in close proximity to the moderating influence of the river. Land use characteristics vary markedly within a 7-mile radius of the temperature sensor at MDT, from open farmland and forested ridge lines to residential and urbanized corridors. (Courtesy of DigitalGlobe and Google Earth.)

changes in surface character within a 7-mile radius of the instrumentation at Harrisburg International Airport (MDT) are visible in this aerial image. This small separation, the distance between MDT and downtown Harrisburg, would generally constitute a source of excellent proximity data. However, a range of microclimates clearly exists within this image and would necessarily complicate the application of MDT data to a nearby location of interest.

The surface-based sampling of air temperature is now affordable and accurate enough to be obtained with the use of field-tested data loggers. The benefits to a meteorologist are obvious: the opportunity to create a higher-resolution sampling network of sufficient quality to document microclimatic temperature variations associated with specific environments (Whiteman et al. 2000). The acquired data can be used in conjunction with soil parameterizations, and near-surface wind and moisture measurements to construct numerical models of surface heat and moisture fluxes.

Of equal importance is the application of this technology to crime scene temperature observations for comparison with quality-controlled proximity data. Data regression efforts are often employed to correlate the *in situ* data with nearby official observations. A statistically robust relationship may be difficult to resolve if instrument errors, microclimatic effects, or both are not accounted for in the regression model. Recognizing these deficiencies, the courtroom testimony of a forensic entomologist would gain credibility by accounting for potential differences between archived climate data and *in situ* observations.

Legal Considerations

Perhaps, the greatest obstacle to a daily or even hourly weather reconstruction is the lack of accurate and representative proximity data. Crime scenes are rarely located in the vicinity of a quality-controlled suite of meteorological instruments. Fortunately, the recent NWS modernization program and the increasingly accessible climate databases available on the World Wide Web have placed terabytes of meteorological information at the fingertips of the forensic entomologist. These data include not only the traditional surface and upper-air observations, but also satellite and radar imagery, local storm reports, numerical model forecasts, NWS statements and summaries, ocean and lake water conditions from anchored buoys, snow and ice depth, liquid water content, and a host of hydrometeorological and geophysical databases. It remains the challenge of the forensic entomologist to sift through this array of information and piece together the weather pertinent to the litigation.

The precision of a forensic entomological evaluation is strongly linked to the accuracy of the derived temperature record and its application to the death scene. Forensic entomologists can access daily maximum and minimum temperatures, pressure, precipitation, and wind data from a growing list of cooperative (COOP) observers numbering more than 11,000 (Potter 2004). The Road Weather Information System (RWIS), Federal Aviation Administration (FAA) automated sites, and military reporting stations, in addition to the COOP reports, enhance the national network of ASOS sites maintained by the NWS. The mixing of weather information from a variety of sources improves the spatial resolution of the combined data set, yet there is concern that a portion of this archived data may be contaminated by a temperature bias due to poor site selection or poor maintenance.

The national repository for the vast array of meteorological observations is the National Climatic Data Center (NCDC) in Asheville, North Carolina. The NCDC will certify that the requested climate data are a true and authentic copy of the original record, enhancing its portability into the courtroom (Potter 2004). Individual Offices of the State Climatologist (available in all fifty states as well as Guam and Puerto Rico) also offer a variety of services beneficial to the forensic entomologist, including seasonal weather summaries and certified temperature and precipitation data. Unofficial data, though useful in closing the gaps in the automated network, must be viewed carefully and with proper consideration for potential errors (see previous discussion on USHCN). Recognizing the potential for deficiencies in the temperature record prior to discovery and deposition will minimize errors in the reconstruction of key weather elements and their subsequent use in the determination of a minimum PMI.

The forensic meteorologist, when consulted, must communicate the inherent uncertainty present in the reconstruction to the forensic entomologist, effectively placing "error bars" on the temperature data to account for the absence of proximity observations. Legal challenges often arise from the application of the "closest" reliable data. Is it representative of the crime scene? To what extent will daily temperature fluctuations observed at nearby NWS instrumentation reflect the environment of the crime scene? Can the crime scene temperatures be reasonably estimated by non-*in situ* observations? What is the uncertainty of the minimum PMI using this temperature reconstruction? An incorporation of temperature lapse rates (i.e., how unsaturated air changes temperature with height), linear interpolation, or more sophisticated regression techniques to account for the effects of terrain, elevation, and exposure will strengthen the use of closest proximity data in a crime scene investigation, and defend against the toughest scrutiny.

Aside from certified data, eyewitness accounts can improve the credibility of the weather conditions determined by the forensic meteorologist at the exact time of a slip and fall, motor vehicle accident, roof failure, or questionable death. Specific attributes of the larger-scale weather regime in which the local scale environment is embedded can be evaluated through the use of satellite and radar imagery and numerical model simulations. A detailed analysis of an additional set of reports that might include accident summaries, news descriptions, and pretrial depositions aids the weather expert in developing the reconstruction. A legal decision may hinge, ultimately, on some element of the weather utilized by the forensic entomologist. The complex interaction described herein among numerous weather attributes, the temperature reconstruction, data verification and certification, as well as the microclimate impacts yet to be resolved through empirical studies suggests that a forensic entomologist should consult a meteorologist to ascertain the importance of each in a case-by-case basis.

The initial sections of this chapter focused on the limitations of the temperature record, particularly when the retrieved data contain periods of questionable sampling. Of equal concern is the accurate interpretation and application of the local meteorological and climatological data to the case at hand. The burden remains with the forensic meteorologist to interpret the findings and convey the inherent uncertainty in those conclusions to the forensic entomologist, who will rely heavily on that data to develop a reasonable minimum PMI. Important controls on temperature, instrument design, and measurement, and the influence of microclimates were developed in earlier discussions. The remainder of this chapter utilizes actual case studies to illustrate the concepts presented.

Case 1: Applying a Temperature Record to a Remote Site in the Case of Suspected Suicide

The skeletal remains of a 19-year-old male were discovered on December 20 just north of a hiking trail along a north-facing slope in steep terrain in the Furnace Hills region of the J. Edward Mack Scout Reservation in northern Lancaster County, Pennsylvania (Figure 17.9a). The dense forest canopy (Figure 17.9b) contributed substantial shade to the recovery site during the summer months, as well as limb and tree branches to a ground cover characterized by several inches of decomposing leaf debris (Figure 17.9c). A preliminary study of the recovery site indicated that the northern exposure, steep elevation, and deciduous tree cover would exert a cooling influence on ambient temperatures.

The NWS maintains three ASOS observation sites within a 25-mile radius of the discovery site: Harrisburg International Airport (MDT), Lancaster Airport (LNS), and Reading Regional Airport (RDG). Table 17.1 lists the proximity reporting sites, elevation above mean sea level, and heading and distance relative to the estimated latitude and longitude of the recovery site.

Case #1:

Figure 17.9 (a) U.S. Geological Survey topographic map illustrating the location where the body was discovered in the Furnace Hills region of the J. Edward Mack Scout Reservation, northern Lancaster County, Pennsylvania. The red arrow points to the body recovery site along steep, north-facing terrain 100 m north of the Horseshoe Trail at an approximate elevation of 890 feet, and below the local maximum of 1,004 feet. (b) Dense forest along the north-facing slope typical of the vegetative cover in the vicinity of the body (c) found beneath tree branches and still attached to the rope.

Table 17.1 Local NWS Reporting Stations: Harrisburg International Airport (MDT), Lancaster Airport (LNS), and Reading Regional Airport (RDG)

Name	Coordinates	Elevation (ft)	Heading/Distance from (40°14.39N, 76°19.03W)
MDT	40°11.37N, 76°45.48W	310	264°/24.3 miles
LNS	40°.07.13N, 76°17.40W	400	173°/8.6 miles
RDG	40°22.24N, 76°57.34W	339	69°/20.5 miles

A fourth reporting site, Muir Army Airfield, located approximately 19 miles to the northwest at Fort Indiantown Gap Military Reservation (MUI), was not considered in this study due to the numerous gaps present in the data archive. The climate record obtained from LNS was relied upon to provide the most accurate weather reconstruction given its proximity to the discovery site. Archived observations were reviewed from late August, when the young man was considered missing, until the discovery of the remains nearly 3 months later. *In situ* data were not available. The primary entomological evidence used in this case included the presence of eclosed *Calliphora vicina* puparia collected from the body in the morgue.

The LNS ASOS instrumentation (utilized in this case) is located on flat, unobstructed terrain on the grounds of the Lancaster Airport. The difference in elevation between the recovery site (~890 feet above mean sea level; Figure 17.9a) and LNS (400 feet above mean sea level) could be responsible for a warm bias of approximately 2–3°F in the vicinity of the accident if the LNS temperature observations were applied uncorrected. Recall that quality-controlled temperature observations are reported at a height of between 1.2 and 2.0 m, as directed by the World Meteorological Organization, to ensure uniformity in measurement. Daily maximum and minimum temperatures were adjusted to account for the observed differences in elevation between the LNS ASOS instrumentation and the actual death scene. Steep terrain and a northern exposure also exerted a microclimatic influence in addition to the cooling effect provided by the dense forest. Consequently, overnight and daily mean temperatures would be expected to be lower in the scout reservation relative to LNS. These adjustments were instrumental in deriving a minimum PMI that differed from a missing person's report by only a few weeks.

Case 2: Artificial Temperature and Exposure in a Homicide Investigation

The homicide occurred 2 years earlier in a small, single-floor cottage constructed on a rectangular concrete slab within a residential neighborhood (Figure 17.10). On April 12, the remains of a 17-year-old male were recovered from a shallow grave located behind a detached garage at the rear of the property. The deceased was covered following the attack and placed in a closet at the south end of the cottage for 1 to 2 months prior to burial. Subsequent examination at the morgue revealed Heleomyzidae larvae and hundreds of *Phormia regina* eclosed puparia on the body. The crime scene phototgraphs, the collected eclosed *Phormia regina* puparia, and the absence of any blow fly larvae or adults from the body, from the wrappings or the burial site, indicated that the adult flies emerged from the puparia while the body was concealed in the cottage.

Case #2:

Figure 17.10 External view of the small cottage in north-central Lancaster County, Pennsylvania, where the remains were stored for approximately 1 to 2 months. The body was covered with sheets and blankets and placed in a closet prior to being removed and placed in a shallow grave behind the garage at the rear of the property.

The rectangular cottage is surrounded on three sides (east, south, and west) by mature hardwoods (estimated 30 to 50 feet in height) and adjacent homes, with an open exposure to the north and northwest (Figure 17.10). The structural orientation produced a long dimension along a 10°–190° heading, and a short dimension along a 100°–280° heading. Nearby dwellings to the southeast, south, and southwest of the cottage, as well as two mature hardwoods (located at 100° and 266° relative to the short dimension of the cottage), would have cast shadows over the cottage during the early morning and late afternoon. A site survey and review of appropriate topographic maps indicated an approximate elevation for the incident scene as 330 feet above mean sea level. The same three NWS ASOS (MDT, LNS, RDG) observation sites noted in case 1 were located within a 50 km (31-mile) radius of the crime scene (Table 17.2). The observed difference in elevation of 73 feet alone would account for 0.4°F cooling relative to LNS. This value is likely within the range of error of the instrument, suggesting the LNS values may be used without adjusting for elevation differences.

The altitude and azimuth of the sun at solar noon for Lancaster, Pennsylvanina (LNS), between March 15 and May 31 are shown in Table 17.3. It is reasonable to conclude from the orientation of the cottage, its close proximity to adjacent buildings, and the solar altitude data that the cottage experienced a reduced diurnal temperature range relative to the nearby LNS observations.

The determination of the minimum PMI in this case was complicated by markedly different controls on temperature. The initial postmortem period was dictated by the small

Table 17.2 Local NWS Reporting Stations: Harrisburg International Airport (MDT), Lancaster Airport (LNS), and Reading Regional Airport (RDG)

Name	Coordinates	Elevation (ft)	Heading/Distance from Crime Scene
MDT	40°11.37N, 76°45.48W	310	280°/31.3 miles
LNS	40°07.13N, 76°17.40W	400	267°/5.0 miles
RDG	40°22.24N, 76°57.34W	339	40°/22.5 miles

Table 17.3   Solar Azimuth and Altitude at Local Noon, the Highest Elevation of the Sun during Its Daily Transit across the Sky for Lancaster, Pennsylvania

Date	Azimuth (E of N)	Altitude
March 15, 2003	178.5°	47.8°
April 20, 2003	177.9°	61.4°
May 31, 2003	178.0°	71.8°

Source: Obtained from the U.S. Naval Observatory.

temperature fluctuations within the warm and dry environment of the cottage; however, this interior temperature history was unavailable for study. The meteorological and geophysical data provided an important perspective on the influence of elevation and exposure at the scene. The forensic entomologist utilized this information to obtain the most accurate developmental temperature to estimate the minimum PMI.

Summary

A forensic entomologist relies on species identification and a determination of the developmental age of insects recovered from a death scene to estimate the minimum PMI. The degree of colonization reflects the exposure of the body to the local weather; thus, daily or hourly temperature data from *in situ* sampling or the nearest quality-controlled NWS site are critical to estimating the accumulated degree-days/accumulated degree-hours (ADD/ADH) prior to corpse discovery. The accuracy of the minimum PMI estimate rests on the precision of the temperature reconstruction. The validity of that estimate can be compromised if environmental controls on temperature (e.g., elevation, exposure, latitude, season, surface characteristics, vegetative cover, and proximity to water) or the sources of error (instrument quality, precision and design, calibration, instrument siting, poor quality control) are not considered.

The forensic entomologist that performs *in situ* sampling of a crime scene should be knowledgeable of the response time, accuracy, and sources of error associated with the instrument of choice. The application of nearby NWS quality-controlled sites can be invalidated if the local environment is strongly dissimilar or if the NWS data are biased in some manner. It is the responsibility of the forensic entomologist to communicate with a forensic meteorologist to better understand the limitations of the climate record as well as the uncertainty in the temperature reconstruction. It cannot be overstated that permanent damage can be inflicted on a forensic entomological study by a poor application of temperature data, thereby compromising the minimum PMI estimation.

Acknowledgments

The authors thank the assistance of the Ephrata Police Department in Ephrata, Pennsylvania, and the East Lampeter Township Police Department, Lancaster, Pennsylvania, for help with background information on some of these cases.

References

Davey, C. A., and R. A. Pielke, Sr. 2005. Microclimate exposures of surface-based weather stations. *Bull. Am. Meteorol. Soc.* 86:479–504.

Davies, L. 2006. Lifetime reproductive output of *Calliphora vicina* and *Lucilia sericata* in outdoor caged and field populations; flight vs. egg production? *Med. Vet. Entomol.* 20:1–6.

DeFelice, T. P. 1998. *An introduction to meteorological instrumentation and measurement.* Upper Saddle River, NJ: Prentice Hall.

Donovan, S. E., M. J. R. Hall, B. D. Turner, and C. B. Moncrieff. 2006. Larval growth rates of the blowfly *Calliphora vicina*, over a range of temperatures. *Med. Vet. Entomol.* 20:106–14.

Gallo, K. P., D. R. Easterling, and T. C. Petersen. 1996. The influence of land use/land cover on climatological values of the diurnal temperature range. *J. Climate* 9:2941–44.

Hart, A. J., and A. P. Whitaker. 2006. Forensic entomology: Insect activity and its role in the decomposition of human cadavers. *Antenna* 30:159–64.

Higley, L. G., L. P. Pedigo, and K. R. Ostlie. 1986. DEGDAY: A program for calculating degree days, and assumptions behind the degree day approach. *Environ. Entomol.* 15:999–1016.

Karl, T. R., P. J. Young, and W. M. Wendland. 1986. A model to estimate the time of observation bias associated with monthly mean maximum, minimum and mean temperatures for the United States. *J. Climate Appl. Meteorol.* 25:145–60.

Lutgens, F. K., and E. J. Tarbuck. 2004. *The atmosphere.* 9th ed. Upper Saddle River, NJ: Prentice Hall.

Nabity, P. D., L. G. Higley, and T. M. Heng-Moss. 2006. Effects of temperature on development of *Phormia regina* (Diptera: Calliphoridae) and use of developmental data in determining time intervals in forensic entomology. *J. Med. Entomol.* 43:1276–86.

Peterson, T. C. 2006. Examination of potential biases in air temperature caused by poor station locations. *Bull. Am. Meteorol. Soc.* 87:1073–80.

Potter, S. 2004. Pieces of evidence. *Weatherwise* 57:28–33.

Slone, D. H., and S. V. Gruner. 2007. Thermoregulation in larval aggregations of carrion-feeding blow flies (Diptera: Calliphoridae). *J. Med. Entomol.* 44:516–23.

Solomon, S., J. S. Daniel, and D. L. Druckenbrod. 2007. Revolutionary minds. *Am. Scientist* 95:430–37.

Sublette, M. S. 2007. The effect of sensor placement on climatological records. Paper presented at 16th Conference on Applied Climatology, San Antonio, TX, January 13–18, 2007.

Tomberlin, J. K., J. R. Wallace, and J. H. Byrd. 2006. Forensic entomology: Myths busted! *Forensic Mag.* 3:10–14.

Walsh, J. E. 2000. *Moonlight: Abraham Lincoln and the Almanac Trial.* New York: St. Martin's Press.

Whiteman, C. D., J. M. Hubbe, and W. J. Shaw. 2000. Evaluation of an inexpensive temperature datalogger for meteorological applications. *J. Atmos. Oceanic Technol.* 15:157–73.

World Meteorological Organization (WMO). 2006. *Guide to meteorological instruments and methods of observations.* 7th ed. WMO 8. Geneva, Switzerland: Secretariat of the World Meteorological Organization.

Entomological Alteration of Bloodstain Evidence

18

M. ANDERSON PARKER
MARK BENECKE
JASON H. BYRD
ROGER HAWKES
ROSEMARY BROWN

Contents

Introduction

The key to the proper interpretation and analysis of physical evidence is for the investigator to possess a combination of formal training within a particular forensic discipline, the experience of working within the chosen field, and a good dose of common sense. This statement is especially true for the analysis of bloodstain patterns. This chapter will address how bloodstain patterns may appear at a crime scene and explore some of their possible interpretations. It will also include examples of the manner in which they are examined or analyzed, and identify the entomological factors that can lead to their alteration and subsequent misinterpretation.

Physical evidence is defined as any item that may be identified as having been associated or involved with a crime scene. The examination, identification, and interpretation of physical evidence are key components of most criminal investigations. To ensure that the value and integrity of this evidence is maintained, it is essential that the proper steps be taken in its documentation, collection, and preservation. One such item that is present at most violent crime scenes is blood evidence. The correct analysis of bloodstains at a scene or on an item of evidence is in many cases the determining factor between guilt and innocence.

The science of forensic entomology is devoted to the study of insects and their arthropod relatives serving in the capacity as evidence in criminal, medical, or civil proceedings. In many cases, entomological evidence is recovered in the form of living or preserved insects that have colonized human remains. Under such circumstances, the arthropod fauna can be of great importance to investigators in determining the time since colonization, or period of insect activity, which can indicate a portion of the total postmortem interval estimation. The advanced technological procedures available in a modern crime laboratory or molecular biology laboratory permit insects and insect fragments to be analyzed for DNA and toxicology when tissue from the deceased is unavailable.

Insects may also make the analysis and interpretation of the events occurring during the commission of a crime more difficult. The removal of flesh and soft tissue from bodies by flies and beetles, and the alteration of sites of trauma are some of the most obvious modifications of evidence. The consumption of skin or hair by ant and cockroach feeding can also cause postmortem artifacts on human remains that might mislead investigators (Figure 18.1). These artifacts may resemble burns or chemical scarring to the uninitiated, but are usually readily apparent to the experienced investigator. One form of evidential alteration that is not usually considered and often overlooked is that the insects present at a crime scene can alter bloodstain patterns and can readily transport blood to new locations.

Relatively few modern textbooks on bloodstains mention artifacts produced by insects. Where it is mentioned in the literature, the reference is most commonly to artifacts produced by flies.

> The activity of flies at the scene where blood has been shed is another possible source of small stains of blood that may be confused with medium- to high-velocity impact spatter.... An understanding of the mechanics of flies feeding on blood and decomposing bodies is essential for proper interpretation of these bloodstains. The horse fly is characterized as a biter, while the common house fly is specialized as a lapper and sucker. Flies ingest blood and regurgitate it onto a surface to allow enzymes to break down the blood. At a later time, the flies return to the areas of regurgitated blood and consume a portion of the blood. The surfaces upon which these activities have taken place will contain small spots of blood material which are often a millimeter or less in diameter with no definite point of convergence or origin. Some of the stains will exhibit dome shaped craters due to the sucking process and others may show swiping due to defecation. These stains may be observed on many surfaces at the scene especially lamp shades, blinds and ceilings as well as on the victim and clothing. Their locations may be inconsistent with blood spatter associated with injuries sustained by the victim. (S. H. James and T. P. Sutton, *Interpretation of Bloodstain Evidence,* quoted after Benecke and Barksdale 2003, p. 152)

The analysis of bloodstain evidence, and the interpretation of bloodstain patterns, has evolved into a forensic discipline of its own. Unfortunately, relatively few investigators receive more than a minimum of training in this highly specialized field. Furthermore, the training that is received is often limited to the protocol for the recovery of samples used for chemical analyses and presumptive tests. This is slowly changing, as the importance of bloodstain analysis is becoming widely recognized and respected. To a bloodstain pattern expert, the size, sequence, and array of blood drops, which may be overlooked by an untrained individual, can hold a wealth of information. The recognition of insect-

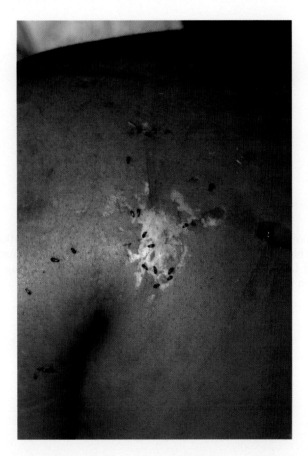

Figure 18.1 Postmortem scavenging by ants, roaches, bees, and wasps on human skin produces patterns that are often confused with perimortem trauma by some investigators. (Photo courtesy of Dr. Jason H. Byrd.)

produced artifacts, and their distinction from unaltered blood present at a crime scene, is essential for the correct interpretation of bloodstain evidence.

Any insect or arthropod may leave bloody tracks of a diminutive nature simply by walking through wet blood and tracking it onto a nearby surface (Figure 18.2). In the case of flies, the most common artifact is in the form of regurgitated or defecated blood, and this common pattern is known as fly specks (Figure 18.3). Fleas can also be responsible for defecating partially digested blood that originated from the victim or others who were present at the scene. Characteristics such as size, shape, and pattern are used by bloodstain analysts to distinguish insect artifacts from legitimate or unaltered blood spatter, and will be discussed in the following sections.

Although not a true spatter pattern, the fly speck or fly spot is often confused with bloodstains created during the commission of the crime. This pattern is created by flies present within the scene that feed on blood, body fluids, and exudates produced during decomposition. The blood and fluids present are tracked about (mechanical transmission), regurgitated, and defecated by the flies. In the instance of the tracking pattern, the marks are extremely small, but a pattern may be evident on close examination (Figure 18.4).

In the case of fly regurgitation, the specks are remarkably symmetrical and are sometimes lighter in color than surrounding blood drops. Most often, the analyst finds these

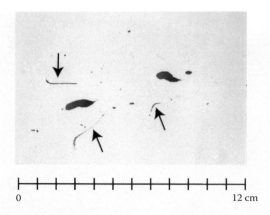

0   12 cm

Figure 18.2 Mechanical transmission of blood produced by a roach walking through pooled blood and tracking blood onto a clean surface. Arrows indicate tracks made by the rear tarsi of a roach. Areas of larger stain are produced by the abdomen contacting the surface. (Photo courtesy of Dr. Jason H. Byrd.)

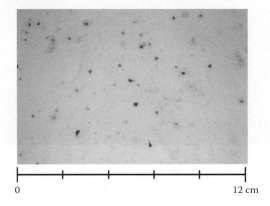

0   12 cm

Figure 18.3 Typical pattern of fly specks produced after flies have contacted free blood and fluids of decomposition. These patterns are from regurgitate and fecal material produced by flies present at the death scene. (Photo courtesy of Dr. Jason H. Byrd.)

patterns in brightly lighted and warm areas where the flies rest, such as high in window corners, along walls where the sun strikes, the back side of blinds and curtains closest to the window, and on objects nearby windows (Figure 18.5). Such stains will usually test positive for blood with a presumptive test. It is critically important for the investigator to understand that the artifacts caused by flies include the victim's blood. Therefore, presumptive blood tests like Hemastix (2190)/Heglostix (Bayer 028165A; hemoglobin catalyzes oxidation of 3,3',5,5'-tetramethylbenzidine [color reagent] by diisopropylbenzole dihydroperoxide from green to blue), Sangur (Merck), or Luminol will not differentiate between the two types of stains. Additionally, DNA typing will not differentiate between the two types of stains. Since these common laboratory tools are useless, this leaves recognition of stain patterns and other physical information as the relevant criteria. Obviously, care should be exercised in evaluating any abnormal patterns that meet these criteria. It is unfortunate that no integrated approach by either the natural sciences or forensic sciences has been used until actual casework (see case examples later in chapter), and presentation in courts made it necessary to develop a method to safely distinguish

Figure 18.4 On close examination, the regurgitate pattern is roughly symmetrical, and the fecal material is an elongated comma shape. The randomness of the pattern and contrasting directionality of the stain indicates that this pattern is produced by insect activity and is not a bloodstain resulting from a violent act committed during the crime. (Photo courtesy of Dr. Mark Benecke.)

between blood spatter caused during the commission of the crime and blood spatter-like patterns produced by flies.

The investigator must have a firm understanding of the important information to be gained in the analysis of bloodstain evidence. This information can include what did (or did not) take place, and answer the question of who may have been involved in these actions. It is an unfortunate truth that in most crime scene investigations it is highly unlikely that a bloodstain pattern expert will actually enter the scene. Therefore, most analysis is done by the expert while reviewing the documentation of another investigator's work. Within our experiences, the proper documentation of bloodstains at the scene usually does not take place.

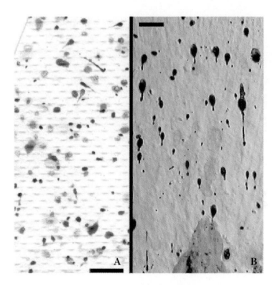

Figure 18.5 Differences between bloodstains caused by violence in contrast to artifacts caused by flies. (a) Stains produced by flies on a vertical surface (sheet of kitchen paper). Note random orientation of tails, the difference in the width of the body and tail is indistinct, and the tails do not end in the satellite spot. (b) Complex bloodstain pattern caused by blood that exited a punctured blood vessel of a man who was standing next to a vertical plane (wall). Larger stains show orientation, body, and tail of droplets, which are easy to distinguish; tail ends in satellite dot. (Photo courtesy of Dr. Mark Benecke.)

It is beyond the scope of this chapter to address all of what bloodstain patterns can provide in the way of information in a criminal investigation. However, it is hoped that this chapter will be a tool to assist in the interpretation of bloodstain evidence when it is present at a crime scene, especially in death investigations where decomposition has occurred and insects are present. This information will be presented as follows: a brief history of bloodstain analysis, basic terminology, techniques, entomological artifacts, and a practical approach to the use of this information. This chapter will conclude with a case review on how bloodstain evidence was documented and analyzed, and how entomological evidence influenced the analysis of bloodstains in an actual death investigation.

> Violent crimes usually result in bloodshed. When liquid blood is acted upon by physical forces, patterns will be deposited on the various items at the crime scene as well as the clothing of the individuals present during the activity. These patterns can yield valuable information concerning the events which lead to their creation. The information can then be used for reconstruction of the incident, as well as to evaluate the statements of the witnesses and the crime participants. Bloodstain pattern evidence can be of value in the investigation of any violent crime in which bloodshed occurs. (Wolson and Johnson, 1994)

History of Bloodstain Analysis

Bloodstain interpretation and analysis is an important tool that may be used by crime scene investigators to explain what actions took place during the act of a crime. It is also just as important in the establishment of what could not have happened at any particular crime scene. However, in order to gain the maximum benefit of blood evidence, the accurate recreation of a crime scene is a basic requirement. This allows those trained in this forensic discipline to answer many of the questions that must be addressed when processing a crime scene.

Although this discipline is considered by many to be new science, its use dates back to the mid-1800s. In fact, even nonscientific references may be found in eleventh-century European law, with the term *red handed* being used as a way to identify a criminal who had blood on his hand. More popular literary references are those of Shakespeare in *Macbeth* and Sir Arthur Conan Doyle in *A Study in Scarlet*. In both stories blood was used as evidence in a crime.

A notable early scientific contributor to this field was Dr. Eduard Piotrowski of the University of Vienna. During the late 1800s, he recognized bloodstain patterns and their causes by analyzing them in laboratory settings. Another of the earlier known studies was by Dr. Paul Jeserich, a chemist who lived in Berlin. He examined the bloodstain patterns that were present at homicide scenes just after the turn of the twentieth century. More recently, Dr. Vincent Balthazard researched bloodstains in the 1930s. His contribution of the impact angle formula to determine the point of origin is perhaps the most significant in the field of blood pattern analysis. Other notable pioneers in the study of bloodstain evidence are Ernest Ziemke and Dr. W. F. Hesselink.

In 1856, J. L. Lassaigne wrote "Neue Untersuchungen zur Erkennung von Blutflecken auf Eisen und Stahl" ["New Examination to Differentiate Bloodspots on Iron and Steel"]. This paper seems to be the first account of marks that appear similar to bloodstains, but were attributed to insects. From his writing, it is clear that he believed that such marks were caused by the "crushing" of dead flies. However, he did imply having found such stains at various scenes, and he did attribute their presence to insects. It is not known if he directly associated the presence of these marks with the regurgitation of flies present at the scene or their semiliquid fecal deposits. He never stated any references to his familiarity to their behavior and biology. However, based on his descriptive work, these stains seem similar to what we today recognize as fly specks. It is generally believed that Lassaigne had indeed found fly specks and had attempted to account for fly activity in his investigations. Lassaigne's detailed observations and effort in differentiating such stains are the first documented reference to insect alteration of bloodstains.

One of the first known cases of bloodstain pattern interpretation in the United States was during the Dr. Sam Sheppard homicide case in 1955. In the 1966 retrial, Dr. Paul Kirk, a professor in Berkeley, California, testified for the defense on several critical points. During the original trial, the prosecution presented information that the suspect holding the murder weapon, dripping with the victim's blood, made the blood trail at the scene. Dr. Kirk testified that the trail could not have been made in this manner, but rather from an open wound. Since there were no such injuries to Dr. Sheppard, the wound had to be present on the suspect. Dr. Kirk's testimony also negated other points concerning bloodstain evidence on Dr. Sheppard's watch, and in the victim's bedroom for the same reason. However, certain aspects of this case are again being presented in court, and many bloodstain experts currently disagree with the original findings of Dr. Kirk. In addition to his pioneering work on this case, Dr. Kirk authored *Crime Investigations*, in which bloodstain pattern interpretation is addressed. Since that time, numerous articles on this discipline have been published that explain the science and how conclusions can be made from blood evidence with a reasonable degree of scientific certainty.

In 1971, *Flight Characteristics and Stain Patterns of Human Blood* was published as the result of the research and studies of Dr. Herbert MacDonell and Lorraine Bialousz. Many researchers in this field have taken the groundwork established by MacDonell and Bialousz and incorporated it into their respective studies, publications, and teachings.

Questions to be answered by the examination of evidence at a crime scene will fit into two basic categories: what is fact and what can be determined within a degree of scientific certainty. Determining factual information is usually the easier of the two equations, but it is the portion that is most often overlooked. Such things as "there lies a dead body" or the fact that "blood is present throughout the scene" are available to us immediately as we enter a scene and are indisputable. In a majority of cases, but not always, the "fact" is not questioned. Therefore, the more difficult aspect of the examination of physical evidence is the second half of the equation. This requires the investigator to answer the what, when, why, and where of all the circumstances surrounding the investigation. With bloodstain patterns, these are exactly the questions that are brought forth as the analysis of the evidence begins. Once fact is established, it then becomes a question of how to determine the answers to the what, when, why, and where. There are many aspects involved with the various avenues of reaching these conclusions, but as with all forensic scientific findings, they should only be based on accepted and proven practices.

With all facets of forensic sciences, there is a protocol and sequence of questions and deductions that must take place to eliminate certain possibilities, and to reach a point of identification or threshold degree of scientific certainty. In many of the cases involving bloodstain patterns, this degree of scientific certainty is the best standard for case analysis. Also, a certain percentage of deductions made from the review of bloodstains will be scenarios of what could not have happened. This protocol of questions and deductions can be a long, detailed, and arduous process. At other times, it is a split-second thought that has become second nature from years of experience.

Before a technique becomes an accepted protocol, it should be used consistently to ensure proper review of the evidence. No matter how time-consuming or detailed the process may be, it is important that proper protocol be practiced and followed without error, and such protocol should be consistent in both its application and its simplest form before being utilized to derive any conclusions. It is beyond the scope of this chapter to provide a complete crime scene protocol. It does not include all that could, or should, be done with bloodstain evidence recovered from a crime scene. What this protocol will cover is general steps in crime scene documentation, the examination and analysis of the blood as it pertains to the pattern interpretation, and the entomological factors that may lead to its alteration.

It must be understood that each investigation and crime scene is unique. There can never be an exact order, or series of steps in any protocol, that can be universally applied to the processing of all crime scenes. There are many extenuating circumstances that will affect the decisions and priorities of the investigation. With that said, there are some guidelines that can be used to establish a crime scene protocol that will ensure that bloodstain evidence is handled properly.

Protocol and Techniques

The key to any investigation is communication among all of the parties involved. No individual or unit can work successfully within a communication void. Very little can be accomplished, and much can be destroyed, if the information from the investigative team and the crime scene unit is not shared. Additionally, an outline of crime scene processes must be created to ensure evidence is documented properly, and that evidence is not destroyed or rendered useless. The outline will serve as the action plan for processing the scene. This is a critical step in the crime scene processing so that the documentation, collection, or analysis of one type of physical evidence will not interfere with another.

Listed are five basic steps that provide a general outline of actions to be taken at all crime scenes. Such basic and elemental steps provide for the proper documentation, collection, and preservation of physical evidence. However, one or more of these steps are frequently overlooked by crime scene investigators.

1. Initial visual inspection, notes, and information gathering
2. Photography and videography
3. Scene sketch and measurements
4. Scene search and collection of evidence
5. On-scene processing (latent prints, bloodstain patterns, casting, etc.)

Within these five areas, the process of bloodstain pattern interpretation has its own guidelines to be used to determine the value of bloodstain pattern evidence. However, this is only in the area of bloodstain pattern interpretation and does not include the processing of an entire scene or the preservation of blood for serological examination.

Utilization of the team approach to processing a scene with bloodstain evidence is preferable. If a team approach is not possible, the procedures will remain the same, but a team allows the various tasks to be done in a timely manner. Also, it is better to have another set of eyes and ears, and another investigative mind, to help eliminate the chance of missing critical evidence.

Bloodstain analysis begins with a visual examination of the overall scene. As obvious as this seems, the importance of a thorough visual examination is often overlooked. This step will determine all subsequent processes within the scene, and will provide the primary direction of the investigation. It is within this step that fact versus the unknown may begin to be established. It must be noted and emphasized that conclusions should not be drawn at this time, although in many cases this information is the foundation for the final interpretation.

Written notes are crucial, and such documentation will be an ongoing process during the entire investigation. This information can come from personal observations, others on the investigative team, victims, witnesses, and sometimes, even the suspects. In addition to the basic crime scene notes, the first observations required for the proper analysis of bloodstain evidence should include environmental conditions, identification of the areas where there is apparent bloodshed, and a record of the blood's condition (i.e., wet or dry, color, consistency, etc.).

The phrase "apparent bloodshed" is often used in this area of forensic science. Determining whether a substance is, or is not, blood will depend on several factors: the level of the analyst's training and experience, the appearance of the substance, and the presence or known presence of a blood source. A blood source may be a person who has been injured, a person who has blood on him or her, or an object with trace amounts of blood. If there is any question that the substance is blood, a presumptive blood test should be performed. This may be done at different times during the analysis. The scene and its components dictate when this will be done. If the test result is negative, the remainder of the analytical process is eliminated, and all such information should be included in the scene notes. Eventually, all of the bloodstains analyzed will be collected and examined for various properties, including a determination whether it is or is not human blood.

In our experience, the following suggestions and techniques are offered for use in differentiating between fly artifacts and human bloodstain patterns (Figure 18.5a and b).

1. Document any fly activity at a scene with photographs and written notations. Flies will be at a scene if access to the scene is available. They will stay at the scene as long as a food source is available to them or as long as they are trapped. Therefore, check for and document any dead flies that may be present. If evidence of flies is present at the scene, assume that fly artifacts will be at the scene. Follow standard protocols for insect collection as outlined in Chapter 3.
2. Document the areas in which stains are found both photographically and with written notations. Fly activity will often be found to concentrate near light sources, on light-colored walls, and near windows and mirrors. It is also common for them to be present in rooms located well away from the body. Make a photographic comparison between stains away from the body and stains near the body.
3. It is of critical importance for the investigator to compare stains at the scene with known fly artifact patterns. In many instances, the patterns produced by insects are remarkably consistent.
4. Identify suspected human bloodstain patterns that are of the spot or teardrop pattern that offer a potential for use in reconstruction, and eliminate the following indicators of entomological origin:
   a. Stains that have a tail/body (Ltl/Lb) ratio greater than 1.
   b. Stains with a distinct head and tail shape (tadpole-like in silhouette).
   c. Stains with a tadpole-like shape that do not end in a small dot.
   d. Any small stains (less than 4 mm) without a distinguishable tail and body, with a clear, white, or yellow central area.
   e. Any stains with a wavy and irregular linear structure.
   f. Any stains that do not have principal directionality consistent with other stains, which suggest a common point of convergence, or a common point of origin. Within a large grouping, bloodstain fly artifacts will point in all directions. However, cast-off blood from human activity will produce stains that within a group have a common general convergence point.

Probably the most important observation for an investigator to make is to note the absence of known human bloodstain pattern characteristics. For instance, the absence of misting around a concentrated mass would suggest the stains might not be from impact or cast-off. Within a group, human cast-off patterns often leave secondary wave cast-off patterns and runoff patterns. As with any investigation involving bloodstain evidence, it must be kept in mind that one or two stains do not make a case. Stains that could be fly artifacts should be eliminated, and an evaluation made based upon stains that can be explained in terms of origin and relevance to the reconstruction. As with general crime scene investigation techniques, always use a high-resolution camera and a macro lens capable of 1:1 reproduction. Always take duplicate photos with and without a photographic scale shown with the evidence.

Other ongoing documentation will be general scene photography, videography (when applicable), and sketches. Proper lighting and light angles are important in scene photography, as the photographs may be used for further analysis and scene reconstruction.

Figure 18.6 Typical cast-off pattern produced by a human hand covered with blood. (Photo courtesy of Melvin R. Bishop.)

Sketches are also useful for visual interpretation of blood evidence notes without the distractions of nonevidentiary objects that may be present in the photographs.

The first step in the actual analysis is to determine, if possible, the bloodstain patterns from any created artifacts. In order to accomplish this, it is necessary to be familiar with the terminology used in this discipline. Some of the most important and basic terminology used to define bloodstain patterns is included in this chapter. The following are some of the most common pattern definitions, while more complete lists may be found in Bevel and Gardner (1997) and MacDonell (1993).

- **Cast-off pattern:** The pattern created when blood is flung from an object in motion (Figure 18.6).
- **Drip pattern:** The pattern created when blood drips into liquid blood (Figure 18.7).

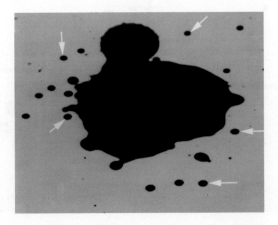

Figure 18.7 Typical drip pattern created when blood from a stationary source drips into previously pooled blood, producing satellite spatter (indicated by arrows). (Photo courtesy of Melvin R. Bishop.)

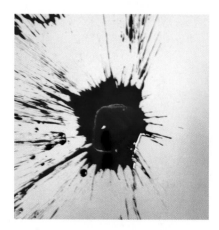

Figure 18.8 Pattern produced when stationary pooled blood is impacted by an external force. In this image, a hammer has impacted a pool of blood. (Photo courtesy of Melvin R. Bishop.)

- **Impact pattern:** The pattern created when stationary blood is put into motion by an external force (Figure 18.8).
- **Passive drop or flow pattern:** The pattern created when the force of gravity alone acts upon blood (Figure 18.9).
- **Projected blood pattern:** The pattern created when blood under pressure and in volume lands on another surface (Figure 18.10).
- **Transfer pattern:** The pattern created when a wet, bloody object comes in contact with another surface (Figure 18.11).
- **Wipe pattern:** The pattern created when a nonbloodied object moves through wet blood and changes the appearance of the blood (Figure 18.12).

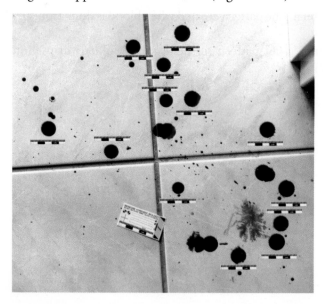

Figure 18.9 Bloodstain pattern produced by the passive dripping of blood from a source with slight movement. (Photo courtesy of Dr. Mark Benecke.)

Figure 18.10 This radiating pattern is produced when blood in motion from a source impacts a stationary surface. (Photo courtesy of Dr. Mark Benecke.)

Figure 18.11 A transfer pattern is produced when a blood object comes in contact with a clean surface. This pattern may repeat itself many times, becoming less pronounced each time it is repeated, due to smaller amounts of blood remaining on the shoe surface after each contact. (Photo courtesy of Melvin R. Bishop.)

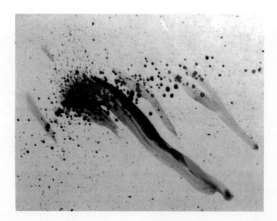

Figure 18.12 A wipe pattern is produced when a clean object contacts and moves through pooled blood, distributing the blood over a greater surface area. (Photo courtesy of the Tallahassee Police Department.)

As stated previously, a single drop of blood does not identify the pattern. The entire area of bloodstains must be evaluated. However, a single drop must not be overlooked, as it could be an important piece of bloodstain evidence. The overall pattern may indicate what happened where, how, and when, but a single drop may provide the who part of the who, what, when, where, and how equation.

In 1995, the Tallahassee Police Department investigated a homicide in which an unknown suspect attacked a local college student in her apartment. During the stabbing assault, there was a violent struggle that began in the rear bedroom and moved throughout the apartment. It ended at the front doorway, where the victim was eventually found. The scene was covered in blood with a multitude of patterns showing the movement and stabbing strikes (Figures 18.13 to 18.17).

From this entire scene, it was not any of the patterns that specifically identified the attacker, but rather one lone drop of blood on the bathroom floor (Figure 18.16). This bloodstain was separate from all other patterns, and it was not consistent with the attack.

Figure 18.13 Rear bedroom where incident started. Blood and hair were found on the floor in several areas, and on the side of the shoe by the bed. (Photo courtesy of the Tallahassee Police Department.)

Figure 18.14 Kitchen area where the stabbing and struggling continued from the rear bedroom. (Photo courtesy of the Tallahassee Police Department.)

Figure 18.15 Victim in doorway of apartment. Bloodstain evidence is clearly visible on the floor, the block wall by the door, and the half-wall near the kitchen area. Bloodstain evidence was also found on the back of the open door. (Photo courtesy of the Tallahassee Police Department.)

The suspect, who was picked up later the same day, had minor wounds on his hands (Figure 18.17). He had apparently transferred this one blood drop as he traveled through the hall adjacent to the bathroom. A blood test confirmed that this drop matched that of the suspect. In this case it was the observation of where the blood did not come from, rather than where it did, that made it stand out and eventually defined its importance.

It is not always possible to accurately identify a pattern, due to conditions and circumstances too numerous to list here. An analyst examines all aspects of the blood evidence available in order to make a pattern determination. The results of these examinations should be included in the scene notes and reports, even if a pattern cannot be identified.

Since each pattern at a scene reveals certain unique information, a brief discussion of each of the previously listed patterns is warranted. The cast-off pattern when made by the instrument of impact attack will tell the minimum number of blows struck. The first blow

Figure 18.16 The lone drop of blood on the bathroom floor that led to the identification and apprehension of the suspect. (Photo courtesy of the Tallahassee Police Department.)

Figure 18.17 The injuries to the suspect's left hand. (Photo courtesy of the Tallahassee Police Department.)

will draw blood that adheres to the instrument, and the subsequent blows will create the pattern. The blood drops in this pattern will generally be linear in appearance (Figure 18.18). A drip pattern may indicate the location of a blood source that did not move for a period of time (although the timeframe is difficult or impossible to determine), and a pool of blood with satellite blood drops characterizes this pattern (see Figures 18.7 and 18.19).

There are three types of impact patterns characteristic of high-, medium-, and low-velocity impact. Each is typically defined by the diameter size of the majority of stain drops, although stains of all sizes may be present. There are certain, but not always inclusive, actions that may be associated with each type. High-impact stains have the smallest diameter (0.1 mm or less) and are mist-like in appearance. Such patterns are produced when blood is exposed to a force of 100 feet per second or greater. However, bloodstains from high-velocity impacts are also produced in the medium-velocity size range. High-velocity impacts are typically associated with gunshot wounds, and sneezing and coughing of fluid

Figure 18.18 Cast-off stain on wall at the scene of a stabbing. These and other types of stains will often be linear in appearance. (Photo courtesy of Dr. Mark Benecke.)

Figure 18.19 Drip pattern produced by movement of blood source showing some satellite stains. Repetitive transfer patterns and smear patterns are also present. (Photo courtesy of Dr. Mark Benecke.)

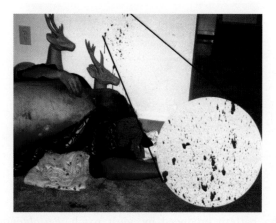

Figure 18.20 Impact stains and other patterns produced by coughing blood. (Photo courtesy of the Tallahassee Police Department.)

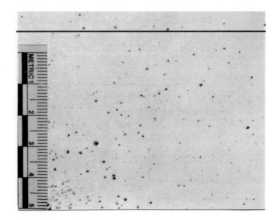

Figure 18.21 High-velocity impact pattern produced by a gunshot wound. (Photo courtesy of the Tallahassee Police Department.)

that contains blood (Figures 18.20 and 18.21). However, such patterns may also be produced when trauma is induced by either heavy, blunt objects, or rapidly moving objects. This type of trauma and the resulting blood patterns are often produced in industrial accidents. Insects present at the crime scene can also create droplet-sized stains such as these. Often the mechanical transmission of blood from a pooled source to a blood-free surface is accomplished by the tarsi (feet) of flies, or the fecal material of fleas. Roaches can also produce apparent blood droplets by their blood-contaminated tarsi, but the droplet size is larger than that produced by fly tarsi.

In the case of roaches, such artifacts are usually easily distinguished, as the overall droplet pattern reveals a series of "tracks" that often have a center drag mark or smear caused by the roach dragging its undersurface or the abdomen tip as it walks (Figures 18.22 and 18.23). It is this intermittent smear that first draws attention, and the imprints made by the roach's tarsi are usually apparent upon closer inspection (Figures 18.24 and 18.25). In many cases, such details are best viewed with a hand lens. With flies, such an event is not as easily distinguished since the pattern produced is more isolated, smaller, and not as uniform. Flies with blood-soaked tarsi may alight on any surface, vertical, horizontal, or

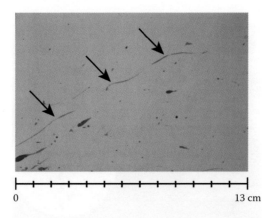

Figure 18.22 Transfer pattern produced by adult roach, *Periplaneta americana*, running through pooled blood. Note the elongated trail produced by the roach dragging its abdomen. (Photo courtesy of Dr. James L. Castner.)

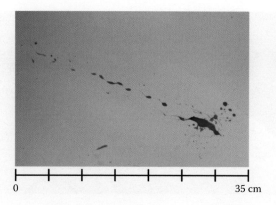

0   35 cm

Figure 18.23 Transfer pattern produced by adult roach, *Periplaneta americana*, slowly walking through pooled blood. (Photo courtesy of Dr. James L. Castner.)

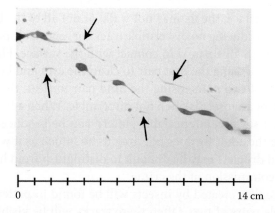

0   14 cm

Figure 18.24 Enlargement of Figure 18.23, showing the drag pattern produced by the roach's abdomen and tarsal imprints. (Photo courtesy of Dr. James L. Castner.)

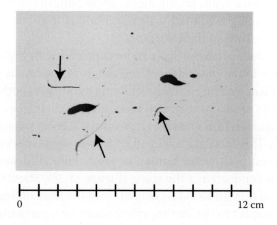

0   12 cm

Figure 18.25 Bloodstain artifacts produced by a roach moving through pooled blood. Note imprints of abdomen and distinct tarsal tracks (indicated by black arrows). (Photo courtesy of Dr. James L. Castner.)

Figure 18.26 All legs of an insect are not always in contact with the surface. Insects groom frequently, and do so by rubbing the body with their legs. This behavior often produces irregular artifacts in bloodstain evidence. Here, *Sarcophaga bullata* (Parker) cleans its compound eyes, leaving only four legs in contact with the surface. (Photo courtesy of Dr. James L. Castner.)

inverted. Once on the surface, the fly may not walk, or not all of the legs may be contaminated with blood, thus producing no discernible tracks or repeatable patterns. Additionally, the fly may not alight with all six tarsi in contact with the surface. Flies often rest on only the hind two pairs of legs, using the first pair to clean the eyes and head (Figure 18.26), or they may rest on the first two pairs, using the hind pair to clean the abdomen. Thus, the typical six-point track of an insect is often not discernible. When a fly with blood-contaminated tarsi walks on a surface, a repeatable pattern can be produced, but no drag mark or smear occurs since the adult fly does not drag its abdomen as it walks. Therefore, such insect-produced blood droplets may be difficult to distinguish from high-velocity droplets produced during the commission of the crime.

Many of these droplets created by insects will be found in patterns of three pairs, or readily distinguishable pairs of two. Often these tracks will be visible as a linear pattern of three pairs of droplets with each row separated by 0.5 to 1.75 cm with flies, and 2.0 to 3.0 cm with roaches. Such tracks will only span a short distance due to the small amounts of blood that typically cling to the small tarsi of flies, and due to the fact that flies do not usually walk far before once again taking flight. These patterns most commonly occur on vertical surfaces (such as light-colored walls) and on ceilings because of the behavioral propensity of flies for landing on these surfaces.

Fleas may also produce artifacts that can be confused with high-velocity blood spatter. Fleas are commonly found in residences and can form enormous populations in a very short period of time. Fleas feed on the blood of living mammals (both human and animal) and pass large quantities of their undigested bloodmeal in their feces. The fecal material of fleas is at first a semiliquid, which dries into a powder if it is not passed on to an absorptive surface (Figure 18.27). This fecal material will appear as small specks that are mist-like in appearance, and will test positive for human or animal blood (depending on the source) if a presumptive test is conducted at the scene (Figure 18.28). Standard laboratory testing can be accomplished on human blood recovered from the feces of fleas, and such evidence can be used to link a suspect to the victim, crime scene, or a particular and relatively confined area. A suspect would not have to linger long before fleas present at the scene would begin to feed and pass human blood onto substrates at the scene. The closely spaced and small droplets of flea fecal material often have the appearance of a fine spray coating over

Figure 18.27 Another common pattern produced by insect activity that is often confused with the high velocity patterns produced from gunshot wounds is called "flea specks." Fleas produce small droplets (less than 1mm) of blood when they feed. These droplets quickly dry and serve as the food source for their larvae. (Photo courtesy of Dr. Nancy Hinkle, University of Georgia, Athens.)

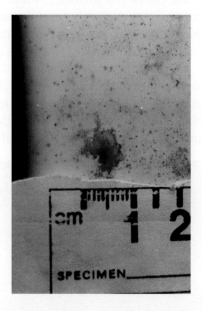

Figure 18.28 "Flea specks" found on an apartment wall at the scene of a bludgeoning. The large stain in the center was produced from the attack. The smaller droplets, initially thought to be high velocity stains from a gunshot wound, are flea specks. (Photo courtesy of the Tallahassee Police Department.)

a surface. This coating of small droplets is found predominantly on white or light-colored surfaces (not only because it is more apparent, but also because fleas are attracted to lighter colors), and is typically confined at a height below 24 inches. The most common areas in which to search for this evidence are along baseboards and floor trim. They are also common on carpets and rugs, but may be difficult to detect with the unaided eye on many fabrics, depending on fabric color and texture.

Medium-velocity impact stains range in size from 1 to 4 mm diameter (Figure 18.29) and are produced by a force other than gravity of 5 to 100 feet per second. They are most

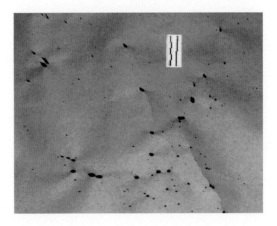

Figure 18.29 A typical medium-velocity impact pattern. (Photo courtesy of Melvin R. Bishop.)

commonly associated with blunt trauma injuries. Many artifacts produced by the presence of insects overlap with this droplet size class as well as the low-velocity impact size class. Low-velocity impact stains have the largest diameter, 4 mm and larger, and are formed when blood is exposed to a force of approximately 5 feet per second or less. These are usually associated with gravitational force and may form part of another pattern, as illustrated in Figure 18.30. Within the medium-velocity size class exists certain patterns that can be immediately recognizable as produced by (or associated with the presence of) flies. Fly specks are the term entomologists give to the deposited liquid fecal material of flies. This material contains large volumes of partially digested and undigested blood if the adult flies have recently fed on free blood, or a body at a crime scene. Thus, a fly feeding on human blood will pass a large amount of partially digested blood that will test positive as human

Figure 18.30 A low-velocity impact pattern (with other patterns) on a wall and baseboard. (Photo courtesy of the Tallahassee Police Department.)

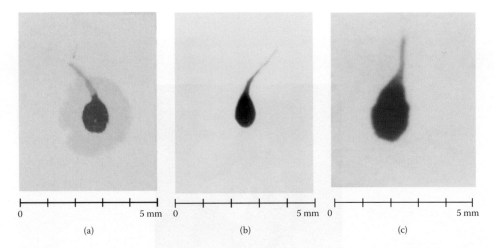

(a)   (b)   (c)

Figure 18.31 Characteristic blood imprints of fly fecal spots, or fly specks, with tail curved to the left (a), right (b), or center (c). Swipe patterns such as these can be easily recognizable to the trained analyst. (Photos courtesy of Dr. James L. Castner.)

if a presumptive test is conducted at the scene. Blood contained within these droplets can be collected and used in all standard laboratory molecular and blood serology tests. Drops such as these can usually be distinguished from those created during the commission of a crime by their overall shape. Fly specks are a swipe pattern that typically exhibits a comma shape, with the tail of the drop trailing to the left, right, or straight of center, depending on the movement of the fly abdomen (Figure 18.31a–c). The fly touching the tip of the abdomen to the surface as it defecates and walks about produces this type of bloodstain artifact. It has the same basic characteristics as many other swipe patterns, and can be easily recognized by the trained analyst.

Flies also will produce medium to large droplets due to their natural feeding behavior and digestive habits. As an adult fly digests its liquefied meal, it frequently regurgitates its gut contents. This regurgitation accumulates as a medium to large droplet at the tip of its sponging mouthparts (Figure 18.32). Often this drop either falls to the surface or is accidentally touched to a surface. Upon contact with a surface, the drop usually adheres or is absorbed and is only partially reconsumed by the fly. This drop is usually symmetrical and has circular borders (Figure 18.33). If the regurgitate is touched to a nonabsorptive vertical surface, it may produce a tail descending straight down the midline and lower edge of the droplet, but this is not always the case due to the viscous nature of the regurgitate.

When an impact pattern is created, the blood drops generally form a radiating pattern, with the highest concentration in the center, but seldom in a complete circle. From this pattern, the analyst may calculate the impact angle, which is defined as the internal angle at which blood impacts a surface, and use the results to determine the point of origin of the blood drops. The mathematical formula utilized to calculate the impact angle is relatively simple. First, select representative and well-formed stains within the pattern. Of these stains, each one will be calculated individually. Measure the width (W) of the stain and divide by the length (L) of the stain. Using any scientific calculator, simply compute the arc sine of the resultant value (angle of impact = arc sine W/L). The result will be the impact angle, and these measurements can then be utilized to calculate the point of origin. The results are used to form a three-dimensional view that provides a range of height and

Figure 18.32 A female *Cochliomyia macellaria* blow fly with a drop of regurgitate on the sponging mouthpart. This is a common feeding behavior found in all species of Calliphoridae, Muscidae, and Sarcophagidae. (Photo courtesy of Steve Grasser, www.bugwood.org.)

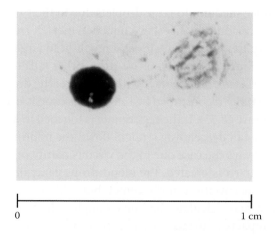

0   1 cm

Figure 18.33 Typical symmetrical bloodstain produced by the regurgitation of blood by an adult fly. This regurgitate spot was deposited on a vertical plaster surface by *Phormia regina* (Meigen). (Photo courtesy of Dr. James L. Castner.)

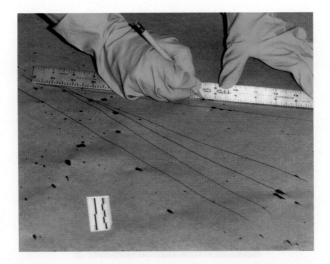

Figure 18.34 Blood drops being charted to determine the point of convergence. This step is an integral component of determining the point of origin. (Photo courtesy of the Tallahassee Police Department.)

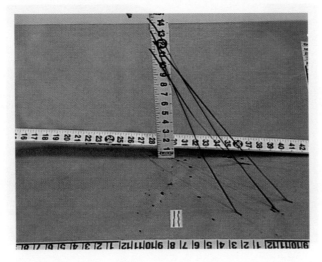

Figure 18.35 The three-dimensional view revealing the height and distance ranges of the blood source from the impact surface. This is the final step in determining the point of origin to reveal the height and distance ranges. (Photo courtesy of the Tallahassee Police Department.)

distance of the blood source from the impact surface. Some of the steps used to determine this are depicted in Figures 18.34 and 18.35.

The passive pattern may be used to show movement. A trail of passive drops will indicate the blood source's direction of travel, or the flow may show movement of a body after death. The most common instance occurs when the subject bleeds from the nose or mouth, as seen in Figure 18.36. The blood will flow from the orifice to the lowest point, independent of the body position.

Projected blood patterns are commonly associated with arterial spurting or gushing. Such patterns may also be created by the vomiting of blood, and are not usually duplicated through insect activity. Movement of a person or object through pooled blood may cause the blood to project, which is characterized by a large volume of blood with spines. On a

Figure 18.36 Victim was shot through the nose and projected blood on the wall and door when he exhaled. (Photo courtesy of the Tallahassee Police Department.)

Figure 18.37 Passive flow from the nose, ear, and head wound in a shooting suicide case. Note the void in the blood pool created by the victim's face and chest. (Photo courtesy of the Tallahassee Police Department.)

vertical surface, it appears as a large upside-down drop with a drip. This phenomenon is illustrated in Figure 18.37, from a shooting case and a dismemberment/arson case, respectively. Larger insects (such as roaches) walking through areas of pooled blood will alter the projected pattern, and typically produce pronounced tracks. Therefore, it is important to note what types of insects were observed at the crime scene and to collect the insects properly. Chapter 3 addresses the proper insect collection protocol at a crime scene.

A transfer pattern may be a smudge or a swipe that could show movement. A flat, bloody hand swiped across a surface will leave its form and show the direction of travel of the hand. The transfer may identify an object such as a weapon, shoe, or fingerprint, and the blood may be in the form of an outline, or a partial or solid pattern. In some cases it is possible to match the object or print when the original is obtained (Figures 18.38 and 18.39).

Figure 18.38 Transfer of bloody axe head on carpet in assault case. (Photo courtesy of the Tallahassee Police Department.)

Figure 18.39 Axe head that created the transfer in Figure 18.38. (Photo courtesy of the Tallahassee Police Department.)

Figure 18.40 Wipe and repetitive transfer stains created after an attempt to clean up a bloodstain. (Photo courtesy of the Tallahassee Police Department.)

Wipe patterns may be used to establish a timeline of events when it can be determined what caused the blood at a particular location. Such bloodstains and patterns will be obviously altered if they interact with another object (Figure 18.40). Another condition that significantly changes the appearance of bloodstains is a void. This is a gap in an otherwise continuous pattern, and it is observable in Figure 18.37. Such patterns sometimes show an identifiable outline of the object that intervened between the blood source and the impact surface. However, pattern identity alone does not provide enough information to reconstruct the scene and incident. Patterns illustrate the actions that took place, and it is necessary to define all of the other bloodstain evidence. There are always general topics to consider, but each scene and its components will dictate which topics need more detailed deliberations.

The analyst considers if and how the observable patterns are connected. This could be through contact, movement, or travel. These examples are detailed below. In Figure 18.41, a cast-off pattern associated with an impact pattern is indicative of blunt trauma. A fabric transfer pattern, such as the one in Figure 18.42, found on several different surfaces, shows movement within the scene. The blood trail with a dripped pattern in Figure 18.43 points to a blood source moving and stopping at various intervals.

Figure 18.41 Cast-off and impact patterns created by hammer blows. (Photo courtesy of the Tallahassee Police Department.)

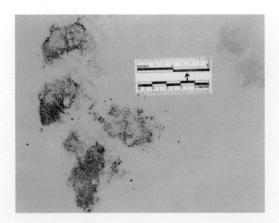

Figure 18.42 Fabric transfer patterns created by a bloody gloved hand. (Photo courtesy of the Tallahassee Police Department.)

Figure 18.43 A shooting suicide victim whose death did not occur immediately left this trail. (Photo courtesy of the Tallahassee Police Department.)

It is important to establish how many persons could be a blood source. The presence, or known presence, of a person or persons at a scene with bleeding injuries indicates the minimum, but does not eliminate an unknown presence. An indicator of such an unknown may be possible from the bloodstain evidence at the scene. A trail of arterial spurts leading away from a body indicates another injured person, although there is a good possibility that a second body will be found at the end of the trail. Such circumstances make it imperative that the collection of blood samples include every area that could have come from an unknown source. Only a specific blood test will confirm or refute the presence of an unknown person.

The process of reconstruction proceeds by linking the bloodstain evidence with the other physical evidence and components of the scene. This starts at the scene and continues at the lab. Items such as furniture may be transported for further analysis in better lighting conditions. Clothing may provide explanations or be an extension of patterns at the scene. Documentation of this evidence is best done in the controlled environment of the forensic laboratory. It is sometimes possible and usually desirable to transport enough of the items related to the evidence to physically reconstruct the scene at the laboratory.

Figure 18.44 The height range from the floor was 12 to 15 inches, and the distance range from the wall was 4 to 6 inches. The foot of the bed is seen in the lower-left corner of the photograph. (Photo courtesy of the Tallahassee Police Department.)

The ultimate goal of reconstruction is to answer the questions of who, what, when, where, and how. Bloodstain evidence analysis plays an important role in those answers. In reference to a specific case, an example of this process would be the analysis of a bloodstain pattern from a reported accidental injury by gunshot to the head. In this case, it was reported that the victim, while leaning over his bed, was shot in the temple area of his head with a 9 mm handgun. An examination of the bloodstain pattern revealed that at the time of the shooting, not only was he not leaning over the bed, but the fatal shot was fired from the opposite side of the room, which contrasted with the original statements provided to the investigators. Figure 18.44 illustrates that not only is the direction of the bloodstain pattern from left to right, to indicate the correct position of the shooter, but from the height of the pattern, the victim had to have been sitting on the floor rather than leaning over the bed.

This case is an example of positive bloodstain evidence that was found not to be consistent with the witness statement. It gave the investigators the ability to eliminate all but a couple of possible scenarios, and greatly assisted in the questioning of the suspect and ultimately determining the truth. In many cases, there are no witnesses, and it is the evidence alone that supplies the information of what did or did not take place. However, many times there are other factors that affect the appearance of the bloodstain evidence at a scene, and all possibilities should be considered.

Entomological Artifacts

Case 1: Homicide in Tallahassee (Parker)

Case Scenario
The following case study will reveal how the physical evidence was consistent throughout the scene except for one area. It is an example of how the unknown, through examination of the evidence and what was known as fact, led to the positive conclusion of the investigation. In January 1998, the Tallahassee Police Department Crime Scene Unit was called to

Figure 18.45 Stairway and landing outside the victim's apartment. The numbers mark blood-stain evidence. (Photo courtesy of the Tallahassee Police Department.)

investigate the scene of a death at an apartment complex on the west side of Tallahassee. Neighbors who lived in the adjacent apartment reported the disappearance of a man last seen in mid-December. They had also reported a foul smell that had grown considerably stronger in the last week.

The residence was a second-floor apartment with an entrance foyer, living area, kitchen, hall, and one bedroom with an adjoining bathroom. Upon the first responding officers' arrival, the front door and all of the windows were secured. The officers were unable to generate a response from anyone inside the apartment. Bloodstains were observable on the outside staircase, which had fourteen steps that led to the landing at the front door (Figure 18.45). The blood was present on the top four steps and the landing. With all of these circumstances, the officers gained immediate entry into the apartment and found the resident's body.

Upon entering the apartment, crime scene unit personnel observed bloodstain evidence throughout the residence. This included blood drops on the floors, as well as projected and transfer blood on the walls. In the areas of the foyer, living area, kitchen, and hall, the bloodstain patterns were consistent with movement throughout the apartment, but not with any violent actions (Figure 18.46). The reasoning behind this assessment was that these patterns did not show any characteristics of impact force or cast-off. The appearance of the bloodstain evidence was also consistent with one bleeding source. The examination of the bedroom and bathroom revealed that the patterns in these areas were consistent with the rest of the apartment (Figure 18.47). The victim was found in a bathtub lying on his left side. He was positioned with his feet at the drain end of the tub and was in an advanced stage of decomposition (Figure 18.48). There were also additional patterns, such as pooled blood, and large areas of transfers. These patterns were produced from the migratory third-instar fly larvae exiting the bathtub. These impressions are characteristic of the long linear bloodstain artifacts in Figure 18.49. Items of clothing that contained blood evidence were also present.

There were very large numbers of both living and dead flies throughout the apartment. These flies were mainly concentrated on the windows, windowsills, and the floors under the windows. Flies and fly pupal cases were also present in large numbers on the bathroom floor. Pupal cases as well as maggots were also found associated with the body.

Figure 18.46 Bedroom floor and bathroom entrance, both displaying bloodstain evidence. (Photo courtesy of the Tallahassee Police Department.)

Figure 18.47 Part of bathroom where the body was found. Note the dead flies on the floor. (Photo courtesy of the Tallahassee Police Department.)

Figure 18.48 The victim's body in the bathtub displaying both bloodstain and entomological evidence. (Photo courtesy of the Tallahassee Police Department.)

Figure 18.49 View of foot showing characteristic maggot trails (linear stains), which are sometimes confused with antemortem movement. (Photo courtesy of Dr. Mark Benecke.)

Along with the bloodstain evidence, there were several items indicating that this case was a suicide. These items included several empty medical prescription bottles and a few empty razor blade packages next to the bathtub. There was a notepaper tablet containing a suicide note in the bathtub with the victim. The only aspect of the evidence that was not consistent within the scene was one particular bloodstain pattern. This concentrated pattern was on the bathroom closet doors (Figures 18.50 and 18.51), mirrors, and portions of the walls. The stains appeared to be all at a 90° angle to the surface, and had a very small diameter. They were consistent in appearance with the patterns that would be indicative of high-velocity impact trauma to the victim. However, there was no evidence of gunfire, and there were no guns found at the scene.

After considerable analysis of these bloodstain patterns, a forensic entomologist was contacted. Through proper documentation, examination, and consultation with

Figure 18.50 Bathroom closet doors where transfer and apparent high-impact patterns were observed. (Photo courtesy of the Tallahassee Police Department.)

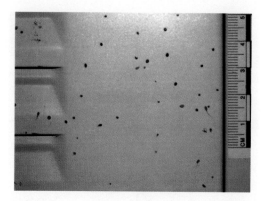

Figure 18.51 Close-up of one area of the doors in Figure 18.50. (Photo courtesy of the Tallahassee Police Department.)

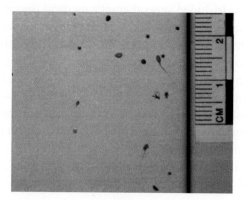

Figure 18.52 Enlargement of stains on closet door suspected of being entomological in origin. (Photo courtesy of the Tallahassee Police Department.)

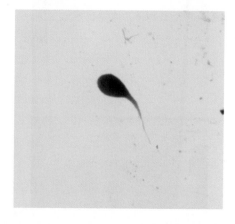

Figure 18.53 Stain produced as fly fecal droplet under laboratory conditions. (Photo courtesy of Dr. James L. Castner.)

the forensic entomologist as part of the forensic science team, it was determined that the patterns originated from the insects present, and that no violent force created these patterns. Proper microphotography of the blood droplets found at the scene was essential in documenting the patterns for later comparison with known samples produced by entomological activity. The scene photographs of different droplet sizes and flow patterns (Figure 18.52) were compared with blood droplets produced by the mechanical transmission and digestion of flies (Figure 18.53) contained in colonies at a forensic entomology laboratory. These latter samples were found to be consistent with the photographs and observations made at the scene.

Case 2: Double Homicide in Nebraska (Benecke)

Case Scenario

The remains of two men were discovered at 14:25 hours on June 14, 1997 in a third-level apartment in a five-plex apartment building in urban Lincoln, Nebraska. Both victims were fully clothed; both had a gunshot wound to the head and gunshot wounds to the torso. One victim was found face down in the kitchen area, and the other victim was prone on the living room floor on carpeting. Pools of a reddish substance were observed around the bodies of the victims. The bodies were in the active decay stage with black putrefaction only just beginning; the skin was intact everywhere except where there were gunshot wounds. At the wound sites, there was dried blood and body fluids and in the areas around the bodies, flies, maggots, and pupae as well as some adult flies were present. The temperature registered 30°C on the wall thermostat.

Entomological Evidence

On June 15, 1997, at 06:45 hours, the police collected six adult flies, three third-instar fly larvae, and several hundred first-instar fly larvae in ethanol from the scene. Adult flies had probably gained access to the apartment by an open space beneath the front door. Both adults and larvae were identified as *Phormia regina* (Meigen) (Diptera: Calliphoridae), the black blow fly, and it was concluded from the presence of small, first-instar larvae and large, third-instar larvae that two distinctly separate periods of egg laying by adult flies had taken place.

Bloodstains

Initial observation of the scene gave the appearance of extensive low-, medium-, and high-velocity blood spatters. Above the position of one of the victims numerous stains of the low- to high-velocity type were found (Figure 18.54). Similar areas were found on a living room hanging lamp (see Figure 18.55), the interior and exterior of the entry door, the bathroom, the two bedrooms, and the walls around the victims. The stains tested positive for blood with a quick test for hemoglobin (Hemastix/Heglostix).

The first assumption to be made was that there had been slinging of a lot of blood around the kitchen and living room. This would suggest not only gunshot wounds, but considerable movement of the victim and suspect(s). It could suggest a motive of robbery, burglary, assault, or a surprise attack. Examination of the kitchen and living room did not indicate struggling or fighting to any great amount. In the bedrooms and bathroom there were flies, but no signs of bloodstain patterns. There were no maggots in these rooms. The conclusion we made was that not much activity had taken place in the bedrooms or the

Figure 18.54 Pattern showing an extensive combination of low-, medium-, and high-velocity bloodstains located above the position of one of the victims. (Photo courtesy of Dr. Mark Benecke.)

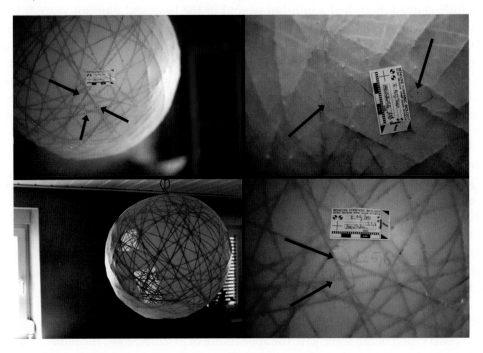

Figure 18.55 Stains produced by insect activity on hanging lamp in living room. (Photos courtesy of Dr. Mark Benecke.)

bathroom of an assaultive nature, and the bloodshed had taken place in the kitchen and living room.

Reconstructing the angle of impact of many of those stains, however, led nowhere. There was no indication that the bodies had been moved, and there were no signs of a struggle in the bedrooms or bathroom. Smaller, round type spatters were mostly <3 mm in length and >1 mm in diameter. Furthermore, stains of a sperm-like shape (irregular, uneven form with tail much longer than the body) as well as a missing systematic directionality were observed. Since all stains were composed of blood, (1) how did they get into all of the rooms and (2) how were they transferred to the walls?

Experiments and Measurements at the Scene of the Crime

It is known that after feeding, flies regurgitate and defecate. Hence, flies could have caused stains, containing blood of the victims, by regurgitation, defecation, and transference. In such cases, it is expected that presumptive blood tests would indicate the presence of blood. To prove that the unusual bloodstain patterns originated from the bodies of the victims, whereas the mechanism of transfer was provided by adult flies, the ratio Ltl/Lb (length of tail/length of body) was calculated as 3.3 ± 2.4 (Table 18.1). Such a high value will not be reached under most case conditions, especially since not only the calculated angles of impact, but also the directionality of the stains never point into one direction, as would be

Table 18.1   Original Data from Crime Scene in Case 2, n = 32 (Suspected Fly Stains, Tear-Drop Shape Only)

Stain Number	Ltot	Lb	Ltl	W	Ltl/Lb	A
1	5.0	2.5	2.5	1.0	1.0	23
2	7.0	2.0	5.0	0.5	2.5	14
3	4.0	2.0	2.0	1.0	1.0	30
4	6.0	2.0	4.0	1.5	2.0	48
5	10.0	2.0	8.0	1.5	4.0	48
6	8.0	2.0	6.0	2.0	3.0	90[a]
7	9.0	2.0	7.0	2.0	3.5	90[a]
8	7.0	1.0	6.0	1.2	6.0	—
9	11.0	2.0	9.0	2.0	4.5	90[a]
10	9.0	4.0	5.0	3.0	1.3	48
11	6.0	1.0	5.0	1.0	5.0	90[a]
12	8.0	2.0	6.0	1.5	3.0	48
13	11.0	2.0	9.0	1.0	4.5	30
14	15.0	2.0	13.0	1.5	6.5	48
15	4.5	1.5	2.5	1.2	1.7	53
16	21.0	3.0	18.0	2.0	6.0	41
17	9.0	2.0	7.0	0.5	3.5	14
18	12.0	1.0	11.0	1.0	11.0	90
19	6.0	1.0	5.0	1.5	5.0	—
20	10.5	1.5	8.5	0.7	5.6	46
21	20.0	3.0	17.0	2.0	5.6	41
22	5.0	2.0	3.0	1.0	1.5	30
23	5.0	3.0	2.0	0.5	0.7	9
24	15.0	5.0	10.0	1.2	2.0	13
25	12.0	2.0	10.0	1.2	5.0	36
26	4.5	2.5	2.0	0.9	0.8	21
27	3.0	2.3	0.7	0.7	0.3	18
28	6.0	5.4	0.6	1.0	0.1	11
29	5.0	2.0	3.0	2.0	1.5	—
30	6.0	4.0	2.0	1.0	0.5	14

Notes: Ltot = total length of stain in mm; Lb = length of body in mm; Ltl = length of tail in mm; W = width of body in mm; Ltl/Lb = ratio of length of tail/length of body; A = impact angle (W/Lb) (°; standard ratio of any bloodstain). [a]Equal body length and width with a tail.

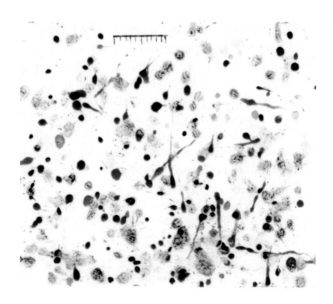

Figure 18.56 Characteristic stain produced by fly activity at a scene. (Photo courtesy of Dr. Mark Benecke.)

expected from a violent impact on a source of blood. Also, a mixture of round, symmetrical and teardrop-like stains was found to be highly suspicious for fly activity (Figure 18.56; see also Figure 18.4). Additional tests under laboratory conditions on vertical paper surfaces with adult *Calliphora vicina* Robineau-Desvoidy (Diptera: Calliphoridae) blow flies matured and maintained at room temperature (20 to 25°C) and supplied with a reddish brown food mixture led to the following results:

- After day in a breeding cage, of 304 stains, 112 (36.8%) had a round shape, whereas 192 (63.2%) had a tear- or sperm-like shape.
- The directionality measured along the longer axis of the tear- and sperm-like stains did show a random distribution of stain orientations with an artificial preference for the top left, where a window (light source) was situated. In this experiment, orientation was in the following directions: 42.8% upwards left, 19.8% upwards right, 19.3% downwards left, and 18.2% downwards right.
- The ratio Ltl/Lb was 1.5 ± 1.6 (n = 80) (i.e., tendency toward ratios > 1). However, single stains will not provide results that are statistically sound. Although the ratio does not conclusively identify a stain as a fly artifact, it provides a tool to eliminate suspect stains.

Conclusion

With information that stains appearing as human blood spatters were fly artifacts, coupled with other scene evidence, we felt confident that the possibility of an execution or revenge slaying could be put into the mix of suspect behaviors at our crime scene.

Case 3: Corpse of the Lonely Woman (Benecke)

In summer 2001, a dead female person was found in her bedroom in an urban apartment in Cologne, Germany. The body had entered the dried-out state of decay, with severe undernourishment during lifetime and an underlying minimal greenish discoloration of the face and the abdominal area after death. In the anal region of the corpse, a few blow fly maggots were found, with the oldest being third instars. As soon as the windows were opened, adult *Lucilia* sp. entered the room. Because numerous small specks appearing to be blood were noted in the face of the decedent, the police asked if blow flies had been present, or if those artifacts could be attributed to another source, in which case a subsequent investigation would be needed.

The windows were closed before the police entered, which explained the presence of only a few flies, mostly pupae of phorid flies (Diptera: Phoridae). Apart from piles of empty pizza delivery cardboard boxes and cigarette butts, which did not provide food sources for blow flies, the apartment was very clean and expensively furnished. The bathtub was half filled with discolored water that was most likely used to wash clothing.

Since the entrance door was regularly locked and no signs of a violent struggle were present, a reddish spatter stain at the end of a ceiling fan chain in the kitchen became of interest (Figure 18.57). The kitchen was located two rooms away from the sleeping room, and there was no visible evidence that linked the kitchen to any violent event. Closer examination led to the conclusion that the stains were fly artifacts. Since the eyes of the corpse were still intact and not used as a food resource by maggots, it was concluded that only very few adult individuals of a smaller fly species had been living in the apartment at some point before, or at the time of death. Those few individuals used the fan chain as a resting place and deposited reddish material with a typical preference for the lower surface margin. The same effect is present in Figure 18.58 under laboratory conditions, yet on a much larger scale.

Figure 18.57 The odd location of this reddish stain came into question during the death investigation. This stain was not a result of a crime committed, but from the natural behavior for flies to rest on the lower margins of hanging objects. (Photo courtesy of Dr. Mark Benecke.)

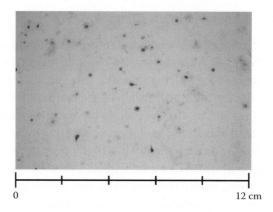

0   12 cm

Figure 18.58 Fly specks produced under controlled laboratory conditions using blood as a food source. (Photo courtesy of Dr. James L. Castner.)

Conclusion

Because of the nature of the stains, no further police investigation was warranted, and the stains found were consistent with the reconstruction of the events at the scene. This was considered to be a case of self-neglect in contrast to homicide or neglect by another person.

Case 4: Slaying of Mother and Child (Benecke)

On January 30, 2001, the dead bodies of a mother and her child were found in the living room of their house on the border of the city of Cologne, Germany. They had been dead for approximately 6 hours. Another child, who had been sleeping upstairs, was alive and not hurt. Bloodstain pattern interpretation was used to determine the course of events.

The crime scene reconstruction based on blood spatter became important to check the statements of an accused man who owned a knife that was used for the stabbing. For legal technicalities (rights of inheritance), it also became important to determine whether the woman or her child had been killed first. Thirdly, the defense lawyer wanted to prove that his client had stabbed the child with brutal force to make clear that his client had no mental control at the moment he performed the stabbing. Apart from medicolegal considerations, it was thought that the velocity of the blood spatter might help to address theses questions.

Among numerous other reddish stains in the house (in this case, due to a local police procedure, all stains were determined as originating from the victims by DNA typing), a few very small stains on a lamp were observed. This lamp was located approximately 1.8 m over ground and had been hanging directly between the locations where the two bodies were found. The police asked if these stains were caused by the impact of violence or by flies. As in most cases, the presence of flies was not looked at by the first investigative team. This team also opened a window in the dwelling, after which any flies in the home may have flown out. Therefore, a combined blood spatter and forensic entomology expert statement was asked for by the police, and later again requested by the judge during the trial.

The tiny, round stains on the lamp were distributed over the complete surface (Figure 18.55). Genetic fingerprinting led to one conclusive DNA type out of six stains (DNA of the child was found in one stain, no result in the other stains). It was discussed

Figure 18.59 Vase with multiple stains. The tiny round stains are produced by natural fly activity at the scene. (Photo courtesy of Dr. Mark Benecke.)

that the stains might have originated from the offender's knife that got stuck in the vertebra of the child (as documented by the forensic pathologist). When the offender took the knife out of the bone with a jerk, a few tiny droplets of blood may have been distributed with a relatively high initial velocity, but got slowed down due to the resistance of the air.

On the other hand, since a possible patterning as in fly artifacts could not be ruled out with certainty (see Figure 18.55), no absolute statement could be made about the nature of these stains. As with the vase in Figure 18.59, bloodstains produced by flies on some objects are difficult to distinguish from those produced by human activity during the commission of the offense. Therefore in court, we reported that because of the season of the year (winter) and the state of the house (no rotting organic material present), it was less likely that flies had produced the stains and more likely that the blood was actually shed off the knife during the stabbing. Droplets then reached the lamp at least at two separate events while the lamp was rotating around its axis. Since the lamp was located around the height of an adult Central European person's head, rotation was most likely induced by the people moving and maybe fighting inside of the room during the crime.

Conclusion

Blood has long been used as evidence to substantiate a crime and to link suspects with a particular scene or violent event, and the recent developments of blood recovery techniques have made it almost impossible to erase or hide traces of bloodstain evidence. However, much information can be obtained from blood relating to the occurrence and sequence of events in a violent crime without chemical analysis, based solely on the expert interpretation of bloodstain patterns. Bloodstain interpretation has developed into a distinct discipline among the forensic sciences, and it emphasizes analysis of the size, shape, and pattern of bloodstains observed at a crime scene. However, insects present at the scene may alter the blood evidence and create artifacts that are easily misinterpreted. Careful examination of the shape and pattern of such blood spots will usually reveal the particular characteristics that indicate their entomological origin. The information in this chapter was presented to help identify common characteristics, and to familiarize the forensic investigator with the basics of bloodstain pattern analysis.

Suggested Readings

Benecke, M., and L. Barksdale. 2003. Distinction of bloodstain patterns from fly artifacts. *Forensic Sci. Int.* 137:152–59.

Benecke, M., L. Barksdale, J. Sundermeier, S. Reibe, and B. C. R. Ratcliffe. 2000. Forensic entomology in a murder case: Blood spatter artifacts caused by flies, and determination of post mortem interval (PMI) by use of blowfly maggots. *Zoology* 103 (Suppl. III): 106.

Bevel, T., and R. M. Gardner. 1997. *Bloodstain pattern analysis*. Boca Raton, FL: CRC Press.

Eckert, W. G., and S. H. James. 1989. *Interpretation of bloodstain evidence at crime scenes*. New York: Elsevier Science Publishing Co.

James, S. H., Ed. 1999. *Scientific and legal applications of bloodstain pattern interpretation*. Boca Raton, FL: CRC Press.

Laber, T. L., and B. Epstein. 1994. *Experiments and protocol exercises in bloodstain pattern analysis*. St. Paul, MN: Midwestern Association of Forensic Scientists.

MacDonell, H. L. 1993. *Bloodstain patterns*. Corning, NY: Laboratory of Forensic Science.

Wolson, T. L., and J. R. Johnson. 1994. *Bloodstain pattern analysis workshop manual*. Miami, FL: Metropolitan Police Institute, 44 pp.

Keys to the Blow Fly Species (Diptera Calliphoridae) of America, North of Mexico

19

TERRY WHITWORTH

Contents

Introduction

The most recent revision of North American blow flies was by Hall (1948); however, Hall's keys have proven to be difficult to use because of his heavy reliance on proportional measurements of characters and the fact that he measured no more than five to ten specimens per species. He also chose specimens representing size extremes rather than "average" individuals (Sabrosky et al., 1989). Subsequently, James (1953, 1955) and Hall and Townsend (1977) provided revised keys, which clarified the identification of regionally selected species. James addressed the Western species of blow flies, while Hall and Townsend provided keys to blow flies found in Virginia. Shewell (1987) provided a key to the genera of North American calliphorids, but did not key species. Rognes (1991) reviewed Palearctic and Holarctic species and recommended numerous changes in blow fly taxonomy. Many of his name changes affected taxa found in North America.

Downes (1965) reduced the North American genera *Angioneura* Brauer and Bergenstamm and *Opsodexia* Townsend to subgenera under the Palearctic genus *Melanomya* Rondani. Later Downes (1986) revised species he had placed within *Melanomya*, describing one new species. Shewell (1987) resurrected *Angioneura* and *Opsodexia* as genera. Dear's (1985) revision of the New World Chrysomyini resulted in *Paralucilia wheeleri* (Hough) being synonymized with *Compsomyiops callipes* Bigot, and *Chloroprocta fuscanipennis* (Macquart) with *Chloroprocta idioidea* (Robineau-Desvoidy).

Sabrosky et al. (1989) revised the genus *Protocalliphora* Hough in North America and described fifteen new species. Subsequently, Whitworth (2002, 2003a) described three additional species. Rognes (1985) synonymized the North American *Protocalliphora hirudo* (Shannon and Dobrosky) and the Palearctic *Trypocalliphora lindneri* Peus with *Trypocalliphora braueri* (Hendel). Sabrosky et al. (1989) agreed with this, but reduced *Trypocalliphora* Peus to a subgenus of *Protocalliphora*. Whitworth (2003b) reevaluated the status of *Trypocalliphora* and agreed with Rognes (1985) that it should be a separate genus.

Rognes (1991) combined Hall's (1948) tribes Phormiini and Chrysomyini under the subfamily Chrysomyinae. Rognes (1991) also proposed the following synonymies: *Boreellus* Aldrich and Shannon = *Protophormia* Townsend (Chrysomyinae); *Phaenicia* Robineau-Desvoidy, *Bufolucilia* Townsend, and *Francilia* Shannon = *Lucilia* Robineau-Desvoidy (Luciliinae); *Acrophaga* Brauer and Bergenstamm, *Acronesia* Hall, *Aldrichina* Townsend, and *Eucalliphora* Townsend = *Calliphora* Robineau-Desvoidy; and *Bellardia agilis* (Meigen) = *B. vulgaris* (Robineau-Desvoidy) (Calliphorinae). He disagreed with Shewell's (1987) revival of the genus *Acrophaga*, which Zumpt (1965) had synonymized with *Calliphora*. Shewell had included three species in this genus: *genarum*, *stelviana* and the Palearctic *subalpina*. Rognes (1991) retained these species in *Calliphora*. He followed Shewell (1987) in retaining *Angioneura* and *Opsodexia* (Melanomyiinae) as genera. Thus, *Melanomya* is a Palearctic genus that does not occur in North America. *Angioneura* is a Holarctic genus represented by five species in North America, while *Opsodexia* is a Nearctic genus with four species in North America. I have adopted all of Rognes' (1991) changes.

Recently, interest in blow flies has increased, along with studies in forensic entomology. Smith (1986) published keys to adult blow flies of Britain, while Greenberg and Kunich (2002) provided keys for Oriental, Australian, South American, and Holarctic species. These keys include some species shared with the Nearctic region, but omit species found only in North America. The lack of any comprehensive species keys for North American blow flies prompted this study.

Materials and Methods

Specimens for this project were obtained from various entomological collections throughout North America and my personal collection. A complete list of my sources for specimens is included in the acknowledgments.

Characters used in my keys are usually visible with the aid of a quality stereomicroscope and a good light. An ocular micrometer will assist in making proportional measurements. It should be noted that fiber-optic lights tend to "wash out" colors, such as yellow and orange, so workers using incandescent lights should consider that when interpreting color characters. Keys were created with specimens killed in cyanide and pinned. Specimens kept in ethyl alcohol or other preservatives may become distorted or discolored making some key characters unreliable. Be especially cautious when using color characters in such specimens. Some specimens need to be relaxed to reveal certain characters. Flies were relaxed over wet sand in shallow plastic containers with tight-fitting lids. Most specimens were sufficiently relaxed after 48 hours in a relaxing chamber, so they could be handled without damage. Those left too long in the relaxing chamber were susceptible to mold or rot and could be ruined. If removed too soon, they were brittle and prone to breakage. Older specimens often had to be relaxed longer before they could be manipulated safely. The shape of the male genitalia proved useful to confirm species when external characters were not distinctive. Male cerci and surstyli were drawn into view following techniques described by Hall (1948). I encountered problems using a bent insect pin, as recommended by Hall, because it tended to flex and sometimes would snap off the genitalia, damaging them. I had better results exposing genitalia using half of a pair of fine-point tweezers bent to a 45° angle. Rognes (1991) has shown that female terminalia have characters useful to identify female specimens to species; however, I did not rely on them in the keys.

Terminology differences in the calliphorid literature can be confusing. For North American terminology, workers should see the *Manual of Nearctic Diptera* (McAlpine, 1981), while for European terminology they should refer to the *Manual of Palearctic Diptera* (Papp and Darvis, 1998). I have primarily followed McAlpine (1981), except as noted below.

Figures 1 to 5 from Rognes (1991) detail many of the characters used for blow fly identification. The names of some characters vary from common North American usage. The following are equivalent terms, with North American terms listed first: postpronotal lobe = humeral callus; postpronotal setae = posthumeral setae (inner and outer); posterior presutural supra-alar seta = presutural seta; propleuron = proepisternal depression; reclinate orbital seta = lateroclinate orbital seta (all are shown in Figures 19.1 and 19.2).

Some changes to older terms are as follows, with the preferred term listed second: parafrontal = fronto-orbital plate; bucca = genal dilation; third antennal segment = first flagellomere (Figures 19.3 to 19.5); inner and outer forceps = cerci and surstyli, respectively (Figures 19.9 and 19.10); and hypopleuron = meron (Figure 19.2).

Some variation will be noted in spelling of the following, with the preferred spelling given second: acrostical = acrostichal, and intraalar = intra-alar. For hyphenated species names, such as *terrae-novae*, a species of both *Calliphora* and *Protophormia*, the hyphen is dropped, as a result of a ruling by the International Commission on Zoological Nomenclature (ICZN, 1999, Article 32.5.2.3.).

Several useful taxonomic characters are available on the wings (Figure 19.6), abdomen (Figures 19.7 and 19.8), and genitalia (males, Figures 19.9 and 19.10; females, Figures 19.11 to 19.13). Important characters that are species specific are illustrated separately. The ratios of head to frons widths are used throughout keys; see Figures 19.23 and 19.24 for how to measure. The average ratio is followed by the range and the total number of specimens measured.

Historically, many terms have been used to describe the hairs and fine dusting observed in adult flies. I use the following convention: macrotrichia are larger hairs with nerves and sockets; microtrichia are cuticular extensions or dusting without sockets. Macrotrichia can be described as setae, setulae, hairs, or bristles. I will avoid the terms *hairs* and *bristles*, and consider larger macrotrichia as setae and smaller macrotrichia as setulae. The term *vestiture* sometimes is used to describe patterns of macrotrichia. Microtrichia patterns have been called dusting, pubescence, pollinosity, microtomentum, or microtrichia. For purposes of this publication, the term *microtomentum* is used to describe this condition.

The scientific names used herein follow Rognes (1991). Where possible, characters used are readily observed with a good microscope and without dissection. The first character listed in a couplet is generally the most distinctive; characters listed after may not be as reliable or may be more difficult to distinguish. See Table 19.1 for a list of species in the order they are addressed and the names used by Hall (1948). The only synonyms given are for Hall's publication. This is not a complete list of synonyms for each species.

Partial keys to adult species of *Protocalliphora* are provided to be integrated with existing keys in Sabrosky et al. (1989). The keys include three new species I have described (Whitworth, 2002, 2003a). I also have added revised illustrations and information to assist in the separation of species of this genus.

Species keys are not provided for *Melanodexia*, *Opsodexia*, or *Angioneura*. Both Hall (1948) and James (1955) provided species keys for *Melanodexia*, but they are difficult to use, and in any case, the genus needs revision, a task that is beyond the scope of this study. Downes (1986) provided keys to species of *Opsodexia* and *Angioneura*, which are effective (N. Woodley, personal communication), but few specimens were available for this study and species keys are therefore not included.

Separating Families

Most calliphorids are readily distinguished from other families by their metallic blue, green, or bronze color and the relatively large size of adults. Metallic muscids and tachinids are frequently found under Calliphoridae in collections because of these shared characters. Metallic muscids are readily separated from calliphorids by the absence of a row of setae on the meron. Metallic tachinids can be distinguished by the prominent subscutellum and bare arista.

The nonmetallic calliphorid genus *Pollenia* Robineau-Desvoidy is common in North America. It can be recognized by a row of setae on the meron and an abundance of silky, crinkly hairs on the thorax. Other nonmetallic calliphorids include the relatively rare *Angioneura*, *Opsodexia*, and *Melanodexia*, which are more or less dull colored. Characters

Table 19.1   Species in the Order They Are Discussed and Comparison of Names Used in the Current Chapter with Names Used by Hall (1948)

Whitworth (2005)	Hall (1948)
Calliphorinae	
Bellardia bayeri	Not included
Bellardia vulgaris	Not included
Calliphora alaskensis	*Acronesia alaskensis*
Calliphora aldrichia	*Acronesia aldrichia*
Calliphora coloradensis	*Calliphora coloradensis*
Calliphora genarum	*Acronesia collini, A. popoffana*
Calliphora grahami	*Aldrichina grahami*
Calliphora latifrons	*Eucalliphora arta, E. lilaea*
Calliphora livida	*Calliphora livida*
Calliphora loewi	*Calliphora mortica*
Calliphora montana	*Acronesia montana*
Calliphora stelviana	*Acronesia abina, A. anana*
Calliphora terraenovae	*Calliphora terrae-novae*
Calliphora vicina	*Calliphora vicina*
Calliphora vomitoria	*Calliphora vomitoria*
Cyanus elongata	*Cyanus elongata*
Cynomya cadaverina	*Cynomyopsis cadaverina*
Cynomya mortuorum	*Cynomya mortuorum, C. hirta*
Chrysomyinae	
Chloroprocta idioidea	*Chloroprocta-idioidea, C. fuscanipennis*
Chrysomya megacephala	Not included
Chrysomya rufifacies	Not included
Cochliomyia aldrichi	*Callitroga aldrichi*
Cochliomyia hominivorax	*Callitroga americana*
Cochliomyia macellaria	*Callitroga macellaria*
Cochliomyia minima	*Callitroga minima*
Compsomyiops callipes	*Paralucilia wheeleri*
Phormia regina	*Phormia regina*
Protocalliphora	*Apaulina*
Protophormia atriceps	*Boreellus atriceps*
Protophormia terraenovae	*Protophormia terrae-novae*
Trypocalliphora braueri	*Apaulina hirudo*
Luciliinae	
Lucilia cluvia	*Phaenicia cluvia*
Lucilia coeruleiviridis	*Phaenicia caeruleiviridis*
Lucilia cuprina	*Phaenicia pallescens*
Lucilia elongata	*Bufolucilia elongata*

Continued

Table 19.1   Species in the Order They Are Discussed and Comparison of Names Used in the Current Chapter with Names Used by Hall (1948) (*Continued*)

Whitworth (2005)	Hall (1948)
Lucilia eximia	*Phaenicia eximia*
Lucilia illustris	*Lucilia illustris*
Lucilia magnicornis	*Francilia alaskensis*
Lucilia mexicana	*Phaenicia mexicana*
Lucilia sericata	*Phaenicia sericata*
Lucilia silvarum	*Bufolucilia silvarum*
Lucilia thatuna	*Phaenicia thatuna*
Polleniinae	
Melanodexia	*Melanodexia*
	Melanodexiopsis
Pollenia angustigena	Not included
Pollenia griseotomentosa	Not included
Pollenia labialis	Not included
Pollenia pediculata	Not included
Pollenia rudis	*Pollenia rudis*
Pollenia vagabunda	Not included
Melanomyinae	
Angioneura	Not included
Opsodexia	Not included

provided in the key should distinguish these genera. Nonmetallic calliphorids are often found in collections with similar-looking muscids, sarcophagids, and tachinids.

Key to Separate Calliphorids from Similar Families

1. Meron without row of setae, sometimes scattered weak setulae.
   .... Muscidae, Anthomyidae, Scathophagidae
   Meron with distinct row of setae (Figures 19.2 and 19.16).
  2

2. Subscutellum strongly developed; arista often bare (not in Dexiini and some other taxa).
   .... Tachinidae
   Subscutellum absent or weak; arista usually setose (Figure 19.3) (except *Angioneura* and most Miltogramminae (Sarcophagidae]).
  3

3. Abdomen, and usually thorax, shining metallic blue, green, or bronze luster, sometimes with darker vittae (e.g., *Cochliomyia*).
   ....most Calliphoridae
   Abdomen and thorax dull gray, brown, or black, never shining metallic.
  4

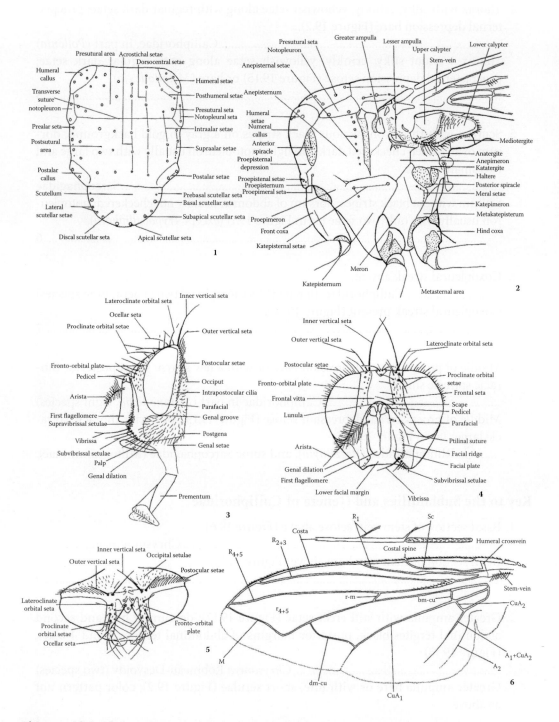

Figures 19.1–19.6 *Calliphora subalpina* (Ringdahl, 1931). 1. Dorsal view of thorax. 2. Left lateral view of thorax. 3. Left lateral view of head. 4. Anterior view of head. 5. Dorsal view of head; from Rognes (1991). 6. *Trypocalliphora braueri*, dorsal view of wing. Inset, portion of costa showing setulae on underside; from Rognes (1991).

4. Thorax with silky, crinkly, yellowish setae along with regular dark setae; proepisternal depression bare (Figure 19.2).

....Calliphoridae, in part (*Pollenia*)

Thorax without silky, crinkly, yellowish setae along with regular dark setae; proepisternal depression setose (Figure 19.15) or bare (*Melanodexia*).

....5

5. Scutum with three conspicuous black stripes on a gray to gold background; dorsum of abdomen checkered dark and light; notopleuron usually with two large and two smaller setae.

....Sarcophagidae

Scutum without black stripes; dorsum of abdomen usually not checkered; notopleuron usually with only two setae (as in Figure 19.1) (except *Trypocalliphora*).

.... 6

6. Coxopleural streak absent.

..............................Calliphoridae, in part (Melanomyinae, two genera, nine species)

Coxopleural streak present (Figure 19.16).

....7

7. Middle of proepisternal depression bare or with a few sparse setae; posterior thoracic spiracle small.

....Calliphoridae in part (*Melanodexia*, eight species)

Middle of proepisternal depression setose (Figure 19.15), posterior thoracic spiracle larger (as in Figure 19.16).

..............Rhinophoridae (*Bezzimyia*), and some Sarcophagidae (Miltogramminae).

Key to the Subfamilies and Genera of Calliphoridae

1. Basal section of stem vein setose above (Figure 19.6).

.... Chrysomyinae 2

Basal section of stem vein bare above (Figure 19.14).

....9

2. Greater ampulla with stiff erect setae (Figure 19.17); dorsum of first and second abdominal tergites black, posterior margins of abdominal tergites 3 and 4 black (Figure 19.18).

.... *Chrysomya* Robineau-Desvoidy (two species)

Greater ampulla bare or with fine, short setulae (Figure 19.2); color pattern not as above.

....3

3. Genal dilation yellow or orange with mostly yellow setae; head with predominantly yellow vestiture; posterior margin of hind coxa setose.

....4

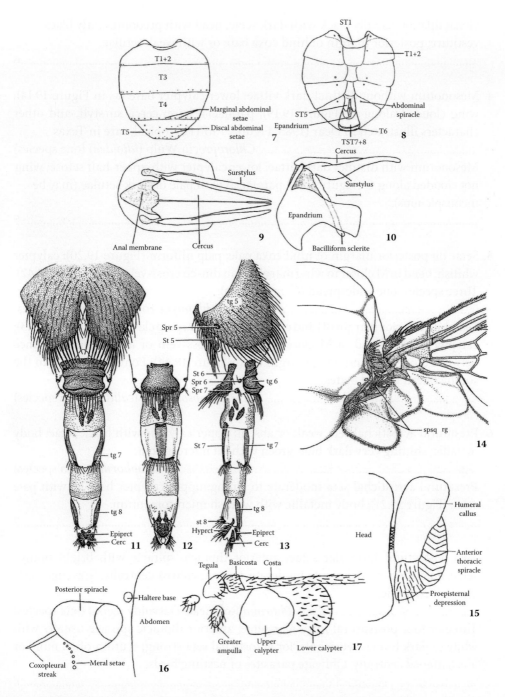

Figures 19.7–19.17 7 and 8. *Pollenia rudis*. 7. Dorsal view of abdomen. 8. Ventral view of abdomen; from Rognes (1991). 9 and 10. *Lucilia magnicornis*, male postabdomen. 9. Posterior view. 10. Left lateral view; from Rognes (1991). 11–13. *Calliphora stelviana*. Female postabdomen. 11. Dorsal view. 12. Ventral view. 13. Left lateral view; from Shewell (1987). cerc = cercus; epiprct = epiproct; hyprct = hypoproct; spr = spiracle; st = sternite; tg = tergite. 14. *Lucilia sericata*. Dorsal view of wing base showing suprasquamal ridge (spsq rg); from Shewell (1987). 15. Diagrammatic left lateral view of anterior portion of thorax, showing setose proepisternal depression. 16. *Lucilia coeruleiviridis*. Left lateral view of posterior thoracic spiracle and coxopleural streak. 17. *Chrysomya rufifacies*. Left lateral view of wing base showing setose greater ampulla.

Genal dilation usually black with dark setae; head with predominantly black vestiture; posterior margin of hind coxa bare or with weak setulae.
....6

4. Mesonotum without distinct dark vittae; lower calypter bare (as in Figure 19.14); wing clouded along C (Figure 19.19); parafacial bare; cerci, surstyli, and other characters illustrated in Dear (1985, Figures 7 to 11); tropical, rare in Texas.
....*Chloroprocta* Wulp (*idioidea*) (one species)
Mesonotum with distinct dark vittae; lower calypter with upper-half setose; wing not clouded along C (Figure 19.6); parafacial with pale or dark setulae (may be inconspicuous).
....5

5. Setae on posterior margin of hind coxa pale; palp filiform (Figure 19.20); calypter whitish; bend in M closer to wing margin than dm-cu cross-vein (as in Figure 19.42). Three species, one widespread (*C. macellaria*).
....*Cochliomyia* Townsend (three species)
Setae on posterior margin of hind coxa long and dark; palp clavate (as in Figure 19.3); calypter brown; bend in M closer to dm-cu cross-vein or about equal distance between cross-vein and wing margin (as in Figure 19.44). Found primarily in the southwestern United States.
....*Compsomyiops* (*callipes*) (one species)

6. Presutural acrostichal seta weak or absent; upper calypter with black setae; body metallic, shining very dark blue-green without microtrichia.
....*Protophormia* (two species)
Presutural acrostichal seta moderate to strong; upper calypter bare or with pale setae (Figure 19.21); body metallic with whitish microtomentum.
....7

7. Two postsutural intra-alar setae; anterior thoracic spiracle with bright orange setae; anterior acrostichal seta moderate; scutum convex centrally. Scavenger species, not parasitic.
....*Phormia* Robineau-Desvoidy *regina* (one species)
Three or four postsutural intra-alar setae; anterior thoracic spiracle usually with white to dark brown setae; anterior acrostichal seta strong; scutum often more or less flattened centrally. Obligate parasites of nestling birds.
....8

8. One or more accessory notopleural setae between the usual anterior and posterior notopleural seta (Figure 19.22); calypter yellowish to brown; frons of male narrow, at narrowest 0.05 (0.04 to 0.07/25) head width (see Figure 19.23 for how to measure); fronto-orbital plates touching, or nearly so; surstylus, cercus, and aedeagus distinctive (see figures in Sabrosky et al., 1989, 272, 273). Female lateroclinate orbital setae absent (see Figures 19.3 to 19.5 for location); thorax and abdomen bronze-green; frons to head ratio 0.22 (0.21 to 0.25/25) (see Figure 19.24 for how

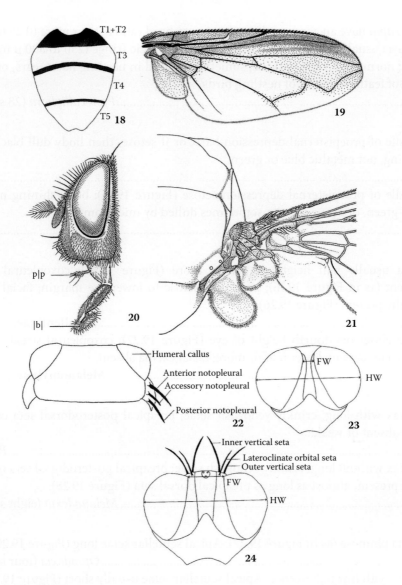

Figures 19.18–19.24 18. *Chrysomya megacephala*. Dorsal view of abdominal, tergites. 19. *Chloroprocta idioidea*. Dorsal view of wing; from Shewell (1987). 20. *Cochliomyia macellaria*. Left lateral view of head; from Shewell (1987). plp = palp; lbl = labellum. 21., *Phormia regina*. Dorsal view of right wing base; from Shewell (1987). 22. *Trypocalliphora braueri*. Dorsal view of prothorax. 23. Measuring male head to frons ratio. fw = frons width at narrowest; hw = head width at widest. 24. Measuring female head to frons ratio. fw = frons width at narrowest; hw = head width at narrowest.

to measure). Puparia appear bare, with sparse spines; prothoracic fringe absent; larvae are obligate subcutaneous parasites of nestling birds.

....*Trypocalliphora braueri* (one species)

Two notopleural setae (as in Figure 19.1); calypter usually whitish, if brown, other characters vary; male frons usually broader, fronto-orbital plates well separated, frons at narrowest 0.06 or more head width (0.06 to 0.16, one species 0.34); surstylus, cercus, and aedeagus variable. Female, with lateroclinate orbital setae present (Figures 19.3 to 19.5); thorax bluish (female *Protocalliphora aenea* and some *P.*

interrupta have an aeneous thorax); frons to head ratio 0.24 or more (0.25 to 0.35). Puparia usually heavily spined; distinct prothoracic fringe (250 to 800 μ in diameter); normally an ectoparasite, sometimes found in nestling nares, ears, or at the base of feather sheaths of nestling birds.

....*Protocalliphora* (28 species)

9. Middle of proepisternal depression bare, or if setose, then body dull black, sub-shining, not metallic blue or green.

....10

Middle of proepisternal depression setose (Figure 19.15); body shining metallic blue-green, or bronze, sheen sometimes dulled by microtomentum.

....13

10. Gena usually half height of eye or more (Figure 19.25); coxopleural streak present (as in Figure 19.16); parafacial setose to lower eye margin; facial carina usually present (Figure 19.26).

....**Polleniinae**............11

Gena about one-fourth height of eye (Figure 19.27); coxopleural streak absent; parafacial bare on lower half or more; facial carina absent.

.... **Melanomyinae**............12

11. Thorax with long, crinkly yellowish setae; preapical posterodorsal seta on hind tibia absent or weak.

....*Pollenia*

Thorax without long, crinkly yellowish setae; preapical posterodorsal seta on hind tibia present, almost as long as preapical dorsal seta (Figure 19.28).

.... *Melanodexia* (eight species)

12. Arista plumose (as in Figure 19.27). Apical scutellar setae long (Figure 19.29).

....*Opsodexia* (four species)

Arista with fine pubescence. Apical scutellar setae usually short (Figure 19.30).

....*Angioneura* (five species)

13. Thorax and abdomen shining green, blue, or bronze. Suprasquamal ridge with conspicuous cluster of setae near the base of scutellum (Figure 19.14); lower calypter bare above (Figure 19.14).

.... **Luciliinae** (one genus, *Lucilia*)

Thorax dull, microtomentose; abdomen usually metallic blue with whitish microtomentum (except Cynomya). Suprasquamal ridge bare or with inconspicuous fine setae (Figure 19.31); lower calypter setose above (Figure 19.31).

....**Calliphorinae**............14

14. Bend of M obtuse (as in Figure 19.32), curvature of apical section even; first flagellomere, at most, twice the length of pedicel; costa usually setulose below only

to junction with subcosta; abdomen blue or olive green; known only from north-eastern North America in the Nearctic region.
....*Bellardia* Robineau-Desvoidy (two species)
Bend of M acute or right angled (Figure 19.42), curvature of apical section greatest just beyond bend; first flagellomere more than twice length of pedicel; costa usually setulose below to junction with R1; abdomen bluish.
....15

15. Upper and lower calypter white (note rim of upper calypter in Cynomya is tan).
....16
Upper and lower calypter light to dark brown, margin may be white.
.... *Calliphora*, in part (11 species)

16. Presutural intra-alar seta absent; abdomen shining, no microtomentum visible when viewed posteriorly.
.... *Cynomya* Robineau-Desvoidy (two species)
Presutural intra-alar seta present (Figure 19.1), abdomen microtomentose when viewed posteriorly.
....17

17. Orange basicosta; abdomen elongate, longer than length of dorsum of thorax; abdomen with light microtomentum when viewed from rear. California to Washington, Colorado to Alberta, usually at higher elevations.
.... *Cyanus* Hall (*elongata*) (one species)
Black basicosta; abdomen no longer than dorsum of thorax; abdomen with heavy microtomentum when viewed from rear. Northern Canada, Alaska, or high elevation only.
....*Calliphora*, in part (two species)

Calliphorinae

This subfamily includes *Bellardia*, *Calliphora*, *Cyanus*, and *Cynomya*. It can be recognized by the following characters: stem vein bare above; lower calypter setose above; proepisternal depression setose; thorax dull, microtomentose; abdomen more or less shining blue; suprasquamal ridge bare or with only a few inconspicuous setae.

Key to Species
Bellardia *Robineau-Desvoidy, 1863*

This Palearctic genus is a recent immigrant to North America and known only from the northeastern United States. It was very rare in my search of collections. The genus can be identified by the obtuse bend in vein M (Figure 19.32). The species are believed to be earthworm parasites. They are the only North American blow flies that are viviparous (Shewell, 1987). Shewell noted that the terminalia of females are very short; he provided two illustrations (Figures 19.38 and 19.39), and he labeled them *Bellardia agilis* (Meigen), which is a synonym of *B. vulgaris* (Robineau-Desvoidy).

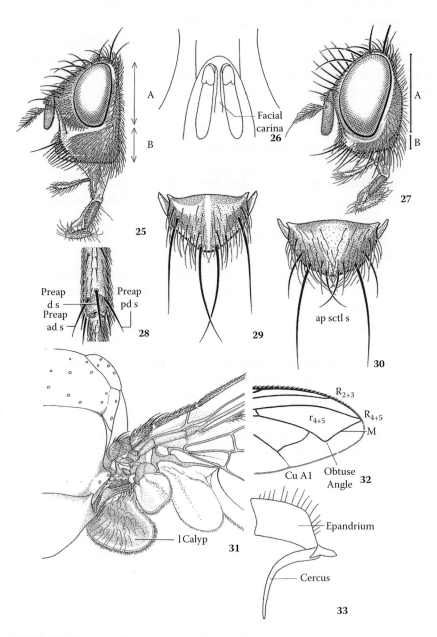

Figures 19.25–19.33 25. Pollenia sp. Female, left lateral view of head: (a) eye height; (b) gena height; after Shewell (1987). 26. Pollenia rudis. Anterior view of antennae and facial carina. 27. Opsodexia sp. Female, left lateral view of head: (a) eye height; (b) gena height; after Shewell (1987). 28. Melanodexia grandis. Distal end of hind tibia; preap d s = preapical dorsal seta; preap ad s = preapical anterodorsal seta; preap pd s = preapical posterodorsal seta; from Shewell (1987). 29. Opsodexia sp. Female, scutellar setae; from Shewell (1987). 30. Angioneura obscura. Female, scutellar setae; ap sctl s = apical scutellar setae; after Shewell (1987). 31. Calliphora sp. Female, dorsal view of right wing base; l calyp = lower calypter; after Shewell (1987). 32. Bellardia vulgaris male. Dorsal view of right wing; after Shewell (1987). 33. Calliphora grahami. Male, left lateral view of postabdomen.

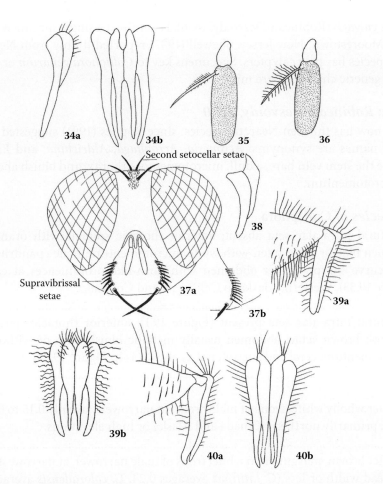

Figures 19.34–19.40 34. *Calliphora vomitoria*. Male, cerci and surstyli: (a) left lateral view; (b) posterior view. 35. *Calliphora genarum*. Male, left lateral view of antenna. 36. *Calliphora stelviana*. Male, left lateral view of antenna. 37. *Calliphora latifrons*. Male: (a) anterior view of head; (b) inset. *Calliphora terraenovae*, supravibrissal setae. 38. *Calliphora latifrons*. Male cercus and surstylus, left lateral view. 39. *Calliphora coloradensis*. Male cerci and surstyli: (a) left lateral view; (b) posterior view. 40. *Calliphora livida*. Male cerci and surstyli: (a) left lateral view; (b) posterior view.

1. Upper parafacial with dark brown spots that do not disappear when viewed from above; lower calypter evenly darkened, light tan; male genitalia tiny, cercus longer than surstylus, as in Rognes (1991; Figures 19.79 and 19.80); a small fly.
   ....*bayeri*
   Upper parafacial without spots; lower calypter white, margins yellowish or light tan; male genitalia larger, cercus shorter than surstylus, as in Rognes (1991; Figures 19.68 and 19.69); a larger fly.
   ....*vulgaris*

Bellardia bayeri (Jacentkovsky, 1937). I examined specimens from Strafford Co., New Hampshire, and Middlesex Co., Massachusetts. This species has dark calypters, and if generic characters are missed, it will tend to key to *Calliphora terraenovae*.

Bellardia vulgaris (Robineau-Desvoidy, 1830). I examined a single specimen collected from Moorestown, New Jersey. Shewell (1987) reported it only from New Jersey. This species has pale calypters. Specimens key to *Calliphora genarum* or *C. stelviana* if generic characters are missed.

Calliphora *Robineau-Desvoidy, 1830*

This genus now has thirteen Nearctic species, since Rognes (1991) suggested the following generic names are synonyms: *Acronesia, Acrophaga, Aldrichina*, and *Eucalliphora*. Species have the stem vein bare, a dull, microtomentose thorax, and bluish abdomen with whitish microtomentum.

Key to Species of Calliphora

1. Presutural intra-alar seta absent; anterior thoracic spiracle with orange setae; abdomen blue or dark green with white microtomentum; male epandrium large, cerci curve sharply under abdomen with horn-like prominences at each base (Figure 19.33). Western, Alaska to California and Colorado.
   ....*grahami*
   Presutural intra-alar seta present (Figure 19.1); anterior thoracic spiracle usually with brown setae; abdomen usually metallic bluish with or without white microtomentum; cerci and surstyli not as above (as in Figure 19.34).
  2

2. Calypter wholly white; frons of male broad, at narrowest, usually 0.15 to 0.21 head width; primarily northern Canada and Alaska or high elevations.
  3
   Calypter brown, margin often white; frons of male narrower, at narrowest, usually 0.14 head width or less (*C. latifrons* averages 0.24, *C. coloradensis* averages 0.15); usually not restricted to northern or high-elevation areas.
  4

3. Arista with short setae above, very short below (Figure 19.35); parafacial with dark chestnut or black ground color; broad undusted stripe between presutural acrostichals usually extending past transverse suture; third abdominal tergite with long median marginal setae, usually more than half the length of those on the fourth tergite; tip of surstylus rounded (see Rognes, 1991, Figure 150). Alaska, northern Canada, to Quebec and Labrador.
   ....*genarum*
   Arista with long setae above and below (Figure 19.36); parafacial orange in ground color on lower half or more; undusted stripe faint, usually narrower and stopping at the transverse suture; third abdominal tergite with shorter median marginal setae, always less than half the length of those on fourth tergite; tip of surstylus pointed (see Rognes, 1991, Figure 173). Alaska, Quebec, high elevations in Colorado.
   ....*stelviana*

4. Facial ridge with row of short, stout, supravibrissal setae, ascending from the vibrissae to a point almost halfway to antennal base (Figure 19.37a); a second set of strong divergent ocellar setae about two-thirds the length of the anterior ocellars,

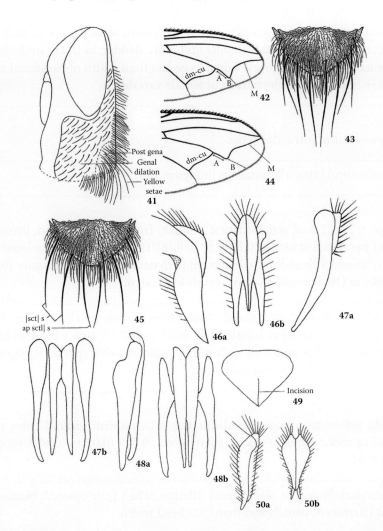

Figures 19.41–19.50 41. *Calliphora vomitoria*. Female, left lateral view of head. 42. *Calliphora vomitoria*. Female, dorsal view of right wing: (a) bend in M to cross-vein; (b) bend to wing margin; after Shewell (1987). 43. *Calliphora terraenovae*. Male, scutellar setae; from Shewell (1987). 44. *Calliphora genarum*. Female, dorsal view of right wing: (a) bend in M to cross-vein; (b) bend to wing margin; after Shewell (1987). 45. *Calliphora montana*. Male, scutellar setae; l sctl s = lateral scutellar setae; ap sctl s = apical scutellar setae; after Shewell (1987). 46. *Calliphora aldrichia*. Male, cerci and surstyli: (a) left lateral view; (b) posterior view. 47. *Calliphora terraenovae*. Male, cerci and surstyli: (a) left lateral view; (b) posterior view. 48. *Calliphora loewi*. Male, cerci and surstyli: (a) left lateral view; (b) posterior view. 49. *Calliphora loewi*. Female, dorsal view of tergite 5. 50. *Calliphora alaskensis*. Male, cerci and surstyli: (a) left lateral view; (b) posterior view.

surrounded by only a few sparse setae (Figure 19.37a). Anterior thoracic spiracle with orange setae. Male genitalia shorter, with a chisel-shaped point (Figure 19.38). Frons of male broad, at narrowest, almost two times the width of parafacial at lunule, frons 0.24 (0.22 to 0.26)/12 head width; female frons 0.37 (0.36 to 0.39)/8 head width. Primarily a Western species, east to Colorado and Wisconsin.

....*latifrons*

Facial ridge with row of slender supravibrissal setae (Figure 19.37b); second set of ocellar setae weak or absent, if stronger (females of some species), surrounded by dense fine setae; anterior spiracle usually with dark setae (except *C. vicina* is also orange); male genitalia usually longer and more slender, as in Figure 19.34a; frons of male narrower, at narrowest, equal to or less than width of parafacial at lunule, frons 0.15 head width or less; frons of female variable.

....5

5. Three postsutural intra-alar setae.

....6

Two postsutural intra-alar setae (as in Figure 19.1).

....7

6. Anterior 1/2 to 2/3 of genal dilation reddish; frons of male broad, broader than width of parafacial at lunule, 0.14 (0.12 to 0.16)/11 times head width; lower portion of surstylus and fifth abdominal segment with dense wavy setae (Figure 19.39a and b). Alaska to Ontario and Indiana, south to Mexico.

.... *coloradensis*

Genal dilation, when fully colored, black; frons of male much narrower, less than half width of parafacial at lunule, 0.06 (0.05 to 0.07)/7 times head width; lower portion of surstylus and fifth abdominal segment with sparse, straighter setae (Figure 19.40a and b). Widespread.

....*livida*

7. Basicosta yellow to orange; genal dilation with reddish ground color on anterior half or more; frons of male, at narrowest, 0.075 (0.07 to 0.08)/4 head width. Widespread.

....*vicina*

Basicosta dark brown or black; genal dilation, when fully colored, black; frons of male, at narrowest, usually less than 0.07 head width.

....8

8. Postgena and lower posterior corner of genal dilation and back of head with long yellow-orange setae, sometimes extending forward along edge of subgena and lower genal dilation (Figure 19.41); vestiture of the occiput below postocular setae primarily pale setae; genal groove reddish or orange. Frons of male, at narrowest, 0.044 (0.04 to 0.05)/10 head width; long, slender, curved surstyli (Figure 19.34a and b); frons of female at narrowest 0.34 (0.31 to 0.35)/10 head width. Widespread.

....*vomitoria*

Postgena and genal dilation with mostly dark or black setae, back of head and rear edge of postgena may have yellow setae; vestiture of the occiput below postocular setae with three or more rows of black setae; genal groove usually black or dark brown (except *C. terraenovae*); other characters vary.

....9

9. Bend in M much closer to wing margin than length of M from cross-vein dm-cu to bend (as in Figure 19.42); usually four (three to five) strong lateral scutellar setae

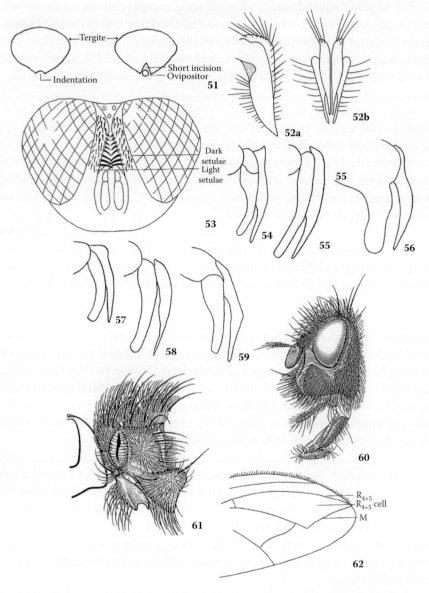

Figures 19.51–19.62 51. *Calliphora alaskensis*. Female, dorsal view of tergite 5, two possible views of condition of tergite above ovipositor: (a) slight indentation; (b) small incision. 52. *Calliphora montana*. Male, cerci and surstyli: (a) left lateral view; (b) posterior view. 53. *Cochliomyia*. Anterior view of head; left side, *C. hominivorax*; right side, *C. macellaria*. 54–59. *Protocalliphora*. Male surstylus and cercus. 54. *P. beameri*. 55. *P. bicolor*. 56. *P. hirundo*. 57. *P. interrupta*. 58. *P. metallica*. 59. *P. parorum*. 60. *Protophormia atriceps*. Female, left lateral view of head; from Shewell (1987). 61. *Protophormia atriceps*. Female, left lateral view of prothorax; after Shewell (1987). 62. *Protophormia atriceps*. Female, dorsal view of right wing.

besides apical pair (Figure 19.43); male frons, at narrowest, 0.04 to 0.07 head width; surstylus curves anteriorly or is straight (Figures 19.34, 19.38, and 19.39).

....10

Bend in M about equal distance between wing margin and dm-cu cross-vein (varies from slightly closer to wing margin to slightly closer to cross-vein) (Figure 19.44 shows the latter condition); usually two or three (occasionally a fourth on one side) strong lateral scutellar setae besides apical pair (Figure 19.45); male frons, at narrowest, 0.06 to 0.14 head width; surstylus curves posteriorly (Figure 19.46a).

....12

10. Genal groove usually reddish or orange; anterior one-third to one-half of genal dilation usually reddish when viewed from above; first flagellomere often orange along lower inside edge; parafacial golden or silvery tan when viewed from above; surstylus long and slender with gentle anterior curve (Figure 19.47). Widespread from Alaska to Newfoundland, south to southern California, and Texas; usually found at higher elevations in the west.

....*terraenovae*

Genal groove usually black or dark brown (rarely reddish or reddish brown); anterior portion of genal dilation entirely black or dark brown; first flagellomere uniformly gray; parafacial silvery black; surstylus not as above (Figure 19.48).

....11

11. Conspicuous polished strife between presutural acrostichals when viewed from rear; surstylus long, straight, parallel sided tapering to an obtuse point (Figure 19.48a); cerci appearing slender when viewed posteriorly (Figure 19.48b); frons of male, at narrowest, 0.043 (0.035 to 0.055)/12 head width; fronto-orbital plates touching at narrowest; fifth abdominal tergite of female with posterior incision one-third to one-half length of segment (Figure 19.49), lateral profile of tergite often tent-like. Alaska, British Columbia, Yukon Territories.

.... *loewi*

No distinct stripe, area uniformly microtomentose. Surstylus curved anteriorly at tips (Figure 19.50a); cerci pear shaped when viewed posteriorly (Figure 19.50b); frons of male wider, at narrowest, 0.06 (0.05 to 0.07)/16 head width, fronto-orbital plates not touching; fifth abdominal tergite of female with short incision, or at most only a slight indentation (Figure 19.51a and b), profile not tent-like. Widespread from Alaska to Oregon and east to North Carolina, primarily at high elevation in southern locations.

....*alaskensis*

12. Two or three lateral scutellar setae in addition to apical pair (Figure 19.45). When a third seta is present, in the prebasal position (see Figure 19.1), it is usually weak. Bend in M vein usually farther from wing margin and closer to cross-vein dm-cu (as in Figure 19.44). Sometimes the bend is equal distance between wing margin and cross-vein. Male frons, at narrowest, 0.11 (0.08 to 0.14)/15 head width. Cercus of male usually shorter (Figure 19.46b). A Western species ranging south from Alaska through British Columbia to California and Colorado.

....*aldrichia*

Three or sometimes four lateral scutellar setae in addition to the apical pair, the seta in the prebasal position usually stronger (as in Figure 19.43). Bend in M vein usually slightly closer to wing margin or equal distance between wing margin and dm-cu cross-vein (similar to Figure 19.42, but the bend is shown much closer to the wing margin). Male frons at narrowest, 0.07 (0.06 to 0.08)/18 head width. Cercus of male usually longer (Figure 19.52b). Primarily east of the Rockies, ranging southeast from northwestern Canada (where it overlaps with *C. aldrichia*) east through the Canadian provinces to Ontario and Labrador.
....*montana*

Calliphora alaskensis Shannon, 1923. This species is widespread but rare, found only at high elevations in the southern portions of its range. Hall (1948) listed specimens from Alaska, Wyoming, and Colorado. I also found this species from seven locations in Canada (three in British Columbia and four in Quebec) and from ten locations in the United States (two in Oregon, two in Utah, four in Colorado, and surprisingly, one each from mountains in Tennessee and North Carolina). This species is normally rare, but I examined about twenty-five females of this species that were bait trapped in the vicinity of Vancouver, British Columbia. These flies were attracted to both beef and chicken liver in second growth timber (K. Needham, in litt.). In seven of fifty specimens examined (six males, one female), the genal groove was reddish to reddish brown, which would place them with *C. terraenovae*. Genitalia will separate males, but females with reddish genal grooves will be difficult to separate.

Calliphora aldrichia Shannon, 1923. Hall (1948) gave records from British Columbia and Quebec, Alaska, Colorado, Wyoming, Montana, and Washington. (I believe the Quebec record was likely *C. Montana*.) I also examined specimens from western Alberta, California, and Oregon. This species is morphologically very close to *C. montana*; the two appear to be sibling species. Males share distinctive surstyli, which curve posteriorly (Figure 19.46a), unlike other *Calliphora*. The most obvious difference between the two species is that males of *C. aldrichia* have, on average, a much wider frons (0.11 of head width, at narrowest) than *C. montana* (0.07 of head width, at narrowest). The term *lateral scutellar setae* is used in the key in a broad sense to include all stronger setae on the margin of the scutellum other than the apical pair (as in Figures 19.43 and 19.45). Other authors assign separate names to these setae, as in Figure 19.1.

The two species appear to have formed as a result of geographical isolation. *Calliphora aldrichia* is found west of the Rocky Mountains, from Alaska to California and Colorado, while *C. montana* is found primarily east of the Rockies, from Northwest Territories and Alberta east to Labrador. Their ranges overlap in northern British Columbia and southern Yukon, where specimens with intermediate characters were found. Since both species are associated with mountains or northern latitudes, the mechanism of isolation is unclear. It appears that *C. aldrichia* is associated with higher elevations, while *C. montana* is found through lower elevations to the east. The area where the species converge is lower elevation at the northern edge of the Rockies. Separating females of these species, based on morphology, will be problematic. Outside the zone where populations converge, distribution appears to be the best way to separate females. As more specimens become available, species distinctions should be reevaluated.

Calliphora coloradensis Hough, 1899. This species is generally rare, but appeared to be locally abundant in areas around Flagstaff, Arizona, and Uvalde, Texas. I also examined specimens from California, New Mexico, Oregon, Wyoming, and South Dakota. Hall (1948) reported its range from Mexico north to Alaska, and east to Ontario and Indiana. Most specimens that I examined were from the southwestern United States. This species has three postsutural intra-alar setae, a character it shares only with *C. livida* and some *C. latifrons*. However, it has a reddish genal dilation, which separates it from *C. livida*. The character is good in fully sclerotized specimens, but can be confusing in teneral individuals, which are fairly common in this species.

Calliphora genarum Zetterstedt, 1838. This species and *Calliphora stelviana* (Brewer and Bergenstamm, 1891) would key to *Acrophaga* in Shewell (1987). I saw few specimens of this species. Hall (1948) gave its range as Alaska and northern Canada, Manitoba, and Labrador. I examined specimens from Yukon, Northwest Territories, and Manitoba. It shares white calypters with *C. stelviana*, which separates both species from other *Calliphora*. The differences in seta length on the arista are used to separate these two species from each other, but the fact that the setae are often damaged makes positive identification more difficult in some cases. Other useful characters to distinguish this species from *C. stelviana* include dark parafacials; a broad undusted stripe between presutural acrostichals; and long median setae on the rear margin of the third abdominal tergite. Characters are illustrated in Rognes (1991, Figures 149 to 158).

Calliphora grahami Aldrich, 1930. This species is indigenous to Asia and an immigrant to the western United States. I saw specimens from California to Alaska. James (1953) also reported it from Colorado and New Mexico. It lacks a presutural intra-alar seta, which distinguishes it from other *Calliphora*, and is a character shared with the genus *Cynomya*. The large, curved cerci (Figure 19.33) of the male are unlike those of any other *Calliphora* in North America.

Calliphora latifrons Hough, 1899. I examined specimens from California to Washington, and Colorado to Wisconsin. Hall (1965) reported that he found the species in the north, from Alaska to Ontario. It is primarily a Western species, though found occasionally in the East. It can be recognized by a combination of several characters, including short, stout supravibrissal setae and a second set of strong divergent ocellar setae. Most *Calliphora* have much finer supravibrissal setae, although *C. coloradensis* can be similar. In most *Calliphora* the second set of ocellar setae is weak or absent, but females of some species, like *C. alaskensis*, have stronger setae. However, the area around the second set of ocellars in *C. latifrons* is mostly bare, while in *C. alaskensis* it is setose. *Calliphora latifrons* sometimes has a small third postsutural intra-alar seta in front of the first strong postsutural intra-alar on one or both sides, which can cause confusion with *C. coloradensis* or *C. livida*.

Calliphora livida Hall, 1948. Widespread in North America. This species is similar to *C. coloradensis*, but the genal dilation is black when fully sclerotized. It can be confused with *C. coloradensis* if the specimen is teneral, a fairly common condition.

Calliphora loewi Enderlein, 1903. Hall (1948) reported this species only in Alaska, but I examined specimens from Kulane Lake in the Yukon, from Terrace and the Queen Charlotte Islands in central British Columbia, Kootenay National Park in southeastern British Columbia, and the Vancouver area in southwestern British Columbia. It can be confused with *C. terraenovae* because an occasional specimen may have a reddish genal groove. I examined many *C. loewi* from the Kola Peninsula in Russia, and several had a bright orange genal groove. In males genitalia are distinctive; in females the shape of the fifth tergite and the presence of a posterior incision separate them from similar species. Characters are illustrated in Rognes (1991, Figures 159 to 168).

Calliphora montana (Shannon, 1926). Hall (1948) commented that he could not find the type of this species in the U.S. National Museum. I borrowed a male-labeled type and a female-labeled allotype from the Canadian National Collection. The male genitalia are similar to those of *C. aldrichia*, but the characters are variable enough that the two species can be difficult to separate. Normally the narrower frons will distinguish males. Because Shannon's description included few details, I redescribe this species below:

- *Calliphora montana* (Shannon, 1926) (Figures 19.45 and 19.52)
- *Steringomyia montana* Shannon, 1926: 135 (♂, ♀)
- *Acronesia montana* Hall, 1948: 280

Diagnosis. Bend in M usually closer to wing margin or equal distance between wing margin and cross-vein dm-cu (Figure 19.44). Usually three, or occasionally four, lateral scutellar setae in addition to the apical pair (see Figure 19.45 for location). Male frons, at narrowest, about 0.07 head width; cerci long (Figure 19.52), longer than in *C. aldrichia* (Figure 19.46).

Male. Head ground color black, with silvery microtomentum, genal groove black, preocellar area triangular, shining to subshining black. Thorax subshining black with white microtomentum. Abdomen metallic blue with silvery microtomentum when viewed from an angle. Frons narrow, at narrowest, 0.072 (0.06 to 0.08)/18 head width; the holotype male ratio is 0.065. Usually three pair lateral scutellar setae in addition to apical pair, sometimes four on one or both sides. Bend in M usually slightly closer to wing margin, occasionally equidistant between wing margin and dm-cu cross-vein or closer to cross-vein. Surstyli curve posteriorly, as in *C. aldrichia*, and unlike all other North American *Calliphora*. Cerci longer than those of *C. aldrichia*.

Female. Color of head as in male, preocellar area unmarked. Thorax and abdomen as in male. Frons, at narrowest, 0.34 (0.32 to 0.36)/8 head width; other characters as in male.

Types. Type male, allotype female, no paratypes designated. Both specimens labeled *Steringomyia montana*, from Edmonton, Alberta, Canada, August 19, 1923, collector E. H. Strickland, label 2444. Male genitalia had been dissected and are in a vial with the specimen.

Additional specimens examined in this study:

CANADA, BRITISH COLUMBIA: Haines Highway, Mile 45, 7/25/1963, G. C. and D. M. Wood, one male; Stone Mountain Provincial Park, 8/2/1989, Paul Arnaud, one male.

LABRADOR: Cartwright, 8/21/1955, E. F. Cashman, one male; Churchhill Road, 10 miles from Goose Bay, 7/28/1987, K. F. Paraday, one male.

MANITOBA: Churchhill, 8/5/1955, D. M. Wood, one male.

NOVA SCOTIA: Cape Breton Highlands National Park, Mackenzie Mountain, 8/29/1983, M. Sharkey, one female.

NORTHWEST TERRITORIES: Norman Wells, 8/16/1969, G. E. Shewell, one female.

ONTARIO: Ogoki, 8/28/1952, J. B. Wallis, one female, 8/18/52, one male; 8/20/1963, D. M. Wood, one male; Temagami, 8/20/1963, G. Taylor and M. Wood, one male.

QUEBEC: Cascapedia River, Gaspe, 30 miles north of New Richmond, 8/1/1983, W. Middlekauff, one male; Grand Valley, 7/30/1963, G. S. Walley, one male; Laurentides Park, Barriere Ste. Anne, 8/15/1971, D. M. Wood, one male; 8/16/1956, R. W. Hodges, two males; La Verendrye Provincial Park, 8/19, 8/20, 8/21/1965, D. M. Wood, two males, one female.

SASKATCHEWAN: Prince Albert, 8/14/1953, W. J. Turnock, one female.

YUKON TERRITORIES: Gravel Lake, 63° 48'N 137° 53'W, 6/16/1981, C. S. Guppy, one male; Kulane Lake, 8/1/1963, G. C. and D. M. Wood, two males; Wolf Creek, Mile 907, Alaska Highway, 8/24/1963, G. C. and D. M. Wood, one female.

Calliphora stelviana (Brauer and Bergenstamm, 1891). This species would key to *Acrophaga* in Shewell (1987). Hall (1948) listed it from Alaska to northern Quebec, and Labrador, also at high elevations in Colorado. I examined specimens from Alaska, Yukon, Northwest Territories, and Quebec. This species and *C. genarum* are the only North American *Calliphora* with white calypters. Characters illustrated in Rognes (1991, Figures 169 to 182).

Calliphora terraenovae Macquart, 1851. This species is widespread from Alaska south to California, and east to Greenland, also known from Wisconsin, Colorado, and New Mexico. Hall (1965) reported it from Florida, and James (1955) reported it from New York. I recently examined a specimen of this species from a 5300 ft. elevation in North Carolina. The Florida record is likely a misidentification. This species lacks any single distinctive character, but can be recognized by a combination of characters.

Calliphora vicina Robineau-Desvoidy, 1830. This species is widespread and common. It is easily recognized, with a yellow to orange basicosta and the anterior half of genal dilation yellowish to reddish. Characters are illustrated in Rognes (1991, Figures 139 to 148).

Calliphora vomitoria (Linnaeus, 1758). This is a common species throughout North America. It is one of the largest *Calliphora*, with bright yellow to orange setae on the rear and lower portion of the postgena, genal dilation, and back of the head. Characters are illustrated in Rognes (1991, Figures 207 to 216).

Cyanus *Hall, 1948*
Represented by a single species.

Cyanus elongata (Hough, 1898). This species is rarely found in collections. Hall (1948) lists it from South Dakota, Colorado, Oregon, and Alberta. James (1953) examined specimens from North Dakota, Nebraska, Colorado, Montana, Utah, Idaho, Washington, Oregon, and California, usually from higher elevations. A collecting trip to southeastern Oregon in August 2005 near the Malheur Wildlife Refuge in Harney County yielded nine specimens of this species. Six came to a trap baited with a dead rabbit set in a swamp at around 4,000 feet elevation. One of each of the other specimens was caught in Malaise traps on Stein's mountain at 4,500, 6,000, and 8,500 feet elevation. It is a large fly with a long shining abdomen and bright orange basicosta. Male genitalia are illustrated in Shannon (1923, Figure 5a and b).

Cynomya *Robineau-Desvoidy, 1830*
The genus has two species, which have white calypters, lack the presutural intra-alar seta, and have a brilliant, shining blue abdomen.

Key to Species of Cynomya
1. Parafacial with bright yellow to orange ground color and golden microtrichia; portions or all of fronto-orbital plate, frontal vitta, antenna, and genal dilation with bright yellow ground color and golden microtrichia; usually one postacrostichal seta; female with center of fifth abdominal tergite distinctly concave and with dense, stout setae. A Holarctic species in North America found only north of the Arctic Circle.
   .... *mortuorum*

2. Parafacial with black to reddish brown ground color and yellowish microtrichia when viewed from above; fronto-orbital plate, frontal vitta, antenna, and genal dilation with black to reddish brown ground color and yellowish microtrichia; usually two postacrostichal setae; female with fifth abdominal tergite more or less straight in profile, setae sparser and weaker; widespread in North America.
   .... *cadaverina*

Cynomya cadaverina Robineau-Desvoidy, 1830. This species is fairly common and widespread throughout North America. Hall (1948) found it from northern Quebec to southern Texas, being most abundant along the Canadian-U.S. border. I rarely found it from the southern United States. The parafacials and genal dilation are black or reddish brown. Male and female abdomens and male genitalia are illustrated in Hall (1948, Figure 29C–F).

Cynomya mortuorum (Linnaeus, 1761). This species is found only in the far north in Alaska near the Arctic Circle. I did not see this species in the unidentified material that I examined from North America, but it was common in a group of blow flies I examined from the Kola Peninsula in Russia. The parafacials and genal dilation are bright yellow. Characters are illustrated in Rognes (1991, Figures 217 to 228).

Chrysomyinae

This subfamily is recognized by a setose stem vein and includes eight genera: *Chloroprocta*, *Chrysomya*, *Cochliomyia*, *Compsomyiops*, *Phormia*, *Protocalliphora*, *Protophormia*, and *Trypocalliphora*.

Chloroprocta *Wulp, 1896*

The genus has a single species.

Chloroprocta idioidea (Robineau-Desvoidy, 1830). This species is occasionally found in southern Texas. It is a small fly that resembles *Cochliomyia*, but it lacks mesonotal vittae, and has dusky wings.

Chrysomya *Robineau-Desvoidy, 1830*

Species of this Old World genus have recently become established in South America and the southern United States (Greenberg and Kunich, 2002), and populations apparently are expanding their distribution. Two other species, *C. albiceps* and *C. putoria* are established in the Neotropical Region and likely will spread to the Neoartic Region in the near future. The genus is recognized by a setose greater ampulla (Figure 19.17).

Key to Species of Chrysomya

1. Vestiture of anterior thoracic spiracle dark brown or dark orange; genal dilation with orange ground color with orange setae; eye of male with upper facets enlarged and sharply demarcated from facets in lower third, as in Zumpt (1965, Figure 113); male frons very narrow, eyes nearly touching, frons, at narrowest, 0.01/5 head width; female frons, at narrowest, 0.32/6 (0.31 to 0.33) head width.
   .... *megacephala*

2. Vestiture of anterior thoracic spiracle pale or white; genal dilation with pale dusting and pale setae; eye of male with upper facets not enlarged, no demarcation in lower third; frons broader, at narrowest, 0.046/5 (0.04 to 0.05) head width; female frons, at narrowest, 0.29/6 (0.28 to 0.30) head width.
   .... *rufifacies*

Chrysomya megacephala (Fabricus, 1794). This species is rarely found in the southern United States, and I examined specimens from Florida only. Woodley (in litt.) reports the U.S. National Museum has this species from Florida, Texas, and California. The vestiture of the anterior thoracic spiracle is dark, and the genal dilation has an orange ground color. Males have the upper facets of the eyes much enlarged, with lower facets of the eyes being much smaller.

Chrysomya rufifacies (Macquart, 1843). Widespread but uncommon in southern California, Arizona, New Mexico, Louisiana, Florida, Illinois, and Michigan (Shahid et al., 2000). Facets of eyes are uniform in size, vestiture of the anterior thoracic spiracle is pale in color, and the genal dilation is pale.

Cochliomyia *Townsend, 1915*

This genus has three species in North America. The genal dilation has orange ground color and yellow setae, with pale setae on posterior margin of hind coxa; palp filiform.

Keys to Species of Cochliomyia

1. Upper anterior portion of genal dialation with few to many short black setulae; fifth tergite cupreous, contrasting in color with preceeding tergites; dorsum of thorax with predominantly metallic black and gray colors; postgenal setulae white. Rare in southern Florida.

  2

   Genal dialation with setulae entirely yellow; fifth tergite blue to green, concolorous with preceeding tergites; dorsum of thorax with predominantly metallic blue or green colors; postgenal setulae yellow.

  3

2. Fifth tergite with a pair of median dorsal silvery microtomentose spots; occiput with few to numerous dark setulae above, just below postocular setae; frons of male narrower, at narrowest, 0.06 (0.05 to 0.065/8) head width; surstylus and cercus long and slender, similar to those in Figure 19.55 (also see Dear, 1985, Figures 39 and 40; Hall, 1948, Figure 17E and F).

   .... *aldrichi*

   Fifth tergite with uniform dusting of microtomentum; occiput with pale setulae only below postocular setae; frons of male broader, at narrowest, 0.083 (0.075 to 0.09/2) head width; surstylus and cercus short, surstylus digitate, similar to those in Figure 19.56 (also see Dear, 1985, Figures 37 and 38; Hall, 1948, Figure 8C–E).

   .... *minima*

3. Fronto-orbital plate with dark setulae outside row of frontal setae (Figure 19.53, right side); lateral areas of fifth tergite without pronounced silvery microtomentum; postgenal setae usually golden yellow; female with dark basicosta; proclinate orbital setae absent. Not found in North America since 1966 due to eradication efforts; found in parts of Mexico and Central and South America.

   .... *hominivorax*

   Lower one-half to one-third of fronto-orbital plate with pale setulae outside row of frontal setae (Figure 19.53, left side); fifth tergite usually with pronounced lateral areas of silvery microtomentum; postgenal setae usually pale yellow; female usually with yellowish basicosta; usually with two pairs of proclinate orbital setae (sometimes one or both sides have only one); widespread in North America.

   .... *macellaria*

Cochliomyia aldrichi Del Ponte, 1938. This species is found occasionally in southern Florida. It is similar to *C. minima*; see discussion under that species.

Cochliomyia hominivorax (Coquerel, 1858). This species is difficult to separate from *C. macellaria* (see comments under that species). Male genitalia are illustrated in Hall (1948, Figure 17G–I). Not in North America north of Mexico; original range was the area south of central California east through Iowa and Indiana to South Carolina (Hall, 1948). This species has been the subject of an intensive eradication effort; most specimens collected in the United States are pre-1960. It was considered eradicated from North America by 1966 (Catts and Mullen, 2002). Specimens collected in North America north of Mexico at later dates may be released sterile males. Overall color usually bluish, lower half of fronto-orbital plate with mostly dark setulae outside row of frontal setae, versus pale setulae in *C. macellaria*. Some specimens have pale setulae mixed with dark in the lower frontal plate. If any dark setulae are present, the specimen is *C. hominivorax*.

Cochliomyia macellaria (Fabricius, 1775). This is the most common *Cochliomyia* in North America, from the southern United States to southern Canada. In good specimens this species can be readily identified by the presence of pale setulae outside the row of frontal setae, and pronounced silvery microtomentum on the lateral areas of the fifth tergite. These characters may be difficult to see in old or damaged specimens. For females, the yellowish basicosta is distinctive. The number of proclinate orbital setae is variable, in a group of sixteen females, eleven had two on each side, while five had only one on each side. Male genitalia are illustrated in Hall (1948, Figure 18A and B).

Cochliomyia minima Shannon, 1926. Dear (1985) identified two females from the Florida Keys; one was from Key West and one was from Stock Island. I examined many *Cochliomyia* from the Keys and never found this species. Dear (1985) also listed this species from Cuba, the Dominican Republic, Jamaica, Puerto Rico, and the Virgin Islands. Male specimens are readily separated from the similar *Cochliomyia aldrichi* by the broader frons and distinctive genitalia. Characters for females are reliable for good specimens, but they are easily damaged and problematic in poor specimens. The pattern of microtomentum on the fifth tergite is sometimes readily visible but is somewhat subjective in many specimens. The color of setulae on the occiput can be difficult to interpret. Some *C. aldrichi* have only a few dark setulae to separate them from *C. minima* with all pale setulae.

Compsomyiops *Townsend, 1918*

A single species.
Compsomyiops callipes (Bigot, 1877) is found primarily in the southwestern United States. I examined specimens from California, Arizona, New Mexico, and Texas. It can be separated from *Cochliomyia* by the clavate palps, long dark setae on the hind coxa, and dark calypter. It is a large bluish fly. Male genitalia are illustrated in Hall (1948, Figure 19A–D); female ovipostitor illustrated in Dear (1985, Figures 47 and 48).

Phormia *Robineau-Desvoidy, 1830*

A single species.
Phormia regina (Meigen, 1826) is very common throughout North America. It is a shining metallic blue or green fly with bright orange setae around the anterior thoracic spiracle. Characters are illustrated by Rognes (1991, Figures 247 to 258).

Protocalliphora *Hough, 1899*

Protocalliphora is a large genus with twenty-eight species known in North America. It is most diverse in temperate regions of the Intermountain West, less common farther south. Sixteen species are found only in the West, six only in the East, while six are widespread in both areas. This genus has been found in forty-six of the lower forty-eight states and

Alaska, but it has not been recorded from Florida or Louisiana. It is uncommon in collections, but common in the nests of many altricial birds. Species of this genus are bird nest parasites whose larvae suck the blood of nestling birds. Characters include three or four postsutural intra-alar setae, two notopleural setae, strong anterior acrostichals, scutum usually flattened on center, puparium usually heavily spined, with a strong prothoracic fringe. This genus is closest to *Trypocalliphora*.

The keys to species relying on adult and puparial *Protocalliphora* in Sabrosky et al. (1989) work well for reared series with matched males, females, and puparia for the twenty-six North American species known at the time of publication (*Trypocalliphora braueri* was included under *Protocalliphora*). The key to males permits the identification of lone males in good condition, and the key to females permits the identification of about fifteen species of lone females in good condition. For males, the shape of the surstyli is a critical character, and some of the sketches provided in Sabrosky et al. (1989) are misleading. I have redrawn the surstyli for *P. beameri* (Sabrosky et al., 1989), *P. bicolor* (Sabrosky et al., 1989), *P. hirundo* (Shannon and Dobroscky, 1924), *P. interrupta* (Sabrosky et al., 1989), *P. metallica* (Townsend, 1919), and *P. parorum* (Sabrosky et al., 1989) (Figures 19.54 to 19.59) to better reflect distinctions for each species. Lone females are often difficult to identify because they have few distinctive characters. Perhaps a detailed study of female genitalia will produce some distinguishing characters in the future, but preliminary examinations have not provided any good characters.

Since the publication of Sabrosky et al. (1989), I have identified three additional North American species of *Protocalliphora* (Whitworth, 2002, 2003a). The former publication describes two new species (*P. bennetti* Whitworth and *P. rugosa* Whitworth); the latter splits *P. sialia* into an Eastern and Western component. *Protocalliphora sialia* Shannon and Dobroscky is the form found in the Midwest and East, while *P. occidentalis* Whitworth is the Western form.

I have provided a key to assist with identification of these new species, and it can be integrated with the adult and puparial key in Sabrosky et al. (1989) starting at couplet 7, p. 77. The unifying character in this group is the digitate surstyli in males. Lone adult females will be difficult to key, but the key is useful to separate females of species in mixed infestations in bird nests. Common mixes in the West include *P. bennetti*, *P. occidentalis*, *P. rugosa*, and occasionally *P. hirundo*. In the East, *P. sialia* and *P. bennetti* are commonly found in the same nest.

Adults in this genus are difficult to collect, though their empty puparia are relatively easy to find in old bird nests. I developed a revised key (Whitworth, 2003b) for the puparia of twenty-seven North American species. To date I have examined over 8,000 bird nests, about half of which were infested with one or more of twenty-seven of the twenty-eight known species of this genus. One species, *P. sapphira* Hall, has not been collected from a nest and is known primarily from a single distinctive male. Three females matched to the male may not be the same species. I have examined many *Protocalliphora* from the same area in Alaska where *P. sapphira* was collected and have found nothing resembling the male holotype. Until recently, *P. beameri* had never been collected from a bird nest. However, in 2004 I received an adult *P. beameri* and matched puparium from a black-throated gray warbler nest. The specimens were provided by Piotr Jablonski, who found the nest in the Chiricahua Mountains in Arizona. The puparium matches those of a previously unidentified species I examined from a barn swallow nest near Fort Davis, Texas.

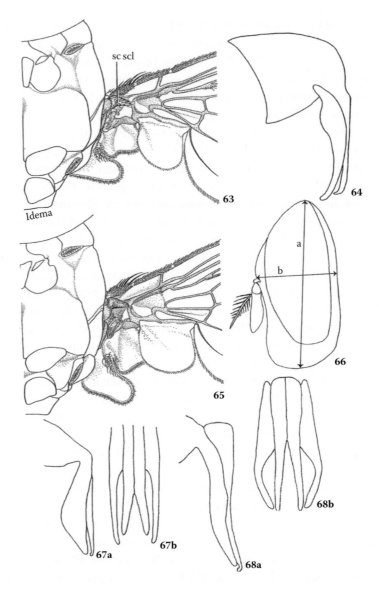

Figures 19.63–19.68 63. *Lucilia illustris*. Female, ventral view of right wing base; sc scl = subcostal sclerite; from Shewell (1987). 64. *Lucilia illustris*. Male, left lateral view of cercus and surstylus. 65. *Lucilia sericata*. Female, ventral view of right wing base; from Shewell (1987). 66. How to measure head proportions: (a) head height; (b) head length. 67. *Lucilia elongata*. Male, cerci and surstyli: (a) left lateral view; (b) posterior view. 68. *Lucilia silvarum*. Male, cerci and surstyli: (a) left lateral view; (b) posterior view.

Key to Male Protocalliphora *with Digitate Surstyli, and White Calypters, with Notes on Females and Puparia*

1. Male surstylus digitate (as in Figure 19.56), not appreciably curved.
  2

   Male surstylus distinctly curved (Figures 19.54, 19.55, 19.57 to 19.59), usually slender, or short and broad, as in Sabrosky et al. (1989, Figures 7 to 9).
   .... 15 species of *Protocalliphora*

2. Calypter white in both sexes, primarily parasites of birds that nest in cavities.
....3

Calypter brown, except calypter white in female *P. cuprina* and *P. halli*. Female *P. cuprina* have fifth tergite cupreous, female *P. halli* are found almost exclusively in barn swallow and phoebe nests. Usually in birds with open nests.
.... *P. cuprina, P. halli, P. hesperia, P. hesperioides*

3. Male and female with postalar wall and tympanic pit bare or with a few pale setae; see Sabrosky et al. (1989, Figure 3b) for location; fore tibia usually with one posterior seta.
....4

Male and female with postalar wall and tympanic pit with a conspicuous tuft of black setae; fore tibia with two posterior setae. Found almost exclusively in bank swallow nests.
.... *P. rognesi*

4. Male and female with preocellar area polished just anterior to median ocellus, polished area varies from small to large and irregular (see Sabrosky et al., 1989, Figures 1 and 2a for location); frons of female, at narrowest, averages 0.25 (0.22 to 0.28) head width. Dorsum of puparium with cuticular ridges faint or absent, or if pronounced, found only east of a line from Alaska to Kentucky.
....5

Male and female with preocellar area dull colored, microtomentose (rarely a small polished area); frons of female, at narrowest, averages 0.28 (0.26 to 0.31) head width. Dorsum of puparium with pronounced cuticular ridges (see Whitworth [2003b] for an explanation of puparial characters). Found primarily in the West, one species, *P. hirundo*, may be found in the East primarily in bank and cliff swallow nests.
....7

5. Male and female with parafacial relatively broad, width at lunule obviously much wider than width of first flagellomere; male frons wider, at narrowest, averaging 0.075 to 0.10 head width. Puparium with hyperstigmatal spines longer, averaging 25 µ or more; posterior ventral spine bands not reduced to rear.
....6

Male and female with parafacial narrow, equal to or barely wider than first flagellomere; male frons narrower, at narrowest, averaging 0.06 (0.05 to 0.07) head width. Puparium with hyperstigmatal spines short, averaging 12.5 µ in length; posterior ventral spine bands reduced to rear.
.... *shannoni*

6. Male frons, at narrowest, 0.075 (0.065 to 0.08) head width; about equal to width of first flagellomere; female with large, triangular polished area encompassing ocellar triangle, which tapers to a point in preocellar area when viewed from below. Puparium with shorter prothoracic fringe averaging 350 µ; dorsal cuticular folds faint. Alaska to northern Idaho, east to northern Minnesota, and southeast to Virginia.
.... *bennetti*

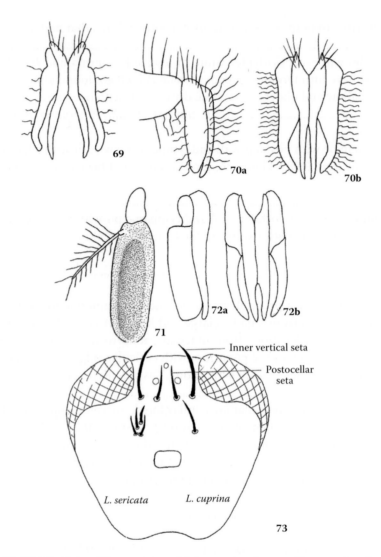

Figures 19.69–19.73 69. *Lucilia mexicana*. Male, cerci and surstyli, posterior view. 70. *Lucilia eximia*. Male, cerci and surstyli: (a) left lateral view; (b) posterior view. 71. *Lucilia thatuna*. Female, antenna, left lateral view. 72. *Lucilia thatuna*. Male, cerci and surstyli: (a) left lateral view; (b) posterior view. 73. *Lucilia sericata*. Posterior view of head showing setae below inner vertical setae, left side; *Lucilia sericata, L. cuprina*.

Male frons broader, at narrowest, 0.10 (0.09 to 0.12) head width; frons of male clearly wider than first flagellomere; female with smaller irregular polished preocellar area, not encompassing ocellar triangle, or if extending upward, not uniformly shining when viewed from below. Puparium with exceptionally long prothoracic fringe, 500 μ or more in diameter; dorsal cuticular folds pronounced. East of a line from Alaska through Saskatchewan and Minnesota to Kentucky.
.... *sialia*

7. Basicosta orange to reddish brown.
....8

Basicosta black or dark brown.

.... *occidentalis*

8. Upper portion of fronto-orbital plate in male significantly narrowed, as in Whitworth (2002, Figure 7). Dorsal cuticular ridges of puparium narrower, less than 50 μ wide, ridges abundant and close together. Primarily in tree, and violet-green swallows and purple martins.

.... *rugosa*

Upper portion of fronto-orbital plate in male not significantly narrowed as in Whitworth (2002, Figure 8). Dorsal cuticular ridges of puparium broader, 50 μ or more, ridges usually widely spaced. Primarily in cliff and bank swallows.

.... *hirundo*

Protophormia *Townsend, 1908*

This genus is represented by only two species in North America. Both have a flattened scutum, like *Protocalliphora*, but the anterior acrostichals are weak or absent.

Key to Species of Protophormia

1. Lower part of face strongly protruding (Figure 19.60); arista almost bare below (Figure 19.60); antenna entirely black; two pairs of marginal scutellar setae in addition to the apical pair; anterior spiracle much enlarged (Figure 19.61), almost as large as humeral callus in lateral view; cell r4+5 closed, or nearly so at wing margin (Figure 19.62); eye small, about two-thirds of head height (Figure 19.60). Rare, found only north of 80° N.

.... *atriceps*

Lower part of face not strongly protruding (as in Figure 19.3); arista plumose (as in Figure 19.3); tip of pedicel and basal part of first flagellomere reddish; three or four pairs of marginal scutellar setae in addition to the apical pair; anterior spiracle smaller, much smaller than humeral callus (as in Figure 19.2); cell r4+5 open at wing margin; eye larger, three-fourths of head height. Common, in the northern United States, Canada, and Alaska.

.... *terraenovae*

Protophormia atriceps (Zetterstedt, 1845). This is a rare species found north of 80°N in North America (Rognes, 1991). It can be recognized by its protruding lower face (Figure 19.60) and large anterior spiracle (Figure 19.61). Various characters are illustrated in Rognes (1991, Figures 311, 313, 315 to 326).

Protophormia terraenovae (Robineau-Desvoidy, 1830). This species is common throughout the northern United States, Canada, and Alaska. I examined specimens from Washington to Ohio, and Alaska to California. The face is not protruding, and it has a smaller anterior spiracle. Various characters are illustrated in Rognes (1991, Figures 310, 312, 314, 327 to 337).

Trypocalliphora *Peus 1960*

Rognes (1985) considered *Trypocalliphora* a valid genus, while Sabrosky et al. (1989) considered it a subgenus of *Protocalliphora*. As a result of my studies of puparia (Whitworth,

2003b), I concluded that *Trypocalliphora* deserves generic status. It is represented by a single Holarctic species.

　　Trypocalliphora braueri (Hendel, 1901) widespread, but uncommon throughout most of the United States, Canada, and Alaska; relatively common in the Northwest (Whitworth, 2003b). Closest to *Protocalliphora*, this species has one or more accessory notopleural setae (Figure 19.22). Larvae are obligate subcutaneous parasites of nesting birds. Puparia have very few spines and lack a prothoracic fringe. Various characters are illustrated in Rognes (1991, Figures 338 to 349).

Luciliinae

This subfamily includes one genus, *Lucilia* (Robineau-Desvoidy, 1830). The genera *Phaenicia, Bufolucilia, and Francilia* were synonymized with *Lucilia* by Rognes (1991). It can be recognized by its shining, green, blue, or bronze thorax and abdomen, suprasquamal ridge with a cluster of setae, and bare lower calypter. The genus includes eleven species in North America. When measuring the head to frons ratios in females, note that the frons is not narrowest at the vertex, as in most female calliphorids.

Lucilia *Robineau-Desvoidy, 1830*
Key to Lucilia *species*

1. Subcostal sclerite on venter of wing with wiry black setulae (Figure 19.63); basicosta tan, dark brown, or black; palp orange; surstylus and cercus of male as in Figure 19.64; ocellar triangle of female large, reaching at least halfway to lunule. Widespread in the northern United States and Canada.

   ....*illustris*

   Subcostal sclerite on venter of wing with pubescence only (Figure 19.65); basicosta orange or black; palp orange or black; surstylus and cercus of male not as above; ocellar triangle of female small, not reaching halfway to lunule.

  2

2. Palp black or brown (note: orange palps may darken in alcohol or if left in kill jars too long); length of head at level of lunule more than half head height (see Figure 19.66 for how to measure); third abdominal tergite with one or two pairs of long, erect median marginal setae (see Figure 19.7 for location); basicosta dark brown or black.

  3

   Palp orange to yellow, not darkened apically; length of head at level of lunule less than half head height (except in *L. thatuna* and some *L. sericata*); third abdominal tergite with marginal setae not especially strong or erect (except male *L. thatuna*); basicosta usually yellow or orange (*L. mexicana* and *L. eximia* have brown basicostas).

  5

3. Three postsutural intra-alar setae with anterior one weak; presutural intra-alar setae absent; arista with short setae, usually much shorter than width of first flagellomere, as in Rognes (1991, Figure 411); first flagellomere long, more than half eye length in profile; male cercus parallel sided, tip of surstylus straight (Figures 19.9 and 19.10); Northern, Alaska to Labrador.

   .... *magnicornis*

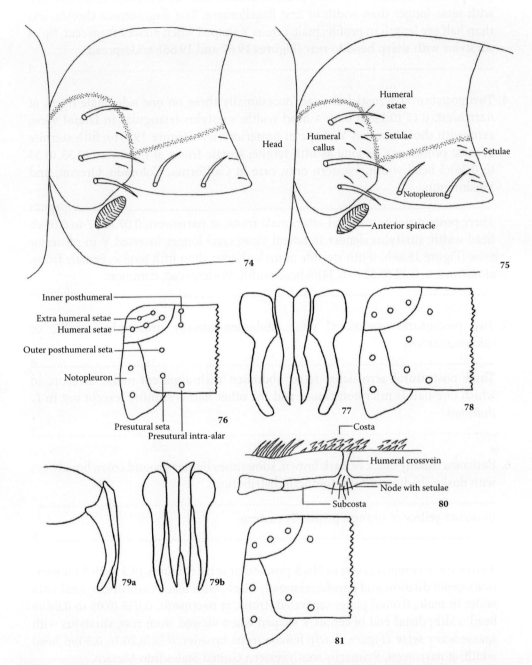

Figures 19.74–19.81 74. *Lucilia cuprina*. Setae at rear of humeral callus and notopleuron. 75. *Lucilia sericata*. Setae at rear of humeral callus and notopleuron. 76. *Pollenia vagabunda*. Dorsal view of left side of prothorax. 77. *Pollenia vagabunda*. Male, cerci and surstyli, posterior view. 78. *Pollenia rudis*. Dorsal view of left side of prothorax. 79. *Pollenia griseotomentosa*. Male, cerci and surstyli: (a) left lateral view; (b) posterior view. 80. *Pollenia pediculata*. Ventral view of junction of humeral cross-vein and subcosta showing bundle of pale setulae. 81. *Pollenia griseotomentosa*. Dorsal view of left side of thorax.

Two postsutural intra-alar setae; presutural intra-alar setae present; arista normal, with setae longer than width of first flagellomere; first flagellomere shorter, less than half eye length in profile; male cercus Y shaped when viewed from rear, tip of surstylus with sharp bend to rear (Figures 19.67 and 19.68); widespread.

....4

4. Two postsutural acrostichal setae, occasionally three on one side; male frons, at narrowest, 0.13 (0.12 to 0.14)/4 head width; surstylus triangular in lateral view; cerci with short inverted V shape in posterior view (Figure 19.67b); fifth sternite of male prominent, as long as fifth tergite; female frons, at narrowest, 0.35 (0.33 to 0.36)/5 head width. Western only, rare in California, Colorado, Oregon, and Washington.

....*elongata*

Three postsutural acrostichal setae; male frons, at narrowest, 0.07 (0.07 to 0.09)/6 head width; surstylus slender in lateral view; cerci longer inverted V in posterior view (Figure 19.68b). Fifth sternite of male shorter than fifth tergite. Female frons, at narrowest, 0.32 (0.32 to 0.34)/5 head width. Widespread, common.

.... *silvarum*

5. Two postsutural acrostichal setae; abdomen usually uniformly metallic or microtomentose.

....6

Three postsutural acrostichal setae; abdomen with apparent mesal division, in which one-half is microtomentose and the other half is shining (except not in *L. thatuna*).

....9

6. Basicosta usually black or dark brown, sometimes lighter ground color, but always with dusky shading; mostly southern distribution.

....7

Basicosta yellow or orange; primarily eastern.

....8

7. Two or more complete rows of black postocular setae (Figures 19.3 to 19.5 for location); genal dilation and parafacial mostly black, with black vestiture; frontal vitta wider in male, frontal plates separated, frons, at narrowest, 0.055 (0.05 to 0.06)/8 head width; distal end of cercus Y shaped when viewed from rear, surstylus with sparse wavy setae (Figure 19.69); female frons broader, 0.28(0.26 to 0.30)/6 head width, at narrowest. Primarily southwestern United States into Mexico.

....*mexicana*

One complete row of black postocular setae; genal dilation and parafacial mostly tan to orange, with vestiture reddish to light brown; frontal vitta in male very narrow, frontal plates touching, or nearly so, frons, at narrowest, 0.035 (0.03 to 0.04)/10 head width; distal end of cercus almost parallel when viewed from rear, surstylus with dense wavy setae (Figure 19.70); female frons narrower 0.25 (0.24 to 0.28)/9 head width, at narrowest. Subtropical, occasionally found in Texas and Florida.

....*eximia*

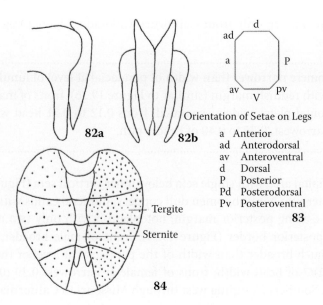

Orientation of Setae on Legs

a   Anterior
ad Anterodorsal
av Anteroventral
d   Dorsal
P   Posterior
Pd Posterodorsal
Pv Posteroventral

Figures 19.82–19.84 82. *Pollenia rudis*. Male, cerci, and surstyli: (a) left lateral view; (b) posterior view. 83. Diagrammatic sketch of seta orientation on legs; a = anterior; ad = antero-dorsal; av = anteroventral; d = dorsal; p = posterior; pd = posterodorsal; pv = posteroventral; v = ventral. 84. Ventral view of vestiture on abdomen. *Pollenia angustigena*, left side, *Pollenia rudis*, right side.

8. Setae on gena dark (note: pale setae on postgena). Frons of male with frontal plates almost touching, frons width, at narrowest, much less than breadth of first flagellomere, frons 0.023 (0.015 to 0.030)/8 head width; male with one laterocli-nate orbital seta slightly anterior to median ocellus (see Figures 19.3 to 19.5 for seta location); female with black setulae outside row of frontal setae on frontal plate; fifth abdominal tergite highly polished, tinged with red or purple in both sexes; mature specimens usually larger, 8.0 to 9.5 mm in length. Maryland south to Florida, north to Michigan and Wisconsin, most common in the southeast, less abundant west of the Mississippi River, but westward to California.

....*coeruleiviridis*

Setae on lower 1/3 to 1/2 of gena pale; frons of male with frontal plates well sepa-rated, frons width, at narrowest, more than width of first flagellomere, frons 0.11 (0.10 to 0.12)/7 head width; male with lateroclinate orbital seta opposite median ocellus, or seta absent; female with pale setulae outside row of frontal setae on fron-tal plate; fifth abdominal tergite generally not more polished than other tergites, usually without reddish or purple cast; mature specimens usually smaller, 8.0 mm length or less. Florida north to North Carolina and west to southern Mississippi.

....*cluvia*

9. First flagellomere broader than width of parafacial at level of lunule, often cupped inward, inner margin often reddish, especially in female (Figure 19.71), sometimes not so cupped in male, still broader than parafacial; frons of male, at narrowest, with frontal plates almost touching, frons 0.044 (0.04 to 0.05)/9 of head width; male surstylus and cercus as in Figure 19.72; frons of female, at narrowest, 0.30 (0.27 to

0.32)/12; rare, known only from California, Colorado, Idaho, Oregon, Utah, and Washington.

....*thatuna*

First flagellomere narrower than width of parafacial at level of lunule, usually not cupped or with reddish margin (similar to Figure 19.36); frons of male, at narrowest, with frontal plates widely separated, frons 0.12 to 0.21 head width; frons of female, at narrowest, 0.35 to 0.40 of head width.

....10

10. Central occipital area with single seta below inner vertical seta (Figure 19.73, right side); metasternum bare; abdomen dull coppery; humeral callus with two or three small setulae along posterior margin; notopleuron with only two or three small setulae on posterior border (Figure 19.74); frons of male broader, at narrowest obviously much broader than width of the parafacial at level of the lunule, 0.20 (0.19 to 0.21)/7 of head width; frons of female at narrowest 0.39 (0.38 to 0.40)/5 head width. Southern, Virginia west through Missouri to California.

....*cuprina*

Central occipital area with group of two to five setae below inner vertical seta (Figure 19.73, left side); metasternum setose; abdomen usually bright green, occasionally shining coppery; humeral callus with six to eight small setulae along posterior margin; notopleuron usually with five or more setulae on rear border (Figure 19.75); frons of male narrower, at narrowest about equal to width of parafacial at level of lunule, 0.13 (0.12 to 0.14)/6 of head width; frons of female at narrowest 0.37 (0.35 to 0.40)/8 head width. Widespread.

....*sericata*

Lucilia cluvia (Walker, 1849). This species is found primarily in the Southeast; I examined specimens from Arkansas to Florida to South Carolina. It is uncommon and very close to *L. coeruleiviridis* in appearance, but *L. cluvia* males can be distinguished by their much broader frons. Females of *L. cluvia* are difficult to separate with confidence from those of *L. coeruleiviridis*, the primary distinction being the color of the fine setulae outside the row of frontal bristles. This character is often variable, damaged, or difficult to see. The difference in the shininess of the fifth tergite is subjective and variable, but can be useful with practice. Based on the material I examined *L. cluvia* tends to be smaller than *L. coeruleiviridis*. Excluding three obviously undersized specimens, twenty specimens of the former were 7.5 to 8 mm in length. For *L. coeruleiviridis*, excluding five undersized specimens, fifty-one ranged from 7.75 to 9.5 mm in length. Better characters are needed for female distinctions; perhaps a study of ovipositors would reveal useful characters to separate the two species.

Lucilia coeruleiviridis Macquart, 1855. This species is in the southeastern United States, but it may be found in the Northeast and Midwest. It is uncommon in the West, and is generally much more commonly encountered than *L. cluvia*. I examined specimens from California to Florida, from Nebraska and Wisconsin to Pennsylvania, and most states south. I did not find it in the Northwest.

Lucilia cuprina Wiedemann, 1826. This species is uncommon throughout the south, from Virginia to Florida west to Missouri and Texas and California. It is usually recognized by its dull coppery sheen, but color alone is not reliable. Some *L. sericata* are quite coppery, though usually more shining. The wider frons in *L. cuprina* readily separates males of each species. A single seta below the inner vertical seta (Figure 19.73, right side) versus two to five setae in *L. sericata* (Figure 19.73, left side) will distinguish specimens of both sexes. This character sometimes varies, or can be hard to see due to the condition of the specimen. The presence or absence of setae on the metasternum (absent in *L. cuprina*) is also useful, but often is difficult to see.

Lucilia elongata Shannon, 1924. This species is rarely found in collections. I examined specimens from California, including San Mateo County, Mendocino County, Tehama County, and Yolo County; Washington, Pierce County; and Oregon, Washington County and Klamath County. James (1955) recorded this species from Orcas Island, Washington, as well as California and possibly Colorado. This species is close to *L. silvarum*, but it normally has only two postsutural acrostichal setae, while *L. silvarum* has three. The frons of males, at narrowest, is much broader, averaging 0.13 of head width in *L. elongata* versus 0.07 in *L. silvarum*. A few females and one male were seen with two setae on one side and three on the other. J. O'Hara (in litt.) reports the Canadian National Collection has fifteen specimens of *L. elongata*; six from various areas in British Columbia have two postsutural acrostichals one each side, and nine from Whatcom County, Washington, include several males with three postsutural acrostichals on one side.

Lucilia eximia (Wiedemann, 1819). This species is rare, found occasionally only in Texas, Oklahoma, and Florida. Hall (1948) noted that this is a common fly in market places of Central America. This is one of only two species with yellow palps and a dark basicosta. It has only one row of black postocular setae, and an orange genal dilation. By contrast, its close relative, *L. mexicana*, has two complete rows of postocular setae and a dark genal dilation. The postocular seta character can be confusing; the row of black postocular setae may be incomplete in *L. eximia*.

Lucilia illustris (Meigen, 1826). This species is widespread and common in the northern United States and Canada. I examined specimens from as far south as southern California and Arizona in the West, but in the Midwest I did not find it south of Missouri to Indiana, while on the East Coast I did not find it south of South Carolina. Various characters are illustrated in Rognes (1991, Figures 371, 411 to 422).

Lucilia magnicornis (Siebke, 1863). This species is uncommon in the far north, from Alaska to Labrador. I examined specimens from Alaska, Northwest Territories, and northern Manitoba. This species has brown palps, three postsutural intraalar setae, and the length of the first flagellomere is more than half the eye length. Various characters are illustrated in Rognes (1991, Figures 371, 411 to 422).

Lucilia mexicana Macquart, 1843. This species is common in the southwestern United States. I examined specimens from California to Texas, Utah, and Oklahoma. Hall

(1948) stated that this species extends as far south as Brazil. It has a brown basicosta, like *L. eximia*, but two complete rows of postocular setae and a dark genal dilation. Its range overlaps with that of *L. eximia* in Texas.

Lucilia sericata (Meigen, 1826). This species is one of the most common *Lucilia*, and is widespread in the United States and southern Canada. It is one of three species with three postsutural setae. It can be separated from *L. cuprina* by the presence of two to five setae on the central occipital area below the inner vertical setae. Specimens tend to be green, but some are so coppery that they can be confused with *L. cuprina*. It also has a setose metasternum, which is often hidden and very difficult to see. This species can be separated from *L. thatuna* by the width of the first flagellomere and the much broader frons of the male. Various characters are illustrated in Rognes (1991, Figures 375, 455 to 465).

Lucilia silvarum (Meigen, 1826). I examined specimens from Washington to California in the West, and Maine to South Carolina and Louisiana in the East. A common, widespread species, Hall (1948) also recorded it from southern Canada. Specimens have three postacrostichal setae and black palps; the male frons is much broader than in the similar *L. elongata*. Various characters are illustrated in Rognes (1991, Figures 376, 466 to 476).

Lucilia thatuna Shannon, 1926. This is an uncommon species; I examined specimens from ten counties in California, most in the northern coastal areas; also Pullman, Washington; Baker County, Oregon; and Cache County, Utah. James (1955) recorded it from many localities in California, and also Montana, Idaho, and Colorado. The presence of three postacrostichal setae and the first flagellomere being broader than the parafacials separate it from *L. cuprina* and *L. sericata*. Specimens are often bluish, which separates them from the green or coppery *L. sericata* or the coppery *L. cuprina*. Males are distinctive, as their frons is much narrower than those of *L. cuprina* and *L. sericata*.

Polleniinae

Species in this subfamily are dull colored, unlike most calliphorids, and there are two genera, *Melanodexia* and *Pollenia*.

Melanodexia *Williston, 1893*

This genus is uncommon in the West; few specimens were encountered in this study, and no attempt was made to sort them to individual species. Both Hall (1948) and James (1955) studied this genus, but species distinctions are difficult, and the genus needs further study. Hall listed three species under this genus and five more under the name *Melanodexiopsis*, a synonym of *Melanodexia*.

Pollenia *Robineau-Desvoidy, 1830*

This genus is widespread in North America. It was thought to be represented by a single species, *P. rudis*, until recently (Rognes, 1991). Six species are now recognized in North

America. Species of this genus are dull-colored calliphorids with distinctive, crinkly yellow setae on the thorax. The key herein was adapted from the one developed by Knut Rognes for Greenberg (1998).

Key to Species of Pollenia

1. Thorax with dark median undusted vitta between presutural acrostichal setae, usually extending forward to extreme anterior slope of thorax and back to the rear of the scutum; usually two or more extra setae in front of the regular row of three (two to four) humeral setae (Figure 19.76); two inner posthumeral setae (Figure 19.76); male cercus broad and flattened when viewed from rear (Figure 19.77). Northeastern and northwestern United States and Canada.
   ....*P. vagabunda*
   No median undusted vitta; no extra humeral setae in front of the regular setae (Figure 19.78); one inner posthumeral seta (Figure 19.78); cercus not as above (Figure 19.79b).
  2

2. Lappets of posterior thoracic spiracle dark brown; facial carina reduced and indistinct; basicosta usually dark brown to black. Northern United States and southeastern Canada.
   ....*P. labialis*
   Lappets of posterior thoracic spiracle pale yellow to orange; facial carina usually distinct (Figure 19.26) (except *P. griseotomentosa*); basicosta yellowish, orange, or light tan.
  3

3. Node at junction of humeral cross-vein and subcosta of wing with a bundle of pale setulae below (Figure 19.80); palpus dark brown or black. Widespread in northern portions of the United States.
   ....*P. pediculata*
   Node without setulae; palpus usually lighter brown or orange (except some *P. rudis*).
  4

4. Outer posthumeral seta absent (Figure 19.81); femur of mid- and hind leg with mostly black vestiture on posteroventral surface (see Figure 19.83 for orientation); facial carina absent or much reduced; male surstyli distinctly curved and slender (Figure 19.79a); frons of male exceptionally narrow, at narrowest, 0.032 (0.025 to 0.04)/3 head width. Rare, northeastern and northwestern United States and Wisconsin.
   ....*P. griseotomentosa*
   Outer posthumeral seta usually present (Figure 19.78); if absent (some *P. angustigena*), femur of mid- and hind leg with mostly yellow vestiture on posteroventral surface; facial carina distinct (Figure 19.27); male surstyli less curved and broader (Figure 19.82); frons of male slightly to much broader.
  5

5. Tibia of mid-leg with one anterodorsal seta (see Figure 19.83 for orientation); mid- and hind femora with yellow or orange posteroventral vestiture; ventral abdominal vestiture in males normal, not particularly fine, dense, or erect (Figure 19.84, left side); male frons narrower, frons 0.038 (0.03 to 0.04)/6 head width. Locally common in Washington, California, Wisconsin, and northeastern United States. .... *P. angustigena*
Tibia of mid-leg usually with two or three anterodorsal setae; mid- and hind femora with black or dark posteroventral vestiture; ventral abdominal vestiture in males fine, erect, and dense (Figure 19.84, right side); male frons broader, frons 0.06 (0.055 to 0.065)/7 head width. Widespread in North America. .... *P. rudis*

Pollenia angustigena Wainwright, 1940. Until this study, this species was known only from northeastern North America. I have examined specimens from California to Washington, Idaho to Wisconsin, Ohio to New Jersey, and south to Virginia. It is similar to *P. rudis*, but males are usually distinctive. Females are difficult to distinguish since the only good character known is the number of anterodorsal setae on the midtibia. If legs are missing or the setae damaged, then identification of females is difficult. Various characters are illustrated in Rognes (1991, Figures 562, 579, 594 to 603).

Pollenia griseotomentosa (Jacentkovsky, 1944). Rognes (personal communication) listed this species from Ontario, Canada. I have seen a specimen from Green County, PA. It is the only North American *Pollenia* lacking an outer posthumeral seta. Various characters are illustrated in Rognes (1991, Figures 563, 604 to 611).

Pollenia labialis Robineau-Desvoidy, 1863. Rognes (1991) recorded this species from Ontario, and Greenberg (1998) listed it from Indiana. I found it from Michigan, Maine, New Hampshire, Oregon, and Washington. The lappets of the posterior spiracle are dark brown, which distinguishes it from other species. Discolored specimens of other species can be confused with it, although the reduced facial carina separates it from most similar species. Various characters are illustrated in Rognes (1991, Figures 565, 622 to 628).

Pollenia pediculata Macquart, 1834. I examined specimens of this species from Washington to Wisconsin and New York, and south to North Carolina. I also found it in Utah, Oregon, and California. Rognes (1991) recorded it from New Mexico. I did not find it in the southeastern United States. This is the second most common *Pollenia* I found, next to *P. rudis*. It is readily identified by a distinctive bundle of setae on the venter of the wing, at the junction of the humeral cross-vein and subcosta. Various characters are illustrated in Rognes (1991, Figures 557, 559, 581, 583, 640 to 650).

Pollenia rudis (Fabricus, 1794). This species is widespread in North America and was once thought to be the only *Pollenia* species present. *Pollenia* specimens in most collections are identified as this species, but I have found that half or more are other species. It is similar to *P. angustigena*, but males have a broader frons and a denser vestiture on the venter of the abdomen. Female characters are limited to setae on the mid tibia. Various characters are illustrated in Rognes (1991, Figures 582, 651 to 661).

Pollenia vagabunda (Meigen, 1826). Rognes (1991) listed this species from British Columbia, Nova Scotia, and Prince Edward Island, and Greenberg (1998) listed it from New York. I also examined specimens from Massachusetts, New Hampshire, and Washington. In specimens in good condition, a dark median stripe between the presutural acrostichal setae is a distinctive character. Accessory humeral setae and two inner posthumeral setae will further confirm its identity. Various characters are illustrated in Rognes (1991, Figures 569, 662 to 669). In one sample of twelve *Pollenia* collected from a home in Tacoma, Washington, on April 1, 2005, eight were *P. vagabunda*, three were *P. angustigena*, and one was *P. rudis*.

Melanomyinae

Downes (1986) synonymized *Angioneura* and *Opsodexia* under *Melanomya*. Shewell (1987) concluded these should be separate genera, an opinion with which Rognes (1991) concurred. Species of both genera are rarely encountered in collections, so keys to species are not provided here. The keys in Downes (1986) are useful in making accurate identifications (N. Woodley, in litt.). Species of both genera are dull colored and nondescript. Their biology is poorly known, but Downes (1986) suspects all might be snail parasites.

Angioneura *Brauer and Bergenstamm, 1893*

This genus includes five species that have relict populations primarily in the East and Midwest.

Opsodexia *Townsend, 1915*

This genus includes four species that apparently have habits and distributions similar to *Angioneura*.

Acknowledgments

This study was made possible with the financial support of my firm, Whitworth Pest Solutions, Inc. I thank my employees for understanding my late arrivals and early departures from work to pursue this study in my lab in the basement of my home. I especially appreciate the help of my general manager, Belinda Bowman, whose diligence has given me the long, undisturbed blocks of time needed to complete this study. Thanks also to my wife, Faye, who has accepted my obsession with blow flies and their intrusion into her home.

I am especially indebted to James O'Hara of the Canadian National Collection, Agriculture and Agri-Food Canada, Ottawa, Ontario, and Knut Rognes of the University of Stavanger, Stavanger, Norway, who provided detailed answers to my many questions and helped inspire me to complete this study. This work would not have been possible without the cooperation of many museum curators who sent me materials for study. Special thanks to Rich Zack and Will Hanson, curators at Washington State University and Utah State University, respectively, who sent me many specimens and were always willing to help.

Rich also acted as a liaison to enable me to get specimens that are not normally loaned to private individuals.

Other curators who sent materials include: from the University of California, Berkeley, Cheryl Barr; University of California, Davis, Steve Heydon; University of California, Riverside, Doug Yanega; California Academy of Sciences, Keve Ribardo; Natural History Museum of Los Angeles, Brian Brown; Florida State Collection of Arthropods, Gary Steck; University of Idaho, Frank Merickel; University of Missouri, Kris Simpson; Montana State University, Richard Hurley; University of New Hampshire, Don Chandler; New Mexico State University, David Richman; Oregon State University, Darlene Judd; National Museum of Natural History, Smithsonian Institution, Norm Woodley; Spencer Museum, University of British Columbia, Karen Needham; University of Wisconsin, Madison, Steven Krauth. Others who sent materials include Eric Eaton, private collector; Neil Haskell, St. Joseph's College, Rensselaer, Indiana; and Jeff Wells, West Virginia University. Thanks to all.

Thanks also to all who reviewed this manuscript; already mentioned are Knut Rognes, James O'Hara, Rich Zack, Norm Woodley, Gary Steck, and Neil Haskell. Other reviewers include James Wallman, University of Wollongong, New South Wales, Australia; Greg Dahlem, Northern Kentucky University, Highland Heights, Kentucky; Gail Anderson, Simon Fraser University, Burnaby, British Columbia; and Bruce Cooper, Canadian National Collection.

Also, thanks to Dawn Nelson, scientific illustrator, who helped me produce quality illustrations to make the keys more understandable. Finally, my Figures 19.1 to 19.10 from Rognes (1991) are reproduced with permission of E.J. Brill/Scandinavian Science Press and the author, Knut Rognes. Figures 19.11 to 19.13 are from McAlpine (1981); Figures 19.14, 19.19 to 19.21, 19.25, 19.27 to 19.32, 19.42, 19.43, 19.60, 19.61, 19.63, and 19.65 are from Shewell (1987); all are reproduced with permission of the Minister of Public Works and Government Services Canada, 2004.

References

Catts, P. E., and G. R. Mullen. 2002. Myiasis (Muscoidea, Oestroidea). In Mullen, G. R., and L. A. Durden, Eds., *Medical and veterinary entomology*. New York: Academic Press, 318–47.

Dear, J. P. 1985. A revision of the New World Chrysomyini (Diptera: Calliphoridae). *Revista Brasileira de Zoologia* 3:100–69.

Downes, W. L. 1965. Tribe Melanomyini. In Stone, A., C. W. Sabrosky, W. W. Wirth, R. H. Foote, and J. R. Coulson, Eds., *A catalog of the Diptera of North America north of Mexico*. Agricultural Handbook 276. Washington, DC: United States Department of Agriculture, Agricultural Research Service, 93233.

Downes, W. L. 1986. *The Nearctic Melanomya and relatives (Diptera: Calliphoridae), a problem in calypterate classification*. Bulletin of the New York State Museum 460.

Greenberg, B. 1998. Reproductive states of some overwintering domestic flies (Diptera: Muscidae and Calliphoridae) with forensic implications. *Arthropod Biology* 91:818–20.

Greenberg, B., and Kunich, J., Eds. 2002. *Entomology and the law: Flies as forensic indicators*. Cambridge, UK: Cambridge University Press.

Hall, D. G. 1948. *The blowflies of North America*. Lafayette, IN: Thomas Say Foundation.

Hall, D. G. 1965. Family Calliphoridae. In *A catalog of the diptera of North America north of Mexico*, Stone, A., C. W. Sabrosky, W. W. Wirth, R. H. Foote, and J. R. Coulson, Eds. Agricultural Handbook 276. Washington, DC: United States Department of Agriculture, Agricultural Research Service, 922–32.

Hall, R. D., and L. H. Townsend. 1977. *The blowflies of Virginia: No. 11.* Research Division Bulletin 123. Virginia Polytechnic Institute and State University.

International Commission on Zoological Nomenclature. 1999. *International code of zoological nomenclature.* 4th ed. London: International Trust for Zoological Nomenclature.

James, M. T. 1953. Notes on the distribution, systematic position, and variation of some Calliphoridae, with particular reference to the species of western North America. *Proceedings of the Entomological Society of Washington* 55:143–48.

James, M. T. 1955. The blowflies of California (Diptera: Calliphoridae). *Bulletin of the California Insect Survey* 4(1).

McAlpine, J. F. 1981. Morphology and terminology—Adults. In *Manual of Nearctic diptera*, McAlpine, J. F., B. V. Peterson, G. E. Shewell, H. J. Teskey, J. R. Vockeroth, and D. M. Wood, Eds. Vol. 1. Agriculture Canada. Monograph 27. Quebec, Canada: Canadian Government Publishing Centre, pp. 9–63.

Papp, L., and B. Darvis, Eds. 1998. *Manual of Palearctic diptera.* Vol. 3. *Higher Brachycera.* Budapest: Science Herald.

Rognes, K. 1985. Revision of the bird-parasitic blowfly genus *Trypocalliphora*, Peus, 1960 (Diptera: Calliphoridae). *Entomologica Scandinavica* 15:371–82.

Rognes, K. 1991. *Blowflies (Diptera, Calliphoridae) of Fennoscandia and Denmark.* E.J. Brill/Scandinavian Science Ltd.

Sabrosky, C. W., Bennett, G. F., and Whitworth, T. L. 1989. *Bird blowflies (Protocalliphora) in North America (Diptera: Calliphoridae), with notes on Palearctic species.* Washington, DC: Smithsonian Institute Press.

Shahid, S. A., R. D. Hall, N. H. Haskell, and R. W. Merritt. 2000. *Chrysomya rufifacies* (Macquart) (Diptera: Calliphoridae) established in the vicinity of Knoxville, Tennessee, USA. *Journal of Forensic Sciences* 45:896–97.

Shannon, R. C. 1923. Genera of Nearctic Calliphoridae, blowflies, with revision of the Calliphorini. *Insecutor Inscitiae Menstruus* 11:101–19.

Shannon, R. C. 1926. Synopsis of the American Calliphoridae (Diptera). *Proceedings of the Entomological Society of Washington* 28:115–19.

Shewell, G. E. 1987. Calliphoridae. In *Manual of Nearctic diptera*, McAlpine, J. F., B. V. Peterson, G. E. Shewell, H. J. Teskey, J. R. Vockeroth, and D. M. Wood, Eds. Vol. 2. Agriculture Canada Monograph 28. Quebec, Canada: Canadian Government Publishing Centre, pp. 1133–45.

Smith, K. G. V. 1986. *A manual of forensic entomology.* London: British Museum (Natural History).

Whitworth, T. L. 2002. Two new species of North American *Protocalliphora* (Diptera: Calliphoridae) from bird nests. *Proceedings of the Entomological Society of Washington* 104:801–11.

Whitworth, T. L. 2003a. A new species of North American *Protocalliphora* (Diptera: Calliphoridae) from bird nests. *Proceedings of the Entomological Society of Washington* 105:664–73.

Whitworth, T. L. 2003b. A key to the puparia of 27 species of North American *Protocalliphora* Hough (Diptera: Calliphoridae) from bird nests and two new puparial descriptions. *Proceedings of the Entomological Society of Washington* 105:995–1033.

Zumpt, F. 1965. Myiasis in man and animals in the Old World. A textbook for physicians, veterinarians and zoologists. London: Butterworth and Company.

Cases of Neglect Involving Entomological Evidence

20

MARK BENECKE

Contents

Introduction

As the many chapters in this text illustrate, forensic entomology has now gained widespread acceptance within the forensic science community. However, much of this acceptance has occurred in recent years despite the fact that applications of entomology in the legal system have been documented in the thirteenth century. Currently, many investigators and legal professionals are not fully aware of the scope and application of forensic entomology. They may be aware of this forensic discipline from death investigations in the criminal context, where entomological evidence can serve to support estimations of the postmortem interval by providing a scientific documentation of the period of insect activity (PIA), or by the determination of the time of colonization (TOC). Most investigators are not aware of the broad scope of forensic entomology in other criminal and civil investigations. While forensic entomologists are commonly called upon to render an expert opinion on cases of child and elder neglect, and in animal neglect and cruelty cases, with increasing frequency, the majority of these requests come from physicians and those in allied health professions. In order to realize the full potential of forensic entomology in these investigations, it is necessary for the forensic investigator to become aware of the utility of entomological evidence (Benecke 2004).

In most cases, forensic entomology has an intricate relationship with other items of physical evidence at the scene, and in many cases, the context in which the entomological evidence is found may be of more importance than the specimens themselves. This can be for many reasons, but one common issue is that the biological variation of the arthropod, its expected growth rate, or the influence of environmental factors on development may never be fully known under field conditions. Therefore, it is often more important to focus on the actual, and sometimes simple, questions that are asked by the court, forensic investigators, physicians, or private persons involved in a case, instead of quarreling over data that cannot be recovered or determined in the first place. This chapter gives some examples for cases in which insects assisted investigators in determining the extent of neglect (i.e., colonization interval) but not the postmortem interval. The following cases

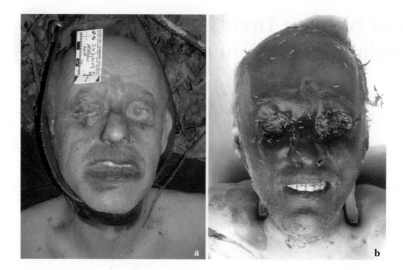

Figure 20.1 Effects of storage on body. (a) Body found at a crime scene during the very hot summer of 2003 in western Germany. Note that the eyes are intact and maggot length did not exceed 5 mm. (b) Three days later, at autopsy, the eyes are no longer intact, and the skin has dried, now bearing no resemblance to the state of the remains at the crime scene. (Photos courtesy of Dr. Mark Benecke.)

also demonstrate that applied forensic entomology starts at the scene of crime instead of at the crime laboratory or morgue, where many insects already have left the corpse and the environmental conditions have drastically changed (Figures 20.1 and 20.2).

Neglect of Elderly and Children

The use of insects as indicators of abuse, or alleged abuse, has been long known. However, in practice, it has often been overlooked by investigators because the insects will alter the physiological state of the tissue, and thus be considered of inconsequence (Benecke 2001; Klingelhöffer 1898). In other common instances they may simply be overlooked, or they may be observed and deemed repulsive by the observer and quickly washed away or discarded. Such cases are of critical importance for the health and medical care professional due to the fact that wounds of living persons are a potential target for the same flies that colonize corpses and decomposing tissues. This can lead to complications in estimation of colonization, determinations as to when the wounds were received, and possibly the overall sequence of events at a death scene. Correctly providing this additional, and sometimes essential, information may be a critical element during the investigation or subsequent trial process. With forensic entomology gaining mainstream acceptance within the forensic science community, and forensic entomologists being more readily available to investigators, even allegedly lower-profile cases like the neglect of elderly people (without violence being used against them, i.e., natural death) and abuse of animals often comes to our attention (Anderson 1999; Anderson and Huitson 2004; Benecke et al. 2004; Benecke and Lessig 2001). Furthermore, life expectancy has increased dramatically within the past few decades, and as a result, the elderly population has grown significantly. With such a burgeoning population, there is an increased awareness for the malpractice of caregivers

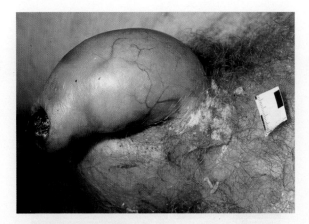

Figure 20.2 The need to make the entomological collections at the crime scene due to the problems with transportation of corpses. Many animals leave the corpse once disturbed, such as these maggots (bottom right), while others may arrive after transport to the morgue or autopsy suite, such as these egg masses. Oviposition was observed by the author. None of the arthropods present indicate neglect, even though they may in other circumstances. (Photo courtesy of Dr. Mark Benecke.)

in private and institutional care. In Germany, an evaluation of 17,000 cases of nonviolent deaths in hospitals showed that around 80% of the elderly suffered from malnutrition, and around 1% from large pressure ulcers (decubitus) (DPA, 2003). Both malnutrition and pressure ulcers are hard to avoid in very old persons, but the extent of problems in care for one of the richest countries of the world should at least raise suspicion. Like DNA typing, forensic entomology may help to exonerate individuals. Often it is shown that caregivers were not at fault for neglect due to the fact that the larvae were deposited on the individual's wounds during the normal and expected interval of caretaker visits. In many extended care facilities, the patient may be allowed and encouraged to spend periods of time outdoors. It is not uncommon for flies to be attracted to the dressed and medicated wounds of patients sitting outside. Wounds such are these are readily colonized and the larvae noted when the facility's staff changes the wound dressing.

From forensic casework, one gets the impression that mistreatment of the elderly is becoming a severe problem in countries with an aging population. From a purely judicial standpoint, it is currently, and likely always will be, very difficult to judge if the caregiver is guilty of misconduct. Therefore, it is imperative that the forensic investigator seek to utilize every contextual clue available to him or her in order to properly interpret the case at hand. Forensic entomology can give important insights into the dynamics of the circumstances surrounding the death, including the frequency and amount of care, as well as details as to the physiological condition of the body and wounds at the time of death of the neglected person. In the following cases, forensic entomology helped to provide a better understanding of the circumstances of death, and especially the circumstances of events preceding death.

Case 1: Clean Apartment with Dead Stable Flies

An elderly woman was found dead in October 2002 in her third-floor apartment in urban Cologne, Germany. The skin had desiccated and the eyes were intact (Figure 20.3a and b).

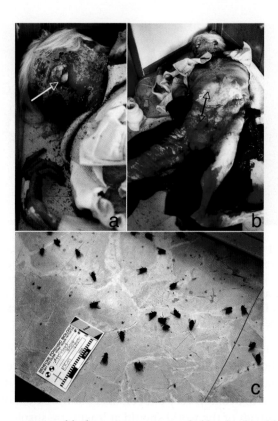

Figure 20.3 Neglect case 1: An elderly woman surrounded by dead stable flies *Muscina stabulans* (note concentration on windowsill). Arrow in (a) points to intact eyes as an indication for absence of blow fly maggots, which are usually common on corpses at this stage of decomposition. Arrow in (b) indicates differential decomposition caused by clothing (which has been cut open by investigators). The presence of dead *M. stabulans* strongly hinted toward neglect, as indicated by the larvae feeding on the excrements of the woman, and not on her body tissue. Note that the windows were closed; therefore, blow flies could not readily enter the apartment, which was generally very clean. (Photos courtesy of Dr. Mark Benecke.)

It was noted that the apartment was very clean except for the bathroom, in which a bathtub had been filled with water and clothing. Apart from the larvae, only dead adult flies of the species *Muscina stabulans* were found spread on the floor, and on a windowsill facing toward northwest (the apartment had no windows on the south) (Figure 20.3c). No insects from the family Calliphoridae (i.e., blow flies) were present in any developmental stage. The conclusion of the forensic entomologist was based on the fact that all adult flies had already emerged from the pupae. Developmental data were available for a range of room temperatures that seemed possible and within reason.

Determination of the minimum interval was of critical importance. An approximate minimum interval of around 3 weeks would have been a misconduct of the paid professional caregiver, who was supposed to check on the well-being of the woman every week. The caregiver, however, claimed that she had called the woman approximately 2 weeks prior to check on her. The well-being checks had been reduced to telephone contact because the now deceased woman allegedly rejected any visits. This possibility could not be ruled out since the old woman was known to be healthy, yet mentally unstable, exhibiting a behavior that was quite "difficult" for everyone that dealt with her.

Table 20.1 Development Information Utilized for *Muscina stabulans* under Room Conditions

Marchenko	19°C	22.8 days
	20°C	21.0 days
	21°C	19.5 days
Nuorteva	ca. 16°C	26–28 days

Source: Nuorteva, P., *Forensic Sci.*, 3:89–94, 1974; Marchenko, M. I., *Forensic Sci. Int.* 120:89–109, 2001.

This case illustrates how very important a crime scene visit is for the forensic entomologist. Without the assistance of a forensic entomologist, the insects would not have been collected by the police since they did not appear to be feeding on the corpse, but seemed to be "by chance on the windowsill." In clear contrast to the entomological findings, it was assumed that the caregiver tried her best, and no charges by the prosecutor's office followed.

Case 2: Deep Tissue Loss at Wrapped Foot

In September 2002, an elderly woman was found dead in her apartment in an urbanized town in western Germany. Her body was in the early stages of decay with lividity present, and no injuries noted (Figure 20.4a). However, the right foot was wrapped in a plastic bag that was closed with a shoelace. After removal of the bag, numerous larvae (maggots) of *Lucilia sericata* were found feeding on her right foot (Figure 20.4b). Police explicitly stated that no adult flies were found inside of the apartment. However, the apartment was in bad hygienic condition, and the landlord had noted in January 2002 that renovations were urgently necessary due to wet spots in the walls. The landlord also noted the presence of "small flies," which this author believes were of the Muscid type, probably *Musca domestica*, or fruit flies (*Drosophila*). The main contributing factor to the unsanitary conditions and fly problem was that the woman did not clean her toilet appropriately (Figure 20.4). Additionally, wet clothing was found in the washbowl. Therefore, a fly population had established in the apartment even without injuries to the woman.

Surprisingly, the caregiver openly stated that it was possible that the foot of the person was wrapped in a plastic bag and that maggots may have been present inside of the plastic bag while the woman was alive. The general practitioner estimated the postmortem interval as 2 days. The age of the maggots was estimated by this author from their size (11 mm) as approximately 4 days old under the recorded outside temperatures of 20°C. However, judging from the deep tissue loss at the foot, it was discussed that most likely, the maggots had been feeding on the living woman for at least a week while she was still alive, but then migrated from the bag to pupate elsewhere. Unfortunately, the apartment could not be checked for pupae. Therefore, the entomological investigation concluded that statements on the possible total interval of neglect had to be restricted to, and proven by, the age of the maggots. Note that the longer interval was more likely, but it could not be proven (Goff and Benecke 2004).

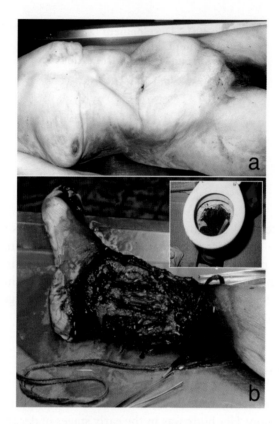

Figure 20.4 Deep tissue loss on foot (b), but very little tissue loss on remainder of body (a). The maggots did feed on the wound while the woman was alive. A shoelace was used to close a plastic bag around the foot "so that the maggots could not crawl out any more" (statement of roommate). (Photo courtesy of the Institute of Legal Medicine, Dortmund, Germany.)

Case 3: Deceased Mother on Couch

In March 2002, the corpse of an elderly woman was found in her urban apartment in a western German city (Figure 20.5a). Because of reddish stains (suspected to be blood) on the sofa, the homicide squad was called. The source of the blood was determined to be pressure ulcers (Figure 20.5b). The apartment was untidy, but otherwise clean. No rotten organic matter was found. Collected from the deceased woman on the sofa were the following insect species:

- Larval *Fannia canicularis* (house flies)
- Larval *Muscina stabulans* (stable flies)
- Adult *Dermestes lardarius* (larder beetles)

These insect species are known to build up populations inside of human dwellings, but the presence of *Fannia* sp. frequently hints toward the presence of feces and urine in cases of neglect (see also case 4). In this case, further evidence for neglect of the living person was found in the fact that the skin of the corpse was not fed on by the larvae, and that pressure ulcers (arrows in Figure 20.5c) had formed even in the region of the neck because the chin was resting on the clavicle. The eyes of the corpse were intact. Pupae of an unknown

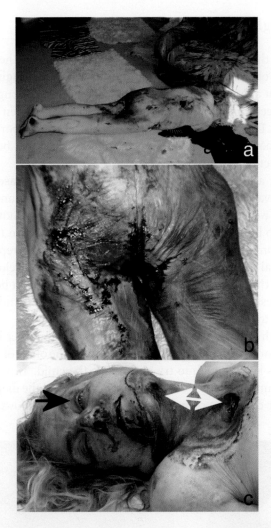

Figure 20.5 Eyes are intact with matching pressure ulcers on the chin and clavicle. Presence of larval *Fannia canicularis* (lesser house fly), larval *Muscina stabulans* (false stable flies), and adult *Dermestes lardarius* (larder beetle) strongly hints toward neglect and colonization of the woman while she was alive. (Photo courtesy of City of Dortmund Police, Germany.)

species were reported but not collected. This led to the conclusion that the foot was not inhabited postmortem. If the eggs had been deposited postmortem, there should have been at least a minimal presence of eggs, or larvae, in the region of the eyes, ears, or nose, since these are (together with wounds) preferred sites of colonization.

The son of the woman was accused of misconduct in the care of his mother. He claimed that he fed his mother the evening before she died, and that she was well at this point. Referring to the entomological findings, and the pressure spots, his statement was not believed by the judge. The court asked the general question if the woman suffered from pain by larvae living on her body; however, that question could not be answered by this author as an entomologist since it is well within the realm of medical science and pain perception. However, from statements of medical doctors performing maggot therapy it is known that blow fly larvae inside of wounds may cause patients to report severe pain, mild discomfort, or no pain at all.

Case 4: Child Neglect

In the following case this author was originally asked to assist in the establishment of the postmortem interval of a child that was found dead in its bed (Figure 20.6). The mother had left the apartment at an unknown time. Due to very severe drug use and an unstable lifestyle (reported street prostitute), she had no recollection of when she was in the apartment for the last time. During the investigation, it also became important to determine the total time the child may have been left alone, and thus neglected, because social workers were accused of neglect by the prosecution.

Under the diaper and on the surface of the skin of the deceased child (anal-genital area), third-instar larvae of the false stable fly *Muscina stabulans* and the lesser house fly *Fannia canicularis* L. were found (Figure 20.6b). It is commonly known that *F. canicularis* adults are attracted to both feces and urine (see also case 3). From the face, larvae of the bluebottle fly *Calliphora vomitoria* were collected. It is commonly reported that *C. vomitoria* maggots are typical early inhabitants of corpses. From the developmental times of the flies, it was estimated that the anal-genital area of the child had not been cleaned for about 14 days, with a total estimated range of time from 7 to 21 days, and that death occurred only 6 to 8 days prior to discovery of the body.

In the first trial, this led to a conviction not only of the mother but also of one of the social workers. The court ruled that the time span between the onset of neglect and the actual death of the child was long enough to try and rescue the child. This is, however, a judicial assumption since it is unknown how often the mother changed the diapers. From a scientific viewpoint, this author could not comment on this assumption. However, it is possible that

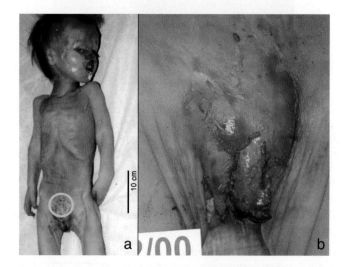

Figure 20.6 On the skin surface under the diaper (anal-genital area), third-instar larvae of the false stable fly *Muscina stabulans* and the lesser house fly *Fannia canicularis* were found. The adults of *F. canicularis* are attracted to both feces and urine. *C. vomitoria* maggots were also found, and they are typical early inhabitants of corpses. From the developmental times of these flies, it was estimated that the anal-genital area of the child had not been cleaned for about 14 days (7- to 21-day range), and that death occurred only 6 to 8 days prior to discovery of the body. This is the first case known to us in which examination of the maggot fauna on a human corpse proved neglect that had occurred prior to death. (Photo courtesy of Institute of Legal Medicine, Leipzig, Germany.)

the mother did not leave the child until a few hours before it died. The first verdict was later overruled; however, the structure of the welfare offices was changed as a result of this case.

This case could only be worked on because of good photographic documentation, the proper collection and preservation of the insects in ethanol (not formalin) by the medicolegal doctors after consultation with the forensic entomologist, and open discussions with the investigators about perspectives beyond determination of PMI.

References

Note to readers: For an extensive reference list, see Benecke (2004).

Anderson GS. (1999). Wildlife forensic entomology: Determining time of death in two illegally killed black bear cubs. *J Forensic Sci* 44:856–59.

Anderson GS, Huitson NR. (2004). Myiasis in pet animals in British Columbia: The potential of forensic entomology for determining duration of possible neglect. *Can Vet J* 45:993–98.

Benecke M. (2001). A brief history of forensic entomology. *Forensic Sci Int* 120:2–14.

Benecke M. (2004). Forensic entomology: Arthropods and corpses. In Tsokos M (Ed.), *Forensic pathology revised*. Vol II. Totowa, NJ: Humana Press, 207–40.

Benecke M, Josephi E, Zweihoff R. (2004). Neglect of the elderly: Forensic entomology cases and considerations. *Forensic Sci Int* 146(Suppl 1):S195–99.

Benecke M, Lessig R. (2001). Child neglect and forensic entomology. *Forensic Sci Int* 120:155–59.

DPA (German Press Agency). (2003). Studie an 17000 Leichen: Jeder Siebte vor Tod falsch gepflegt. dpa#051402Jan03, January 5, 2003. Scientific paper not yet published due to political pressure on the research team.

Goff ML, Benecke M. (2004). "I don't know" can be the best answer. *Forensic Entomol Anil Aggrawal's Internet J Forensic Med Toxicol* 5:58–59.

Klingelhöffer D. (1898). Zweifelhafte Leichenbefunde durch Benagung von Insekten [Misinterpretation on the cause of death as a result of insects feeding on corpses]. *Vierteljahresschrift Gerichtliche Medizin* 25:58–63.

Marchenko MI. (2001). Medicolegal relevance of cadaver entomofauna for the determination of the time of death. *Forensic Sci Int* 120:89–109.

Nuorteva P. (1974). Age determination of a blood stain in a decaying shirt by entomological means. *Forensic Sci* 3:89–94.

Acarology in Criminolegal Investigations
The Human Acarofauna during Life and Death

21

M. ALEJANDRA PEROTTI
HENK R. BRAIG

Contents

Introduction

Most healthy humans will obtain one or more species of mites from their mother and carry these with them for the rest of their life. The sebaceous glands and hair follicles of humans and most mammals are inhabited by symbiotic follicular mites of the family Demodecidae (Acari: Prostigmata) (Krantz, 1978; Evans, 1992; Walter and Proctor, 1999; Desch, 2009). The follicular mites develop in synchrony with their human hosts. They accomplish their complete life cycle inside the human skin, being restricted to the pilose-baceous glands.

Outside the glands, on the skin, freely mobile acarines are also present, and undetected by the public, an army of so-called house dust mites crawl over our skin, move up and down, in, out, and through the fabric of the clothing we wear (Colloff, 1988; Arlian, 1989; Kniest et al., 1989; de Boer and van der Geest, 1990; Colloff et al., 1996; Horak et al., 1996; Miller and Miller, 1996; Mollet and Robinson, 1996a, 1996b; Bischoff et al., 1998; Franjola and Rosinelli, 1999; Arlian et al., 2003; Mahakittikun et al., 2006). Clothing itself constitutes an important piece of evidence in almost every forensic investigation. Clothes shed fibers and collect fibers, hairs, food debris, and several forms of dust; of the dust, the main living components are yeasts, pollen, and dust mites. The amount of mites supported by a used trouser or by our bed clothing or the cover of our pillow is of such importance that special methods had to be developed to eliminate the mites from the fabric worn by or in contact with people allergic to them (Miller and Miller, 1996; Bischoff et al., 1998; Arlian et al., 2003; Cieslak et al., 2007).

Mites are omnipresent in our domestic environment. Practically every kind of food microparticle left on any surface or hidden in any miniscule corner of the kitchen will support dust mites in vast numbers and diversity. But the fauna of the kitchen will vary from that of the living room and the bedroom, places where our skin flakes dominate the dust (Hughes, 1976; Frost et al., 2009). The diversity per microarea, occurrence, and abundance make mites the most available organisms to be analyzed for trace or transfer evidence in most forensic investigations, including illegal trade of protected animal and plant species, cultural artifacts, drugs, and contaminated food.

Forensic Acarology

Forensic entomology has become a well-established discipline within the forensic sciences, and its value relies on the rapid colonization of the remains by insects. Blow flies or flesh flies are often the first insects colonizing a dead body. The flies will seize any available orifice or wound by depositing eggs or larvae in large amounts. The infestation, if not interrupted, will be followed by a complete life cycle of the flies on the corpse (Anderson and Cervenka, 2002; Greenberg and Kunich, 2005), and a more or less continuous succession of arthropods and other consumers will reduce the remains to a skeleton. Life history data for blow flies and flesh flies have been well documented and have been proven to be accurate enough to estimate the time of death in many cases.

However, there are environments where insects are absent or rare, or even circumstances of death that impede the access of flies or their development on or off the corpse. Mites are almost always present and can colonize a body without the help of insects, but are systematically overlooked. Probably because of their micrometrical size or the lack of knowledge about their biology, mites are easily missed by the untrained eye.

Pierre Mégnin was the first acarologist using mites to estimate time of death in a forensic investigation. He also described eight distinct waves of arthropods colonizing corpses. He realized that the very first colonizers are not only insects but also mites. The sixth wave of arthropods was composed exclusively of mites (Mégnin, 1894, 1895). After Mégnin, only a few researchers have specifically studied the Acari in forensic investigations (Goff et al., 1986; Goff, 1989, 1991; Leclercq and Verstraeten, 1993). Recently, new ideas have been proposed to incorporate mites in forensic investigations, and dedicated projects in forensic acarology are pursued. Research is being conducted under environmental conditions where insects are rare, absent, or access to corpses is restricted (Perotti et al., 2006). The challenge is to identify how the acarine fauna might contribute to and underpin the estimation of time of death, circumstances of death, and might serve as evidential data on movement or relocation of bodies. It is encouraging to see how fast the more than 100-year-old idea of forensic acarology has been imbued with new life, vitality, and international recognition (Rasmy, 2007; Perotti et al., 2009).

Although mites are already present in and on the human body during life, once death takes place, new mite species will arrive and colonize the body as fast and as well as insects do. They will arrive by walking or will be carried by currents of air (airborne), or as phoretics on insects visiting the corpse but not necessarily colonizing it. Insects might rest on a body, or investigate a body only to return to the corpse to colonize it at a much later stage. The initial contact with the corpse might only leave mites as trace. Mites do not have wings; therefore, they do not fly but are able to use other animals as carriers. Blow flies, filth flies, scuttle flies, carrion beetles, rove beetles, carpet beetles, moths, myriapods, and even mites themselves will carry phoretic mites to a corpse. The mechanisms that mites of forensic importance utilize to get to corpses are described by Perotti (Perotti, 2009; Perotti et al., 2009). The diversity of the mite fauna on and around corpses is amazing (Mégnin, 1894; Baker, 2009; Braig and Perotti, 2009). An embalmed body of a 28-year-old female with a gunshot wound to the head that was recently exhumed after spending 28 years inside an unsealed casket in an unsealed cement vault underground yielded large numbers of live springtails and glycyphagid mites (Merritt et al., 2007). Indoor mites and arthropods have a great potential to provide forensically important information (Solarz, 2009; Frost et al., 2009). After more than 100 years, mites again are used to estimate the time of death (Goff, 1989, 1991, 1993). Astigmatid mites are associated with forensic cases and have recently been found in masses on an insect-free corpse (Russell et al., 2004; O'Connor, 2009). However, submerged bodies do not seem to link with freshwater mites (Proctor, 2009). In this chapter, we introduce the living acarological fauna of the human skin and clothes, and comment on their potential relevance and role as forensic indicators and trace evidence.

Mites in Healthy Skin of Living and Dead People

Pairs of follicular mites can be found in several mammal hosts; among them, the most common are humans with *Demodex folliculorum* and *Demodex brevis* (Desch and Nutting, 1972). Combinations of more than two mite species are also possible for different hosts, and triplets and quartets have been isolated from the brown rat, *Rattus norvegicus* (Bukva, 1995). Interestingly, water seems not to be an obstacle for follicular mites, as they have also been demonstrated to live in the skin of marine mammals such as seals and sea lions (Desch et al., 2003).

Demodex are small mites (Figure 21.1). The adults average 200 μm in length; the females are wider and longer than the males. Eggs and the immature stages such as larvae are inconspicuous forms that can be easily confounded with particles or microfibers. The two human *Demodex* species can occur in separate or the same individuals. *Demodex folliculorum* is the biggest of the two species (averaging 279.7 μm for male mites and 294.0 μm for female mites) (Desch and Nutting, 1972). Its shape is elongated in comparison with the shorter forms of *D. brevis* (with males averaging 165.8 μm and females 208.3

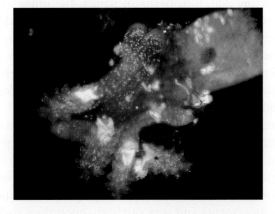

Figure 21.1 Fluorescent microscopy picture of four specimens of the human follicle mite *Demodex* sp. adjacent to a follicle obtained from the skin (forehead) of an adult male. The four pairs of legs of each adult mite are easily recognizable. DAPI stains the individual nuclei of the mites. (Merged image: UV channel [DAPI stain] and green channel [LP 560 nm]). 600×. Length of body ~ 200 μm.

µm) (Desch and Nutting, 1972). Modern microscopical techniques can help to localize and rapidly identify these acarines from minute skin samples or skin products. For example, a recent study on the two human *Demodex* species performed with environmental scanning electron microscopy (ESEM) shows in detail the ultrastructure of the 1 to 2 µm piercing mouthparts of the mites (Jing et al., 2005). The use of fluorescent dyes also proved to improve detection of mites accompanying removed eyelashes when using fluorescent instead of light microscopy (Kheirkhah et al., 2007).

D. folliculorum inhabits the opening of the duct of the follicles, sometimes exhibiting half of the body outside the pore. *D. brevis* prefers to live deep inside the glands. Both species can be found simultaneously in the same pilosebaceous complex, but *D. folliculorum* tends to inhabit the follicles of simple hairs and *D. brevis* is more often associated with the vellus hairs (Desch and Nutting, 1972). As it happens with other demodecids, *D. folliculorum* feeds off cell contents and secretions (Nutting and Rauch, 1963; Desch and Nutting, 1972). *D. brevis* are found in the glandular acini with the mouthparts directed toward the fundus of the acinus, where they seem to ingest individual gland cells (Desch and Nutting, 1972); however, damage to the glands as a whole has not been reported. Follicular mites are easily screened for on the face, external ear, eyelashes, upper chest, nipples, and more rarely, on the scalp, foreskin, penis, mons veneris, buttocks, and ectopic sebaceous gland in the buccal mucosa (Breckenridge, 1953; Goihman-Yahr, 1998). Originally, *Demodex* was discovered in the ear canal and in earwax of humans. Earwax might contain more forensically important information than previously thought.

There is not a defined method to precisely measure *Demodex* prevalence in the human population, as it would require the screening of the complete area covered by the skin of each individual of the population sample (Forton, 2007). However, estimations from unrelated sources suggest a homogeneous prevalence in humans from all regions of the globe, independent of ethnicity but directly correlated with the age of their host (Wilson, 1844; Breckenridge, 1953; Desch and Nutting, 1972; Rufli and Mumcuoglu, 1981; Nutting, 1985; Hellerich and Metzelder, 1994; Mihaescu and Weber-Chappuis, 1996; Mathiesen and Settnes, 1997; Forton and Song, 1998; Goihman-Yahr, 1998; Zhao et al., 1998; Jiang et al., 2003; Ding and Huang, 2005; Dolenc-Voljc et al., 2005; Gao et al., 2005; Gotia et al., 2006; Okyay et al., 2006; Forton, 2007). In newborn babies, the mites are difficult to localize and commonly not found. Mites increase in numbers during childhood, enlarging the mite population in synchrony with the development of the pilosebaceous glands of their host, being able to reach 100% prevalence in healthy adult humans worldwide (Rufli and Mumcuoglu, 1981; Okyay et al., 2006; Forton, 2007). There is a tendency of increased abundance of mites on the scalp of people of advanced age, dark hair color, or with a bald head (Hellerich and Metzelder, 1994).

Although two recent reports on human *Demodex* recovered from corpses attempted to characterize the mites as parasites and potential vectors of pathogenic organisms (Ozdemir et al., 2003, 2005), the transmission of a mite from one individual to another has never been reported in the extensive literature on human *Demodex*. *Demodex* cannot switch hosts. The mites present mobile legs in nymphal and adult stages. They display restricted, swimming movements and very rarely abandon their shelter. Special attention has been dedicated to the fact that if the mites are forced to leave their glandular shelter, they soon die of desiccation on the surface of the skin and make no attempt to reenter any follicles (Nutting and Rauch, 1963). Horizontal transmission between lifelong partners has not been documented, but anecdotal evidence from acarologists specializing in *Demodex* biology for many decades suggests that horizontal transmission is likely absent during a human's lifetime. Early work on other mammal species has revealed that transfer between female and male animals is extremely rare, even during copula. Transfer from mother to offspring is the rule (Fisher, 1973; Bukva, 1990). *Demodex* are transmitted from mother to offspring by close physical contact during birth, nursing, breast-feeding, and throughout infancy. Cross-infection is not even known between closely related host species (Drager and Paine, 1980).

   D. folliculorum happily lives in our skin, exposed to a wide range of temperatures varying naturally between 8 and 30°C. In laboratory cultures, the optimal temperature was 26 to 27°C. Survival is drastically reduced when temperatures decrease to 0°C and increase above 37°C. The lethal temperature is around 54 to 54.5°C (Sheard and Hardenbergh, 1927; Zhao et al., 2005). Progressive rise of temperature stimulates movement of the mites, and they become very active at 45°C, while at 53.5°C they lose motility, and die after 5 minutes of heating to 54.5°C (Sheard and Hardenbergh, 1927). These studies suggest two temperature ranges at which the survival of the mites is impaired, one between 0 and 8°C and another between 30 and 54°C. Mites should be found alive on a corpse between 0 and 53°C. Raising the temperature of the skin up to 45°C should very much increase the chances of visualizing living mites moving actively on the dead skin of a corpse. Raising the temperature in specific areas of the skin mirrors the maternal transfer of mites. Maternal transmission is likely to happen by localized increased temperature between mother and offspring during birth, caressing, and breast-feeding.

   The use of fluorescent microscopy and the application of fluorescent dyes like DAPI (4′,6-diamidino-2-phenylindole) on removed, detached, or plucked off facial hairs or any hairs rooted in pilosebaceous glands will dramatically help to obtain evidential information because of the rapid visualization of mites still attached to the root of the hair, independently of the presence or absence of follicular cells of the host (Kheirkhah et al., 2007). The mitochondrial and nuclear DNA of individuals of one or both *Demodex* species might act as a proxy for the host individual.

   During the first 48 to 72 hours after death, medical parameters are of most use to estimate the postmortem interval. After this period, *Demodex* specimens might be still alive for some hours or up to 14 or 15 days, depending on the circumstances (Wilson, 1844; Rufli and Mumcuoglu, 1981). In almost every microscopical survey of the skin during autopsies, follicular mites have been visualized since more than a century ago (Wilson, 1844; Gmeiner, 1908). In an early study in Basel and Zürich in Switzerland, Henle, who discovered *Demodex* at the same time as Berger in France, found eleven out of twelve human corpses carrying *Demodex* in the skin (Henle, 1845). In Germany, all eight studied corpses of children and adults showed *Demodex* in the skin of the face; only two newborn babies were negative (Simon, 1842). A more extensive study recovered mites from 97 of 100 corpses in the skin of the face (Gmeiner, 1908). The three negative samples originated from a 2- and an 8-day-old baby and from a 9-year-old girl, the last of which the author regarded as a false negative. From the same corpses, eyelashes were also recovered and screened for *Demodex*. The eyelashes only carried *Demodex* in 49 of the 100 corpses on average, but a strong age relation was evident. The incidence ranged from 10% in the 1- to 10-year-old category to around 60% in the age range 20 to 90. Another study in Germany recovered mites from 79% of the eyelashes of corpses (Hunsche, 1900). Gmeiner repeated his original study with an additional 100 corpses and obtained an incidence of 100% for *Demodex* in skin samples and 50% for eyelashes (Gmeiner, 1908). A French study on skin samples confirmed the 100% incidence for 100 corpses 1 to 82 years of age (Fuss, 1933). In Denmark, eyelashes from another 100 corpses in a more recent study were positive for mites in 89% (Norn, 1970).

The survival of *Demodex* mites could be linked to the environmental conditions at the time of death and thereafter. Then, estimation of the postmortem time interval may also be based upon survival of the mites, and might well become important for those circumstances where flies or insects cannot or have not yet colonized the body. Baseline and reference data need to be established on the survival of Demodex on dead bodies to facilitate estimates of time of death. This area of research is awaiting development and might perfectly complement and link unrelated data provided by medical or physiological parameters.

Mites on and in Clothes

While Demodex will not survive outside our skin and are not prepared to easily leave our follicles, the dust mites do crawl on our skin and make their way through our clothes in search of food, mates, and places to lay eggs or to abandon our bodies and clothes in search of new habitats elsewhere. Decisions to disembark will depend upon changes in the environmental conditions.

Within the huge mite order of Astigmata, which includes free-living species, predators, commensals, and parasites, only one family has mites specialized in feeding on human skin flakes: the Pyroglyphidae. Pyroglyphids are also called house dust mites, but this is a term that involves other families belonging to different orders. It means that our home will likely be inhabited by a rich acarofauna. Members of the orders Astigmata, Mesostigmata, Prostigmata, and Oribatida will be more or less represented in any of the rooms of the house (Hughes, 1976; van Bronswijk, 1979; Frost et al., 2009). All these house acarines will have specific habits and habitats. That is, pyroglyphids would be present in any habitat where our skin flakes or dander are found, because our skin debris is their favorite food. Habitats where dander is abundant include our skin, clothes, bed, sofa, carpet, and curtains, which might be plagued by skin mites. About 80% of the material seen floating in a sunbeam can actually be human dander. The same applies to the other mite species and their different niches. Generally, the mesostigmatid mites of the house are predators, and together with predatory prostigmatids, they will feed on the pyroglyphids and other astigmatids that are more represented in the kitchen and living room, where they have access to our food reserves. The mites in the kitchen are known as the stored food mites, mostly of the families Glycyphagidae and Acaridae (Astigmata) (Hughes, 1976; van Bronswijk, 1979). The prominent genus in the family Pyroglyphidae, as far as humans are concerned, is *Dermatophagoides*. From an evolutionary point of view, the original habitats of most *Dermatophagoides* mites (Pyroglyphidae, Astigmata) were likely the nests and lairs of birds and mammals (habitats where they are still found) (Kniest, 1994; Walter and Proctor, 1999). As we humans evolved and moved from open space to caves to houses, our domestic mite fauna followed.

A worldwide analysis has indicated that three of the species, *Dermatophagoides pteronyssinus*, *Dermatophagoides farinae*, and *Euroglyphus maynei*, constitute the majority of our skin and clothing arthropod fauna (Blythe et al., 1974; Crowther et al., 2000). Original, early work on these mites identified *D. farinae* as the American dust mite and *D. pteronyssinus* as the European species, most common in temperate climates. In the tropics and

Figure 21.2 Light microscopy picture of adult females of the mite *Dermatophagoides* sp. collected on the inside of pajama trousers worn by a male. From left to right, setae of the idiosoma, the genital opening that is Y-shaped, and four pairs of legs are discernable in each specimen; ventral view. The cryptic appearance of the mites is evident. Length of body = 350 μm.

subtropical areas, *Blomia tropicalis* (Glycyphagidae, Astigmata) is perhaps the dominant species (Chew et al., 1999).

Pyroglyphidae mites are very well known, not just because they are around our skin, but because they have built a unique reputation as one of the most notorious human allergen producers, being responsible for varied chronic respiratory diseases or respiratory disorders. The biology of Pyroglyphidae mites is probably one of the best characterized in the Acari, and there is a huge body of information in the medical literature regarding their ecological parameters and life history traits, always linked to allergy research. These data will facilitate attempts to estimate their behavior in forensic settings.

Skin mites are still small, but very agile and more robust than the follicular mites (Figure 21.2). The length of the idiosoma, the large body region that carries the legs but excludes the mouthparts, of an adult *Dermatophagoides* ranges from 300 μm for the male to 370 μm for the female, while for *Euroglyphus* it reaches ~200 μm (male) and ~280 μm (female) in adulthood. *D. pteronyssinus* females and males present similar length distributions, and both sexes are able to disperse or move to new habitats. *D. farinae* as well as *D. pteronyssinus* females exhibit gregarious behavior, grouping in breeding sites where they deposit the eggs in clusters (Figure 21.3), while the males of *D. farinae* maintain a distance from the gravid females and only follow them when females become receptive. The dispersal and distribution of the skin mites in a substrate under optimal conditions (e.g., clothing or skin) seems to respond to the release of varied pheromones, some attracting the females, some repelling the males (van Asselt et al., 1996). According to van Bronswijk (1979), the most stable microclimate provided by the human environment is found in the bed. There, our body provides the perfect nesting environment. The stratification of the *Dermatophagoides* species can be mapped on the mattress of our bed. Independent studies show that the most populated areas are at the edges of the mattresses, then the bottom of the bed surrounding the feet area, but if the humidity is adequate, the abundance of mites could be surprisingly high in blankets or pillow covers on top (Dusbabek, 1979; Yoshikawa and Bennett, 1979). Inside a typical mattress, the number of mites might vary between 100,000 and 10 million. Dead mites and mite droppings may constitute up to 10% of the weight of a 2-year-old pillow. Human sweat, fluids, and secretions are also attractants for the bed acarofauna. At the edges, where

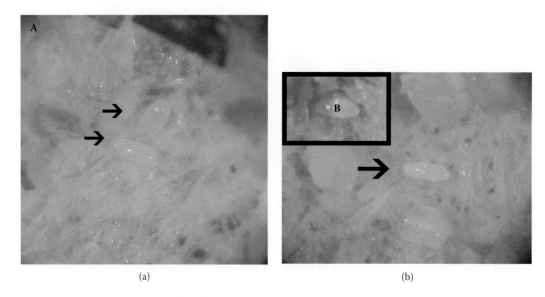

(a) (b)

Figure 21.3 Eggs of *Dermatophagoides farinae*. (a) Two eggs were deposited together on the rearing substrate (bed linen and tissue paper covered with human dander). The eggs are glued by the mother to the substrate (clothing, mattresses, etc.). (b) In the upper-left insert, a freshly deposited egg is shown, while in the main field a several-day-old egg is ready to hatch. *Dermatophagoides* eggs measure ~40 μm.

the population crowds, *D. pteronyssinus* is more frequent than *D. farinae*; otherwise, both species present similar abundances (Dusbabek, 1979). Pyroglyphids osmoregulate through the cuticle, which allows them to withstand humidity changes associated with a wide range of temperatures; nevertheless, humidity is the most critical factor for survival (Hart, 1998). The protonymphal stage is more resistant to desiccation. At over 60% relative humidity (RH) at room temperature, the mites can easily complete their life cycle in 1 month. A study on *D. farinae* indicates that the time required for development is inversely related to the amount of moist air given daily (Arlian et al., 1999); in a regime of constant temperature and 75% RH, the life cycle was completed in 41.1 ± 0.50 days. Many human-associated mites require humidity for water uptake only for about 1 to 3 hours a day, during which the mites can take up enough water to survive for the rest of the day under quite desiccating conditions (Schei et al., 2002).

House dust mites are not found in newly built houses or in houses before they become inhabited; they are inoculated by us once we move in (Warner et al., 1999; Crowther et al., 2000). Our clothing carries the mites (Bischoff and Kniest, 1998). Dust mites inhabiting our skin and clothes are able to switch hosts as long as humidity levels are acceptable. The mites will likely move to another human or his or her clothing by physical contact between the hosts or by sharing clothing, including clothed or upholstered furniture. Two individuals can present different mite communities with distinct species compositions and ratios, based on their hygienic habits. The diversity of the personal acarofauna can be studied by classical morphological methods or by molecular analyses. Modern tools for molecular genetics can provide insights even at the population level within a single species. Unfortunately, not much knowledge has been developed yet. The link of a suspect to the scene of a crime might be confirmed by just analyzing the diversity and kinship of the mites present.

Environmental Contact with Mites

Away from our skin, but yet associated with our bodies and our movements, there is still a different habitat for mites: our shoes. Everywhere we walk, we collect dust and soil. In fact, shoe prints and the materials our shoes collect constitute a body of important evidence in modern forensic investigations. Soil mites are adhered to the shoes together with the soil they belong to. Oribatida or hard soil mites can be identified from fragments of their carcasses. Their strong cuticula will preserve in good shape for extensive periods of time, including archaeological time spans (Chepstow-Lusty et al., 2007). In soil, all orders of mites are represented. The richness or diversity of soil mites is microhabitat specific; they also exhibit a microbiogeographical distribution. A species of mite might only be present in an area of soil, in an open field, in a playground, around the roots of a tree, in a backyard, in a grave, and only there and nowhere else. Almost every systematic acarologist has found and described new species of mites from his or her garden! Soil mites are also strongly linked to any activities of humans. Archaeologists use this link extensively to reconstruct human activities and habitats.

Aside from soil traces, particular habitats might be characterized by the activity of mites, particularly of blood-sucking mites. At a crime scene of a 24-year-old female victim in California, people investigating the scene were bitten by chigger mites of the species *Eutrombicula belkini* (Trombiculidae, Prostigmata). Previous biological studies on this species had shown that the distribution of the mite was very restricted (Webb et al., 1983; Prichard et al., 1986). Interestingly, only one of the suspects exhibited a similar biting pattern as the people scrutinizing the scene. The bites of the mites developed a characteristic swelling pattern on the skin, which provided information to estimate the time of the original bites. This allowed placing the suspect at the scene of the crime at the time of the crime. The suspect later confessed (Webb et al., 1983; Prichard et al., 1986).

Acknowledgments

The authors appreciate the funding of research on forensic acarology by Leverhulme Trust.

References

Anderson, G. S., and V. J. Cervenka. 2002. Insects associated with the body: Their use and analyses. In *Advances in forensic taphonomy*, Haglund, W. D., and M. H. Sorg, Eds. Boca Raton, FL: CRC Press, 173–200.

Arlian, L. G. 1989. Biology and ecology of house dust mites, *Dermatophagoides* spp. and *Euroglyphus* spp. *Immunol. Allergy Clin. North Am.* 9:339–56.

Arlian, L. G., J. S. Neal, and D. L. Vyszenski-Moher. 1999. Fluctuating hydrating and dehydrating relative humidities effects on the life cycle of *Dermatophagoides farinae* (Acari: Pyroglyphidae). *J. Med. Entomol.* 36:457–61.

Arlian, L. G., D. L. Vyszenski-Moher, and M. S. Morgan. 2003. Mite and mite allergen removal during machine washing of laundry. *J. Allergy Clin. Immunol.* 111:1269–273.

Baker, A. S. 2009. Mummies and mites. *Exp. Appl. Acarol.*, in press.

Bischoff, E. R. C., A. Fischer, B. Liebenberg, and F. M. Kniest. 1998. Mite control with low temperature washing. II. Elimination of living mites on clothing. *Clin. Exp. Allergy* 28:60–65.

Bischoff, E. R. C., and F. M. Kniest. 1998. Differences in the migration behaviour of two most widespread species of house dust mites (HDM). *J. Allergy Clin. Immunol.* 101:S28.

Blythe, M. E., J. D. Williams, and J. M. Smith. 1974. Distribution of pyroglyphid mites in Birmingham with particular reference to *Euroglyphus maynei*. *Clin. Exp. Allergy* 4:25–33.

Braig, H. R., and M. A. Perotti. 2009. Carcasses and mites. *Exp. Appl. Acarol.*, in press.

Breckenridge, R. L. 1953. Infestation of the skin with *Demodex folliculorum*. *Am. J. Clin. Pathol.* 23:348–52.

Bukva, V. 1990. Transmission of *Demodex flagellurus* (Acari: Demodicidae) in the house mouse, *Mus musculus*, under laboratory conditions. *Exp. Appl. Acarol.* 10:53–60.

Bukva, V. 1995. *Demodex* species (Acari: Demodecidae) parasitizing the brown rat, *Rattus norvegicus* (Rodentia): Redescription of *Demodex ratti* and description of *D. norvegicus* sp. n. and *D. rotticola* sp. n. *Folia Parasitol.* 42:149–60.

Chepstow-Lusty, A. J., M. R. Frogley, B. S. Bauer, M. J. Leng, A. B. Cundy, K. P. Boessenkool, and A. Gioda. 2007. Evaluating socio-economic change in the Andes using oribatid mite abundances as indicators of domestic animal densities. *J. Archaeol. Sci.* 34:1178–86.

Chew, F. T., L. Zhang, T. M. Ho, and B. W. Lee. 1999. House dust mite fauna of tropical Singapore. *Clin. Exp. Allergy* 29:201–6.

Cieslak, M., I. Kaminska, S. Wrobel, K. Solarz, E. Szilman, and P. Szilman. 2007. Effects of modified textile floor coverings on house dust mites. *Pol. J. Environ. Stud.* 16:35–42.

Colloff, M. J. 1988. Mite ecology and microclimate in my bed. In *Mite allergy. A world-wide problem*, de Weck, A. L., and A. Todt, Eds. Brussels: UCB Institute of Allergy, 51–54.

Colloff, M. J., C. Taylor, and T. G. Merrett. 1996. *Killing dust mites (Pyroglyphidae) with steam*. Acarology IX, Vol. 1. Paper presented at the Proceedings of the 9th International Congress of Acarology, Columbus, OH.

Crowther, D., J. Horwood, N. Baker, D. Thomson, S. Pretlove, I. Ridley, and T. Oreszczyn. 2000. *House dust mites and the built environment: A literature review*. London: University College London.

de Boer, R., and L. P. S. van der Geest. 1990. House-dust mite (Pyroglyphidae) populations in mattresses, and their control by electric blankets. *Exp. Appl. Acarol.* 9:113–22.

Desch, C. E. 2009. Human hair follicle mites, the most frequently found parasites, and forensic acarology. *Exp. Appl. Acarol.*, in press.

Desch, C. E., M. D. Dailey, and P. Tuomi. 2003. Description of a hair follicle mite (Acari: Demodecidae) parasitic in the earless seal family Phocidae (Mammalia: Carnivora) from the harbor seal *Phoca vitulina* Linnaeus, 1758. *Int. J. Acarol.* 29:231–35.

Desch, C. E., and W. B. Nutting. 1972. *Demodex folliculorum* (Simon) and *D. brevis* Akbulatova of man: Redescription and reevaluation. *J. Parasitol.* 58:169–77.

Ding, Y., and X. Huang. 2005. [Investigation of external auditory meatus secretion *Demodex folliculorum* and *Demodex brevis* infection in college students]. *J. Clin. Otorhinolaryngol. Head Neck Surg.* 19:176–77.

Dolenc-Voljc, M., M. Pohar, and T. S. Lunder. 2005. Density of *Demodex folliculorum* in perioral dermatitis. *Acta Derm. Venereol.* 85:211–15.

Drager, N., and G. D. Paine. 1980. Demodicosis in African buffalo (*Syncerus caffer caffer*) in Botswana. *J. Wildl. Dis.* 16:521–24.

Dusbabek, F. 1979. Dynamics and structure of mixed populations of *Dermatophagoides farinae* and *D. pteronyssinus*. 1. Paper presented at the V International Congress of Acarology, East Lansing, MI.

Evans, G. O. 1992. *Principles of acarology*. Wallingford, UK: CAB International.

Fisher, W. F. 1973. Natural transmission of *Demodex bovis* Stiles in cattle. *J. Parasitol.* 59:223–24.

Forton, F. 2007. Standardized skin surface biopsy: Method to estimate the *Demodex folliculorum* density, not to study the *Demodex folliculorum* prevalence. *J. Eur. Acad. Dermatol. Venereol.* 21:1301–2.

Forton, F., and M. Song. 1998. Limitations of standardized skin surface biopsy in measurement of the density of *Demodex folliculorum*. A case report. *Br. J. Dermatol.* 139:697–700.

Franjola, T. R., and D. Rosinelli M. 1999. Acaros del polvo de habitaciones en la ciudad de Punta Arenas, Chile [Mites in house dust in the city of Punta Arenas, Chile]. *Bol. Chil. Parasitol.* 54:82–88.

Frost, C. L., H. R. Braig, and M. A. Perotti. 2009. Indoor arthropods of forensic importance. In *Forensic entomology: New trends and technologies*, Amendt, J., C. P. Campobasso, M. Grassberger, and M. L. Goff, Eds. Amsterdam: Springer, in press.

Fuss, F. 1933. La vie parasitaire du *Demodex folliculorum hominis* [The parasitic life of *Demodex folliculorum hominis*]. *Ann. Dermatol. Syphiligr.* 4:1053–62.

Gao, Y.-Y., M. A. Di Pascuale, W. Li, D. T.-S. Liu, A. Baradaran-Rafii, A. Elizondo, T. Kawakita, V. K. Raju, and S. C. G. Tseng. 2005. High prevalence of Demodex in eyelashes with cylindrical dandruff. *IOVS* 46:3089–94.

Gmeiner, F. 1908. *Demodex folliculorum* des Menschen und der Tiere [*Demodex folliculorum* of humans and animals]. *Arch. Dermatol. Syphil.* 92:25–96.

Goff, M. L. 1989. Gamasid mites as potential indicators of postmortem interval. In *Progress in acarology*, Channabasavanna, G. P., and C. A. Viraktamath, Eds. Vol. 1. New Delhi: Oxford & IBH Publishing, 443–50.

Goff, M. L. 1991. Use of acari in establishing a postmortem interval in a homicide case on the island of Oahu, Hawaii. In *Modern acarology*, Dusbábek, E., and V. Bukva, Eds. Vol. 1. The Hague: SPB Academic Publishing, 439–42.

Goff, M. L. 1993. Estimation of postmortem interval using arthropod development and successional patterns. *Forensic Sci. Rev.* 5:81–94.

Goff, M. L., M. Early, C. B. Odom, and K. Tullis. 1986. A preliminary checklist of arthropods associated with exposed carrion in the Hawaiian Islands, USA. *Proc. Hawaiian Entomol. Soc.* 26:53–57.

Goihman-Yahr, M. 1998. Demodecidosis manifested on the external genitalia. *Int. J. Dermatol.* 37:634–36.

Gotia, S. R., P. Ghitulescu, C. Solovan, S. L. Gotia, C. Fira-Mladinescu, and M. Crestescu. 2006. Seasonal fluctuation in the incidence of human *Demodex folliculorum* infection. *Cent. Eur. J. Occup. Environ. Med.* 12:37–42.

Greenberg, B., and J. C. Kunich. 2005. *Entomology and the law: Flies as forensic indicators.* New ed. Cambridge, UK: Cambridge University Press.

Hart, B. J. 1998. Life cycle and reproduction of house-dust mites: Environmental factors influencing mite populations. *Allergy Asthma Proc.* 53:13–17.

Hellerich, U., and M. Metzelder. 1994. The frequency of scalp infestation by the ectoparasite *Demodex folliculorum* Simon in pathological and forensic autopsy material. *Arch. Kriminol.* 194:111–18.

Henle, F. 1845. Bericht über die Arbeiten im Gebiet der rationellen Pathologie [Report on work in the area of rational pathology]. *Z. Ration. Med.* 3:28.

Horak, B., J. Dutkiewicz, and K. Solarz. 1996. Microflora and acarofauna of bed dust from homes in Upper Silesia, Poland. *Ann. Allergy Asthma Immunol.* 76:41–50.

Hughes, A. M. 1976. *The mites of stored food and houses.* London: Her Majesty's Stationary Office.

Hunsche, K. 1900. Das Vorkommen des *Demodex folliculorum* am Augenlide und seine Beziehung zu Liderkrankungen [The occurence of *Demodex folliculorum* on eyelids and its relationship to diseases of the eyelid]. *Münch. Med. Wochensch.* 45:1563.

Jiang, S. F., T. J. Li, and S. P. Chi. 2003. [Investigation on the prevalence of *Demodex folliculorum* and *Demodex brevis*]. *Chin. J. Vector Biol. Control* 14:294–95.

Jing, X., G. Shuling, and L. Ying. 2005. Environmental scanning electron microscopy observation of the ultrastructure of Demodex. *Microsc. Res. Tech.* 68:284–89.

Kheirkhah, A., G. Blanco, V. Casas, and S. C. G. Tseng. 2007. Fluorescein dye improves microscopic evaluation and counting of Demodex in blepharitis with cylindrical dandruff. *Cornea* 26:697–700.

Kniest, F., B. Liebenberg, and E. Bischoff. 1989. Presence and transporting of dust mites in clothing. *J. Allergy Clin. Immunol.* 83:262.

Kniest, F. M. 1994. Are storage mites different from house-dust mites? *Atemswegs. Lungenkrankh.* 20:40–45.

Krantz, G. W. 1978. *A manual of acarology*. 2nd ed. Corvallis: Oregon State University.

Leclercq, M., and C. Verstraeten. 1993. Entomologie et médecine légale. L'entomofaune des cadavres humains: sa succession par son interprétation, ses résultats, ses perspectives [Entomology and forensic medicine. The entomofauna of human corpses: Its succession and interpretation, its results, its prospects]. *J. Méd. Légale Droit Méd.* 36:205–22.

Mahakittikun, V., J. J. Boitano, E. Tovey, C. Bunnag, P. Ninsanit, T. Matsumoto, and C. Andre. 2006. Mite penetration of different types of material claimed as mite proof by the Siriraj chamber method. *J. Allergy Clin. Immunol.* 118:1164–68.

Mathiesen, P. R., and O. P. Settnes. 1997. *Demodex folliculorum* and *Demodex brevis* (Acari: Demodicidae) in a population sample of Ojobi, Ghana. *Bull. Scand. Soc. Parasitol.* 7:83.

Mégnin, P. 1894. *La Faune des Cadavres. Application de l'Entomologie à la Médecine Lég*ale *[The fauna of corpses. Application of entomology to forensic medicine]*. Paris: G. Masson and Gauthier-Villars et Fils.

Mégnin, P. 1895. La faune des cadavres [The fauna of carcasses]. *Ann. Hyg. Publ. Méd. Lég.* 33:64–67.

Merritt, R. W., R. Snider, J. L. de Jong, M. E. Benbow, R. K. Kimbirauskas, and R. E. Kolar. 2007. Collembola of the grave: A cold case history involving arthropods 28 years after death. *J. Forensic Sci.* 52:1359–61.

Mihaescu, A., and K. Weber-Chappuis. 1996. Demodex in a fine-needle aspiration of the cheekbone area: A case report with short description of the parasite. *Arch. Anat. Cytol. Pathol.* 44:113–16.

Miller, J. D., and A. Miller. 1996. Ten minutes in a clothes dryer kills all mites in blankets. *J. Allergy Clin. Immunol.* 97:423.

Mollet, J. A., and W. H. Robinson. 1996a. Evaluating the dispersal of the American house dust mite (*Dermatophagoides farinae* Hughes (Pyroglyphidae)) using marked mites. Acarology IX, Vol. 1. Paper presented at the Proceedings of the 9th International Congress of Acarology, Columbus, OH.

Mollet, J. A., and W. H. Robinson. 1996b. Dispersal of American house dust mites (Acari: Pyroglyphidae) in a residence. *J. Med. Entomol.* 33:844–47.

Norn, M. S. 1970. *Demodex folliculorum*, incidence and possible pathogenic role in the human eyelid. *Acta Ophthalmol. Suppl.* 108:1–85.

Nutting, W. B. 1985. *Prostigmata-Mammalia: Validation of coevolutionary phylogenies*. New York: Wiley-Interscience.

Nutting, W. B., and H. Rauch. 1963. Distribution of *Demodex aurati* in the host (*Mesocricetus auratus*) skin complex. *J. Parasitol.* 49:323–29.

O'Connor, B. M. 2009. Astigmatid mites (Acari: Sarcoptiformes) of forensic interest. *Exp. Appl. Acarol.*, in press.

Okyay, P., H. Ertabaklar, E. Savk, and S. Ertug. 2006. Prevalence of *Demodex folliculorum* in young adults: Relation with sociodemographic/hygienic factors and acne vulgaris. *J. Eur. Acad. Dermatol. Venereol.* 20:474–76.

Ozdemir, M. H., U. Aksoy, C. Akisu, E. Sonmez, and M. A. Cakmak. 2003. Investigating Demodex in forensic autopsy cases. *Forensic Sci. Int.* 135:226–31.

Ozdemir, M. H., U. Aksoy, E. Sonmez, C. Akisu, C. Yorulmaz, and A. Hilal. 2005. Prevalence of Demodex in health personnel working in the autopsy room. *Am. J. Forensic Med. Pathol.* 26:18–23.

Perotti, M. A. 2009. Phoretic mites associated with animal and human decomposition. *Exp. Appl. Acarol.*, in press.

Perotti, M. A., H. R. Braig, and M. L. Goff. 2009. Phoretic mites and carcasses. In *Forensic entomology—New trends and technologies*, Amendt, J., C. P. Campobasso, M. Grassberger, and M. L. Goff, Eds. Amsterdam: Springer, in press.

Perotti, M. A., M. L. Goff, A. S. Baker, B. D. Turner, and H. R. Braig. 2009. Forensic acarology. *Exp. Appl. Acarol.*, in press.

Perotti, M. A., B. D. Turner, A. S. Baker, and H. R. Braig. 2006. Forensic acarology. Paper presented at the 4th Meeting of the European Association for Forensic Entomology, Bari, Italy.

Prichard, J. G., P. D. Kossoris, R. A. Leibovitch, L. D. Robertson, and F. W. Lovell. 1986. Implications of trombiculid mite bites: Reports of a case and submission of evidence in a murder trial. *J. Forensic Sci.* 31:301–6.

Proctor, H. C. 2009. Can freshwater mites act as forensic tools? *Exp. Appl. Acarol.*, in press.

Rasmy, A. H. 2007. Forensic acarology: A new area for forensic investigation. *ACARINES J. Egypt. Soc. Acarol.* 1:3.

Rufli, T., and Y. Mumcuoglu. 1981. The hair follicle mites *Demodex folliculorum* and *Demodex brevis*: Biology and medical importance. A review. *Dermatologica* 162:1–11.

Russell, D. J., M. M. Schulz, and B. M. O'Connor. 2004. Mass occurrence of astigmatid mites on human remains. *Abh. Ber. Naturkundemus. Görlitz* 76:51–56.

Schei, M. A., J. O. Hessen, and E. Lund. 2002. House-dust mites and mattresses. *Allergy* 57:538–42.

Sheard, C., and J. G. Hardenbergh. 1927. The effects of ultraviolet and infra-red irradiation on *Demodex folliculorum. J. Parasitol.* 14:36–42.

Simon, G. 1842. Über eine in den kranken und normalen Haarsäcken des Menschen lebende Milben [About mites living in diseased and healthy hair follicles of humans]. *Arch. Anat. Physiol. Wiss. Med.* 9:218–37.

Solarz, K. 2009. Indoor and dust mites. *Exp. Appl. Acarol.*, in press.

van Asselt, L., G. Wauthy, and P. Grootaert. 1996. Spatial distributions of *Dermatophagoides farinae* Hughes and *Dermatophagoides pteronyssinus* (Trouessart) (Pyroglyphidae). Acarology IX, Vol. 1. Paper presented at the Proceedings of the 9th International Congress of Acarology, Columbus, OH.

van Bronswijk, J. E. M. H. 1979. House-dust as an ecosystem. 1. Paper presented at the V International Congress of Acarology, East Lansing, MI.

Walter, D. E., and H. C. Proctor. 1999. *Mites—Ecology, evolution and behaviour*. Wallingford, UK: CABI Publishing.

Warner, A., S. Bostrom, C. Moller, and N. I. M. Kjellman. 1999. Mite fauna in the home and sensitivity to house-dust and storage mites. *Allergy* 54:681–90.

Webb, J. P., Jr., R. B. Loomis, M. B. Madon, S. G. Bennett, and G. E. Greene. 1983. The chigger species *Eutrombicula belkini* Gould (Acari: Trombiculidae) as a forensic tool in a homicide investigation in Ventura County, California. *Bull. Soc. Vector Ecol.* 8:141–46.

Wilson, E. 1844. Researches into the structure and development of a newly discovered parasitic animalcule of the human skin—The *Entozoon folliculorum. Philos. Trans. R. Soc. Lond.* 134:305–19.

Yoshikawa, M., and P. H. Bennett. 1979. House dust mites in Columbus, Ohio, USA. *Ohio J. Sci.* 79:280–82.

Zhao, R., Y. Hou, and Y. Liang. 1998. [Relationship between Demodex infection and age as well as skin disease]. *Chin. J. Vector Biol. Control* 10:50–54.

Zhao, Y. E., N. Guo, X. Zheng, S.-F. Yang, M.-X. Zhang, L.-M. Zhang, and K.-J. Wu. 2005. [Observations on morphology and the survival temperature range of *Demodex folliculorum*]. *Acta Entomol. Sin.* 48:754–58.

Index

A

abdomen, *26, 28,*
ABFE, *see* American Board of Entomology (ABFE)
acarology, *see also* Mites
    clothing, 642–644
    environmental contact, 645
    forensic applications, 638–639
    fundamentals, 637–638
    healthy skin of living and dead people
        comparison, 639–642
access control, 336–338
accidents, *see* Traffic accidents
accumulated degree-days (ADD), 397, 460, 521, 523,
    537
accumulated degree-hours (ADH), 459–460, 537
acrobat ants, *120, see also Crematogaster lineolata*
Acronesia, see Calliphora sp.
Acrophaga, see Calliphora sp.
active decay, 408
Acyrthosiphon pisum, 390
adaptive responses, 500
adipocere fromation, 411
admissibility, 454–455
adult emergence, laboratory rearing, 189–192
advanced decay, 409–410
advanced floating decay decomposition stage, *277,*
    278–279
aedeagus, 32
aerial collection of evidence, 140, *141–146, 152*
affidavits, 466–467
AFLP, *see* Amplified fragment length polymorphism
    (AFLP)
Africa, global review, 261–262
Africanized bees, 113, *see also Apis mellifera*
    scutellata
agricultural species, 396–397
Agyrtidae family, 211
albedo, 527
Aldrichina, see Calliphora sp.
Aldrichina grahami, 375
Aleochura curtula, 229
alimemazine detection, 431
alkalinity effects, 413
allergic reactions, 478
Alphitobius diaperinus, 488, 488–489

alteration, bloodstain evidence
    case studies, 568–579
    cast-off pattern, 549, *549, 555, 565*
    cockroach transfer pattern, 556, *556–557*
    double homicide in Nebraska case study,
        573–576
    drip pattern, 549, *549, 555*
    fleas, 558–559, *559*
    flow pattern, 550, *550, 564, 564*
    fly impact on patterns, 560–561, *561–562*
    fundamentals, 539–544, 579
    grooming impact, 558, *558*
    high-impact patterns, 554, *556*
    historical developments, 544–546
    homicide, Tallahassee case study, 568–572
    impact pattern, 550, *550, 555, 565*
    lonely woman case study, 577–578
    low-velocity impact stains, 560, *560*
    medium-velocity impact stains, 559–560, *560*
    passive drop pattern, 550, *550*
    projected pattern, 550, *551,* 563–564, *564*
    protocols and techniques, 546–568
    reconstruction, 567–568
    regurgitation by flies, 560–561, *561–562*
    transfer pattern, 550, *551, 556, 564, 565–566*
    wipe pattern, 550, *552, 565, 565*
ambient air temperature, 135, *136–137*
Amblyomma americanum, 117, see also Lone star
    tick
ambush bugs, 27
Ameletidae family, 288
American Academy of Forensic Sciences, 465
American Board of Entomology (ABFE), 8, 12, 456
American burying beetles, 89, *90, see also* Burying
    beetles; *Nicrophorus americanus*
American carrion beetles, 88–89, *89, see also*
    Carrion beetles; *Necrophila americana*
American cockroaches, *see also Periplaneta*
    americana
    damage similar to abrasions or burns, *120*
    description, 119, *120*
    segmented cerci, *30*
American dog tick, *117, see also Dermacentor*
    variabilis
American Registry of Professional Entomologists
    (ARPE), 5–6

651